FOOD
Finder

Food Sources of Vitamins & Minerals

SECOND EDITION

(Previously SECTION D
of the FOOD PROCESSOR II Manual)

Elizabeth S. Hands

ESHA Research
PO Box 13028, Salem, Oregon 97309, USA

COPYRIGHT (c) 1985   1st edition, 1st printing: Betty & Bob Geltz, ESHA Research
COPYRIGHT (c) 1987   1st edition, 2nd printing: Elizabeth Hands Geltz, ESHA Research
COPYRIGHT (c) 1990   2nd edition, 1st printing: Elizabeth S. Hands (Geltz), ESHA Research
                                1193 Royvonne SE, Suite 23   (zip code 97302)
                                PO Box 13028
                                Salem, Oregon 97309, USA

Library of Congress Catalog Card Number 89-82411

ISBN 0-940071-06-1

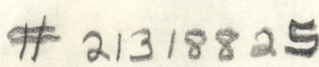

Acknowledgements

Special acknowledgement to Nancy S. Belleque for her extensive work, dedication and competence. This document could not have been completed without her careful and patient eye for numbers and accuracy.

Further thanks go to Ann Marchant MS, RD for her scientific review of the narratives; to Patti Bishop for the cover design and graphics; and to Ken Parksion, Lori Walling and Steve Davison for their assistance with this project.

Contents

Preface & Background

As the second, and expanded edition of *Food Sources of Vitamins & Minerals*, it has been renamed *The Food Finder*. It has evolved from being Section D of the computer manual for The Food Processor software systems, to a book that many of our customers requested. This special compilation of nutrients in foods is organized so you can look up the nutrient you want, and find a variety of food sources for it. The information in this book is the result of several years of initial research, which now continues on a regular basis. The approximately 450 scientific sources of information are listed at the end of the book.

Information is presented for 24 nutrients: 12 vitamins, 10 minerals, cholesterol and dietary fiber. It is almost twice as large as the previous edition. We are pleased to add calories, and nutrient density for each food item (nutrient per 100 calories) so you can select foods for the nutrient level you want, and also relate it to the calories it will contribute. We have doubled the number of foods, and added more nutrients. We have also added more food items that are currently popular. And, the nutrient information is the latest from the USDA and other scientific sources.

It took us much longer than we thought, and for those who waited, we want you to know that the extra time also provides you with the newest RDAs (Recommended Dietary Allowances) and some major updates on nutrient data for grains and vegetables that were only released in November of 1989.

Not all the foods listed have a high amount of the selected nutrient. Since cooked and processed foods retain varying proportions of nutrients, we have listed foods in different forms so you can see the difference. Also, some foods may be a nutrient contributor because they are consumed in larger volume than other foods. Some foods are listed because they have *none* of the selected nutrient.

This document and The Food Processor nutrition software systems were originally researched and prepared for personal use. Because the authors have scientific backgrounds, it was considered unacceptable to have missing values when calculating and evaluating diets and menus. So the research began, and several years later the first "ESHA" database was used in the first Food Processor nutrition system. Now it is used in major research efforts in locations across the United States and in 30 other countries. ... And the research continues.

Considerable effort has been made to report the most accurate data available, as discussed on the next page. There will always be changes in the future, and we welcome any suggestions or comments.

About the Nutrient Data

This data is compiled from over 450 scientific sources of information, including the most recent USDA data. These additional sources include scientific journal articles; food composition tables from England, Canada, and Europe; information from other nutrient data banks and publications; unpublished scientific data; and manufacturers' data. It also includes the valuable assistance of, and many conversations with, the professional staff members of the USDA Human Nutrition Information Service in Hyattsville, Maryland.

Estimates of nutrient amounts for foods and nutrients include all possible adjustments in the interest of accuracy. When multiple values were reported for a nutrient, the numbers were averaged and weighted with consideration for the original number of samples in the separate sources, as well as the type of analytical technique used. Whenever water percentages were available, estimates of nutrient amounts were adjusted for water content. Whenever a reported weight appeared inconsistent (e.g. cooked eggplant and collards), many kitchen tests were made, and the average weight of the typical product was given as tested.

When estimates of nutrient amounts in cooked foods were derived from reported amounts in raw foods, published retention factors were applied. Some reported data for combination foods were modified to include newer data available for major ingredients. For example, since "pies" were analyzed and reported, newer data on fruits (and pumpkin) have been published. Older reported data on certain bakery items were updated for new enrichment levels for certain nutrients.

It is Important to Know

. . . that there can be many different nutrient values reported for the same food. Many factors influence the amounts of nutrients in foods: the mineral content of the soil, fertilizer used, genetics of the plant or animal, the diet of the animal, the method of processing, season of the year, methods of analysis, difference in moisture content of the samples analyzed, length of storage, method of storage, and methods of cooking the food.

As a result, different values for the same food item are reported even by reliable sources. Although each nutrient from USDA government data is presented as a single number, it is actually an average of a range of numbers. USDA data will have different reported values for foods as well, as older information is replaced with newer data in the more recent publications. *Therefore, nutrient data should be viewed and used only as a guide, a close approximation of nutrient content.*

The sources and references are listed at the end of this book.

About the Tables

Each nutrient has its own chapter. An informative summary starts with a discussion about the vitamin or mineral -- why the body needs it; how deficiencies or excesses may affect your health; and other basic information. Then there are brief discussions about food losses, recommended nutrient intake (American and Canadian), and food sources. Tables of information follow.

Foods are listed in the table alphabetically within food groups. Each food is listed with the following information: the gram weight and serving size; calories per serving; the nutrient per 100 calories (a measure of nutrient density); and the nutrient per serving is listed last, in the shaded column.

When the serving is 1 ounce, it refers to the "weight" of one ounce. The gram equivalent weight of one ounce has been abbreviated from 28.35g to 28 g for the tables. Weights and the other information are *for the edible portion only*.

Interpreting the Information

Because analytical results are often reported averages from several tests, it will be possible to get small variations between foods that are "mathematical" in nature. Small variations do not signify that one food is better nutritionally than another. Keep in mind that the nutrient data should be viewed and used only as a guide, a close approximation of nutrient content.

Sometimes frozen foods will show a larger amount of a nutrient compared to the same portion of the fresh, or cooked from fresh item. Because freezing breaks down the cell walls of the "plants," more food can be compressed into the same portion size when it is thawed. Also, when items are cooked they can lose water content. This can result in a higher density of nutrient, simply because of the water loss.

Dietary Fiber deserves a special word. It is important to know that it is emerging data, and to expect additional research and refinement of analytical techniques to modify the data in the future. In this table, either an estimate of the total dietary fiber (a specific analytical technique); or a combination of analytical results for the insoluble and soluble fractions are used. The combination measures for the insoluble components (NDF method) plus the values of the soluble fractions (pectin & gums) are used when the information is available.

Niacin values are for preformed niacin and do not include additional niacin that may form in the body from the conversion of tryptophan.

Legumes are listed in the vegetable section. Soybeans enjoy multiple status, however, and products can be found in Dairy (soy milk), Grains (flour), Nuts & Seeds (roasted soybeans), and Other (miso, natto, tempeh, soy sauce). Cooked soybeans and tofu are in the Vegetable section .

The energy (Calories) and nutrients in recipes and combination foods vary widely, depending on the ingredients. In addition, manufacturers modify their formulations and recipes, and values will change accordingly.

Cuts of meat are the common cuts. Expanded descriptions of chops, and raw to cooked forms are listed in the descriptions for Thiamin and Riboflavin (vitamins B1 and B2). A single breast of chicken is referred to as a half breast in research data. In this document, we use the more popular term of 1 breast of chicken (two per bird).

Expanded food descriptions, where they existed, are listed in the sections for Thiamin and Riboflavin.

Methods of Improving your Diet, Menus or Recipes

To <u>Increase</u> the Content of a Specific Nutrient

1. To get the information you need, scan the shaded column which is the "Nutrient/Serving". Look for the larger numbers, and then identify the foods that are practical additions for you. Be sure to see if the serving size is appropriate as well.

2. Or, scan the foods or food groups first, and see if the item you want to add is going to have enough of the nutrient to meet your needs.

To <u>Increase</u> the Nutrient Content, with the <u>Lowest</u> Additional Calories

1. To get the information you need, first scan the "food names" and/or the shaded "Nutrient/Serving" column to determine the food items you want to consider.

2. Then, look at the middle column which are the Calories (Cal) for the serving size stated. The numbers in the column indicate the relative "Calories per serving" for the foods you want to consider. If the calories are pretty similar, it won't make a difference which items you choose. But, as an example, a cup of whole milk will have more calories than a cup of skim milk.

3. Another way is to look at the "Nutrient/100 Cal" column. This is to the left of the shaded "Nutrient/serving". This column will show at a glance which items have the highest *relative* concentration of "nutrient" for the same amount of calories (100 calories). This is especially helpful when the two serving sizes are different.

To <u>Decrease</u> the Content of a Selected Nutrient

You will most likely want to *decrease* such nutrients as sodium and cholesterol. In that case the process is just a variation of the above. Instead of looking for the larger numbers, you should focus on the smaller numbers of the nutrient.

Example of the Thought Process

In this example, we want to increase the niacin content of a menu. We want to select the highest amount of niacin for the fewest calories, between the items listed. The thought process might go something like this . . .

	Wt (g)	Svg	Cal	Niacin 100 Cal	Niacin Svg
Apple, whole, 2.75" diam	138	1 ea	80	.13	.106
Blueberries, fresh	145	1 c	82	.85	.700
Peach half, canned, juice pack	77	1 half	34	1.3	.448
Peach half, canned, heavy syrup	81	1 half	60	.80	.497

"The blueberries have the highest amount of niacin at .700 mg per serving. However, a whole cup of fresh blueberries is more than we want to add. The apple has a low amount, so we will look at the canned peaches."

"A single peach half (juice pack) has .45 mg of niacin (.448 rounded), and .50 mg (.497 rounded) if it's canned in heavy syrup. The "heavy syrup" canned peach half has more niacin, but it also has almost twice the calories. The nutrient density is higher with the juice pack peach half (1.3 compared to .8). So, to add the most niacin for the least number of calories, the juice-pack canned peach half is the best choice in this example."

Recommended Dietary Allowances Revised 1989 *

"The Allowances are expressed as average daily intakes over time, and are intended to provide for individual variations among most normal persons under usual environmental stresses in the United States ."**

Age (years) & gender	Reference Weight lbs	kg	Reference Height cm	in	Protein g	Vitamin A RE	Thiamin mg	Riboflavin mg	Niacin NE	Vitamin B6 mg	Folacin µg	Vitamin B12 µg	Vitamin C mg	Vitamin D µg	Vitamin E αTE	Vitamin K µg	Calcium mg	Iodine µg	Iron mg	Magnesium mg	Phosphorus mg	Selenium µg	Zinc mg
Infants																							
0.0 - 0.5	6	13	60	24	13	375	0.3	0.4	5	0.3	25	0.3	30	7.5	3	5	400	40	6	40	300	10	5
0.5 - 1.0	9	20	71	28	14	375	0.4	0.5	6	0.6	35	0.5	35	10	4	10	600	50	10	60	500	15	5
Children																							
1 - 3	13	29	90	35	16	400	0.7	0.8	9	1.0	50	0.7	40	10	6	15	800	70	10	80	800	20	10
4 - 6	20	44	112	44	24	500	0.9	1.1	12	1.1	75	1.0	45	10	7	20	800	90	10	120	800	20	10
7 - 10	28	62	132	52	28	700	1.0	1.2	13	1.4	100	1.4	45	10	7	30	800	120	10	170	800	30	10
Males																							
11 - 14	45	99	157	62	45	1000	1.3	1.5	17	1.7	150	2.0	50	10	10	45	1200	150	12	270	1200	40	15
15 - 18	66	145	176	69	59	1000	1.5	1.8	20	2.0	200	2.0	60	10	10	65	1200	150	12	400	1200	50	15
19 - 24	72	160	177	70	58	1000	1.5	1.7	19	2.0	200	2.0	60	10	10	70	1200	150	10	350	1200	70	15
25 - 50	79	174	176	70	63	1000	1.5	1.7	19	2.0	200	2.0	60	5	10	80	800	150	10	350	800	70	15
51 +	77	170	173	68	63	1000	1.2	1.4	15	2.0	200	2.0	60	5	10	80	800	150	10	350	800	70	15
Females																							
11 - 14	46	101	157	62	46	800	1.1	1.3	15	1.4	150	2.0	50	10	8	45	1200	150	15	280	1200	45	12
15 - 18	55	120	163	64	44	800	1.1	1.3	15	1.5	180	2.0	60	10	8	55	1200	150	15	300	1200	50	12
19 - 24	58	128	164	65	46	800	1.1	1.3	15	1.6	180	2.0	60	10	8	60	1200	150	15	280	1200	55	12
25 - 50	63	138	163	64	50	800	1.1	1.3	15	1.6	180	2.0	60	5	8	65	800	150	15	280	800	55	12
51+	65	143	160	63	50	800	1.0	1.2	13	1.6	180	2.0	60	5	8	65	800	150	10	280	800	55	12
Pregnant					60	800	1.5	1.6	17	2.2	400	2.2	70	10	10	65	1200	175	30	320	1200	65	15
Lactating																							
1st 6 mo.					65	1300	1.6	1.8	20	2.1	280	2.6	95	10	12	65	1200	200	15	355	1200	75	19
2nd 6 mo.					62	1200	1.6	1.7	20	2.1	260	2.6	90	10	11	65	1200	200	15	340	1200	75	16

* *Recommended Dietary Allowances*, 10th revised edition (c) 1989, by the National Academy of Sciences, National Academy Press, Washington DC. The RDAs are designed for the maintenance of good nutrition of practically all healthy people in the United States.

This table is prepared by ESHA Research, Salem, Oregon, for the users of The Food Processor nutrition system, with permission from the National Academy Press. The Recommended amounts are related to the reference heights and weights listed here. Weights and heights are the medians for the U.S. Population as reported in NHANES II ; the median weights of those under 19 years of age from Hamill et al, 1979. *The Food Processor*® nutrition systems use the formulas themselves, so if you are not a "reference" person, it will be more accurate for you.

DEFINITIONS:

mcg or µg = micrograms; 1000 mcg = 1 mg; 1000 mg = 1 gram. Thiamin = Vit B1; Riboflavin = Vit B2; Niacin = Vit B3.
RE (Retinol equivalents) = 1 µg Vitamin A from animal sources, or 6 µg of Vitamin A from β-carotene (plant sources).
Vitamin D: 10 µg of Vitamin D (as cholecalciferol) = 400 IU (International Units) IUs are an older measure.
Vitamin E: 1 mg of d-α tocopherol = 1 α-TE (TE = tocopherol equivalent).
Niacin (Vitamin B3): NE (niacin equivalent) is 1 mg of niacin or 60 mg of dietary tryptophan. Also referred to as mg-NE.

ESHA Research, 1193 Royvonne SE, #23 (zip code = 97302), PO Box 13028, Salem, OR 97309.
Phone (503) 585-6242 Fax (503) 585-5543.

Estimated Safe and Adequate Daily Dietary Intakes
of Selected Vitamins & Minerals, 1989

Because there is less information on which to base allowances, these figures are provided in the form of ranges of recommended amounts. Also, since toxic levels for many trace elements may only be several times the usual intakes, the upper levels for the elements in this table should not be habitually exceeded.

Age (years)	Biotin (mcg)	Pantothenic Acid (mg)	Copper (mg)	Manganese (mg)	Fluoride (mg)	Chromium (mcg)	Molybdenum (mcg)
0.0-0.5	10	2	0.4 - 0.6	0.3 - 0.6	0.1 - 0.5	10-40	15-30
0.5-1.0	15	3	0.6 - 0.7	0.6 - 1.0	0.2 - 1.0	20-60	20-40
1 - 3	20	3	0.7 - 1.0	1.0 - 1.5	0.5 - 1.5	20-80	25-50
4 - 6	25	3 - 4	1.0 - 1.5	1.5 - 2.0	1.0 - 2.5	30-120	30-75
7-10	30	4 - 5	1.0 - 2.0	2.0 - 3.0	1.5 - 2.5	50-200	50-150
11+	30-100	4 - 7	1.5 - 2.5	2.0 - 5.0	1.5 - 2.5	50-200	75-250
Adults	30-100	4 - 7	1.5 - 3.0	2.0 - 5.0	1.5 - 4.0	5 -200	75-250

Source: Information from *Recommended Dietary Allowances,* 10th revised edition (c) 1989, by the National Academy of Sciences

Estimated <u>Minimum</u> * Requirements for Healthy Persons
for Sodium, Chloride, and Potassium

The 1989 Recommended Dietary Allowances suggest *Minimum* requirements for these elements, rather than the "ranges" recommended in previous editions.

Age	Weight (kg)	Weight (lb)	Sodium (mg)	Chloride (mg)	Potassium (mg)
up to 5 months	4.5	10	120	180	500
6 - 11 months	8.9	20	200	300	700
1 year	11	24	225	350	1,000
2 - 5	16	35	300	500	1,400
6 - 9	25	55	400	600	1,600
10 - 18	50	110	500	750	2,000
over 18 years	70	154	500*- 2400	750	2,000

Source: Information from *Recommended Dietary Allowances,* 10th revised edition (c) 1989, by the National Academy of Sciences

Regarding Sodium *

A <u>committee of the Food and Nutrition Board recently recommended that daily intakes of sodium chloride (salt) be limited to 6 grams or less.</u> The <u>Sodium "part" would be 2.4 grams, or 2400 mg. — 4.8 times the minimum requirement.</u> The *minimum* requirement -- a different concept from the "recommended range" in previous recommendations, is 500 mg for adults. This "minimum" is substantially exceeded in the average American diet, largely from added sodium products in processed foods. The 500 mg "minimum" does not include an allowance for prolonged losses from the skin through sweat. For those that are susceptible to hypertension (about 1 in 5 persons), reduced sodium intake is clearly beneficial. For others, there may not be such an obvious benefit.

Regarding Potassium *

Desirable intakes of potassium may considerably exceed these minimum values ($\approx$ 3,500 mg for adults). The National Research Council recommends increased intake of fruits and vegetables because of the considerable evidence that dietary potassium exerts a beneficial effect in hypertension.

Regarding Chloride * Intakes and losses of Chloride parallel those of sodium.

ESHA Research

US RDA -- Used for Food Labels

The U.S. RDAs are standards set by the Food and Drug Administration in 1973 using the Recommended Dietary Allowances of the National Academy of Sciences, National Research Council. As of 1989, the numbers used on food labels, are those used from the 1968 RDA, which makes them about 20 years old. However, there has not been a great need to update them because there have not been significant changes in the base recommendations. Also, the highest needs in the RDA table for age and gender were generally selected for use as the U.S. RDA, so the values are generous. The nutrition information on food labels is expressed as a precent of the U. S. RDA.

The 1989 Recommended Dietary Allowances have modified previous RDAs however, and are listed on page vi.

The US RDA has different recommendations for infants, children under 4 years, adults & children over 4 years, and for pregnant or nursing women. They are all listed here for reference. The Food Processor program(s) come with the Adult Label recommendation already entered into the system.

	Unit of Measure	Infants (birth-12 months)	Children under 4 years	Adults & children 4 or more years	Pregnant or Nursing
Protein (veg)*	g	25	28	65	no change
Protein (meat)**	g	18	20	45	no change
Vitamin A ***	RE	300	500	1000	1600
Vitamin D	IU	400	400	400	400
Vitamin E	IU	5	10	30	30
Vitamin C	mg	35	40	60	60
Folacin / Folic acid	mg (mcg)	0.1 (100)	0.2 (200)	0.4 (400)	0.8 (400)
Thiamin (B1)	mg	0.5	0.7	1.5	1.7
Riboflavin (B2)	mg	0.6	0.8	1.7	2.0
Niacin (B3)	mg	8	9	20	20
Vitamin B6	mg	0.4	0.7	2.0	2.5
Vitamin B12	mcg	2	3	6	8
Biotin	mg	0.05	0.15	0.3	0.3
Pantothenic Acid	mg	3	5	10	10
Calcium	g (mg)	0.6 (600)	0.8 (800)	1.0 (1000)	1.3 (1300)
Phosphorus	g (mg)	0.6 (600)	0.8 (800)	1.0 (1000)	1.3 (1300)
Iodine	mcg	45	70	150	150
Iron	mg	15	10	18	18
Magnesium	mg	70	200	400	450
Copper	mg	0.6	1.0	2.0	2.0
Zinc	mg	5	8	15	15

Source: adopted from *Food Technology* 28 (7):5, 1974

* The protein efficiency ratio is less than casein, which means a basicly vegetarian diet.

** The protein efficiency ratio is greater than or equal to casein. This is for diets that include meat.

*** Vitamin A was originally expressed in IUs: 1000 RE = 5000 IU/

Dietary Guidelines & Goals

In the 1940s there was a prevalence of under nutrition in the United States. The four food group plan was originated at that time to give Americans a better idea of how to prepare an adequate diet. A far different situation began to emerge in the mid 1960s, and the guidelines here are focused on the results of *over-nutrition*. Excessive intakes of fat, salt, sugar, and even protein, have contributed to the diseases of today -- heart disease, cancer, diabetes, liver disease and obesity, and probably more. The suggestions for better food choices are made with these Dietary Guidelines.

Dietary Guidelines for Americans

1.   <u>Eat a variety of foods daily</u>. Include these foods every day: fruits and vegetables; whole grain and enriched breads and cereals and other products made from grains; milk and milk products; meats, fish, poultry and eggs; and dried peas and beans.

2.   <u>Maintain desirable weight</u>. Increase physical activity; control overeating by eating slowly, taking smaller portions, and avoiding "seconds"; eat fewer fatty foods and sweets and less sugar, drink fewer alcoholic beverages, and eat more foods that are low in calories and high in nutrients.

3.   <u>Avoid too much fat, saturated fat, and cholesterol.</u> Choose low-fat protein sources such as lean meats, fish, poultry, and dry peas and beans; use eggs and organ meats in moderation; limit intake of fats on and in foods; trim fats from meats; broil, bake, or boil--don't fry; limit breaded and deep-fried foods; read food labels for fat contents.

4.   <u>Eat foods with adequate starch and fiber.</u> Substitute starchy foods for foods high in fats and sugars; select whole-grain breads and cereal, fruits and vegetables, and dried beans and peas to increase fiber and starch intake.

5.   <u>Avoid too much sugar.</u> Use less sugar, syrup, and honey; reduce concentrated sweets like candy, soft drinks, cookies, and the like; select fresh fruits or fruits canned in light syrup or their own juices; read food labels-- sucrose, glucose, dextrose, maltose, lactose, fructose, syrups and honey are all sugars; eat sugar less often to reduce dental caries.

6.   <u>Avoid too much sodium.</u> Learn to enjoy the flavors of unsalted foods; flavor foods with herbs, spices, and lemon juice; reduce salt in cooking; add little or no salt at the table; limit salty foods like potato chips, pretzels, salted nuts, popcorn, condiments (soy sauce, steak sauce, and garlic salt), some cheeses, pickled foods and cured meats, and some canned vegetables and soups. read food labels for sodium or salt contents, especially in processed and snack foods; use lower-sodium products when available.

7.   <u>If you drink alcoholic beverages, do so in moderation.</u> For individuals who drink, limit all alcoholic beverages (including wine, beer, liquors and so on) to one or two drinks per day. "One drink" means 12 ounces of beer, 3 oz of wine, or 1 1/2 oz of distilled spirits. Pregnant women should refrain from the use of alcohol. If you drink, do not drive.

> Source: U.S. Department of Agriculture, U.S. Department of Health and Human Services, Nutrition and Your Health, dietary Guidelines for Americans, 2nf Ed (Washington DC: Government Printing Office, 1985).

Dietary Goals

The Dietary Goals recommend that our daily calories should come from the following food energy sources. Protein shows 12% "by default" because protein needs are actually measured by your personal body weight.

Carbohydrates:	58% of calories
Fats	30% of calories
Protein	12% of calories
Total	100%

Total fats should be equally divided between saturated (10%, mono-unsaturated (10%) and poly-unsaturated (10%). Cholesterol intake is recommended at less than 300 mg a day.

> Source: Dietary Goals for the United States, 2nd edition: Select Committee on Nutrition and Human Needs, United States Senate, Washington D.C., 1977.

ESHA Research

The Canadian RNI -- Recommended Nutrient Intake

All recommended intakes are designed to cover individual variations in essentially all of a healthy population subsisting upon a variety of common foods available in Canada. All the figures are examples because the primary units are expressed per kilogram of body weight. The recommended intakes during periods of growth are taken as appropriate for individuals representative of the mid-point in each age group.

Many of the Canadian recommendations are driven by formulas, and this table shows examples of Recommended Nutrient Intakes for the age and reference weights shown. Nutrients not shown in the table are indicated by formulas listed here, and are referenced in each nutrient chapter.

Thiamin = 0.4 mg/1000 kcal   Niacin = 7.2 NE/1000kcal   Phosphorus = same as Calcium
Riboflavin = 0.5 mg/1000kcal   Vitamin B6 = 15 µg as pyridoxine / gram of protein.

Summary <u>Examples</u> of Recommended Nutrient Intakes for Canadians

Age & Gender	Weight (kg)	Protein g/day	Vit A RE/day	Vit D µg/day	Vit E mg/day	Vit C mg/day	Folacin µg/day	Vit B12 µg/day	Calcium mg/day	Magnesium mg/day	Iron mg/day	Iodine µg/day	Zinc ng/day
Months													
0-2	4.5	11	400	10	3	20	50	0.3	350	30	0.4	25	2
3-5	7.0	14	400	10	3	20	50	0.3	350	40	5	35	3
6-8	8.5	17	400	10	3	20	50	0.3	400	50	7	40	3
9-11	9.5	18	400	10	3	20	55	0.3	400	50	7	45	3
Years													
1	11	19	400	10	3	20	65	0.3	500	55	6	55	4
2-3	14	22	400	5	4	20	80	0.4	500	70	6	65	4
4-6	18	26	500	5	5	25	90	0.5	600	90	6	85	5
7-9													
Males	25	30	700	2.5	7	35	125	0.8	700	110	7	110	6
Females	25	30	700	2.5	6	30	125	0.8	700	110	7	95	6
10-12													
Males	34	38	800	2.5	8	40	170	1.0	900	150	10	125	7
Females	36	40	800	2.5	7	40	180	1.0	1000	160	10	110	7
13-15													
Males	50	50	900	2.5	9	50	150	1.5	1100	210	12	160	9
Females	48	42	800	2.5	7	45	145	1.5	800	200	13	160	8
16-18													
Males	62	55	1000	2.5	10	55	185	1.9	900	250	10	160	9
Females	53	43	800	2.5	7	45	160	1.9	700	215	14	160	8
19-24													
Males	71	58	1000	2.5	10	60	210	2.0	800	240	8	160	9
Females	58	43	800	2.5	7	45	175	2.0	700	200	14	160	8
25-49													
Males	74	61	1000	2.5	9	60	220	2.0	800	250	8	160	9
Females	59	44	800	2.5	6	45	175	2.0	700	200	14	160	8
50-74													
Males	73	60	1000	2.5	7	60	220	2.0	800	250	8	160	9
Females	63	47	800	2.5	6	45	190	2.0	800	210	7	160	8
75+													
Males	69	57	1000	2.5	6	60	205	2.0	800	230	8	160	9
Females	64	47	800	2.5	5	45	190	2.0	800	220	7	160	8
Pregnancy (additional)													
1st trimester	15	100	2.5	2	0	305	1.0	500	15	6	25	0	
2nd trimester	20	100	2.5	2	20	305	1.0	500	20	6	25	1	
3rd trimester	25	100	2.5	2	20	305	1.0	500	25	6	25	2	
Lactation (additional)	20	400	2.5	3	30	120	0.5	500	80	0	50	6	

Source: *Recommended Nutrient Intakes for Canadians*, Health and Welfare Canada), Ottawa, Canada, 1984

Recommended Energy Intake*
For U. S. Individuals, (Median Heights and Weights)
Assuming Light - Medium Activity Level

Individuals look to the recommended energy intake (Calories) for guidance on "how many calories" their diets or menus should contain. It is really important to know that the tables and figures on the subject are the results of averages, and the recommended amounts for an individual can vary from 20% in both directions. If a person has undergone fasting or a series of low-calorie diets from time to time, the metabolism changes further. To offset the reduced food intake, the metabolism becomes more efficient, and the average calories per day to maintain weight will be even lower.

Age (years) or Condition	Weight (kg)	(lb)	Height (cm)	(in)	REE (kcal/day)	Multiples of REE **	Per kg	Calories* Per day
Infants								
0.0 - 0.5	6	13	60	24	320		108	650
0.5 - 1.0	9	20	71	28	500		98	850
Children								
1 - 3	13	29	90	35	740		102	1,300
4 - 6	20	44	112	44	950		90	1,800
7 - 10	28	62	132	52	1,130		70	2,000
Males								
11 - 14	45	99	157	62	1,440	1.70	55	2,500
15 - 18	66	145	176	69	1,760	1.67	45	3,000
19 - 24	72	160	177	70	1,780	1.67	40	2,900
25 - 50	79	174	176	70	1,800	1.60	37	2,900
51+	77	170	173	68	1,530	1.50	30	2,300
Females								
11 - 14	46	102	157	62	1,310	1.67	47	2,200
15 - 18	55	120	163	64	1,370	1.60	40	2,200
19 - 24	58	128	164	65	1,350	1.60	38	2,200
25 - 50	63	138	163	64	1,380	1.55	36	2,200
51+	65	143	160	63	1,280	1.50	30	1,900
Pregnant								
1st trimester								+ 0
2nd trimester								+ 300
3rd trimester								+ 300
Lactating								+ 500

* These figures represent the *AVERAGE* needs of individuals, and <u>can vary + or - 20% *within the light-to-moderate activity level*</u>, depending on the individual's personal metabolism. The numbers listed for Calories per day are calculated based on FAO equations and then rounded.

** <u>REE</u> is Resting Energy Expenditure. The <u>Multiples</u> are the activity factors suggested by WHO, which represent the additional Caloric needs for persons with light-to-moderate activity. Keep in mind, that these numbers can vary plus or minus 20% in individuals.

ESHA Research

Food for Thought, continued

o **High Blood Pressure and Salt.** There is a lot of evidence associating high blood pressure (hypertension) and high sodium intake. . . but, lowering sodium intake doesn't always lower blood pressure in all people. It is estimated that about 1 in 5 persons are sodium/salt sensitive. There is a fairly equal relationship of sodium and potassium in the body, and some studies suggest that the balance of these two elements may be more important in managing high blood pressure than we currently realize.

o **Sodium in Your Water?** Do you have a water softener in your home? Hard water is high in calcium and magnesium, which is very good for you, by the way, but these minerals are taken out to make it "soft." They are exchanged for sodium. In some homes, water softeners are attached to the hot water lines only, and then the cold water is used for cooking with its good-for-you minerals left intact.

o **Sodium Additives in our Food.** Although sodium chloride (table salt) is the most common contributor of sodium in the diet, there are other food additives that do it too. Monosodium glutamate (MSG, Accent) and disodium guanylate or inosinate are called flavor enhancers. Sodium phosphate is added to frozen vegetables in sauces to prevent separation.

Other sodium compounds may be added as preservatives, thickeners, sweeteners (yes, sweeteners), and binders. Even if you don't add salt to your food; even if it doesn't taste "salty;" if you eat a lot of processed foods you are getting a lot of sodium.

o **Another Excess in the Making?** It will be interesting to see the developments of the increased use of salt substitutes, which are primarily potassium-based rather than sodium-based. If not an excess, certainly another imbalance could be in the making. Generally our bodies need a balance between potassium and sodium.

o **No Bugs in Sugar.** Did you ever notice that sugar stored in the cupboard never has bugs in it? that is because there are no nutrients in it for them to eat!

o **Few Bugs in White Flour.** One of the reasons white wheat flour was so popular was because it stayed on the shelf so much longer than whole wheat flour. It was many years before the reason was discovered—the bugs didn't bother the white flour because most of the nutrients were processed out.

It was many years before some of those nutrients were replaced in the flour, and called "enrichment." Enrichment is done for Vitamins B-1 (thiamin), B-2 (riboflavin), B-3 (niacin), and iron. It affects white wheat flours, cornmeal, and their products—breads, noodles, and pasta. Polished white rice has enrichment for these nutrients except for riboflavin. Even with this enrichment, however, "white" flour is not as nutritious as its former self—whole wheat flour. There are too many other nutrients that are partially processed out and not replaced.

o **High Protein . . . Low Calcium?** Diets high in protein promote the excretion of calcium, making it necessary to consume more calcium. An interesting association, since there is an increased concern about osteoporosis in our older adult population and a typical American diet includes twice the protein needed.

Osteoporosis, a bone-thinning disease, can be influenced by several factors beyond high protein diets—lack of exercise, especially weight bearing exercise, and a lack of calcium in the diet in the growth years. The lack of sunshine (or vitamin D) or of certain other nutrients that help absorb and balance calcium in the body (magnesium and phosphorus) will also have an effect. Women are at higher risk after menopause or a complete hysterectomy, because the cessation of the female hormone is known to have an effect. Many physicians prescribe estrogen for women in this situation, as a preventative measure against osteoporosis .

Food for Thought, continued

o **Carbonated Beverages... Low Calcium?** There has been a 300% increase in the consumption of carbonated drinks in the last 30 years, and many carbonated beverages contain phosphoric acid. There has been some question whether too much phosphorus intake can cause an imbalance, and force excretion of calcium from the body. An interesting study in the July, 1989 Journal of Orthopedic Research gives more information. Over 2,600 female former athletes were surveyed. Those who now drink carbonated beverages showed a 2.28 greater incidence rate of bone fractures after age 40 than those who do not drink soda pop. There was a significant relationship between the amount of soda pop and the number of fractures in women over age 40. At that age, bone mass starts declining, as well as estrogen levels. A low milk diet was also found to increase fracture risk. More studies are needed, of course, but this bears noting.

o **Where is the Flavor in Foods?** Most of the good flavors and smells we associate with foods are primarily found in the fat portions of the food.

o **What do Fat and Fiber have in Common?** Both provide a feeling of "fullness" that is associated with satisfaction. Because there are a lot of different components to fiber, some will provide more feeling of fullness than others.

o **Alter your Diet Slowly.** If you have decided to alter your diet and eat foods with more fiber. . . change slowly. The body adjusts to patterns of eating. If you have not eaten much fiber, suddenly adding copious amounts will find your system unprepared, upset, and unable to handle it well. Start gently and increase amounts over time. Your body will tell you what is right for you.

o **Another High Rating for Dietary Fiber.** A high fiber diet apparently can influence the metabolism of certain nutrients. Reports indicate that persons with mild cases of diabetes have been able to reduce their insulin intakes when on high fiber diets. And studies also show a relationship with reduced serum cholesterol levels.

o **When it Floats...** What is the right amount of dietary fiber? One scientist summed it up simply by saying that when your stool floats, you are getting the right amount of fiber.

o **Satisfying that Hungry Feeling . . . Are All Calories Equal?** Well, maybe not. Some recent studies indicate that different foods **can** make a difference. Foods with table sugar in them—sucrose—(that morning doughnut?) will make you hungrier later than foods without sugar or foods with fruit sugar (fructose). Fructose isn't absorbed as quickly as glucose and produces a much slower rise in blood sugar. As a result, the rise in insulin is also slower and more moderate, and the appetite is more satisfied for a longer period of time.

Foods that cause a **rapid** rise in blood sugar cause a rapid rise in insulin, which can stay at a high level long after the body has absorbed the sugar (glucose). It appears from the studies that the insulin may have more to do with appetite than with the level of blood sugar. (Dr. Judith Rodin, Yale). This may also explain why some diabetics who receive large amounts of insulin complain that they are hungry all the time.

o **The _New_ Carbohydrate.** Because of convenience foods, the mass-scale processing of many foods, and our high sugar consumption, we now have a **new** (to this century) type of carbohydrate. . . THE REFINED CARBOHYDRATE. Where complex carbohydrates are high in nutrients and fiber, and low in calories; the new refined carbohydrates are just the opposite. They are low in nutrients and fiber, and high in calories—and they now provide almost **one-fourth** of the calories in the average American diet.

o **When is an Unsaturated Fat no Longer Unsaturated ?** When it is hydrogenated. Unsaturated fats are usually liquid at room temperature (cooking oils) and can be perishable. Thus, they are often refrigerated and tightly wrapped to keep fresh (margarine). But the stability of the product can be really increased through a process called "hydrogenation," which makes it more solid at room temperature. Unfortunately, this process also reduces the original polyunsaturated fat content, which possibly reduces its health value.

o **About Hydrogenated Fats.** In the late 1970s a significant discovery was made about hydrogenation; a fact whose consequence is still not clearly understood—hydrogenation can create fatty acids that are not normally found in nature. Almost all unsaturated fatty acids occur in nature in a form that resembles a horseshoe (called a cis-fatty acid). When partial hydrogenation occurs, the bend in the horseshoe can change, forcing one side of the "U" to extend the other way. These are called trans-fatty acids.

Frankly, it is not known if the different molecular shape matters are not. But there are questions—can our bodies handle these unusual molecules?. . . in large quantities? Such fatty acids are not made by the body, and they are rare in foods. However, for a better part of this century we have been feeding ourselves a lot of them.

From 1910 to 1978 there was a 27% increase in the total dietary fat intake in the United States, with most of the increase through vegetable oils. For example, butter was largely replaced by margarine, and lard and meat drippings were largely replaced with vegetable shortening. The majority of these oils are modified by partial hydrogenation.

o **Cholesterol — Our body makes it.** From 70% to 85% of the cholesterol in our bodies is made by our liver. If you ate no cholesterol at all, the liver would make sure you had enough, because cholesterol is so important to the body. It is the basis of many of our important hormones.

There is a lot of controversy and confusion about cholesterol, partially because it can be found in the arteries (along with fat) of people with heart disease (arteriosclerosis, or hardening of the arteries). But, blood serum cholesterol is not the same as "dietary cholesterol," and the presence of cholesterol in the artery walls may more likely be the **indicator** of the problem rather than the cause.

Some predictors of heart disease are smoking, high blood pressure, high blood triglycerides (certain fatty acids), and high blood cholesterol. One real culprit may simply be a high fat intake. It has been shown that blood cholesterol will be reduced when saturated fat intake is reduced.

o **On Birth Control Pills?** Studies indicate that birth control pills may cause extra requirements for such nutrients as Vitamin B-6, folacin, zinc, and others. Check with your physician or dietitian for the latest information.

o **Tea and Coffee** have an inhibiting effect on the body's absorption of iron. Scientific studies have shown that a cup of tea with a meal can decrease iron absorption by 64%, and a cup of coffee within one hour of a meal can reduce iron absorption of a hamburger meal by 39%. No decrease in iron absorption occurred if coffee was consumed one hour before the meal, however.

o **Is Fat more Fattening?** It appears so!! Not only does fat contain more than twice the calories of carbohydrates (<u>nine</u> calories per gram for fat versus <u>four</u> calories per gram for carbohydrates), but fat is more easily stored in the body than carbohydrates. For every 100 calories consumed as <u>carbohydrate,</u> it takes about 23 calories to convert it to fat, and about 77 calories get stored as fatty tissue. However, for every 100 calories consumed as <u>fat</u>, it takes only about three calories to convert it to storable form, leaving a whopping 97 calories available for body fat.

Food for Thought, continued

o   **Carbonated Beverages... Low Calcium?**   There has been a 300% increase in the consumption of carbonated drinks in the last 30 years, and many carbonated beverages contain phosphoric acid. There has been some question whether too much phosphorus intake can cause an imbalance, and force excretion of calcium from the body. An interesting study in the July, 1989 Journal of Orthopedic Research gives more information. Over 2,600 female former athletes were surveyed. Those who now drink carbonated beverages showed a 2.28 greater incidence rate of bone fractures after age 40 than those who do not drink soda pop.   There was a significant relationship between the amount of soda pop and the number of fractures in women over age 40. At that age, bone mass starts declining, as well as estrogen levels. A low milk diet was also found to increase fracture risk. More studies are needed, of course, but this bears noting.

o   **Where is the Flavor in Foods?**   Most of the good flavors and smells we associate with foods are primarily found in the fat portions of the food.

o   **What do Fat and Fiber have in Common?**   Both provide a feeling of "fullness" that is associated with satisfaction. Because there are a lot of different components to fiber, some will provide more feeling of fullness than others.

o   **Alter your Diet Slowly.**   If you have decided to alter your diet and eat foods with more fiber. . . change slowly. The body adjusts to patterns of eating. If you have not eaten much fiber, suddenly adding copious amounts will find your system unprepared, upset, and unable to handle it well. Start gently and increase amounts over time. Your body will tell you what is right for you.

o   **Another High Rating for Dietary Fiber.**   A high fiber diet apparently can influence the metabolism of certain nutrients. Reports indicate that persons with mild cases of diabetes have been able to reduce their insulin intakes when on high fiber diets. And studies also show a relationship with reduced serum cholesterol levels.

o   **When it Floats...** What is the right amount of dietary fiber? One scientist summed it up simply by saying that when your stool floats, you are getting the right amount of fiber.

o   **Satisfying that Hungry Feeling . . . Are All Calories Equal?**   Well, maybe not. Some recent studies indicate that different foods **can** make a difference. Foods with table sugar in them—sucrose—(that morning doughnut?) will make you hungrier later than foods without sugar or foods with fruit sugar (fructose). Fructose isn't absorbed as quickly as glucose and produces a much slower rise in blood sugar. As a result, the rise in insulin is also slower and more moderate, and the appetite is more satisfied for a longer period of time.

Foods that cause a **rapid** rise in blood sugar cause a rapid rise in insulin, which can stay at a high level long after the body has absorbed the sugar (glucose). It appears from the studies that the insulin may have more to do with appetite than with the level of blood sugar. (Dr. Judith Rodin, Yale). This may also explain why some diabetics who receive large amounts of insulin complain that they are hungry all the time.

o   **The _New_ Carbohydrate.**   Because of convenience foods, the mass-scale processing of many foods, and our high sugar consumption, we now have a **new** (to this century) type of carbohydrate. . . THE REFINED CARBOHYDRATE. Where complex carbohydrates are high in nutrients and fiber, and low in calories; the new refined carbohydrates are just the opposite. They are low in nutrients and fiber, and high in calories—and they now provide almost **one-fourth** of the calories in the average American diet.

Food for Thought, continued

o **When is an Unsaturated Fat no Longer Unsaturated ?** When it is hydrogenated. Unsaturated fats are usually liquid at room temperature (cooking oils) and can be perishable. Thus, they are often refrigerated and tightly wrapped to keep fresh (margarine). But the stability of the product can be really increased through a process called "hydrogenation," which makes it more solid at room temperature. Unfortunately, this process also reduces the original polyunsaturated fat content, which possibly reduces its health value.

o **About Hydrogenated Fats.** In the late 1970s a significant discovery was made about hydrogenation; a fact whose consequence is still not clearly understood—hydrogenation can create fatty acids that are not normally found in nature. Almost all unsaturated fatty acids occur in nature in a form that resembles a horseshoe (called a cis-fatty acid). When partial hydrogenation occurs, the bend in the horseshoe can change, forcing one side of the "U" to extend the other way. These are called trans-fatty acids.

Frankly, it is not known if the different molecular shape matters are not. But there are questions—can our bodies handle these unusual molecules?. . . in large quantities? Such fatty acids are not made by the body, and they are rare in foods. However, for a better part of this century we have been feeding ourselves a lot of them.

From 1910 to 1978 there was a 27% increase in the total dietary fat intake in the United States, with most of the increase through vegetable oils. For example, butter was largely replaced by margarine, and lard and meat drippings were largely replaced with vegetable shortening. The majority of these oils are modified by partial hydrogenation.

o **Cholesterol — Our body makes it.** From 70% to 85% of the cholesterol in our bodies is made by our liver. If you ate no cholesterol at all, the liver would make sure you had enough, because cholesterol is so important to the body. It is the basis of many of our important hormones.

There is a lot of controversy and confusion about cholesterol, partially because it can be found in the arteries (along with fat) of people with heart disease (arteriosclerosis, or hardening of the arteries). But, blood serum cholesterol is not the same as "dietary cholesterol," and the presence of cholesterol in the artery walls may more likely be the **indicator** of the problem rather than the cause.

Some predictors of heart disease are smoking, high blood pressure, high blood triglycerides (certain fatty acids), and high blood cholesterol. One real culprit may simply be a high fat intake. It has been shown that blood cholesterol will be reduced when saturated fat intake is reduced.

o **On Birth Control Pills?** Studies indicate that birth control pills may cause extra requirements for such nutrients as Vitamin B-6, folacin, zinc, and others. Check with your physician or dietitian for the latest information.

o **Tea and Coffee** have an inhibiting effect on the body's absorption of iron. Scientific studies have shown that a cup of tea with a meal can decrease iron absorption by 64%, and a cup of coffee within one hour of a meal can reduce iron absorption of a hamburger meal by 39%. No decrease in iron absorption occurred if coffee was consumed one hour before the meal, however.

o **Is Fat more Fattening?** It appears so!! Not only does fat contain more than twice the calories of carbohydrates (nine calories per gram for fat versus four calories per gram for carbohydrates), but fat is more easily stored in the body than carbohydrates. For every 100 calories consumed as carbohydrate, it takes about 23 calories to convert it to fat, and about 77 calories get stored as fatty tissue. However, for every 100 calories consumed as fat, it takes only about three calories to convert it to storable form, leaving a whopping 97 calories available for body fat.

Conversions for Weights & Measures

Weight Measures

1 kg	=	1000 grams = 2.21 pounds
1 gram	=	1000 mg (milligrams) = .035 ounce = .001 kg
1 mg	=	1000 mcg (μg)
1 oz	=	28.35 grams (rounded to 28 grms, and sometimes 30 grms)
1 pound	=	1 kilogram divided by 2.21 = 454 grams = 16 ounces (weight ounces)

Volume and Household measures

1 tsp	=	5 grams dry weight (eg, sugar, salt, some powders) = 4.9 ml
1 Tbsp	=	3 tsp (teaspoons) = about 15 ml
1 fluid oz	=	2 Tbsp (tablespoons) = 29.6 ml (milliliters) = about 30-31 grams depending on the weight of the liquid
1 cup	=	16 Tbsp (tablespoons) = 8 fluid ounces = about 237g of water = 237 ml
1 pint	=	2 cups =16 fluid ounces
1 quart	=	4 cups = 32 fluid ounces = .946 liter
1 liter	=	1000 ml = 1.06 quarts or .85 imperial quart = 4.24 cups = 33.9 fluid ounces = .264 gallion
1 ml	=	1 cc (cubic centimeter) = .001 liter = .034 fluid ounce

Abbreviations

avg	average		mg	milligrams
c	cup		micro	microwave cooked
cal	calories or kilocalories		oz	ounce
Calif	California		pkg	package
choc	chocolate		pkt	packet
ckd	cooked		pnut	peanut
cnd	canned (commercially)		prep	prepared
conc	concentrate		RE	Retinol Equivalents
cond	condensed		reg	regular
crm	cream		RTS	Ready to serve
dm/diam	diameter		sce	sauce
enr	enriched		sm/sml	small
fl oz	fluid ounce		std	standard
f/fr	from		svg	serving
fz/fzn/frzn	frozen (commercially)		t	teaspoon
g	gram(s)		T	tablespoon
imit	imitation		unenr	unenriched
inst	instant		veg	vegetable
IU	International Unit		vol	volume
jce	juice		whpd	whipped
kcal	calorie(s)		wt	weight
lb	pound		w/	with
lg, lrg	large		w/o	without
liq	liquid		≈	about/approximately
marg	margarine		&	and
mcg	micrograms		/	per/or
med	medium			

o   **What do Craving for Sweets and Pre-Menstrual Syndrome (PMS) have in common?**

Lack of Sunlight. Some fascinating connections are being revealed in research by Judith J. Wurtman at the Department of Nutrition and Food Science at the Massachusetts Institute of Technology. For those that are sensitive, the *lack of sunlight* can be connected to a variety of seemingly unconnected items: Carbohydrate craving; Pre-menstrual Syndrome (PMS); and depression.   There are many factors involved in these conditions, but our society's increased employment levels in offices (less sunlight), and increased use of sunglasses, may contribute another set of issues. We'll look forward to future research on this one.

o   **The Protein of Meat
            At a Fraction of the Cost
            At a Fraction of the Fat
            With lots of Dietary Fiber...**

Protein is actually a combination of a variety of "pieces" called amino acids. The protein in meats and milk and eggs are the most complete for all the amino acids and are of the highest quality. People who eat meat needn't worry about protein intake, as long as calories are adequate.

Because the protein in meat is considered complete, it is often used as a standard to compare the protein in plants. Protein in plants is limited in one or more of the essential amino acids, but by combining proteins from different plants, the amino acids can be complemented to more closely resemble the high quality of meat protein. Recent research indicates that the combinations do not have to be at the same meal --protein adequacy can be achieved even if the foods are taken at 5 hour intervals. Children may do better when complementary proteins are taken in the same meal.

More research needs to be done in this area, but it appears even easier than we thought to get enough protein and be well nourished on a vegetarian or vegan diet.

o   **Dark Leaves** of leafy vegetables have more vitamin A, vitamin E, iron and other nutrients-- than the pale leaves.

o **When is an Unsaturated Fat no Longer Unsaturated ?** When it is hydrogenated. Unsaturated fats are usually liquid at room temperature (cooking oils) and can be perishable. Thus, they are often refrigerated and tightly wrapped to keep fresh (margarine). But the stability of the product can be really increased through a process called "hydrogenation," which makes it more solid at room temperature. Unfortunately, this process also reduces the original polyunsaturated fat content, which possibly reduces its health value.

o **About Hydrogenated Fats.** In the late 1970s a significant discovery was made about hydrogenation; a fact whose consequence is still not clearly understood—hydrogenation can create fatty acids that are not normally found in nature. Almost all unsaturated fatty acids occur in nature in a form that resembles a horseshoe (called a cis-fatty acid). When partial hydrogenation occurs, the bend in the horseshoe can change, forcing one side of the "U" to extend the other way. These are called trans-fatty acids.

Frankly, it is not known if the different molecular shape matters are not. But there are questions—can our bodies handle these unusual molecules?. . . in large quantities? Such fatty acids are not made by the body, and they are rare in foods. However, for a better part of this century we have been feeding ourselves a lot of them.

From 1910 to 1978 there was a 27% increase in the total dietary fat intake in the United States, with most of the increase through vegetable oils. For example, butter was largely replaced by margarine, and lard and meat drippings were largely replaced with vegetable shortening. The majority of these oils are modified by partial hydrogenation.

o **Cholesterol — Our body makes it.** From 70% to 85% of the cholesterol in our bodies is made by our liver. If you ate no cholesterol at all, the liver would make sure you had enough, because cholesterol is so important to the body. It is the basis of many of our important hormones.

There is a lot of controversy and confusion about cholesterol, partially because it can be found in the arteries (along with fat) of people with heart disease (arteriosclerosis, or hardening of the arteries). But, blood serum cholesterol is not the same as "dietary cholesterol," and the presence of cholesterol in the artery walls may more likely be the **indicator** of the problem rather than the cause.

Some predictors of heart disease are smoking, high blood pressure, high blood triglycerides (certain fatty acids), and high blood cholesterol. One real culprit may simply be a high fat intake. It has been shown that blood cholesterol will be reduced when saturated fat intake is reduced.

o **On Birth Control Pills?** Studies indicate that birth control pills may cause extra requirements for such nutrients as Vitamin B-6, folacin, zinc, and others. Check with your physician or dietitian for the latest information.

o **Tea and Coffee** have an inhibiting effect on the body's absorption of iron. Scientific studies have shown that a cup of tea with a meal can decrease iron absorption by 64%, and a cup of coffee within one hour of a meal can reduce iron absorption of a hamburger meal by 39%. No decrease in iron absorption occurred if coffee was consumed one hour before the meal, however.

o **Is Fat more Fattening?** It appears so!! Not only does fat contain more than twice the calories of carbohydrates (nine calories per gram for fat versus four calories per gram for carbohydrates), but fat is more easily stored in the body than carbohydrates. For every 100 calories consumed as carbohydrate, it takes about 23 calories to convert it to fat, and about 77 calories get stored as fatty tissue. However, for every 100 calories consumed as fat, it takes only about three calories to convert it to storable form, leaving a whopping 97 calories available for body fat.

o **Carbonated Beverages... Low Calcium?** There has been a 300% increase in the consumption of carbonated drinks in the last 30 years, and many carbonated beverages contain phosphoric acid. There has been some question whether too much phosphorus intake can cause an imbalance, and force excretion of calcium from the body. An interesting study in the July, 1989 Journal of Orthopedic Research gives more information. Over 2,600 female former athletes were surveyed. Those who now drink carbonated beverages showed a 2.28 greater incidence rate of bone fractures after age 40 than those who do not drink soda pop. There was a significant relationship between the amount of soda pop and the number of fractures in women over age 40. At that age, bone mass starts declining, as well as estrogen levels. A low milk diet was also found to increase fracture risk. More studies are needed, of course, but this bears noting.

o **Where is the Flavor in Foods?** Most of the good flavors and smells we associate with foods are primarily found in the fat portions of the food.

o **What do Fat and Fiber have in Common?** Both provide a feeling of "fullness" that is associated with satisfaction. Because there are a lot of different components to fiber, some will provide more feeling of fullness than others.

o **Alter your Diet Slowly.** If you have decided to alter your diet and eat foods with more fiber. . . change slowly. The body adjusts to patterns of eating. If you have not eaten much fiber, suddenly adding copious amounts will find your system unprepared, upset, and unable to handle it well. Start gently and increase amounts over time. Your body will tell you what is right for you.

o **Another High Rating for Dietary Fiber.** A high fiber diet apparently can influence the metabolism of certain nutrients. Reports indicate that persons with mild cases of diabetes have been able to reduce their insulin intakes when on high fiber diets. And studies also show a relationship with reduced serum cholesterol levels.

o **When it Floats...** What is the right amount of dietary fiber? One scientist summed it up simply by saying that when your stool floats, you are getting the right amount of fiber.

o **Satisfying that Hungry Feeling . . . Are All Calories Equal?** Well, maybe not. Some recent studies indicate that different foods **can** make a difference. Foods with table sugar in them—sucrose—(that morning doughnut?) will make you hungrier later than foods without sugar or foods with fruit sugar (fructose). Fructose isn't absorbed as quickly as glucose and produces a much slower rise in blood sugar. As a result, the rise in insulin is also slower and more moderate, and the appetite is more satisfied for a longer period of time.

Foods that cause a **rapid** rise in blood sugar cause a rapid rise in insulin, which can stay at a high level long after the body has absorbed the sugar (glucose). It appears from the studies that the insulin may have more to do with appetite than with the level of blood sugar. (Dr. Judith Rodin, Yale). This may also explain why some diabetics who receive large amounts of insulin complain that they are hungry all the time.

o **The __New__ Carbohydrate.** Because of convenience foods, the mass-scale processing of many foods, and our high sugar consumption, we now have a **new** (to this century) type of carbohydrate. . . THE REFINED CARBOHYDRATE. Where complex carbohydrates are high in nutrients and fiber, and low in calories; the new refined carbohydrates are just the opposite. They are low in nutrients and fiber, and high in calories—and they now provide almost **one-fourth** of the calories in the average American diet.

Vitamin A

From Plants: Precursor vitamin A, Carotenoids
From Animal Sources: Preformed Retinol

Vitamin A plays a vital role in the eye's function, and prevents (or sometimes cures) night blindness. However, only one-thousandth of the Vitamin A in our bodies is in the retina of the eye. The rest is in the blood stream, most body tissues, and stored in the liver.

Vitamin A is essential for healthy skin and the inside surfaces of the mucous membranes—the linings of the mouth, lungs, stomach, intestines, and urinary tract. It plays a role in the body's immunity reactions. It is involved in the growth of bones; maintains the stability of the cell membranes; helps manufacture red blood cells and maintain nerve cell sheaths; prevents some forms of impotence; helps ensure a normal output of thyroxin from the thyroid gland, and more.

The body converts carotenes from plant sources into vitamin A. Carotenes are found primarily in vegetables of the yellow-orange-red color and where there is the dark green pigment chlorophyll that masks the orange color. When the body gets more than it needs from plant sources, it will slow down its conversion of the vitamin, so you cannot overdose on vitamin A from plant sources.

However, when the body gets vitamin A from animal sources, it has already been converted and the body will take it all in and store it. Because it is fat soluble, excess amounts are not routinely flushed out of the system by the kidneys. There is a lot of discussion about overdosing on "A" which we will not repeat here, but basically you really have to work hard to overdose on this vitamin *from food*. It would most likely happen from taking supplements or eating something like polar bear liver. Toxicity from supplements has been reported at intake levels of only 10 times the RDA for one month.

Vitamin Losses

Vitamin A is stable during mild cooking, but is destroyed at high temperatures in the presence of oxygen (air). Although it is a fat soluble vitamin and needs some oil for absorption into the body, mineral oil (sold as a laxative) is not absorbed by the body, so it will absorb the vitamin A and carry it right out of the body. Other vitamin A destroyers are tannic acid from tea; nitrates and benzoate from preserved food; aspirin; barbituates; an artificial lemon flavoring called citral; and ferrous sulfate, an iron supplement.

Recommended Dietary Allowances for Adults

USA RDA: Men, 1000 RE; Women, 800 RE (pregnant, 800 RE; Nursing, 1300 RE)
Canadian RNI: Men, 1000 RE; Women, 800 RE (pregnant, 900 RE; Nursing, 1200 RE)
    The recommended amounts for vitamin A are somewhat related to body weight, which is why the average female recommendation is slightly less than the average male.

The recommended amounts are given in RE (Retinol Equivalents) where 1 RE equals 1 mcg of Retinol (from animal sources) or 6 mcg of Beta-Carotene (from plant sources). Older listings may use international units for Vitamin A, which can lead to confusion. The IU was based on studies that did not allow for poor absorption and bioavailability of the carotenoids. One RE equals 3.3 IU of vitamin activity from Retinol (animal sources) and 10 IU of vitamin activity from Carotenoids (plant sources). This would be 5000 IU for the adult man and 4000 IU for the adult woman.

Food Sources

Vitamin A is widely distributed in dark green leafy or yellow-orange-red vegetables, and some fruits (apricots, cantaloupe, mango, papaya). It is present in fairly large quantities in liver, and many margarines and milks are fortified with vitamin A. It is found in small amounts in fat tissue, egg yolks and cream.

1

	Wt (g)	Svg	Cal	Vit A (RE) 100 Cal	Vit A (RE) Svg

Baked Goods

Grains and flours are not sources of vitamin A. It is present in baked goods from egg, milk or fruit ingredients.

	Wt (g)	Svg	Cal	100 Cal	Svg
Apple crisp, 3" x 3"	78	1 pce	146	45	65
Bagel, egg, 3.5" diam.	68	1 ea	180	4	8
CAKES, piece = 1/16th cake unless otherwise stated:					
Boston cream pie, 1/8th	120	1 pce	260	27	70
Carrot cake, w/cream cheese frosting, 2.5" x 3"	112	1 pce	406	163	663
Cheesecake, 1/8th	103	1 pce	300	37	110
Chocolate w/choc frstng	69	1 pce	235	13	31
Coffee cake, 2.4" x 2.8"	72	1 pce	230	14	32
Gingerbread, recipe, 3" x 3"	110	1 pce	351	38	134
Pound cake, 1/2" slice	30	1 pce	115	44	51
Sheet cake, plain, 3" x 3"	86	1 pce	315	13	41
Sponge cake, 1/12 cake	66	1 pce	194	20	39
White, commercial, w/frosting	74	1 pce	278	4	12
Yellow, chocolate frosting	69	1 pce	240	12	30
Cheese puffs (Cheetos)	28	1 oz	158	16	26
Cherry crisp, 3" x 3"	138	1 pce	157	92	144
Chips: see corn & tortilla this section; potato chips under Vegetables.					
COOKIES:					
Butter cookies	25	5 ea	115	36	41
Lady fingers	44	4 ea	158	47	75
Oatmeal raisin	52	4 ea	245	5	12
Shortbread, homemade, large	28	2 ea	145	61	89
Snickerdoodles	20	1 ea	110	39	43
Vanilla wafer cookies	40	10 ea	185	8	14
Cornbread, muffin (yellow)	45	1 ea	145	11	16
Corn chips	28	1 oz	155	7	11
Crackers, cheese	10	10 ea	50	10	5
Croissant, 4.5" x 4" x 2"	57	1 ea	235	6	13
Cream puff, custard filling	110	1 ea	280	72	202
Danish pastry, average	61	1 ea	228	6	14
Eclair, custard filled, choc icing	94	1 ea	262	61	161
MUFFINS:					
Blueberry, from mix	45	1 ea	138	8	11
Bran, wheat or oat:					
Recipe	45	1 ea	125	24	30
From mix	45	1 ea	140	10	14
Cornbread, muffin, made w/yellow cornmeal	45	1 ea	145	11	16
PIES, piece = 1/6 of 9" pie:					
Apple pie	158	1 pce	405	1	5
Banana cream, recipe	198	1 pce	319	12	37
Blueberry	158	1 pce	380	4	14
Cherry	158	1 pce	410	17	70
Chocolate cream	175	1 pce	311	11	34
Coconut cream	172	1 pce	343	10	34
Coconut custard	165	1 pce	384	25	96
Cream pie, commercial	152	1 pce	455	14	65
Custard pie	152	1 pce	293	33	96
Lemon meringue	140	1 pce	355	19	66
Peach pie	158	1 pce	405	28	115
Pecan pie	138	1 pce	583	7	41
Pumpkin	200	1 pce	367	507	1861
Strawberry chiffon, recipe	162	1 pce	372	35	129
Peach crisp, 3" x 3"	139	1 pce	166	63	104
PopTart type toaster pastry, fortified	54	1 ea	210	71	150
Tortilla, corn, 6" diam, fried	30	1 ea	87	9	8
Tortilla chips, flavored	28	1 oz	140	10	14

WAFFLES

	Wt (g)	Svg	Cal	100 Cal	Svg
Recipe, 7" diam.	75	1 ea	245	16	39
From mix, 7" diam.	75	1 ea	205	24	49
Frozen, 4" diam.	35	1 ea	98	134	131

Dairy & Dairy Products

	Wt (g)	Svg	Cal	100 Cal	Svg
CHEESE (1.5" cube ≈ 1 oz):					
American processed cheese	28	1 oz	106	77	82
American cheese food:					
Cold pack	28	1 oz	94	51	48
Jar (pasteurized)	28	1 oz	93	67	62
American cheese spread	28	1 oz	82	66	54
Blue cheese	28	1 oz	100	65	65
Brick cheese	28	1 oz	105	82	86
Brie cheese	28	1 oz	95	60	57
Camembert	28	1 oz	85	84	71
Caraway	28	1 oz	107	77	82
Cheddar cheese	28	1 oz	114	75	86
Colby	28	1 oz	112	70	78
Cottage cheese:					
Lowfat 1%	226	1 c	205	22	45
Creamed (4.5% fat)	225	1 c	225	46	104
Dry curd (<.5% fat)	145	1 c	123	10	12
Cream cheese (1 T = 15g)	28	1 oz	99	125	124
Edam cheese	28	1 oz	101	71	72
Feta cheese	28	1 oz	75	48	36
Fontina	28	1 oz	110	91	100
Gjetost	28	1 oz	132	57	75
Gorgonzola	28	1 oz	111	93	103
Gouda	28	1 oz	101	49	49
Gruyere	28	1 oz	117	84	98
Parmesan, grated (1 T = 5g)	28	1 oz	129	38	49
Limburger	28	1 oz	93	124	115
Monterey jack	28	1 oz	106	76	81
Mozzarella, part skim, low moisture	28	1 oz	80	68	54
Muenster	28	1 oz	104	87	90
Neufchatel	28	1 oz	74	101	75
Port du salut	28	1 oz	100	105	105
Provolone	28	1 oz	100	75	75
Ricotta, part skim	246	1 c	340	82	278
Ricotta, whole milk	246	1 c	428	77	330
Romano	28	1 oz	110	36	40
Roquefort	28	1 oz	105	90	95
Swiss cheese	28	1 oz	107	67	72
CREAM, SWEET, fluid:					
Table or coffee cream	15	1 T	30	90	27
Half and half	15	1 T	20	80	16
Light whipping cream	15	1 T	44	100	44
Heavy whipping cream	15	1 T	51	124	63
CREAM, SWEET, whipped:					
Cream, sweet, whipped	119	1 c	410	122	501
Pressurized	60	1 c	154	81	124
CREAM, SOUR:					
Cultured, dairy	14	1 T	30	90	27
Half and half, dairy	15	1 T	20	85	17
Cream, sour, Imitation, non-dairy	230	1 c	479	0	0
CREAM SUBSTITUTES, non-dairy:					
Coffee whitener, liquid	15	1 T	20	7	1.4
Coffee whitener, powdered	2	1 tsp	11	4	0.4
Dessert toppings, non-dairy:					
Frozen (like Coolwhip)	75	1 c	239	27	65
Dessert powder, dry	43	1.5 oz	245	19	46
Pressurized topping	70	1 c	185	18	33

	Wt (g)	Svg	Cal	Vit A (RE) 100 Cal	Vit A (RE) Svg
MILK (cow):					
Skim	245	1 c	86	173	149
Lowfat 1%	244	1 c	102	142	145
Lowfat 2%	244	1 c	121	116	140
Whole (3.3%)	244	1 c	150	51	76
Buttermilk	245	1 c	99	20	20
Canned:					
Skim, evaporated	255	1 c	200	150	300
Whole, evaporated	252	1 c	340	40	136
Sweetened, condensed	306	1 c	982	25	248
Dried instant nonfat	68	1 c	244	198	483
Milk (other):					
Goat milk	244	1 c	168	82	137
Human milk	246	1 c	171	92	157
Soy milk	240	1 c	79	10	8
MILK BEVERAGES &					
beverage mixes:					
Eggnog, commercial	254	1 c	342	59	203
Instant Breakfast powder	37	1 env	130	135	175
Milkshakes, 10 fl oz = 1.25 c:					
Chocolate	283	1.25 c	360	18	64
Strawberry	283	1.25 c	319	26	83
Vanilla	283	1.25 c	314	29	90
MILK DESSERTS:					
Custard, baked	265	1 c	305	48	146
Ice cream, vanilla:					
Regular	133	1 c	269	49	133
Rich	148	1 c	349	63	219
Soft serve	173	1 c	377	53	199
Ice milk, regular	131	1 c	184	28	52
Ice milk, soft serve	175	1 c	223	20	44
Puddings, prepared:					
Assorted flavors, regular	260	1 c	300	22	67
Canned (5 oz can ≈ 1/2 c)	142	1 can	205	15	31
Pudding pops, average	57	1 ea	97	26	25
Sherbet (2%)	193	1 c	270	14	39
Yogurt, frozen, average	174	1 c	220	16	36
YOGURT:					
Lowfat, plain	227	1 c	144	25	36
Lowfat, with fruit	227	1 c	231	11	25
Nonfat	227	1 c	127	4	5
Whole milk	227	1 c	138	49	68

Eggs

	Wt (g)	Svg	Cal	100 Cal	Svg
EGG, chicken, raw or cooked					
(Vitamin A is in the yolk):					
Whole egg	50	1 ea	75	127	97
Yolk only	16.6	1 ea	59	164	97

Fruits & Fruit Juices

	Wt (g)	Svg	Cal	100 Cal	Svg
Acerola juice, fresh	242	1 c	51	241	123
APPLES w/peel, 2.75" dm	138	1 ea	80	9	7
Applesauce, unsweetened	244	1 c	106	7	7
APRICOTS:					
Fresh, whole, pitted	106	3 ea	51	325	166
Canned halves, juice pack	84	3 ea	40	355	142
Canned halves, heavy syrup	85	3 ea	70	150	105
Dried halves	35	10 ea	83	305	253
Apricot nectar, canned	251	1 c	141	234	330
Avocado, whole:					
California	173	1 ea	305	35	106
Florida	304	1 ea	340	55	186
Banana, 8.75", 176g w/peel	114	1 ea	105	9	9

	Wt (g)	Svg	Cal	Vit A (RE) 100 Cal	Vit A (RE) Svg
Blackberries:					
Fresh berries	144	1 c	74	32	24
Frozen, unthawed	151	1 c	97	18	17
Canned	256	1 c	236	24	56
Blueberries:					
Fresh berries	145	1 c	82	18	15
Frozen, unthawed	155	1 c	78	16	13
Canned	256	1 c	225	7	16
Boysenberries:					
Frozen, unthawed	132	1 c	66	13	9
Canned	256	1 c	225	5	10
Cantaloupe: see Melon.					
Cherries, sour:					
Frozen	155	1 c	72	188	135
Canned	244	1 c	90	204	184
CHERRIES, SWEET:					
Fresh, whole, pitted	68	10 ea	49	31	15
Frozen, thawed	259	1 c	232	21	49
Canned	257	1 c	213	19	40
Cranberry sauce, canned	277	1 c	419	1	6
Currants:					
Fresh, black	112	1 c	71	37	26
Fresh, red or white	112	1 c	63	21	13
Dried, Zante	144	1 c	407	3	10
Dates, chopped, pitted	178	1 c	489	2	9
Figs, fresh, medium	50	1 ea	37	19	7
Figs, dried	19	1 ea	48	5	3
Fruit cocktail, canned:					
Juice pack	248	1 c	115	66	76
Heavy syrup	255	1 c	185	28	52
GRAPEFRUIT, 241g w/rind:					
Pink or red half	123	1 ea	37	86	32
White half	118	1 ea	39	3	1
Grapes, Thompson 10=50g	160	1 c	114	10	12
Kiwi fruit	76	1 ea	46	28	13
Lemon juice, fresh	244	1 c	60	8	5
Loganberries, fresh	150	1 c	105	11	12
Mandarin oranges, canned	252	1 c	155	137	212
Mango, sliced	165	1 c	108	594	642
Melons: also see Watermelon.					
Cantaloupe cubes	160	1 c	57	905	516
Casaba cubes	170	1 c	45	11	5
Honeydew cubes	170	1 c	60	11	7
Frozen, mixed melon balls	173	1 c	55	558	307
Nectarine, fresh	136	1 med	67	149	100
ORANGE, 2-5/8" (180g w/peel)	131	1 ea	60	45	27
Orange juice:					
Fresh juice	248	1 c	111	45	50
From frozen conc.	249	1 c	110	17	19
Canned	249	1 c	105	42	44
Orange grapefruit jce, canned	247	1 c	105	28	29
Papaya nectar, canned	250	1 c	142	20	28
Papaya, whole, 454g w/refuse	304	1 ea	117	523	612
Passion fruit juice:					
Purple	247	1 c	126	140	177
Yellow	247	1 c	149	399	595
PEACHES, peeled:					
Fresh, pitted, 2.5" dm	87	1 ea	37	127	47
Frozen, thawed	250	1 c	235	30	71
Cnd halves, juice pack	77	1 ea	34	86	29
Cnd halves, heavy syrup	81	1 ea	60	45	27
Dried halves	130	10 ea	311	90	281
Peach nectar, canned	249	1 c	134	48	64
PEAR, Bartlett	166	1 ea	98	3	3
Persimmon, fresh	168	1 ea	118	308	364

ESHA Research

	Wt (g)	Svg	Cal	Vit A (RE) 100 Cal	Svg
Fruits & Fruit Juices continued:					
PINEAPPLE pieces,					
canned, juice pack	250	1 c	150	6	10
Plantain:					
Fresh slices	148	1 c	181	92	167
Cooked	154	1 c	179	78	140
Plums:					
Fresh, 2-1/8" diam	66	1 ea	36	59	21
Cnd, juice pack	95	3 ea	55	174	96
Cnd, heavy syrup	110	3 ea	98	29	29
Prunes, dried, whole, pitted	84	10 ea	201	83	167
Raisins, packed measure	165	1 c	494	1	1
Raspberries:					
Fresh berries	123	1 c	60	27	16
Frozen, thawed	250	1 c	255	6	15
Canned with liquid	256	1 c	234	4	9
Raspberry juice, fresh	240	1 c	98	24	24
Rhubarb, cooked	240	1 c	279	6	17
STRAWBERRIES:					
Fresh berries	160	1 c	48	9	4.5
Frozen, unsweetened	149	1 c	52	13	7
Frozen, thawed, sweetened	255	1 c	245	2	6
Tangerine	84	1 ea	37	209	77
Tangerine juice:					
From frozen	241	1 c	110	125	138
Canned	249	1 c	125	84	105
Watermelon, diced	160	1 c	50	117	59

Grains & Grain Products

Vitamin A is not present in grains (except for corn). Breakfast cereals are often fortified with vitamin A. Check label.

	Wt (g)	Svg	Cal	100 Cal	Svg
Corn flour	117	1 c	428	9	40
Corn grits, yellow, cooked (white grits have no vit. A)	242	1 c	146	10	15
Cornmeal, dry:					
Degermed, enriched, dry	138	1 c	502	12	61
Nearly whole, bolted	122	1 c	440	14	62
Egg noodles, cooked	160	1 c	200	17	34
Soy flour, full fat, stirred	42	1/2 c	183	3	5
Spinach noodles, cooked	160	1 c	178	4	8
Whole wheat, sprouted	106	1 c	236	3	6

Meats: Fish & Shellfish

	Wt (g)	Svg	Cal	100 Cal	Svg
Bluefish:					
Baked/broiled	100	3.5 oz	159	80	127
Fried in crumbs	100	3.5 oz	205	59	120
Catfish, cornmeal fried	100	3.5 oz	229	4	8
Clams:					
Canned, drained	160	1 c	236	116	274
Breaded, fried small	188	20 ea	379	45	170
Steamed	90	20 ea	133	116	154
Cod, baked or broiled	85	3 oz	89	13	12
CRAB meat:					
Alaska King, ckd leg	134	1 ea	129	9	12
Blue crab, ckd	135	1 c	138	14	20
Blue crab, canned	135	1 c	133	11	14
Dungeness crab, ckd meat	101	3/4 c	85	16	14
Crab, imitation, fr/Surimi	85	3 oz	87	7	6
Eel:					
Baked or broiled	100	3.5 oz	236	481	1136
Smoked	100	3.5 oz	330	227	750
Fish cakes, fried fr/frzn	100	3.5 oz	213	9	20
Fish sticks, heated fr/frzn	57	2 ea	155	12	18

	Wt (g)	Svg	Cal	Vit A (RE) 100 Cal	Svg
Gefiltefish, commercial	42	1 pce	35	31	11
Haddock:					
Baked/broiled	85	3 oz	95	17	16
Breaded, fried	85	3 oz	175	11	20
Smoked	100	3.5 oz	116	19	22
HALIBUT:					
Baked/broiled	85	3 oz	119	39	46
Smoked	100	3.5 oz	224	20	45
Herring:					
Baked/broiled	100	3.5 oz	203	15	31
Canned w/liquid	100	3.5 oz	208	19	39
Smoked or kippered	100	3.5 oz	217	18	39
Pickled, 1 piece ≈ 15g	100	3.5 oz	262	98	258
Lobster, meat only, cooked	145	1 c	142	27	38
Mackerel:					
Cooked	100	3.5 oz	262	21	54
Canned, Jack, #1 tall can	361	1 can	563	83	470
Mullet, baked/broiled	85	3 oz	127	28	36
Ocean Perch:					
Baked or broiled	100	3.5 oz	121	12	14
Breaded, fried	85	3 oz	185	11	20
OYSTERS:					
Raw, fresh, Eastern	248	1 c	170	131	223
Raw, fresh, Pacific	248	1 c	200	112	223
Breaded, fried	88	6 ea	173	50	86
Pike, Northern, baked/broiled	100	3.5 oz	113	21	24
Pollock, baked/broiled	100	3.5 oz	99	9	9
Rockfish, baked /broiled	100	3.5 oz	121	55	66
SALMON, cooked:					
Average, baked/broiled	85	3 oz	183	29	53
Chinook, smoked	85	3 oz	99	22	22
Coho, poached	100	3.5 oz	185	51	95
Sockeye, cooked fr/fresh	100	3.5 oz	216	29	63
Canned, Atlantic, small can	220	1 can	281	9	24
Chum, #1 can	369	1 can	521	13	67
Pink, #1 can	454	1 can	631	12	75
Sockeye, #1 can	369	1 can	566	34	194
Sardines, canned, drained:					
In oil, Atlantic (2 = 24g)	92	1 can	192	32	62
In sauce, Pacific (1 = 38g)	100	3.5 oz	178	39	70
Scallops:					
Breaded, fried	93	6 ea	200	11	21
Steamed	100	3.5 oz	113	19	21
Sea Bass, baked/broiled	100	3.5 oz	124	52	64
Seatrout (Steelhead), cooked	100	3.5 oz	131	32	42
Shark, batter-fried	85	3 oz	194	24	46
SHRIMP:					
Boiled	100	4.5 ea	99	18	18
Breaded, fried	90	12 ea	218	11	24
Canned	85	2/3 c	102	15	15
Snapper, baked/broiled	100	3.5 oz	128	9	12
SOLE (Flounder):					
Baked or broiled	85	3 oz	99	10	10
Fried in batter	100	3.5 oz	214	6	12
Surimi: see Crab, imitation.					
Sturgeon:					
Ckd from fresh	85	3 oz	115	179	206
Smoked	85	3 oz	147	112	165
Swordfish, baked/broiled	100	3.5 oz	155	26	41
Trout, baked or broiled	85	3 oz	129	15	19
TUNA:					
Bluefin, baked or broiled	85	3 oz	157	409	642
Canned, small can, drained:					
Light, oil pack	171	1 can	339	12	40
Light, water pack	165	1 can	216	29	62

Meats
Beef, Lamb, Pork, Veal

	Wt (g)	Svg	Cal	Vit A (RE) 100 Cal	Vit A (RE) Svg
Beef kidney, simmered	85	3 oz	122	255	312
Beef liver:					
Braised	85	3 oz	137	6577	9011
Pan-fried	85	3 oz	184	4956	9119
Lamb liver, braised	85	3 oz	187	3404	6366
Pork liver, braised	85	3 oz	141	3255	4589
Veal liver:					
Braised	85	3 oz	140	4887	6842
Pan-fried	85	3 oz	208	2300	4784

Meats: Poultry

	Wt (g)	Svg	Cal	Vit A (RE) 100 Cal	Vit A (RE) Svg
CHICKEN:					
All types:					
Fried	140	1 c	307	8	25
Roasted	140	1 c	266	8	22
Stewed	140	1 c	248	8	21
Canned, with broth	142	5 oz	235	20	48
Dark meat:					
Fried	85	1 oz	203	10	20
Roasted	85	1 oz	174	10	18
Stewed	85	1 oz	163	11	18
Light meat:					
Fried	85	3 oz	163	5	8
Roasted	85	3 oz	147	5	7
Stewed	85	3 oz	135	5	7
Breast, meat & skin:					
Batter-fried	140	1 ea	364	8	28
Flour-fried	98	1 ea	218	7	15
Roasted	98	1 ea	193	13	26
Breast meat, cooked:					
Batter-fried	72	1 ea	193	10	19
Flour-fried	49	1 ea	120	10	12
Roasted	52	1 ea	112	13	15
Drumstick, meat, cooked	43	1 ea	79	10	8
Thigh, meat & skin:					
Batter-fried	86	1 ea	238	11	25
Flour-fried	62	1 ea	162	11	18
Roasted	62	1 ea	153	20	30
Thigh meat, cooked	52	1 ea	111	10	10
Wing, meat & skin:					
Batter-fried	49	1 ea	159	11	17
Flour-fried	32	1 ea	103	12	12
Roasted	34	1 ea	99	16	16
Chicken gizzard, cooked	22	1 ea	34	35	12
Chicken liver, cooked	20	1 ea	31	3171	983
DUCK, domestic, roasted:					
Meat and skin	85	3 oz	286	19	54
Meat only	85	3 oz	171	11	20
GOOSE, domestic, roasted:					
Meat and skin	85	3 oz	259	7	18
Meat only	85	3 oz	202	5	10
Liver, raw	94	1 ea	125	6982	8728
TURKEY:					
Gizzard	67	1 ea	109	34	37
Liver	75	1 ea	127	2209	2806

Meats: Sausages & Lunchmeats

	Wt (g)	Svg	Cal	Vit A (RE) 100 Cal	Vit A (RE) Svg
Braunschweiger sausage	57	2 oz	205	1174	2406
Chicken roll, light meat	57	2 oz	90	16	14

	Wt (g)	Svg	Cal	Vit A (RE) 100 Cal	Vit A (RE) Svg
FRANKFURTERS (hotdogs):					
Beef/pork, 10/package	45	1 link	145	0	0
Chicken, 10/package	45	1 link	115	15	17
Turkey, 10/package	45	1 link	102	17	17
Ham and cheese roll	57	2 oz	147	28	41
Liverwurst, pork	18	1 pce	59	1236	729

Mixed Dishes & Fast Foods

	Wt (g)	Svg	Cal	Vit A (RE) 100 Cal	Vit A (RE) Svg
Beef, macaroni & tomato sauce, recipe	226	1 c	189	59	111
Beef pot pie, fr/frozen	234	1 pie	426	44	187
BURRITO:					
Bean burrito	174	1 ea	322	18	58
Beef burrito	177	1 ea	463	12	55
Deluxe, beef and bean	198	1 ea	424	19	82
Cheese soufflé, recipe	112	1 c	221	84	185
Chicken & noodles, recipe	240	1 c	365	36	130
Chicken a la king, recipe	245	1 c	470	58	272
Chicken chow mein, recipe	250	1 c	255	20	50
Chicken curry, recipe	337	1.5 c	305	50	153
Chicken salad	78	1/2 c	266	12	31
Chili w/beans, canned	255	1 c	286	30	86
Chop suey w/ beef and pork	250	1 c	300	20	60
Cole slaw	120	1 c	84	117	98
Corn pudding	250	1 c	271	33	89
Egg roll, chicken	100	1 ea	242	10	23
Egg salad	183	1 c	438	68	300
ENCHILADA:					
Beef enchilada	120	1 ea	292	28	82
Chicken	120	1 ea	269	32	87
Cheese	120	1 ea	330	51	168
French toast, recipe	65	1 pce	123	46	57
LASAGNA:					
With meat, recipe	245	1 pce	398	42	168
Without meat, recipe	218	1 pce	316	53	168
Macaroni & cheese:					
Recipe	200	1 c	430	54	232
Canned	240	1 c	230	31	72
Macaroni salad, w/o cheese	141	1 c	371	11	40
Manicotti, w/meat & tomato sauce, recipe	233	1 ea	320	60	191
Meat loaf	85	1 pce	200	12	26
Moussaka, lamb & eggplant	250	1 c	250	50	125
PIZZA, cheese:					
Regular crust, 1/8 of 15"	120	1 pce	290	37	106
Thick crust, 1/2 of 10"	208	1 pce	519	28	146
Potato salad, w/mayo & egg	250	1 c	358	23	83
Quiche Lorraine, 1/8 pie	176	1 pce	600	76	454
Ravioli, beef	28	2 ea	33	55	18
SANDWICHES, Fast food items:					
Cheeseburger, 3 oz beef	112	1 ea	300	22	65
Cheeseburger, 4 oz beef	194	1 ea	524	24	128
Chicken patty sandwich	157	1 ea	436	4	16
English muffin, w/egg, cheese & bacon	138	1 ea	360	44	160
Fish sandwich:					
Large, without cheese	170	1 ea	470	3	15
Regular, with cheese	140	1 ea	420	6	25
Hamburger, 3 oz beef	98	1 ea	245	6	14
Hamburger, 4 oz beef	174	1 ea	445	6	28

ESHA Research

Mixed Dishes & Fast Foods, continued:	Wt (g)	Svg	Cal	Vit A (RE) 100 Cal	Svg
SANDWICHES on					
part whole wheat bread:					
Avocado, cheese, sprouts					
& tomato	195	1 ea	432	37	160
Grilled cheese	117	1 ea	393	54	214
Egg salad sandwich	111	1 ea	319	28	90
SPAGHETTI, pasta &					
tomato sauce w/cheese:					
Recipe	250	1 c	260	54	140
Canned	250	1 c	190	63	120
Spinach soufflé, recipe	245	1 c	220	259	569
Taco:					
Beef	78	1 ea	207	13	27
Chicken	78	1 ea	172	19	34
Tostada:					
With beans and beef	192	1 ea	332	40	132
With beans and chicken	157	1 ea	249	32	81
Refried beans only	157	1 ea	212	35	74
Tuna noodle casserole, recipe	202	1 c	251	14	34
Tuna salad	205	1 c	383	14	54
Turkey pot pie, from frozen	233	1 ea	416	35	145

Nuts & Seeds

	Wt (g)	Svg	Cal	Vit A (RE) 100 Cal	Svg
Filberts (hazelnuts) chpd	115	1 c	727	1	8
Pecans, dried, chpd	119	1 c	794	2	15
Pistachios, dry (no shells)	128	1 c	739	4	30
Pumpkin /squash seed					
kernels, dried	138	1 c	747	7	53
Sunflower seeds, dry/oil rsted	140	1 c	825	1	7
Walnuts, chopped:					
Black	125	1 c	759	5	37
English	120	1 c	770	2	15

Soups, Sauces & Gravies

	Wt (g)	Svg	Cal	Vit A (RE) 100 Cal	Svg
Gravy, chicken:					
Recipe	130	1/2 c	163	13	21
Canned	119	1/2 c	95	140	132
SAUCES:					
Spaghetti sauce, plain:					
Recipe	220	1 c	179	131	235
Canned	249	1 c	272	113	306
White sauce, medium:					
Recipe	250	1 c	395	86	340
From mix, w/milk	264	1 c	240	38	92
SOUPS: Soups are canned unless					
otherwise stated. For soups prep.					
w/milk, assume whole milk.					
RTS = Ready To Serve.					
Cheese soup, prep. w/milk	251	1 c	230	64	147
Chicken broth,					
Dry mix packet	6	1 ea	16	56	9
Chicken, cream of					
prepared w/milk	248	1 c	191	49	94
CHICKEN NOODLE:					
From canned	240	1 c	114	107	122
Canned, chunky, RTS	251	1 c	178	73	130
From dry	252	1 c	53	12	6
Chicken & rice	241	1 c	60	110	66
Chicken & vegetable					
Canned	241	1 c	74	359	266
Canned, chunky, RTS	240	1 c	167	359	599
Chili beef soup	250	1 c	169	298	503

Clam chowder:	Wt (g)	Svg	Cal	Vit A (RE) 100 Cal	Svg
Manhatten style	244	1 c	78	118	92
New England style	248	1 c	163	25	40
Minestrone	241	1 c	80	293	234
Mushroom, cream of, cond	248	1 c	205	19	38
Split pea soup w/ham:					
Canned	253	1 c	189	23	44
Chunky, RTS, w/ham	240	1 c	184	265	487
Prepared fr/dry	255	1 c	133	4	5
TOMATO, cream of:					
Prepared w/milk	248	1 c	160	68	109
Prepared w/water	244	1 c	86	80	69
Prepared fr/dry	265	1 c	102	81	83
Turkey vegetable	241	1 c	74	330	244
Vegetable beef, canned	244	1 c	79	239	189
Vegetable beef, fr/dry	253	1 c	53	45	24
Vegetarian vegetable	241	1 c	70	430	301

Vegetables & Legumes

	Wt (g)	Svg	Cal	Vit A (RE) 100 Cal	Svg
Alfalfa sprouts	33	1 c	10	51	5
Amaranth leaves:					
Fresh, chopped	28	1 c	7	1122	83
Cooked	132	1 c	28	1307	366
Artichoke, globe, cooked					
300g with refuse	120	1 ea	53	32	17
Artichoke hearts:					
Ckd from frozen-pkg	240	9 oz	108	36	39
Marinated-jar	170	6 oz	168	17	28
Asparagus, pieces:					
Fresh, uncooked	67	1/2 c	15	407	61
Ckd from fresh	90	1/2 c	23	332	75
Ckd from frozen	180	1 c	50	292	147
Canned, drained	121	1/2 c	16	425	68
BEANS: also see garbanzos,					
lentils, soybeans.					
Baked beans (dry white					
beans w/ spices & sauce):					
Canned, plain/vegetarian	254	1 c	235	18	43
Canned w/pork, sweet sce	253	1 c	282	10	29
Canned w/pork, tomato sce	253	1 c	247	13	31
Broadbeans, canned	256	1 c	183	1	3
Green (snap) beans:					
Fresh, uncooked	110	1 c	34	216	74
Ckd from fresh	125	1 c	44	189	83
Ckd from frozen					
Canned, drained	135	1 c	26	181	47
Lima beans:					
Ckd from frozen	88	1/2 c	90	17	16
Canned, drained	170	1 c	164	20	32
Winged, ckd f/fresh	62	1 c	23	23	5
Yardlong, ckd fr/fresh	104	1 c	49	96	47
Yellow wax beans:					
Ckd fr/frozen	135	1 c	36	42	15
Canned, drained	135	1 c	26	55	14
Beet greens:					
Fresh	19	1/2 c	4	2900	116
Cooked, drained	144	1 c	40	1835	734
Borage:					
Fresh/raw	44	1/2 c	9	2056	185
Cooked	100	3.5 oz	25	1752	438
Broccoli, chopped:					
Fresh chopped	88	1 c	24	567	136
Ckd from fresh	156	1 c	46	478	220
Ckd from frozen	184	1 c	51	533	272
Brussels sprouts:					
Ckd from fresh	156	1 c	60	187	112
Ckd from frozen	155	1 c	65	140	91

	Wt (g)	Svg	Cal	Vit A 100 Cal	(RE) Svg		Wt (g)	Svg	Cal	Vit A 100 Cal	(RE) Svg
Vegetables & Legumes, continued:						Mustard greens:					
CABBAGE:						Fresh greens	56	1 c	15	2034	297
Common, fresh, shredded	70	1 c	16	55	9	Ckd from fresh	140	1 c	21	2019	424
Common, cooked	150	1 c	32	41	13	Ckd from frozen	150	1 c	29	2354	671
Bok-choy, fresh, shredded	70	1 c	9	2333	210	Okra, cooked:					
Bok-choy, cooked	170	1 c	20	2185	437	Pods, ckd fr/fresh	85	8 ea	27	180	49
Savoy, fresh, shredded	70	1 c	20	350	70	Sliced, ckd fr/frozen	92	1/2 c	34	138	47
Savoy, cooked	145	1 c	35	369	129	Onion, spring, chopped	50	1/2 c	13	1923	250
						Parsley:					
CARROTS:						Fresh, sprigs	10	10 ea	3	1576	52
Fresh, 7.5" x 1-1/8"	72	1 ea	31	6532	2025	Freeze-dried	1.4	1/4 c	4	2225	89
Fresh, grated	55	1/2 c	24	6446	1547						
Ckd fr/ fresh, slices	78	1/2 c	35	5471	1915	**PEAS:**					
Ckd fr/ frozen	73	1/2 c	26	4969	1292	Black-eyed peas:					
Canned, w/ liquid	123	1/2 c	28	5786	1620	Ckd from fresh	165	1 c	179	59	105
Canned, drained	73	1/2 c	17	5918	1006	Ckd from frozen	170	1 c	224	6	13
Carrot juice	123	1/2 c	49	6447	3159	Green peas, edible-pods:					
Celery:						Fresh, uncooked	145	1 c	61	49	30
Fresh, large outer stalk	40	1 ea	6	78	5	Ckd from fresh	160	1 c	67	45	30
Cooked, diced	150	1 c	22	74	16	Ckd from frozen	80	1/2 c	42	36	15
Chard, Swiss:						Green peas:					
Fresh	36	1 c	7	3787	259	Fresh, uncooked	146	1 c	118	114	134
Cooked	175	1 c	35	3423	1198	Ckd from fresh	160	1 c	134	102	137
Chicory greens, chopped	90	1/2 c	21	1714	360	Ckd from frozen	80	1/2 c	63	122	77
Chrysanthemum garland:						Canned, drained	85	1/2 c	59	110	65
Fresh	25	1 c	4	9175	367	Canned, with liquid	124	1/2 c	61	77	47
Cooked	100	1 c	20	2525	505	Split peas, ckd fr/dry	196	1 c	231	.6	1
Collards:						Peas & carrots, fr/frozen	70	1/2 c	37	1797	665
Fresh collards	36	1 c	7	1754	120	Peas, mature, sprouted:					
Ckd from fresh	145	1 c	20	1586	322	Fresh sprouts	120	1 c	154	13	20
Ckd from frozen	170	1 c	61	1667	1017	Cooked	100	3.5 oz	118	9	11
CORN, yellow only (white corn has little Vitamin A):						**PEPPERS, HOT, green** chili:					
Fresh kernels, uncooked	77	1/2 c	66	33	22	Fresh, chopped	75	1/2 c	30	193	58
Ckd from fresh	82	1/2 c	89	20	18	Canned, hot green	68	1/2 c	17	247	42
Ckd from frozen	82	1/2 c	67	30	20	Jalapeño, canned, chopped	68	1/2 c	17	682	116
Canned, drained	82	1/2 c	66	19	13	**PEPPERS, HOT, red** chili:					
Canned, vacuum pack	210	1 c	166	31	51	Fresh, chopped	75	1/2 c	30	2687	806
Canned, cream style	128	1/2 c	93	13	12	Canned, w/liquid	68	1/2 c	17	4759	809
Dandelion greens:											
Fresh, chopped	55	1 c	25	3080	770	**PEPPERS, SWEET, green:**					
Cooked	105	1 c	35	3511	1229	Fresh, chopped	50	1/2 c	12	221	27
Dock (sorrel greens):						Ckd from fresh, chopped	68	1/2 c	12	220	26
Fresh	133	1 c	29	1834	532	Ckd from frozen	68	1/2 c	12	161	20
Cooked	100	1 c	20	1735	347	**PEPPERS, SWEET, red:**					
Eggplant, cooked cubes	160	1 c	45	23	10	Fresh, chopped	50	1/2 c	12	2375	285
Endive, fresh, chopped	25	1/2 c	4	1283	51	Ckd from fresh, pod	73	1 ea	13	2115	275
Escarole (curly endive)	50	1 c	9	1212	103	Pimento, canned	57	2 oz	15	873	131
Garbanzo (chickpeas):						Poi, two finger	240	1 c	269	2	5
Cooked from dry	164	1 c	269	2	4						
Canned (solids and liquid)	240	1 c	285	2	6	**POTATOES** have no Vitamin A.					
Garden cress:						Pumpkin, mashed:					
Fresh	25	1/2 c	8	279	22	Cooked from fresh	245	1 c	50	530	265
Cooked	135	1 c	31	335	104	Canned	123	1/2 c	42	6433	2702
Kale, chopped:						Purslane:					
Fresh, chopped	67	1 c	33	1806	596	Fresh, chopped	43	1 c	7	811	57
Ckd fr/fresh	65	1/2 c	21	2290	481	Boiled	115	1 c	21	1014	213
Ckd fr/ frozen	65	1/2 c	20	2065	413	Seaweed:					
Lambquarters, chopped						Fresh, Lavar	28	1 oz	10	1485	147
Fresh	56	1 c	24	2708	650	Dried, Spirulina	28	1 oz	82	19	16
Cooked	180	1 c	58	3010	1746						
Leeks, fresh, chopped	104	1 c	63	16	10	**Soybeans:**					
Lentils, cooked from dry	99	1/2 c	115	1	1	Ckd fr/fresh veg	90	1/2 c	127	11.0	14
						Ckd fr/dry legume	86	1/2 c	149	0.5	.8
LETTUCE, fresh, chopped:											
Butterhead	56	1 c	7.3	746	54	**SOYBEAN PRODUCTS:** see tofu this					
Iceberg	56	1 c	7.3	254	19	section; miso, and tempeh					
Looseleaf	56	1 c	10	1050	106	in Other; roasted soybeans in					
Romaine	56	1 c	9	1629	146	Nuts and Seeds; soy milk in Dairy					
						& soy flour in Grains.					

ESHA Research

	Wt (g)	Svg	Cal	Vit A (RE) 100 Cal	Svg
Vegetables & Legumes, continued:					
Spinach:					
Fresh, chopped	56	1 c	12	3642	448
Cooked from fresh	180	1 c	41	4268	1750
Cooked from frozen	190	1 c	53	3301	1756
Canned, drained	214	1 c	50	3756	1878
SQUASH, SUMMER, sliced:					
Crookneck, fresh	130	1 c	24	183	44
Crookneck, cooked	180	1 c	36	144	52
Scallop, fresh	130	1 c	24	60	14
Scallop, cooked	90	1/2 c	14	55	8
Zucchini, fresh	130	1 c	19	233	44
Zucchini, cooked	180	1 c	29	149	43
SQUASH, WINTER, mashed:					
Acorn (Danish), baked	245	1 c	137	77	105
Acorn (Danish), boiled	245	1 c	83	76	63
Butternut, baked	245	1 c	99	1732	1715
Butternut, ckd fr/frozen	240	1 c	94	852	801
Hubbard, baked	240	1 c	120	1208	1450
Hubbard, boiled	236	1 c	70	1350	945
Spaghetti, baked or boiled	155	1 c	45	38	17
Succotash:					
Cooked from fresh	192	1 c	222	25	56
Ckd from frozen	170	1 c	158	25	39
Sweet potato:					
Baked in skin, then peeled	114	1 ea	118	2108	2488
Boiled, peeled	151	1 ea	160	1609	2575
Canned, mashed	128	1/2 c	129	1495	1929
Canned, vacuum pack	255	1 c	233	874	2036
Candied, recipe	105	1 pce	144	306	440
Tofu (soybean curd)	124	1/2 c	94	11	11
TOMATOES:					
Fresh, whole	123	1 ea	24	579	139
Fresh, chopped	180	1 c	35	583	204
Cooked from fresh	240	1 c	60	542	325
Canned, whole	240	1 c	47	309	145
Tomato juice, canned	244	1 c	42	328	136
Tomato paste, canned	262	1 c	220	294	647
Tomato puree, canned	250	1 c	102	333	340
Tomato sauce, canned	245	1 c	74	324	240
Turnip greens:					
Ckd from fresh	86	1/2 c	15	2731	396
Ckd from frozen	82	1/2 c	24	2725	654
Vegetable juice cocktail	242	1 c	46	615	283
VEGETABLES, MIXED,					
cooked from frozen:					
Broccoli, carrots, pasta	95	2/3 c	88	886	780
Broccoli, carrots &					
water chestnuts	91	2/3 c	32	1600	512
Broccoli, cauliflower					
& red pepper	95	2/3 c	25	308	77
Broccoli, water chestnuts	95	1/2 c	33	361	119
Mixed vegetables (corn, peas,					
carrots, green beans & limas):					
Cooked from frozen	182	1 c	107	728	779
Canned, drained	163	1 c	77	2466	1899
Peas, carrots, onions					
& butter sauce	71	2/3 c	100	110	110
Peas, mushrooms	95	1/2 c	73	102	74
Peas, onions	95	1/2 c	71	90	64
Peas, pasta, onion, corn					
in cream sauce	95	1/2 c	132	40	53
Peas, pasta, mushrooms					
& cream sauce	95	1/2 c	129	53	69
Peas, rice & mushrooms	66	.66 c	108	39	42
Watercress, fresh	17	1/2 c	2	3995	80

Yam, orange: see Sweet potatoes.
Zucchini: see Squash, summer

Other
Beverages, Fats, Condiments, Sweets, Spices

	Wt (g)	Svg	Cal	Vit A (RE) 100 Cal	Svg
BEVERAGES:					
Alcoholic drinks, (vitamin A					
is present in the mixer):					
Bloody mary, 5 fl.oz drink	148	1 ea	116	44	51
Screwdriver, 7 fl. oz drink	213	1 ea	174	8	13
Tequila sunrise, 5.5 fl. oz	172	1 ea	189	9	17
Non-alcoholic:					
Clam and tomato juice	166	2/3 c	77	47	36
Lemonade, from frozen	248	1 c	100	5	5
Pineapple grapefruit drink	250	1 c	117	8	9
Pineapple orange drink	250	1 c	125	106	133
Barbecue sauce	16	1 T	10	140	14
Butter	141	1 T	100	106	106
Butter oil (ghee)	13	1 T	112	106	119
Catsup	171	1 T	18	132	24
Chicken fat	13	1 T	115	16	18
Chili sauce, hot red pepper	31	2 T	6	4900	294
CHOCOLATE:					
Baking chocolate, unsweetened	28	1 oz	145	.7	1
Hot fudge topping	75	1/4 c	255	10	26
Milk chocolate candy	28	1 oz	145	7	10
Cocoa powder	86	1 c	224	3	7
Cod liver oil	14	1 T	126	2833	3570
Miso (soybean product)	138	1/2 c	284	4	12
Olives:					
Green	39	10 ea	45	27	12
Ripe, large	47	10 ea	78	7	5
Pickle, dill	65	1 ea	5	140	7
Pickle relish	15	1 T	20	10	2
Salsa:					
Picante by Tostitos	85	6 T	40	158	63
Recipe	108	1/2 c	46	190	150
SPICES:					
Basil, dried	4.5	1 T	11	384	42
Cayenne, red pepper	5.3	1 T	17	1300	221
Chili powder	7.5	1 T	24	1092	262
Chives, fresh	3	1 T	1	1400	14
Chives, freeze-dried	.8	1/4 c	2	2750	55
Coriander leaf, fresh	4	1/4 c	1	1100	11
Marjoram, dried	1.7	1 T	5	274	14
Oregano, ground	4.5	1 T	14	222	31
Paprika	6.9	1 T	20	2090	418
Tarragon, ground	4.8	1 T	14	144	20
Thyme, ground	4.3	1 T	12	136	16
Tempeh (soybean product)	83	1/2 c	165	35	57

Thiamin   (Vitamin B1)

Thiamin is essential for making the energy in food available to the body. It works with the other "B" vitamins in the many stages of metabolism to convert carbohydrates into a usable form of energy — glucose. More thiamin is needed if you are very active and use lots of calories, or if you have a high percentage of calories coming from carbohydrate.

Since the brain and nervous system rely on glucose for energy, they are sensitive to a shortfall of this vitamin. Mild deficiencies result in the inability to concentrate, irritability, depression, and muscle weakness. Major deficiencies produce far more severe symptoms: edema, atrophy of leg muscles, peripheral nerve changes, paralysis, and heart failure. The clinical condition of severe, prolonged deficiency is usually called Beriberi, and it led to the discovery of the vitamin.

Beriberi became epidemic in the Orient when whole grain rice was replaced with polished white rice brought in by Western traders. The hull that was polished off was rich in thiamin along with other "B" vitamins and several more nutrients. It took a long time for researchers to discover that Beriberi was the result of something *NOT* in the food, rather than by some bacteria or substance *IN* the food.

Thiamin is one of several nutrients that is required to be added to flour and cereals in the process called "enrichment." This is to compensate for the loss of the vitamin during the milling and processing of the grain.

Thiamin deficiency is seen in alcoholics, and for those who consume large quantities of unenriched white rice,white flour and baked goods made with unenriched flour. A continual diet of only unenriched dry snacks (most corn chips, pretzels), soft drinks and candy (a high calorie, low nutrient diet) may create a deficiency.

Vitamin Losses

Thiamin is one of the most easily destroyed vitamins. It is removed from grains when they are processed or milled. It is also one of the vitamins replaced in white wheat flour, cornmeal, and white rice (enrichment). The heat from baking bread results in a 15-30% loss, mostly in the crust, and toasting a slice can destroy an additional 10-30 % depending on how long it is toasted.

Thiamin is water soluble and some will leach into cooking water. If the cooking water is retained and used, there should be fairly complete retention.   Sulfur dioxide, which is sometimes used to preserve vitamin C, also destroys thiamin. Absorption will be reduced by alcohol, barbituates and large quantities of tea.

Recommended Dietary Allowances for Adults

USA RDA: Men, 1.5 mg; Women, 1.1 mg (pregnant, 1.5 mg; nursing, 1.6 mg) with a minimum of 1.0 mg/day.
Canadian RNI: 0.4 mg of Thiamin for every 1000 kcalories of intake, with a minimum of 0.8 mg/day for persons 19 or
   older.
Because Thiamin is essential for metabolism of energy (calories), the recommended amount of thiamin is related to the food consumed in the diet. The USA recommendation is about 0.5 mg of Thiamin for every 1000 kcalories; Canadian recommendations are for 0.4 mg of Thiamin for every 1000 kcalories of intake. The Food Processor® nutrition systems automatically calculate the more accurate amount (0.5 mg or 0.4 mg of thiamin per 1000 calories) when computing your personal RDA and RNI.

Food Sources

Thiamin is present in many foods, especially in unrefined whole grains, legumes (beans and peas), seeds, pork, organ meats (liver, heart, kidney), and brewers yeast. Enriched and fortified white flour, cornmeal, rice, cereals and bakery products also provide Thiamin in the diet, compensating for loss in processing. Not all refined grain products have been enriched, however, so it is worth reading the labels.

ESHA Research

	Wt (g)	Svg	Cal	Thiamin (mg) 100 Cal	Svg

Baked Goods

All purpose flour is enriched with vitamin B1.
Baked goods containing <u>unenriched</u> white
flour will have much less Vitamin B1 than shown.

	Wt (g)	Svg	Cal	100 Cal	Svg
Bagels, 3.5" diam, egg or plain	68	1 ea	180	.1	.258
Biscuits:					
Homemade	28	1 ea	100	.1	.080
From refrigerator dough	20	1 ea	65	.1	.082
From mix	28	1 ea	94	.1	.122
BREADS: Vitamin B1 is added to all-purpose flour (enrichment).					
Banana nut bread, 1/2"	50	1 pce	161	.1	.095
Cracked wheat	25	1 pce	65	.1	.095
Cornbread muffin	45	1 ea	145	.1	.100
French, 5" x 2.5" x 1" piece	35	1 pce	100	.2	.160
Italian, 4.75" x 4" x 1/2"	30	1 pce	83	.1	.123
Mixed grain	25	1 pce	65	.2	.100
Oatmeal bread	25	1 pce	65	.2	.115
Pita pocket bread, enr. 6.5"	60	1 pce	165	.2	.274
Pumpernickel, 5" x 4" x 3/8"	32	1 pce	80	.1	.109
Raisin bread	25	1 pce	68	.1	.082
Rye, light 5" x 3.5" x 7/16"	25	1 pce	65	.2	.102
Vienna bread	25	1 pce	70	.2	.115
Wheat bread (white & whole wheat flour)	28	1 pce	72	.2	.129
White bread	28	1 pce	75	.2	.131
Whole wheat bread	28	1 pce	70	.1	.100
Bread pudding, with raisins	165	1 c	349	.1	.176
CAKES, piece = 1/16th cake unless stated otherwise:					
Carrot cake, cream cheese frosting 2.5" x 3"	112	1 pce	406	<.1	.110
Chocolate, choc frosting	69	1 pce	235	<.1	.070
Coffee cake, 2.4" x 2.8"	72	1 pce	230	.1	.137
Gingerbread, 3" x 3"	63	1 pce	174	.1	.095
Pound cake, 1/2" slice	30	1 pce	115	<.1	.055
Sponge cake, 1/12 cake	66	1 pce	194	.1	.100
White cake, with frosting	73	1 pce	260-290	.1	.180
Yellow cake, choc frosting	69	1 pce	240	<.1	.063
Chips: see corn & tortilla this section; potato chips under Vegetables.					
COOKIES, average, 1-2 per oz	28	1 oz	varies	<.1	.02-.08
Corn chips	28	1 oz	155	<.1	.040
CRACKERS:					
Cheese crackers	10	10 ea	50	.1	.050
Cheese, w/peanut butter fill	30	4 ea	150	.1	.160
Graham cracker	14	2 ea	60	<.1	.020
Rye wafers, whole grain	14	2 ea	55	.1	.060
Saltines	12	4 ea	50	.1	.060
Sesame	12	4 ea	60	.1	.060
Round, like Ritz	9	3 ea	45	.1	.030
Wheat, thin	8	4 ea	35	.1	.040
Whole wheat crackers	8	2 ea	35	.1	.020
Croissant, 4.5" x 4" x 2"	57	1 ea	235	.1	.170
Croutons, dry bread cubes	30	1 c	111	.1	.105
Danish pastry, avg.	61	1 ea	228	.1	.160
Doughnut, yeast raised, plain	60	1 ea	235	.1	.280
English muffins:					
Plain	57	1 ea	140	.2	.262
Sourdough	56	1 ea	129	.2	.220
With raisins	56	1 ea	146	.2	.220
MUFFINS:					
Blueberry, from recipe	45	1 ea	135	.1	.100
Blueberry, from mix	45	1 ea	140	.1	.112
Bran, wheat, from recipe	45	1 ea	125	.1	.110
Bran, wheat, from mix	45	1 ea	140	.1	.080

	Wt (g)	Svg	Cal	100 Cal	Svg
Pancakes, 4" diam:					
Buckwheat, from mix	27	1 ea	55	.1	.040
Plain, homemade	27	1 ea	60	.1	.060
Plain, from mix	27	1 ea	60	.2	.090
Whole wheat, from mix.	52	1 ea	94	.1	.093
PIES: each piece is 1/6 of a 9" pie.					
Apple pie	158	1 pce	405	<.1	.175
Banana cream	198	1 pce	319	<.1	.128
Blueberry	158	1 pce	380	<.1	.173
Cherry	158	1 pce	410	<.1	.188
Chocolate cream pie	175	1 pce	311	<.1	.151
Coconut cream	172	1 pce	343	<.1	.130
Cream, commercial	152	1 pce	455	<.1	.060
Custard pie	152	1 pce	293	<.1	.137
Lemon meringue	140	1 pce	355	<.1	.100
Mincemeat	160	1 pce	395	<.1	.157
Peach pie	158	1 pce	405	<.1	.175
Pecan pie	138	1 pce	583	<.1	.217
Pumpkin pie	200	1 pce	367	<.1	.137
Strawberry chiffon, recipe	162	1 pce	372	<.1	.123
PopTart-type pastry, fortified	54	1 ea	210	.1	.173
Pretzels, dutch twist	16	1 ea	65	.1	.050
Pretzel, thin twist	60	10 ea	240	.1	.190
ROLLS:					
Dinner rolls, 2.5" x 2"	28	1 ea	85	.2	.140
Hamburger bun	45	1 ea	129	.2	.220
Hard roll, white	50	1 ea	155	.1	.200
Hotdog bun	40	1 ea	115	.2	.196
Rye roll, dark	28	1 ea	79	.1	.117
Rye roll, light	28	1 ea	76	.2	.114
Submarine roll (hoagie)	135	1 ea	400	.1	.540
Whole wheat roll	35	1 ea	88	.1	.120
Tortillas:					
Corn, enr, 6" diam., fried	30	1 ea	87	.1	.050
Flour, 10.5" diam.	57	1 ea	168	.1	.205
Flour, 8" diam.	35	1 ea	105	.1	.127
Tortilla chips:					
Plain	28	1 oz	139	<.1	.010
Doritos, Nacho flavored	28	1 oz	139	<.1	.040
Doritos, Taco flavored	28	1 oz	140	.1	.080
Waffles, from mix, 7" diam.	75	1 ea	205	.1	.140
Waffles, frozen, 4" diam.	35	1 ea	98	.2	.154

Dairy & Dairy Products

	Wt (g)	Svg	Cal	100 Cal	Svg
CHEESE (1.5" cube ≈ 1 ounce):					
American cheese	28	1 oz	106	<.1	.008
American cheese spread	28	1 oz	82	<.1	.014
Brie cheese	28	1 oz	95	<.1	.020
Cheddar cheese	28	1 oz	114	<.1	.008
Cheshire cheese	28	1 oz	110	<.1	.013
Cottage cheese:					
Lowfat, 1% or 2%	113	1/2 c	92	.1	.051
Creamed (4.5% fat), avg.	108	1/2 c	113	<.1	.024
Cream cheese (1 T = 15g)	28	1 oz	99	<.1	.005
Feta cheese	28	1 oz	75	<.1	.040
Mozzarella, part skim, low moisture	28	1 oz	80	<.1	.006
Parmesan, grated (1T = 6g)	28	1 oz	129	<.1	.013
Ricotta, part skim	123	1/2 c	170	<.1	.026
Swiss cheese	28	1 oz	107	<.1	.006
CREAM, SWEET, fluid:					
Coffee or table cream	240	1 c	469	<.1	.077
Half and half	242	1 c	315	<.1	.085
Light whipping cream	239	1 c	699	<.1	.057

Thiamin (Vitamin B1) mg

	Wt (g)	Svg	Cal	Thiamin (mg) 100 Cal	Svg
Dairy & Dairy Products, continued:					
CREAM, SWEET, whipped:					
Heavy cream, unsweetened	119	1 c	410	<.1	.026
Pressurized	60	1 c	154	<.1	.022
CREAM, SOUR, dairy:					
Cultured, dairy	230	1 c	493	<.1	.081
Half and half, dairy	15	1 T	20	<.1	.005
Cream, sour, imitation, non-dairy	230	1 c	479	0	0
CREAM SUBSTITUTES, non-dairy:					
Coffee whitener, liquid/frzn	120	1/2 c	163	0	0
Coffee whitener, powder	94	1 c	514	0	0
Dessert Toppings, non-dairy or part dairy:					
Frozen (e.g. Coolwhip)	75	1 c	239	0	0
Dessert powder, dry mix	43	1.5 oz	245	0	0
Pressurized, non-dairy	70	1 c	185	0	0
Kefir beverage	233	1 c	160	.3	.450
MILK (cow):					
Buttermilk (< 1% fat)	245	1 c	99	.1	.083
1% or 2% lowfat	244	1 c	102	.1	.095
Skim	245	1 c	86	.1	.088
Whole (3.3% fat)	244	1 c	150	.1	.093
Canned, whole, evaporated	252	1 c	340	<.1	.120
Canned, skim, evaporated	255	1 c	200	.1	.114
Dried, instant nonfat, env.	91	1 ea	326	<.1	.094
Milk (other):					
Goat milk	244	1 c	168	<.1	.117
Human breast milk	246	1 c	171	<.1	.034
Soy milk	240	1 c	79	.5	.386
MILK BEVERAGES and mixes:					
Chocolate flavored, to be mixed with water:					
Powder (includes dry milk)	28	1 oz	100	<.1	.031
Drink	206	3/4 c	100	<.1	.031
Chocolate flavored, to be mixed with whole milk:					
Powder	22	3/4 oz	75	<.1	.032
Drink	266	1 c	226	<.1	.101
Instant Breakfast, dry mix	37	1 env.	130	.2	.300
Milkshakes (10 fl oz = 1.25 c):					
Chocolate	283	1.25 c	360	<.1	.164
Strawberry	283	1.25 c	319	<.1	.127
MILK DESSERTS:					
Custard, baked	265	1 c	305	<.1	.110
Ice cream, vanilla:					
Regular	133	1 c	269	<.1	.052
Rich	148	1 c	349	<.1	.044
Soft serve	173	1 c	377	<.1	.080
Ice milk, vanilla:					
Regular	131	1 c	184	<.1	.076
Soft serve	175	1 c	223	.1	.117
Puddings, prepared:					
Assorted flavors, instant	149	1/2 c	145-155	<.1	.040
Assorted, instant, low cal	130	1/2 c	69	<.1	.050
Chocolate, from mix, all	130	1/2 c	155	<.1	.045
Chocolate, 5 oz can ≈ 1/2 c	142	1 can	205	<.1	.040
Pudding pops, all	57	1 ea	99	<.1	.030
Sherbet	193	1 c	270	<.1	.033
Yogurt, frozen, average	174	1 c	220	<.1	.055
YOGURT:					
Lowfat, plain	227	1 c	144	.1	.100
Lowfat, fruit	227	1 c	231	<.1	.084
Lowfat, coffee or vanilla	227	1 c	193	<.1	.095
Nonfat	227	1 c	127	.1	.109
Whole	227	1 c	138	<.1	.066

Eggs

	Wt (g)	Svg	Cal	Thiamin (mg) 100 Cal	Svg
Cooked (all ways)	50	1 ea	77.5	<.1	.026
Raw, whole	50	1 ea	75	<.1	.031
Raw, white	33.4	1 ea	17	<.1	.002
Raw, yolk	16.6	1 ea	59	<.1	.028

Fruits & Fruit Juices

	Wt (g)	Svg	Cal	Thiamin (mg) 100 Cal	Svg
Apple, 2.75" diam., with or without peel	138	1 ea	80	<.1	.023
Applesauce, unsweetened	244	1 c	106	<.1	.032
Apple juice, canned, bottled	248	1 c	116	<.1	.052
Apple juice, from frozen conc.	239	1 c	111	<.1	.007
APRICOTS:					
Fresh, pitted	106	3 ea	51	.1	.032
Canned, juice pack	84	3 ea	40	<.1	.015
Canned, heavy syrup	85	3 ea	70	<.1	017
Dried halves	35	10 ea	83	<.1	.003
Apricot nectar, canned	251	1 c	141	<.1	.023
Avocado, fresh, whole:					
California (227g with refuse)	173	1 ea	305	.1	.187
Florida (454g with refuse)	304	1 ea	340	.1	.328
Banana, 8.75" long, 176g w/peel	114	1 ea	105	<.1	.051
Blackberries:					
Fresh berries	144	1 c	74	<.1	.043
Frozen, unthawed	151	1 c	97	<.1	.044
Canned	256	1 c	236	<.1	.069
Blueberries:					
Fresh berries	145	1 c	82	.1	.070
Frozen, unthawed	155	1 c	78	.1	.050
Canned	256	1 c	225	<.1	.087
Boysenberries:					
Canned	256	1 c	225	<.1	.067
Frozen, unthawed	132	1 c	66	.1	.070
Cantaloupe: see Melon.					
Cherries, sour, frozen, unthawed	155	1 c	72	.1	.068
Cherries, sour, canned	244	1 c	90	<.1	.041
CHERRIES, SWEET:					
Fresh, pitted	68	10 ea	49	.1	.030
Frozen	259	1 c	232	<.1	.070
Canned	257	1 c	213	<.1	.054
Cranberry juice cocktail	253	1 c	145	<.1	.020
Currants:					
Fresh, Black	112	1 c	71	.1	.056
Fresh, Red or white	112	1 c	63	.1	.045
Dried (Zante)	144	1 c	407	<.1	.230
Dates, whole, pitted	83	10 ea	228	<.1	.075
Figs, dried	18	1 ea	48	<.1	.013
Fig, fresh, medium	50	1 ea	37	.1	.030
Fruit cocktail, heavy syrup	255	1 c	185	<.1	.045
Fruit cocktail, juice pack	248	1 c	115	<.1	.045
Grapes, fresh:					
Thompson, seedless	50	10 ea	35	.1	.046
Tokay, Emperor	57	10 ea	40	.1	.052
Grape juice, from frozen	250	1 c	128	<.1	.038
Grape juice, bottled, canned	253	1 c	155	<.1	.066
GRAPEFRUIT (half = 241g w/rind):					
Half, white - pink/red	118-123	1 ea	38	.1	.043
Canned with liquid	254	1 c	152	.1	.097
Grapefruit Juice:					
Fresh juice	247	1 c	96	.1	.099
From frozen concentrate	247	1 c	102	.1	.101
Canned, unsweetened	247	1 c	93	.1	.104
Kiwi fruit, medium	76	1 ea	46	<.1	.020
Mandarin oranges, canned	252	1 c	155	.1	.130

Values are for edible portion of foods

Fruits & Fruit Juices, continued:	Wt (g)	Svg	Cal	100 Cal	Svg
Mango, sliced	165	1 c	108	.1	.096
Melon: also see Watermelon.					
Cantaloupe, fresh cubes	160	1 c	57	.1	.036
Casaba, fresh cubes	170	1 c	45	.2	.102
Honeydew, fresh cubes	170	1 c	60	.2	.131
Frozen, melon balls, mixed	173	1 c	55	.1	.030
Nectarine	136	1 ea	67	<.1	.023
ORANGE, 2-5/8", 180g whole	131	1 ea	60	.2	.114
Orange juice:					
Fresh	248	1 c	111	.2	.223
Chilled	249	1 c	110	.3	.276
From frozen	249	1 c	110	.2	.200
Canned, unsweetened	249	1 c	105	.1	.149
Orange grapefruit juice, canned	247	1 c	105	.1	.138
Papaya, (454g whole)	304	1 ea	117	.1	.082
PEACHES: peeled (we couldn't find any data with skin).					
Fresh, whole, pitted, peeled	87	1 ea	37	<.1	.015
Fresh slices	170	1 c	73	<.1	.029
Canned halves, juice pack	77	1 half	34	<.1	.006
Canned slices, juice pack	248	1 c	109	<.1	.020
Canned slices, heavy syrup	250	1 c	190	<.1	.028
PEARS:					
Bartlett (1 ea = 1 cup slices)	166	1 ea	98	<.1	.033
Canned, juice pack	77	1 half	38	<.1	.010
Canned, heavy syrup	79	1 half	59	<.1	.010
PINEAPPLE:					
Fresh pieces or chunks	155	1 c	76	.19	.143
Canned slice, juice pack	58	1 slice	35	.15	.055
Canned pieces, juice pack	250	1 c	140	.17	.238
Canned pieces, heavy syrup	255	1 c	199	.10	.230
Pineapple juice:					
Prepared from frozen conc.	250	1 c	129	.1	.175
Canned, unsweetened	250	1 c	140	.1	.138
Plantain, slices (cooked/fresh)	154	1 c	179	<.1	.074
Plums, medium, 2-1/8" diam.	66	1 ea	36	.1	.028
Plums, canned, juice pack	95	3 ea	55	<.1	.022
Prunes, dried, pitted	34	4 ea	80	<.1	.027
Prune juice, bottled	256	1 c	181	<.1	.041
Raisins, packed measure	165	1 c	494	.1	.25
Raspberries, fresh	123	1 c	60	.1	.037
Raspberries, frozen or canned	253	1 c	245	<.1	.049
Rhubarb, cooked with sugar	240	1 c	279	<.1	.043
STRAWBERRIES:					
Fresh berries	149	1 c	45	.1	.030
Frozen, unthawed	149	1 c	52	.1	.033
Frozen, thawed, sweetened	255	1 c	245	<.1	.041
Tangerine, whole	84	1 ea	37	.2	.088
Tangerine juice, from frozen	241	1 c	110	.1	.125
Tangerine juice, canned	249	1 c	125	.1	.149
Watermelon, diced	160	1 c	50	.3	.128

Grains & Grain Products
Cereals, Flour, Grains, Noodles, Pasta, and Popcorn

	Wt (g)	Svg	Cal	100 Cal	Svg
Amaranth grain, dry	195	1 c	729	<.1	.096
Barley, pearled, cooked	157	1 c	193	<.1	.130
Bran: see Oat, Rice, Wheat.					
Buckwheat, flour, dark	98	1 c	338	.2	.578
Buckwheat, flour, light	98	1 c	340	<.1	.091
Bulgar wheat, cooked	182	1 c	152	<.1	.104

CEREALS, COLD: "Ready to eat" cereals may be fortified and enriched withThiamin. Brands vary. See the label.

CEREALS, HOT, cooked:	Wt (g)	Svg	Cal	100 Cal	Svg
Corn grits, cooked, enriched	242	1 c	145	.2	.240
Corn grits, cooked, unenrich.	242	1 c	146	<.1	.048
Cream of Rice	244	1 c	126	.1	.100
Cream of Wheat	244	1 c	140	.2	.240
Farina, enriched	233	1 c	116	.2	.190
Farina, unenriched	233	1 c	116	<.1	.023
Malt-O-Meal	240	1 c	122	.4	.480
Maypo cereal	180	3/4 c	128	.4	.500
Oatmeal, from rolled oats-all (regular, quick or instant)	234	1 c	145	.2	.260
Oatmeal, fortified instant, prepared from packet:					
Plain	177	3/4 c	104	.5	.530
With apples	149	3/4 c	135	.4	.480
With bran and raisin	195	3/4 c	158	.4	.560
With cinnamon and spice	161	3/4 c	177	.3	.560
With maple and raisins	155	3/4 c	162	.3	.520
Ralston cereal	253	1 c	134	.1	.200
Roman Meal, cooked	181	3/4 c	111	.2	.180
Whole wheat cereal	242	1 c	150	.1	.170
Corn grits (hominy): see Cereals, Hot.					
Cornmeal:					
Degermed, enriched, cooked	240	1 c	120	.1	.140
Degermed, enriched, dry	138	1 c	505	.2	.987
Bolted, nearly whl, dry, unenr.	122	1 c	441	.1	.370
Flour: see specific grain, nut or vegetable.					
MACARONI:					
Cooked, enriched	140	1 c	197	.1	.286
Cooked whole wheat	140	1 c	174	.1	.151
Cooked vegetable, enriched	134	1 c	172	.1	.150
Masa Harina, enriched	110	1 c	407	.4	1.56
Millet, cooked	120	1/2 c	143	.1	.127
Noodles:					
Chow mein noodles, dry	45	1 c	237	.1	.260
Egg noodles, cooked	160	1 c	213	.1	.298
Spinach noodles, cooked	140	1 c	182	.1	.136
Oat bran, 1T ≈ 6 g	94	1 c	132	.8	1.1
Oats, rolled, dry	81	1 c	311	.2	.590
Oatmeal: see Cereals, hot.					
PASTA: see Macaroni, Noodles, Spaghetti.					
Quinoa grain, dry	170	1 c	635	.1	.337
RICE, cooked:					
Brown rice	195	1 c	217	.1	.187
White, regular (enriched)	205	1 c	264	.1	.334
White, converted (enriched)	175	1 c	200	.2	.438
White, instant (enriched)	165	1 c	162	.1	.124
Wild rice	82	1/2 c	83	.1	.043
Rice bran	83	1 c	262	.9	2.29
Rice flour	158	1 c	578	<.1	.218
Rye flour, dark	128	1 c	415	.1	.404
Rye flour, medium	102	1 c	374	.1	.338
Rye flour, light	102	1 c	361	.1	.293
Soy flour, low fat	88	1 c	326	.1	.334
Soy flour, full fat, stirred	85	1 c	370	<.1	.17-.24
Spaghetti, pasta:					
Enriched	140	1 c	197	.1	.286
Whole wheat spaghetti	140	1 c	174	.1	.151
WHEAT:					
Bran, dry	30	1/2 c	65	.2	.157
FLOURS are enriched with vitamin B1, except for whole wheat:					
All-purpose, white, unsifted	125	1 c	455	.2	.981
Cake, sifted	96	1 c	348	.2	.856
Self-rising, enriched	125	1 c	442	.2	.843
Semolina, enriched	167	1 c	601	.2	1.35
Whole wheat	120	1 c	407	.1	.536

Values are for edible portion of foods

Thiamin (Vitamin B1) mg

	Wt (g)	Svg	Cal	Thiamin (mg) 100 Cal	Svg
Grains & Grain Products, continued:					
WHEAT, continued:					
Wheat germ, raw	100	1 c	360	.5	1.88
Wheat germ, toasted	113	1 c	432	.4	1.89
Wheat, rolled, cooked	240	1 c	142	.1	.170
Wheat, rolled, dry	85	1 c	289	.1	.310
Wheat, sprouted	106	1 c	236	.1	.246
Whole grain wheat (wheat berries), cooked	50	1/3 c	28	.1	.04

Meats: Fish & Shellfish

	Wt (g)	Svg	Cal	100 Cal	Svg
Bass, freshwater, baked/broiled	100	3.5 oz	125-140	<.1-.1	.1-1.5
Bluefish, steamed	100	3.5 oz	148	<.1	.057
Catfish, fried in cornmeal	100	3.5 oz	229	<.1	.073
Clams, meat only, steamed	45	10 ea	67	<.1	.005
Cod, baked or broiled	100	3.5 oz	105	.1	.088
Cod, smoked	100	3.5 oz	79	.1	.080
CRAB, meat only:					
Alaskan King leg, cooked	134	1 ea	129	.1	.071
Blue crab, cooked, unpacked	135	1 c	138	.1	.145
Blue crab, canned, unpacked	135	1 c	133	.1	.110
Dungeness crab, cooked	101	3/4 c	85	<.1	.042
Crab, imitation from surimi	85	3 oz	87	<.1	.027
Eel, baked or broiled	100	3.5 oz	236	<.1	.183
Eel, smoked	100	3.5 oz	330	<.1	.140
Fish cakes, fried from frozen	100	3.5 oz	213	<.1	.060
Fish sticks, heated from frozen	57	2 ea	155	<.1	.072
Gefiltefish, sweet, commercial	42	1 pce	35	<.1	.027
HADDOCK:					
Baked or broiled	85	3 oz	95	<.1	.034
Breaded, fried	85	3 oz	175	<.1	.060
Smoked	100	3.5 oz	116	<.1	.047
Halibut, baked or broiled	85	3.5 oz	119	<.1	.059
Halibut, smoked	100	3.5 oz	224	<.1	.050
Herring:					
Baked or broiled	100	3.5 oz	203	.1	.112
Canned with liquid	100	3.5 oz	208	<.1	.025
Smoked or kippered	100	3.5 oz	217	.1	.126
Pickled, 1 piece ≈15 g	100	3.5 oz	262	<.1	.036
Lobster, meat only, cooked	145	1 c	142	<.1	.100
Mackerel:					
Baked or broiled, average	100	3.5 oz	210	<.1	.145
Canned, Jack, tall can	361	1 can	563	<.1	.144
Ocean perch, baked or broiled	100	3.5 oz	121	<.1	.090
Ocean perch, breaded, fried	85	3 oz	185	<.1	.100
OYSTERS:					
Raw, Eastern	124	1/2 c	85	.2	.170
Raw, Pacific	124	1/2 c	100	.1	.083
Breaded, fried, medium	88	6 ea	173	.1	.130
Simmered, Eastern or Atlantic	100	3.5 oz	137	.1	.170
Pollock, baked or broiled	100	3.5 oz	99	.1	.062
Pompano, baked or broiled	100	3.5 oz	211	.2	.500
Rockfish, baked or broiled	100	3.5 oz	121	<.1	.044
Roe, raw, mixed species	28	1 oz	39	.5	.210
SALMON, cooked:					
Average, baked or broiled	85	3 oz	183	.1	.183
Chinook, smoked	85	3 oz	99	<.1	.020
Canned, Atlantic, small can	220	1 can	281	.2	.440
Canned, Chum, #1 can	369	1 can	521	<.1	.251
Canned, Pink, #1 can, drained	454	1 can	631	<.1	.104
Sardines, canned, drained:					
Atlantic, 2 sardines = 24g	92	1 can	192	<.1	.074
Pacific, 1 sardine = 38g	100	3.5 oz	178	<.1	.044

	Wt (g)	Svg	Cal	100 Cal	Svg
Seatrout/Steelhead, cooked	100	3.5 oz	131	.1	.085
Shark, batter-fried	85	3 oz	194	<.1	.061
SHRIMP:					
Boiled, 2 large ≈ 11g	100	3.5 oz	99	<.1	.031
Breaded, fried, 2 large ≈ 15g	90	12 ea	218	.1	.116
Canned, with liquid	100	3.5 oz	102	<.1	.028
Canned, drained	128	1 c	154	<.1	.035
Shrimp, imitation from surimi	85	3 oz	86	<.1	.020
Smelt, Rainbow, cooked	85	3 oz	106	.1	.056
Snapper, baked or broiled	100	3.5 oz	128	<.1	.053
SOLE (Flounder):					
Baked or broiled	85	3 oz	99	.1	.068
Fried in batter	85	3 oz	250	.1	.170
Breaded, fried	100	3.5 oz	188	.1	.130
Steamed	100	3.5 oz	92	.1	.060
Squid, fried in flour	85	3 oz	149	<.1	.048
Surimi, processed walleye pollock:					
see also imitation Crab, Shrimp	85	3 oz	84	<.1	.017
Swordfish, baked or broiled	100	3.5 oz	155	<.1	.043
Trout, baked or broiled	85	3 oz	129	<.1	.072
TUNA:					
Baked or broiled, Bluefin	100	3.5 oz	159	.1	.076
Canned, drained, oil pack	171	1 can	339	<.1	.065
Canned, drained, water pack	165	1 can	216	<.1	.058

Meats
Beef, Pork, Lamb etc.

	Wt (g)	Svg	Cal	100 Cal	Svg
BEEF:					
Breakfast strips, cured beef	34	3 ea	153	<.1	.031
Chuck blade, pot roasted, all grades:					
Lean and fat (5.4 oz raw)	85	3 oz	325	<.1	.059
Lean only	85	3 oz	230	<.1	.071
Ground beef, average of baked, broiled, pan-fried:					
Extra Lean, 17% fat, raw	85	3 oz	215	<.1	.051
Lean, 21% fat, raw	85	3 oz	231	<.1	.043
Regular, 26.6% fat, raw	85	3 oz	250	<.1	.026
Rib, choice, oven roasted:					
Lean and fat (5 oz raw)	85	3 oz	324	<.1	.056
Lean only	85	3 oz	204	<.1	.070
Round steak, broiled, choice:					
Lean and fat (4.5 oz raw)	85	3 oz	233	<.1	.077
Lean only	85	3 oz	165	.1	.086
Round tip, oven roasted:					
Lean and fat	85	3 oz	213	<.1	.077
Lean only	85	3 oz	162	.1	.083
Sirloin steak, broiled, all grades:					
(11.3 oz raw steak = 8.2 oz lean and fat cooked, 6.9 oz lean)					
Lean and fat	85	3 oz	238	<.1	.068
Lean only	85	3 oz	172	<.1	.077
T-bone steak, choice, broiled:					
(16 oz raw = 9.7 oz lean and fat cooked, 7.4 oz lean)					
Lean and fat	85	3 oz	276	<.1	.077
Lean only	85	3 oz	182	.1	.094
Beef heart, pieces, simmered	85	3 oz	140	.1	.119
Beef kidney, cooked pieces	140	1 c	201	.1	.267
Beef liver, cooked	85	3 oz	184	.1	.179
Beef, dried, cured	28	1 oz	47	.1	.050
FROG LEGS, flour-fried	144	6 ea	418	<.1	.170

HAM: See Pork, cured; Turkey Ham; and Sausages and Lunchmeats group.

13

ESHA Research

	Wt (g)	Svg	Cal	Thiamin (mg) 100 Cal	Svg		Wt (g)	Svg	Cal	Thiamin (mg) 100 Cal	Svg
LAMB:						**Meats: Poultry**					
Arm chop, braised (5.6 oz raw with bone):											
Lean and fat (2.5 oz ckd)	70	1 chop	244	<.1	.050	**CHICKEN:** A 3 lb chicken					
Lean only (1.9 oz ckd)	55	1 chop	152	<.1	.040	yields ≈ 1.45 lbs raw meat;					
Loin chop, broiled (4.2 oz raw with bone):						≈ 1.1 lbs cooked.					
Lean & fat (2.3 oz ckd)	64	1 chop	201	<.1	.064	**All types:**					
Lean only (1.6 oz ckd)	46	1 chop	100	.1	.051	Fried	140	1 c	307	<.1	.119
Cutlet, lean, cooked average	85	3 oz	175	<.1	.090	Roasted	140	1 c	266	<.1	.097
Leg of lamb, roasted:						Stewed	140	1 c	248	<.1	.069
Lean & fat	85	3 oz	219	<.1	.085	Canned with broth	142	5 oz	235	<.1	.021
Lean only	85	3 oz	162	<.1	.093	**Dark meat only:**					
Liver, braised	85	3 oz	187	.1	.200	Fried	85	3 oz	203	<.1	.079
						Roasted	85	3 oz	174	<.1	.062
PORK:						Stewed	85	3 oz	163	<.1	.047
Bacon, cooked:						**Light meat only:**					
Regular	19	3 pce	109	.1	.131	Fried	85	3 oz	163	<.1	.062
Canadian style	47	2 pce	86	.4	.383	Roasted	85	3 oz	147	<.1	.055
Breakfast strips	34	3 pce	156	.2	.251	Stewed	85	3 oz	135	<.1	.036
Center loin chop: cut 3 per lb,						**Breast*, meat and skin** =145g raw:					
which = 151g (5.3 oz) raw with						Batter-fried	140	1 ea	364	<.1	.161
bone; 124g (4.4 oz) without						Flour-fried	98	1 ea	218	<.1	.080
bone. Cooked values follow.						Roasted	98	1 ea	193	<.1	.065
Braised, lean and fat	75	1 ea	266	.22	.581	Stewed	110	1 ea	202	<.1	.045
Braised, lean only	61	1 ea	166	.45	.748	**Breast*, meat only** = 118g raw:					
Broiled, lean and fat	82	1 ea	284	.3	.870	Fried	86	1 ea	161	<.1	.068
Broiled, lean only	72	1 ea	166	.5	.827	Roasted	86	1 ea	142	<.1	.060
Pan-fried, lean and fat	89	1 ea	334	.3	.909	Stewed	95	1 ea	144	<.1	.040
Pan-fried, lean only	67	1 ea	178	.5	.838	* two pieces per bird					
Roasted, lean and fat	88	1 ea	268	.3	.727	**Drumstick, meat & skin** = 73g					
Roasted, lean only	72	1 ea	180	.4	.681	raw, ≈ 110g with bone:					
Center rib chop: cut 3 per lb,						Batter-fried	72	1 ea	193	<.1	.081
which = 151g (5.3 oz) raw with						Flour-fried	49	1 ea	120	<.1	.040
bone; ≈ 112g (3.9 oz) without						Roasted	52	1 ea	112	<.1	.036
bone. Cooked values follow.						Stewed	57	1 ea	116	<.1	.029
Braised, lean and fat	67	1 ea	246	.1	.355	**Drumstick, meat only** = 62g raw:					
Braised, lean only	53	1 ea	147	.2	.310	Fried	42	1 ea	82	<.1	.032
Broiled, lean and fat	77	1 ea	264	.2	.604	Roasted	44	1 ea	76	<.1	.033
Broiled, lean only	63	1 ea	162	.3	.563	Stewed	46	1 ea	78	<.1	.025
Pan-fried, lean and fat	88	1 ea	343	.2	.560	**Thigh, meat & skin** = 94g raw,					
Pan-fried, lean only	62	1 ea	160	.3	.476	≈ 120g with bone:					
Roasted, lean and fat	79	1 ea	252	.2	.465	Batter-fried	86	1 ea	238	<.1	.102
Roasted, lean only	66	1 ea	162	.3	.422	Flour-fried	62	1 ea	162	<.1	.058
Pork roast:						Roasted	62	1 ea	153	<.1	.042
Leg, lean & fat	85	3 oz	250	.2	.539	Stewed	68	1 ea	158	<.1	.039
Leg, lean only	85	3 oz	187	.3	.587	**Thigh, meat only** = 69g raw:					
Loin/rib, lean & fat	85	3 oz	265	.2	.601	Flour fried	52	1 ea	113	<.1	.046
Loin/rib, lean only	85	3 oz	206	.3	.658	Roasted	52	1 ea	109	<.1	.038
Shoulder, braised from 6.8 oz						Stewed	55	1 ea	107	<.1	.035
raw meat with bone:						Chicken liver, simmered	20	1 ea	30	.1	.030
Lean and fat	85	3 oz	293	.2	.460						
Lean only	67	2.4 oz	166	.2	.402	**DUCK**, domestic, roasted:					
Spareribs, ckd fr/ 1 lb raw	177	6.25 oz	703	.1	.722	Meat and skin	85	3 oz	286	<.1	.148
						Meat only	85	3 oz	171	.1	.221
PORK, CURED — HAM (also see Lunchmeat											
section, turkey ham, and Bacon under Pork)						**GOOSE**, domestic, roasted:					
Roasted, lean and fat	85	3 oz	207	.3	.551	Meat and skin	85	3 oz	259	<.1	.065
Roasted, lean only	85	3 oz	133	.4	.578	Meat only	85	3 oz	202	<.1	.078
Canned, average	85	3 oz	142	.6	.817						
						TURKEY:					
RABBIT, roasted	85	3 oz	175	<.1	.060	Ground, cooked	100	3.5 oz	229	<.1	.054
						All types, roasted	140	1 c	238	<.1	.087
VEAL (calf):						Dark meat, roasted	85	1 c	159	<.1	.053
Cutlet, braised/broiled	100	3.5 oz	181	<.1	.050	Light meat, roasted	85	1 c	133	<.1	.052
Rib roast	85	3 oz	151	<.1	.050	Slices with gravy fr/frozen	142	5 oz	95	<.1	.034
Liver, pan-fried	85	3 oz	208	.1	.213	Turkey liver, cooked	75	1 ea	127	<.1	.039
						Turkey patty, breaded, fried	64	1 ea	181	<.1	.064
VENISON (deer), roasted	85	3 oz	134	.4	.510						

Thiamin (Vitamin B1) mg

Meats: Sausages & Lunchmeats

	Wt (g)	Svg	Cal	Thiamin 100 Cal	Thiamin Svg
Beef lunchmeat, loaf or roll	28	1 oz	87	<.1	.031
Beerwurst, beer salami, beef	23	1 pce	75	<.1	.026
Beerwurst, beer salami, pork	23	1 pce	55	.2	.127
Berliner sausage	23	1 pce	53	.2	.087
Bologna:					
Beef and pork	28	1 oz	89	.1	.049
Pork, cured	23	1 pce	57	.2	.120
Braunschweiger sausage	57	2 oz	205	<.1	.143
Brotwurst, link	70	1 ea	226	<.1	.175
Cheesefurter (cheese smokies)	43	1 ea	141	<.1	.106
Chicken roll, light meat	57	2 oz	90	<.1	.037
FRANKFURTER (hotdog):					
Beef, 8 per package	57	1 ea	184	<.1	.029
Beef and pork, 8 per package	57	1 ea	183	.1	.113
Chicken, 10 per package	45	1 ea	115	<.1	.030
Turkey, 10 per package	45	1 ea	102	<.1	.037
HAM, cured pork:					
Lunchmeat, extra lean	57	2 oz	75	.7	.528
Lunchmeat, regular	57	2 oz	103	.5	.489
Chopped	42	2 pce	98	.3	.246
Minced	42	2 pce	110	.3	.300
Ham and cheese roll or loaf	57	2 oz	147	.2	.341
Ham salad spread	60	1/4 c	130	.2	.260
Italian sausage link, cooked	67	1 ea	216	.2	.417
Knockwurst, link	68	1 ea	209	.1	.233
Luncheon meat, canned	21	1 pce	70	.1	.077
Luxury loaf	57	2 oz	80	.5	.401
Peppered loaf	57	2 oz	198	<.1	.054
Pickle and pimento loaf	57	2 oz	149	.1	.166
Polish sausage	28	1 oz	92	.2	.142
PORK SAUSAGE, cooked:					
Link	13	1 ea	48	.2	.096
Patty (yield from 2 oz raw)	27	1 pce	100	.2	.200
Brown and serve, links	13	1 ea	50	.1	.050
Salami:					
Beef	23	1 pce	58	<.1	.029
Beef and pork, dry	20	2 pce	85	.1	.120
Pork and beef (1 piece = 1 oz)	57	2 oz	143	.1	.136
Turkey (1 piece = 1 oz)	57	2 oz	111	.1	.059
Smoked link sausage:					
Beef and pork	68	1 ea	229	.1	.177
Pork	68	1 ea	265	.2	.476
Turkey ham or pastrami	57	2 oz	73	.1	.042
Turkey roll, light and dark	57	2 pce	84	.1	.052
Turkey roll, light meat	57	2 pce	83	.1	.050

Mixed Dishes & Fast Foods

	Wt (g)	Svg	Cal	Thiamin 100 Cal	Thiamin Svg
Beef & vegetable stew, recipe	245	1 c	220	<.1	.150
Beef & vegetable stew, canned	245	1 c	194	<.1	.067
Beef, macaroni, tomato sauce					
recipe	226	1 c	189	.1	.191
Beef pot pie, from frozen	234	1 ea	426	<.1	.180
Burrito, bean, average	182	1 ea	341	.1	.46
Burrito, beef	177	1 ea	463	<.1	.260
Chicken egg roll	100	1 ea	242	<.1	.120
Chicken pot pie:					
Recipe, 1/3 pie	232	1 pce	545	.1	.320
From frozen	230	1 ea	430	<.1	.167
Chili with beans, canned	255	1 c	286	<.1	.122
Chop suey with beef and pork	250	1 c	300	.1	.280
Cole slaw	120	1 c	84	.1	.079
Corn dog	111	1 ea	330	.1	.280
Corn pudding	250	1 c	271	.4	1.03

	Wt (g)	Svg	Cal	Thiamin 100 Cal	Thiamin Svg
Corned beef hash, canned	220	1 c	382	<.1	.123
Egg salad	183	1 c	438	<.1	.120
Enchilada	230	1 ea	235	.1	.180
Enchirito	213	1 ea	382	.1	.388
Fajita (steak taco)	142	1 ea	235	.2	.405
French toast, recipe	65	1 pce	123	.1	.153
LASAGNA:					
With meat, recipe	245	1 pce	398	<.1	.216
Without meat, recipe	218	1 pce	316	.1	.206
Frozen entree	205	1 pce	275	.1	.190
Macaroni and cheese	200-240	1 c	230	<.1	.10-.12
Macaroni salad, without cheese	141	1 c	371	<.1	.098
Manicotti, frozen entree	225	1 ea	271	.1	.240
Meat loaf, beef only	87	1 pce	193	<.1	.058
Meat loaf, beef and 1/3 pork	87	1 pce	212	.1	.189
Moussaka, lamb and eggplant	250	1 c	250	.1	.253
Pies, fried, commercial:					
Apple pie	85	1 ea	255	<.1	.090
Cherry pie	85	1 ea	250	<.1	.060
PIZZA, cheese:					
Regular crust, 1/8 of 15"	120	1 pce	290	.1	.340
Thick crust, 1/2 of 10"	208	1 pce	519	.1	.678
Potato salad w/mayo and eggs	250	1 c	358	<.1	.193
Quiche Lorraine, 1/8 pie	176	1 pce	600	<.1	.110
SANDWICHES, Fast food items:					
Cheeseburger, 3 oz meat	112	1 ea	300	.1	.260
Cheeseburger, 4 oz meat	194	1 ea	524	.1	.325
Chicken patty sandwich	157	1 ea	436	<.1	.289
English muffin with egg,					
cheese & bacon	138	1 ea	360	.1	.460
Fish sandwich:					
Large, without cheese	170	1 ea	470	.1	.350
Regular, with cheese	140	1 ea	420	.1	.322
Hamburger, 3 oz meat patty	98	1 ea	245	.1	.230
Hamburger, 4 oz meat patty	174	1 ea	445	.1	.382
Hotdog (frankfurter) and bun	85	1 ea	260	.1	.286
Roast beef on bun	150	1 ea	345	.1	.390
SANDWICHES (on part whole wheat bread, except when stated as rye):					
Bacon, lettuce, and tomato	135	1 ea	327	.1	.418
Ham & cheese (American)	151	1 ea	363	.2	.806
Ham on rye bread	116	1 ea	242	.3	.744
Ham and swiss, on rye	145	1 ea	350	.2	.750
Ham salad sandwich	125	1 ea	339	.2	.520
Ham sandwich	122	1 ea	256	.3	.798
Peanut butter and jam	100	1 ea	341	<.1	.296
SPAGHETTI, pasta & tomato sauce w/cheese:					
Homemade	250	1 c	260	<.1	.250
Canned	250	1 c	190	.2	.350
Taco, beef	78	1 ea	207	<.1	.030
Taco, chicken	78	1 ea	172	<.1	.039
Soft taco, beef	92	1 ea	228	.2	.387
Tostada:					
With refried beans	157	1 ea	243	.2	.10-.40
With beans and beef	192	1 ea	332	<.1	.078
With beans and chicken	157	1 ea	249	<.1	.072
Turkey pot pie, frozen	233	1 ea	416	<.1	.170

ESHA Research

Thiamin (Vitamin B1)   mg

	Wt (g)	Svg	Cal	Thiamin (mg) 100 Cal	Thiamin (mg) Svg
Nuts & Seeds					
Almonds, whole, dried	28	1 oz	167	<.1	.060
Almond butter:	16	1 T	101	<.1	.021
Brazil nuts, dry (about 7)	28	1 oz	186	.2	.284
Cashew nuts, oil roasted	28	1 oz	163	.1	.120
Chestnuts, roasted	36	1/4 c	88	.1	.087
Coconut:					
Raw, grated	80	1 c	283	<.1	.053
Flaked, sweetened	76	1 c	346	<.1	.023
Dried, unsweetened	78	1 c	515	<.1	.047
Coconut water, raw	240	1 c	46	.2	.072
Filberts (hazelnuts), whole	135	1 c	853	.1	.675
Macadamias, oil roasted	134	1 c	962	<.1	.285
MIXED NUTS w/ peanuts (almonds, brazil nuts, cashews, filberts, pecans):					
Dry roasted	137	1 c	814	<.1	.274
MIXED NUTS w/o peanut (almonds, brazil nuts, cashews, filberts and pecans): oil roasted	144	1 c	886	.1	.726
PEANUTS:					
Dried	146	1 c	827	.1	.969
Oil roasted	144	1 c	837	<.1	.364
Peanut butter, smooth style	32	2 T	190	<.1	.044
Peanut flour, defatted	60	1 c	196	.2	.420
Pecans, dried halves	28	1 oz	190	.1	.241
Pine nuts, dried, Pignola	28	1 oz	146	.2	.230
Pine nuts, dried, Pinyon	28	1 oz	161	.2	.353
Pistachios (no shells):					
Dried nuts	128	1 c	739	.1	1.05
Dry roasted	128	1 c	776	.1	.541
Poppyseed	8.8	1 T	47	.2	.075
Pumpkin/Squash seed kernels:					
Dried kernels	138	1 c	747	<.1	.290
Roasted kernels	227	1 c	1185	<.1	.250
Sesame seeds:					
Kernels, dried	38	1/4 c	221	.1	.270
Whole, dried	36	1/4 c	206	.1	.285
Soybeans, dry roasted	86	1/2 c	405	<.1	.086
Sunflower seeds:					
Dry kernels	36	1/4 c	205	.4	.825
Oil roasted	34	1/4 c	387	.1	.367
Tahini, sesame butter	15	1 T	91	.3	.238
Walnuts, chopped:					
Black	28	1 oz	172	<.1	.062
English	28	1 oz	182	.1	.108
Vegetables & Legumes					
Alfalfa sprouts	33	1 c	10	.3	.025
Amaranth leaves, fresh, chopped	28	1 c	7	.1	.008
Amaranth leaves, cooked	132	1 c	28	.1	.026
Artichoke globe, cooked from fresh, 300g with refuse	120	1 ea	60	.1	.078
Artichoke hearts:					
Cooked from frozen-pkg	240	9 oz	108	.1	.149
Marinated, jar	170	6 oz	168	<.1	.060
Asparagus:					
Fresh spears, uncooked	58	4 ea	13	.5	.066
Cooked from fresh, pieces	90	1/2 c	23	.4	.089
Cooked from frozen, pieces	90	1/2 c	25	.2	.059
Canned spears	80	4 ea	15	.3	.047
Bamboo shoots, sliced:					
Fresh, uncooked	151	1 c	41	.6	.227
Cooked from fresh	120	1 c	15	.2	.024
Canned	131	1 c	25	.1	.034

	Wt (g)	Svg	Cal	Thiamin (mg) 100 Cal	Thiamin (mg) Svg
BEANS: see also Garbanzos, Lentils, Soybeans.					
Baked beans (dry white beans with spices and sauce):					
Home prepared	253	1 c	382	.1	.344
Canned, plain/vegetarian	254	1 c	235	.2	.389
Canned with franks	257	1 c	366	<.1	.149
Canned w/pork, sweet sce	253	1 c	282	<.1	.119
Canned w/pork, tomato sce	253	1 c	247	<.1	.132
Black beans, cooked fr/dry	172	1 c	227	.2	.420
Broadbeans:					
Cooked fr/dry legume	170	1 c	186	.1	.165
Cooked fr/fresh veg	100	3.5 oz	56	.2	.128
Great northern, cooked fr/dry	177	1 c	210	.1	.280
Green (snap) beans:					
Raw, uncooked	110	1 c	34	.3	.092
Cooked from fresh	125	1 c	44	.2	.093
Cooked from frozen	135	1 c	36	.2	.065
Canned, drained	135	1 c	26	.1	.020
Canned with liquid	240	1 c	36	.2	.060
Hyacinth, cooked fr/dry	194	1 c	228	.3	.524
Kidney:					
Cooked fr/dry	177	1 c	225	.1	.283
Canned	256	1 c	208	.1	.279
Lima beans:					
Cooked from fresh	170	1 c	208	.1	.238
Cooked from frozen, all	88	1/2 c	90	.1	.063
Cooked from dry	188	1 c	217	.1	.303
Canned, drained	170	1 c	164	<.1	.060
Canned with liquid	241	1 c	191	.1	.133
Navy beans, cooked fr/dry	182	1 c	259	.1	.368
Pinto beans:					
Canned	240	1 c	186	.1	.242
Cooked fr/dry	171	1 c	235	.1	.318
Refried beans, canned	253	1 c	270	<.1	.124
White beans, cooked fr/dry	179	1 c	253	.2	.422
Winged beans:					
Cooked fr/raw veg, sliced	62	1 c	23	.2	.053
Cooked fr/dry legume	172	1 c	252	.2	.507
Yardlong, cooked fr/dry	171	1 c	202	.2	.363
Yellow wax beans: see green beans.					
Bean sprouts (mung beans):					
Fresh sprouts	104	1 c	31	.3	.087
Cooked from fresh, stir fried	124	1 c	62	.3	.174
Cooked, boiled	124	1 c	26	.2	.062
Canned, drained	125	1 c	16	.2	.038
Beet greens, cooked from fresh	144	1 c	40	.4	.168
Beets, diced:					
Cooked from fresh	85	1/2 c	26	.1	.026
Canned, drained	85	1/2 c	27	<.1	.010
BROCCOLI chopped:					
Fresh, uncooked	88	1 c	24	.2	.058
Cooked from fresh	156	1 c	44	.2	.086
Cooked from frozen	184	1 c	51	.2	.101
Brussels sprouts, cooked:					
From fresh	156	1 c	60	.3	.166
From frozen	155	1 c	65	.2	.160
CABBAGE:					
Common, fresh, shredded	70	1 c	16	.2	.036
Common, cooked	150	1 c	32	.3	.086
Bok choy, fresh, shredded	70	1 c	9	.3	.028
Bok choy, cooked	170	1 c	20	.3	.054
Red, fresh, chopped	70	1 c	19	.2	.045
Red, cooked	75	1/2 c	16	.2	.026
Savoy, fresh, chopped	70	1 c	20	.2	.049
Savoy, cooked	145	1 c	35	.2	.074
CARROTS:					
Whole, 7.5" x 1.5"	72	1 ea	31	.2	.070
Grated, fresh	55	1/2 c	24	.2	.053

Thiamin (Vitamin B1) mg

Vegetables & Legumes, continued:	Wt (g)	Svg	Cal	Thiamin (mg) 100 Cal	Thiamin (mg) Svg		Wt (g)	Svg	Cal	Thiamin (mg) 100 Cal	Thiamin (mg) Svg
CARROTS, continued:						Onions, Spring, chopped, all	50	1/2 c	16	.2	.028
Cooked from fresh, sliced	78	1/2 c	35	.1	.027	Parsley, fresh, chopped	30	1/2 c	10	.2	.024
Cooked from frozen, sliced	73	1/2 c	26	.1	.020	Parsnips, cooked, sliced	156	1 c	125	.1	.130
Canned, drained	73	1/2 c	17	.1	.013						
Canned with liquid	123	1/2 c	28	.1	.023	**PEAS:**					
Carrot juice	123	1/2 c	49	.2	.113	Black-eyed peas:					
Cauliflower:						Cooked from fresh	165	1 c	160	.1	.167
Fresh, uncooked	50	1/2 c	12	.3	.038	Cooked from frozen	170	1 c	224	.2	.422
Cooked from fresh	62	1/2 c	15	.3	.039	Cooked from dry	171	1 c	198	.2	.345
Cooked from frozen	90	1/2 c	17	.2	.034	Canned	240	1 c	184	.1	.182
Celery:						Green peas:					
Raw, chopped	60	1/2 c	10	.3	.033	Fresh, uncooked	73	1/2 c	59	.33	.194
Cooked fr/fresh	150	1 c	27	.2	.065	Cooked from fresh	80	1/2 c	67	.31	.207
Chard, Swiss, cooked fr/fresh	175	1 c	35	.2	.060	Cooked from frozen	80	1/2 c	63	.35	.226
Collards:						Canned, drained	85	1/2 c	59	.17	.103
Cooked fr/fresh	128	1 c	35	.1	.027	Canned w/liquid	124	1/2 c	61	23	.139
Cooked fr/frozen	170	1 c	63	.1	.080	Green peas, edible-pods:					
						Fresh, uncooked	145	1 c	61	.4	.218
CORN:						Cooked from frozen	80	1/2 c	42	.1	.051
Fresh, kernels	77	1/2 c	66	.2	.154	Cooked from fresh	160	1 c	67	.3	.205
Cooked from fresh, kernels	82	1/2 c	89	.2	.176	Split peas, cooked fr/dry	196	1 c	231	.16	.372
Cooked from frozen, kernels	82	1/2 c	67	.1	.057	Peas & carrots:					
Canned, drained	82	1/2 c	66	<.1	.025	Cooked from frozen	80	1/2 c	38	.5	.180
Canned with liquid	128	1/2 c	79	<.1	.033	Canned with liquid	128	1/2 c	48	.2	.095
Canned, vacuum pack	105	1/2 c	83	.1	.043	Peas, sprouted, mature:					
Cream style corn, canned	128	1/2 c	93	<.1	.032	Fresh peas	120	1 c	154	.2	.270
Cucumber, whole, 8" x 2+"	301	1 ea	39	.2	.090	Cooked from fresh	100	3.5 oz	118	.2	.216
Dandelion greens:											
Raw, chopped	55	1 c	25	.4	.105	**PEPPERS, HOT** (1 pod ≈ 45g):					
Cooked	105	1 c	35	.4	.137	Chili peppers, fresh, chopped	75	1/2 c	30	.2	.068
Eggplant, cooked from fresh	160	1 c	45	.3	.122	Jalapenos, canned, chopped	68	1/2 c	17	.1	.020
Garbanzo beans (chickpeas):											
Cooked from dry	164	1 c	269	.1	.190	**PEPPERS, SWEET,** green or					
Canned	240	1 c	285	<.1	.07	red, chopped (1 pod ≈ 73g):					
Jerusalem artichoke, slices	150	1 c	114	.3	.300	Fresh, chopped	50	1/2 c	14	.2	.033
Jicama, sliced	100	3.5 oz	20	.1	.020	Cooked from fresh	68	1/2 c	19	.2	.040
Kale, chopped:						Poi, two finger	240	1 c	269	.1	.312
Fresh leaves, chopped	67	1 c	33	.2	.074						
Cooked fr/fresh	130	1 c	42	.2	.069	**POTATOES:**					
Cooked fr/frozen	130	1 c	39	.1	.056	Baked in oven:					
Kohlrabi, fresh chopped	140	1 c	38	.2	.070	(4.75" long, 2-1/3" diam)					
Kohlrabi, cooked fr/fresh	165	1 c	48	.1	.066	Flesh and skin	156	1 ea	145	.1	.164
Lambquarters:						Flesh only	202	1 ea	220	<.1	.216
Chopped, fresh	56	1 c	24	.4	.090	Potato skin	58	1 ea	115	<.1	.071
Cooked fr/fresh	180	1 c	58	.3	.180	Baked in microwave:					
Leeks, chopped, fresh	104	1 c	63	.1	.062	Flesh and skin	156	1 ea	156	.1	.201
Lentils, cooked from dry	198	1 c	231	.1	.335	Flesh only	202	1 ea	212	.1	.242
Lentils, sprouted:						Boiled, peeled afterward	136	1 ea	119	.1	.144
Fresh	77	1 c	81	.2	.176	Boiled without skin	135	1 ea	116	.1	.132
Stir-fried	100	3.5 oz	101	.2	.220	Canned, 1" diam.	70	2 ea	42	.1	.048
						French fries, from frozen:					
LETTUCE, chopped:						Fried in oil	50	10 strips	158	.1	.089
Butterhead	56	1 c	7.3	.5	.034	Oven heated	50	10 strips	111	.1	.061
Iceberg	56	1 c	7.3	.4	.026	Hash browns, from frozen	156	1 c	340	.1	.173
Loose leaf	56	1 c	10	.3	.028	Mashed potatoes:					
Romaine	56	1 c	9	.6	.056	Prepared with milk	210	1 c	162	.1	.185
Mushrooms:						Prepared from instant	215	1 c	239	.1	.336
Fresh, sliced	35	1/2 c	9	.4	.036	Potato puffs (tater tots)	62	1/2 c	138	.1	.122
Cooked from fresh	78	1/2 c	21	.3	.057	Potatoes au gratin:					
Canned, drained	78	1/2 c	19	.3	.048	Recipe	245	1 c	322	<.1	.157
Mustard greens:						From dry mix	245	1 c	228	<.1	.049
Fresh, chopped	56	1 c	15	.3	.045	Potatoes, scalloped:					
Cooked from frozen	150	1 c	29	.2	.060	Recipe	245	1 c	210	.1	.169
Cooked from fresh	140	1 c	21	.3	.057	From dry mix	245	1 c	228	<.1	.047
Okra pods, cooked fr/fresh	85	8 ea	27	.4	.112	Potato chips	28	14 ea	148	<.1	.041
Okra slices, cooked fr/frozen	92	1/2 c	34	.3	.091	Pumpkin:					
Onions:						Cooked from fresh	245	1 c	50	.2	.076
Fresh, chopped	160	1 c	61	.1	.068	Canned (can be a mixture of pumpkin & squash)	245	1 c	84	.1	.060
Cooked from fresh	105	1/2 c	46	.1	.044	Radish seeds, sprouted	38	1 c	16	.3	.039

	Wt (g)	Svg	Cal	Thiamin (mg) 100 Cal	Svg
Vegetables & Legumes, continued:					
Rutabaga:					
Fresh, cubed	140	1 c	51	.2	.126
Cooked from fresh	85	1/2 c	29	.2	.061
Sauerkraut, canned w/liquid	236	1 c	44	.1	.050
Seaweed, lavar, fresh	28	1 oz	10	.3	.028
Seaweed, spirulina, dried	28	1 oz	82	.8	.675
Soybeans:					
Fresh, green, uncooked	256	1 c	376	.3	1.12
Cooked from fresh	90	1/2 c	127	.2	.234
Cooked from dry	172	1 c	298	.1	.267
Soybeans, sprouted:					
Fresh beans	35	1/2 c	45	.3	.119
Cooked, steamed	94	1/2 c	76	.3	.193
Stir fried	100	3.5 oz	125	.3	.420
SOYBEAN PRODUCTS: see tofu in this section; miso, natto, and tempeh in Other; roasted soybeans in Nuts and Seeds; soy flour in Grains, and soy milk in Dairy.					
Spinach:					
Fresh, chopped	56	1 c	12.3	.4	.044
Cooked from fresh	180	1 c	41	.4	.171
Cooked from frozen, leaf	190	1 c	53	.2	.114
Canned, drained	214	1 c	50	.1	.034
SQUASH, SUMMER, sliced:					
Crookneck, fresh slices	130	1 c	24	.3	.068
Crookneck, cooked	180	1 c	36	.2	.088
Scallop, fresh slices	130	1 c	24	.4	.091
Scallop, cooked	90	1/2 c	14	.3	.046
Zucchini, fresh slices	130	1 c	19	.5	.091
Zucchini, cooked	180	1 c	29	.3	.074
SQUASH, WINTER, mashed:					
Acorn (Danish), baked	245	1 c	137	.3	.409
Acorn (Danish), boiled	245	1 c	83	.3	.245
Butternut, baked	245	1 c	99	.2	.177
Butternut, cooked from frozen	240	1 c	94	.1	.120
Hubbard, baked	240	1 c	120	.1	.178
Hubbard, boiled	236	1 c	70	.1	.099
Spaghetti, baked or boiled	155	1 c	45	.1	.059
Succotash, cooked from frozen	170	1 c	158	.1	.126
Sweet potato, 5" x 2":					
Baked in skin, then peeled	114	1 ea	118	.1	.083
Boiled (peeled first)	151	1 ea	160	.1	.080
Canned, mashed, regular	128	1/2 c	129	<.1	.034
Canned, vacuum pack	128	1/2 c	117	<.1	.047
Tofu, raw:					
Firm	126	1/2 c	183	.1	.199
Regular	124	1/2 c	94	.1	.100
TOMATOES:					
Fresh, whole, 2.6" diam.	123	1 ea	26	.3	.073
Fresh, chopped	180	1 c	38	.3	.108
Cooked from fresh	240	1 c	65	.3	.168
Canned, whole	240	1 c	47	.2	.108
Tomato juice, canned	244	1 c	42	.3	.115
Tomato paste, canned	262	1 c	220	.2	.406
Tomato puree, canned	250	1 c	102	.2	.178
Tomato sauce, canned	245	1 c	74	.2	.162
Turnips:					
Fresh cubes	130	1 c	35	.15	.052
Cooked from fresh	78	1/2 c	14	.2	.021
Turnip greens, cooked fr/fresh	72	1/2 c	14	.2	.032

	Wt (g)	Svg	Cal	Thiamin (mg) 100 Cal	Svg
VEGETABLES, MIXED:					
Broccoli, cauliflower, and red pepper, fr/frzn	95	2/3 c	25	.2	.060
Mixed vegetables (corn, peas, carrots, green beans and lima beans):					
From frozen	182	1 c	107	.1	.129
Canned, drained	163	1 c	77	.1	.075
Peas, carrots and onions	91	1/2 c	54	.3	.170
Peas and mushrooms	95	1/2 c	73	.3	.250
Peas and onions	95	1/2 c	71	.3	.240
Vegetable juice cocktail	242	1 c	46	.2	.104
Yam, orange: see Sweet potato.					
Yam, white, cooked fr/fresh	136	1 c	158	.1	.129
Zucchini: see Squash, summer.					

Other

Cooking ingredients, some beverages, candy, condiments, flavorings, sweeteners, spices, etc.

	Wt (g)	Svg	Cal	Thiamin (mg) 100 Cal	Svg
BEVERAGES: also see milk beverages & fruit juices.					
Beer (12 fl oz = 1.5 c)	356	1.5 c	146	<.1	.021
Light beer (12 fl oz = 1.5 c)	354	1.5 c	100	<.1	.032
Grape drink, non-carbonated	250	1 c	112	.1	.079
Pineapple grapefruit drink	250	1 c	121	.1	.075
Pineapple orange drink	250	1 c	121	.1	.075
CANDY:					
Chocolate coated almonds	165	1 c	935	<.1	.300
Chocolate coated peanuts	170	1 c	954	.1	.515
Chocolate coated raisins	187	1 c	733	<.1	.225
Catsup	15	1 T	16	.1	.013
Chocolate, baking, unsweetened	28	1 oz	145	<.1	.015
Cocoa powder	86	1 c	224	<.1	.070
Hummous, Humous	123	1/2 c	210	<.1	.113
Miso (soybean product)	138	1/2 c	284	<.1	.134
Molasses, blackstrap	40	2 T	85	<.1	.040
Natto (soybean product)	88	1/2 c	187	.1	.141
Poppyseed	9	1 T	47	.2	.075
SPICES:					
Caraway seed	6.7	1 T	22	.1	.026
Chili powder	7.5	1 T	24	.1	.026
Coriander leaf, dried	1.8	1 T	5	.5	.023
Cumin seed	6	1 T	2	2.2	.038
Fenugreek seed	11	1 T	3	6.1	.036
Garlic cloves	12	4 ea	18	.1	.024
Garlic powder	8.4	1 T	28	.1	.039
Onion powder	6.5	1 T	15	.2	.027
Paprika	6.9	1 T	2	.2	.045
Thyme, ground	4.3	1 T	1	2.2	.022
Tempeh (soybean product)	83	1/2 c	165	.1	.109
Yeast:					
Brewer's	8	1 tbsp	25	5.0	1.25
Dry active, package	7	1 pkg	20	.8	.165

Riboflavin   (Vitamin B2)

Riboflavin is needed to release stored energy for use, and is essential to the functioning of vitamins B6 and niacin. As a component of several different enzymes, riboflavin is essential to many steps in the metabolism of carbohydrates, fats, and protein. People who are more active need more riboflavin.

Riboflavin is also necessary for building and maintaining body tissues, for making red blood cells, for helping the body protect itself from common skin and eye disorders, and in the synthesis of corticosteroids. Deficiencies may cause an inflamed mouth with cracks in the corners, scaly, dry facial skin, or similar symptoms.

Vitamin Losses

It is relatively stable to heat if it is dry or in an acid medium. However, it decomposes in a heated alkaline solution, such as soda. Small amounts will leach into cooking water since it is water soluble, but if the cooking water is used there should be fairly complete retention.

Riboflavin is most sensitive to light, and is destroyed by the ultraviolet rays of the sun. Ever wonder why milk is seldom sold in clear glass bottles? Fifty percent of the riboflavin in a clear bottle is destroyed in two hours if exposed to direct sunlight, and about 20% if it is an overcast day.

Riboflavin is one of the vitamins that is processed out of the whole grain when it is milled into flour, and rice when it is polished. It is put back into white flour and cornmeal through "enrichment," but it is not put back into white rice.

Recommended Dietary Allowances for Adults
USA RDA: Men, 1.7 mg; Women, 1.3 mg (pregnant, 1.6 mg; nursing, 1.8 mg), and a minimum of 1.2 mg/day.
   The recommended amounts of riboflavin are related to calorie intake — 0.6 mg per 1000 calories of food
   consumed.
Canadian RNI: 0.5 mg/1000 kcalories of daily intake, with a minimum of 1.0 mg/day for age 19 and older.

Food Sources

Riboflavin is widely distributed in animal protein (meat, poultry, fish), eggs, and especially dairy products. Other good sources are fortified cereals, and baked goods made with enriched white wheat flour; and green vegetables (broccoli, asparagus, turnip greens, spinach).

Baked Goods

Breads, Cakes, Cookies, Crackers, Muffins, Pancakes, Pastries, Pies, Rolls, some desserts. Recipes here use flour enriched according to U.S. standards. Baked goods containing <u>unenriched</u> white flour will have much less vitamin B-2 than shown.

Food	Wt (g)	Svg	Cal	100 Cal	Svg
Apple crisp, 3" x 3"	78	1 pce	146	<.1	.037
Bagel, 3.5" dm, plain or egg	68	1 ea	180	.1	.197
Biscuits, from mix	28	1 ea	94.4	.1	.108
Biscuits, from refrig dough	20	1 ea	65.2	.1	.052
BREADS (vitamin B2 is added to white all-purpose flour in USA):					
Cornbread muffin, avg	45	1 ea	145	.1	.100
Cracked wheat	25	1 pce	65	.1	.095
French, 5" x 2.5" x 1"	35	1 pce	100	.1	.120
Mixed grain	25	1 pce	65	.2	.100
Oatmeal bread	25	1 pce	65	.1	.066
Pita pocket bread, enr. 6.5"	60	1 pce	165	.1	.130
Pumpernickel, 5" x 4" x 3/8"	32	1 pce	80	.2	.166
Raisin bread	25	1 pce	68	.2	.155
Rye, light, 5" x 3.5" x 7/16"	25	1 pce	65	.1	.080
Vienna bread	25	1 pce	70	.1	.089
Wheat (white & whole wheat flour)	28	1 pce	72	.1	.089
White bread	28	1 pce	75	.1	.087
Whole wheat bread	28	1 pce	70	.1	.059
CAKES, pce = 1/16th cake unless otherwise stated:					
Angel food, 1/12 tube	53	1 pce	125	.1	.106
Boston cream pie, 1/8	120	1 pce	260	.1	.180
Carrot cake, cream cheese frosting 2.5" x 3"	112	1 pce	406	<.1	.130
Cheesecake:					
From recipe, 1/12	92	1 pce	278	<.1	.119
From mix, 1/8	103	1 pce	300	.1	.270
Chocolate, choc. frosting	69	1 pce	235	<.1	.100
Snack cake, filled:					
Chocolate (Ding Dongs)	28	1 ea	105	.1	.090
Spongecake (Twinkies)	42	1 ea	155	<.1	.060
Coffee cake, 2.4" x 2.8"	72	1 pce	230	.1	.151
Gingerbread, f/mix, 3" x 3"	63	1 pce	174	.1	.114
Pound cake, 1/2", f/mix	30	1 pce	115	.1	.060
Sheet cake:					
Plain, 3" x 3"	86	1 pce	315	<.1	.150
W/white frosting	121	1 pce	445	<.1	.160
Sponge cake, 1/12 tube	66	1 pce	194	.1	.107
White cake:					
W/chocolate frosting	77	1 pce	291	.1	.151
W/coconut or white frosting	70	1 pce	265	<.1	.131
Yellow, choc. frosting	69	1 pce	245	.1	.140
Cheese puffs (Cheetos)	28	1 oz	158	<.1	.030
Chips: see corn & tortilla this section; potato chips under Vegetables.					
COOKIES:					
Animal (1 box = 1 oz)	28	27 ea	120	.1	.130
Chocolate chip:					
Commercial	42	4 ea	180	.1	.230
From refrigerator dough	48	4 ea	225	<.1	.101
Recipe	40	4 ea	185	<.1	.060
Fig bars	56	4 ea	210	<.1	.073
Lady fingers	44	4 ea	158	<.1	.060
Oatmeal raisin	52	4 ea	245	<.1	.080
Peanut butter, recipe	48	4 ea	245	<.1	.070
Sandwich type, all	40	4 ea	195	<.1	.070
Shortbread cookies, comm.	32	4 ea	155	.1	.090
Sugar, from chilled dough	48	4 ea	235	<.1	.060
Snickerdoodle	20	1 ea	110	<.1	.047
Vanilla wafer cookies	40	10 ea	185	.1	.100
Corn chips, like Fritos	28	1 oz	155	<.1	.050
CRACKERS:					
Armenian cracker bread	28	4 pce	117	<.1	.040
Cheese cracker	10	10 ea	50	.1	.040
Cheese cracker with peanut butter filling	30	4 ea	150	.1	.120
Graham cracker	14	2 ea	60	.1	.030
Oyster (10 = 7 g)	28	1 oz	120	.1	.140
Round, like Ritz	9	3 ea	45	.1	.030
Rye wafers, whole grain	14	2 ea	55	.1	.030
Saltines	12	4 ea	50	.1	.050
Sesame	12	4 ea	60	.1	.040
Wheat, thin	8	4 ea	35	.1	.030
Whole wheat cracker	8	2 ea	35	.1	.030
Croissant, 4.5" x 4" x 2"	57	1 ea	235	<.1	.130
Croutons:					
Dry bread cubes	30	1 c	111	.1	.105
Herb seasoned	28	1 oz	99	.2	.200
Danish pastry, plain	57	1 ea	220	<.1	.170
Danish pastry with fruit	65	1 ea	235	<.1	.140
Doughnuts:					
Cake type, medium	50	1 ea	210	.1	.120
Yeast-raised, plain	60	1 ea	235	<.1	.120
Yeast-raised, jelly filled	65	1 ea	226	<.1	.100
English muffin:					
Plain	57	1 ea	140	.1	.182
Sourdough	56	1 ea	129	.1	.140
With raisins	56	1 ea	146	.1	.170
MUFFINS:					
Blueberry, recipe	45	1 ea	135	.1	.110
Blueberry, from mix	45	1 ea	140	.1	.126
Bran, wheat, recipe	45	1 ea	125	.1	.130
Bran, wheat, from mix	45	1 ea	140	.1	.120
Cornmeal, recipe or mix	45	1 ea	145	.1	.100
Pancakes:					
Buckwheat, f/mix, 4" dm.	27	1 ea	55	.1	.050
Plain, f/mix, 4" dm.	27	1 ea	60	.2	.120
Whole wheat, 5" dm.	52	1 ea	94	.1	.088
PIES: pce = 1/6 of 9" pie.					
Apple	158	1 pce	405	<.1	.128
Banana cream	198	1 pce	319	.1	.257
Blueberry	158	1 pce	380	<.1	.142
Cherry	158	1 pce	410	<.1	.142
Chocolate cream	175	1 pce	311	.1	.300
Coconut cream	172	1 pce	343	.1	.257
Coconut custard	165	1 pce	384	.1	.318
Cream, commercial	152	1 pce	455	<.1	.150
Custard	152	1 pce	293	.1	.267
Lemon meringue	140	1 pce	355	<.1	.140
Mincemeat	160	1 pce	395	<.1	.136
Peach	158	1 pce	405	<.1	.155
Pecan	138	1 pce	583	<.1	.165
Pumpkin	200	1 pce	367	.1	.293
PopTart-type pastry, fortified	54	1 ea	210	.1	.184
Pretzels, dutch twist	16	1 ea	65	.1	.040
Pretzels, thin twist	60	10 ea	240	.1	.150
ROLLS:					
Dinner rolls, 2.5" x 2"	28	1 ea	85	.1	.090
Hamburger bun	45	1 ea	129	.1	.148
Hard roll, white	50	1 ea	155	.1	.120
Hotdog bun	40	1 ea	115	.1	.132
Rye roll, dark	28	1 ea	79	.2	.121
Rye roll, light	28	1 ea	76	.1	.090
Submarine roll (Hoagie)	135	1 ea	400	.1	.330
Whole wheat roll	35	1 ea	88	.1	.073

Riboflavin (Vitamin B2) mg

	Wt (g)	Svg	Cal	Riboflavin (mg) 100 Cal	Riboflavin (mg) Svg
Baked Goods, continued:					
Tortillas:					
Corn, enr., fried, 6" dm.	30	1 ea	87	<.1	.030
Flour, large, 10.5" dm.	57	1 ea	168	.1	.125
Flour, regular, 8" dm.	35	1 ea	105	.1	.078
Tortilla chips:					
Plain	28	1 oz	139	<.1	.020
Doritos, nacho flavor	28	1 oz	139	<.1	.030
Doritos, taco flavor	28	1 oz	140	.1	.090
Waffles:					
Homemade, 7" dm.	75	1 ea	245	.1	.240
From mix, 7" dm.	75	1 ea	205	.1	.230
From frozen, 4" dm.	35	1 ea	98	.2	.185

Dairy & Dairy Products

	Wt (g)	Svg	Cal	Riboflavin (mg) 100 Cal	Riboflavin (mg) Svg
CHEESE (1.5" cube ≈ 1 ounce):					
American, processed	28	1 oz	106	.1	.100
American cheese food,					
cold pack	28	1 oz	94	.1	.126
American cheese spread	28	1 oz	82	.1	.122
Blue	28	1 oz	100	.1	.108
Brick	28	1 oz	105	.1	.100
Brie	28	1 oz	95	.2	.147
Camembert	28	1 oz	85	.2	.138
Caraway	28	1 oz	107	.1	.128
Cheddar cheese	28	1 oz	114	.1	.106
Colby	28	1 oz	112	.1	.106
Cottage cheese:					
Lowfat 1%	226	1 c	164	.2	.373
Lowfat 2%	226	1 c	205	.2	.418
Creamed (4.5% fat):					
Large curd	225	1 c	235	.2	.370
Small curd	210	1 c	215	.2	.342
Dry curd (<.5% fat)	145	1 c	123	.2	.206
Cream cheese 1 T = 15g	28	1 oz	99	.1	.056
Edam	28	1 oz	101	.1	.110
Feta	28	1 oz	75	.3	.230
Fontina	28	1 oz	110	.1	.058
Gjetost	28	1 oz	132	.1	.100
Gouda	28	1 oz	101	.1	.095
Monterey jack	28	1 oz	106	.1	.111
Mozzarella, part skim,					
low moisture	28	1 oz	80	.1	.097
Muenster	28	1 oz	104	.1	.091
Parmesan, grated (1 T = 5g)	28	1 oz	129	.1	.109
Pimento, processed	28	1 oz	106	.1	.100
Provolone	28	1 oz	100	.1	.091
Ricotta, with part skim	246	1 c	340	.1	.455
Romano, grated (1 oz = 5.6 T)	28	1 oz	128	.1	.122
Roquefort	28	1 oz	105	.2	.166
Swiss	28	1 oz	107	.1	.103
Swiss, processed	28	1 oz	95	.1	.078
Swiss cheese food	28	1 oz	92	.1	.113
CREAM, SWEET, fluid:					
Coffee or table (19.3% fat)	15	1 T	30	.1	.022
Half and half (11.5%)	15	1 T	20	.1	.023
Light (30.9%)	239	1 c	699	<.1	.299
Heavy (37% fat)	238	1 c	821	<.1	.262
CREAM SWEET, whipped:					
Heavy, unsweetened (37% fat)	119	1 c	410	<.1	.131
Pressurized (22% fat)	60	1 c	154	<.1	.039
CREAM, SOUR, dairy:					
Cultured (21% fat)	14	1 T	30	<.1	.021
Half & half (12% fat)	15	1 T	20	.1	.022
Cream, sour, Imitation					
non-dairy (19.5% fat)	14	1 T	29	0	0

	Wt (g)	Svg	Cal	Riboflavin (mg) 100 Cal	Riboflavin (mg) Svg
CREAM SUBSTITUTES, non-dairy:					
Coffee whitener (check label):					
Liquid/frzn (10% fat)	120	1/2 c	163	0	0
Powder (35.5% fat)	94	1 c	514	<.1	.155
Dessert Toppings, non-dairy:					
Frozen (Coolwhip), 25.3% fat	75	1 c	239	0	0
Dessert mix, dry, 39.5% fat	43	1.5 oz	245	0	0
Pressurized, 22.3% fat	70	1 c	185	0	0
Kefir, beverage	233	1 c	160	.3	.440
MILK (cow):					
Skim	245	1 c	86	.4	.343
Lowfat 1% fat	244	1 c	102	.4	.407
Lowfat 2% fat	244	1 c	121	.3	.403
Whole (3.3% fat)	244	1 c	150	.3	.395
Buttermilk (< .01% fat)	245	1 c	99	.4	.377
Canned:					
Skim, evaporated	255	1 c	200	.4	.796
Whole (7.6% fat) evap.	252	1 c	340	.2	.800
Sweetened, cond (8.7% fat)	306	1 c	982	.1	1.27
Dried, nonfat, instant	68	1 c	244	.5	1.19
Dried, buttermilk	120	1 c	464	.4	1.90
Milk (other):					
Goat milk (4.1% fat)	244	1 c	168	.2	.337
Human breast (4.4% fat)	246	1 c	171	.1	.089
Soy milk (1.9% fat)	240	1 c	79	.2	.168
MILK BEVERAGES & mixes:					
Chocolate beverage mix,					
to be mixed w/water:					
Powder (includes dry milk)	28	1 oz	100	.2	.170
Drink, prep w/water	206	3/4 c	100	.2	.170
Chocolate beverage mix,					
to be mixed w/milk:					
Powder	21.6	3/4 oz	75	<.1	.032
Drink, prep w/whole milk	266	1 c	226	.2	.428
Eggnog, commercial	254	1 c	342	.1	.483
Instant Breakfast, powder	37	1 env	130	.1	.068
Malt drink, from dry mix:					
Chocolate flavor	265	1 c	229	.2	.437
Natural flavor	265	1 c	237	.2	.588
Milkshakes (1.25 c = 10 fl oz):					
Chocolate	283	1.25 c	360	.2	693
Strawberry	283	1.25 c	319	.2	552
Vanilla	283	1.25 c	314	.2	.515
MILK DESSERTS:					
Custard, baked	265	1 c	305	.2	.500
Ice cream, vanilla:					
Regular (11% fat)	133	1 c	269	.1	.329
Rich (16% fat)	148	1 c	349	.1	.283
Soft serve (≈ 13% fat)	173	1 c	377	.1	.448
Ice milk, regular	131	1 c	184	.2	.347
Ice milk, soft serve	175	1 c	223	.2	.541
Puddings (5 oz can ≈ 1/2+ c):					
Assorted, cooked f/mix:					
Regular	130	1/2 c	150	.1	.18-.20
Low calorie	130	1/2 c	69	.3	.200
Chocolate, instant/cooked	130	1/2 c	155	.1	.188
Chocolate, canned	142	1 can	205	.1	.170
Coconut, instant	149	1/2 c	184	.1	.200
Rice, instant	149	1/2 c	175	.1	.200
Tapioca:					
Recipe	165	1 c	220	.1	.300
From mix	130	1/2 c	145	.1	.180
Canned	142	1 can	160	.1	.140
Vanilla, instant/cooked	130	1/2 c	148	.1	.175
Vanilla, canned	142	1 can	220	<.1	.120
Pudding pops, all flavors	57	1 ea	93-99	.1	.110
Sherbet (2% fat)	193	1 c	270	<.1	.089

Riboflavin (Vitamin B2) mg   Values are for edible portion of foods

Dairy, Milk Desserts, continued:	Wt (g)	Svg	Cal	100 Cal	Svg		Wt (g)	Svg	Cal	100 Cal	Svg
Yogurt, frozen, avg	174	1 c	220	.1	.276	Fruit cocktail, canned:					
YOGURT:						Juice pack	248	1 c	115	<.1	.040
Lowfat, plain (1.5% fat)	227	1 c	144	.3	.486	Heavy syrup	255	1 c	185	<.1	.048
Lowfat with fruit (1.1% fat)	227	1 c	231	.2	.404	Gooseberries:					
Lowfat, coffee/vanilla (1.2% fat)	227	1 c	193	.2	.456	Fresh	150	1 c	67	.1	.045
Nonfat	227	1 c	127	.4	.531	Canned, w/liquid	252	1 c	185	.1	.134
Whole (3.25% fat)	227	1 c	138	.2	.322	Grapes, Thompson seedless:					
						Fresh	160	1 c	114	.1	.091
Eggs						Canned, heavy syrup	256	1 c	187	<.1	.056
						Grape juice:					
Egg, chicken, whole:						From frozen	250	1 c	128	.1	.065
Cooked, average	50	1 ea	77.5	.3	.256	Bottled, canned	253	1 c	155	.1	.094
Raw	50	1 ea	75	.3	.254	**GRAPEFRUIT** (half = 241g w/rind):					
White, raw	33.4	1 ea	17	.9	.151	Pink or red half	123	1 ea	37	.1	.025
Yolk, raw	16.6	1 ea	59	.2	.106	White half	118	1 ea	39	.1	.024
						Canned sections	254	1 c	152	<.1	.051
Fruits & Fruit Juices						Grapefruit juice:					
						Fresh	247	1 c	96	.1	.049
Acerola juice, fresh	242	1 c	51	.3	.150	From frozen conc.	247	1 c	102	.1	.054
APPLE, 2.75" diam.:						Canned, unsweetened	247	1 c	93	.1	.049
With peel	138	1 ea	80	<.1	.019	Kiwi fruit, medium	76	1 ea	46	.1	.040
Without peel	128	1 ea	72	<.1	.013	Lemon juice:					
Apple juice:						Fresh	244	1 c	60	<.1	.024
Canned, bottled	248	1 c	116	<.1	.042	Bottled	244	1 c	52	<.1	.022
From frozen	239	1 c	111	<.1	.036	Lime juice, fresh	246	1 c	65	<.1	.025
Applesauce, unsweetened	244	1 c	106	<.1	.006	Loganberries:					
						Fresh berries	150	1 c	105	<.1	.051
APRICOTS:						Frozen, unthawed	147	1 c	80	.1	.050
Fresh, pitted (114g w/refuse)	106	3 ea	51	.1	.042	Mandarin oranges, canned	252	1 c	155	.1	.110
Canned, juice pack	248	1 c	119	<.1	.047	Mango, fresh slices	165	1 c	108	.1	.094
Canned, heavy syrup	258	1 c	214	<.1	.057	Melon: also see Watermelon.					
Dried, halves	35	10 ea	83	.1	.053	Cantaloupe cubes	160	1 c	57	.1	.034
Apricot nectar, canned	251	1 c	141	<.1	.035	Casaba cubes	170	1 c	45	.1	.034
Avocado (1c mashed = 230g):						Honeydew cubes	170	1 c	60	.1	.031
California (227g w/refuse)	173	1 ea	305	.1	.211	Frozen melon balls, mixed	173	1 c	55	.1	.038
Florida (454g w/refuse)	304	1 ea	340	.1	.371	Nectarine, med. = 1 c slices	136	1 ea	67	.1	.056
Bananas, 8.75" long, 176g w/peel	114	1 ea	105	.1	.114						
Blackberries:						**ORANGE** 2-5/8", 180g w/rind	131	1 ea	60	.1	.052
Fresh berries	144	1 c	74	.1	.058	Orange juice:					
Frozen, unthawed	151	1 c	97	.1	.069	Fresh	248	1 c	111	.1	.074
Canned	256	1 c	236	<.1	.100	Chilled	249	1 c	110	<.1	.052
Blackberry juice, fresh	250	1 c	93	.1	.080	Prepared f/frozen	249	1 c	110	<.1	.040
Blueberries:						Canned, unsweetened	249	1 c	105	.1	.070
Fresh berries	145	1 c	82	.1	.073	Orange grapefruit juice, cnd	247	1 c	105	.1	.074
Frozen, unthawed	155	1 c	78	.1	.057	Papaya (454g w refuse)	304	1 ea	117	.1	.097
Canned	256	1 c	225	.1	.136	Passion fruit juice:					
Boysenberries:						Purple	247	1 c	126	.3	.324
Frozen, unthawed	132	1 c	66	.1	.049	Yellow	247	1 c	149	.2	.249
Canned	256	1 c	225	<.1	.074						
Cantaloupe: see Melon.						**PEACHES:**					
Cherries, sour:						Fresh, whole, 2.5" dm.	87	1 ea	37	.1	.036
Frozen, unthawed	155	1 c	72	.1	.053	Frozen, thawed	250	1 c	235	<.1	.088
Canned	244	1 c	90	.1	.100	Canned, heavy syrup	256	1 c	190	<.1	.061
						Canned, juice pack	249	1 c	109	<.1	.042
CHERRIES, SWEET:						Peach nectar, canned	249	1 c	134	<.1	.035
Fresh	145	1 c	104	.1	.087	**PEARS:**					
Frozen, unthawed	259	1 c	232	.1	.122	Bartlett, 180g w/refuse	166	1 ea	98	.1	.066
Canned with liquid	257	1 c	213	<.1	.103	Canned, juice pack	77	1 half	38	<.1	.008
Cranberry apple juice	253	1 c	169	<.1	.051	Canned, heavy syrup	79	1 half	58	<.1	.017
Cranberry juice cocktail	253	1 c	145	<.1	.020	Pear nectar, canned	250	1 c	149	<.1	.033
Currants:						Persimmon, fresh, large	168	1 ea	118	<.1	.034
Fresh, red/white avg	112	1 c	67	.1	.056						
Dried, Zante	144	1 c	407	<.1	.204	**PINEAPPLE:**					
Dates, whole, pitted	83	10 ea	228	<.1	.083	Fresh pieces	155	1 c	76	.1	.056
Fig, fresh, medium	50	1 ea	37	.1	.025	Cnd pieces, juice pack	250	1 c	150	<.1	.048
Fig, dried, large	19	1 ea	48	<.1	.017	Cnd pieces, heavy syrup	255	1 c	199	<.1	.064
						Cnd slice, juice pack	58	1 ring	35	<.1	.015
						Pineapple juice	250	1 c	135	<.1	.053
						Plantain slices, ckd f/fresh	151	1 c	180	<.1	.080

ESHA Research

22

PLUMS:	Wt (g)	Svg	Cal	100 Cal	Svg
Fresh, 2-1/8" diam.	66	1 ea	36	.2	.063
Cnd w/peel, juice pack	95	3 ea	55	.1	.056
Cnd w/peel, heavy syrup	110	3 ea	98	<.1	.042
Pomegranate juice, fresh	63	1/4 c	28	.1	.037
Prunes, dried, pitted	84	10 ea	201	.1	.136
Prune juice, bottled	256	1 c	181	.1	.179
Raisins, dark, packed meas	165	1 c	494	<.1	.145
Raspberries:					
Fresh berries	123	1 c	60	.2	.111
Frozen, thawed	250	1 c	255	<.1	.113
Cnd with liquid	256	1 c	234	<.1	.079
Raspberry juice, fresh	240	1 c	98	.4	.400
Rhubarb, fresh, diced	122	1 c	26	.1	.037
STRAWBERRIES:					
Fresh	149	1 c	45	.2	.098
Frozen, thawed, sweetened	255	1 c	245	.1	.130
Tangerine	84	1 ea	37	<.1	.018
Tangerine juice, f/frozen	241	1 c	110	<.1	.046
Watermelon, diced pieces	160	1 c	50	.1	.032

Grains & Grain Products
Cereals, Flour, Grains, Noodles, Pasta, Popcorn

	Wt (g)	Svg	Cal	100 Cal	Svg
Amaranth grain, dry	195	1 c	729	<.1	.300
Barley, cooked, pearled	157	1 c	193	<.1	.097
Buckwheat flour:					
Dark	98	1 c	338	<.1	.155
Light	98	1 c	340	<.1	.047
Bulgar wheat, cooked	182	1 c	151	<.1	.051
Carob flour	103	1 c	185	.3	.475

CEREALS, COLD (Ready to eat): Most cereals are enriched with vitamin B2. Brands vary. See the label.

CEREALS, HOT (Cooked): Some cereals are enriched with vitamin B2.

Corn grits:	Wt (g)	Svg	Cal	100 Cal	Svg
Enriched	242	1 c	145	.1	.150
Unenriched	242	1 c	146	<.1	.024
Cream of Wheat	244	1 c	140	.1	.070
Farina:					
Enriched	233	1 c	116	.1	.117
Unenriched	233	1 c	116	<.1	.023
Malt-O-Meal	240	1 c	122	.2	.240
Maypo	180	3/4 c	128	.5	.600
Oatmeal:					
From rolled oats, unfortified	234	1 c	145	<.1	.047
Instant, fortified, packet:					
Plain	177	3/4 c	104	.3	.290
With bran and raisins	195	3/4 c	158	.4	.630
Other flavors, averaged	156	3/4 c	159	.2	.325
Ralston	253	1 c	134	.1	.180
Roman Meal	181	3/4 c	111	.1	.090
Wheatena	243	1 c	135	<.1	.050
Whole wheat cereal	242	1 c	150	.1	.120
Corn flour:					
Regular	117	1 c	422	<.1	.094
Masa Harina, enriched	114	1 c	416	.2	.86
Cornmeal, dry:					
Degermed, enriched	138	1 c	505	.1	.562
Whole (not enriched)	122	1 c	435	<.1	.130

FLOUR: see specific grain, nut, or vegetable.

MACARONI, cooked:	Wt (g)	Svg	Cal	100 Cal	Svg
Enriched	140	1 c	197	.1	.137
Whole wheat	140	1 c	174	<.1	.063
Vegetable, enriched	134	1 c	172	<.1	.082
Millet, cooked from dry	120	1/2 c	143	.1	.098
Noodles:					
Chow mein, dry	45	1 c	237	.1	.189
Egg noodles, cooked	160	1 c	213	.1	.133
Spinach noodles, cooked	160	1 c	182	.1	.144
Oat bran (1T ≈ 6g)	94	1 c	132	.2	.206
Popcorn, popped in oil	11	1 c	55	<.1	.020
Quinoa grain, dry	170	1 c	635	.1	.673
RICE, cooked:					
Brown rice	195	1 c	217	<.1	.049
White, converted, enriched	175	1 c	200	<.1	.031
White, instant, enriched	165	1 c	162	<.1	.076
White, regular, enriched	205	1 c	264	<.1	.027
Wild rice	82	1/2 c	83	.1	.072
Rice bran	83	3.5 oz	262	.1	.236
Rice flour	158	1 c	578	<.1	.033
Rye flour:					
Dark	128	1 c	415	.1	.321
Light	102	1 c	361	<.1	.092
Soy flour:					
Full fat, raw	42	1/2 c	182	.3	.487
Low fat	44	1/2 c	163	.1	.125
Spaghetti noodles, cooked:					
Enriched	140	1 c	197	.1	.137
Whole wheat spaghetti	140	1 c	174	<.1	.063
Tapioca, dry	152	1 c	518	<.1	.152
WHEAT:					
Bran	30	1/2 c	65	.3	.173
FLOURS (white flour is enriched w/vitamin B2):					
All purpose, white, unsifted	125	1 c	455	.1	.618
Cake, sifted, enriched	96	1 c	348	.1	.413
Self-rising	125	1 c	442	.1	.518
Semolina	167	1 c	601	.2	.954
Whole wheat flour	120	1 c	402	.1	.258
Wheat germ:					
Raw	100	1 c	360	.1	.510
Toasted	113	1 c	432	.2	.930
Wheat, rolled:					
Cooked	240	1 c	142	<.1	.070
Dry	85	1 c	289	<.1	.100
Whole grain wheat (wheatberries) cooked	50	1/3 c	28	<.1	.120
Whole wheat, sprouted	108	1 c	214	.1	.177

Meats: Fish & Shellfish

	Wt (g)	Svg	Cal	100 Cal	Svg
Anchovies, canned in oil, drained	45	11 ea	95	.2	.163
Bluefish, fried in crumbs	100	3.5 oz	205	<.1	.080
Catfish, fried in cornmeal	100	3.5 oz	229	.1	.133
CLAMS:					
Steamed, meat only	45	10 ea	66	.3	.192
Canned, drained	160	1 c	236	.3	.682
Breaded, fried	188	20 ea	379	.1	.459
COD:					
Baked/broiled	100	3.5 oz	105	.1	.079
Batter-fried	100	3.5 oz	199	<.1	.040
Dried, salted	28	1 oz	81	.1	.067
Smoked	100	3.5 oz	79	.1	.070

ESHA Research

Riboflavin (Vitamin B2) mg

	Wt (g)	Svg	Cal	100 Cal	Svg
Fish, continued:					
CRAB:					
Alaska King leg, cooked	134	1 ea	129	.1	.074
Blue crab meat, cooked	135	1 c	138	.1	.119
Blue crab, canned	135	1 c	133	.1	.113
Dungeness crab, cooked	101	3/4 c	85	.2	.155
Crab, imitation from surimi	85	3 oz	87	<.1	.023
Crayfish, cooked, moist heat	85	3 oz	97	.1	.065
Eel:					
Baked/broiled	100	3.5 oz	236	<.1	.051
Smoked	100	3.5 oz	330	.1	.350
Fish cakes, fried f/frzn	100	3.5 oz	213	<.1	.060
Fish sticks, heated f/frzn	57	2 ea	155	.1	.101
Gefiltefish, sweet, comm.	42	1 pce	35	.1	.025
Haddock:					
Baked/broiled	85	3 oz	95	<.1	.038
Breaded, fried	85	3 oz	175	.1	.100
Smoked	100	3.5 oz	116	<.1	.049
HALIBUT, baked/broiled	85	3 oz	119	.1	.077
Herring:					
Baked/broiled	100	3.5 oz	203	.1	.299
Canned in oil	100	3.5 oz	208	.1	.180
Pickled (1 piece ≈ 15g)	100	3.5 oz	262	.1	.139
Smoked, kippered	100	3.5 oz	217	.1	.319
Lobster meat, cooked	145	1 c	142	.1	.096
Mackerel:					
Baked/broiled	100	3.5 oz	262	.2	.412
Baked/broiled, Spanish	100	3.5 oz	158	.1	.210
Canned, Jack Mackerel	361	1 can	563	.1	.765
Ocean perch:					
Baked/broiled	100	3.5 oz	121	.1	.134
Breaded, fried	85	3 oz	185	.1	.110
OYSTERS:					
Raw, Eastern	248	1 c	170	.2	.412
Raw, Pacific	248	1 c	200	.3	.578
Breaded, fried	88	6 med	173	.1	.178
Simmered	42	6 med	58	.2	.139
Pollock, baked/broiled	100	3.5 oz	99	.2	.200
Pompano, baked/broiled	100	3.5 oz	211	.1	.270
Roe, raw, mixed species	28	1 oz	39	.5	.190
SALMON, cooked:					
Average, baked/broiled	85	3 oz	183	.1	.145
Chinook, smoked	85	3 oz	99	.1	.086
Canned:					
Atlantic, small can	220	1 can	281	.3	.752
Chinook, #1 tall can	369	1 can	521	.1	.567
Pink, #1 can, drained	454	1 can	631	.1	.844
Sockeye, #1 can, drained	369	1 can	566	.1	.712
Sardines, canned, drained:					
Atlantic, 2 = 24g	92	1 can	192	.1	.209
Pacific, 1 = 38g	100	3.5 oz	174	.1	.233
Scallops:					
Breaded, fried	93	6 ea	200	.1	.102
Steamed	100	3.5 oz	113	.1	.075
Shad, bkd w/bacon	100	3.5 oz	201	.1	.260
Shark, batter-fried	85	3 oz	194	<.1	.082
SHRIMP:					
Breaded, fried (2 large = 15g)	90	12 ea	218	.1	.122
Boiled (2 large = 11g)	100	3.5 oz	99	<.1	.032
Canned, drained	128	1 c	154	<.1	.047
Shrimp, imitation f/surimi	85	3 oz	86	<.1	.029
Snapper, baked/broiled	100	3.5 oz	128	.1	.075
Sole (Flounder):					
Baked/broiled, no added fat	85	3 oz	99	.1	.097
Batter-fried	85	3 oz	250	.1	.127
Squid, flour-fried	85	3 oz	149	.3	.389
Steelhead (seatrout), cooked	100	3.5 oz	131	.2	.225
Surimi, processed walleye pollock: see imitation crab or shrimp	85	3 oz	84	<.1	.018
Swordfish, baked/broiled	85	3 oz	132	<.1	.099
Trout, baked/broiled	85	3 oz	129	.1	.191
TUNA:					
Baked/broiled, Bluefin	100	3.5 oz	159	.1	.105
Canned, drained:					
Light, oil pack	171	1 can	339	.1	.181
Light, water pack	165	1 can	216	.1	.194

Meats

Beef, Lamb, Pork, Ham, Rabbit, Frog legs, Venison and Veal

	Wt (g)	Svg	Cal	100 Cal	Svg
BEEF:					
Breakfast strips, cured beef	34	3 strips	153	.1	.088
Chuck blade, pot roasted, all:					
Lean & fat (5.4 oz raw)	85	3 oz	325	.1	.196
Lean only	85	3 oz	230	.1	.241
Ground beef, cooked average:					
Extra lean, 17% fat raw	85	3 oz	215	.10	.218
Lean, 21% fat raw	85	3 oz	231	.08	.176
Regular, 26.6% fat raw	85	3 oz	250	.06	.156
Frozen patty, broiled, 23% fat raw	85	3 oz	240	.07	.170
Rib roast, oven roasted:					
Lean & fat (5 oz raw)	85	3 oz	324	<.1	.146
Lean only	85	3 oz	204	.1	.179
Round steak, broiled, choice:					
Lean & fat (4.5 oz raw)	85	3 oz	233	.1	.175
Lean only	85	3 oz	165	.1	.195
Round tip, oven roasted:					
Lean & fat	85	3 oz	213	.1	.208
Lean only	85	3 oz	162	.1	.228
Sirloin steak, broiled, all: 11.3 oz raw = 8.2 oz lean & fat cooked; 6.9 oz lean:					
Lean & fat	85	3 oz	238	.1	.153
Lean only	85	3 oz	172	.1	.168
T-bone steak, broiled, choice: 16 oz raw = 9.7 oz lean & fat cooked; 7.4 oz lean:					
Lean & fat	85	3 oz	276	.1	.178
Lean only	85	3 oz	182	.1	.209
Variety meats:					
Heart, simmered	85	3 oz	140	.9	1.31
Kidney, cooked	140	1 ea	201	2.8	5.68
Liver, fried	85	3 oz	184	1.9	3.52
Tongue, cooked	85	3 oz	241	.1	.298
Corned beef, canned	85	3 oz	213	.1	.125
Dried beef, cured	28	1 oz	47	.5	.231
FROG LEGS, flour- fried	144	6 ea	418	.1	.350
HAM: see Cured Pork, Turkey ham, Sausages & Lunchmeats section.					
LAMB:					
Arm chop, braised: (Raw w/bone = 160g or 5.6 oz)					
Lean & fat	70	1 ea	244	.1	.18
Lean part	55	1 ea	152	.1	.15
Loin chop, broiled: (Raw, w/bone = 120g or 4.2 oz)					
Lean & fat	64	1 ea	201	.08	.16
Lean only	46	1 ea	100	.13	.13
Cutlet, lean, ckd average	85	3 oz	175	.14	.24

	Wt (g)	Svg	Cal	Riboflavin (mg) 100 Cal	Svg
LAMB, continued:					
Leg of lamb, roasted:					
Lean & fat	85	3 oz	219	.10	.23
Lean only	85	3 oz	162	.15	.25
PORK:					
Bacon, cooked:					
Regular	19	3 pce	109	<.1	.054
Canadian style	47	2 pce	86	.1	.092
Breakfast strips	34	3 pce	156	<.1	.125
Center loin chop: 3 per lb = 151g (5.3 oz) raw w/bone.					
This ≈ 124g (4.4 oz) w/o bone. Cooked values follow.					
Braised, lean & fat	75	1 ea	266	.1	.179
Braised, lean only	61	1 ea	166	.1	.165
Broiled, lean & fat	82	1 ea	284	.1	.217
Broiled, lean only	72	1 ea	166	.1	.222
Pan-fried, lean & fat	89	1 ea	333	.1	.245
Pan-fried, lean only	67	1 ea	178	.1	.221
Roasted, lean & fat	88	1 ea	268	.1	.210
Roasted, lean only	72	1 ea	180	.1	.196
Center rib chop: 3 per lb = 151g (5.3 oz) raw w/bone.					
This ≈ 112g (3.9 oz) w/o bone. Cooked values follow.					
Braised, lean & fat	67	1 ea	246	.1	.184
Braised, lean only	53	1 ea	147	.1	.169
Broiled, lean & fat	77	1 ea	264	.1	.217
Broiled, lean only	63	1 ea	162	.1	.203
Pan-fried, lean & fat	88	1 ea	343	.1	.245
Pan-fried, lean only	62	1 ea	160	.1	.217
Roasted, lean & fat	79	1 ea	252	.1	.220
Roasted, lean only	66	1 ea	162	.1	.206
Pork roast, leg:					
Lean & fat	85	3 oz	250	.1	.265
Lean only	85	3 oz	187	.2	.297
Pork roast, loin & rib:					
Lean & fat	85	1 pce	265	.1	.220
Lean only	85	1 pce	206	.1	.244
Shoulder, braised (yield fr/ 6.8 oz raw w/bone):					
Lean & fat	85	3 oz	293	.1	.261
Lean only	67	2.4 oz	166	.1	.241
Spareribs, ckd fr/1 lb raw	177	6.25 oz	703	.1	.676
PORK, CURED - HAM: see Sausage & Lunchmeat Group, Turkey Ham, and Bacon under Pork.					
Roasted, lean & fat	85	3 oz	207	.1	.188
Roasted, lean only	85	3 oz	133	.2	.216
Canned, roasted	85	3 oz	140	.2	.213
RABBIT, roasted	85	3 oz	175	.1	.14
VEAL (calf):					
Cutlet, lean, ckd average	85	3 oz	166	.2	.290
Rib roast	85	3 oz	151	.2	.250
Liver, simmered	85	3 oz	222	.7	1.65
VENISON (deer) roasted	85	3 oz	134	.4	.51

Meats: Poultry

	Wt (g)	Svg	Cal	Riboflavin (mg) 100 Cal	Svg
CHICKEN: A 3 lb chicken is ≈1.45 lbs raw; ≈ 1.1 lbs cooked.					
All types of meat:					
Fried	140	1 c	307	.1	.277
Roasted	140	1 c	266	.1	.249
Stewed	140	1 c	248	.1	.228
Canned w/ broth, 5 oz	142	1 can	235	.1	.183
Dark meat:					
Fried	85	3 oz	203	.1	.212
Roasted	85	3 oz	174	.1	.193
Stewed	85	3 oz	163	.1	.172

	Wt (g)	Svg	Cal	Riboflavin (mg) 100 Cal	Svg
CHICKEN, continued:					
Light meat:					
Fried	85	3 oz	163	.1	.107
Roastd	85	3 oz	147	.1	.098
Stewed	85	3 oz	135	.1	.100
Breast*, meat & skin (145g raw ≈ 180g w/bone):					
Batter-fried	140	1 ea	364	.1	.204
Flour-fried	98	1 ea	218	.1	.128
Roasted	98	1 ea	193	.1	.117
Stewed	110	1 ea	202	.1	.127
Breast* meat: (raw ≈ 118g):					
Stewed	95	1 ea	144	.1	.113
Roasted	86	1 ea	142	.1	.098
Fried	86	1 ea	161	.1	.108
*2 pieces per bird					
Drumstick, meat & skin (73g raw ≈110g w/bone):					
Batter-fried	72	1 ea	193	.1	.155
Flour-fried	49	1 ea	120	.1	.110
Roasted	52	1 ea	112	.1	.112
Stewed	57	1 ea	116	.1	.108
Drumstick meat (raw ≈ 62g):					
Roasted	44	1 ea	76	.1	.103
Stewed	46	1 ea	78	.1	.098
Thigh, meat & skin (94g raw ≈120g w/bone):					
Batter-fried	86	1 ea	238	.1	.195
Flour-fried	62	1 ea	162	.1	.151
Roasted	62	1 ea	153	.1	.131
Thigh meat (raw ≈ 69g):					
Fried	52	1 ea	113	.1	.133
Roasted	52	1 ea	109	.1	.120
Stewed	55	1 ea	107	.1	.120
Wing, meat & skin (49g raw ≈ 90g w/bone):					
Batter fried	49	1 ea	159	<.1	.074
Flour fried	32	1 ea	103	<.1	.044
Roasted	34	1 ea	99	<.1	.044
Stewed	68	1 ea	158	.1	.130
Wing meat (raw ≈ 29g):					
Roasted	21	1 ea	43	.1	.027
Fried	20	1 ea	42	.1	.026
Stewed	24	1 ea	43	.1	.026
Chicken gizzard	22	1 ea	34	.2	.054
Chicken heart	3.3	1 ea	6	.4	.024
Chicken liver	20	1 ea	30	1.2	.350
DUCK, domestic, roasted:					
Meat & skin	85	3 oz	286	.1	.229
Meat only	85	3 oz	171	.2	.385
GOOSE, domestic, roasted:					
Meat & skin	85	3 oz	259	.1	.275
Meat only	85	3 oz	202	.2	.332
Liver, raw	94	1 ea	125	.7	.838
TURKEY:					
Breast meat:					
Barbecued	28	1 oz	40	.1	.030
Hickory smoked	28	1 oz	35	.1	.030
Ground turkey, cooked	100	3.5 oz	229	.1	.168
Roasted (1 cup = 140g):					
All types	85	3 oz	145	.1	.155
Light meat	85	3 oz	133	.1	.115
Dark meat	85	3 oz	159	.1	.211
Frozen slices w/gravy	142	5 oz	95	.2	.180
Turkey gizzard	67	1 ea	109	.2	.219
Turkey heart	16	1 ea	28	.5	.141
Turkey liver	75	1 ea	127	.8	1.07
Turkey patty, breaded, fried	64	1 ea	181	.1	.122

ESHA Research

	Wt (g)	Svg	Cal	Riboflavin (mg) 100 Cal	Svg		Wt (g)	Svg	Cal	Riboflavin (mg) 100 Cal	Svg
Meats: Sausages & Lunchmeats						Corn pudding, recipe	250	1 c	271	.1	.320
						Corned beef hash, canned	220	1 c	382	.1	.403
						Egg roll, chicken & vegetable	100	1 ea	242	<.1	.120
Barbecue loaf, pork/beef	23	1 pce	40	.1	.057	Egg salad	183	1 c	438	.1	.451
Beef lunchmeat, thin sliced	28	1 oz	50	.1	.054	**Enchiladas:**					
Braunschweiger	28	1 oz	102	.4	.435	Beef	120	1 ea	292	.1	.163
Dutch brand loaf	28	1 oz	68	.1	.076	Cheese	120	1 ea	330	.1	.210
FRANKFURTERS (hotdogs):						Chicken	120	1 ea	269	.1	.154
Beef, 8/package	57	1 ea	184	.03	.058	Enchirito	207	1 ea	441	.1	.273
Beef & pork, 8/package	57	1 ea	183	.04	.068	French toast, recipe	65	1 pce	123	.1	.173
Chicken, 10/package	45	1 ea	115	.05	.052						
Turkey, 10/package	45	1 ea	102	.08	.080	**LASAGNA:**					
						With meat, recipe	245	1 pce	398	.1	.333
HAM, lunchmeat:						Without meat, recipe	218	1 pce	316	.1	.280
Thin sliced	28	3 pce	37	.2	.063	Macaroni & cheese:					
Extra lean	28	1 oz	37	.2	.063	Canned	240	1 c	230	.1	.240
Regular	28	1 oz	51	.1	.071	Recipe	200	1 c	430	.1	.400
Liverwurst, pork	18	1 pce	59	.3	.185	Macaroni salad, w/o cheese	141	1 c	371	<.1	.068
Luxury loaf	57	2 oz	80	.2	.168	Manicotti w/meat, tomato sce	233	1 ea	320	.1	.355
Olive loaf	57	2 oz	133	.1	.147	Meat loaf:					
Pastrami, beef	57	2 oz	198	<.1	.096	Beef and 1/3 pork	87	1 pce	212	.1	.190
Pastrami, turkey	57	2 oz	74	.2	.151	Beef only	87	1 pce	193	.1	.182
Peppered loaf	28	1 oz	42	.2	.086	Moussaka, lamb & eggplant	250	1 c	250	.1	.317
Polish sausage	28	1 oz	92	<.1	.042	Potato salad w/mayo. & eggs	250	1 c	358	<.1	.150
						Pizza, cheese:					
PORK SAUSAGE:						Regular crust 1/8 of 15"	120	1 pce	290	.1	.290
Brown & serve links	13	1 link	50	<.1	.020	Thick crust 1/2 of 10"	208	1 pce	519	.1	.539
Link, small, cooked	13	1 link	48	.1	.033	Quiche Lorraine, 1/8th pie	176	1 pce	600	.1	.320
Salami:						Ravioli, beef, canned	226	1 c	220	.1	.165
Beef	23	1 pce	58	.1	.059						
Beef/Pork	57	2 oz	143	.1	.213	**SANDWICHES,** fast food:					
Turkey	57	2 oz	111	.1	.150	Cheeseburger, 3 oz patty	112	1 ea	300	.1	.240
Smoked link sausage:						Cheeseburger, 4 oz patty	194	1 ea	524	.1	.485
Beef/Pork	68	1 link	229	.1	.116	Chicken patty sandwich	157	1 ea	436	.1	.258
Pork	68	1 link	265	.1	.175	English muffin, egg,					
Turkey	28	1 oz	55	.1	.060	cheese and bacon	138	1 ea	360	.1	.500
Summer sausage:						Fish sandwich:					
Beef/Pork	23	1 pce	80	.1	.069	Large, w/o cheese	170	1 ea	470	.1	.238
Turkey	28	1 oz	50	.2	.120	Regular, w/cheese	140	1 ea	420	.1	.266
Turkey breakfast sausage	28	1 oz	65	.1	.080	Hamburger, 3 oz patty	98	1 ea	245	.1	.240
Turkey ham	57	2 oz	73	.2	.150	Hamburger, 4 oz patty	174	1 ea	445	.1	.382
Turkey roll, light & dark	57	2 oz	84	.2	.161	Hotdog (frankfurter) w/bun	85	1 ea	260	.1	.186
Turkey roll, light meat	57	2 oz	83	.2	.128	Roast beef on bun	150	1 ea	345	.1	.330
Mixed Dishes & Fast Foods						**SANDWICHES** (on part whole wheat bread, except when stated as rye):					
						Ham & cheese	151	1 ea	363	.1	.419
Beef & vegetable stew:						Ham & swiss on rye	145	1 ea	350	.1	.404
Homemade	245	1 c	220	.1	.170	Ham on rye	116	1 ea	242	.1	.300
Canned	245	1 c	194	.1	.123	Ham sandwich	122	1 ea	256	.1	.318
Beef, macaroni, tomato						Roast beef sandwich	122	1 ea	280	.1	.290
sauce, recipe	226	1 c	189	.1	.169	Peanut butter & jam	100	1 ea	341	.1	.213
Beef pot pie, f/frozen	234	1 ea	426	<.1	.150	Tuna salad sandwich	116	1 ea	303	.1	.217
BURRITOS:						Turkey sandwich	122	1 ea	271	.1	.242
Bean	174	1 ea	322	.1	.226	Turkey ham	122	1 ea	253	.1	.340
Beef	177	1 ea	463	.1	.364						
Beef & bean	175	1 ea	390	.1	.293	**SPAGHETTI** (pasta & tomato sauce w/cheese):					
Deluxe combination	198	1 ea	424	.1	.320	Home recipe	250	1 c	260	.1	.180
Cheese soufflé, recipe	112	1 c	221	.1	.288	Canned	250	1 c	190	.1	.280
Chicken & noodles, recipe	240	1 c	365	<.1	.170	Spinach soufflé	136	1 c	218	.1	.305
Chicken a la king, recipe	245	1 c	470	.1	.420	Taco:					
Chicken chow mein:						Beef	78	1 ea	207	.1	.134
Canned	250	1 c	95	.1	.100	Chicken	78	1 ea	172	.1	.119
Recipe	250	1 c	255	.1	.230	**Tostada,** with:					
Chicken pot pie, f/frozen	230	1 ea	430	<.1	.167	Refried beans	157	1 ea	212	.1	.138
Chicken salad w/celery	78	1/2 c	266	<.1	.078	Beans & beef	192	1 ea	332	.1	.243
Chili with beans, canned	255	1 c	286	.1	.268	Beans & chicken	157	1 ea	249	.1	.190
Chop suey w/beef & pork	250	1 c	300	.1	.380	Tuna noodle casserole, recipe	202	1 c	251	.1	.169
Cole slaw	120	1 c	84	.1	.074	Tuna salad	205	1 c	383	<.1	.140
Corn dog	111	1 ea	330	.1	.170	Turkey pot pie, frozen	233	1 ea	416	<.1	.170

Values are for edible portion of foods

Riboflavin (Vitamin B2) mg

Nuts & Seeds

Food	Wt (g)	Svg	Cal	100 Cal	Svg
ALMONDS:					
Dried, whole	142	1 c	837	.1	1.11
Dry roasted	138	1 c	810	.1	.827
Almond butter	16	1 T	101	.1	.098
Brazil nuts, dry (about 7)	28	1 oz	184	<.1	.035
Cashews:					
Dry roasted	137	1 c	787	<.1	.274
Oil roasted	130	1 c	748	<.1	.228
Chestnuts, roasted	143	1 c	350	.1	.250
Coconut, dried unsweetened	78	1 c	515	<.1	.078
Coconut water, raw	240	1 c	46	.3	.137
Filberts (hazelnuts), whole	135	1 c	853	<.1	.149
Macadamias, oil roasted or dry	28	1 oz	204	<.1	.031
MIXED NUTS with peanuts (cashews, peanuts, brazil nuts, filberts, almonds, pecans):					
Dry roasted	137	1 c	814	<.1	.274
Oil roasted	142	1 c	876	<.1	.315
MIXED NUTS without peanuts (cashews, brazil nuts, filberts, almonds, pecans):					
Oil roasted	144	1 c	886	.1	.700
PEANUTS:					
Dried	146	1 c	827	<.1	.191
Oil roasted	144	1 c	837	<.1	.156
Peanut butter, smooth style	32	2 T	190	<.1	.032
Peanut flour, defatted	60	1 c	196	.1	.288
Pecans, chopped	30	1/4 c	199	<.1	.038
Pine nuts, dried:					
Pignola	28	1 oz	146	<.1	.054
Pinyon	28	1 oz	161	<.1	.063
Pistachios, shelled, dried	28	1 oz	164	<.1	.049
Poppyseed	8.8	1 T	47	<.1	.015
Pumpkin/squash seeds:					
Kernels, dried	138	1 c	747	.1	.442
Kernels, roasted	227	1 c	1185	.1	.663
Whole, roasted	64	1 c	285	.1	.155
Sesame seeds:					
Kernels, dried	37	1/4 c	221	<.1	.032
Whole, dried	36	1/4 c	206	<.1	.089
Soybeans, dry roasted	86	1/2 c	405	<.1	.125
Sunflower seed					
kernels, dry/oil roasted	35	1/4 c	206	<.1	.095
Walnuts, chopped:					
Black	125	1 c	759	<.1	.136
English	120	1 c	770	<.1	.178

Soups, Sauces & Gravies

Food	Wt (g)	Svg	Cal	100 Cal	Svg
GRAVIES:					
Beef gravy:					
Recipe	135	1/2 c	151	<.1	.056
Canned	233	1 c	124	.1	.084
Chicken gravy:					
Recipe	130	1/2 c	163	<.1	.066
Canned	238	1 c	189	.1	.103
From dry mix	260	1 c	85	.2	.150
Mushroom gravy, canned	238	1 c	120	.1	.150
Pork gravy, f/dry mix	258	1 c	76	.1	.059
Turkey gravy:					
Canned	238	1 c	122	.2	.191
From dry mix	261	1 c	87	.1	.112

SAUCES (also see Other):

Food	Wt (g)	Svg	Cal	100 Cal	Svg
Au Jus, canned	238	1 c	38	.4	.143
Cheese sce, recipe	101	1/2 c	216	.1	.221
Cheese sce, fr/dry mix	139	1/2 c	153	.2	.280
White sce, recipe	250	1 c	395	.1	.430
White sce, fr/dry mix	264	1 c	240	.2	.450

SOUPS: soups are canned unless otherwise stated. RTS = ReadyTo Serve. For soups prep. w/milk, assume whole milk.

Food	Wt (g)	Svg	Cal	100 Cal	Svg
Beef broth:					
From canned	240	1 c	16	.3	.050
From dry mix	244	1 c	19	.1	.020
Beef noodle:					
From canned	244	1 c	84	.1	.059
From dry mix	251	1 c	41	.1	.060
Chicken broth:					
From canned	244	1 c	39	.2	.071
From dry cube	243	1 c	13	.2	.024
From dry mix	244	1 c	21	.2	.034
Chicken, cream of:					
Condensed, undiluted	251	1 c	233	.1	.120
Prep with milk	248	1 c	191	.1	.258
Prep with water	244	1 c	115	.1	.061
Prep from dry	261	1 c	107	.2	.204
Chicken noodle:					
Prep with water	241	1 c	75	.1	.060
Chunky, RTS	240	1 c	114	.1	.168
Prep from dry mix	252	1 c	53	.1	.058
Chili beef soup	250	1 c	169	<.1	.075
Clam chowder, New England	248	1 c	163	.1	.236
Clam chowder, tomato based:					
Manhattan w/water	244	1 c	78	.1	.039
Manhattan, chunky, RTS	240	1 c	133	<.1	.062
Mushroom, cream of:					
Condensed, undiluted	251	1 c	257	.1	.166
Prep with milk	248	1 c	205	.1	.280
Prep with water	244	1 c	130	.1	.090
Onion soup:					
From canned	241	1 c	57	<.1	.024
Fr/dry packet	184	3/4 c	20	.2	.042
Oyster stew, with milk	245	1 c	134	.2	.233
Split pea soup:					
With or without ham	253	1 c	189	<.1	.078
Chunky, RTS	240	1 c	184	.1	.094
Prepared from dry	255	1 c	133	.1	.152
Tomato soup, w/milk	248	1 c	160	.2	.248
Tomato vegetable, f/dry	189	3/4 c	41	.1	.034
Turkey noodle soup	244	1 c	69	.1	.063
Vegetable, cream of, f/dry	260	1 c	105	.1	.107

Vegetables & Legumes

Food	Wt (g)	Svg	Cal	100 Cal	Svg
Alfalfa sprouts	33	1 c	10	.4	.042
Amaranth leaves:					
Fresh, chopped	28	1 c	7.4	.6	.045
Cooked, boiled	132	1 c	28	.6	.177
Artichoke:					
Globe, ckd (300g w/refuse)	120	1 ea	60	.1	.079
Hearts, marinated-jar	170	6 oz	168	.1	.174
Asparagus:					
Fresh, uncooked	58	4 ea	13	.6	.072
Ckd from fresh, pces	90	1/2 c	22.5	.5	.109
Ckd from frozen, pces	180	1 c	50.4	.4	.185
Canned, pces w/liquid	121	1/2 c	16	.8	.123
Bamboo shoots, sliced:					
Cooked from fresh	120	1 c	15	.4	.060
Canned	131	1 c	25	.1	.034

Riboflavin (Vitamin B2) mg

Left column

	Wt (g)	Svg	Cal	Riboflavin (mg) 100 Cal	Riboflavin (mg) Svg
Vegetables & Legumes, continued:					
BEANS (also see Garbanzo, Lentils, Soybeans):					
Baked beans (dry white beans w/spices & sauce):					
Home prepared	253	1 c	382	<.1	.124
Cnd, plain/vegetarian	254	1 c	235	.1	.152
Cnd with franks	257	1 c	366	<.1	.144
Cnd with pork	253	1 c	268	<.1	.096
Cnd w/pork, sweet sce	253	1 c	282	.1	.154
Cnd w/pork, tomato sce	253	1 c	247	<.1	.116
Black beans, ckd from dry	172	1 c	227	<.1	.101
Broadbeans:					
Ckd from dry legumes	170	1 c	186	.1	.151
Ckd from fresh veg.	100	3.5 oz	56	.2	.09
Great northern, ckd f/dry	177	1 c	210	<.1	.104
Green (snap) beans:					
Fresh, uncooked	110	1 c	34	.3	.116
Ckd from fresh	125	1 c	44	.3	.121
Ckd from frozen	135	1 c	36	.3	.100
Canned, drained	135	1 c	26	.3	.076
Canned w/liquid	240	1 c	36	.3	.122
Navy beans, ckd f/dry	182	1 c	259	.1	.111
Pinto beans, ckd f/dry	171	1 c	235	.1	.156
Refried beans, canned	253	1 c	270	<.1	.139
Winged beans:					
Ckd from fresh veg.	62	1 c	23	.2	.045
Ckd from dry legume	172	1 c	252	.1	.222
Yardlong, cooked	104	1 c	49	.2	.103
Yellow wax: see green beans.					
Bean sprouts, (Mung beans):					
Fresh, uncooked	104	1 c	31.2	.4	.129
Boiled, drained	124	1 c	26	.5	.126
Stir fried	124	1 c	62	.4	.223
Canned, drained	125	1 c	16	.5	.088
Beets, diced, cnd, drained	85	1/2 c	27	.1	.035
Beet greens:					
Fresh pieces	19	1/2 c	4	1.1	.042
Cooked, drained	144	1 c	40	1.0	.416
BROCCOLI, chopped:					
Fresh, uncooked	88	1 c	24	.4	.104
Ckd from fresh	156	1 c	44	.4	.176
Ckd from frozen	184	1 c	51	.3	.149
With cheese sauce	142	1/2 c	166	.1	.180
Brussels sprouts:					
Ckd from fresh	156	1 c	60	.2	.124
Ckd from frozen	155	1 c	65	.3	.175
CABBAGE, chopped:					
Common, fresh	70	1 c	16	.1	.022
Common, cooked	150	1 c	32	.3	.083
Bok choy, fresh	70	1 c	9	.5	.049
Bok choy, cooked	170	1 c	20	.5	.107
Red cabbage, fresh	70	1 c	19	.1	.021
Red cabbage, cooked	75	1/2 c	16	.1	.015
Pe-tsai, fresh	76	1 c	11	.3	.038
Pe-tsai, cooked	119	1 c	16	.3	.052
Savoy, fresh	70	1 c	20	.1	.021
Savoy, cooked	145	1 c	35	.1	.029
CARROT:					
Whole 7.5" x 1-1/8"	72	1 ea	31	.1	.042
Fresh, grated	55	1/2 c	24	.1	.032
Ckd from fresh, slices	78	1/2 c	35	.1	.044
Ckd from frozen, slices	73	1/2 c	26	.1	.027
Canned with liquid	123	1/2 c	28	.1	.033
Canned, drained	73	1/2 c	17	.1	.022
Carrot juice	123	1/2 c	49	.1	.067

Right column

	Wt (g)	Svg	Cal	Riboflavin (mg) 100 Cal	Riboflavin (mg) Svg
Cauliflower:					
Fresh pieces	50	1/2 c	12	.2	.029
Ckd from fresh	62	1/2 c	15	.3	.040
Ckd from frozen	180	1 c	34	.3	.095
Celery:					
Fresh outer stalk, 8" x 1.5"	40	1 ea	6.4	.3	.018
Cooked, diced	150	1 c	27	.3	.071
Chard, Swiss:					
Fresh, chopped	36	1 c	7	.5	.032
Ckd from fresh	175	1 c	35	.4	.150
Collards:					
Fresh, chopped	36	1 c	7	.3	.023
Ckd from fresh	128	1 c	35	.2	.067
Ckd from frozen	170	1 c	63	.3	.196
CORN:					
Fresh kernels, uncooked	77	1/2 c	66	.1	.046
Ckd from fresh, kernels	82	1/2 c	89	.1	.059
Ckd from fresh, 5" cob	77	1 ea	83	.1	.055
Ckd from frozen	82	1/2 c	67	.1	.060
Canned kernels:					
Drained	82	1/2 c	66	.1	.041
With liquid	128	1/2 c	79	.1	.078
Vacuum pack	105	1/2 c	83	.1	.077
Cream style	128	1/2 c	93	.1	.068
Cucumber, whole, 8" x 2+"	301	1 ea	39	.2	.060
Dandelion greens:					
Fresh	55	1 c	25	.6	.143
Cooked	105	1 c	35	.5	.184
Dock (sorrel greens):					
Fresh	133	1 c	29	.5	.133
Cooked	100	1 c	20	.4	.086
Eggplant, cooked cubes	160	1 c	45	.1	.032
Escarole (curly endive)	50	1 c	8.5	.4	.038
Garbanzo beans (chickpeas):					
Ckd from dry	164	1 c	269	<.1	.103
Canned	240	1 c	285	<.1	.079
Garden cress:					
Fresh, chopped	25	1/2 c	8	.8	.065
Cooked	135	1 c	31	.7	.216
Jerusalem artichoke, fresh	150	1 c	114	.1	.090
Kale:					
Fresh, chopped	67	1 c	33	.3	.087
Ckd from fresh	130	1 c	41.6	.2	.090
Ckd from frozen	130	1 c	39	.4	.148
Kohlrabi:					
Fresh slices	140	1 c	38	.1	.028
Ckd fr/fresh	165	1 c	48	.1	.033
Lambquarters, chopped:					
Fresh vegetable	56	1 c	24	1.0	.246
Cooked	180	1 c	58	.8	.468
Lentils, ckd from dry	198	1 c	231	.1	.145
Lentils, sprouted, fresh	77	1 c	81	.1	.099
Lentils, sprouted, stir-fried	100	3.5 oz	101	.1	.090
LETTUCE, chopped:					
Butterhead	56	1 c	7.3	.5	.034
Iceberg	56	1 c	7.3	.2	.017
Loose leaf	56	1 c	10	.4	.045
Romaine	56	1 c	9	.6	.056
Mushrooms:					
Fresh slices	35	1/2 c	9	1.8	.157
Ckd from fresh	78	1/2 c	21	1.1	.234
Canned, drained	78	1/2 c	19	.9	.172
Mustard greens:					
Fresh, chopped	56	1 c	15	.4	.062
Ckd from fresh	140	1 c	21	.4	.088
Ckd from frozen	150	1 c	29	.3	.080
Okra:					
Pods, ckd f/fresh	85	8 ea	27	.2	.047
Slices, ckd f/frozen	92	1/2 c	34	.3	.113

Vegetables & Legumes, continued:

	Wt (g)	Svg	Cal	Riboflavin (mg) 100 Cal	Svg
Onions, spring, chopped	50	1/2 c	16	.3	.040
Parsley:					
Fresh, chopped	30	1/2 c	10	.3	.033
Freeze- dried	1.4	1/4 c	4	.8	.032
Parsnips, sliced, cooked	156	1 c	125	.1	.080
PEAS:					
Black-eyed:					
Cooked from frozen	170	1 c	224	<.1	.109
Cooked from dry	171	1 c	198	<.1	.094
Canned	240	1 c	184	.1	.178
Green peas:					
Fresh, uncooked	73	1/2 c	59	.2	.096
Ckd from fresh	80	1/2 c	67	.2	.119
Ckd from frozen	80	1/2 c	63	.1	.080
Canned, drained	85	1/2 c	59	.1	.066
Canned with liquid	124	1/2 c	61	.1	.091
Green peas, edible-pods:					
Fresh, uncooked	145	1 c	61	.2	.116
Ckd from fresh	160	1 c	67	.2	.122
Ckd from frozen	80	1/2 c	42	.2	.095
Split peas, cooked f/dry	196	1 c	231	<.1	.110
Peas & carrots:					
Ckd from frozen	80	1/2 c	38	.1	.056
Canned with liquid	128	1/2 c	48	.1	.068
Peas, mature, sprouted:					
Fresh	120	1 c	154	.1	.186
Cooked	100	3.5 oz	118	.2	.285
PEPPERS, HOT (1 pod ≈ 45g):					
Fresh, chopped	75	1/2 c	30	.2	.068
Canned, chopped	68	1/2 c	17	.2	.034
Jalapeno peppers, canned	68	1/2 c	17	.2	.034
PEPPERS, SWEET, green or red, chopped (1 pod ≈ 74g)					
Fresh	50	1/2 c	14	.1	.015
Ckd from fresh	68	1/2 c	19	.1	.020
Pimento, canned	57	2 oz	13	.3	.034
Poi, two finger	240	1 c	269	<.1	.096
POTATOES:					
Baked in oven, 4.75 x 2.3" dm:					
Flesh and skin	202	1 ea	220	<.1	.067
Flesh only	156	1 ea	145	<.1	.033
Potato skin	58	1 ea	115	.1	.069
Boiled, 2.5" diam.	136	1 ea	119	<.1	.027
French fried	50	10 ea	158	<.1	.014
Hash browns, f/frozen	156	1 c	340	<.1	.031
Mashed potatoes:					
Prep with milk	210	1 c	162	.1	.084
Prep from instant	220	1 c	239	.1	.139
Potato dishes, prepared:					
Au gratin, recipe	245	1 c	322	<.1	.284
Au gratin, from mix	245	1 c	228	<.1	.199
Scalloped, recipe	245	1 c	210	.1	.225
Scalloped, from mix	245	1 c	228	<.1	.137
Pumpkin:					
Cooked from fresh	245	1 c	50	.4	.191
Canned (can be mixture of pumpkin & squash)	123	1/2 c	42	.2	.066
Purslane:					
Fresh, chopped	43	1 c	7	.7	.048
Cooked	115	1 c	21	.5	.104
Radish seeds, sprouted	38	1 c	15.6	.3	.039
Radishes, red	45	10 ea	7	.3	.020
Rutabaga, cubes:					
Fresh	140	1 c	51	.1	.056
Cooked	85	1/2 c	29	.1	.031
Salsify, cooked slices	135	1 c	92	.3	.234
Sauerkraut, canned w/liquid	236	1 c	44	.1	.052

	Wt (g)	Svg	Cal	Riboflavin (mg) 100 Cal	Svg
Seaweed:					
Irish moss, fresh	28	1 oz	13.9	.9	.132
Kelp, fresh	28	1 oz	12.2	.4	.043
Lavar, fresh	28	1 oz	9.9	1.3	.126
Wakame, fresh	28	1 oz	12.8	.5	.065
Spiralina, dried	28	1 oz	82.2	1.3	1.04
Soybeans:					
Ckd from fresh veg	180	1 c	255	.1	.279
Ckd from dry legume	172	1 c	298	.2	.490
Soybeans, mature, sprouted:					
Fresh	35	1/2 c	45	.09	.041
Steamed	94	1 c	76	.07	.050
Stir- fried	100	3.5 oz	125	.15	.190
SOY PRODUCTS: see tofu in this section; miso, natto, tempeh in Other; roasted soybeans in Nuts & Seeds; soy milk in Dairy; soy flour in Grains.					
Spinach, chopped:					
Fresh, chopped	56	1 c	12	.9	.106
Ckd from fresh	180	1 c	41	1.0	.425
Ckd from frozen	190	1 c	53	.6	.319
Canned, drained	214	1 c	50	.6	.295
SQUASH, SUMMER, sliced:					
Crookneck, fresh slices	130	1 c	24	.2	.056
Crookneck, cooked	180	1 c	36	.2	.088
Scallop, fresh slices	130	1 c	24	.2	.039
Scallop, cooked	90	1/2 c	14	.2	.023
Zucchini, fresh slices	130	1 c	19	.2	.039
Zucchini, cooked	180	1 c	29	.3	.074
SQUASH, WINTER, mashed:					
Acorn / Danish, baked	245	1 c	137	<.1	.032
Acorn / Danish, boiled	245	1 c	83	<.1	.020
Butternut, baked	245	1 c	99	<.1	.042
Butternut, ckd f/frozen	240	1 c	94	.1	.094
Hubbard, baked	240	1 c	120	.1	.113
Hubbard, boiled	236	1 c	70	.1	.066
Spaghetti, baked/boiled	155	1 c	45	.1	.034
Succotash:					
Cooked from fresh	192	1 c	222	.1	.184
Cooked from frozen	170	1 c	158	.1	.116
Sweet potatoes (5" x 2"dm):					
Baked in skin, peeled	114	1 ea	118	.1	.145
Boiled (peeled before)	151	1 ea	160	.1	.210
Canned, vacuum pack	255	1 c	233	.1	.145
Candied (recipe) pce = 2.5 x 2"	105	1 pce	144	<.1	.044
Tofu (soybean curd):					
Firm, raw	126	1/2 c	183	.07	.129
Regular, raw	124	1/2 c	94	.07	.064
TOMATOES:					
Fresh, 2.4" diam.	123	1 ea	26	.2	.059
Fresh, chopped	180	1 c	38	.2	.086
Cooked from fresh	240	1 c	65	.2	.137
Canned, whole	240	1 c	47	.2	.074
Tomato juice, canned	244	1 c	41.5	.2	.076
Tomato paste, canned	262	1 c	220	.2	.498
Tomato puree, canned	250	1 c	102	.1	.135
Tomato sauce, canned	245	1 c	74	.2	.142
Turnips, cubed:					
Fresh	130	1 c	35	.1	.039
Cooked	156	1 c	28	.1	.018
Turnip greens:					
Ckd from fresh	144	1 c	29	.4	.104
Ckd from frozen	82	1/2 c	24	.3	.061

ESHA Research

	Wt (g)	Svg	Cal	Riboflavin (mg) 100 Cal	Svg
VEGETABLES, MIXED, cooked from frozen:					
Broccoli, cauliflower, red pepper	95	2/3 c	25	.3	.080
Broccoli, carrots, pasta	95	2/3 c	88	.1	.070
Mixed vegetables (corn, carrots, peas, green beans, lima beans):					
Ckd from frozen	182	1 c	107	.2	.218
Canned, drained	163	1 c	77	.1	.078
Peas, carrots, onions	91	1/2 c	54	.1	.070
Peas, carrots, onions, pasta	95	1/2 c	122	.1	.080
Peas, cauliflower, cream sce	95	1/2 c	118	.1	.130
Peas & mushrooms	95	1/2 c	73	.2	.110
Peas & onions	95	1/2 c	71	.1	.080
Peas, onions, cheese sce	142	1/2 c	165	.1	.150
Peas, potatoes, cream sce	76	1/2 c	140	.1	.130
Vegetable juice cocktail	242	1 c	46	.1	.068
Water chestnuts:					
Fresh	62	1/2 c	66	.2	.124
Canned whole	28	4 ea	14	.1	.007
Watercress, fresh chopped	17	1/2 c	2	1.0	.020
Yam, orange: see sweet potato.					
Zucchini: see Squash.					

Other

Beverages, condiments, fats, sweets, flavorings, spices, etc.

	Wt (g)	Svg	Cal	Riboflavin (mg) 100 Cal	Svg
Beer (12 fluid oz = 1/5 c):					
Regular	356	1.5 c	146	.1	.093
Light	354	1.5 c	100	.1	.106
Butterscotch topping	50	3 T	156	<.1	.040
Coffee, brewed	240	1 c	2	1.0	.020
FATS:					
Butter	14	1 T	102	<.1	.005
Margarine, hard or soft:					
40% fat	14	1 T	49	<.1	.003
60% fat	14	1 T	76	<.1	.004
80% fat	14	1 T	102	<.1	.005
CANDY and CANDY BARS:					
Almonds, sugar coated	28	7 ea	146	.1	.156
Caramel, plain or chocolate	28	1 oz	115	<.1	.050
Chocolate kisses	28	6 pce	154	.1	.080
Chocolate coated:					
Almonds	165	1 c	935	.1	1.08
Mints	28	1 oz	116	<.1	.020
Peanuts	170	1 c	954	<.1	.258
Raisins	187	1 c	733	<.1	.165
KIT KAT	43	1 ea	210	.1	.110
KRACKLE	34	1 ea	179	.1	.090
M & M's plain candies	48	1 pkg	237	.1	.123
M & M's peanut candies	47	1 pkg	240	<.1	.094
MARS bar	50	1 ea	240	.1	.164
MILKY WAY	60	1 ea	260	.1	.148
Milk chocolate:					
Plain	28	1 oz	145	.1	.100
With almonds	28	1 oz	150	.1	.130
With peanuts	28	1 oz	155	<.1	.065
With crispy rice cereal	28	1 oz	140	.1	.080
MR. GOODBAR	47	1 ea	250	<.1	.120
REESE's peanut butter cup	45	2 ea	240	<.1	.050
SNICKERS, 2.2 oz	61	1 bar	290	<.1	.107
Caramel topping	50	3 T	155	<.1	.050
Carob flour	103	1 c	185	.3	.475
Catsup	15	1 T	16	.1	.011

	Wt (g)	Svg	Cal	Riboflavin (mg) 100 Cal	Svg
Chili sauce, hot red pepper	31	2 T	6	.5	.028
CHOCOLATE:					
Baking, unsweetened	28	1 oz	145	.1	.099
Bittersweet	28	1 oz	141	<.1	.050
Chocolate chips, semi sweet	170	1 c	860	<.1	.140
Dark, sweet	28	1 oz	150	<.1	.040
Chocolate syrup, thin type	38	2 T	85	<.1	.019
Chocolate fudge topping	38	2 T	125	.1	.082
Cocoa powder	86	1 c	224	.2	.456
Fruit punch:					
From frozen	247	1 c	113	<.1	.032
Canned	253	1 c	118	<.1	.058
Lemonade, from frozen	248	1 c	100	.1	.052
Mayonnaise	14	1 T	99	<.1	.006
Miso (soybean product)	138	1/2 c	284	.1	.345
Molasses, blackstrap	40	2 T	85	.1	.080
Mustard, prepared	16	1 T	12	.3	.031
Natto (soybean product)	88	1/2 c	187	.1	.167
Pineapple grapefruit drink	250	1 c	117	<.1	.040
Pineapple orange drink	250	1 c	125	<.1	.048
Salad dressings:					
Blue cheese	15	1 T	75	<.1	.020
Ranch style dressing	15	1 T	54	<.1	.021
Salsa: Picante by Tostitos	85	6 T	40	.2	.070
Soy sauce:					
Regular (wheat + soy) shoyu	18	1 T	9	.25	.023
Tamari (soy)	18	1 T	11	.25	.027
Fr/hydrolyzed veg protein	18	1 T	7	.27	.020
SPICES:					
Caraway seed	6.7	1 T	22	.1	.025
Cayenne or red pepper	5.3	1 T	17	.3	.049
Chili powder	7.5	1 T	24	.3	.060
Coriander leaf, dried	1.8	1 T	5	.5	.027
Cumin seed	6	1 T	22	.1	.020
Fenugreek seed	11.1	1 T	36	.1	.041
Garlic powder	8.4	1 T	28	<.1	.013
Mace, ground	5.3	1 T	25	.1	.024
Paprika	2.1	1 tsp	6.1	.6	.037
Tarragon, ground	4.8	1 T	14	.5	.064
Tea, brewed	240	1 c	2	1.7	.034
Tempeh (soybean product)	83	1/2 c	165	.1	.092
Tobasco sauce	15	1 T	1.6	.65	.011
Wine:					
Red	118	1/2 c	85	<.1	.033
White	118	1/2 c	79	<.1	.006
Yeast:					
Brewer's	7.5	1 T	25	1.4	.342
Dry active	7.5	1 T	20	2.1	.410

Niacin   (Vitamin B3)

Niacin is essential to almost every biochemical link in the metabolism of carbohydrates, proteins, and fats for energy. If there is a shortfall, the energy reactions are blocked. As a result, the amount of niacin needed is proportional to the calories eaten. It also protects the skin, nervous tissues, and the digestive tract from disorders.

Mild deficiencies can occur when the diet is low in niacin and protein. Symptoms include digestive disorders, nervous irritability, insomnia, headaches, and sometimes a swollen, red, sore tongue. Children can be weak and have poor growth.

With a severe deficiency, there are severe symptoms of skin and gastro-intestinal lesions, diarrhea, inflammation of the mucous membranes, and mental disorders leading to dementia and death. This condition is called Pellagra, and at the turn of the century it was a severe problem in the South, where the diet of the poor was very low in protein and was dependent primarily on cornmeal. The milling process takes a lot of the niacin out of the corn; thus, the cornmeal and salt pork (really just fat) diet was nutritionally insufficient. Niacin played a big role in "curing" a large number of mental patients in the South after the turn of the century. . . nearly half of the cases in insane asylums at that time were cleared up when niacin was given.

The niacin in food tables only represents a part of the niacin available to the body -- the niacin already preformed in foods. It does not count the amount that is converted from the amino acid Tryptophan. 60 milligrams of Tryptophan is considered, by convention, to be equivalent to 1 milligram of niacin. The actual amount ranges from 39 to 86 mg, and appears to be influenced by hormones (larger amounts are converted during pregnancy or when contraceptive pills are taken).   When this additional form of niacin is considered, they are combined and called Niacin Equivalents (NEs) or milligrams of equivalent (mg equiv) of Niacin.

Vitamin Losses

The milling process takes niacin out of grains, however, it is one of the vitamins that is put back into wheat flour, polished white rice, and cornmeal as "enrichment." Niacin is a very stable nutrient otherwise, and most of the small losses are from leaching into the cooking water (it is water soluble). If the cooking water is retained and used, there should be fairly high retention.

Recommended Dietary Allowances for Adults   (NE = mg equivalents of Niacin)

USA RDA: Men, 19 NE; Women, 15 NE; (pregnant: 17 NE; nursing: 20 NE), with a minimum of 13 NE age 10 and
   over. Because niacin is involved with metabolizing foods, the recommendation is based on the number of calories
   consumed at a rate of 6.6 NE per 1000 kcalories (with a minimum required of 13 NE per day for adults).
Canadian RNI: 7.2 NE per 1000 kcal, with a minimum of 14.4 NE/day for people 19 years of age and older.

The recommendations assume sources of niacin from both preformed niacin, and the amount that can be converted from tryptophan -- niacin equivalents.   For adults aged 19-50 in the United States, average diets supply about 700 mg of tryptophan for women and 1,100 mg for men. This converts to about 11.7 to 18.3 mg of niacin for adult women and men, over and above the amount of preformed niacin already in the food. Therefore, diets may contain more niacin than analysis of the preformed niacin will indicate.

Food Sources

The best sources of niacin are legumes (dried, mature beans and peas), fish, and organ meats. Whole grains contain good amounts of niacin, but as much as 70% of the niacin may be unavailable. Because niacin is enriched in flour and cereals, breads, baked goods and cereals made with enriched flour are good sources of niacin. Some niacin in grains is bound up in a way that makes it unavailable to the body. In Mexico, the traditional way of grinding corn with limewater releases much of this bound niacin, so that it is available to the body. Eggs and milk contain very little pre-formed niacin, but have sufficient tryptophan to compensate.

	Wt (g)	Svg	Cal	Niacin (mg) 100 Cal	Niacin (mg) Svg

The values for niacin are for preformed only, and do not include niacin that could be contributed by tryptophan, a niacin precursor.

Baked Goods

Recipes here use white (wheat) flour enriched with niacin, according to U.S. standards.

	Wt (g)	Svg	Cal	100 Cal	Svg
Bagel, egg or plain, 3.5" dm.	68	1 ea	180	1.3	2.40
Biscuits:					
Homemade	28	1 ea	100	.8	.80
From mix	28	1 ea	94	.9	.85
From refrig dough	20	1 ea	65	1.0	.67
BREADS (all-purpose flour is enriched with niacin):					
Banana nut bread, 1/2" slice	50	1 pce	161	.5	.776
Boston brown bread, canned	45	1 pce	95	.7	.700
Cornbread muffin	45	1 ea	145	.6	.850
French, 5" x 2.5" x 1"	35	1 pce	100	1.4	1.40
Cracked wheat bread	25	1 pce	65	1.3	.843
Italian, 4.75" x 4" x 1/2"	30	1 pce	84	1.2	1.00
Mixed grain bread	25	1 pce	65	1.7	1.10
Oatmeal bread	25	1 pce	65	1.3	.850
Pumpernickel, 5" x 4" x 3/8"	32	1 pce	80	1.3	1.06
Raisin bread	25	1 pce	68	1.5	1.02
Rye bread, light 5" x 3.5"	25	1 pce	65	1.3	.828
Wheat bread (white & whole wheat flour)	28	1 pce	72	1.8	1.26
White bread	28	1 pce	75	1.4	1.05
Whole wheat bread	28	1 pce	70	1.6	1.09
CAKES (pce = 1/16th cake unless otherwise stated):					
Boston cream pie, 1/8 cake	120	1 pce	260	.3	.700
Carrot cake w/cream cheese frosting 2.5" x 3"	112	1 pce	406	<.1	.110
Cheesecake from mix, 1/8	103	1 pce	300	.2	.600
Chocolate, chocolate frosting	69	1 pce	235	.3	.600
Coffeecake, 2.4" x 2.8"	72	1 pce	230	.6	1.29
Fruitcake, dark, 2/3" arc	43	1 pce	165	.3	.500
Gingerbread, 1/9 of 8" square	63	1 pce	174	.5	.818
Pound cake, 1/2" slice	30	1 pce	120	.4	.500
Sheet cake:					
Plain, 3" x 3"	86	1 pce	315	.3	1.10
W/white frosting	121	1 pce	445	.2	1.10
Snack cake, cream filled:					
Chocolate, like Ding dongs	28	1 pce	105	.6	.700
Sponge cake, like Twinkies	42	1 ea	155	.4	.600
Sponge cake, 1/12 tube cake	66	1 pce	194	.4	.802
White cake with frosting:					
Chocolate frosting	77	1 pce	291	.6	1.78
Coconut or white frosting	70	1 pce	265	.6	1.71
Yellow cake, chocolate frosting	69	1 pce	235	.3	.691
Cherry crisp, 3" x 3" pce	138	1 pce	157	.4	.563
Chips: see corn & tortilla this section; potato chips under vegetable.					
COOKIES:					
Animal cookies	28	27 ea	120	.9	1.10
Chocolate chip cookies:					
Home recipe	40	4 ea	185	.3	.584
From refrigerated dough	48	4 ea	225	.4	.893
Commercial	42	4 ea	180	.5	.900
Fig bars	56	4 ea	210	.3	.728
Oatmeal raisin	52	4 ea	245	.4	1.00
Peanut butter, recipe	48	4 ea	245	.8	1.90
Sandwich type cookies, all	40	4 ea	195	.4	.800
Snickerdoodles	20	1 ea	110	.5	.507
Shortbread, commercial	32	4 ea	155	.6	.900
Sugar cookies f/chilled dough	48	4 ea	235	.5	1.10
Vanilla wafers	40	10 ea	185	.5	1.00

	Wt (g)	Svg	Cal	100 Cal	Svg
CRACKERS:					
Armenian cracker bread	28	4 pce	117	.9	1.05
Cheese, peanut butter filling	30	4 ea	150	1.6	2.40
Graham crackers	14	2 ea	60	1.0	.600
Oyster crackers (10 = 7g)	28	1 oz	120	1.3	1.50
Round, like Ritz	9	3 ea	45	.7	.300
Rye wafers, whole grain	14	2 ea	55	.9	.500
Saltine crackers	12	4 ea	50	1.2	.600
Sesame crackers	12	4 ea	60	.5	.300
Wheat crackers, thin	8	4 ea	35	1.1	.400
Whole wheat crackers	8	2 ea	35	1.1	.400
Croissant 4.5" x 4" x 2"	57	1 ea	235	.6	1.30
Croutons (dry bread cubes)	30	1 c	111	1.3	1.44
Danish pastry, plain	57	1 ea	220	.6	1.40
Doughnuts:					
Cake type, medium	50	1 ea	210	.5	1.10
Yeast raised, plain	60	1 ea	235	.8	1.80
English muffin:					
Plain muffin	57	1 ea	140	1.5	2.14
Sourdough	56	1 ea	129	1.7	2.20
With raisins	56	1 ea	146	1.1	1.60
MUFFINS, from mix:					
Blueberry muffin	45	1 ea	140	.8	1.17
Bran muffin	45	1 ea	140	1.4	1.90
Cornbread muffin	45	1 ea	145	.6	.800
Pancakes:					
Buckwheat, 4" dm, f/mix	27	1 ea	55	.4	.200
Plain, 4" dm, homemade	27	1 ea	60	.8	.500
Plain, 4" dm, f/mix	27	1 ea	60	1.3	.800
Whole wheat, 5" dm, f/mix	52	1 ea	94	.8	.726
Peach crisp, 3" x 3"	139	1 pce	166	.6	1.01
PIES: pce = 1/6 of 9" pie.					
Apple pie	158	1 pce	405	.4	1.60
Banana cream	198	1 pce	319	.4	1.17
Blueberry	158	1 pce	380	.5	1.73
Cherry	158	1 pce	410	.4	1.58
Chocolate cream	175	1 pce	311	.3	1.06
Coconut cream	172	1 pce	343	.3	1.08
Coconut custard	165	1 pce	384	.3	1.10
Cream pie, commercial	152	1 pce	455	.2	1.10
Custard pie	152	1 pce	293	.3	.917
Lemon meringue	140	1 pce	355	.2	.833
Mincemeat	160	1 pce	395	.4	1.57
Peach	158	1 pce	405	.6	2.30
Pecan	138	1 pce	583	.2	1.10
Pumpkin	200	1 pce	367	.3	1.22
Strawberry chiffon, recipe	162	1 pce	372	.3	1.28
Poptart type pastry, fortified	54	1 ea	210	1.1	2.27
Pretzels, dutch twist	16	1 ea	65	1.1	.701
Pretzels, thin twists	60	10 ea	240	1.1	2.60
ROLLS:					
Cinnamon bun, small	50	1 ea	158	.8	1.20
Dinner rolls, commercial	28	1 ea	85	1.3	1.10
Hamburger bun	45	1 ea	129	1.4	1.78
Hard roll, white	50	1 ea	155	1.1	1.70
Hotdog bun	40	1 ea	115	1.4	1.58
Rye, dark	28	1 ea	79	1.3	1.02
Rye, light	28	1 ea	76	1.2	.927
Submarine (hoagie) roll	135	1 ea	400	1.1	4.50
Whole wheat roll	35	1 ea	88	1.5	1.31
Stuffing, prep.w/enr. bread:					
From dry bread cubes	140	1 c	500	.5	2.50
Stove Top Stuffing	108	1/2 c	176	.9	1.50
Tortilla:					
Corn, uncooked	30	1 ea	65	.6	.400
Flour, 10.5" dm.	57	1 ea	168	1.1	1.93
Flour, 8" dm.	35	1 ea	105	1.1	1.20

	Wt (g)	Svg	Cal	Niacin (mg) 100 Cal	Svg
Waffles:					
Homemade, 7" dm.	75	1 ea	245	.6	1.50
Prepared from mix, 7" dm.	75	1 ea	205	.4	.900
From frozen, 4" dm.	35	1 ea	98	1.9	1.85

Beverages

Also see Milk beverages, Fruit juices, and Vegetable juices

	Wt (g)	Svg	Cal	Niacin (mg) 100 Cal	Svg
Alchoholic:					
Beer, 12 fluid oz = 1.5 c	356	1.5 c	146	1.1	1.61
Beer, light, 12 fl oz	354	1.5 c	100	1.4	1.39
Bloody mary, 5 fl oz	148	1 ea	116	.6	.642
Wine, red	118	1/2 c	85	.1	.095
Wine, white	118	1/2 c	80	.1	.080
Non-alcoholic:					
Clam & Tomato jce, cnd	120	1/2 c	56	.1	.042
Coffee, brewed	240	1 c	2	27.0	.533
Lemon lime soda	149	1.5 c	149	<.1	.055
Lemonade, prep. f/frozen	248	1 c	100	<.1	.040
Limeade, prep. f/frozen	247	1 c	102	.1	.054
Pineapple grapefruit drink	250	1 c	117	.6	.668
Pineapple orange drink	250	1 c	125	.4	.518
Tea, brewed	240	1 c	2	5.0	.100

Dairy & Dairy Products

	Wt (g)	Svg	Cal	Niacin (mg) 100 Cal	Svg
CHEESE (1.5" cube ≈ 1 oz):					
American cheese	28	1 oz	106	<.1	.020
Blue cheese	28	1 oz	100	.3	.288
Cheddar cheese	28	1 oz	114	<.1	.023
Cottage cheese, lowfat 1%	57	1/4 c	41	.2	.075
Cottage cheese, creamed	54	1/4 c	56	.1	.071
Cream cheese (1 T = 15g)	28	1 oz	99	<.1	.029
Gjetost cheese	28	1 oz	132	.2	.230
Parmesan, grated, (1 T = 5g)	28	1 oz	129	.1	.089
Swiss cheese	289	1 oz	107	<.1	.026
CREAM, SWEET, fluid:					
Coffee or table cream	240	1 c	469	<.1	.137
Half and half	242	1 c	315	.1	.189
Light whipping cream	239	1 c	699	<.1	.100
Heavy whipping cream	238	1 c	821	<.1	.093
CREAM, SWEET, whipped:					
Whipped cream, unsweetened	119	1 c	410	<.1	.047
Pressurized	60	1 c	154	<.1	.042
CREAM, SOUR:					
Cultured, dairy	29	2 T	62	<.1	.019
Half & half, dairy	15	1 T	20	.1	.010
Cream, sour, Imitation, non-dairy	230	1 c	479	0	0
CREAM SUBSTITUTES, non-dairy:					
Coffee whitener, frzn/liquid	120	1/2 c	163	0	0
Coffee whitener, powder	94	1 c	514	0	0
Dessert Toppings, non-dairy:					
Frozen (like Coolwhip)	75	1 c	239	0	0
Dessert powder, dry mix	43	1.5 oz	245	0	0
Pressurized, non-dairy	70	1 c	185	0	0
Kefir, beverage	233	1 c	160	.2	.300
MILK (cow):					
Skim	245	1 c	86	.25	.216
Lowfat 1%	244	1 c	102	.2	.21
Lowfat 2%	244	1 c	121	.2	.21
Whole (3.3% fat)	244	1 c	150	.1	.205
Canned, evaporated, whole	252	1 c	340	.14	.488
Dry, instant nonfat	23	1/3 c	81	.2	.202

	Wt (g)	Svg	Cal	Niacin (mg) 100 Cal	Svg
Milk (other):					
Goat milk	244	1 c	168	.4	.676
Human breast milk	246	1 c	171	.25	.435
Soy milk	240	1 c	79	.45	.353
MILK BEVERAGES and mixes:					
Chocolate flavored, to be mixed with water:					
Powder w/dry milk	28	1 oz	100	.18	.179
Drink	206	3/4 c	100	.18	.179
Chocolate flavored, to be mixed with milk:					
Powder only	22	3/4 oz	75	.15	.110
Drink	266	1 c	226	.14	.317
Instant Breakfast, powder only	37	1 env	130	3.8	5.00
Milkshakes (10 fl oz = 1.25 cup):					
Chocolate	283	1.25 c	360	.13	.460
Strawberry	283	1.25 c	319	.15	.495
Vanilla	283	1.25 c	314	.2	.524
MILK DESSERTS:					
Ice cream, regular	133	1 c	269	<.1	.134
Puddings, prepared:					
Canned, choc or vanilla, 5 oz can ≈ 1/2 cup	142	1 can	205	.3	.600
Chocolate, fr/ mix-ckd or inst	260	1 c	305	.1	.235
Rice pudding, fr/ instant	149	1/2 c	175	.3	.600
Vanilla, cooked or instant	260	1 c	295	.1	.200
Yogurt, frozen, average	174	1 c	220	.1	.138
YOGURT:					
Coffee/Vanilla flavored	227	1 c	193	.13	.243
Fruit flavored	227	1 c	231	.09	.216
Lowfat, plain	227	1 c	144	.18	.259
Nonfat, plain	227	1 c	127	.22	.281
Whole milk yogurt	227	12 c	138	.12	.170

Eggs

	Wt (g)	Svg	Cal	Niacin (mg) 100 Cal	Svg
Whole egg, large, cooked	50	1 ea	77	<.1	.033
Whole egg, large, raw	50	1 ea	75	<.1	.037
White only, raw	33.4	1 ea	17	.2	.031
Yolk only, raw	16.6	1 ea	59	<.01	.002

Fruits & Fruit Juices

	Wt (g)	Svg	Cal	Niacin (mg) 100 Cal	Svg
Acerola juice, fresh	242	1 c	51	2.0	1.00
APPLES, 2.75" diam.	138	1 ea	80	.13	.106
Apple juice, canned/bottled	248	1 c	116	.2	.248
Applesauce, canned, unswtnd	244	1 c	106	.43	.459
APRICOTS:					
Fresh, pitted	106	3 ea	51	1.2	.636
Canned, juice pack	84	3 ea	40	.7	.289
Canned, heavy syrup	85	3 ea	70	.5	.320
Dried halves	35	10 ea	83	1.3	1.05
Apricot nectar, canned	251	1 c	141	.5	.653
Avocado, whole:					
California	173	1 ea	305	1.1	3.32
Florida	304	1 ea	340	1.7	5.84
Banana, 8.75" long, 176g w/peel	114	1 ea	105	.6	.616
Blackberries:					
Fresh berries	144	1 c	74	.8	.576
Frozen, unthawed	151	1 c	97	.6	.604
Canned	256	1 c	236	.3	.735
Blueberries:					
Fresh berries	145	1 c	82	.85	.700
Frozen, unthawed	155	1 c	78	1.0	.806
Canned	256	1 c	225	.13	.289

Values are for edible portion of foods

	Wt (g)	Svg	Cal	Niacin (mg) 100 Cal	Svg
Fruits, continued:					
Boysenberries:					
Frozen	132	1 c	66	1.5	1.01
Canned	256	1 c	225	.3	.589
Cantaloupe: see Melon.					
Cherries, sour:					
Frozen, unthawed	155	1 c	72	.3	.212
Canned	244	1 c	90	.5	.432
CHERRIES, SWEET:					
Fresh cherries	145	1 c	104	.6	.580
Frozen, sweetened, thawed	259	1 c	232	.2	.458
Canned with liquid	257	1 c	213	.5	1.02
Cranberries, whole, raw	95	1 c	46	.2	.095
Cranberry juices:					
Cranberry-Apple	253	1 c	169	.1	.150
Cranberry-Apricot Juice	184	3/4 c	118	.2	.221
Cranberry-Grape Juice	184	3/4 c	103	.2	.221
Cranberry juice cocktail	253	1 c	145	.1	.090
Dates, whole, pitted	8.3	1 ea	23	.8	.183
Elderberries, fresh	145	1 c	105	.7	.725
Figs:					
Fresh, medium	50	1 ea	37	.5	.200
Dried	19	1 ea	48	.3	.130
Fruit cocktail, canned:					
Juice pack	248	1 c	115	.9	.999
Heavy syrup	255	1 c	185	.5	.954
Grapes, fresh:					
Thompson seedless	160	1 c	114	.42	.480
Tokay/Emperor	57	10 ea	40	.43	.171
Grape juice:					
Bottled or canned	253	1 c	155	.4	.663
Prepared from frozen	250	1 c	128	.24	.310
GRAPEFRUIT:					
(halves = 241g w/rind)					
Half, pink or red	123	1 ea	37	.65	.240
Half, white	118	1 ea	39	.8	.320
Canned sections	254	1 c	152	.4	.617
Grapefruit juice:					
Prepared from frozen	247	1 c	102	.5	.536
Canned, unsweetened	247	1 c	93	.6	.571
Canned, sweetened	250	1 c	115	.7	.800
Kiwi fruit	76	1 med	46	.9	.400
Lemon juice:					
Fresh	244	1 c	60	.4	.244
Canned or bottled	244	1 c	52	.9	.481
Lime juice:					
Fresh	246	1 c	66	.37	.246
Canned or bottled	246	1 c	51	.78	.401
Loganberries:					
Fresh	100	2/3 c	70	1.2	.840
Frozen, unthawed	147	1 c	80	1.6	1.24
Mangos, fresh slices	165	1 c	108	.9	.964
Melon: also see Watermelon.					
Cantaloupe, cubes	160	1 c	57	1.6	.918
Casaba, cubes	170	1 c	45	1.5	.680
Honeydew, cubes	170	1 c	60	1.7	1.02
Frozen, mixed melon balls	173	1 c	55	2.0	1.10
Nectarine, 1 med ≈ 1 c sliced	138	1 c	68	2.0	1.37
ORANGE, 2-5/8",180g w/peel	131	1 ea	60	.6	.369
Orange juice:					
Fresh juice	248	1 c	111	.9	.992
Prep. from frozen conc.	249	1 c	110	.5	.500
Chilled	249	1 c	110	.6	.697
Canned, unsweetened	249	1 c	105	.7	.782
Orange grapefruit jce, canned	247	1 c	105	.8	.830
Papaya, whole	304	1 ea	117	.9	1.03
Papaya nectar, canned	250	1 c	142	.3	.375

	Wt (g)	Svg	Cal	Niacin (mg) 100 Cal	Svg
PEACHES:					
Fresh, whole, 2.5" dm.	87	1 ea	37	2.3	.861
Frozen slices, thawed	250	1 c	235	.7	1.63
Canned half, juice pack	77	1 ea	34	1.3	.448
Canned half, heavy syrup	81	1 ea	60	.8	.497
Peach nectar, canned	249	1 c	134	.5	.717
PEARS:					
Fresh, Bartlett,181g w/refuse	166	1 ea	98	.2	.170
Canned, juice pack	77	1 half	38	.4	.154
Canned, heavy syrup	79	1 half	59	.3	.191
PINEAPPLE:					
Fresh chunks	155	1 c	76	.9	.651
Canned pieces, juice pack	250	1 c	150	.5	.710
Canned pieces, heavy syrup	255	1 c	199	.4	.732
Pineapple juice:					
Fr/frozen	250	1 c	129	.4	.500
Canned, unsweetened	250	1 c	140	.5	.643
Plantain, cooked slices	154	1 c	179	.6	1.16
PLUMS:					
Fresh, medium, 2-1/8" dm.	66	1 ea	36	.9	.330
Canned, pitted, juice pack	95	3 ea	55	.8	.449
Canned, pitted, heavy syrup	110	3 ea	98	.3	.320
Prunes, dried, pitted	84	10 ea	201	.8	1.65
Prune juice, bottled	256	1 c	181	1.1	2.0
Raisins, packed measure	165	1 c	494	.3	1.35
Raspberries:					
Fresh berries	123	1 c	60	1.9	1.11
Frozen, thawed	250	1 c	255	.6	1.50
Canned with liquid	256	1 c	234	.5	1.13
STRAWBERRIES:					
Fresh, whole, capped	149	1 c	45	.75	.340
Frozen, unsweetened	149	1 c	52	1.3	.688
Frozen, thawed, sweetened	255	1 c	245	.4	1.02
Tangerine	84	1 ea	37	.4	.134
Watermelon, cubes	160	1 c	50	.6	.320

Grains & Grain Products

Cereals and most processed wheat flour & corn products are enriched with niacin to compensate for what is processed out.

	Wt (g)	Svg	Cal	Niacin (mg) 100 Cal	Svg
Amaranth grain, dry	195	1 c	729	.34	2.51
Barley, pearled, cooked	157	1 c	193	1.7	3.24
BRAN: see Oat, Rice, Wheat.					
Buckwheat flour, dark	98	1 c	338	.8	2.75
Buckwheat flour, light	98	1 c	340	.12	.47
Bulgar wheat, cooked	182	1 c	151	1.2	1.82
CEREALS, COLD (Ready to eat): Cereals are enriched with niacin. Brands will vary. See label.					
CEREALS, HOT (cooked): Can be enriched w/niacin. See label.					
Corn grits:					
Enriched	242	1 c	146	1.3	1.96
Unenriched	242	1 c	146	.3	.484
Cream of Rice	244	1 c	126	.8	1.00
Cream of Wheat	244	1 c	140	1.1	1.50
Farina:					
Enriched	233	1 c	116	1.1	1.28
Unenriched	233	1 c	116	.2	.233
Oatmeal, cooked rolled oats (Regular, Instant, Quick)	234	1 c	145	.21	.300
Oatmeal, fortified, Instant:					
Plain (one packet)	177	3/4 c	104	5.3	5.49
With bran and raisin	195	7/8 c	158	5.1	8.12
Average of all flavors	164	3/4 c	160	3.7	5.90

Values are for edible portion of foods

	Wt (g)	Svg	Cal	Niacin (mg) 100 Cal	Svg
Grains: Cereals, continued:					
Malt-O-Meal	240	1 c	122	4.8	5.80
Maypo	180	3/4 c	128	5.5	7.00
Ralston	253	1 c	134	1.5	2.05
Roman Meal, cooked	181	3/4 c	111	2.1	2.32
Wheatena cereal	243	1 c	135	1.0	1.34
Whole wheat cereal	242	1 c	150	1.4	2.13
Corn flour:					
Regular, unenriched	117	1 c	422	.5	2.22
Masa harina, enriched	114	1 c	416	2.7	11.2
Cornmeal. dry:					
Degermed, enriched, dry	138	1 c	505	1.4	6.95
Nearly whole, bolted	122	1 c	440	.5	2.30
FLOUR: see specific grain, vegetable, or nut.					
MACARONI, cooked:					
Regular, enriched	140	1 c	197	1.2	2.34
Vegetable, enriched	134	1 c	172	.8	1.44
Whole wheat	140	1 c	174	.6	.99
Millet, cooked	120	1/2 c	143	1.1	1.6
Noodles:					
Chow mein noodles	45	1 c	237	1.1	2.68
Egg noodles, cooked	160	1 c	213	1.1	2.38
Spinach noodles, cooked	140	1 c	182	1.2	2.14
Oat bran, 1T ≈ 6 g	94	1 c	132	.7	.878
Oats, rolled, dry, uncooked	81	1 c	311	.2	.630
Oats, cooked: see Cereals, hot.					
PASTA: see macaroni, noodles, spaghetti.					
Quinoa grain, dry	170	1 c	635	.8	4.98
RICE, cooked:					
Brown rice	195	1 c	217	1.4	2.98
White, converted, enriched	175	1 c	200	1.2	2.45
White, regular, enriched	205	1 c	264	1.1	3.03
White, instant, enriched	165	1 c	162	.9	1.45
Wild rice	82	1 c	83	1.3	1.06
Rice bran	83	1 c	262	10.8	28.2
Rice flour	158	1 c	578	10.7	4.09
Rye flour:					
Dark	128	1 c	415	1.3	5.47
Light	102	1 c	361	.2	.816
Soy flour, stirred:					
Low fat flour	44	1/2 c	163	.58	.95
Full fat raw	42	1/2 c	182	.99	1.81
Spaghetti noodles, cooked:					
Enriched	140	1 c	197	1.2	2.34
Whole wheat spaghetti	140	1 c	174	1.0	.99
WHEAT:					
Bran	30	1/2 c	65	6.3	4.07
Flour (enriched w/niacin, except for whole wheat):					
All-purpose white, unsifted	125	1 c	455	1.6	7.38
Cake flour, sifted	96	1 c	348	1.9	6.52
Self-rising flour	125	1 c	442	1.6	7.29
Semolina flour	167	1 c	601	1.7	10.0
Whole wheat flour	120	1 c	407	1.9	7.64
Wheat germ:					
Raw	100	1 c	360	1.9	6.81
Toasted	113	1 c	432	1.6	6.81
Wheat, rolled:					
Cooked	240	1 c	142	1.5	2.20
Dry	85	1 c	289	1.2	3.50
Whole grain wheat (wheat berries) cooked	50	1/3 c	28	1.4	.40
Whole grain wheat, sprouted	108	1 c	214	1.6	3.35

	Wt (g)	Svg	Cal	Niacin (mg) 100 Cal	Svg
Meats: Fish & Shellfish					
Anchovies, canned, drained	45	11 ea	95	9.4	8.96
Bluefish:					
Baked or broiled	100	3.5 oz	159	4.9	7.78
Fried in crumbs	100	3.5 oz	205	2.7	5.50
Catfish, fried in cornmeal	100	3.5 oz	229	1.0	2.28
CLAMS, meat only:					
Breaded, fried, small	188	20 ea	379	1.0	3.88
Canned, small can, drained	160	1 can	236	2.3	5.37
Steamed, whole, meat only	90	20 ea	133	2.3	3.02
COD:					
Baked or broiled	100	3.5 oz	105	2.4	2.51
Batter-fried	100	3.5 oz	199	1.1	2.20
CRAB, cooked meat:					
Alaska King crab leg	134	1 leg	129	1.4	1.80
Blue crab meat, cooked	135	1 c	138	2.2	3.00
Blue crab, canned	135	1 c	133	1.4	1.85
Dungeness crab	101	3/4 c	85	3.4	2.92
Crab, imitation from surimi	85	3 oz	87	.18	.153
Crayfish, cooked, moist heat	85	3 oz	97	2.6	2.49
Eel:					
Baked or broiled	100	3.5 oz	236	1.9	4.49
Smoked	100	3.5 oz	330	1.2	3.80
Fish cakes, fried	100	3.5 oz	213	.6	1.20
Fish sticks, heated fr/frozen	57	2 ea	155	.8	1.21
HADDOCK:					
Baked or broiled	85	3 oz	95	4.1	3.94
Breaded, fried	85	3 oz	175	1.7	2.90
Smoked	100	3.5 oz	116	4.4	5.07
Halibut:					
Baked or broiled	85	3 oz	119	5.1	6.06
Smoked	100	3.5 oz	224	2.6	5.80
Herring:					
Baked or broiled	100	3.5 oz	203	2.0	4.12
Canned with liquid	100	3.5 oz	208	1.8	3.80
Pickled, 1 piece ≈ 15 g	100	3.5 oz	262	1.1	2.80
Smoked or kippered	100	3.5 oz	217	2.0	4.40
Lobster, meat, cooked	145	1 c	142	1.1	1.55
Mackerel:					
Atlantic, baked or broiled	100	3.5 oz	262	2.6	6.85
Spanish, baked or broiled	100	3.5 oz	158	3.2	5.00
Jack, #1 tall can	361	1 can	563	4.0	22.3
Ocean perch:					
Baked or broiled	100	3.5 oz	121	2.0	2.44
Breaded, fried	85	3 oz	185	1.1	2.00
OYSTERS:					
Raw, Eastern	248	1 c	170	1.9	3.25
Raw, Pacific	248	1 c	200	2.5	4.98
Fried, breaded, med. (Eastern)	88	6 ea	173	.8	1.45
Simmered (Eastern)	100	3.5 oz	137	1.8	2.49
Pike, Northern, baked or broiled	100	3.5 oz	113	1.9	2.18
Pollock, baked / broiled:					
Mixed species	100	3.5 oz	99	3.8	3.80
Walleye (Alaska)	100	3.5 oz	113	1.5	1.65
Rockfish, baked or broiled	100	3.5 oz	121	3.2	3.92
SALMON, cooked:					
Baked or broiled, average	85	3 oz	183	3.1	5.67
Smoked, Chinook	85	3 oz	99	4.1	4.01
Canned, Atlantic, small can	220	1 can	281	5.6	15.6
Pink, boned, #1 can drained	454	1 can	631	4.7	29.7
Sockeye, #1 can drained	369	1 can	566	3.6	20.2
Sardines, canned, drained:					
Atlantic, 2 ea = 24g	92	1 can	192	2.5	4.83
Pacific, 1 ea = 38g	100	3.5 oz	174	2.4	4.20

ESHA Research

Values are for edible portion of foods

Fish & Shellfish, continued:	Wt (g)	Svg	Cal	Niacin (mg) 100 Cal	Svg
Scallops:					
Breaded, fried	93	6 ea	200	.7	1.40
Steamed	100	3.5 oz	113	1.2	1.41
Scallops, imitation from surimi	85	3 oz	84	.31	.264
Seatrout (Steelhead), cooked	100	3.5 oz	131	7.4	9.67
Shark, batter-fried	85	3 oz	194	1.2	2.37
SHRIMP:					
Boiled (2 large ≈ 11g)	100	3.5 oz	99	2.6	2.59
Fried in bread crumbs (2 ≈ 15g)	90	12 ea	218	1.3	2.76
Shrimp, canned, drained	128	1 c	154	2.3	3.53
Shrimp, imitation from surimi	85	3 oz	86	.17	.145
Snapper, baked or broiled	100	3.5 oz	128	2.7	3.46
SOLE (Flounder):					
Baked or broiled	85	3 oz	99	1.9	1.85
Fried in batter	85	3 oz	250	.8	2.12
Fried in bread crumbs	100	3.5 oz	188	1.1	2.10
Squid, fried in flour	85	3 oz	149	1.5	2.20
Surimi, processed walleye(Alaska)					
pollock: also see imitation					
crab, scallops, shrimp.	85	3 oz	84	.2	.187
Swordfish, broiled or baked	85	3 oz	132	7.5	10.0
Trout, baked or broiled	85	3 oz	129	1.8	2.30
TUNA:					
Baked or broiled from fresh	85	3 oz	157	5.7	8.96
Canned, drained, small can:					
Light, oil pack	171	1 can	339	6.0	20.3
Light, water pack	165	1 can	216	12.0	26.0

Meats
Beef, Pork, Ham, etc.

BEEF:	Wt (g)	Svg	Cal	Niacin (mg) 100 Cal	Svg
Breakfast strips, cured					
beef, cooked	34	3 ea	153	1.4	2.20
Chuck blade, pot roasted:					
Lean and fat	85	3 oz	325	.6	2.01
Lean only	85	3 oz	230	1.0	2.27
GROUND BEEF, cooked average:					
Extra lean (17% fat raw)	85	3 oz	215	1.9	4.11
Lean (21% fat raw)	85	3 oz	231	1.8	4.23
Regular (26.6% fat raw)	85	3 oz	250	2.0	4.94
Frozen patty, broiled (23%)	85	3 oz	240	1.9	4.48
Rib, oven roasted, choice:					
Lean and fat	85	3 oz	324	.9	2.80
Lean only	85	3 oz	204	1.7	3.50
Round steak, broiled, choice:					
Lean and fat	85	3 oz	233	1.4	3.18
Lean only	85	3 oz	165	2.2	3.55
Round tip, roasted, all grades:					
Lean and fat	85	3 oz	213	1.4	2.95
Lean only	85	3 oz	162	2.0	3.18
Sirloin steak, broiled:					
Lean and fat	85	3 oz	238	1.7	4.02
Lean only	85	3 oz	172	2.6	4.54
T-bone steak, broiled:					
Lean and fat	85	3 oz	276	1.2	3.32
Lean only	85	3 oz	182	2.2	3.94
Variety meats:					
Heart, simmered	85	3 oz	140	2.5	3.46
Kidney, cooked	140	1 ea	201	4.2	8.43
Liver, fried	85	3 oz	184	6.7	12.3
Dried beef, cured (6-7 pieces)	28	1 oz	47	5.7	2.70
Corned beef, canned	85	3 oz	213	1.0	2.07

	Wt (g)	Svg	Cal	Niacin (mg) 100 Cal	Svg
FROG LEGS, flour-fried	144	6 ea	418	.4	1.80
HAM: see Pork, cured and Sausages & Lunchmeats group.					
LAMB:					
Arm chop, braised:					
Lean and fat	70	1 ea	244	1.9	4.69
Lean only	55	1 ea	152	2.3	3.46
Loin chop, broiled:					
Lean and fat	64	1 ea	201	2.3	4.53
Lean only	46	1 ea	100	3.2	3.18
Cutlet, lean, cooked average	85	3 oz	175	3.1	5.37
Leg of lamb, roasted:					
Lean and fat	85	3 oz	219	2.6	5.60
Lean only	85	3 oz	162	3.3	5.39
Liver, pan-fried	85	3 oz	202	7.0	14.2
PORK:					
Bacon, cooked:					
Regular	19	3 pce	109	1.3	1.39
Breakfast strips	34	3 pce	156	1.7	2.58
Canadian style	47	2 pce	86	3.7	3.22
Center loin chop:					
Braised, lean and fat	75	1 ea	266	1.7	4.49
Braised, lean only	61	1 ea	166	2.5	4.15
Broiled, lean and fat	82	1 ea	284	1.6	4.35
Broiled, lean only	72	1 ea	166	2.4	3.99
Pan-fried, lean and fat	89	1 ea	333	1.4	4.58
Pan-fried, lean only	67	1 ea	178	2.3	4.03
Roasted, lean and fat	88	1 ea	268	1.7	4.44
Roasted, lean only	72	1 ea	180	2.3	4.10
Center rib chop:					
Braised, lean and fat	67	1 ea	246	1.5	3.78
Braised, lean only	53	1 ea	147	2.3	3.43
Broiled, lean and fat	77	1 ea	264	1.4	3.63
Broiled, lean only	63	1 ea	162	2.0	3.30
Pan-fried, lean and fat	88	1 ea	343	1.1	3.86
Pan-fried, lean only	62	1 ea	160	2.0	3.20
Roasted, lean and fat	79	1 ea	252	1.5	3.87
Roasted, lean only	66	1 ea	162	2.2	3.53
Pork roast, average-loin & rib:					
Lean and fat	85	1 pce	265	1.6	4.23
Lean only	85	1 pce	206	2.2	4.60
Shoulder, braised:					
(from 6.8 oz raw w/bone):					
Lean and fat	85	3 oz	293	1.5	4.43
Lean only	67	2.4 oz	166	2.4	4.00
Spareribs, ckd f/1 lb raw	177	6.25 oz	703	1.4	9.69
PORK, CURED - HAM: also see Turkey ham & Lunchmeat section.					
Roasted, lean and fat	85	3 oz	207	1.8	3.80
Roasted, lean only	85	3 oz	133	3.2	4.27
Canned, regular, roasted	85	3 oz	140	3.0	4.27
RABBIT, roasted	85	3 oz	175	3.2	5.62
VEAL (calf):					
Cutlet, lean, cooked average	85	3 oz	166	4.3	7.16
Rib, roasted	85	3 oz	151	4.2	6.37
Liver, pan fried	85	3 oz	208	6.9	14.4
VENISON (deer) roasted	85	3 oz	134	4.3	5.70

Values are for edible portion of foods

Niacin (Vitamin B3) mg

Meats: Poultry

	Wt (g)	Svg	Cal	Niacin per 100 Cal	Niacin Svg
CHICKEN:					
All types of meat:					
Fried	140	1 c	307	4.4	13.5
Roasted	140	1 c	266	4.8	12.8
Stewed	140	1 c	248	3.5	8.56
Canned with broth	142	5 oz	235	3.8	8.99
Dark meat:					
Fried	85	3 oz	203	3.0	6.01
Roasted	85	3 oz	174	3.2	5.57
Stewed	85	3 oz	163	2.5	4.03
Light meat:					
Fried	85	3 oz	163	7.0	11.4
Roasted	85	3 oz	147	7.2	10.6
Stewed	85	3 oz	135	4.9	6.6
Breast*, meat & skin:					
Batter-fried	140	1 ea	364	4.0	14.7
Flour-fried	98	1 ea	218	6.2	13.5
Roasted	98	1 ea	193	6.5	12.5
Stewed	110	1 ea	202	4.3	8.6
Breast*, meat only:					
Fried	86	1 ea	161	7.9	12.7
Roasted	86	1 ea	142	8.3	11.8
Stewed	95	1 ea	144	5.9	8.50
*2 pieces per bird.					
Drumstick, meat & skin:					
Batter-fried	72	1 ea	193	1.9	3.67
Flour-fried	49	1 ea	120	2.5	2.96
Roasted	52	1 ea	112	2.8	3.12
Stewed	57	1 ea	116	2.1	2.40
Drumstick, meat only:					
Fried	42	1 ea	82	3.1	2.58
Roasted	44	1 ea	76	3.5	2.67
Stewed	46	1 ea	78	2.5	1.98
Thigh, meat & skin:					
Batter-fried	86	1 ea	238	2.1	4.92
Flour-fried	62	1 ea	162	2.7	4.31
Roasted	62	1 ea	153	2.6	3.95
Stewed	68	1 ea	158	2.1	3.33
Thigh, meat only:					
Fried	52	1 ea	113	3.3	3.70
Roasted	52	1 ea	109	3.1	3.39
Stewed	55	1 ea	107	2.7	2.86
Wing, meat & skin:					
Batter-fried	49	1 ea	159	1.6	2.58
Flour-fried	32	1 ea	103	2.1	2.14
Roasted	34	1 ea	99	1.9	2.26
Stewed	40	1 ea	100	1.9	1.85
Wing, meat only:					
Fried	20	1 ea	42	3.5	1.45
Roasted	21	1 ea	43	3.6	1.54
Stewed	24	1 ea	43	2.9	1.25
Chicken gizzard, simmered	22	1 ea	34	2.6	.87
Chicken liver, simmered	20	1 ea	30	3.0	.89
DUCK, domestic, roasted:					
Meat and skin	85	3 oz	286	1.4	4.1
Meat only	85	3 oz	171	2.5	4.3
GOOSE, domestic, roasted:					
Meat and skin	85	3 oz	259	1.4	3.5
Meat only	85	3 oz	202	1.7	3.5
Liver, raw	94	1 ea	125	4.9	6.11
TURKEY:					
Breast meat:					
Barbecued	28	1 oz	40	6.8	2.73
Hickory smoked	28	1 oz	35	7.9	2.75
Ground meat, cooked	100	3.5 oz	229	2.1	4.82

	Wt (g)	Svg	Cal	Niacin per 100 Cal	Niacin Svg
TURKEY, continued:					
Roasted:					
All types	85	3 oz	145	3.2	4.63
Dark meat	85	3 oz	159	1.9	3.10
Light meat	85	3 oz	133	4.4	5.81
Breast meat	85	1 pce	93	7.6	7.08
Turkey liver	75	1 ea	127	3.5	4.46
Turkey patty, breaded, fried	64	1 ea	181	.8	1.47

Meats: Sausages & Lunchmeats

	Wt (g)	Svg	Cal	Niacin per 100 Cal	Niacin Svg
Beef lunchmeat:					
Loaf or roll	28	1 oz	87	1.2	1.04
Thin sliced	28	1 oz	50	3.0	1.49
Beerwurst (beer salami):					
Beef	23	1 pce	75	.9	.660
Pork	23	1 pce	55	1.4	.748
Berliner sausage	23	1 pce	53	1.3	.715
Bologna:					
Beef bologna	23	1 pce	72	.8	.605
Beef and pork	28	1 oz	89	.8	.731
Pork bologna	23	1 pce	57	1.6	.897
Turkey	28	1 oz	56	1.9	1.04
Braunschweiger sausage	57	2 oz	205	2.3	4.78
Brotwurst, link	70	1 ea	226	1.0	2.31
Cheesefurter (cheese smokies)	43	1 ea	141	.9	1.25
Chicken roll, light meat	57	2 oz	90	3.3	3.00
Dutch brand loaf	28	1 oz	68	1.0	.677
FRANKFURTERS (hotdogs):					
Beef, 8/package	57	1 ea	184	.8	1.44
Beef and pork, 10/package	45	1 ea	145	.8	1.18
Chicken, 10/package	45	1 ea	115	1.2	1.39
Turkey, 10/package	45	1 ea	102	1.7	1.70
HAM, cured Pork:					
Chopped, packaged	42	2 pce	98	1.4	1.40
Lunchmeat, regular	57	2 oz	103	2.9	2.98
Thin sliced, 3 slices= 1 oz	28	1 oz	37	3.7	1.37
Minced	21	1 pce	55	1.6	.874
Ham and cheese roll (loaf)	57	2 oz	147	1.3	1.96
Italian sausage link, cooked	67	1 ea	216	1.3	2.79
Liverwurst, pork	18	1 pce	59	2.3	1.36
Luncheon meat, canned	21	1 pce	70	.9	.656
Luxury loaf	57	2 oz	80	2.5	1.97
Peppered loaf	28	1 poz	42	2.1	.873
Pepperoni sausage, sliced	22	4 pce	109	1.0	1.09
Polish sausage	28	1 oz	92	1.1	.976
Pork sausage:					
Cooked link	13	1 link	48	1.2	.587
Cooked patty	27	1 pce	100	1.2	1.22
Salami:					
Beef salami	23	1 pce	58	1.4	.785
Pork and beef	57	2 oz	143	1.4	2.01
Turkey	57	2 oz	111	2.0	2.23
Smoked sausage:					
Beef & pork, link	68	1 link	229	1.0	2.19
Pork, link	68	1 link	265	1.2	3.08
Turkey	28	1 oz	55	2.1	1.16
Summer sausage	23	1 pce	80	1.2	.940
TURKEY, additional lunchmeats:					
Breakfast sausage	28	1 oz	65	2.2	1.42
Ham	57	2 oz	73	3.8	2.75
Loaf, breast meat only	43	2 pce	46	7.7	3.54
Pastrami	57	2 oz	74	3.4	2.48
Roll, light and dark meat	57	2 oz	84	3.2	2.72
Roll, light meat only	57	2 oz	83	4.8	3.97
Summer sausage	28	1 oz	50	2.8	1.40

ESHA Research

Mixed Dishes and Fast Food

	Wt (g)	Svg	Cal	Niacin (mg) 100 Cal	Niacin (mg) Svg
Beef & vegetable stew:					
Recipe	245	1 c	220	2.1	4.70
Canned	245	1 c	194	1.3	2.43
Beef, macaroni, & tomato					
sauce, recipe	226	1 c	189	1.9	3.51
Beef pot pie from frozen	234	1 pie	426	.7	3.05
BURRITOS:					
Bean burrito	174	1 ea	322	.7	2.40
Beef burrito	177	1 ea	463	1.4	6.42
Beef and bean	175	1 ea	390	1.1	4.36
Deluxe combination	198	1 ea	424	1.0	4.44
Chicken a la king, recipe	245	1 c	470	1.1	5.40
Chicken and noodles, recipe	240	1 c	365	1.2	4.30
Chicken chow mein:					
Recipe	250	1 c	255	1.7	4.30
Canned	250	1 c	95	1.1	1.00
Chicken pot pie from frozen	230	1 ea	430	.9	3.93
Chicken salad with celery	78	1/2 c	266	1.2	3.25
Chili with beans, canned	255	1 c	286	.3	.913
Chop suey, with beef and pork	250	1 c	300	1.7	5.00
Corn dog	111	1 ea	330	1.0	3.27
Corn pudding	250	1 c	271	.9	2.47
Corned beef hash, canned	220	1 c	382	1.2	4.60
ENCHILADA:					
Beef enchilada	120	1 ea	292	.7	2.09
Cheese	120	1 ea	330	.2	.646
Chicken	120	1 ea	269	1.2	3.22
French toast, recipe	65	1 pce	123	.9	1.09
LASAGNA:					
With meat, recipe	245	1 pce	398	.9	3.64
Without meat, recipe	218	1 pce	316	.6	2.01
Macaroni & cheese:					
Recipe	200	1 c	430	.4	1.80
Canned	240	1 c	230	.4	1.00
Macaroni salad, no cheese	141	1 c	371	.2	.669
Meat loaf:					
Beef and 1/3 pork	87	1 pce	212	1.4	3.07
Beef only	87	1 pce	193	1.7	3.19
Moussaka, recipe	250	1 c	250	1.9	4.78
Pies, fried, commercial:					
Apple pie	85	1 ea	255	.4	1.00
Cherry pie	85	1 ea	250	.2	.600
PIZZA, cheese:					
Regular crust, 1/8 of 15"	120	1 pce	290	1.4	4.20
Thick crust, 1/2 of 10"	208	1 ea	519	1.4	7.46
Potato salad w/ mayo & eggs	250	1 c	358	.6	2.23
Quiche Lorraine, 1/8th pie	176	1 pce	600	.2	1.20
Ravioli, beef, canned	226	1 c	220	1.6	3.43
SANDWICHES, fast food items:					
Cheeseburger, 3 oz meat	112	1 ea	300	1.2	3.70
Cheeseburger, 4 oz meat	194	1 ea	524	1.4	7.37
Chicken patty sandwich	157	1 ea	436	2.1	9.21
English muffin, with egg,					
cheese & bacon	138	1 ea	360	1.0	3.71
Fish sandwich:					
Regular, w/ cheese	140	1 ea	420	.8	3.30
Large, without cheese	170	1 ea	470	.7	3.52
Hamburger, 3 oz meat	98	1 ea	245	1.6	3.80
Hamburger, 4 oz meat	174	1 ea	445	1.8	7.85
Hotdog (frankfurter) with bun	85	1 ea	260	1.0	2.48

	Wt (g)	Svg	Cal	Niacin (mg) 100 Cal	Niacin (mg) Svg
SANDWICHES on part whole wheat bread unless noted on rye:					
Bacon, lettuce & tomato	135	1 ea	327	1.3	4.13
Chicken salad	100	1 ea	294	1.4	4.15
Egg salad	111	1 ea	319	.8	2.56
Grilled cheese	117	1 ea	393	.7	2.56
Ham sandwich	122	1 ea	256	2.1	5.36
Ham on rye	116	1 ea	242	1.9	4.50
Ham & cheese	151	1 ea	363	1.5	5.38
Ham & swiss on rye	145	1 ea	350	1.3	4.52
Peanut butter & jam	100	1 ea	341	1.7	5.81
Roast beef	122	1 ea	280	2.0	5.52
Tuna salad	116	1 ea	303	1.9	5.85
Turkey ham	122	1 ea	253	2.1	5.32
Turkey ham on rye	116	1 ea	239	1.9	4.46
SPAGHETTI, pasta & tomato sauce w/cheese:					
Homemade	250	1 c	260	.9	2.30
Canned	250	1 c	190	2.4	4.50
Taco, beef	78	1 ea	207	1.2	2.49
Taco, chicken	78	1 ea	172	2.4	4.17
Tostadas:					
W/beans & beef	192	1 ea	332	.9	2.94
W/beans & chicken	157	1 ea	249	1.8	4.53
W/refried beans	157	1 ea	212	.4	.845
Tuna noodle casserole, recipe	202	1 c	251	3.4	8.59
Tuna salad, f/recipe	205	1 c	383	3.5	13.3
Turkey pot pie, frozen	233	1 ea	416	.9	3.80

Nuts & Seeds

	Wt (g)	Svg	Cal	Niacin (mg) 100 Cal	Niacin (mg) Svg
ALMONDS:					
Dried, whole	142	1 c	837	.6	4.77
Dry roasted	138	1 c	810	.5	3.89
Brazil nuts, dry	140	1 c	919	.2	2.27
Cashews:					
Dry roasted	137	1 c	787	.2	1.92
Oil roasted	130	1 c	748	.3	2.34
Coconut:					
Fresh, grated	80	1 c	283	.15	.432
Dried, unsweetened	28	1 oz	187	.1	.171
Flaked, sweetened, pkgd	74	1 c	351	.1	.222
Filberts (hazelnuts), whole	135	1 c	853	.2	1.54
Macadamia nuts:					
Dry roasted	137	1 c	814	.8	6.44
Oil roasted	134	1 c	962	.3	2.71
MIXED NUTS w/o peanuts (cashews, almonds, brazil nuts, pecans & filberts):					
Oil roasted	144	1 c	886	.3	2.83
MIXED NUTS w/peanuts (almonds, brazil nuts, cashews, filberts, & pecans):					
Oil roasted	142	1 c	876	.8	7.19
Dry roasted	137	1 c	814	.8	6.44
PEANUTS, roasted	144	1 c	837	2.5	20.6
Peanut butter, smooth	32	2 T	190	2.2	4.23
Pecans, dried, chopped	119	1 c	794	.1	1.06
Pine nuts, dried:					
Pignola	28	1 oz	146	.7	1.01
Pinyon	28	1 oz	161	.8	1.24
Pistachio nuts:					
Dried	128	1 c	739	.2	1.38
Dry roasted	128	1 c	776	.2	1.80
Poppyseed	8.8	1 T	47	.2	.086

	Wt (g)	Svg	Cal	Niacin (mg) 100 Cal	Niacin (mg) Svg
Pumpkin/squash seeds:					
Kernels, dried	138	1 c	747	.3	2.41
Kernels, roasted	227	1 c	1185	.3	3.60
Whole, roasted	64	1 c	285	.3	.846
Sesame seeds, dried:					
Kernels	150	1 c	882	.8	7.02
Whole	144	1 c	825	.8	6.50
Soybeans, dry roasted	86	1/2 c	405	.3	1.21
SUNFLOWER seed kernels:					
Dried	36	1/4 c	205	.8	1.62
Oil roasted	135	1 c	830	.7	5.58
Tahini (sesame butter)	15	1 T	91	.9	.847
Walnuts, chopped:					
Black	125	1 c	759	.1	.863
English	120	1 c	770	.2	1.25

Soups, Sauces & Gravies

	Wt (g)	Svg	Cal	100 Cal	Svg
GRAVY:					
Beef gravy, homemade	135	1/2 c	151	.9	1.34
Beef gravy, canned	233	1 c	124	1.2	1.54
Chicken gravy:					
Homemade	130	1/2 c	163	1.3	2.08
Canned	238	1 c	189	.6	1.06
From dry mix	260	1 c	85	.9	.80
SAUCES:					
Au Jus, canned	238	1 c	38	5.1	1.93
Spaghetti sauce, plain:					
Recipe	220	1 c	179	1.7	3.00
Canned	249	1 c	272	1.4	3.75
White sauce:					
Recipe	250	1 c	395	.2	.800
From mix with milk	264	1 c	240	.2	.500

SOUPS: soups are prepared from canned unless otherwise stated. RTS=Ready To Serve. For soups prepared with milk, assume whole milk.

	Wt (g)	Svg	Cal	100 Cal	Svg
Beef broth:					
Fr/canned	240	1 c	16	11.7	1.87
Fr/dry mix	244	1 c	19	1.9	.356
Fr/dry cube	241	1 c	8	2.0	.162
Beef noodle soup:					
Prepared w/water	244	1 c	84	1.3	1.07
Fr/dry mix	251	1 c	41	1.7	.690
Chicken broth:					
Fr/canned	244	1 c	39	8.6	3.35
Fr/dry mix	244	1 c	21	.9	.195
Chicken soup, chunky, RTS	251	1 c	178	2.5	4.42
Chicken, cream of:					
Prepared w/milk	248	1 c	191	.5	.923
Prepared w/water	244	1 c	115	.7	.820
Chicken noodle soup:					
Prepared w/water	241	1 c	75	1.9	1.39
Prepared w/milk	240	1 c	114	3.8	4.32
Fr/dry mix	252	1 c	53	1.7	.883
Chicken rice soup:					
Prepared w/water	241	1 c	60	1.9	1.13
Chunky style, RTS	240	1 c	127	3.2	4.10
Chicken vegetable:					
Chunky-RTS	240	1 c	167	2.0	3.30
Fr/dry mix	251	1 c	49	1.4	.687
Clam chowder:					
New England style	248	1 c	163	.6	1.03
Manhatten style	244	1 c	78	1.2	.817
Lentil and ham soup	248	1 c	140	1.0	1.35
Minestrone	241	1 c	80	1.2	.942
Onion soup	241	1 c	57	1.1	.600

	Wt (g)	Svg	Cal	100 Cal	Svg
Split pea w/ham:					
Prepared w/water	253	1 c	189	.8	1.48
Chunky style, RTS	240	1 c	184	1.4	2.52
Tomato soup:					
Prepared w/milk	248	1 c	160	1.0	1.52
Prepared w/water	244	1 c	86	1.7	1.42
From dry	265	1 c	102	.8	.782
Tomato vegetable, f/dry	253	1 c	55	1.4	.789
Turkey noodle	244	1 c	69	2.0	1.40
Turkey vegetable	241	1 c	74	1.4	1.01
Vegetable soup, chunky, RTS	240	1 c	122	1.0	1.20
Vegetable beef	244	1 c	79	1.3	1.03
Vegetable, cream of, f/dry	260	1 c	105	.5	.520
Vegetarian vegetable	241	1 c	70	1.3	.916

Vegetables & Legumes

	Wt (g)	Svg	Cal	100 Cal	Svg
Alfalfa sprouts	33	1 c	10	1.6	.159
Amaranth leaves:					
Fresh, chopped	28	1 c	7	2.6	.184
Cooked, boiled	132	1 c	28	2.6	.738
Artichoke globe, cooked (300g w/refuse)	120	1 ea	60	2.0	1.2
Artichoke hearts:					
Cooked f/ frozen, pkg	240	9 oz	108	2.0	2.20
Marinated, jar	170	6 oz	168	.8	1.38
Asparagus:					
Fresh spears, uncooked	58	4 ea	13	5.1	.660
Ckd fr/ fresh pieces	90	1/2 c	23	4.2	.947
Ckd fr/ frozen pieces	180	1 c	50	3.7	1.87
BEANS: also see garbanzo, lentils, soybeans.					
Baked beans (dry white beans w/ spices & sauce):					
Home prepared	253	1 c	382	.3	1.03
Canned, plain or vegetarian	254	1 c	235	.5	1.09
Canned w/franks	257	1 c	366	.6	2.32
Canned w/pork	253	1 c	268	.4	1.13
Canned w/pork, sweet sce	253	1 c	282	.3	.888
Canned w/pork, tomato ce	253	1 c	247	.5	1.26
Black beans, ckd f/dry	172	1 c	227	.4	.869
Broadbeans-vegetable:					
Fresh, sliced = 8g	109	1 c	79	2.1	1.6
Ckd from fresh	100	3.5 oz	56	2.1	1.2
Broadbeans-mature legumes:					
Ckd from dry	170	1 c	186	.7	1.21
Canned with liquid	256	1 c	183	1.3	2.46
Great northern:					
Canned	262	1 c	300	.4	1.21
Ckd fr/dry	177	1 c	210	.6	1.21
Green (snap) beans:					
Fresh, uncooked	110	1 c	34	2.4	.827
Ckd from fresh	125	1 c	44	1.7	.768
Ckd from frozen	135	1 c	36	1.6	.563
Canned, drained	135	1 c	26	1.0	.272
Canned with liquid	240	1 c	36	1.3	.480
Kidney beans:					
Canned	256	1 c	208	.6	1.29
Ckd fr/dry	177	1 c	225	.5	1.02
Lima beans:					
Ckd fr/frozen, small	90	1/2 c	94	.7	.693
Ckd fr/frozen, large	85	1/2 c	85	1.1	.909
Ckd fr/dry, avg	185	1 c	223	.3	.696
Canned, drained	170	1 c	164	.5	.800
Canned with liquid	241	1 c	191	.3	.629

39

	Wt (g)	Svg	Cal	Niacin (mg) 100 Cal	Niacin (mg) Svg
Vegetables & Legumes, continued:					
BEANS, continued:					
Navy beans, cooked fr/dry	182	1 c	259	.4	.966
Pinto beans:					
Canned	240	1 c	186	.38	.701
Ckd fr/dry	171	1 c	235	.3	.684
Refried beans, canned	253	1 c	270	.5	1.23
White beans, cooked fr/dry	179	1 c	253	.2	.487
Winged beans:					
Ckd from fresh veg	31	1/2 c	12	1.7	.202
Ckd from dry legume	172	1 c	252	.6	1.43
Yardlong beans:					
Ckd fr/fresh veg	104	1 c	49	1.3	.655
Ckd fr/dry legume	171	1 c	202	.5	.942
Yellow wax: see green beans.					
Bean sprouts (mung beans):					
Fresh sprouts	104	1 c	31	2.5	.779
Stir fried	124	1 c	62	2.4	1.49
Boiled	124	1 c	26	4	1.01
Canned	125	1 c	16	1.7	.275
Beets:					
Ckd fr/ fresh, 2" dm.	100	2 ea	31	.90	.273
Canned, drained, diced	85	1/2 c	27	.56	.150
Beet greens, ckd fr/fresh	144	1 c	40	1.8	.719
Broccoli, chopped:					
Fresh pieces	88	1 c	24	2.3	.562
Ckd from fresh	156	1 c	44	2.0	.896
Ckd from frozen	184	1 c	51	1.7	.843
Brussels sprouts:					
Ckd from fresh	156	1 c	60	1.6	.946
Ckd from frozen	155	1 c	65	1.3	.832
CABBAGE, shredded:					
Common, fresh	70	1 c	16	1.3	.210
Common, cooked	150	1 c	32	1.1	.345
Bok choy, cooked	170	1 c	20	3.6	.728
Pe-tsai, fresh	76	1 c	11	2.8	.304
Pe-tsai, cooked	119	1 c	16	3.7	.595
Red cabbage, fresh	70	1 c	19	1.1	.210
Red cabbage, cooked	75	1/2 c	16	.94	.150
CARROTS:					
Fresh, whole (7.5" x 1-1/8)	72	1 ea	31	2.2	.668
Fresh, grated	55	1/2 c	24	2.1	.510
Cooked from fresh, sliced	78	1/2 c	35	1.1	.395
Cooked from frozen	73	1/2 c	26	1.2	.320
Carrot juice	123	1/2 c	49	1.0	.473
Cauliflower:					
Fresh pieces	50	1/2 c	12	2.6	.317
Cooked from fresh	62	1/2 c	15	2.3	.342
Cooked from frozen	90	1/2 c	17	1.6	.279
Celery:					
Fresh-outer stalk, 8" x 1.5"	40	1 pce	6	2.2	.129
Cooked pieces	75	1/2 c	11	2.2	.240
Chard, Swiss, cooked	175	1 c	35	1.8	.630
Collards:					
Fresh, chopped	36	1 c	11	1.2	.135
Ckd from fresh	128	1 c	35	1.1	.371
Ckd from frozen	170	1 c	63	1.8	1.08
CORN:					
Fresh kernels, uncooked	77	1/2 c	66	2.0	1.31
Ckd fr/fresh, kernels	82	1/2 c	89	1.5	1.32
Ckd fr/frozen, kernels	82	1/2 c	67	1.6	1.05
Canned, drained	82	1/2 c	66	1.1	.720
Canned with liquid	128	1/2 c	79	1.5	1.20
Canned, vacuum pack	105	1/2 c	83	1.5	1.23
Cream style, canned	128	1/2 c	93	1.3	1.23
Cucumber, whole (8" x 2" diam)	301	1 ea	39	2.3	.903
Dandelion greens:					
Fresh	55	1 c	25	1.6	.390
Cooked	105	1 c	35	1.4	.500
Dock (sorrel greens):					
Fresh	133	1 c	29	2.3	.665
Cooked	100	1 c	20	2.1	.411
Eggplant, cooked	160	1 c	45	2.1	.960
Escarole, curly endive	50	1 c	8.5	2.3	.200
Garbanzo beans (chickpeas):					
Canned	240	1 c	285	.12	.331
Cooked from dry	164	1 c	269	.3	.863
Garden cress, cooked	135	1 c	31	3.5	1.08
LETTUCE, chopped:					
Butterhead	56	1 c	7	2.3	.168
Iceberg	56	1 c	7	1.4	.105
Looseleaf	56	1 c	10	2.2	.224
Romaine	56	1 c	9	3.1	.280
Jerusalem artichoke, fresh	150	1 c	114	1.7	1.95
Jicama, fresh	100	3.5 oz	20	.6	.110
Kale, chopped:					
Fresh, chopped	67	1 c	33	2.0	.670
Ckd from fresh	130	1 c	42	1.7	.700
Ckd from frozen	130	1 c	39	2.2	.874
Kohlrabi:					
Fresh slices	140	1 c	38	1.5	.560
Ckd from fresh	165	1 c	48	1.3	.644
Lambquarters:					
Fresh, chopped	56	1 c	24	2.8	.672
Cooked	180	1 c	58	2.8	1.62
Lentils, ckd from dry	198	1 c	231	.9	2.10
Lentils, sprouted:					
Fresh	77	1 c	81	1.1	.869
Stir fried	100	3.5 oz	101	1.2	1.20
MUSHROOMS:					
Fresh slices (1 avg = 18g)	35	1/2 c	9	16.5	1.44
Ckd from fresh	78	1/2 c	21	16.6	3.48
Canned, drained	78	1/2 c	19	6.7	1.25
Mustard greens:					
Fresh, chopped	56	1 c	15	3.1	.448
Ckd from fresh	140	1 c	21	2.9	.606
Ckd from frozen	150	1 c	28	1.4	.387
Okra:					
Pods ckd fr/fresh	85	8 pods	27	2.7	.740
Slices ckd fr/ frozen	92	1/2 c	34	2.1	.722
Onions, spring, chopped	50	1/2 c	16	1.6	.263
Parsley:					
Fresh, chopped	30	1/2 c	10	2.1	.210
Freeze-dried	1.4	1.4 c	4	3.8	.146
Parsnips, cooked fr/fresh	156	1 c	125	.9	1.10
PEAS:					
Black-eyed peas:					
Ckd from fresh	165	1 c	160	1.4	2.32
Ckd from frozen	170	1 c	224	.6	1.24
Ckd from dry	171	1 c	198	.4	.846
Canned	240	1 c	184	.5	.850
Green peas:					
Fresh, uncooked	145	1 c	118	2.6	3.05
Ckd from fresh	160	1 c	134	2.4	3.23
Ckd from frozen	80	1/2 c	63	1.9	1.18
Canned, drained	85	1/2 c	59	1.1	.622
Canned with liquid	124	1/2 c	61	1.7	1.04
Green, edible-pod peas:					
Fresh, uncooked	145	1 c	61	1.4	.870
Ckd from fresh	160	1 c	67	1.3	.862
Split peas, cooked f/dry	196	1 c	231	.8	1.74
Peas, green, sprouted:					
Fresh sprouts	120	1 c	154	2.4	3.71
Ckd from fresh	100	3.5 oz	118	.9	1.07

Values are for edible portion of foods

	Wt (g)	Svg	Cal	Niacin (mg) 100 Cal	Svg
Vegetables & Legumes, continued:					
Peas & carrots:					
Ckd from frozen	80	1/2 c	38	2.4	.923
Canned with liquid	128	1/2 c	48	1.6	.744
PEPPER, HOT, green or red:					
Fresh, chopped	75	1/2 c	30	2.4	.713
Chili pepper, canned	68	1/2 c	17	3.2	.544
Jalapenos, chopped, canned	68	1/2 c	17	2.0	.340
PEPPER, SWEET, green or red:					
Fresh, chopped (1 pod = 74g)	50	1/2 c	14	1.8	.255
Cooked from fresh	68	1/2 c	19	1.7	.324
Poi, two finger	240	1 c	269	1.0	2.64
POTATOES:					
Baked in oven:					
(4.75" long, 2-1/3" dm)					
Flesh & skin	202	1 ea	220	1.5	3.32
Flesh only	156	1 ea	145	1.5	2.18
Potato skin	58	1 ea	115	1.5	1.78
Baked in microwave:					
Flesh & skin	202	1 ea	212	1.6	3.46
Flesh only	156	1 ea	156	1.6	2.54
Boiled, 2-1/2" diam:					
Cooked without skin	135	1 ea	116	1.5	1.77
Cooked in skin, peeled after	136	1 ea	119	1.6	1.96
Canned potatoes, 1" diam.	70	2 ea	42	1.5	.640
French fries fr/frozen:					
Fried in oil	50	10 ea	158	1.0	1.63
Oven heated	50	10 ea	111	1.0	1.15
Hash browns	156	1 c	340	1.1	3.78
Mashed:					
Prepared w/milk	210	1 c	162	1.5	2.35
Prepared fr/instant	220	1 c	239	.8	1.91
Potatoes au gratin:					
From recipe	245	1 c	322	.8	2.43
From mix	245	1 c	228	1.0	2.30
Potatoes, scalloped:					
From recipe	245	1 c	210	1.2	2.58
From mix	245	1 c	228	1.1	2.52
Potato chips, plain	28	14 ea	148	.8	1.19
Potato pancakes	76	1 ea	237	.7	1.61
PUMPKIN, mashed:					
Cooked from fresh	245	1 c	50	2.0	1.01
Canned	246	1 c	84	1.1	.900
Purslane, boiled	115	1 c	21	2.5	.529
Radishes, red	45	10 ea	7	1.9	.135
Radish seeds, sprouted	38	1 c	15.6	6.9	1.08
Rutabaga:					
Fresh cubes	140	1 c	51	1.9	.980
Cooked	85	1/2 c	29	1.8	.536
Salsify, cooked slices	135	1 c	92	.6	.529
Sauerkraut, canned w/liquid	236	1 c	44	.8	.337
Seaweed:					
Irish moss, fresh	28	1 oz	14	1.2	.168
Kelp, fresh	28	1 oz	12	1.1	.133
Lavar, fresh	28	1 oz	10	4.2	.417
Wakame, fresh	28	1 oz	12.8	3.5	.454
Spirulina, dried	28	1 oz	82.2	4.4	3.63
Soybeans, ckd from dry	172	1 c	298	.2	.686
Soybeans, sprouted:					
Fresh sprouts	35	1/2 c	45	.9	.402
Steamed	94	1 c	76	1.4	1.03

SOYBEAN PRODUCTS: see tofu, this section; miso, natto, tempeh in Other; roasted soybeans in Nuts & Seeds; soy milk in Dairy; & soy flour in Grains.

	Wt (g)	Svg	Cal	Niacin (mg) 100 Cal	Svg
Spinach:					
Fresh, chopped	56	1 c	12	3.3	.405
Ckd from fresh	180	1 c	41	2.2	.882
Ckd from frozen	190	1 c	53	1.5	.796
Canned, drained	214	1 c	50	1.7	.830
SQUASH, SUMMER, sliced:					
Crookneck, cooked	180	1 c	36	2.6	.923
Scallop, fresh slices	130	1 c	24	3.3	.780
Scallop, cooked	90	1/2 c	14	3.0	.418
Zucchini, fresh slices	130	1 c	19	2.7	.520
Zucchini, cooked	180	1 c	29	2.7	.770
SQUASH, WINTER, mashed:					
Acorn / Danish, baked	245	1 c	137	1.6	2.16
Acorn / Danish, boiled	245	1 c	83	1.6	1.30
Butternut, baked	245	1 c	99	2.4	2.38
Butternut, ckd fr/frozen	240	1 c	94	1.2	1.11
Hubbard, baked	240	1 c	120	1.1	1.34
Hubbard, boiled	236	1 c	70	1.1	.788
Spaghetti, baked or boiled	155	1 c	45	2.8	1.26
Succotash, ckd f/frozen	170	1 c	158	1.4	2.22
Sweet potato, 5" x 2" diam:					
Baked in skin	.114	1 ea	118	.6	.700
Boiled, peeled	151	1 ea	160	.6	1.00
Canned, mashed	128	1/2 c	129	.9	1.20
Canned, vacuum packed	255	1 c	233	.8	1.89
Candied, recipe, 2.5" x 2"	105	1 pce	144	.3	.414
Tofu (soybean curd):					
Firm, raw	126	1/2 c	183	.26	.480
Regular, raw	124	1/2 c	94	.26	.242
TOMATOES:					
Fresh, whole, 2.6" dm.	123	1 ea	26	3.0	.772
Fresh, chopped	180	1 c	38	3.0	1.13
Cooked from fresh	240	1 c	65	2.8	1.80
Canned, whole	240	1 c	47	3.7	1.76
Tomato juice, canned	244	1 c	42	4.0	1.64
Tomato paste, canned	262	1 c	220	3.8	8.44
Tomato puree, canned	250	1 c	102	4.2	4.29
Tomato sauce, canned	245	1 c	74	3.8	2.82
Turnips:					
Fresh cubes	130	1 c	35	1.5	.520
Ckd from fresh	78	1/2 c	14	1.7	.233
Turnip greens:					
Ckd from fresh	144	1 c	29	2.0	.592
Ckd from frozen	164	1 c	48	1.6	.768
VEGETABLES, combinations, cooked from frozen:					
Broccoli, cauliflower, red pepper	95	2/3 c	25	1.6	.400
Broccoli, carrots, pasta	95	2/3 c	88	.7	.600
Cantonese stir fry	95	1/2 c	53	1.1	.600
Chinese stir fry	95	1/2 c	31	1.6	.500
Green beans & spaetzle, Bavarian style	95	1/2 c	108	.3	.300
Japanese stir fry	95	1/2 c	29	2.1	.600
Mixed (carrots, corn, peas, green beans, lima beans):					
From frozen	182	1 c	107	1.4	1.55
Canned, drained	163	1 c	77	1.2	.941
Peas, carrots, onions	91	1/2 c	54	2.4	1.30
Peas, carrots, onions & pasta	95	1/2 c	122	.9	1.10
Peas, cauliflower & crm sce	95	1/2 c	118	1.0	1.20
Peas & mushrooms	95	1/2 c	73	2.9	2.10
Peas & onions	95	1/2 c	71	2.5	1.80
Peas, potatoes & cream sce	76	1/2 c	140	.7	1.00

Values are for edible portion of foods

	Wt (g)	Svg	Cal	Niacin (mg) 100 Cal	Niacin (mg) Svg
VEGETABLE COMBINATIONS, continued:					
Peas, pasta, corn					
& cream sauce	95	1/2 c	132	1.2	1.60
Peas, pasta, mushrooms					
& cream sauce	95	1/2 c	129	1.5	1.90
Vegetable juice cocktail	242	1 c	46	3.8	1.76
Water chestnuts, canned slices	70	1/2 c	35	.7	.252
Watercress, chopped	17	1/2 c	2	1.7	.034
Yams, orange: see Sweet potatoes.					
Zucchini: see Squash, summer.					

Other

	Wt (g)	Svg	Cal	Niacin (mg) 100 Cal	Niacin (mg) Svg
Barbecue sauce	250	1 c	160	.5	.800
CANDY and CANDY BARS:					
Chocolate covered items:					
Almonds	165	1 c	935	.4	4.17
Peanuts	170	1 c	954	1.7	16.0
Raisins	187	1 c	733	.2	1.29
M&M's peanut candies	47	1 pkg	240	.6	1.48
M&M's plain chocolate candies	48	1 pkg	237	.1	.270
Milk chocolate, plain	28	1 oz	145	<.1	.100
Milk chocolate w/peanuts	28	1 oz	155	1.4	2.20
Milk chocolate w/rice cereal	28	1 oz	140	<.1	.100
Reese's peanut butter cup	45	2 ea	240	.9	2.12
Snickers candy bar, 2.2 oz	61	1 bar	290	.6	1.84
Carob flour	103	1 c	185	1.1	1.95
Catsup	15	1 T	16	1.3	.206
Chili sauce:					
Hot red pepper based	31	2 T	6	3.1	.188
Tomato based	273	1 c	284	1.5	4.40
CHOCOLATE:					
Baking, unsweetened	28	1 oz	145	.3	.38
Bittersweet	28	1 oz	141	.2	.318
Chocolate chips, semi-sweet	170	1 c	860	.1	.900
Chocolate syrup:					
Hot fudge topping	300	1 c	1020	<.1	.600
Thin type	300	1 c	680	.1	.900
Cocoa powder	86	1 c	224	.8	1.80
Gelatin salad /dessert	120	1/2 c	70	.3	.200
Ginger root, fresh slices	11	5 ea	8	1.0	.077
Honey	339	1 c	1030	<.1	1.00
Jalapeno bean dip, Fritos	28	1 oz	33	3.3	1.10
Miso (soybean product)	275	1 c	565	.4	2.37
Molasses, blackstrap	40	2 T	85	.9	.800
Mustard, prepared	125	1/2 c	94	1.7	1.56
Natto (soybean product)	100	3.5 oz	167	.7	1.10
Salsa:					
Picante by Tostitos	85	6 T	40	1.3	.500
Homemade	184	7/8 c	79	1.2	.928
Soy sauce:					
Regular (soy and wheat) shoyu	18	1 T	9	6.7	.605
Tamari (soy)	18	1 T	11	6.5	.711
Made f/fhydrolyzed veg. protein	18	1 T	7	7.3	.509
SPICES:					
Basil, dried	4.5	1 T	11	2.8	.313
Caraway seed	6.7	1 T	22	1.1	.242
Cayenne, red pepper	5.3	1 T	17	2.7	.461
Chili powder	7.5	1 T	24	2.5	.592
Cinnamon	6.8	1 T	18	.5	.088
Garlic cloves	12	4 ea	18	.5	.084
Garlic powder	8.4	1 T	28	.2	.058
Ginger, ground powder	5.4	1 T	19	1.5	.278
Nutmeg, ground	7	1 T	37	.24	.091

	Wt (g)	Svg	Cal	Niacin (mg) 100 Cal	Niacin (mg) Svg
SPICES, continued:					
Oregano, ground powder	4.5	1 T	14	2.0	.280
Paprika	6.9	1 T	20	5.3	1.06
Pepper, black	6.4	1 T	16	.45	.073
Sage, ground	2	1 T	6	1.9	.114
Tarragon, ground powder	4.8	1 T	14	3.1	.430
Thyme, ground	4.3	1 T	12	1.8	.212
Turmeric, ground	6.8	1 T	24	1.5	.350
Tempeh (soybean product)	166	1 c	331	2.3	7.7
Teriyaki sauce	18	1 T	15	1.5	.229
Tobasco sauce	15	1 T	1.6	3.0	.050
Tofu: see vegetable section.					
Yeast:					
Brewer's yeast	8	1 T	25	12.6	3.16
Dry active yeast	30	4 T	80	15.0	12.0

Vitamin B6   (Pyridoxine)

Vitamin B-6 plays important roles in the body, especially in the metabolism of protein; therefore dietary needs are increased when the intake of protein is increased. Its coenzymes are part of the process that converts protein into energy, and that converts glycogen into glucose for muscle tissue. It is also required for building some amino acids and for converting others to hormones, including the synthesis of niacin from tryptophan. Pyridoxine is necessary for the production of red blood cells and the functioning of nerve tissue. It may also be involved with the metabolism of polyunsaturated fats.

Deficiencies can result in diseases of the skin, tongue and mouth sores, anemia, insulin sensitivity, nausea, nervousness, and convulsions. Deficiencies rarely occur alone however, and are most likely to be seen in people who are deficient in several B-complex vitamins.

There can be toxic effects from vitamin B6 taken in "gram" quantities over an extended period of time. This can occur when taking supplements that contain high doses of B6, or when prescribed by physicians over extended periods to treat PMS or mental disorders. Reported consequences can vary from permanent nerve damage, to reversal of the symptoms when the overdoses are eliminated. Watch for more research results in this area.

Vitamin Losses

The processing of foods drastically reduces the B6 content.   Processed and refined foods may contain less than 50% of that found in the original food.   From 50 to 70% percent is lost when grains are milled into cereal; about 75% is lost when wheat is milled into white flour, and it is not replaced. From 50 to 70% is lost in processing luncheon meats; frozen vegetables show losses ranging from 37 to 56 percent; canned vegetables show losses of 57 to 77 percent. Fruits show an average of 15 percent loss in freezing and about 38 percent loss in canning. Little is lost in processing dairy products.

Heat also takes its toll, and B-6 content may be reduced by exposure to ultraviolet light. In 1951, vitamin B-6 was found to be an essential vitamin: babies became irritable and convulsive when exclusively fed a particular formula that had been sterilized at a heat high enough to destroy most of its vitamin B-6 (in some foods, however, B-6 is less sensitive to heat and to cooking).

Recommended Dietary Allowances for Adults
USA RDA: Men, 2.0 mg; Women, 1.6 mg (Pregnant, 2.2 mg; Lactation/Nursing, 2.1 mg) These allowances
   assume that the reported average protein intakes are approximately 100g/day for men, and 60 g/day for women.
   They may not be sufficient for those whose habitual protein intake is at or above this amount for extended periods.
Canadian RNI: 15 mcg of B6 per gram of protein. An additional 0.3 mg / day is recommended for lactation.
   For 50 g intake of protein, this would result in a recommended amount of 750 mcg or .75 mg of Vitamin B6.

There are many medicines (approximately 40) known to affect the metabolism or bioavailability of vitamin B6. Vitamin B-6 needs are increased if oral contraceptives and certain drugs are taken (such as penicillamine and isoniazid), if diuretics are taken, or if a lot of alcohol is consumed. Caution is noted for megadoses of this vitamin. Recent evidence indicates potential problems of excessive intake.   Such excessive intake can only happen with excessive supplements, not regular dietary intake.

Food Sources

Vitamin B-6 is widely distributed in unprocessed foods. Good sources include chicken, fish, pork, eggs and liver. whole grains (unmilled rice, oats, whole wheat products) legumes (soybeans, peanuts), fruits, nuts, many vegetables, and especially avocados and bananas. Relatively poor sources are red meat and dairy products.

Values are for edible portion of foods

	Wt (g)	Svg	Cal	Vit B6 (mg) 100 Cal	Svg		Wt (g)	Svg	Cal	Vit B6 (mg) 100 Cal	Svg
Baked Goods						Tortillas:					
						Corn, 6" dm.	30	1 ea	65	.1	.090
						Flour, 10.5" dm.	57	1 ea	168	<.1	.023
BREADS:						Tortilla chips:					
Banana nut, 1/2" slice	50	1 pce	161	.08	.126	Plain	28	1 oz	139	.1	.080
Boston brown bread, cnd	45	1 pce	95	.06	.060	Doritos, flavored	28	1 oz	140	.1	.104
Cornbread, muffin	45	1 ea	145	<.1	.040	Waffle, ckd from mix, 7" diam	75	1 ea	205	<.1	.029
Pumpernickel, 5" x 4" x 3/8"	32	1 pce	80	.06	.051	Waffle, frozen 4" diam	35	1 ea	98	.1	.093
Raisin bread	25	1 pce	68	<.01	.010						
Rye, light, 5" x 4" x 7/16"	25	1 pce	65	.03	.020						
Wheat bread (white &						**Dairy & Dairy Products**					
whole wheat flour)	25	1 pce	65	.05	.030						
White bread	28	1 pce	75	<.01	.009	**CHEESE** (1.5" cube ≈ 1 oz):					
Whole wheat bread	28	1 pce	70	.07	.052	Brie cheese	28	1 oz	95	.1	.067
						Camembert	28	1 oz	85	.1	.064
CAKES, pce = 1/16th cake						Cheddar cheese, shredded	113	1 c	455	<.1	.084
unless otherwise stated:						Cottage cheese:					
Angel food, 1/2 tube cake	53	1 pce	125	<.1	.010	Lowfat 1%	226	1 c	164	.1	.154
Boston cream pie, 1/8	120	1 pce	260	<.1	.050	Lowfat 2%	226	1 c	205	.1	.172
Carrot cake, cream cheese						Creamed, large curd	225	1 c	235	.1	.140
frosting 2.5" x 3"	112	1 pce	406	<.1	.060	Creamed, small curd	210	1 c	215	.1	.141
Cheesecake, average	98	1 pce	290	<.1	.054	Creamed with fruit	226	1 c	279	<.1	.120
Coffee cake, 2.4" x 2.8"	72	1 pce	230	<.1	.020	Dry curd	145	1 c	123	.1	.119
Fruitcake, dark, 2/3" arc	43	1 pce	165	<.1	.054	Gjetost	28	1 oz	132	.1	.077
Gingerbread, recipe, 3" x 3"	110	1 pce	351	<.1	.058	Parmesan, grated, 1 T = 5 g	28	1 oz	129	<.1	.030
Pound cake, 1/2" slice	30	1 pce	115	<.1	.020	Ricotta, made w/whole milk	246	1 c	428	<.1	.106
Sheet cake, 3" x 3" piece:											
Plain	86	1 pce	315	<.1	.030	**CREAM, SOUR, dairy**	230	1 c	493	<.1	.037
W/white frosting	121	1 pce	445	<.1	.030	Cream, Sour, Imitation, non-dairy	230	1 c	479	0	0
Yellow cake, chocolate frosting	69	1 pce	240	<.1	.030						
Cherry crisp, 3" x 3"	138	1 pce	157	<.1	.054	**CREAM, SWEET:**					
Chips: see corn & tortilla this section;						Half and Half	242	1 c	315	<.1	.094
potato chips under Vegetable.						Heavy cream, unsweetened	119	1 c	410	<.1	.031
						Whipped, pressurized	60	1 c	154	<.1	.030
COOKIES:						**Kefir** beverage	233	1 c	160	<.1	.087
Average cookie	40	4 ea	150-250	<.1	.012						
Fig bars	56	4 ea	210	<.1	.070	**MILK** (cow):					
Doughnuts:						Skim	245	1 c	86	.1	.098
Cake type, medium	50	1 ea	210	<.1	.020	Lowfat, 1%	244	1 c	102	.1	.105
Yeast raised, jelly filled	65	1 ea	226	.1	.300	Lowfat, 2%	244	1 c	121	.1	.105
Yeast raised, plain	60	1 ea	235	.1	.277	Whole (3.3% fat)	244	1 c	150	.1	.102
MUFFIN, wheat bran	45	1 ea	140	<.1	.124	Buttermilk	245	1 c	99	.1	.083
Pancakes, from mix:						Canned:					
Buckwheat, 4" diam	27	1 ea	55	.1	.060	Evaporated, skim	255	1 c	200	.1	.140
Plain/sourdough, 4" diam	27	1 ea	60	<.1	.010	Evaporated, whole	252	1 c	340	<.1	.126
Whole wheat, 5" diam	52	1 ea	94	.2	.157	Sweetened, condensed	306	1 c	982	<.1	.156
						Dried, instant non-fat	68	1 c	244	.1	.235
PIES: pce = 1/6 of 9" pie.						Milk (other):					
Apple pie	158	1 pce	405	<.1	.083	Goat	244	1 c	168	.1	.112
Banana cream	198	1 pce	319	.1	.460	Human breast milk	246	1 c	171	<.1	.027
Blueberry	158	1 pce	380	<.1	.072	Soy milk	240	1 c	79	.1	.098
Cherry	158	1 pce	410	<.1	.083						
Chocolate cream	175	1 pce	311	<.1	.089	**MILK BEVERAGES** & mixes:					
Coconut cream	172	1 pce	343	.3	1.01	**Chocolate flavored,**					
Coconut custard	165	1 pce	384	.1	.236	**to be mixed w/water:**					
Cream pie, commercial	152	1 pce	455	<.1	.065	Powder (includes dry milk)	28	1 oz	100	<.1	.040
Custard pie	152	1 pce	293	<.1	.084	Drink	206	3/4 c	100	<.1	.040
Lemon meringue	140	1 pce	355	<.1	.050	**Chocolate flavored,**					
Mincemeat	160	1 pce	395	<.1	.147	**to be mixed w/milk:**					
Peach	158	1 pce	405	<.1	.065	Powder	21.6	3.4 oz	75	<.01	.002
Pecan	138	1 pce	583	<.1	.080	Drink	266	1 c	226	<.1	.104
Pumpkin	200	1 pce	367	<.1	.107	Cocoa, hot, w/whole milk	250	1 c	218	<.1	.107
Poptart-type pastry, fortified	54	1 ea	210	.1	.205	Instant Breakfast, dry	37	1 env	130	.3	.400
						Malted milk, w/whole milk:					
						Chocolate flavor	265	1 c	229	.1	.135
ROLLS:						Natural flavor	265	1 c	237	.1	.188
Dinner roll, commercial	28	1 ea	85	<.1	.012	Milkshakes (10 fl oz = 1.25 c):					
Hamburger bun	45	1 ea	129	<.1	.016	Chocolate	283	1.25 c	360	<.1	.142
Hotdog bun	40	1 ea	115	<.1	.014	Strawberry	283	1.25 c	319	<.1	.125
Whole wheat roll	35	1 ea	88	.1	.064	Vanilla	283	1.25 c	314	<.1	.147

Vitamin B-6 (Pyridoxine) mg

	Wt (g)	Svg	Cal	Vit B6 (mg) 100 Cal	Svg
Dairy, continued:					
MILK DESSERTS:					
Custard, baked	265	1 c	305	<.1	.132
Ice cream, vanilla					
Regular	133	1 c	269	<.1	.061
Rich	148	1 c	349	<.1	.053
Soft serve	173	1 c	377	<.1	.095
Ice milk, regular	131	1 c	184	<.1	.085
Ice milk, soft serve	175	1 c	223	.1	.133
PUDDING, prepared:					
Assorted, low calorie	130	1/2 ea	69	.1	.060
Average pudding, canned					
5 oz can≈1/2 c	142	1 can	260+	<.1	.030
Chocolate, cooked or					
instant, f/mix	260	1 c	305	<.1	.128
Flavors fr/mix, instant	145	1/2 c	175	<.1	.050
Tapioca, ckd fr/ mix	130	1/2 c	145	<.1	.051
Vanilla fr/mix	130	1/2 c	148	<.1	.051
Yogurt, frozen, average	174	1 c	220	<.1	.069
YOGURT:					
Lowfat, plain	227	1 c	144	.1	.111
Lowfat, fruit	227	1 c	231	<.1	.091
Lowfat, coffee / vanilla	227	1 c	193	.1	.102
Nonfat	227	1 c	127	.1	.120
Whole	227	1 c	138	.1	.073

Eggs

	Wt (g)	Svg	Cal	Vit B6 (mg) 100 Cal	Svg
Egg, large, cooked	50	1 ea	77	.1	.060
Egg, large, raw	50	1 ea	77	.1	.069
Egg yolk, raw	16.6	1 ea	59	<.1	.065

Fruits & Fruit Juices

	Wt (g)	Svg	Cal	Vit B6 (mg) 100 Cal	Svg
APPLES, 2.75" diam:					
With peel	138	1 ea	80	<.1	.066
Wthout peel	128	1 ea	72	<.1	.059
Applesauce, unsweetened	244	1 c	106	<.1	.063
Apple juice, bottled	248	1 c	116	<.1	.074
APRICOTS:					
Fresh, whole, pitted	106	3 ea	51	.1	.057
Canned, juice pack, halves	84	3 ea	40	.2	.060
Canned, heavy syrup	85	3 ea	70	.1	.050
Dried, halves	35	10 ea	83	.1	.060
Apricot nectar, canned	251	1 c	141	.1	.160
Avocado, whole:					
California varieties	173	1 ea	305	.2	.484
Florida	304	1 ea	340	.3	.851
Banana, 8.75", 176g w/peel	114	1 ea	105	.6	.659
Blackberries:					
Fresh berries	144	1 c	74	.1	.084
Frozen, unthawed	151	1 c	97	<.1	.092
Canned	256	1 c	236	<.1	.092
Blueberries:					
Fresh berries	145	1 c	82	<.1	.052
Frozen, unthawed	155	1 c	78	.1	.091
Canned	256	1 c	225	<.1	.096
Boysenberries:					
Frozen	132	1 c	66	.1	.074
Canned	256	1 c	225	<.1	.097
Cantaloupe, Casaba: see Melon.					
Cherries, sour:					
Frozen	155	1 c	72	.1	.104
Canned	244	1 c	90	.1	.107

	Wt (g)	Svg	Cal	Vit B6 (mg) 100 Cal	Svg
CHERRIES, SWEET:					
Fresh, pitted (10 = 68g)	145	1 c	104	.1	.052
Frozen, thawed	259	1 c	232	<.1	.078
Canned with liquid	257	1 c	213	.1	.233
Cranberries, whole, fresh	95	1 c	46	.1	.062
Cranberry apple juice	253	1 c	169	<.1	.060
Cranberry juice cocktail	253	1 c	145	<.1	.050
Cranberry sauce,					
canned, strained	277	1 c	419	<.1	.050
Dates, whole, pitted	83	10 ea	228	.1	.159
Elderberries, fresh	145	1 c	105	.3	.334
Figs:					
Fresh, medium	50	1 ea	37	.2	.057
Dried	187	10 ea	477	.1	.419
Fruit cocktail, canned:					
Juice pack	248	1 c	115	.1	.128
Heavy syrup	255	1 c	185	.1	.106
Gooseberries, fresh	150	1 c	67	.2	.120
Grapes, fresh:					
Thompson, seedless	50	10 ea	35	.2	.055
Tokay or Emperor	57	10 ea	40	.2	.063
Grape juice, canned/bottled	253	1 c	155	.1	.164
GRAPEFRUIT, half = 241 g w/peel:					
Half, pink or red	123	1 ea	37	.1	.052
Half, white	118	1 ea	39	.1	.051
Canned, sections	254	1 c	152	<.1	.051
Grapefruit juice:					
From frozen conc.	247	1 c	102	.1	.109
Canned, sweetened	250	1 c	115	<.1	.050
Guava, fresh	90	1 ea	45	.3	.129
Lemon juice:					
Fresh	244	1 c	60	.2	.124
Bottled	244	1 c	52	.2	.105
Lime juice:					
Fresh	15	1 T	4	1.9	.077
Bottled	246	1 c	50	.1	.066
Loganberries:					
Fresh	100	2/3 c	70	.1	.065
Frozen, unthawed	147	1 c	80	.1	.096
Mandarin oranges, canned	252	1 c	155	.1	.172
Mango, fresh, slices	165	1 c	108	.2	.221
Melons: see also Watermelon.					
Cantaloupe cubes	160	1 c	57	.3	.184
Casaba cubes	170	1 c	45	.2	.098
Honeydew cubes	170	1 c	60	.2	.100
ORANGE, 2-5/8", 180g w/peel	131	1 ea	60	.1	.079
Orange juice:					
Fresh juice	248	1 c	111	.1	.099
Chilled	249	1 c	110	.1	.134
Prepared f/frozen	249	1 c	110	.1	.110
Orange grapefruit juice	247	1 c	105	.1	.057
Papaya, whole (454 g w/refuse)	304	1 ea	117	<.1	.058
PEACHES:					
Fresh, peeled slices (1 avg					
peach ≈ 87g)	170	1 c	73	<.1	.030
Canned half, heavy syrup	81	1 ea	60	<.1	.015
Canned half, juice pack	79	1 ea	34	<.1	.015
PEARS:					
Fresh, Bartlett	166	1 ea	98	<.1	.030
Canned half, heavy syrup	79	1 ea	59	<.1	.011
PINEAPPLE:					
Fresh, chunks	155	1 c	76	.2	.135
Cnd pieces, juice pack	250	1 c	150	.1	.189
Cnd pieces, heavy syrup	255	1 c	199	.1	.189

Values are for edible portion of foods

	Wt (g)	Svg	Cal	Vit B6 (mg) 100 Cal	Svg
Fruits & Fruit Juices, continued:					
Pineapple juice:					
From frozen	250	1 c	129	.1	.185
Cnd, unsweetened	250	1 c	140	.2	.240
Plantain slices:					
Fresh	148	1 c	181	.2	.443
Cooked	154	1 c	179	.2	.370
Plums:					
Fresh, 2-1/8" diam.	66	1 ea	36	.1	.053
Canned, juice pack	95	3 ea	55	.1	.040
Canned, heavy syrup	110	3 ea	98	<.1	.030
Prunes, pitted (97g w/pits)	84	10 ea	201	.1	.220
Prune juice, bottled	256	1 c	181	.3	.558
Raisins, seedless, dark, packed	165	1 c	494	.1	.411
Raspberries:					
Fresh berries	123	1 c	60	.1	.070
Frozen, unsweetened, thawed	250	1 c	255	<.1	.085
Canned	256	1 c	234	<.1	.108
STRAWBERRIES:					
Fresh berries	149	1 c	45	.2	.088
Frozen, sweetened, thawed	255	1 c	245	<.1	.077
Tangerine	84	1 ea	37	.2	.056
Tangerine juice:					
From frozen	241	1 c	110	.1	.101
Canned	249	1 c	125	.1	.080
Watermelon, fresh pieces	160	1 c	50	.5	.230

Grains & Grain Products

Cereals, Grains, Flours, Noodles, Pasta, Popcorn

	Wt (g)	Svg	Cal	Vit B6 (mg) 100 Cal	Svg
Amaranth grain, dry	195	1 c	729	.1	.435
Barley, pearled, ckd	157	1 c	193	.1	.181
BRAN: see Oat, Rice, Wheat.					
Buckwheat flour:					
Light	98	1 c	340	<.1	.090
Dark	98	1 c	338	.1	.405
Bulgar wheat, cooked	182	1 c	151	.1	.151
CEREALS, COLD (Ready to eat): Cereals are sometimes fortified w/vitamin B6. Check label.					
CEREALS, HOT (Cooked): Some cereals are fortified w/ B6. Check label.					
Corn grits	242	1 c	145	<.1	.058
Cream of rice	244	1 c	126	.1	.066
Maypo	180	3/4 c	128	.5	.650
Oatmeal, unfortified	234	1/2 c	145	<.1	.047
Oatmeal, instant, fortified:					
Plain	177	3/4 c	104	.7	.742
With bran and raisins	195	7/8 c	158	.5	.761
Flavored, averaged	164	3/4 c	160	.5	.765
Ralston cereal	253	1 c	134	<.1	.114
Roman Meal	181	3/4 c	111	<.1	.085
Whole wheat cereal	242	1 c	150	<.1	.067
Corn flour:					
Regular	117	1 c	422	.1	.433
Masa Harina enriched	114	1 c	416	.1	.422
Corn grits: see Cereals, hot.					
Cornmeal, dry:					
Degermed, enriched	138	1 c	505	.1	.355
Nearly whole, bolted	122	1 c	440	.1	.560
FLOURS: see specific grain, nut, or vegetable.					

	Wt (g)	Svg	Cal	Vit B6 (mg) 100 Cal	Svg
MACARONI, cooked:					
White, enriched	140	1 c	197	<.1	.049
Whole wheat	140	1 c	174	.1	.111
Vegetable, enriched	134	1 c	172	<.1	.032
Millet	120	1/2 c	143	.1	.13
Noodles, cooked:					
Egg noodles	160	1 c	213	<.1	.058
Spinach	160	1 c	182	.1	.134
Oat bran (1T ≈ 6g)	94	1 c	132	.1	.156
Oats, rolled:					
Dry	81	1 c	311	<.1	.097
Cooked: see Cereals, hot.					
PASTA: see Macaroni, Noodles, and Spaghetti.					
RICE, cooked:					
Brown	195	1 c	217	.1	.283
White, regular	205	1 c	264	.1	.191
White, converted	175	1 c	200	<.1	.033
White, instant	165	1 c	162	<.1	.017
Wild rice	164	1 c	166	.1	.221
Rice bran	83	1 c	262	1.3	3.38
Rice polish	158	1 c	578	.1	.689
Rye flour:					
Dark	128	1 c	415	.1	.567
Light	102	1 c	361	.1	.239
Medium	102	1 c	374	.1	.273
Soy flour:					
Full fat, raw	85	1 c	368	.1	.392
Low fat	88	1 c	326	.1	.459
Spaghetti, cooked:					
Enriched	140	1 c	197	<.1	.049
Whole wheat spaghetti	140	1 c	174	.1	.111
WHEAT:					
Bran	30	1/2 c	65	.6	.391
Flours:					
All purpose, white, unsifted	125	1 c	455	<.1	.055
Semolina	167	1 c	601	<.1	.172
Whole wheat flour	120	1 c	407	.1	.409
Wheat germ:					
Raw	100	1 c	360	.4	1.3
Toasted	113	1 c	432	.3	1.11
Wheat, rolled:					
Cooked	240	1 c	142	.1	.078
Dry	85	1 c	289	.1	.404
Whole grain wheat (wheatberries) cooked	50	1/3 c	28	.1	.030
Whole grain wheat, sprouted	108	1 c	214	.1	.289

Meats: Fish & Shellfish

	Wt (g)	Svg	Cal	Vit B6 (mg) 100 Cal	Svg
Bluefish:					
Baked or broiled	100	3.5 oz	159	.3	.526
Fried in crumbs	100	3.5 oz	205	.2	.365
Catfish, fried in cornmeal	100	3.5 oz	229	<.1	.135
CLAMS:					
Breaded, fried small	188	20 ea	379	<.1	.096
Canned, drained	160	1 c	236	.1	.133
Canned with liquid, small can	183	1 ea	145	.1	.081
Steamed clams, meat only	90	20 ea	133	.1	.080
COD:					
Baked or broiled	100	3.5 oz	105	.3	.283
Batter fried	100	3.5 oz	199	.1	.240
Poached or steamed	100	3.5 oz	102	.3	.283
Smoked	100	3.5 oz	79	.3	.270
CRAB meat, cooked:					
Blue crab, unpacked	135	1 c	138	.2	.325
Blue crab, canned	135	1 c	133	.3	.405
Dungenous crab, cooked	101	3/4 c	85	.4	.325

Vitamin B-6 (Pyridoxine) mg

Fish & Shellfish, continued:

	Wt (g)	Svg	Cal	Vit B6 (mg) 100 Cal	Svg
Crayfish, ckd w/moist heat	85	3 oz	97	.2	.170
Eel, baked or broiled	100	3.5 oz	236	<.1	.077
Eel, smoked	100	3.5 oz	330	<.1	.130
Fish cakes, fried f/frozen	100	3.5 oz	213	<.1	.050
HADDOCK:					
Baked or broiled	85	3 oz	95	.3	.294
Breaded, fried	85	3 oz	175	.1	.130
Smoked	100	3.5 oz	116	.3	.400
Halibut:					
Baked or broiled	85	35 oz	119	.3	.337
Smoked	100	3.5 oz	224	.1	.330
Herring:					
Baked or broiled	100	3.5 oz	203	.2	.348
Canned w/liquid	100	3.5 oz	208	.1	.160
Smoked or kippered	100	3.5 oz	217	.2	.413
Pickled (1 piece ≈ 15 g)	100	3.5 oz	262	<.1	.114
Lobster, meat only, cooked	145	1 c	142	.1	.112
Mackerel:					
Baked/broiled, Atlantic	100	3.5 oz	262	.2	.460
Canned, Jack Mackerel	361	1 can	563	.1	.758
Mullet, baked or broiled	85	3 oz	127	.3	.417
Mussels, Blue					
steamed, meat only	85	3 oz	147	.4	.630
Northern pike, baked/broiled	100	3.5 oz	113	.1	.135
Ocean Perch:					
Baked or broiled	100	3.5 oz	121	.3	.330
Breaded, fried	85	3 oz	185	.1	.218
OYSTERS:					
Raw, Eastern	248	1 c	170	.1	.124
Raw, Pacific	248	1 c	200	.1	.120
Breaded, fried, med., Eastern	88	6 ea	173	<.1	.056
Simmered, Eastern	100	3.5 oz	137	.1	.095
Pollock, poached	100	3.5 oz	128	.2	.270
Roe, raw, mixed species	28	1 oz	39	.2	.079
SALMON:					
Baked or broiled, average	85	3 oz	183	.1	.186
Smoked, salmon, Chinook	85	3 oz	99	.2	.236
Canned, Atlantic, small can	220	1 can	281	.5	1.50
Canned, Sockeye, #1 can	369	1 can	566	.2	1.22
Sardines, canned, drained:					
Atlantic, 2 ea = 24 g	92	1 can	192	.1	.154
Pacific, 1 ea = 38 g	76	2 ea	136	.1	.094
Scallops:					
Breaded, fried	93	6 ea	200	.1	.176
Steamed	100	3.5 oz	113	.2	.210
Seatrout/Steelhead, ckd	100	3.5 oz	131	.6	.796
SHRIMP:					
Boiled, 2 lrg ≈ 11g	100	3.5 oz	99	.1	.127
Breaded, fried, 2 lrg ≈ 15 g	90	12 ea	218	<.1	.088
Canned, drained	128	1 c	154	.1	.142
Canned with liquid	100	3.5 oz	102	.1	.096
Smelt, Rainbow, cooked	85	3 oz	106	1.4	1.50
Snapper, baked/broiled	100	3.5 oz	128	.2	.274
SOLE (flounder):					
Baked or broiled	85	3 oz	99	.2	.204
Breaded, fried	100	3.5 oz	188	.1	.174
Fried in batter	85	3 oz	250	.1	.133
Squid, fried in flour	85	3 oz	149	<.1	.050
Swordfish, baked/broiled	100	3.5 oz	155	.2	.381
Trout, baked/broiled	85	3 oz	129	.3	.415
TUNA:					
Baked/broiled, Bluefin	85	3 oz	157	.3	.446
Canned, drained, No. 1/2 can:					
Light, oil pack	171	1 can	339	.1	.188
Light, water pack	165	1 can	216	.3	.624

Meats

Beef, Pork, Ham, etc.

	Wt (g)	Svg	Cal	Vit B6 (mg) 100 Cal	Svg
BEEF:					
Chuck roast, pot roasted:					
Lean & fat	85	3 oz	325	.1	.210
Lean only	85	3 oz	230	.1	.250
Ground beef, cooked average,					
baked, broiled, pan fried:					
Extra lean (17% fat raw)	85	3 oz	215	.11	.230
Lean (21% fat raw)	85	3 oz	230	.09	.210
Regular (26.6% fat raw)	85	3 oz	250	.08	.210
Rib, roast, choice:					
Lean and fat	85	3 oz	324	.1	.210
Lean only	85	3 oz	204	.1	.260
Round steak, broiled, choice:					
Lean and fat	85	3 oz	233	.2	.380
Lean only	85	3 oz	165	.3	.430
Round tip, roasted:					
Lean and fat	85	3 oz	213	.1	.310
Lean only	85	3 oz	162	.2	.340
Sirloin steak, broiled:					
Lean and fat	85	3 oz	238	.1	.320
Lean only	85	3 oz	172	.2	.360
T-bone steak, broiled:					
Lean and fat	85	3 oz	276	.1	.280
Lean only	85	3 oz	182	.2	.330
Beef liver, fried	85	3 oz	184	.7	1.22
Corned beef, canned	85	3 oz	213	.1	.111
Dried beef, cured (6 to 7 pieces)	28	1 oz	47	.3	.144
HAM: see Pork, cured; Turkey ham;					
and Sausages & Lunchmeats.					
LAMB:					
Arm chop, braised:					
Lean and fat	70	1 ea	244	<.1	.080
Lean only	55	1 ea	152	<.1	.070
Loin chop, broiled:					
Lean and fat	64	1 ea	201	<.1	.083
Lean only	46	1 ea	100	.1	.074
Cutlet, cooked average	85	3 oz	175	.1	.140
Leg of lamb, roasted:					
Lean and fat	85	3 oz	219	.1	.130
Lean only	85	3 oz	162	.1	.140
Rib roast:					
Lean and fat	85	3 oz	305	<.1	.100
Lean only	85	3 oz	197	.1	.130
Shoulder roast:					
Lean and fat	85	3 oz	235	<.1	.111
Lean only	85	3 oz	173	.1	.128
PORK:					
Bacon, cooked:					
Regular	19	3 pce	109	<.1	.050
Canadian style	47	2 pce	86	.2	.210
Breakfast strips	34	3 pce	156	.1	.120
Center loin chop:					
Braised, lean & fat	75	1 ea	266	.1	.280
Braised, lean only	61	1 ea	166	.2	.270
Broiled, lean & fat	82	1 ea	284	.1	.348
Broiled, lean only	72	1 ea	166	.2	.340
Pan-fried, lean & fat	89	1 ea	333	.1	.350
Pan-fried, lean only	67	1 ea	178	.2	.340
Roasted, lean & fat	88	1 ea	268	.1	.350
Roasted, lean only	72	1 ea	180	.2	.340
Center rib chop:					
Braised, lean & fat	67	1 ea	246	.1	.210
Braised, lean only	53	1 ea	147	.1	.210
Broiled, lean & fat	77	1 ea	264	.1	.260
Broiled, lean only	63	1 ea	162	.2	.250

ESHA Research

Vitamin B-6 (Pyridoxine) mg

Values are for edible portion of foods

	Wt (g)	Svg	Cal	Vit B6 (mg) 100 Cal	Svg
PORK, continued:					
Center rib chop, continued:					
Pan-fried, lean & fat	88	1 ea	343	.1	.290
Pan-fried, lean only	62	1 ea	160	.2	.270
Roasted, lean & fat	79	1 ea	252	.1	.270
Roasted, lean only	66	1 ea	162	.2	.260
Pork roast, avg loin / rib:					
Lean & fat	85	3 oz	265	.1	.315
Lean only	85	3 oz	206	.2	.360
Shoulder, braised, from 6.8 oz raw w/bone:					
Lean & fat	85	3 oz	293	.1	.230
Lean only	67	2.4 oz	166	.2	.275
Spareribs f/ 1 lb raw	177	6.25 oz	703	.1	.620
PORK, CURED - HAM, roasted:					
Lean & fat	85	3 oz	207	.2	.323
Lean only	85	3 oz	133	.3	.400
Canned, average	85	3 oz	142	.2	.340
RABBIT, roasted, average	85	3 oz	175	.2	.315
VEAL (calf):					
Cutlet, lean, avg	85	3 oz	166	.2	.280
Rib, roasted	85	3 oz	151	.2	.230
Liver, pan fried	85	3 oz	208	.4	.730

Meats: Poultry

	Wt (g)	Svg	Cal	Vit B6 (mg) 100 Cal	Svg
CHICKEN:					
All types of meat:					
Fried	140	1 c	307	.2	.670
Roasted	140	1 c	266	.2	.650
Stewed	140	1 c	248	.1	.370
Canned, boned w/ broth	142	5 oz	235	.2	.500
Dark meat:					
Fried	85	3 oz	203	.2	.316
Roasted	85	3 oz	174	.2	.304
Stewed	85	3 oz	163	.1	.176
Light meat:					
Fried	85	3 oz	163	.3	.540
Roasted	85	3 oz	147	.3	.510
Stewed	85	3 oz	135	.2	.279
Breast*, meat & skin:					
Batter-fried	140	1 ea	364	.2	.600
Flour-fried	98	1 ea	218	.3	.570
Roasted	98	1 ea	193	.3	.540
Stewed	110	1 ea	202	.2	.320
Breast*, meat only:					
Fried	86	1 ea	161	.3	.550
Roasted	86	1 ea	142	.4	.516
Stewed	95	1 ea	144	.2	.320
*2 pieces per bird					
Drumstick, meat & skin:					
Batter-fried	72	1 ea	193	.1	.200
Flour-fried	49	1 ea	120	.1	.170
Roasted	52	1 ea	112	.2	.180
Stewed	57	1 ea	116	.1	.110
Drumstick, meat:					
Fried	42	1 ea	82	.2	.170
Roasted	44	1 ea	76	.2	.170
Stewed	46	1 ea	78	.1	.100
Thigh, meat & skin:					
Batter-fried	86	1 ea	238	.1	.230
Flour-fried	62	1 ea	162	.1	.210
Roasted	62	1 ea	153	.1	.190
Stewed	68	1 ea	158	.1	.120
Thigh, meat only:					
Fried	52	1 ea	113	.2	.200
Roasted	52	1 ea	109	.2	.180
Stewed	55	1 ea	107	.1	.110

	Wt (g)	Svg	Cal	Vit B6 (mg) 100 Cal	Svg
Wing, meat & skin:					
Batter-fried	49	1 ea	159	.1	.150
Flour-fried	32	1 ea	103	.1	.130
Roasted	102	3 ea	297	.1	.420
Stewed	40	1 ea	100	.1	.090
Wing, meat:					
Fried	20	1 ea	42	.3	.120
Roasted	21	1 ea	43	.3	.120
Stewed	24	1 ea	43	.2	.080
Chicken liver, simmered	20	1 ea	30	.4	.117
DUCK, domestic, roasted:					
Meat and skin	85	3 oz	286	.1	.156
Meat only	85	3 oz	171	.1	.212
GOOSE, domestic:					
Meat and skin, roasted	85	3 oz	259	.1	.317
Meat only, roasted	85	3 oz	202	.2	.396
Liver, raw	94	1 ea	125	.6	.720
TURKEY:					
Breast meat:					
Barbecued	28	1 oz	40	.3	.110
Hickory smoked	28	1 oz	35	.3	.110
Ground meat, cooked	100	3.5 oz	229	.2	.39
Roasted:					
All types	85	3 oz	145	.3	.389
Dark meat	85	3 oz	159	.2	.304
Light meat	85	3 oz	133	.3	.455
Turkey heart	16	1 ea	28	.2	.051
Turkey liver	75	1 ea	127	.3	.391
Turkey patty, breaded, fried	64	1 ea	181	.1	.128
Frozen, slices w/gravy	142	5 oz	95	.1	.142

Meats: Sausages & Lunchmeats

	Wt (g)	Svg	Cal	Vit B6 (mg) 100 Cal	Svg
Barbecue loaf, pork or beef	23	1 pce	40	.2	.060
Beef:					
Loaf or roll lunchmeat	28	1 oz	87	.1	.050
Thin sliced beef	28	1 oz	50	.2	.100
Beerwurst (beer salami):					
Beef salami	23	1 pce	75	.1	.050
Pork salami	23	1 pce	55	.1	.080
Berliner sausage	23	1 pce	53	.1	.050
BOLOGNA:					
Beef & pork	28	1 oz	89	.1	.050
Cured pork	23	1 pce	57	.1	.060
Turkey	28	1 oz	56	.1	.050
Braunschweiger sausage	18	1 pce	65	.1	.060
Brotwurst, link	70	1 ea	226	<.1	.090
Cheesefurter (cheese smoki)	43	1 ea	141	<.1	.050
Chicken roll, light meat	57	2 oz	90	.3	.308
Dutch brand loaf	28	1 oz	68	.1	.060
FRANKFURTERS (hotdogs):					
Beef, 8/package	57	1 ea	184	<.1	.060
Beef & pork, 8/package	57	1 ea	183	<.1	.080
Chicken, 10/package	45	1 ea	115	.1	.090
Turkey, 10/package	45	1 ea	102	.1	.095
HAM:					
Chopped, packaged	42	2 pce	98	.1	.134
Ham, thin sliced, 3 pce ≈ 1 oz	28	1 oz	37	.4	.130
Ham, extra lean	57	2 oz	75	.3	.261
Ham, regular	57	2 oz	103	.2	.193
Minced	21	1 pce	55	.1	.060
Ham patty, cooked	60	1 ea	203	<.1	.100
Ham salad spread	240	1 c	518	.1	.359
Ham & cheese roll or loaf	57	2 oz	147	.1	.150

Meats: Sausages & Lunchmeats, continued:

	Wt (g)	Svg	Cal	100 Cal	Svg
Italian sausage link, ckd	67	1 ea	216	.1	.220
Knockwurst, link	68	1 ea	209	.1	.110
Luxury loaf	57	2 oz	80	.2	.180
Olive loaf	57	2 oz	133	.1	.130
Peppered loaf	28	1 oz	42	.2	.080
Pepperoni, small slice	22	4 pce	109	.1	.055
PORK SAUSAGE patty, ckd	27	1 pce	100	.1	.090
Salami:					
Beef	23	1 pce	58	.1	.050
Beef and pork, dry	20	2 pce	85	.1	.100
Pork and beef	57	2 oz	143	.1	.119
Turkey	57	2 oz	111	.1	.140
Smoked link sausage, beef & pork	68	1 ea	229	.1	.120
Smoked link sausage, pork	68	1 ea	265	.1	.240
Summer sausage	23	1 pce	80	.1	.070
TURKEY lunchmeats (other):					
Breakfast sausage	28	1 oz	65	.1	.080
Pastrami	57	2 oz	74	.2	.164
Turkey ham	57	2 oz	73	.2	.159
Turkey roll, light & dark meat	57	2 oz	84	.3	.225
Turkey roll, light meat only	57	2 oz	83	.3	.225
Smoked turkey sausage	28	1 oz	55	.1	.060
Summer sausage	28	1 oz	50	.2	.080

Mixed Dishes & Fast Foods

	Wt (g)	Svg	Cal	100 Cal	Svg
Beef & vegetable stew:					
Recipe	245	1 c	220	.1	.276
Canned	245	1 c	194	.1	.202
Beef, macaroni, & tomato sauce, recipe	226	1 c	189	.2	.299
Beef pot pie, from frozen	234	1 ea	426	.1	.267
BURRITOS:					
Bean burrito	174	1 ea	322	.3	1.01
Beef burrito	177	1 ea	463	.1	.295
Beef and bean	175	1 ea	390	.2	.727
Deluxe combination	198	1 ea	424	.2	.666
Cheese souffle, recipe	112	1 c	221	<.1	.096
Chicken a la king, recipe	245	1 c	470	.1	.233
Chicken & noodles, recipe	240	1 c	365	<.1	.156
Chicken chow mein, recipe	250	1 c	255	.2	.413
Chicken pot pie, fr/frozen	230	1 ea	430	.1	.460
Chicken salad w/celery	78	1/2 c	266	.1	.174
Chili w/beans, canned	255	1 c	286	.1	.337
Chop suey w/beef & pork	250	1 c	300	.1	.320
Cole slaw	120	1 c	84	.2	.175
Corn dog	111	1 ea	330	<.1	.110
Corn pudding	250	1 c	271	.1	.295
Corned beef hash, canned	220	1 c	382	.1	.410
Egg salad	183	1 c	438	<.1	.182
ENCHILADA:					
Beef	120	1 ea	292	.1	.152
Cheese	120	1 ea	330	<.1	.094
Chicken	120	1 ea	269	.1	.205
LASAGNA:					
Recipe w/meat	245	1 pce	398	.1	.351
Recipe w/o meat	218	1 pce	316	.1	.221
Macaroni & cheese, recipe	200	1 c	430	<.1	.050
Macaroni salad, no cheese	141	1 c	371	<.1	.069
Manicotti, frozen entree	225	1 ea	271	.1	.221
Meat loaf:					
Beef only	87	1 pce	193	.1	.182
Beef and 1/3 pork	87	1 pce	212	.1	.192

	Wt (g)	Svg	Cal	100 Cal	Svg
Moussaka, lamb & eggplant	250	1 c	250	.1	.354
PIZZA:					
Thick crust, 1/2 of 10"	208	1 pce	519	<.1	.071
Thin crust, 1/8 of 15"	120	1 pce	290	<.1	.041
Potato salad, w/mayo & eggs	250	1 c	358	.1	.353
Ravioli, beef, canned	226	1 c	220	.1	.281
SANDWICHES, Fast Food items:					
Cheeseburger, 3 oz beef	112	1 ea	300	<.1	.112
Cheeseburger, 4 oz beef	194	1 ea	524	<.1	.233
Chicken patty sandwich	157	1 ea	436	.1	.368
English muffin, with egg, cheese & bacon	138	1 ea	360	<.1	.152
Fish sandwich:					
Large, without cheese	170	1 ea	470	<.1	.119
Regular, with cheese	140	1 ea	420	<.1	.098
Hamburger, 3 oz beef	98	1 ea	245	<.1	.117
Hamburger, 4 oz beef	174	1 ea	445	.1	.278
Hotdog (frankfurter) and bun	85	1 ea	260	<.1	.074
SANDWICHES (on part whole wheat bread except when stated as on rye):					
Bacon, lettuce & tomato	135	1 ea	327	<.1	.146
Egg salad sandwich	111	1 ea	319	<.1	.110
Grilled cheese	117	1 ea	393	<.1	.103
Ham	122	1 ea	256	.1	.330
Ham on rye	116	1 ea	242	.1	.316
Ham & cheese	151	1 ea	363	.1	.350
Ham & swiss on rye	145	1 ea	350	.1	.341
Peanut butter & jam	100	1 ea	341	<.1	.161
Roast beef	122	1 ea	280	.1	.253
Tuna salad sandwich	116	1 ea	303	.1	.186
Turkey	122	1 ea	271	.1	.269
Turkey ham	122	1 ea	253	.1	.225
SPAGHETTI (pasta & tomato sauce with cheese):					
Homemade	250	1 c	260	.1	.197
Canned	250	1 c	190	.1	.131
Spinach souffle	136	1 c	218	.1	.120
Taco, beef	78	1 ea	207	.1	.158
Taco, chicken	78	1 ea	172	.1	.236
Tostada:					
W/beans & beef	192	1 ea	332	.2	.668
W/beans & chicken	157	1 ea	249	.3	.732
W/refried beans	157	1 ea	212	.5	1.01
Tuna noodle casserole, home	202	1 c	251	.1	.243
Tuna salad	205	1 c	383	<.1	.166
Turkey pot pie, frozen	233	1 ea	416	.1	.460

Nuts & Seeds

	Wt (g)	Svg	Cal	100 Cal	Svg
ALMONDS:					
Dried, whole	142	1 c	837	<.1	.160
Dry roasted	138	1 c	810	<.1	.102
Brazil nuts, dry, unsalted	140	1 c	919	<.1	.351
Cashews, dry or oil roasted	130	1 c	748	<.1	.325
Chestnuts, dried	28	1 oz	105	.2	.223
COCONUT:					
Dried, unsweetened	78	1 c	515	.5	2.34
Flaked, sweetened, packaged	74	1 c	351	.1	.213
Coconut water, raw	240	1 c	46	.2	.077
Filberts, hazelnuts, whole	135	1 c	853	.1	.826
Macadamias, dried	134	1 c	940	<.1	.381
MIXED NUTS w/ peanuts (cashews, peanuts, brazil nuts, filberts, almonds & pecans):					
Dry roasted	137	1 c	814	<.1	.406
Oil roasted	142	1 c	876	<.1	.341

Vitamin B-6 (Pyridoxine) mg   Values are for edible portion of foods

	Wt (g)	Svg	Cal	Vit B6 (mg) 100 Cal	Svg
Nuts & Seeds, continued:					
MIXED NUTS, without peanuts					
(cashews, almonds, brazil nuts,					
pecans, filberts):					
Oil roasted	144	1 c	886	<.1	.259
PEANUTS, dry or oil roasted	144	1 c	837	<.1	.367
Peanut butter	32	2 T	188	.1	.124
Peanut flour, defatted	60	1 c	196	.2	.302
Pecans, chopped	119	1 c	794	<.1	.224
Pine nuts, dried:					
Pignola	28	1 oz	146	.1	.080
Pinyon	28	1 oz	161	<.1	.080
Pistachios, dried, shelled	128	1 c	739	<.1	.270
Pumpkin / Squash seed					
kernels, dry	138	1 c	747	<.1	.124
Sesame seed kernels, dried	150	1 c	882	.1	1.18
Soybeans, dry roasted	86	1/2 c	405	<.1	.179
SUNFLOWER SEEDS:					
Dried kernels	144	1 c	821	.2	1.83
Oil roasted	135	1 c	830	.2	1.61
Sunflower seed butter	16	1 T	93	.2	.176
Tahini, sesame butter	15	1 T	91	.1	.056
Walnuts, chopped:					
Black	125	1 c	759	.1	.698
English	120	1 c	770	.1	.670

Soups, Sauces & Gravies

	Wt (g)	Svg	Cal	Vit B6 (mg) 100 Cal	Svg
SAUCES (also see Other):					
Cheese, mix w/ milk	279	1 c	305	<.1	.101
Spaghetti sauce:					
Plain, recipe	220	1 c	179	.2	.395
Plain, canned	249	1 c	272	.1	.395
W/meat, recipe	248	1 c	297	.2	.530
W/meat, canned	206	.8 c	220	.1	.265
White sauce, medium:					
From recipe	250	1 c	395	<.1	.101
From mix, made w/milk	264	1 c	240	<.1	.079

SOUPS, prepared: soups are from canned unless otherwise stated; RTS = ready to serve. For soups prep. with milk, assume whole milk.

	Wt (g)	Svg	Cal	Vit B6 (mg) 100 Cal	Svg
Beef broth/bouillon:					
Canned	240	1 c	16	.4	.071
From dry mix	244	1 c	19	.1	.024
Beef, chunky, RTS	240	1 c	171	.1	.132
Chicken, cream of:					
Prepared with milk	248	1 c	191	<.1	.067
From dry	261	1 c	107	<.1	.052
Chicken & vegetable:					
Chunky, RTS	240	1 c	167	.1	.096
Prepared from dry mix	251	1 c	49	.1	.070
Chicken broth/bouillon:					
Canned	244	1 c	39	.1	.024
From dry mix	244	1 c	21	.1	.024
Chicken noodle soup	241	1 c	75	<.1	.008
Chili beef soup	250	1 c	169	.1	.158
Clam chowder:					
Manhatten, chunky, RTS	240	1 c	133	.2	.264
Manhatten	244	1 c	78	.1	.083
New England, cream style	248	1 c	163	.1	.126
Gazpacho RTS	244	1 c	57	.3	.146
Lentil and ham RTS	248	1 c	140	.2	.223
Minestrone	241	1 c	80	.1	.099

	Wt (g)	Svg	Cal	Vit B6 (mg) 100 Cal	Svg
Split pea soup, canned:					
Without ham	253	1 c	189	<.1	.070
With ham	253	1 c	189	<.1	.068
Tomato, cream of:					
With milk	248	1 c	160	.1	.164
With water	244	1 c	86	.1	.112
From dry	265	1 c	102	.1	.098
Tomato & vegetable, fr/dry	253	1 c	55	.2	.094
Tomato beef noodle	244	1 c	140	.1	.088
Turkey, chunky, RTS	236	1 c	136	.5	.692
Vegetable beef:					
Canned	244	1 c	79	.1	.076
From dry mix	253	1 c	53	.1	.053
Vegetable, chunky, RTS	240	1 c	122	.2	.192
Vegetable, cream of, fr/dry	260	1 c	105	<.1	.052
Vegetarian vegetable	241	1 c	70	.1	.055

Vegetables & Legumes

	Wt (g)	Svg	Cal	Vit B6 (mg) 100 Cal	Svg
Artichoke globe, cooked	120	1 ea	60	.2	.133
Artichoke hearts:					
Cooked fr/ frozen-pkg	240	9 oz	108	.2	.209
Marinated-jar	170	6 oz	168	.1	.150
Asparagus:					
Fresh pieces	67	1/2 c	15	.7	.103
Ckd from fresh	90	1/2 c	23	.6	.127
Ckd from frozen	180	1 c	50	.3	.159
Canned, drained	121	1/2 c	16	.4	.061
Canned, with liquid	122	1/2 c	17	.7	.120
BEANS: also see Garbanzo, Lentils, Soybeans.					
Baked beans (dry white					
beans w/ spices & sauce):					
Home prepared	253	1 c	382	.1	.228
Canned, plain/vegetarian	254	1 c	235	.1	.340
Canned w/franks	257	1 c	366	<.1	.118
Canned w/pork	253	1 c	268	.1	.162
Canned w/pork, sweet sce	253	1 c	282	.1	.215
Canned w/pork, tomato sce	253	1 c	247	.1	.175
Black beans, cooked	172	1 c	227	.1	.119
Broadbeans, cooked fr/dry	170	1 c	186	.1	.122
Great northern, fr/dry	177	1 c	210	.1	.207
Green (snap) beans:					
Fresh	110	1 c	34	.2	.081
Cooked fr/ fresh	125	1 c	44	.2	.070
Cooked fr/ frozen	135	1 c	36	.2	.076
Canned, drained	135	1 c	26	.2	.050
Canned, with liquid	240	1 c	36	.2	.072
Kidney beans:					
Ckd fr/ dry	177	1 c	225	.1	.212
Canned with liquid	256	1 c	208	.1	.177
Lima beans:					
Ckd fr/ fresh	170	1 c	208	.2	.328
Ckd fr/ frozen, large	85	1/2 c	85	.1	.104
Ckd fr/ frozen, baby	90	1/2 c	94	.1	.104
Ckd fr/ dry	188	1 c	217	.1	.303
Canned with liquid	241	1 c	191	.1	.219
Navy, cooked fr/ dry	182	1 c	259	.1	.298
Pinto beans:					
Canned	240	1 c	186	.1	.178
Ckd from dry	171	1 c	235	.1	.265
Refried beans, canned	253	1 c	270	.1	.280
White beans, ckd fr/ dry	179	1 c	253	.1	.227
Winged beans:					
Ckd from fresh veg	62	1 c	23	.2	.051
Ckd from dry legume	172	1 c	252	<.1	.081
Yardlong:					
from dry legume	171	1 c	202	.1	.162
Yellow wax: see green beans.					

ESHA Research   50

Vitamin B-6 (Pyridoxine) mg

	Wt (g)	Svg	Cal	Vit B6 (mg) 100 Cal	Svg
Vegetables & Legumes, continued:					
Bean sprouts (Mung beans):					
Fresh sprouts	104	1 c	31	.3	.092
Cooked fr/ fresh, stir-fried	124	1 c	62	.2	.101
Beet greens, cooked, drained	144	1 c	40	.5	.190
Beets:					
Cooked from fresh	85	1/2 c	26	.1	.026
Canned w/liquid, diced	123	1/2 c	36	.2	.068
Broccoli, chopped:					
Fresh, uncooked	88	1 c	24	.6	.140
Ckd from fresh	156	1 c	44	.5	.224
Ckd from frozen	184	1 c	51	.5	.239
W/cheese sauce	142	1/2 c	166	<.1	.110
Brussels sprouts:					
Ckd from fresh	156	1 c	60	.5	.310
Ckd from frozen	155	1 c	65	.4	.271
CABBAGE, shredded:					
Common, fresh	70	1 c	16	.4	.066
Common, cooked	150	1 c	32	.3	.096
Bok choy, fresh	70	1 c	9	.7	.066
Bok choy, cooked	170	1 c	20	1.5	.300
Pe-tsai, fresh	76	1 c	11	1.5	.176
Pe-tsai, cooked	119	1 c	16	1.3	.211
Red, fresh	70	1 c	19	.8	.147
Red, cooked	75	1/2 c	16	.7	.105
Savoy, fresh	70	1 c	20	.7	.133
Savoy, cooked	145	1 c	35	.6	.220
CARROTS:					
Fresh, whole, 7.5" x 1-1/8"	72	1 ea	31	.3	.106
Fresh, grated	55	1/2 c	24	.3	.081
Ckd fr/fresh, slices	78	1/2 c	35	.5	.192
Ckd fr/frozen, slices	73	1/2 c	26	.4	.094
Canned, drained	73	1/2 c	17	.5	.082
Canned, with liquid	123	1/2 c	28	.5	.138
Carrot juice	123	1/2 c	49	.5	.266
Cauliflower:					
Fresh pieces	50	1/2 c	12	1.0	.116
Cooked from fresh	62	1/2 c	15	.8	.125
Cooked from frozen	180	1 c	34	.5	.158
Celeriac, celery root, cooked	100	3.5 oz	25	.4	.101
Celery, raw, chopped	60	1/2 c	9.6	.5	.052
Celery, cooked, chopped	150	1 c	27	.5	.129
Chard, Swiss chard, cooked	175	1 c	35	.3	.117
Collards:					
Ckd from fresh	128	1 c	35	.2	.067
Ckd from frozen	170	1 c	61	.3	.194
CORN:					
Cooked fr/ fresh	82	1/2 c	89	.2	.138
Cooked fr/ frozen	82	1/2 c	67	.3	.178
Canned, vacuum pack	210	1 c	166	.1	.116
Canned, with liquid	128	1/2 c	79	.3	.256
Cream style	128	1/2 c	93	.1	.081
Cucumber, whole, 8" x 2+"	301	1 ea	39	.4	.156
Daikon radish, slices	44	1/2 c	8	1.9	.153
Eggplant, cooked cubes	160	1 c	45	.3	.138
Garbanzo beans (chickpeas):					
Canned	240	1 c	285	.4	1.14
Cooked from dry	164	1 c	269	.1	.228
Garden cress, fresh	25	1/2 c	8	.8	.062
Jerusalem artichoke, fresh	150	1 c	114	.1	.107
Jicama	100	3.5 oz	20	.5	.100
Kale, chopped:					
Fresh, uncooked	67	1 c	33	.6	.182
Cooked from fresh	130	1 c	42	.4	.179
Cooked from frozen	130	1 c	39	.3	.112
Kohlrabi:					
Fresh slices	140	1 c	38	.6	.210
Cooked fr/fresh	165	1 c	48	.3	.139
Leeks, chopped:					
Fresh	104	1 c	63	.4	.242
Cooked	52	1/2 c	24	.3	.075
Lentils, cooked fr/dry	198	1 c	231	.2	.352
Lentils, sprouted:					
Fresh sprouts	77	1 c	81	.2	.146
Stir-fried	100	3.5 oz	101	.2	.161
LETTUCE, chopped:					
Butterhead	56	1 c	7	.51	.037
Iceberg	56	1 c	7	.30	.022
Looseleaf	56	1 c	10	.31	.031
Romaine	56	1 c	9	.37	.034
Mushrooms, slices:					
Fresh, sliced	35	1/2 c	9	.4	.034
Cooked from fresh	78	1/2 c	21	.4	.074
Canned, drained	78	1/2 c	19	.3	.055
Mustard greens:					
Fresh, chopped	56	1 c	15	.7	.096
Cooked fr/frozen	150	1 c	29	.6	.162
Cooked fr/fresh	140	1 c	21	.9	.180
Okra, cooked:					
Pods, cooked fr/fresh	85	8 ea	27	.6	.159
Sliced, from frozen	92	1/2 c	34	.1	.043
Onion:					
Fresh, chopped	160	1 c	61	.3	.186
Ckd fr/fresh, chopped	105	1/2 c	46	.3	.135
Dehydrated flakes	14	1/4 c	45	.5	.224
Onion, spring, chopped	50	1/2 c	16	.2	.025
Parsnips:					
Fresh slices	133	1 c	100	.1	.120
Cooked	156	1 c	125	.1	.146
PEAS:					
Black-eyed peas:					
Cooked from fresh	165	1 c	160	.1	.107
Cooked from frozen	170	1 c	224	.1	.162
Cooked from dry	171	1 c	198	.1	.171
Canned	240	1 c	184	.1	.108
Green peas, edible-pods:					
Fresh	145	1 c	61	.4	.232
Ckd from fresh	160	1 c	67	.3	.230
Ckd from frozen	80	1/2 c	42	.3	.139
Green:					
Fresh, uncooked	145	1 c	118	.2	.245
Ckd from fresh	160	1 c	134	.2	.250
Ckd from frozen	80	1/2 c	63	.1	.090
Canned, drained	85	1/2 c	59	.1	.054
Canned, with liquid	124	1/2 c	61	.1	.081
Split peas, ckd fr/dry	196	1 c	231	<.1	.094
Peas, mature, sprouted	120	1 c	154	.2	.318
Peas, sprouted, cooked	100	3.5 oz	118	.1	.128
Peas & carrots:					
Cooked from frozen	80	1/2 c	38	.2	.070
Canned	128	1/2 c	48	.2	.113
PEPPERS, HOT, chili, green or red, chopped:					
Fresh	75	1/2 c	30	.7	.209
Canned	68	1/2 c	17	.5	.084
PEPPERS, SWEET, green or red, chopped:					
Fresh	50	1/2 c	14	.9	.124
Cooked from fresh	68	1/2 c	19	.8	.158
Pimento, canned	57	2 oz	13	.9	.123
POTATOES:					
Baked in oven:					
Flesh and skin	202	1 ea	220	.3	.701
Flesh only	156	1 ea	145	.3	.470
Potato skin	58	1 ea	115	.3	.352
Boiled, flesh only	135	1 ea	116	.3	.363

ESHA Research

Vitamin B-6 (Pyridoxine) mg

Values are for edible portion of foods

	Wt (g)	Svg	Cal	Vit B6 (mg) 100 Cal	Svg		Wt (g)	Svg	Cal	Vit B6 (mg) 100 Cal	Svg
Vegetables & Legumes, continued:						Tofu, raw:					
						Rirm	126	1/2 c	183	.1	.116
POTATOES, continued:						Regular	124	1/2 c	94	.1	.058
Canned, 1" diam.	70	2 ea	42	.3	.132	**TOMATOES:**					
French fries, fr/ frozen:						Fresh, whole	123	1 ea	26	.4	.098
Cooked in oil	50	10 ea	158	.1	.118	Fresh, chopped	180	1 c	38	.4	.144
Oven heated	50	10 ea	111	.1	.116	Cooked from fresh	240	1 c	65	.4	.228
Hash browns:						Canned, whole	240	1 c	47	.5	.216
From fresh, homemade	156	1 c	163	.1	.217	Tomato juice, canned	244	1 c	42	.7	.271
From frozen	156	1 c	340	.1	.197	Tomato paste, canned	262	1 c	220	.5	.996
Mashed potatoes:						Tomato puree, canned	250	1 c	102	.4	.380
Prepared w/milk	210	1 c	162	.3	.489	Tomato sauce, canned	245	1 c	74	.5	.333
Prepared fr/Instant	220	1 c	239	.1	.257	Turnips, cubed:					
Potato puffs, (tater tots)	62	1/2 c	138	.1	.143	Fresh cubes	130	1 c	35	.3	.117
Potato dishes:						Cooked fr/fresh	78	1/2 c	14	.4	.052
Au gratin, recipe	245	1 c	322	.1	.426	Turnip greens, cooked:					
Au gratin, fr/mix	245	1 c	228	<.1	.098	From fresh	144	1 c	29	.9	.259
Scalloped, recipe	245	1 c	210	.2	.436	From frozen	82	1/2 c	24	.2	.055
Scalloped, fr/mix	245	1 c	228	<.1	.103	Vegetable juice cocktail	242	1 c	46	.7	.339
Potato chips, plain	28	14 ea	148	.1	.144						
Potato pancakes	76	1 ea	237	.1	.290	**VEGETABLE COMBINATIONS,**					
Pumpkin, cooked, mashed:						Cooked fr/ frozen:					
Cooked from fresh	245	1 c	50	.3	.157	Broccoli, carrots, pasta	95	2/3 c	88	.1	.060
Canned	123	1/2 c	42	.2	.069	Broccoli, carrots					
Radish seeds, sprouted	38	1 c	16	.7	.108	& water chestnuts	91	2/3 c	32	.3	.090
Rutabaga:						Broccoli, cauliflower					
Fresh, cubes	140	1 c	51	.3	.140	& red pepper	95	2/3 c	25	.5	.120
Cooked	85	1/2 c	29	.3	.077	Broccoli, water chestnuts	95	1/2 c	33	.4	.120
Sauerkraut, canned w/liquid	236	1 c	44	.7	.307	Cantonese stir fry	95	1/2 c	53	.2	.100
Seaweed (spirulina), dried	28	1 oz	82	.1	.103	Chinese stir fry	95	1/2 c	31	.3	.080
Shallots, freeze dried-chpd	3.6	1/4 c	13	.5	.060	Green beans & spaetzle,					
Soybeans, cooked fr/dry	172	1 c	298	.1	.402	Bavarian style	95	1/2 c	108	<.1	.050
Soybeans, mature-sprouted:						Japanese style	95	1/2 c	29	.3	.090
Fresh	35	1/2 c	45	.1	.062	**Mixed vegetables** (corn,					
Steamed	94	1/2 c	76	.1	.108	peas, carrots, green					
						beans & lima beans):					
SOYBEAN PRODUCTS: see tofu this section; miso, tempeh in Other; roasted soybeans in Nuts & Seeds; soy milk in Dairy; soy flour in Grains.						Cooked from frozen	182	1 c	107	.1	.135
						Canned, drained	163	1 c	77	.2	.129
						Peas, carrots, onions, pasta	95	1/2 c	122	<.1	.060
SQUASH, SUMMER, sliced:						Peas & mushrooms	95	1/2 c	73	.2	.130
Crookneck, fresh	130	1 c	24	.6	.142	Peas & onions	91	1/2 c	54	.2	.090
Crookneck, cooked	180	1 c	36	.5	.169	Peas & onions & cheese sce	142	1/2 c	165	.1	.130
Scallop, fresh	130	1 c	24	.6	.142	Peas, potatoes & cream sce	76	1/2 c	140	.1	.100
Scallop, cooked	90	1/2 c	14	.5	.077	Peas, rice & mushrooms	66	2/3 c	108	.1	.070
Zucchini, fresh	130	1 c	19	.6	.116	Peas, pasta, corn, cream sce	95	1/2 c	132	.1	.100
Zucchini, cooked	180	1 c	29	.5	.140	Peas, pasta, mushrooms,					
						& cream sauce	95	1/2 c	129	.1	.110
SQUASH, WINTER, mashed:						Spinach & water chestnuts	95	1/2 c	29	.3	.080
Acorn, baked	245	1 c	137	.3	.476	Yam, orange: see sweet potato.					
Acorn, boiled	245	1 c	83	.3	.287	Yam, white, cooked cubes	136	1 c	158	.2	.310
Butternut, baked	245	1 c	99	.3	.303	Zucchini: see Squash.					
Butternut, ckd fr/ frozen	240	1 c	94	.2	.166						
Hubbard, baked	240	1 c	120	.3	.413	**Other**					
Hubbard, boiled	236	1 c	70	.3	.243						
Spaghetti baked or boiled	155	1 c	45	.3	.153						
Spinach:						Barbecue sauce	250	1 c	160	.2	.245
Fresh, chopped	56	1 c	12	.9	.109	Beer (1.5 c = 12 fl oz):					
Ckd from fresh	180	1 c	41	1.1	.436	Regular	356	1.5 c	146	.1	.178
Ckd from frozen	190	1 c	53	.5	.278	Light	354	1.5 c	100	.1	.120
Canned, drained	214	1 c	50	.4	.214	**CANDY:**					
Succotash, cooked:						Chocolate covered:					
From fresh	192	1 c	222	.1	.223	Almonds	165	1 c	935	<.1	.126
From frozen	170	1 c	158	.1	.162	Coconut	28	1 oz	133	.1	.100
Sweet potato:						Peanuts	170	1 c	954	<.1	.419
Baked with peel	114	1 ea	118	.2	.275	Raisins	187	1 c	733	<.1	.340
Boiled, peeled	151	1 ea	160	.2	.360	Milk chocolate w/peanuts	28	1 oz	155	<.1	.053
Canned, mashed	128	1/2 c	129	.2	.243	REESE's peanut butter cup	45	2 ea	240	<.1	.059
Vacuum packed, mashed	255	1 c	233	.2	.485	Carob flour	103	1 c	185	.2	.377
Candied, recipe	105	1 pce	144	.1	.165	Catsup	245	1 c	255	.2	.44
						Chili sauce, tomato based	273	1 c	284	.1	.325

ESHA Research

52

	Wt (g)	Svg	Cal	Vit B6 (mg) 100 Cal	Vit B6 (mg) Svg		Wt (g)	Svg	Cal	Vit B6 (mg) 100 Cal	Vit B6 (mg) Svg
Other, continued:											
CHOCOLATE:											
Baking, unsweetened	28	1 oz	145	<.1	.011						
Choc. chips, semi-sweet	170	1 c	860	<.1	.044						
Milk chocolate	28	1 oz	145	<.1	.016						
Cocoa powder	86	1 c	224	<.1	.060						
Hummous/Humous	246	1 c	420	.2	.979						
Mayonnaise	220	1 c	1577	<.1	.060						
Miso (soybean product)	138	1/2 c	284	.1	.297						
Molasses, blackstrap	40	2 T	85	.1	.108						
Mustard, prepared	125	1/2 c	94	.1	.089						
Pineapple grapefruit drink	250	1 c	117	.1	.105						
Pineapple orange drink	250	1 c	125	.1	.118						
Poppyseed	8.8	1 T	47	.1	.040						
SALAD DRESSINGS:											
Blue cheese dressing	123	1/2 c	618	<.1	.046						
Ranch dressing	119	1/2 c	435	<.1	.053						
1000 island dressing	125	1/2 c	472	<.1	.027						
Salsa, Picante by Tostitos	85	6 T	40	.2	.080						
Salsa, recipe	108	.5 c	46	.2	.095						
SPICES:											
Cinnamon	6.8	1 T	18	.3	.060						
Dill weed, dried	3.1	1 T	8	.6	.045						
Garlic, cloves	12	4 ea	18	2.2	.400						
Garlic, powder	8.4	1 T	28	6.1	1.70						
Onion powder	6.5	1 T	15	.7	.100						
Tempeh (soybean product)	83	1/2 c	165	.2	.248						
Yeast:											
Brewer's	8	1 T	25	1.6	.404						
Dry active, regular	30	4 T	80	.8	.600						

ESHA Research

Folacin (Folate)

Folacin is essential for the formation of nucleic acids, like DNA and RNA, and therefore is especially important where cell turnover is greatest. Examples include the bone marrow, which produces blood cells; the intestinal tract, where regeneration of some cells occurs every several days (which, of course, affects the ability to absorb other nutrients); and increased needs during pregnancy for the growth of the fetus.

Folacin is also required for the synthesis and breakdown of amino acids (protein). A deficiency affects the nervous system and brain function. Deficiencies can occur from alcohol abuse, or any condition that requires increased cell production; such as burns, blood loss, and skin diseases (e.g. measles and chicken pox).

It has recently been recognized that the folacin measures in foods have been understated, or calculated by older methods of analysis. The data in this book and in the data base used by the Food Processor® computer systems reflects the more recent analyses. Even so, food analysis methods present difficulty, and the most recent values may still be as much as 20% low due to incomplete recovery.

Vitamin Losses

Folacin (folate) is found in many foods, especially leafy vegetables, liver, yeast, legumes and some fruits. Food processing can drastically reduce folacin content, however. Normal household preparation can destroy as much as 50% percent in vegetables, and reheating destroys even more, since vitamin C will not be there to protect it. Folacin is somewhat stable to heat in an acidic medium, however.

If vegetables are stored at room temperature, up to 70 percent of their folacin is gone in three days. Store green vegetables in dark, airtight places in the refrigerator, or eat them freshly picked, since loss is accelerated by air and light.

Recommended Dietary Allowances for Adults

USA RDA: Men, 200 mcg (2.0 mg); Women, 180 mcg (1.8mg) Pregnant, 400 mcg; 280 mcg for lactation.
  These 1989 recommendations are less than previous recommendations. The amounts for healthy adults are based on
  approximately 3 mcg per kg of body weight per day.
Canadian RNI: Recommendations are based on 3 mcg per kg of body weight per day.. This generally translates to: Men,
  220 mcg; Women, 175 mcg (Pregnant, 480 mcg; and Lactating/Nursing, 295 mcg).

There is some evidence that Folic acid and the anti-convulsant drug phenytoin inhibit the uptake of each other.

Food Sources

Think of foliage -- leafy green. If eaten fresh, it is one of the richest sources for the vitamin called folacin. It is also in liver, legumes, oranges, peanuts, sunflower seeds, and whole grains (in white flour, large portions are milled out and not replaced).

Baked Goods

	Wt (g)	Svg	Cal	Folacin (mcg) 100 Cal	Svg
Apple crisp, 3 x 3"	78	1 ea	146	2	3
Bagel, egg/plain, 3.5" dm	68	1 ea	180	9	16
Biscuits, average	28	1 ea	93	1.5	1+
BREADS:					
Banana nut, 1/2"	50	1 pce	161	7	11
Boston brown,1/2" slc	45	1 pce	95	8.4	8
Cornbread muffin	45	1 ea	145	3	5
Cracked wheat	25	1 pce	65	19	12
French, 5" x 2.5" x 1"	35	1 pce	100	13	13
Mixed grain	25	1 pce	65	25	16
Oatmeal bread	25	1 pce	65	12	8
Pita pocket, 6.5" dm.	60	1 ea	165	7	12
Pumpernickel, 4" x 5" x 3/8"	32	1 pce	80	20	16
Raisin bread	25	1 pce	68	13	9
Rye, 5" x 3.5"	25	1 pce	65	15	10
Wheat bread (white and whole wheat flour)	28	1 pce	72	18	13
White bread	28	1 pce	75	13	10
Whole wheat	28	1 pce	70	22	16
CAKES, pce = 1/16 cake unless otherwise noted:					
Angel food, 1/12	53	1 pce	125	3	4
Boston cream pie, 1/8	120	1 pce	260	3	7
Carrot w/cream cheese frosting, 2.5"x 3" pce	112	1 pce	406	3	11
Cheesecake f/recipe, 1/12	92	1 pce	278	6	17
Chocolate, choc. frosting	69	1 pce	235	2	4
Coffee cake f/mix, 2.4" x 2.8"	72	1 pce	230	2	5
Gingerbread, 3" x 3"	63	1 pce	174	2	4
Pound cake, 1/2"	30	1 pce	115	3	3
Sheet cake, 3x3":					
Plain	86	1 pce	315	5	15
White frosting	121	1 pce	445	3	12
Spongecake, 1/12 tube	66	1 pce	194	6	11
Snack cake, like Twinkies	42	1 ea	155	3	4
White cake:					
Chocolate frosting	77	1 pce	291	3	8
Coconut frosting	70	1 pce	270	1.5	4
Yellow cake, choc. frosting	69	1 pce	240	2	5
Cherry crisp, 3" x 3"	138	1 pce	157	6	10
Chips: see corn & tortilla this section, potato chips under Vegetable.					
Cookies, average	45	4 ea	180-245	2	4
CRACKERS:					
Armenian cracker bread	28	4 pce	117	10	12
Graham crackers	14	2 ea	60	3	2
Rye wafers, whole grain	14	2 ea	55	18	10
Sesame crackers	12	4 ea	60	8	5
Wheat cracker, thin	8	4 ea	35	9	3
Crepe (no filling)	27	1 ea	47	11	5
Croissant, 4.5" x 4" x 2"	57	1 ea	235	8	18
Danish pastry, average	61	1 ea	228	7	15
Doughnut, yeast raised	60	1 ea	235	5	13
English muffin:					
Plain, enriched	57	1 ea	140	13	18
Sourdough	56	1 ea	129	12	15
MUFFINS, from mix:					
Blueberry	45	1 ea	140	10	14
Bran muffin	45	1 ea	140	14	19
Cornmeal	45	1 ea	145	3	5
Pancakes:					
Buckwheat, f/mix, 4" dm.	27	1 ea	55	10	6
Plain, recipe, 4" dm.	27	1 ea	60	6	4
Whole wheat, 5" dm.	52	1 ea	94	10	9

	Wt (g)	Svg	Cal	Folacin (mcg) 100 Cal	Svg
PIES, pce = 1/6 of 9" pie:					
Apple	158	1 pce	405	2	8
Banana cream	198	1 pce	319	7	22
Blueberry	158	1 pce	380	4	14
Cherry	158	1 pce	410	4	16
Chocolate cream	175	1 pce	311	4	11
Coconut cream	172	1 pce	343	3	11
Coconut custard	165	1 pce	384	7	25
Cream pie, commercial	152	1 pce	455	4	18
Custard pie	152	1 pce	293	5	15
Pecan	138	1 pce	583	3	18
Lemon meringue	140	1 pce	355	4	13
Mincemeat	160	1 pce	395	2	9
Peach	158	1 pce	405	3	12
Pumpkin	200	1 pce	367	5	20
Strawberry chiffon, recipe	162	1 pce	372	6	21
PopTart-type pastry, fortified	54	1 ea	210	21	43
Pretzels, thin twists	60	10 ea	240	4	10
ROLLS:					
Cinnamon, small	50	1 ea	158	11	18
Dinner, 2.5" x 2"	28	1 ea	85	12	10
Hamburger bun	45	1 ea	129	13	17
Hard roll, white	50	1 ea	155	11	17
Hotdog bun	40	1 ea	115	13	15
Rye roll, dark	28	1 ea	79	15	12
Rye roll, light	28	1 ea	76	15	11
Submarine roll (hoagie)	135	1 ea	400	12	49
Whole wheat roll	35	1 ea	88	22	20
Stuffing, w/enr. bread:					
Bread stuffing fr/dry	140	1 c	500	3	14
Stove Top stuffing	108	1/2 c	176	13	22
Tortillas:					
Corn, enr., fried, 6" dm.	30	1 ea	87	6	5
Flour, 10.5" dm.	57	1 ea	168	15	25
Flour, 8" dm.	35	1 ea	105	15	16
Waffles, 7" dm:					
From recipe	75	1 ea	245	5	13
From mix	75	1 ea	205	2	4

Dairy & Dairy Products

	Wt (g)	Svg	Cal	Folacin (mcg) 100 Cal	Svg
CHEESE (1.5" cube ≈ 1 oz):					
American cheese	28	1 oz	106	2	2
Blue cheese	28	1 oz	100	10	10
Brick cheese	28	1 oz	105	6	6
Brie cheese	28	1 oz	95	19	18
Camembert	28	1 oz	85	21	18
Cheddar cheese	28	1 oz	114	4	5
Cheshire	28	1 oz	110	4	4
Colby cheese	28	1 oz	112	5	5
Cottage cheese:					
Lowfat 1%	226	1 c	164	17	28
Lowfat 2%	226	1 c	205	15	30
Creamed, large curd	225	1 c	235	12	27
Creamed, small curd	210	1 c	215	12	26
Creamed, with fruit	226	1 c	279	8	22
Dry curd	145	1 c	123	17	21
Cream, cheese (1 T = 15g)	28	1 oz	99	4	4
Edam cheese	28	1 oz	101	5	5
Gorgonzola	28	1 oz	111	8	9
Gouda cheese	28	1 oz	101	6	6
Liederkranz	28	1 oz	87	39	34
Limburger	28	1 oz	93	17	16
Parmesan, grated (1 T ≈ 5g)	100	1 c	455	2	8
Ricotta cheese, part skim	246	1 c	340	14	34
Roquefort	28	1 oz	105	13	14
Swiss	28	1 oz	92	2	2

ESHA Research

Dairy & Dairy Products, continued:	Wt (g)	Svg	Cal	Folacin (mcg) 100 Cal	Svg
CREAM, SWEET, fluid:					
Coffee or table	240	1 c	469	1	6
Half and half	15	1 T	20	10	2
Light whipping cream	239	1 c	699	1	9
CREAM, SWEET whipped:					
Heavy cream, whipped	119	1 c	410	1	5
Pressurized	60	1 c	154	.6	1
CREAM, SOUR, dairy:					
Cultured, dairy	230	1 c	493	5	25
Sour dressing, dairy	235	1 c	416	7	28
Cream, sour, Imitation, non-dairy	230	1 c	479	0	0
CREAM SUBSTITUTES, non-dairy:					
Coffee whitener, liq/frzn	120	1/2 c	163	0	0
Coffee whitener, powder	94	1 c	541	0	0
Dessert toppings, non-dairy:					
Frozen, like Coolwhip	75	1 c	239	0	0
Dessert powder, dry	43	1.5 oz	245	0	0
Pressurized, non-dairy	70	1 c	185	0	0
Kefir beverage	233	1 c	160	13	20
MILK (cow):					
Skim	245	1 c	86	16	14
Lowfat 1%	244	1 c	102	12	12
Lowfat 2%	244	1 c	121	10	12
Whole (3.3% fat)	244	1 c	150	8	12
Buttermilk (< 1% fat)	245	1 c	99	12	12
Canned:					
Skim, evaporated	255	1 c	200	11	22
Whole, evaporated	252	1 c	340	5	18
Sweetened, condensed	306	1 c	982	4	34
Dry, instant, nonfat	68	1 c	244	14	34
Dry, instant, buttermilk	120	1 c	464	12	57
Milk (other):					
Human breast milk	246	1 c	171	14	24
Soy milk	240	1 c	79	5	3.6
MILK BEVERAGES and mixes:					
Chocolate flavor to be mixed w/milk:					
Powder	21.6	3/4 oz	75	5	4
Drink w/whole milk	266	1 c	226	5	12
Chocolate flavor to be mixed w/water:					
Powder (includes dry milk)	28	1 oz	100	3	3
Drink	206	3/4 c	100	3	3
Cocoa, hot, w/whole milk	250	1 c	218	6	12
Instant Breakfast, fortified, dry	37	1 env	130	77	100
Malted milk, w/whole milk:					
Chocolate flavor	265	1 c	229	7	16
Natural flavor	265	1 c	237	9	22
Milkshakes, 10 fl oz = 1.25 c:					
Chocolate	283	1.25 c	360	3	10
Strawberry	283	1.25 c	319	3	9
Vanilla	283	1.25 c	314	3	9
MILK DESSERTS:					
Custard, baked	265	1 c	305	8	24
Ice cream, vanilla					
Regular	133	1 c	269	1	3
Soft serve	173	1 c	377	2	9
Ice milk, soft serve, vanilla	175	1 c	223	2	5
Puddings (5 oz can ≈ 1/2 c):					
Assorted flavors:					
Low calorie	130	3/4 c	69	10	7
Regular	135	1/2 c	150-175	3-4	6
Chocolate:					
Cooked from mix	260	1 c	300	3	8
From instant	260	1 c	310	3	10
Canned	142	1 can	205	1.5	3

MILK DESSERTS continued:	Wt (g)	Svg	Cal	Folacin (mcg) 100 Cal	Svg
Puddings, continued:					
Lemon or coconut f/inst	149	3/4 c	181	3	6
Vanilla, canned	142	1 can	220	1.4	3
Sherbet (2% fat)	193	1 c	270	5	14
Yogurt, frozen, average	174	1 c	220	6	14
YOGURT:					
Lowfat, plain	227	1 c	144	17	25
Lowfat, fruit	227	1 c	231	9	21
Lowfat, coffee or vanilla	227	1 c	193	12	23
Nonfat yogurt	227	1 c	127	22	28
Whole	227	1 c	138	12	16

Eggs

	Wt (g)	Svg	Cal	Folacin (mcg) 100 Cal	Svg
Whole egg (chicken):					
Cooked	50	1 ea	77.5	30	23
Raw	50	1 ea	75	31	23
White, raw	33.4	1 ea	17	6	1
Yolk, raw	16.6	1 ea	59	41	24

Fruits & Fruit Juices

	Wt (g)	Svg	Cal	Folacin (mcg) 100 Cal	Svg
Apple, 2.75" diam:					
With peel	138	1 ea	80	5.0	4.0
Without peel	128	1 ea	72	0.7	0.5
APRICOTS:					
Fresh, pitted	106	3 ea	51	18	9
Canned, juice pack	248	1 c	119	4	5
Canned, heavy syrup	258	1 c	214	2	4
Dried, halves	35	10 ea	83	4	4
Avocado, whole:					
California	173	1 ea	305	37	113
Florida	304	1 ea	340	48	162
Banana, 8.75", 176g w/peel	114	1 ea	105	23	24
Blackberries:					
Fresh berries	144	1 c	74	66	49
Frozen, unthawed	151	1 c	97	53	51
Canned	256	1 c	236	29	68
Blueberries:					
Fresh berries	145	1 c	82	11	9
Frozen, unsweetened	155	1 c	78	13	10
Canned	256	1 c	225	3	7
Boysenberries:					
Frozen, unthawed	132	1 c	66	127	84
Canned	256	1 c	225	39	88
Cherries, sour:					
Frozen, unthawed	155	1 c	72	10	7
Canned	244	1 c	90	22	20
CHERRIES, SWEET:					
Fresh, pitted,10 = 68g	145	1 c	104	8	8
Canned	257	1 c	213	4	7
Cantaloupe: see Melons.					
Currants:					
Fresh, Black	112	1 c	71	6	4
Fresh, Red or white	112	1 c	63	6	4
Dried, (Zante)	144	1 c	407	4	15
Dates, fresh, pitted	83	10 ea	228	6	14
Figs, fresh, medium	50	1 med	37	4	1.5
Figs, dried	187	10 ea	477	3	16

Fruits & Fruit Juices, continued:	Wt (g)	Svg	Cal	Folacin (mcg) 100 Cal	Svg
Fruit cocktail, canned:					
Juice pack	248	1 c	115	2	1.5
Heavy syrup	255	1 c	185	.6	1.2
Gooseberries, canned w/liq.	252	1 c	185	4	8
GRAPEFRUIT, half=241g w/rind:					
Half, pink or red	123	1 ea	37	41	15
Half, white	118	1 ea	39	30	12
Canned, sections	254	1 c	152	14	22
Grapefruit juice:					
Fresh juice	247	1 c	96	54	52
Prep f/frzn conc.	247	1 c	102	51	52
Canned, unsweetened	247	1 c	93	28	26
Grapes:					
Thompson, seedless	50	10 ea	35	10	4
Tokay or Emperor	57	10 ea	40	10	4
Canned, heavy syrup	256	1 c	187	4	8
Grape juice:					
From frozen	250	1 c	128	3	4
Bottled or canned	253	1 c	155	4	7
Guava, raw	90	1 ea	45	28	13
Kiwi fruit	76	1 ea	46	37	17
Lemon juice:					
Fresh juice	244	1 c	60	53	32
Frozen, standard strength	244	1 c	54	43	23
Bottled	244	1 c	52	47	25
Lime juice:					
Fresh juice	246	1 c	65	32	21
Bottled	246	1 c	50	39	20
Loganberries, fresh	100	2/3 c	70	37	26
Mandarin oranges, canned	252	1 c	155	13	20
Mango, fresh slices	165	1 c	108	29	31
Melon, cubes, also see Watermelon:					
Cantaloupe	160	1 c	57	84	48
Casaba	170	1 c	45	89	40
Honeydew	170	1 c	60	85	51
Frozen, melon balls, mixed	173	1 c	55	81	45
Nectarine (med = 1 c slc)	136	1 med	67	8	5
ORANGE 2-5/8", 180g w/peel	131	1 ea	60	66	40
Orange juice:					
Fresh juice	248	1 c	111	98	109
Chilled	249	1 c	110	41	45
Prep. frzn concentrate	249	1 c	110	99	109
Canned, unsweetened	249	1 c	105	14	15
Orange grapefruit jce, canned	247	1 c	105	19	20
Papaya, 454g w/refuse	304	1 ea	117	41	48
Papaya nectar, canned	250	1 c	142	4	5
PEACHES:					
Fresh, peeled slices	170	1 c	73	8	6
Frozen, thawed slices	250	1 c	235	3	8
Canned, juice pack	77	1 half	34	8	2.6
Canned, heavy syrup	81	1 half	60	4	2.6
PEARS:					
Bartlett, 180g w/refuse	166	1 ea	98	12	12
Canned, heavy syrup	79	1 half	59	1.5	.9
Canned, juice pack	77	1 half	38	4	1.6
Persimmon, Japanese, large	168	1 ea	118	11	13
PINEAPPLE:					
Fresh, chunks	155	1 c	76	22	16
Canned pieces, juice pack	250	1 c	150	8	12
Canned pieces, heavy syrup	255	1 c	199	6	12
Pineapple juice:					
From frozen	250	1 c	129	50	64
Canned, unsweetened	250	1 c	140	41	58
Plantain slices, fresh	148	1 c	181	18	33
Plantain slices, cooked	154	1 c	179	22	40

	Wt (g)	Svg	Cal	Folacin (mcg) 100 Cal	Svg
PLUMS:					
Medium, 2-1/8" dm.	66	1 ea	36	9	3
Canned, juice pack	95	3 ea	55	5	2.8
Canned, heavy syrup	110	3 ea	98	3	2.8
Prunes, dried	84	10 ea	201	2	3.4
Raisins, dark, unpacked meas	145	1 c	435	1	5
Raspberries:					
Fresh berries	123	1 c	60	54	33
Frozen, thawed measure	250	1 c	255	26	65
Canned	256	1 c	234	12	27
Rhubarb:					
Fresh, diced	122	1 c	26	34	9
Cooked with sugar	240	1 c	279	5	13
STRAWBERRIES:					
Fresh berries	149	1 c	45	62	28
Frozen, thawed, sweetened	255	1 c	245	17	42
Frozen, unsweetened	149	1 c	52	54	28
Tangerines, medium	84	1 ea	37	46	17
Tangerine juice:					
From frozen	241	1 c	110	10	11
Canned, sweetened	249	1 c	125	6	8
Watermelon, diced pieces	160	1 cup	50	7	3.4

Grains & Grain Products

	Wt (g)	Svg	Cal	Folacin (mcg) 100 Cal	Svg
Amaranth grain, dry	195	1 c	729	13	95
Barley, pearled, cooked	157	1 c	193	13	25
Bran: see Oat, Rice, Wheat.					
Buckwheat flour:					
Dark	98	1 c	338	37	125
Light	98	1 c	340	29	100
Bulgar wheat, cooked	182	1 c	151	22	33

CEREALS, COLD (Ready To Eat): Cereals can be fortified with folacin. Amounts vary. Check the label.

CEREALS, HOT (cooked):	Wt (g)	Svg	Cal	Folacin (mcg) 100 Cal	Svg
Corn grits	242	1 c	145	1.4	2
Cream of Rice	244	1 c	126	6	8
Cream of Wheat	244	1 c	140	7	9
Farina, cooked	233	1 c	116	4	5
Malt-O-Meal	240	1 c	122	4	5
Maypo, cooked	180	3/4 c	128	6	7
Oatmeal, regular, quick/inst.	234	1 c	145	7	9.4
Oatmeal, fortified instant:					
Plain, from packet	177	3/4 c	104	144	150
Other flavors averaged	164	3/4 c	160	94	150
Ralston	253	1 c	134	13	18
Roman Meal	181	3/4 c	111	16	18
Wheatena	243	1 c	135	13	17
Whole wheat cereal	242	1 c	150	17	25
Corn flour:					
Regular	117	1 c	422	7	29
Masa Harina, enriched	114	1 c	416	6	27
Cornmeal, dry, degermed	138	1 c	505	13	66

FLOUR: see specific grain, nut, or vegetable.

	Wt (g)	Svg	Cal	Folacin (mcg) 100 Cal	Svg
MACARONI, cooked, enriched	140	1 c	197	5	10
Millet, cooked	120	1/2 c	143	16	23
Noodles:					
Chow mein, dry	45	1 c	237	4	10
Egg, cooked	160	1 c	213	5	11
Spinach, cooked	140	1 c	182	9	17

Values are for edible part of foods

	Wt (g)	Svg	Cal	Folacin (mcg) 100 Cal	Svg		Wt (g)	Svg	Cal	Folacin (mcg) 100 Cal	Svg
Grains & Grain Products, continued:						**HALIBUT:**					
Oat bran (1T ≈ 6g)	94	1 c	132	37	49	Baked or broiled	85	35 oz	119	7	8
Oats, rolled, dry	81	1 c	311	8	26	Smoked	100	3.5 oz	224	2	5
PASTA: see Macaroni,						Steamed, pacific	100	3.5 oz	131	8	11
Noodles, Spaghetti.						Herring:					
Popcorn, popped in oil	11	1 c	55	5	3	Baked or broiled	100	3.5 oz	203	3	5
						Canned with liquid	100	3.5 oz	208	2	5
RICE, cooked:						Smoked or kippered	100	3.5 oz	217	2	4
Brown rice	195	1 c	217	4	8	Lobster, meat only, cooked	145	1 c	142	11	16
White, regular, enriched	205	1 c	264	2	6	Mackerel:					
White, converted, enriched	175	1 c	200	4	7	Baked/broiled, Atlantic	100	3.5 oz	262	3	7
White, instant	165	1 c	162	4	7	Baked/broiled, Spanish	100	3.5 oz	158	4	7
Wild rice	164	1 c	166	26	43	Canned, Jack, #300 can-tall	361	1 can	563	3	18
Rice bran	83	1 c	262	20	52	Mullet, baked/broiled	85	3 oz	127	7	8
Rice flour	158	1 c	578	1	6						
Rye flour:						**OYSTERS:**					
Dark	128	1 c	415	19	77	Raw, Eastern	248	1 c	170	14	25
Light	102	1 c	361	6	22	Raw, Pacific	248	1 c	200	12	24
Soy flour, stirred:						Fried, Eastern, medium	88	6 ea	173	7	12
Full fat, raw	85	1 c	368	80	293	Simmered, Eastern	100	3.5 oz	137	13	18
Full fat, roasted	85	1 c	373	52	193	Perch, Ocean:					
Lowfat	85	1 c	326	111	361	Baked or broiled	100	3.5 oz	121	7	9
Defatted	100	1 c	327	93	305	Breaded, fried	85	3 oz	185	3	6
Spaghetti, cooked:						Pike, Northern, baked/broiled	100	3.5 oz	113	27	30
Enriched	140	1 c	197	5	10	Pollock, baked or broiled	100	3.5 oz	99	13	13
Whole wheat spaghetti	140	1 c	174	4	7						
Tapioca, dry	152	1 c	518	1	6	**SALMON, cooked:**					
						Broiled or baked, avg	85	3 oz	183	8	14
WHEAT:						Smoked salmon, Chinook	85	3 oz	99	2	1.6
Wheat bran	30	1/2 c	65	36	23.7	Canned, Atlantic, small can	220	1 can	281	12	35
FLOUR, unbleached:						Pink, #1 can, drained	454	1 can	631	11	70
All purpose, white, unsifted	125	1 c	455	7	32.5	Sockeye, #1 can, drained	369	1 can	566	6	36
Cake, sifted	96	1 c	348	5.5	19	Sardines, canned, drained:					
Semolina	167	1 c	601	20	120	Atlantic, 2 ea =24g	92	1 can	192	6	11
Whole wheat flour	120	1 c	407	13	53	Pacific, 1 ea =38g	100	3.5 oz	178	14	24
Wheat germ, raw	100	1 c	360	78	281	Scallops:					
Wheat germ, toasted	113	1 c	432	92	398	Breaded, fried	93	6 ea	200	5	11
Wheat, rolled:						Steamed	100	3.5 oz	113	16	18
Cooked	240	1 c	142	19	27	Seatrout or Steelhead, cooked	100	3.5 oz	131	7	10
Dry	85	1 c	289	19	54						
Whole grain wheat						**SHRIMP:**					
(wheat berries) cooked	50	1/3 c	86	7	6	Boiled, 2 large ≈ 11g	100	3.5 oz	99	4	4
Whole wheat, sprouted	108	1 c	214	21	44	Breaded, fried, 2 large ≈ 15g	90	12 ea	218	3	7
						Smelt, Rainbow, cooked	85	3 oz	106	15	16
						Snapper, baked or broiled	100	3.5 oz	128	7	9
Meats: Fish & Shellfish											
						SOLE (Flounder):					
						Baked or broiled	85	3 oz	99	10	10
Bass, baked / broiled	100	3.5 oz	125	7	9	Fried in batter	85	3 oz	250	3	7
Carp, baked / broiled	100	3.5 oz	162	6	9	Breaded, fried	100	3.5 oz	188	5	9
Catfish, fried w/cornmeal	100	3.5 oz	229	3	7	Steamed	100	3.5 oz	92	11	10
Clams:						Swordfish, baked/broiled	100	3.5 oz	155	10	16
Breaded, fried, small	188	20 ea	379	1	5	**TUNA,** canned, drained-No.1/2 can:					
Steamed clams, meat only	90	20 ea	133	3	4	Light, oil pack	171	1 can	339	3	9
Canned, drained	160	1 c	236	3	7	Light, water pack	165	1 can	216	4	8
Cod:											
Baked or broiled	100	3.5 oz	105	10	10						
Fried with batter	100	3.5 oz	199	4	9	**Meats:**					
Canned with liquid, 11 oz	312	1 can	327	9	28	Beef, Pork, Ham, etc.					
Crayfish, cooked, moist heat	85	3 oz	97	6	6						
						BEEF:					
CRAB, meat:						**Chuck blade**, pot roasted:					
Blue, canned, unpacked	135	1 c	133	17	22	Lean & fat	85	3 oz	325	1.5	5
Dungeness, cooked	101	3/4 c	85	24	20	Lean only	85	3 oz	230	2.2	5
Eel, smoked	100	3.5 oz	330	2	8	**Ground beef**, baked, broiled,					
Fish cakes, recipe	100	3.5 oz	172	5	8	pan-fried average:					
Fish sticks, heated fr/frozen	57	2 ea	155	7	10	Extra lean, 17% fat raw	85	3 oz	215	3.6	7.7
Haddock:						Lean, 20.7% fat raw	85	3 oz	231	3.3	7.7
Baked, broiled or poached	85	3 oz	95	14	13	Regular, 26.6% fat, raw	85	3 oz	250	3.1	7.7
Breaded, fried	85	3 oz	175	8	14	Frozen patty, broiled, 23% fat	85	3 oz	240	3.2	7.7

BEEF, continued:	Wt (g)	Svg	Cal	Folacin (mcg) 100 Cal	Svg
Rib, oven roasted:					
Lean & fat	85	3 oz	324	2	6
Lean only	85	3 oz	204	3	7
Round steak, broiled:					
Lean & fat	85	3 oz	233	3	8
Lean only	85	3 oz	165	5	9
Round tip, oven roasted:					
Lean & fat	85	3 oz	213	3	6
Lean only	85	3 oz	162	4	7
Sirloin steak, broiled:					
Lean & fat	85	3 oz	238	3	6
Lean only	85	3 oz	172	4	7
T-bone steak, broiled:					
Lean & fat	85	3 oz	276	2	6
Lean only	85	3 oz	182	4	7
Beef kidney, cooked	140	1 ea	201	68	137
Beef liver, fried	85	3 oz	184	102	187
Dried beef, cured (6-7 pieces)	28	1 oz	47	8	4
Corned beef, canned	85	3 oz	213	2	5
HAM: see Pork, cured; Lunch- meats group & Turkey ham.					
LAMB:					
Arm chop, braised:					
Lean & fat	70	1 ea	244	5	13
Lean only	55	1 ea	152	8	12
Loin chop, broiled:					
Lean & fat	64	1 ea	201	6	12
Lean only	46	1 ea	100	11	11
Cutlet, lean, cooked average	85	3 oz	175	11	19
Leg of lamb, roasted:					
Lean & fat	85	3 oz	219	8	17
Lean only	85	3 oz	162	12	20
Shoulder roast:					
Lean & fat	85	3 oz	235	8	18
Lean only	85	3 oz	173	12	21
Liver	85	3 oz	202	168	340
PORK:					
Bacon, regular, cooked	19	3 pces	109	1	1
Center loin chop:					
Braised, lean & fat	75	1 ea	266	1	3
Braised, lean only	61	1 ea	166	2	3
Broiled, lean & fat	82	1 ea	284	2	4
Broiled, lean only	72	1 ea	166	3	4
Fried, lean & fat	89	1 ea	333	1	4
Fried, lean only	67	1 ea	178	2	4
Roasted, lean & fat	88	1 ea	268	<1	1
Roasted, lean only	72	1 ea	180	<1	1
Center rib chop:					
Braised, lean & fat	67	1 ea	246	2	4
Braised, lean only	53	1 ea	147	3	4
Broiled, lean & fat	77	1 ea	264	2	6
Broiled, lean only	63	1 ea	162	3	5
Fried, lean & fat	88	1 ea	343	1	5
Fried, lean only	62	1 ea	160	3	5
Roasted, lean & fat	79	1 ea	252	2	6
Roasted, lean	66	1 ea	162	4	6
Pork roast, average loin/rib:					
Lean and fat	85	3 oz	265	3	7
Lean only	85	3 oz	206	4	8
Spareribs, cooked fr/1 lb raw	177	6.25 oz	703	1	7
PORK CURED - HAM: also see					
Lunchmeat group & Turkey ham.					
Roasted, lean & fat	140	1 c	341	1	4
Roasted, lean only	140	1 c	219	3	6
Canned, roasted, average	85	3 oz	142	3	4

	Wt (g)	Svg	Cal	Folacin (mcg) 100 Cal	Svg
RABBIT, roasted meat	85	3 oz	175	4	7
VEAL (calf):					
Cutlet, lean, cooked avg	85	3 oz	166	8	13
Liver, pan-fried	85	3 oz	208	131	272
Rib roast	85	3 oz	151	8	12

Meats: Poultry

	Wt (g)	Svg	Cal	Folacin (mcg) 100 Cal	Svg
CHICKEN:					
All types of meat:					
Fried	140	1 c	307	3	10
Roasted	140	1 c	266	3	8
Stewed	140	1 c	248	3	8
Canned, boned w/broth	142	5 oz	235	1.6	4
Dark meat:					
Fried	85	3 oz	203	4	7.0
Roasted	85	3 oz	174	4	6.7
Stewed	85	3 oz	163	4	6.0
Light meat:					
Fried	85	3 oz	163	2	3.6
Roasted	85	3 oz	147	2	3.0
Stewed	85	3 oz	135	2	3.0
Breast*, meat & skin:					
Batter-fried	140	1 ea	364	2	8
Flour-fried	98	1 ea	218	2	4
Roasted	98	1 ea	193	2	3
Breast*, meat only:					
Fried	86	1 ea	161	2	4
Roasted	86	1 ea	142	2	3
* two pieces per bird					
Drumstick, meat & skin:					
Batter-fried	72	1 ea	193	3	6
Flour-fried	49	1 ea	120	3	4
Roasted	52	1 ea	112	4	4
Stewed	57	1 ea	116	3	4
Drumstick, meat:					
Fried	42	1 ea	82	5	4
Roasted	44	1 ea	76	5	4
Stewed	46	1 ea	78	5	4
Thigh, meat & skin:					
Batter-fried	86	1 ea	238	3	8
Flour-fried	62	1 ea	162	3	5
Roasted	62	1 ea	153	3	4
Stewed	68	1 ea	158	2.5	4
Thigh, meat:					
Fried	52	1 ea	113	3.5	4
Roasted	52	1 ea	109	4	4
Stewed	55	1 ea	107	4	4
Chicken gizzard	22	1 ea	34	34	12
Chicken liver	20	1 ea	30	513	154
DUCK, domestic, roasted:					
Meat and skin	85	3 oz	286	2	6
Meat only	85	3 oz	171	5	8.5
GOOSE, domestic, roasted:					
Meat and skin	85	3 oz	259	.7	2
Meat only	85	3 oz	202	.9	2
Liver pate, canned	13	1 T	41	19	8
TURKEY, roasted:					
All types	85	3 oz	145	4	6
Dark meat	85	3 oz	159	5	8
Light meat	85	3 oz	133	4	5
Turkey gizzard	67	1 ea	109	33	36
Turkey heart	16	1 ea	28	45	13
Turkey liver	75	1 ea	127	393	499
Ground turkey, cooked	100	3.5 oz	229	3	7

	Wt (g)	Svg	Cal	Folacin (mcg) 100 Cal	Svg

Meats: Sausages & Lunchmeats

	Wt (g)	Svg	Cal	100 Cal	Svg
Braunschweiger	57	2 oz	205	28	57
Italian sausage link, cooked	67	1 ea	216	2	4
Liverwurst, pork	18	1 pce	59	9	5
Salami, turkey	57	2 oz	111	5	5
Turkey ham	57	2 oz	73	6	4
Turkey pastrami	57	2 oz	74	5	4

Mixed Dishes & Fast Foods

	Wt (g)	Svg	Cal	100 Cal	Svg
Beef & vegetable stew:					
Recipe	245	1 c	220	17	37
Canned	245	1 c	194	16	31
Beef, macaroni & tomato sauce, recipe	226	1 c	189	12	23
Beef pot pie, from frozen	234	1 ea	426	4	17
Burrito:					
Bean	174	1 ea	322	17	55
Beef	177	1 ea	463	8	35
Beef and bean	175	1 ea	390	12	48
Deluxe combination	198	1 ea	424	12	51
Cheese souffle, recipe	112	1 c	221	13	29
Chicken a la king, recipe	245	1 c	470	2	11
Chicken & noodles, recipe	240	1 c	365	3	9
Chicken chow mein:					
Recipe	250	1 c	255	7	19
Canned	250	1 c	95	13	12
Chicken egg roll	100	1 ea	242	18	44
Chicken pot pie, from frozen	230	1 ea	430	7	29
Chicken salad w/celery	78	1/2 c	266	2	4
Chili w/beans, canned	255	1 c	286	14	41
Cole slaw	120	1 c	84	38	32
Chop suey w/beef & pork	250	1 c	300	7	22
Corn fritter, recipe	45	1 ea	116	15	17
Corn pudding	250	1 c	271	23	63
Corned beef hash, canned	220	1 c	382	4	15
Egg salad	183	1 c	438	17	74
Enchilada:					
Beef	120	1 ea	292	4	11
Cheese	120	1 ea	330	5	15
Chicken	120	1 ea	269	4	12
French toast, recipe	65	1 pce	123	15	18
LASAGNA, recipe:					
With meat	245	1 pce	398	4	16
Without meat	218	1 pce	316	5	14
Macaroni & cheese:					
Recipe	200	1 c	430	2	10
Canned	240	1 c	230	3	8
Manicotti, frozen entree	225	1 ea	271	10	28
Meat loaf, average	87	1 pce	203	4	8
Moussaka, lamb & eggplant	250	1 c	250	18	44
PIZZA, cheese:					
Regular crust, 1/8 of 15"	120	1 pce	290	14	40
Thick crust, 1/2 of 10"	208	1 pce	519	14	70
Potato salad w/mayo & eggs	250	1 c	358	5	17
Quiche Lorraine, 1/8 Pie	176	1 pce	600	3	17
Ravioli, beef, canned	226	1 c	220	10	21
SANDWICHES, fast food items:					
Cheeseburger, 3 oz beef	112	1 ea	300	7	20
Cheeseburger, 4 oz beef	194	1 ea	524	4	23
Chicken patty sandwich	157	1 ea	436	4	18
English muffin with egg, cheese and bacon	38	1 ea	360	10	35

	Wt (g)	Svg	Cal	100 Cal	Svg
Fish sandwich:					
Large without cheese	170	1 ea	470	9	43
Regular with cheese	140	1 ea	420	6	24
Hamburger, 3 oz beef	98	1 ea	245	7	16
Hamburger, 4 oz beef	174	1 ea	445	5	24
Hotdog (frankfurter)& bun	85	1 ea	260	7	17
SANDWICHES (on part whole wheat bread, except when stated as rye):					
Avocado, cheese, sprouts & tomato	195	1 ea	432	18	76
Bacon, lettuce & tomato	135	1 ea	327	12	41
Chicken salad sandwich	100	1 ea	294	10	28
Egg salad sandwich	111	1 ea	319	14	44
Grilled cheese	117	1 ea	393	8	30
Ham & cheese	151	1 ea	363	9	31
Ham & swiss on rye	145	1 ea	350	7	25
Ham on rye	116	1 ea	242	10	23
Ham sandwich	122	1 ea	256	11	29
Peanut butter & jam	100	1 ea	341	14	47
Roast beef sandwich	122	1 ea	280	10	29
Tuna salad sandwich	116	1 ea	303	12	36
Turkey sandwich	122	1 ea	271	11	28
SPAGHETTI (pasta & tomato sauce w/cheese):					
Recipe	250	1 c	260	3	8
Canned	250	1 c	190	3	6
Spinach souffle	136	1 c	218	28	62
Taco, beef	78	1 ea	207	6	13
Taco, chicken	78	1 ea	172	8	14
Tostadas, with:					
Beans & beef	192	1 ea	332	11	37
Beans & chicken	157	1 ea	249	14	34
Refried beans only	157	1 ea	212	22	47
Tuna noodle casserole, recipe	202	1 c	251	5	13
Tuna salad, without egg	205	1 c	383	4	15
Turkey pot pie, frozen	233	1 ea	416	6	24

Nuts & Seeds

	Wt (g)	Svg	Cal	100 Cal	Svg
ALMONDS, dried, whole	142	1 c	837	10	83
Brazil nuts, dry, unsalted	140	1 c	919	.6	6
Cashew butter	16	1 T	94	12	11
Cashews:					
Dry roasted	137	1 c	787	12	95
Oil roasted	130	1 c	748	12	88
Chestnuts, roasted	143	1 c	350	29	100
Coconut:					
Raw, grated	80	1 c	283	7	21
Shredded, sweetened, pkg	93	1 c	466	2	9
Dried unsweetened	78	1 c	515	1	7
Coconut milk, raw	240	1 c	552	1	6
Filberts (hazelnuts), whole	135	1 c	853	11	97
Macadamias, dried	134	1 c	940	10	91
MIXED NUTS w/peanuts (cashews, peanuts, brazil nuts, filberts, almonds, pecans):					
Dry roasted	137	1 c	814	8	69
Oil roasted	142	1 c	876	13	118
MIXED NUTS w/o peanuts (cashews, almonds, brazil nuts, pecans & filberts) oil roasted	144	1 c	886	9	81
PEANUTS:					
Dry roasted	146	1 c	855	25	212
Oil roasted	144	1 c	837	22	181

Nuts, Seeds & Products, continued:

Food	Wt (g)	Svg	Cal	Folacin 100 Cal	Folacin Svg
Peanut butter:					
Chunky	258	1 c	1520	16	237
Smooth	258	1 c	1517	13	202
Pecans, dried, chopped	119	1 c	794	6	47
Pine nuts, dried:					
Pignola	28	1 oz	146	13	19
Pinyon	28	1 oz	161	12	19
Pistachios, dried, shelled	128	1 c	739	10	74
Pumpkin/squash seeds:					
Kernels, roasted	227	1 c	1185	10	115
Whole, roasted	64	1 c	285	9	27
Sesame seeds:					
Kernels, dried	150	1 c	882	17	150
Whole, dried	144	1 c	825	17	139
Sesame flour:					
High fat	28	1 oz	149	6	9
Low fat	28	1 oz	95	9	8
Soybeans, dry roasted	172	1 c	810	45	364
Sunflower seed kernels:					
Dry roasted	128	1 c	745	37	272
Oil roasted	135	1 c	830	38	316
Sunflower seed butter	16	1 T	93	37	34
Walnuts, chopped:					
Black	125	1 c	759	11	83
English	120	1 c	770	10	79

Soups, Sauces & Gravies

Food	Wt (g)	Svg	Cal	Folacin 100 Cal	Folacin Svg
Beef gravy:					
Recipe	135	1/2 c	151	5	7
Canned	233	1 c	124	6	7
SAUCES (also see Other):					
Cheese sauce:					
Regular	101	1/2 c	216	4	9
From mix w/milk	279	1 c	305	4	12
Hollandaise sauce, recipe	160	1 c	867	7	60
Spaghetti sauce, plain:					
Recipe	220	1 c	179	13	23
Canned	249	1 c	272	14	39
White sauce, recipe, med	250	1 c	395	3	12

SOUPS: All soups are canned unless otherwise stated. For soups prep. w/milk, assume whole milk. RTS = Ready To Serve.

Food	Wt (g)	Svg	Cal	Folacin 100 Cal	Folacin Svg
Bean w/bacon	253	1 c	173	18	32
Beef broth/bouillon	240	1 c	16	10	2
Beef noodle	244	1 c	84	5	4
Beef, chunky, RTS	240	1 c	171	8	13
Black bean soup, prepared	247	1 c	116	21	25
Celery, cream of, prepared with milk	248	1 c	165	5	9
Chicken noodle	241	1 c	75	3	2.2
Chicken rice, chunky, RTS	240	1 c	127	3	4
Chicken, chunky, RTS	251	1 c	178	3	5
Chicken, cream of, w/milk	248	1 c	191	4	8
Chili beef	250	1 c	169	6	10
Clam chowder:					
New England	248	1 c	163	7	12
Manhatten style, RTS	240	1 c	133	7	9
Lentil and ham, RTS	248	1 c	140	35	50
Minestrone	241	1 c	80	20	16
Mushroom, cream of:					
Condensed	251	1 c	257	3	7
Prepared w/milk	248	1 c	205	7	15
Onion soup, canned	241	1 c	57	27	15
Oyster stew, w/milk	245	1 c	134	5	7
Potato, cream of	248	1 c	148	6	9

Food	Wt (g)	Svg	Cal	Folacin 100 Cal	Folacin Svg
Split pea soup:					
With ham, chunky, RTS	240	1 c	184	3	5
From dry mix	255	1 c	133	11	15
Tomato soup:					
Prepared w/milk	248	1 c	160	13	21
Prepared w/water	244	1 c	86	17	15
Prepared fr/dry	265	1 c	102	7	7
Turkey soup, chunky, RTS	236	1 c	136	18	25
Vegetable beef	244	1 c	79	13	11
Vegetable, chunky, RTS	240	1 c	122	14	17
Vegetarian vegetable	241	1 c	70	15	11

Vegetables & Legumes

Food	Wt (g)	Svg	Cal	Folacin 100 Cal	Folacin Svg
Alfalfa sprouts	33	1 c	10	122	12
Amaranth leaves:					
Chopped, fresh	28	1 c	7	327	24
Cooked	132	1 c	28	328	92
Artichoke, globe, cooked	120	1 ea	60	102	61
Artichoke, hearts:					
Cooked from frozen-pkg	240	9 oz	108	264	285
Marinated-jar	170	6 oz	168	89	149
Asparagus, pieces:					
Fresh, pieces	67	1/2 c	15	467	70
Ckd from fresh	90	1/2 c	23	392	88
Ckd from frozen	180	1 c	50	349	176
Canned, drained	121	1/2 c	16	613	98
Canned with liquid	122	1/2 c	17	612	104
Bamboo shoots, sliced, canned	131	1 c	25	160	40
BEANS (see also Garbanzo, Lentils, Soybeans):					
Baked beans (dry white beans w/spices & sauce):					
Home prepared	253	1 c	382	32	122
Canned, plain/vegetarian	254	1 c	235	26	61
Canned w/franks	257	1 c	366	21	77
Canned w/pork	253	1 c	268	34	92
Canned w/sweet sauce	253	1 c	282	34	95
Canned w/tomato sauce	253	1 c	247	23	57
Black beans, cooked fr/dry	172	1 c	227	113	256
Broadbeans:					
Ckd from dry	170	1 c	186	95	177
Canned	256	1 c	183	46	84
Great northern:					
Ckd from dry	177	1 c	210	86	181
Canned	262	1 c	300	71	213
Green beans, snap beans:					
Fresh, uncooked	110	1 c	34	118	40
Ckd from fresh	125	1 c	44	95	42
Ckd from frozen	135	1 c	36	117	42
Canned, drained	135	1 c	26	165	43
Canned with liquid	240	1 c	36	121	44
Kidney beans:					
Ckd fr/dry	177	1 c	225	102	229
Canned with liquid	256	1 c	208	61	126
Lima beans:					
Ckd fr/resh	170	1 c	208	41	86
Ckd fr/frozen, average	88	1/2 c	90	63	57
Ckd fr/dry, large	188	1 c	217	72	156
Ckd fr/dry, small	182	1 c	229	119	273
Canned, drained	170	1 c	164	24	40
Canned with liquid	241	1 c	191	63	121
Navy, ckd fr/dry	182	1 c	259	99	255
Pinto beans:					
Canned	240	1 c	186	78	145
Ckd fr/dry	171	1 c	235	125	294
Refried, canned	253	1 c	270	56	150
White, ckd from dry	179	1 c	253	97	245
Winged, ckd from dry	172	1 c	252	7	18

ESHA Research

Folacin (mcg)

Vegetables & Legumes, continued:

	Wt (g)	Svg	Cal	100 Cal	Svg
BEANS, continued:					
Yardlong, ckd from dry	171	1 c	202	123	249
Yellow wax: see green beans.					
Bean sprouts (Mung beans):					
Fresh sprouts	104	1 c	31	203	63
Boiled, drained	124	1 c	26	135	35
Stir-fried	124	1 c	62	116	72
Canned, drained	125	1 c	16	76	12
Beet greens, cooked, drained	144	1 c	40	118	47
Beets:					
Ckd from fresh, whole	100	2 ea	31	277	86
Canned, drained, diced	85	1/2 c	27	82	22
Pickled, slices	114	1/2 c	74	47	35
BROCCOLI, chopped:					
Fresh, chopped	88	1 c	24	260	62
Ckd from fresh	156	1 c	44	177	78
Ckd from frozen	184	1 c	51	108	55
W/cheese sauce	142	1/2 c	166	66	110
W/hollandaise sauce	95	1/2 c	105	101	106
Brussels sprouts:					
Ckd fr/fresh	156	1 c	60	156	94
Ckd fr/frozen	155	1 c	65	242	157
CABBAGE:					
Common, fresh, shredded	70	1 c	16	248	40
Common, cooked	150	1 c	32	97	31
Bok choy, fresh, shredded	70	1 c	9	633	57
Bok choy, cooked	170	1 c	20	160	32
Pe-Tsai, fresh, shredded	76	1 c	12	498	60
Pe-Tsai, cooked	119	1 c	16	397	64
Red, fresh, shredded	70	1 c	19	100	19
Red, cooked	75	1/2 c	16	59	9
Savoy, fresh, shredded	70	1 c	20	160	32
Savoy cooked	145	1 c	35	106	37
CARROTS:					
Fresh, whole, 7.5" x 1-1/8"	72	1 ea	31	33	10
Fresh, grated	55	1/2 c	24	32	8
Ckd from fresh, sliced	78	1/2 c	35	31	11
Ckd from frozen	73	1/2 c	26	30	8
Canned, drained	73	1/2 c	17	39	7
Canned with liquid	123	1/2 c	28	36	10
Carrot juice	123	1/2 c	49	10	5
Cauliflower:					
Fresh pieces	50	1/2 c	12	276	33
Ckd fr/fresh	62	1/2 c	15	211	32
Ckd fr/frozen	180	1 c	34	217	74
Celery:					
Fresh, large outer stalk	40	1 ea	6	183	11
Ckd, diced	150	1 c	27	122	33
Chard, Swiss, fresh, chopped	36	1 c	7	285	20
Chard, Swiss, cooked	175	1 c	35	163	57
Collards:					
Fresh, chopped	36	1 c	11	36	4
Cooked fr/fresh	128	1 c	35	23	8
Cooked fr/frozen	170	1 c	63	212	129
CORN:					
Fresh kernels, uncooked	77	1/2 c	66	54	35
Ckd from fresh	82	1/2 c	89	43	38
Ckd from frozen	82	1/2 c	67	28	19
Canned, drained	82	1/2 c	66	46	30
Canned, with liquid	128	1/2 c	79	62	49
Canned, vacuum pack	210	1 c	166	63	104
Canned, cream style	128	1/2 c	93	62	57
Cucumber w/peel, 1/8" slices	28	7 pce	4	107	4

	Wt (g)	Svg	Cal	100 Cal	Svg
Dandelion greens:					
Fresh	55	1 c	25	256	64
Cooked	105	1 c	35	234	82
Eggplant, cooked	160	1 c	45	51	23
Endive, fresh, chopped	25	1/2 c	4	888	36
Escarole/curly endive	50	1 c	9	835	71
Garbanzo beans (chickpeas):					
Ckd from dry	164	1 c	269	105	282
Canned	240	1 c	285	56	160
Jerusalem artichoke, slices	150	1 c	114	13	15
Jicama	100	3.5 oz	20	75	15
Kale:					
Fresh, chopped	67	1 c	33	59	20
Ckd fr/raw or frozen	130	1 c	40	75	30
Kohlrabi:					
Fresh, sliced	140	1 c	38	37	14
Ckd from fresh	165	1 c	48	28	13
Leeks, chopped:					
Fresh	104	1 c	63	105	67
Cooked	52	1/2 c	24	67	16
Lentils, cooked from dry	198	1 c	231	155	358
Lentils, sprouted:					
Fresh sprouts	77	1 c	81	95	77
Stir-fried	100	3.5 oz	101	83	84
LETTUCE:					
Butterhead	56	1 c	7.3	563	41
Iceberg	56	1 c	7.3	431	31
Loose leaf	56	1 c	10	594	60
Romaine	56	1 c	9	848	76
Mushrooms:					
Fresh slices	35	1/2 c	9	82	7.4
Cooked from fresh	78	1/2 c	21	68	14
Canned, drained	78	1/2 c	18	51	10
Mustard greens:					
Fresh, chopped	56	1 c	145	226	33
Cooked from fresh	140	1 c	21	95	20
Cooked from frozen	150	1 c	29	70	20
Okra:					
Pods, ckd fr/fresh	85	8 ea	27	143	39
Slices, ckd fr/frozen	92	1/2 c	34	394	134
Onions:					
Fresh, chopped	160	1 c	61	49	30
Cooked fr/fresh, chopped	105	1/2 c	46	35	16
Dehydrated flakes	14	1/4 c	45	52	23
Parsley:					
Fresh, chopped	30	1/2 c	10	549	55
Freeze-dried	1.4	1/4 c	4	538	22
Parsnips:					
Fresh slices	133	1 c	100	89	89
Cooked	156	1 c	125	73	91
PEAS:					
Black-eyed peas:					
Ckd from fresh	165	1 c	160	131	210
Ckd from frozen	170	1 c	224	107	240
Ckd from dry	171	1 c	198	180	356
Canned	240	1 c	184	67	123
Green peas:					
Fresh, uncooked	145	1 c	118	80	94
Ckd from fresh	160	1 c	134	75	101
Ckd from frozen	80	1/2 c	63	74	47
Canned, drained	85	1/2 c	59	64	38
Green peas, edible-pods:					
Fresh, uncooked	145	1 c	61	71	44
Ckd from fresh	160	1 c	67	72	48
Ckd from frozen	80	1/2 c	42	57	24
Split, ckd fr/dry	196	1 c	231	55	127
Peas & carrots:					
Canned	128	1/2 c	48	49	24
Ckd from frozen	80	1/2 c	38	55	21

Folacin (mcg)

	Wt (g)	Svg	Cal	Folacin 100 Cal	Folacin Svg
Vegetables & Legumes, continued:					
Peas, sprouted, mature:					
Fresh sprouts	120	1 c	154	112	173
Ckd from fresh	100	3.5 oz	118	31	36
PEPPERS, HOT, green/red:					
Fresh, chopped, 1 pod ≈ 45g	75	1/2 c	30	58	18
Canned, hot chili or Jalapeno	68	1/2 c	17	206	35
PEPPERS, SWEET, green/red:					
Fresh, chopped, 1 pod ≈ 74g	50	1/2 c	14	79	11
Cooked fr/fresh	68	1/2 c	19	53	10
POTATOES:					
Baked in oven:					
Flesh & skin	202	1 ea	220	10	22
Flesh only	156	1 ea	145	10	14
Potato skin	58	1 ea	115	11	13
Baked in microwave oven:					
Flesh & skin	202	1 ea	212	11	24
Flesh only	156	1 ea	156	12	19
Potato skin	58	1 ea	77	13	10
Boiled, 2.5" diam, flesh only:					
Cooked without skin	135	1 ea	116	10	12
Boiled in skin, then peeled	136	1 ea	119	11	14
Ckd fr/frozen, small	70	1 ea	46	13	6
French fries, fr/frozen:					
Fried in oil	50	10 pces	158	9	15
Oven heated	50	10 pces	111	8	8
Hash browned, fr/frozen	156	1 c	340	8	26
Mashed potatoes:					
Prep. w/milk	210	1 c	162	11	17
Prep. fr/instant	215	1 c	239	6	15
Potato puffs (tater tots)	62	1/2 c	138	7	10
Canned, 1" diam.	70	2 ea	42	11	4
Potato chips	28	14 ea	148	9	13
Potatoes au gratin, recipe	245	1 c	322	6	20
Potatoes scalloped, recipe	245	1 c	210	10	21
Potato pancakes	76	1 ea	237	9	22
Pumpkin:					
Ckd fr/fresh, mashed	245	1 c	50	66	33
Canned	123	1/2 c	42	36	15
Radishes, daikon	44	1/2 c	8	120	10
Radishes, red	45	10 ea	7	174	12
Radish seeds, sprouted	38	1 c	16	231	36
Rutabaga, cubes:					
Fresh	140	1 c	51	56	29
Cooked	85	1/2 c	29	46	13
Sauerkraut, canned w/liquid	236	1 c	44	9	4
Seaweed (kelp), fresh	28	1 oz	12	418	51
Shallots, freeze-dried, chopped	3.6	1/4 c	13	32	4
Spinach:					
Fresh, chopped	56	1 c	12	886	109
Ckd fr/fresh	180	1 c	41	639	262
Ckd fr/frozen	190	1 c	53	384	204
Canned, drained	214	1 c	50	418	209
Soybeans, ckd from dry	172	1 c	298	31	93
Soybeans, mature, sprouted:					
Fresh sprouts	35	1/2 c	45	134	60
Steamed	94	1/2 c	76	101	77
SOYBEAN PRODUCTS: see tofu this section; miso, & tempeh in Other; roasted soybeans in Nuts & Seeds; soy milk in Dairy; soy flour in Grains.					
SQUASH, SUMMER varieties:					
Crookneck, fresh slices	130	1 c	24	124	30
Crookneck, cooked	180	1 c	36	101	36
Scallop, fresh slices	130	1 c	24	163	39
Scallop, cooked	90	1/2 c	14	133	19
Zucchini, fresh slices	130	1 c	19	152	29
Zucchini, cooked	180	1 c	29	104	30
SQUASH, WINTER, mashed:					
Acorn/Danish, baked	245	1 c	137	33	46
Acorn/Danish, boiled	245	1 c	83	33	28
Butternut, baked	245	1 c	99	48	47
Butternut, cooked fr/frozen	240	1 c	94	31	29
Hubbard, baked	240	1 c	120	32	39
Hubbard, boiled	236	1 c	70	33	23
Spaghetti, baked or boiled	155	1 c	45	28	12
Succotash:					
Ckd from fresh	192	1 c	222	34	75
Ckd from frozen	170	1 c	158	36	57
Sweet potatoes:					
Baked in skin	114	1 ea	118	22	26
Boiled, peeled	151	1 ea	160	14	22
Candied, recipe	105	1 pce	144	8	12
Canned, mashed	128	1/2 c	129	16	21
Tofu, raw (soybean product):					
Firm	126	1/2 c	183	20	37
Regular	124	1/2 c	94	20	19
TOMATOES:					
Fresh, whole	123	1 ea	26	73	19
Fresh, chopped	180	1 c	38	71	27
Cooked from fresh	240	1 c	65	48	31
Canned, whole	240	1 c	47	75	35
Tomato juice, canned	244	1 c	41.5	117	49
Tomato paste, canned	262	1 c	220	18	40
Tomato puree, canned	250	1 c	102	38	39
Tomato sauce, canned	245	1 c	74	52	39
Turnip, cubes:					
Fresh cubes	130	1 c	35	54	19
Ckd from fresh	78	1/2 c	14	51	7
Turnip greens, cooked:					
From fresh	144	1 c	29	590	171
From frozen	82	1/2 c	24	135	32
Vegetable juice cocktail	242	1 c	46	83	38
VEGETABLES, MIXED, cooked from frozen:					
Broccoli, carrots & pasta	95	2/3 c	88	68	60
Broccoli, carrots & water chestnuts	91	2/3 c	32	266	85
Broccoli, cauliflower & red pepper	95	2/3 c	25	196	49
Broccoli & water chestnuts	95	1/2 c	33	346	114
Cantonese stir fry vegetables	95	1/2 c	53	87	46
Chinese stir fry vegetables	95	1/2 c	31	32	10
Green beans & spaetzle, Bavarian style	95	1/2 c	108	26	28
Japanese style vegetables	95	1/2 c	29	117	34
Mixed vegetables (corn, peas, carrots, green beans & limas):					
Canned, drained	163	1 c	77	50	39
Cooked from frozen	182	1 c	107	32	35
Peas, carrots and onions	91	1/2 c	54	104	56
Peas, cauliflower & cream sce	95	1/2 c	118	41	48
Peas & mushrooms	95	1/2 c	73	111	81
Peas & onions	95	1/2 c	71	103	73
Peas, onions & pasta	95	1/2 c	122	31	38
Peas, onions & cheese sce	142	1/2 c	165	41	67
Peas, pasta, corn, & cream sce	95	1/2 c	132	40	5
Peas, pasta, mushrooms & cream sauce	95	1/2 c	129	53	68
Peas, potatoes & cream sce	76	1/2 c	140	29	41
Peas, rice & mushrooms	66	2/3 c	108	27	29
Water chestnuts, cnd, slices	70	1/2 c	35	21	8
Watercress, fresh	17	1/2 c	2	1700	34
Yam, orange: see Sweet potato.					
Yam, white, cooked, cubes	136	1 c	158	14	22
Zucchini: see Squash.					

ESHA Research

	Wt (g)	Svg	Cal	Folacin (mcg) 100 Cal	Svg		Wt (g)	Svg	Cal	Folacin (mcg) 100 Cal	Svg

Other

	Wt (g)	Svg	Cal	100 Cal	Svg
1000 Island salad dressing	16	1 T	60	10	6
Barbecue sauce	103	1 c	185	16	30
Blue cheese salad dressing	15	1 T	75	11	8
BEVERAGES (also see Dairy, Fruits & Vegetables)					
Beer (1.5 cup = 12 fl oz)	356	1.5 c	146	15	21
Beer, light	354	1.5 c	100	15	15
Bloody mary, 5 fl oz drink	148	1 ea	116	17	20
Lemonade/Limeade fr/frzn	248	1 c	101	6	6
Pineapple grapefruit drink	250	1 c	117	22	26
Pineapple orange drink	250	1 c	125	22	27
Screwdriver, 7 fl oz drink	213	1 ea	174	43	75
Tequila sunrise, 5.5 oz drink	172	1 ea	189	31	58
Tea, brewed	240	1 c	2	500-620	10-12
Butter	227	1 c	1626	.4	7
CANDY:					
Chocolate covered:					
Almonds	165	1 c	935	14	128
Coconut	28	1 oz	133	4	5
Peanuts	170	1 c	954	18	171
Raisins	187	1 c	733	2	17
Fudge, average, w/nuts	28	1 oz	118	3	4
M & M's plain candies	48	1 pkg	237	2	5
M & M's peanut candies	47	1 pkg	240	2	5
Milk chocolate, w/almonds	28	1 oz	150	3	4
Milk chocolate, w/peanuts	28	1 oz	155	10	16
REESE's peanut butter cup	45	2 ea	240	7	17
SNICKERS, 2.2 oz	61	1 bar	290	2	6
Catsup	245	1 c	255	15	37
Chili sauce, tomato based	273	1 c	284	7	20
CHOCOLATE:					
Baking unsweetened	28	1 oz	145	12	18
Bittersweet	28	1 oz	141	10	14
Chocolate chips, semi-sweet	170	1 c	860	3	22
Hot fudge topping	300	1 c	1020	2	23
Syrup, thin	300	1 c	680	4	24
Cocoa powder	86	1 c	224	15	33
Granola bar	28	1 ea	127	18	23
Honey	339	1 c	1030	3	32
Hummous or Humous	246	1 c	420	35	146
Margarine, 80% fat	227	1 c	1626	.2	2.4-3
Mayonnaise	220	1 c	1577	.4	6
Miso (soybean product)	275	1 c	565	16	91
Molasses, blackstrap	40	2 T	85	7	6
Salsa:					
Picante by Tostitos	85	6 T	40	35	14
Recipe	108	1/2 c	46	36	28
Soy sauce:					
Regular (wheat & soy)	18	1 T	9	31	2.8
Tamari (soy)	18	1 T	11	30	3.3
From hydrolyzed protein	18	1 T	7	33	2.3
SPICES:					
Chili powder	7.5	1 T	24	16	4
Fenugreek seed	11.1	1 T	36	18	6
Garlic cloves	9	3 cloves	13	2	0.3
Garlic powder	8.4	1 T	28	18	5
Onion powder	6.5	1 T	15	60	9
Tempeh (soybean product)	166	1 c	331	26	86
Teriyaki sauce	18	1 T	15	24	4
Tobasco sauce	15	1 T	1.6	12	.2
Yeast:					
Brewer's	8	1 T	25	1252	313
Dry active, regular	30	4 T	80	1425	1140

Vitamin B-12

Many coenzymes use vitamin B-12, and it is needed with folacin to help manufacture red blood cells. Vitamin B-12 plays a role in building and maintaining the sheath that protects the nerve fibers, and since the function of B-12 and folacin are closely related, a deficiency in B-12 can impair cell growth where folacin is involved with DNA.

Deficiencies in B-12 or folacin can cause a form of anemia. But, lack of B-12 will also result in an ever-increasing paralysis of the muscles and nerves. Since the early symptoms are not detectable from a blood test and early detection is necessary to prevent permanent damage, this type of sneaky deficiency damage has earned the name Pernicious Anemia. There is another way to get Pernicious Anemia, however. To absorb vitamin B-12, the body needs a protein that is made in the stomach called Castle's Intrinsic factor. If it is absent, then the B-12 will not be absorbed. This is not a common condition however, and dietary deficiency of B-12 is rare.

Because folacin and B-12 work together, intakes of folacin will clear up the early anemia condition and mask the more serious B-12 deficiency. As a result, the over-the-counter preparations of folacin are limited by law to 400 mcg, an amount too low to have this masking effect over vitamin B-12.

Vitamin Losses

Vitamin B-12 is generally fairly stable, however there are small losses in processing and when cooking foods at high temperature in an alkaline solution (such as soda). Up to 10 percent is lost in milk during pasteurization and another 30 percent can be destroyed by boiling for several minutes. Iron and vitamin B-6 deficiencies and gastritis will decrease absorption of vitamin B-12.

Recommended Dietary Allowances for Adults

USA RDA: Adults, 2.0 mcg (Pregnant, 2.2 mcg; Nursing/lactating, 2.6 mcg)
Canadian RNI: Adults, 2.0 mcg (Pregnant, 3.0 mcg; Nursing/lactating, 2.5 mcg)

Food Sources

Vitamin B12 is found exclusively in animal food — meats, dairy products, and eggs. It is not found in plant food, so insufficient intake of this nutrient may be a concern for strict vegetarians. If dairy products are added to the vegetarian diet, however, sufficient amounts of B12 should be consumed. The body can save B-12 and can store enough in the liver to last two to three years or more, making it unnecessary to consume it regularly.

Values are for edible portion of foods

	Wt (g)	Svg	Cal	Vit B-12 (mcg) 100 Cal	Svg
Dairy & Dairy Products					
CHEESE (1.5" cube ≈ 1 oz):					
American, processed	28	1 oz	106	.2	.197
American cheese food, cold	28	1 oz	94	.4	.363
Blue	28	1 oz	100	.3	.345
Brick	28	1 oz	105	.3	.356
Brie	28	1 oz	95	.5	.468
Camembert	28	1 oz	85	.4	.367
Cheddar cheese	28	1 oz	114	.2	.234
Cheshire	28	1 oz	110	.3	.300
Colby	28	1 oz	112	.2	.234
Cottage cheese:					
Lowfat 1%	226	1 c	164	.9	1.43
Lowfat 2%	226	1 c	205	.8	1.61
Creamed, large curd	225	1 c	235	.6	1.31
Creamed, small curd	210	1 c	215	.6	1.31
Creamed, with fruit	226	1 c	279	.4	1.12
Dry curd	145	1 c	123	1.0	1.20
Cream cheese (1 T = 15g)	28	1 oz	99	.1	.120
Edam	28	1 oz	101	.4	.435
Feta	28	1 oz	75	.4	.300
Fontina	28	1 oz	110	.3	.300
Gjetost	28	1 oz	132	.2	.300
Gorgonzola	28	1 oz	111	.2	.260
Gouda	28	1 oz	101	.4	.435
Gruyere	28	1 oz	117	.4	.454
Liederkranz	28	1 oz	87	.3	.260
Limburger	28	1 oz	93	.3	.295
Monterey jack	28	1 oz	106	.3	.300
Mozzarella, part skim,					
low moisture	28	1 oz	80	.3	.262
Muenster	28	1 oz	104	.4	.418
Parmesan, grated (1 T = 5g)	28	1 oz	129	.3	.420
Pimento processed	28	1 oz	106	.2	.197
Port du salut	28	1 oz	100	.4	.425
Provolone	28	1 oz	100	.4	.415
Romano, grated (1 T = 5g)	28	1 oz	128	.3	.420
Swiss cheese	28	1 oz	107	.4	.475
Swiss, processed	28	1 oz	95	.4	.348
Swiss cheese food	28	1 oz	92	.7	.652
CREAM, SWEET, fluid:					
Coffee or table	240	1 c	469	.1	.528
Half and half (12% fat)	242	1 c	315	.3	.796
Light whipping cream	239	1 c	699	<.1	.466
Heavy whipping cream	238	1 c	821	<.1	.428
CREAM, SWEET, whipped:					
Heavy cream	119	1 c	410	<.1	.214
Pressurized	60	1 c	154	.1	.175
CREAM, SOUR:					
Cultured, dairy	230	1 c	493	.1	.690
Half and half, dairy	15	1 T	20	.23	.045
Sour cream, Imitation, non-dairy	230	1 c	479	0	0
CREAM SUBSTITUTES non-dairy:					
Coffee whitener, liquid/frozen	120	1/2 c	163	0	0
Coffee whitener, powder	94	1 c	514	0	0
Dessert toppings, non-dairy:					
Frozen, like Coolwhip	75	1 c	239	0	0
Dessert powder, dry	43	1.5 oz	245	0	0
Pressurized topping	70	1 c	185	0	0
Kefir beverage	233	1 c	160	.6	.900
MILK (cow):					
Skim	245	1 c	86	1.1	.926
Lowfat 1%	244	1 c	102	.9	.898
Lowfat 2%	244	1 c	121	.7	.888

	Wt (g)	Svg	Cal	Vit B-12 (mcg) 100 Cal	Svg
Milk, continued:					
Whole (3.3% fat)	244	1 c	150	.6	.871
Buttermilk	245	1 c	99	.5	.537
Canned:					
Skim, evaporated	255	1 c	200	.3	.610
Whole, evaporated	252	1 c	340	.1	.410
Sweetened, condensed	306	1 c	982	.1	1.36
Dry, instant nonfat	68	1 c	244	1.1	2.72
Dry, buttermilk	120	1 c	464	1.0	4.59
Milk (other):					
Goat	244	1 c	168	<.1	.159
Human breast milk	246	1 c	171	<.1	.111
Soy milk	240	1 c	79	0	0
MILK BEVERAGES & beverage mixes:					
Chocolate flavored to be					
mixed with water:					
Powder (includes dry milk)	28	1 oz	100	.4	.444
Drink	206	3/4 c	100	.4	.444
Chocolate flavored to be					
mixed with milk:					
Powder	21.6	3/4 oz	75	0	0
Drink	266	1 c	226	.4	.870
Instant Breakfast, dry, fortified	37	1 env	130	.5	.600
Malted milk, prep.w/whole milk:					
Chocolate	265	1 c	229	.4	.910
Natural	265	1 c	237	.4	1.03
Milkshake (10 fl oz = 1.25 c):					
Chocolate	283	1.25 c	360	.3	.970
Strawberry	283	1.25 c	319	.3	.880
Vanilla	283	1.25 c	314	.3	1.01
MILK DESSERTS:					
Custard, prep. fr/mix	143	1/2 c	161	.4	.710
Ice cream, vanilla:					
Regular	133	1 c	269	.2	.625
Rich	148	1 c	349	.2	.537
Soft serve	173	1 c	377	.3	.996
Ice milk, vanilla:					
Regular	131	1 c	184	.5	.875
Soft serve	175	1 c	223	.6	1.36
Puddings:					
Avg, canned (5oz can≈1/2 c)	142	1 can	212	<.1	.070
Avg, low calorie, fr/mix	130	1/2 c	69	.7	.500
Chocolate, ckd or instant/avg	260	1 c	305	.1	.200
Coconut, instant	149	1/2 c	184	.2	.440
Lemon, from instant	149	1/2 c	178	.2	.440
Rice pudding, average	141	1/2 c	165	.3	.420
Tapioca, ckd fr/mix	130	1/2 c	145	.3	.400
Vanilla, ckd or instant	130	1/2 c	148	.3	.400
Pudding pops, average	57	1 ea	94	.2	.164
Sherbet	193	1 c	270	.1	.158
Yogurt, frozen, average	174	1 c	220	.3	.706
YOGURT:					
Lowfat, plain	227	1 c	144	.9	1.28
Lowfat, fruit	227	1 c	231	.5	1.06
Lowfat, coffee or vanilla	227	1 c	193	0.6	1.20
Nonfat	227	1 c	127	1.1	1.40
Whole	227	1 c	138	.6	.844
Eggs					
Chicken egg, raw, large:					
Whole egg	50	1 ea	77	.77	.59
White only	33.4	1 ea	17	.40	.07
Yolk only	16.6	1 ea	59	.88	.52

Values are for edible portion of foods

Meats: Fish & Shellfish

	Wt (g)	Svg	Cal	Vit B-12 (mcg) 100 Cal	Svg
Anchovies, canned in oil	45	11 ea	95	.4	.40
Bass, baked / broiled	100	3.5 oz	125	2.7	3.40
Bluefish:					
Baked or broiled	100	3.5 oz	159	4.4	7.05
Fried in crumbs	100	3.5 oz	205	2.4	4.90
Carp, baked / broiled	100	3.5 oz	162	.9	1.47
Catfish, fried in cornmeal	100	3.5 oz	229	.9	1.97
CLAMS:					
Breaded, fried, small	188	20 ea	379	20	77
Canned, drained	160	1 c	236	67	158
Canned w/liquid, small can	183	1 can	145	68	99
Steamed, meat only	90	20 ea	133	67	89
Clam nectar, canned	240	1 c	6	140	8.4
Cod:					
Baked, broiled, poached	100	3.5 oz	105	1.0	1.05
Fried in batter	100	3.5 oz	199	.5	.90
Canned, w/liquid, 11oz	312	1 can	327	1.0	3.26
Smoked	100	3.5 oz	79	2.2	1.70
CRAB meat only:					
Blue, ckd f/fresh	135	1 c	138	7.2	9.90
Blue, canned	135	1 c	133	.5	.62
Dungeness, cooked	101	3/4 c	85	11.8	10.0
Crab, imitation fr/surimi	85	3 oz	87	1.6	1.43
Crab cakes, fr/recipe	60	1 ea	93	3.8	3.56
Crayfish, cooked, moist heat	85	3 oz	97	3.0	2.94
Eel, smoked	100	3.5 oz	330	1.8	5.80
Fish cakes, fried f/frozen	100	3.5 oz	213	.5	1.00
Fish sticks, frozen, heated	57	2 ea	155	.7	1.02
Gefiltefish, sweet, commercial	42	1 pce	35	1.0	.354
Grouper, cooked	100	3.5 oz	118	.6	.692
Haddock:					
Baked or broiled	85	3 oz	95	1.2	1.18
Breaded, fried	85	3 oz	175	.5	.791
Smoked	100	3.5 oz	116	1.4	1.60
HALIBUT:					
Baked or broiled	85	35 oz	119	1.0	1.16
Smoked	100	3.5 oz	224	.4	.900
Herring:					
Baked or broiled	100	3.5 oz	203	6.5	13.1
Canned in oil, w/liq	100	3.5 oz	208	4.1	8.50
Canned w/tomato sauce	100	3.5 oz	173	4.9	8.50
Smoked or kippered	100	3.5 oz	217	8.6	18.7
Pickled, 1 pce ≈ 15g	100	3.5 oz	262	1.6	4.27
Lobster, meat only, cooked	145	1 c	142	3.2	4.51
Mackerel:					
Baked/broiled, Atlantic	100	3.5 oz	262	7.3	19.0
Baked/broiled, Spanish	100	3.5 oz	158	4.4	7.00
Canned, Jack, tall can	361	1 can	563	4.5	25.1
Mussels, Blue,					
meat only, steamed	85	3 oz	147	1.0	1.40
OYSTERS:					
Raw, Eastern	248	1 c	170	28	48
Raw, Pacific	248	1 c	200	20	40
Breaded, fried, Eastern	88	6 med	173	8	14
Simmered, Eastern	100	3.5 oz	137	28	38
Perch, ocean:					
Baked / broiled	100	3.5 oz	121	1.0	1.15
Fried, breaded	85	3 oz	185	.4	.770
Pike, Northern, baked/broiled	100	3.5 oz	113	1.0	1.15
Pollock, baked/broiled:					
Mixed species	100	3.5 oz	99	3.1	3.10
Walleye	100	3.5 oz	113	3.7	4.20
Rockfish, baked / broiled	100	3.5 oz	121	2.6	3.16
Roe, raw, mixed species	28	1 oz	39	6.2	2.40

	Wt (g)	Svg	Cal	Vit B-12 (mcg) 100 Cal	Svg
SALMON:					
Baked or broiled, average	85	3 oz	183	2.7	4.93
Smoked salmon, Chinook	85	3 oz	99	2.8	2.77
Canned, drained:					
Atlantic, small can	220	1 can	281	2.2	6.30
Pink, #1 can, boned	454	1 can	631	3.3	21.0
Sockeye, #1 can	369	1 can	566	2.9	16.6
Sardines, canned:					
Atlantic, 2 sardines = 24 g	92	1 can	192	4.3	8.23
Pacific, 1 sardine = 38 g	100	3.5 oz	178	5.1	9.00
Scallops:					
Fried, breaded	93	6 ea	200	.6	1.23
Steamed	100	3.5 oz	113	1.7	1.87
Seatrout/Steelhead, cooked	100	3.5 oz	131	4.4	5.70
Shark, batter fried	85	3 oz	194	.5	1.03
SHRIMP:					
Boiled, 2 large ≈ 11 g	100	3.5 oz	99	1.5	1.49
Breaded, fried, 2 lrg ≈ 15g	90	12 ea	218	.8	1.68
Canned with liquid	100	3.5 oz	102	.9	.953
Canned, drained	128	1 c	154	.9	1.44
Smelt, Rainbow, cooked	85	3 oz	106	3.2	3.37
Snapper, baked /broiled	100	3.5 oz	128	1.3	1.60
SOLE (Flounder):					
Baked or broiled	85	3 oz	99	2.2	2.13
Batter-fried	85	3 oz	250	.6	1.39
Breaded, fried	100	3.5 oz	188	1.0	1.82
Steamed	100	3.5 oz	92	2.3	2.08
Squid, flour-fried	85	3 oz	149	.6	.932
Swordfish, baked/broiled	100	3.5 oz	155	1.3	2.02
Trout, baked/broiled	85	3 oz	129	1.9	2.40
TUNA:					
Bluefin, baked	85	3 oz	157	5.9	9.25
Canned, drained (No. 1/2 can):					
Light, canned in oil	171	1 can	339	1.7	5.61
Light, water pack	165	1 can	216	2.6	5.61

Meats
Beef, Pork, Ham, etc.

	Wt (g)	Svg	Cal	Vit B-12 (mcg) 100 Cal	Svg
BEEF:					
Breakfast strips, cured, cooked	34	3 pce	153	.8	1.17
Chuck blade, pot roasted:					
Lean & fat	85	3 oz	325	.6	1.90
Lean only	85	3 oz	230	.9	2.10
GROUND BEEF, broiled/fried:					
Extra lean, 17% fat raw	85	3 oz	215	.8	1.77
Lean, 20.7% fat raw	85	3 oz	231	.9	1.97
Regular, 26.6% fat raw	85	3 oz	250	1.0	2.40
Frozen patty, 23% fat, raw	85	3 oz	240	.9	2.10
Rib, choice, roasted:					
Lean & fat	85	3 oz	324	.7	2.13
Lean only	85	3 oz	204	1.2	2.48
Round steak, broiled:					
Lean & fat	85	3 oz	233	1.0	2.34
Lean only	85	3 oz	165	1.5	2.53
Round tip, roasted:					
Lean & fat	85	3 oz	213	1.1	2.33
Lean only	85	3 oz	162	1.5	2.46
Round, pot roasted/braised:					
Lean & fat	85	3 oz	222	.9	2.04
Lean only	85	3 oz	189	1.1	2.10
Sirloin steak, broiled:					
Lean & fat	85	3 oz	238	.7	1.65
Lean only	85	3 oz	172	1.0	1.70

ESHA Research

Values are for edible portion of foods

	Wt (g)	Svg	Cal	Vit B-12 (mcg) 100 Cal	Svg		Wt (g)	Svg	Cal	Vit B-12 (mcg) 100 Cal	Svg
BEEF, continued:						**PORK, continued:**					
T-bone steak, broiled:						**Pork roast**, rib/loin:					
Lean & fat	85	3 oz	276	.7	1.80	Lean & fat	85	3 oz	265	.19	.495
Lean only	85	3 oz	182	1.1	1.93	Lean only	85	3 oz	206	.24	.490
Variety Meats:						**Pork roast**, leg:					
Heart	85	3 oz	140	9	12	Lean & fat	85	3 oz	250	.2	.595
Kidney, cooked	140	1 ea	201	36	72	Lean only	85	3 oz	187	.3	.612
Liver, fried	85	3 oz	184	52	95	**Shoulder, braised**, from					
Dried beef, cured, 6-7 pieces	28	1 oz	47	3	1.3	6.8 oz raw w/bone:					
Corned beef, canned	85	3 oz	213	.6	1.4	Lean & fat	85	3 oz	293	.2	.587
						Lean only	67	2.4 oz	166	.3	.476
HAM: see Pork, cured; Turkey						**Spareribs,** cooked					
Ham; and Sausages & Lunchmeats.						from 1 lb raw	177	6.25 oz	703	.3	1.91
LAMB:						**HAM, CURED PORK** (also see					
						Turkey ham & Lunchmeat group):					
Arm chop, braised:						Roasted, lean and fat	85	3 oz	207	.3	.550
Lean & fat	70	1 ea	244	.7	1.82	Roasted, lean only	85	3 oz	133	.5	.600
Lean only	55	1 ea	152	1.0	1.45	Canned, roasted	85	3 oz	140	.6	.830
Loin chop, broiled:						**RABBIT**, roasted	85	3 oz	131	4.2	5.53
Lean & fat	64	1 ea	201	.8	1.58	**VEAL**:					
Lean only	46	1 ea	100	1.2	1.17	Cutlet, braised or broiled	85	3 oz	166	.8	1.41
Cutlet, cooked average	85	3 oz	175	1.3	2.22	Rib, roasted	85	3 oz	151	.9	1.34
Leg of lamb, roasted:						Liver, simmered	85	3 oz	208	26.0	54.4
Lean & fat	85	3 oz	219	1.0	2.20						
Lean only	85	3 oz	162	1.4	2.24						
Rib roast:											
Lean & fat	85	3 oz	305	.6	1.90						
Lean only	85	3 oz	197	.9	1.84						
Shoulder roast:						**Meats: Poultry**					
Lean & fat	85	3 oz	235	1.0	2.24						
Lean only	85	3 oz	173	1.3	2.30	**CHICKEN:**					
						All types of meat:					
PORK:						Fried	140	1 c	307	.2	.480
Bacon, cooked:						Roasted	140	1 c	266	.2	.460
Regular	19	3 pce	109	.3	.330	Stewed	140	1 c	248	.1	.310
Breakfast strips	34	3 pce	156	.4	.600	Canned, bonned w/broth	142	5 oz	235	.2	.420
Canadian style	47	2 pce	86	.4	.360	**Dark meat** only:					
Blade chop:						Fried	85	3 oz	203	.1	.279
Braised:						Roasted	85	3 oz	174	.2	.267
Lean & fat	67	1 ea	275	.2	.510	Stewed	85	3 oz	163	.1	.188
Lean only	50	1 ea	156	.3	.410	**Light meat** only:					
Pan-fried:						Fried	85	3 oz	163	.2	.310
Lean & fat	89	1 ea	368	.2	.780	Roasted	85	3 oz	147	.2	.291
Lean only	62	1 ea	175	.4	.620	Stewed	85	3 oz	136	.1	.194
Center loin chop:						**Breast*, meat & skin:**					
Braised:						Batter-fried	140	1 ea	364	.1	.410
Lean & fat	75	1 ea	266	.2	.460	Flour-fried	98	1 ea	218	.2	.340
Lean only	61	1 ea	166	.2	.370	Roasted	98	1 ea	193	.2	.320
Broiled:						Stewed	110	1 ea	202	.1	.230
Lean & fat	82	1 ea	284	.2	.620	**Breast*, meat:**					
Lean only	72	1 ea	166	.3	.530	Fried	86	1 ea	161	.2	.310
Pan-fried:						Roasted	86	1 ea	142	.2	.292
Lean & fat	89	1 ea	333	.2	.690	Stewed	95	1 ea	144	.2	.220
Lean only	67	1 ea	178	.3	.490	*** 2 pieces per bird**					
Roasted:						**Drumstick, meat & skin:**					
Lean & fat	88	1 ea	268	.2	.530	Batter fried	72	1 ea	193	.1	.200
Lean only	72	1 ea	180	.3	.450	Flour fried	49	1 ea	120	.1	.160
Center rib chop:						Roasted	52	1 ea	112	.2	.170
Braised:						Stewed	57	1 ea	116	.1	.120
Lean & fat	67	1 ea	246	.1	.360	**Drumstick, meat:**					
Lean only	53	1 ea	147	.2	.280	Fried	42	1 ea	82	.2	.140
Broiled:						Roasted	44	1 ea	76	.2	.150
Lean & fat	77	1 ea	264	.2	.520	Stewed	46	1 ea	78	.1	.110
Lean only	63	1 ea	162	.3	.440	**Thigh, meat & skin:**					
Pan-fried:						Batter-fried	86	1 ea	238	.1	.240
Lean & fat	88	1 ea	343	.2	.540	Flour-fried	62	1 ea	162	.1	.190
Lean only	62	1 ea	160	.2	.390	Roasted	62	1 ea	153	.1	.180
Roasted:						Stewed	68	1 ea	158	<.1	.130
Lean & fat	79	1 ea	252	.2	.440	**Thigh meat:**					
Lean only	66	1 ea	162	.2	.370	Fried	52	1 ea	113	.2	.170
						Roasted	52	1 ea	109	.2	.160
						Stewed	55	1 ea	107	.1	.120

	Wt (g)	Svg	Cal	Vit B-12 (mcg) 100 Cal	Svg
Chicken, continued:					
Wing, meat & skin:					
Batter-fried	49	1 ea	159	.1	.120
Roasted	34	1 ea	99	.1	.100
Variety meats:					
Heart, simmered	3.3	1 ea	6	4	.240
Liver, simmered	20	1 ea	30	13	3.87
DUCK, domestic, roasted:					
Meat and skin	85	3 oz	286	.1	.250
Meat only	85	3 oz	171	.2	.338
GOOSE, domestic, roasted:					
Meat and skin	85	3 oz	259	.2	.418
Meat only	85	3 oz	202	.3	.548
Liver pate, canned	13	1 T	41	1.0	.420
TURKEY:					
Breast meat:					
Barbecued	28	1 oz	40	.4	.160
Hickory smoked	28	1 oz	35	.5	.190
Frozen, slices w/gravy	142	5 oz	95	.1	.115
Ground, cooked	100	3.5 oz	229	.1	.330
Roasted:					
All types	140	1 c	238	.2	.520
Dark meat	85	3 oz	159	.2	.316
Light meat	85	3 oz	133	.2	.316
Liver, cooked	75	1 ea	127	28.0	35.6
Turkey patty, breaded, fried	64	1 ea	181	.1	.140

Meats: Sausages & Lunchmeats

	Wt (g)	Svg	Cal	Vit B-12 (mcg) 100 Cal	Svg
Beef lunchmeat, thin sliced	28	1 oz	50	1.5	.730
Beerwurst (beer salami):					
Beef	23	1 pce	75	.7	.490
Pork	23	1 pce	55	.4	.200
Berliner sausage	23	1 pce	53	1.2	.610
Bologna:					
Beef	23	1 pce	72	.4	.320
Beef and pork	28	1 pce	89	.4	.380
Cured pork	23	1 pce	57	.4	.210
Turkey	28	1 oz	56	.8	.420
Braunschweiger sausage	57	2 oz	205	5.6	11.5
Brotwurst, link	70	1 ea	226	.6	1.44
Cheesefurter/cheese smokies	43	1 ea	141	.5	.740
Chicken roll, light meat	57	2 oz	90	.2	.195
Dutch brand loaf	28	1 oz	68	.5	.370
FRANKFURTER (hotdog):					
Beef, 8/package	57	1 ea	184	.5	.940
Beef and pork, 8/pkg	57	1 ea	183	.4	.740
Chicken, 10/pkg	45	1 ea	115	.5	.580
Turkey, 10/pkg	45	1 ea	102	.6	.585
HAM, lunchmeat:					
Chopped, packaged	42	2 pce	98	.3	.294
Extra lean	57	2 oz	75	.6	.425
Regular	57	2 oz	103	.5	.471
Thin sliced, 3 pieces = 1 oz	28	1 oz	37	.6	.210
Ham salad spread	240	1 c	518	.4	1.83
Italian sausage link, ckd	67	1 ea	216	.4	.870
Keilbasa sausage	26	1 pce	81	.5	.420
Knockwurst sausage, link	68	1 ea	209	.4	.800
Liverwurst, pork	18	1 pce	59	4.1	2.42
Luncheon meat, canned	21	1 pce	70	1.0	.710
Luncheon sausage, beef/pork	23	1 pce	60	.8	.450
Luxury loaf	57	2 oz	80	1.0	.780
Pastrami:					
Beef, cured	57	2 oz	198	.5	1.0
Turkey, cured	57	2 oz	74	1.6	1.15
Peppered loaf	28	1 pce	42	1.3	.560

	Wt (g)	Svg	Cal	Vit B-12 (mcg) 100 Cal	Svg
Pepperoni, small slice	22	4 pce	109	.5	.552
Pickle & pimento loaf	57	2 oz	149	.4	.670
PORK SAUSAGE, cooked:					
Brown and serve	13	1 link	50	.4	.200
Link, small	13	1 link	48	.5	.220
Patty, cooked	27	1 pce	100	.5	.470
Salami:					
Beef	23	1 pce	58	1.9	1.11
Beef and pork, dry	20	2 pce	85	.4	.380
Pork and beef	57	2 oz	143	1.4	2.07
Turkey	57	2 oz	111	1.8	1.98
Smoked link sausage:					
Beef and pork	68	1 ea	229	.4	1.03
Pork	68	1 ea	265	.4	1.11
Smoked turkey sausage	28	1 oz	55	1.0	.560
Summer sausage	23	1 pce	80	1.3	1.06
Turkey lunchmeats (other):					
Breakfast sausage	28	1 pce	65	.8	.500
Ham	57	2 oz	73	1.8	1.29
Roll, light or dark	57	2 oz	84	.2	.180
Summer sausage	28	1 oz	50	2.3	1.15
Vienna sausage, canned	16	1 ea	45	.4	.160

Mixed Dishes & Fast Foods

	Wt (g)	Svg	Cal	Vit B-12 (mcg) 100 Cal	Svg
Beef & vegetable stew:					
Recipe	245	1 c	220	.7	1.60
Canned	245	1 c	194	.8	1.59
Beef, macaroni & tomato					
sauce, recipe	226	1 c	189	.4	.765
Beef pot pie, from frozen	234	1 ea	426	.3	1.30
Burrito:					
Bean	174	1 ea	322	<.1	.160
Beef	177	1 ea	463	.6	2.72
Beef and bean	175	1 ea	390	.4	1.59
Deluxe combination	198	1 ea	424	.3	1.48
Cheese souffle, recipe	112	1 c	221	.4	.986
Chicken & noodles, recipe	240	1 c	365	.1	.531
Chicken a la king, recipe	245	1 c	470	.1	.307
Chicken chow mein, recipe	250	1 c	255	.1	.224
Chicken pot pie, fr/frozen	230	1 ea	430	.1	.230
Chicken salad w/celery	78	1/2 c	266	.1	.115
Corn dog	111	1 ea	330	.2	.580
Corn pudding	250	1 c	271	19.9	54.0
Corned beef hash, canned	220	1 c	382	.4	1.54
Egg salad	183	1 c	438	.4	1.97
Enchilada:					
Beef	120	1 ea	292	.3	1.02
Cheese	120	1 ea	330	.1	.386
Chicken	120	1 ea	269	.1	.245
Enchirito	207	1 ea	441	.4	1.56
French toast, recipe	65	1 pce	123	.2	.274
LASAGNA:					
Recipe, w/meat	245	1 pce	398	.4	1.45
Recipe, w/o meat	218	1 pce	316	.2	.484
Macaroni & cheese:					
Recipe	200	1 c	430	.1	.298
Canned	240	1 c	230	.1	.200
Manicotti, frozen entree	225	1 ea	271	.2	.540
Meat loaf:					
Beef and 1/3 pork	87	1 pce	212	.8	1.64
Beef only	87	1 pce	193	1.0	1.89
Moussaka, lamb & eggplant	250	1 c	250	.6	1.42
PIZZA, cheese:					
Regular crust, 1/8 of 15"	120	1 pce	290	0.2	.480
Thick crust, 1/2 of 10"	208	1 pce	519	.1	.660

ESHA Research

Mixed Foods & Fast Foods, continued:

Food	Wt (g)	Svg	Cal	Vit B-12 (mcg) 100 Cal	Vit B-12 (mcg) Svg
Potato salad, w/mayo & eggs	250	1 c	358	.1	.385
Quiche Lorraine	176	1 pce	600	<.1	.170
Ravioli, beef, canned	226	1 c	220	.2	.428

SANDWICHES, fast food:

Food	Wt (g)	Svg	Cal	100 Cal	Svg
Cheeseburger, 3 oz meat	112	1 ea	300	.3	.884
Cheeseburger, 4 oz meat	194	1 ea	524	.4	2.33
Chicken patty sandwich	157	1 ea	436	<.1	.198
English muffin, w/egg, cheese & bacon	138	1 ea	360	.2	.828
Fish sandwich:					
Large, without cheese	170	1 ea	470	<.1	.153
Regular, with cheese	140	1 ea	420	.2	.840
Hamburger, 3 oz meat	98	1 ea	245	.3	.774
Hamburger, 4 oz meat	174	1 ea	445	.5	2.26
Hotdog (frankfurter) & bun	85	1 ea	260	.2	.580

SANDWICHES on part whole wheat bread, except when stated as rye:

Food	Wt (g)	Svg	Cal	100 Cal	Svg
Corned beef & swiss on rye	147	1 ea	429	.4	1.57
Egg salad sandwich	111	1 ea	319	.2	.492
Ham sandwich	122	1 ea	256	.2	.425
Ham on rye	116	1 ea	242	.2	.425
Ham & cheese	151	1 ea	363	.2	.625
Ham & swiss on rye	145	1 ea	350	.3	.906
Roast beef	122	1 ea	280	.5	1.46
Tuna salad sandwich	116	1 ea	303	.3	.897
Turkey sandwich	122	1 ea	271	.4	1.15
Turkey ham sandwich	122	1 ea	253	.5	1.28

SPAGHETTI (pasta & tomato sauce w/meat):

Food	Wt (g)	Svg	Cal	100 Cal	Svg
Recipe	248	1 c	330	.4	1.20
Canned	250	1 c	260	.3	.822
Spinach souffle	136	1 c	218	.3	.680
Tacos:					
Beef	78	1 ea	207	.7	1.36
Chicken	78	1 ea	172	.1	.198
Tostada:					
W/beans and beef	192	1 ea	332	.5	1.66
W/beans and chicken	157	1 ea	249	.2	.436
W/refried beans	157	1 ea	212	.1	.317
Tuna noodle casserole, recipe	202	1 c	251	.7	1.78
Tuna salad	205	1 c	383	.9	3.59
Turkey pot pie, frozen	233	1 ea	416	<.1	.200

Soups, Sauces & Gravies

Food	Wt (g)	Svg	Cal	100 Cal	Svg
Beef gravy	233	1 c	124	.2	.230
Chicken gravy	238	1 c	189	.1	.210

SAUCES:

Food	Wt (g)	Svg	Cal	100 Cal	Svg
Cheese sauce:					
Recipe	101	1/2 c	216	.2	.458
From mix w/milk	279	1 c	305	.3	.871
Hollandaise, recipe	160	1 c	867	.3	2.20
Spaghetti sauce with meat:					
Recipe	248	1 c	297	1.2	3.52
Canned	250	1 c	267	.8	2.13
White sauce:					
Recipe, meduim	250	1 c	395	.2	.880
Mix, prepared w/milk	264	1 c	240	.3	.826

SOUPS: soups are canned unless otherwise stated.
RTS = Ready To Serve. For soups prep.w/milk, assume whole milk.

Food	Wt (g)	Svg	Cal	100 Cal	Svg
Beef broth or bouillon	240	1 c	16	1.9	.300
Beef noodle	244	1 c	84	.2	.200
Beef, chunky, RTS	240	1 c	171	.4	.610
Celery, cream of, w/milk	248	1 c	165	.3	.433
Cheese, prep. w/milk	251	1 c	230	.2	.440
Chicken broth	244	1 c	39	.6	.240
Chicken, chunky, RTS	251	1 c	178	.1	.250
Chicken, cream of, w/milk	248	1 c	191	.2	.450
Chili beef	250	1 c	169	.2	.320
Clam chowder:					
Manhatten style, RTS	240	1 c	133	6.0	7.92
New England style	248	1 c	163	6.3	10.3
Lentil and ham, RTS	248	1 c	140	.2	.300
Mushroom, cream of	248	1 c	205	.2	.444
Oyster stew	245	1 c	134	2.0	2.63
Tomato, prepared w/milk	248	1 c	160	.3	.440
Turkey, chunky, RTS	236	1 c	136	3.5	4.79
Turkey vegetable	241	1 c	74	.2	.170
Vegetable beef	244	1 c	79	.4	.310

Grains, Vegetables & Legumes

Vitamin B12 is not found in vegetables or grains. Its presence in vegetable and cereal dishes is due to added ingredients, such as meat, eggs, milk or cheese which contain Vitamin B12.

Other

Food	Wt (g)	Svg	Cal	100 Cal	Svg
Butter	113	1/2 c	813	—	trace
Cheesecake, fr/recipe, 1/12	92	1 pce	278	.2	.455
Margarine:					
Regular, 80% fat	14	1 T	102	<.1	.013
Soft, 40% fat	14	1 T	49	.3	.129
PIES: pce = 1/6 of a 9" pie.					
Coconut custard	165	1 pce	384	.2	.762
Custard	152	1 pce	293	.2	.617
Popovers	51	1 ea	96	.3	.243
Spongecake, 1/12 cake	66	1 pce	194	.2	.328
Miso (soybean product)	275	1 c	565	.1	.570
Noodles, egg, cooked	160	1 c	213	.2	.144
Tempeh (soybean product)	166	1 c	331	.4	1.39

Pantothenic Acid

This vitamin, as a constituent of coenzyme A, plays a central role in the metabolic process for proteins, fats, and carbohydrates; as well as for making fatty acids, cholesterol, acetylcholine, and steroid hormones. It is also essential for the formation of certain nerve-regulating substances and hormones.

"Panto"means everywhere and, indeed, pantothenic acid is found in all cells and tissues, including the ones we eat. If our food is processed, however, the amounts are severely reduced.

Dietary deficiencies are not known. However, experimentally-induced deficiencies can create weakness, reduced antibody production, muscle cramps, and reduced adrenal function.

Vitamin Losses

Milling destroys at least half the pantothenic acid in grains, and it is not replaced by enrichment. Processing of foods (canning and freezing) also takes its toll. Canned fruits and vegetables show average losses of 50 to 62 percent; frozen fruits and vegetables show average losses of 7 and 47 percent, respectively. It is unstable in either alkaline or acid solutions, and in high heat.

Estimated Safe and Adequate Amounts for Adults

USA: Adults: 4 mg to 7 mg

Because there is less information upon which to base allowances, a 4 mg to 7 mg range is given for pantothenic acid by the National Research Council's Food and Nutrition Board. Canada does not have an official recommendation either. However, discussion of this vitamin in "Recommended Nutrient Intakes for Canadians" supports approximately the same amount as the USA suggested range.

Food Sources

Sources include whole grains, legumes, some vegetables and fruits, organ meats (liver, heart, etc.), yeast, and egg yolk.

	Wt (g)	Svg	Cal	Panto (mg) 100 Cal	Panto (mg) Svg

Baked Goods
Breads, Cakes, Cookies, Crackers, Muffins, Pancakes, Pastries, Pies, Rolls, and some desserts.

	Wt (g)	Svg	Cal	100 Cal	Svg
Apple crisp, 3" x 3" piece	78	1 pce	146	<.1	.061
Bagel, 3.5" diam.	68	1 ea	180	.1	.267
BREADS:					
Banana nut bread, 1/2" slice	50	1 pce	161	.1	.209
Cornbread, muffin	45	1 ea	145	.3	.374
Cracked wheat bread	25	1 pce	65	.2	.151
French bread, 5" x 2.5" x 1"	35	1 pce	100	.1	.126
Mixed grain bread	25	1 pce	65	.2	.158
Pita pocket bread, enr, 6.5"	60	1 ea	165	.2	.256
Pumpernickel, 5" x 4" x 3/8"	32	1 pce	80	.2	.153
Raisin bread	25	1 pce	68	.2	.108
Rye, light, 5" x 3.5" x 7/16"	25	1 pce	65	.2	.111
White bread	28	1 pce	75	.2	.122
Wheat bread (white and whole wheat flour)	28	1 pce	72	.2	.119
Whole wheat bread	35	1 pce	86	.3	.258
CAKE:					
Angel food cake, 1/12 cake	53	1 pce	125	.1	.115
Cheesecake, 1/12 cake	92	1 pce	278	.2	.527
Sponge cake, 1/12 cake	66	1 pce	194	.2	.401
Chips: see tortilla chips this section.					
COOKIES:					
Chocolate chip, commercial	42	4 ea	180	.3	.540
Lady fingers	44	4 ea	158	.4	.691
Oatmeal raisin	52	4 ea	245	<.1	.070
Peanut butter	48	4 ea	245	<.1	.067
Cream puff w/custard filling	110	1 ea	280	.2	.698
Danish pastry, w/fruit	65	1 ea	235	.1	.250
Doughnut, yeast raised	60	1 ea	235	.1	.231
Eclair w/custard filling & chocolate icing	94	1 ea	262	.2	.544
English muffin, average	57	1 ea	135	.2	.250
MUFFINS:					
Bran muffin (wheat)	45	1 ea	125	.2	.210
Cornmeal muffin	45	1 ea	145	.3	.374
Pancakes, from mix:					
Buckwheat, 4" diam	27	1 ea	55	.15	.084
Plain, 4" diam	27	1 ea	60	.14	.084
Whole wheat, 5" diam.	52	1 ea	94	.3	.295
PIES: piece = 1/6 of 9" pie.					
Chocolate cream	175	1 pce	311	.1	.457
Coconut cream	172	1 pce	343	.2	.517
Coconut custard	165	1 pce	384	.2	.949
Custard pie	152	1 pce	293	.2	.683
Pumpkin pie	200	1 pce	367	.2	.850
ROLLS:					
Hamburger bun	45	1 ea	129	.2	.238
Hotdog bun	40	1 ea	115	.2	.212
Submarine (hoagie)	135	1 ea	400	.2	.667
Whole wheat roll	35	1 ea	88	.3	.260
Tortillas:					
Corn, 6" diam.	30	1 ea	87	<.1	.034
Flour, 8" diam.	35	1 ea	105	.1	.091
Tortilla chips, plain	28	1 oz	139	<.1	.015
Waffles:					
Plain, recipe or mix, 7" diam.	75	1 ea	225	.1	.242
Frozen, 4" diam	35	1 ea	98	.1	.124

Dairy & Dairy Products

	Wt (g)	Svg	Cal	100 Cal	Svg
CHEESE (1.5" cube ≈ 1 oz):					
American, processed	28	1 oz	106	.1	.137
American cheese food, cold pack	28	1 oz	94	.3	.277
American cheese spread	28	1 oz	82	.2	.194
Blue cheese	28	1 oz	100	.5	.490
Camembert	28	1 oz	85	.5	.387
Cheddar cheese	28	1 oz	114	.1	.117
Cottage cheese:					
Lowfat 1%	226	1 c	164	.3	.486
Lowfat 2%	226	1 c	205	.3	.547
Creamed:					
Large curd (4.5% fat)	225	1 c	235	.2	.447
Small curd (4.5% fat)	210	1 c	215	.2	.447
With fruit (3.4% fat)	226	1 c	279	.1	.380
Dry curd (<.5% fat)	145	1 c	123	.2	.236
Cream cheese	28	1 oz	99	.1	.077
Gorgonzola	28	1 oz	111	.4	.490
Monterey Jack	28	1 oz	106	<.1	.080
Mozzerella, part skim, low moisture	28	1 oz	80	<.1	.026
Neufchatel	28	1 oz	74	.2	.160
Parmesan, grated (1 T = 5g)	28	1 oz	129	.1	.149
Ricotta, with part skim	246	1 c	340	.1	.400
Roquefort cheese	28	1 oz	105	.5	.491
Swiss cheese	28	1 oz	107	.1	.122
CREAM, SWEET, fluid:					
Coffee or table	240	1 c	469	.1	.662
Half and half	242	1 c	315	.2	.699
Light whipping cream	239	1 c	699	.1	.619
Heavy whipping cream	238	1 c	821	.1	.607
CREAM, SWEET, whipped:					
Whipped cream (heavy)	119	1 c	410	.1	.304
Pressurized	60	1 c	154	.1	.183
CREAM, SOUR, dairy:					
Cultured, dairy	230	1 c	493	.2	.828
Half and half, dairy	15	1 T	20	.3	.054
Cream, sour, Imitation, non-dairy	230	1 c	479	0	0
CREAM SUBSTITUTES, non-dairy:					
Coffee whitener, liquid	120	1/2 c	163	0	0
Coffee whitener, powder	94	1 c	514	0	0
Dessert Toppings, non-dairy:					
Frozen (e.g. Coolwhip)	75	1 c	239	0	0
Dessert powder, dry	43	1.5 oz	245	0	0
Pressurized, non-dairy	70	1 c	185	0	0
Eggnog, commercial	254	1 c	342	.3	1.06
Kefir beverage	233	1 c	160	0	0
MILK (cow):					
Skim	245	1 c	86	.9	.806
Lowfat, 1%	244	1 c	102	.8	.788
Lowfat, 2%	244	1 c	121	.6	.781
Whole (3.3% fat)	244	1 c	150	.5	.766
Buttermilk (< 1% fat)	245	1 c	99	.7	.674
Canned:					
Skim, evaporated	255	1 c	200	.9	1.80
Whole, evap. (7.6% fat)	252	1 c	340	.5	1.61
Sweetened, cond. (8.7% fat)	306	1 c	982	.2	2.30
Dried instant non-fat	68	1 c	244	.9	2.20
Dried buttermilk	120	1 c	464	.8	3.80
Milk (other):					
Goat milk	244	1 c	168	.5	.756
Human breast milk	246	1 c	171	.6	1.00
Soy milk	240	1 c	79	.1	.115

	Wt (g)	Svg	Cal	Panto (mg) 100 Cal	Svg		Wt (g)	Svg	Cal	Panto (mg) 100 Cal	Svg
Dairy & Dairy Products, continued:						Apple juice, fr/frzn conc.	239	1 c	111	.14	.151
MILK BEVERAGES & Mixes:						Applesauce, unsweetened	244	1 c	106	.2	.232
Chocolate flavored,						**APRICOTS:**					
to be mixed w/water:						Fresh, whole, pitted	106	3 ea	51	.5	.254
Powder (includes dry milk)	28	1 oz	100	.3	.334	Canned, juice pack	85	3 ea	40	.2	.077
Drink	206	3/4 c	100	.3	.334	Canned, heavy syrup	85	3 ea	70	.1	.078
To be mixed w/ milk:						Dried halves	35	10 ea	83	.3	.264
Powder	21	3/4 oz	75	<.1	.017	Avocado, whole:					
Drink	266	1 c	226	.3	.766	California varieties	173	1 ea	305	.6	1.68
Instant Breakfast, dry mix	37	1 env	130	1.5	2.00	Florida variety	304	1 ea	340	.9	2.95
Malted milk, w/whole milk:						Banana, 8.75" long, 176g w/peel	114	1 ea	105	.3	.296
Chocolate flavor	265	1 c	229	.3	.766	Blackberries:					
Natural flavor	265	1 c	237	.3	.766	Fresh berries	144	1 c	74	.5	.346
Milkshakes (10 fl oz = 1.25 c):						Frozen, unthawed	151	1 c	97	.2	.228
Chocolate	283	1.25 c	360	.3	1.10	Canned	256	1 c	236	.2	.387
Vanilla	283	1.25 c	314	.4	1.18	Blueberries:					
Strawberry	283	1.25 c	319	.4	1.39	Fresh berries	145	1 c	82	.2	.135
						Frozen, sweetened, thawed	230	1 c	185	.2	.288
MILK DESSERTS:						Canned	256	1 c	225	.1	.228
Custard, baked	265	1 c	305	.5	1.59	Boysenberries:					
Ice cream, vanilla:						Frozen, unthawed	132	1 c	66	.5	.330
Regular	133	1 c	269	.2	.654	Canned	256	1 c	225	.1	.335
Rich	148	1 c	349	.2	.562	Cantaloupe/Casaba: see Melon					
Soft serve	173	1 c	377	.3	1.07	Cherries, sour:					
Ice milk:						Frozen, unthawed	155	1 c	72	.4	.276
Regular	131	1 c	184	.4	.662	Canned	244	1 c	90	.3	.256
Soft serve	175	1 c	223	.5	1.03						
PUDDING, prepared:						**CHERRIES, SWEET:**					
(5 oz can ≈ 1/2+ cup)						Fresh, whole, pitted	68	10 ea	49	.2	.086
Chocolate pudding:						Frozen	259	1 c	232	<.1	.074
Cooked from mix	130	1/2 c	150	.3	.375	Canned with liquid	257	1 c	213	<.1	.074
Instant	260	1 c	310	.2	.750	Cranberries, fresh, whole	95	1 c	46	.5	.208
Canned	142	1 can	205	.1	.270	Cranberry sauce, canned	77	1 c	419	.1	.247
Coconut, instant	149	1/2 c	184	.2	.380	Currants, black, fresh	112	1 c	71	.6	.446
Lemon, instant	149	1/2 c	178	.2	.390	Dates, whole, pitted	83	10 ea	228	.3	.648
Low calorie, assorted	130	1/2 ea	69	.7	.480	Figs:					
Rice pudding:						Fresh, medium	50	1 med	37	.4	.150
Cooked from mix	132	1/2 c	155	.2	.367	Dried	187	10 ea	477	.2	.813
Instant	149	1/2 c	175	.2	.410	Fruit cocktail, canned:					
Tapioca:						Juice pack	248	1 c	115	.1	.153
Recipe	165	1 c	220	.2	.495	Heavy syrup	255	1 c	185	.1	.153
Canned	142	1 can	160	.2	.270	Gooseberries:					
Vanilla:						Fresh	150	1 c	67	.6	.429
From mix, instant	130	1/2 c	148	.2	.367	Canned w/liquid	252	1 c	185	.2	.348
Canned	142	1 can	220	.1	.270						
Pudding pops, all flavors	57	1 ea	95	.2	.220	**GRAPEFRUIT** (half = 241g w/rind):					
Yogurt, frozen, average	174	1 c	220	.6	1.34	Fresh halves, pink/red	123	1 ea	37	.9	.348
						Fresh halves, white	118	1 ea	39	.9	.334
YOGURT:						Canned sections	254	1 c	152	.2	.305
Lowfat, plain	227	1 c	144	.9	1.34	Grapefruit juice:					
Lowfat, w/ fruit	227	1 c	231	.5	1.11	Fresh juice	247	1 c	96	.5	.467
Lowfat, coffee or vanilla	227	1 c	193	.6	1.25	From frozen concentrate	247	1 c	102	.5	.467
Nonfat	227	1 c	127	1.1	1.46	Canned, unsweetened	247	1 c	93	.3	.321
Whole	227	1 c	138	.6	.883	Canned, sweetened	250	1 c	115	.3	.325
						Kiwi fruit, fresh	76	1 ea	46	.5	.215
Eggs						Lemon juice:					
						Fresh	244	1 c	60	.4	.251
Cooked, average	50	1 ea	77.5	.8	.600	Bottled	244	1 c	52	.4	.222
Whole, large, raw	50	1 ea	75	.9	.672	Lime juice, fresh	246	1 c	65	.5	.339
Yolk, raw	16.6	1 ea	59	1.1	.632	Loganberries:					
White, raw	33.4	1 ea	17	.24	.040	Fresh berries	150	1 c	105	.3	.366
						Frozen, unthawed	147	1 c	80	.4	.359
						Mandarin oranges, canned	252	1 c	155	.2	.375
Fruits & Fruit Juices						Mango, fresh slices	165	1 c	108	.2	.264
						Melon: also see Watermelon.					
Acerola juice, fresh	242	1 c	51	1.0	.500	Cantaloupe, fresh cubes	160	1 c	57	.4	.205
APPLE, 2.75" diam:						Casaba, fresh cubes	170	1 c	45	.6	.250
with peel	138	1 ea	80	.1	.084	Honeydew, fresh cubes	170	1 c	60	.6	.352
without peel	128	1 ea	72	.1	.073	Frozen melon balls, mixed	173	1 c	55	.5	.282
						Nectarine	136	1 ea	67	.3	.215

ESHA Research

Fruits & Fruit Juices, continued:	Wt (g)	Svg	Cal	Panto (mg) 100 Cal	Svg
ORANGE, 2-5/8 (180g whole)	131	1 ea	60	.5	.328
Orange juice:					
Fresh juice	248	1 c	111	.4	.476
Chilled	249	1 c	110	.4	.476
From frozen concentrate	249	1 c	110	.4	.393
Canned, unsweetened	249	1 c	105	.4	.374
Orange grapefruit juice, canned	247	1 c	105	.3	.346
Papaya (454g whole)	304	1 ea	117	.6	.663
PEACHES:					
Fresh, peeled slices	170	1 c	73	.4	.289
Frozen, thawed slices	250	1 c	235	.1	.330
Canned half, juice pack	77	1 ea	34	1.1	.388
Canned half, heavy syrup	81	1 ea	60	.1	.041
PEARS:					
Bartlett (181g whole)	166	1 ea	98	.1	.116
Canned half, juice pack	77	1 half	38	.1	.042
Canned half, heavy syrup	79	1 half	59	<.1	.017
PINEAPPLE:					
Fresh chunks, pieces	155	1 c	76	.3	.248
Canned pieces, juice pack	250	1 c	150	.5	.734
Canned pieces, heavy syrup	255	1 c	199	.1	.255
Pineapple juice:					
Prepared from frozen	250	1 c	129	.2	.313
Canned, unsweetened	250	1 c	140	.2	.250
Plantain, cooked slices	154	1 c	179	.2	.359
Plums, 2-1/8" diam.	66	1 ea	36	.3	.120
Pomegranate, 3.5" diam.	154	1 ea	104	.9	.918
Prunes, dried, pitted	84	10 ea	201	.2	.386
Prune juice, bottled	256	1 c	181	.1	.240
Raisins, dark	145	1 c	435	<.1	.065
Raspberries:					
Fresh	123	1 c	60	.5	.295
Frozen, thawed	250	1 c	255	.1	.375
STRAWBERRIES:					
Fresh berries	149	1 c	45	1.1	.507
Frozen, sweetened, thawed	255	1 c	245	.1	.275
Tangerine	84	1 ea	37	.5	.168
Tangerine juice:					
From frozen	241	1 c	110	.3	.301
Canned	249	1 c	125	.7	.924
Watermelon, diced pieces	160	1 c	50	.7	.339

Grains & Grain Products

Cereals, Grains, Flours, Noodles, Pasta, Popcorn

	Wt (g)	Svg	Cal	Panto (mg) 100 Cal	Svg
Amaranth grain, dry	195	1 c	729	.3	2.04
Barley, pearled, cooked	157	1 c	193	.1	.212
Bran: see Oat, Rice, Wheat.					
Buckwheat flour, light	98	1 c	340	.3	1.00
Buckwheat flour, dark	98	1 c	338	.4	1.42
Bulgar wheat, cooked	182	1 c	151	.4	.626
CEREAL, COLD (Ready to eat)					
Some cereals may be fortified with Pantothenic acid. Check label.					
CEREAL, HOT, cooked:					
Corn grits, cooked	242	1 c	145	.1	.082
Maypo cereal	180	3/4 c	128	.2	.254
Oatmeal, fr/rolled oats					
Regular, Quick or Instant	234	1 c	145	.3	.468
Ralston cereal	253	1 c	134	.2	.329
Roman Meal, cooked	181	3/4 c	111	.3	.279
Whole wheat cereal	242	1 c	150	.3	.404

Corn flour:	Wt (g)	Svg	Cal	Panto (mg) 100 Cal	Svg
Regular	117	1 c	422	.2	.770
Masa Harina flour	114	1 c	416	.2	.750
Cornmeal, dry:					
Degermed, enriched	138	1 c	505	.1	.431
Bolted, nearly whole	122	1 c	441	.1	.519
FLOUR: see specific grain, nut or vegetable.					
MACARONI, cooked:					
Enriched	140	1 c	197	.1	.137
Whole wheat	140	1 c	174	.3	.587
Vegetable, enriched	134	1 c	172	.3	.469
Millet, cooked	120	1/2 c	143	.1	.205
Noodles:					
Egg noodles, cooked	160	1 c	213	.1	.232
Spinach noodles, cooked	140	1 c	182	.1	.256
Oat bran (1T = 6g)	94	1 c	132	1.1	1.4
Oats, rolled:					
Dry, uncooked	81	1 c	311	.3	1.01
Cooked oatmeal: see Cereals, hot.					
PASTA: see Macaroni, Noodles, Spaghetti.					
RICE, cooked:					
Brown rice	195	1 c	217	.3	.556
White, converted	175	1 c	200	.3	.567
White, regular, enriched	205	1 c	264	.3	.799
White, instant	165	1 c	162	.2	.294
Wild rice	164	1 c	166	.2	.253
Rice bran	83	1 c	262	2.3	6.13
Rice flour	158	1 c	578	.2	1.29
Rye flour:					
Dark	128	1 c	415	.4	1.86
Light	102	1 c	361	.2	.678
Soy flour:					
Full fat, raw	42	1/2 c	182	.4	.668
Full fat, roasted	42	1/2 c	184	.3	.508
Low fat	44	1/2 c	163	.5	.801
Defatted	50	1/2 c	164	.6	.998
Spaghetti noodles, cooked:					
Enriched	140	1 c	197	.1	.157
Whole wheat spaghetti	140	1 c	174	.3	.587
WHEAT:					
Wheat bran	30	1/2 c	65	1.0	.654
Flour:					
All purpose, white, unsifted	125	1 c	455	.1	.548
Cake flour, sifted	96	1 c	348	.1	.439
Semolina flour	167	1 c	601	.1	.969
Whole wheat flour	120	1 c	407	.3	1.30
Wheat germ:					
Raw	100	1 c	360	.6	2.26
Toasted	113	1 c	432	.4	1.57
Wheat, rolled:					
Cooked	240	1 c	142	.3	.44
Dry	85	1 c	289	.4	1.05
Whole grain wheat (wheat berries) cooked	50	1/3 c	28	.3	.09
Whole wheat, sprouted	108	1 c	214	.5	1.13

Meats: Fish & Shellfish

Bluefish:	Wt (g)	Svg	Cal	Panto (mg) 100 Cal	Svg
Baked or broiled	100	3.5 oz	159	.4	.585
Fried in crumbs	100	3.5 oz	205	.2	.505
Steamed	100	3.5 oz	148	.5	.807
Catfish, fried in cornmeal	100	3.5 oz	229	.2	.418

	Wt (g)	Svg	Cal	Panto (mg) 100 Cal	Svg
Fish & Shellfish, continued:					
Clams:					
Breaded, fried, small	188	20 ea	379	.2	.681
Steamed, meat only	90	20 ea	133	.4	.585
Canned, drained	160	1 c	236	.4	.945
COD:					
Baked or broiled	100	3.5 oz	105	.2	.250
Batter fried	100	3.5 oz	199	.1	.215
Canned w/liquid (11 oz can)	312	1 can	327	.2	.690
CRAB:					
Blue crab meat:					
Cooked, unpacked	135	1 c	138	.4	.575
Canned, unpacked	135	1 c	133	.6	.740
Dungeness, cooked meat	101	3/4 c	85	.7	.575
Fish cakes, fried from frozen	100	3.5 oz	213	.1	.250
Haddock, breaded, fried	85	3 oz	175	.1	.213
HALIBUT:					
Baked or broiled	85	3.5 oz	119	.2	.212
Steamed	100	3.5 oz	131	.2	.271
Smoked	100	3.5 oz	224	.1	.250
Herring:					
Baked or broiled	100	3.5 oz	203	.4	.740
Canned in oil	100	3.5 oz	208	.3	.700
Smoked (kippered)	100	3.5 oz	217	.5	.985
Pickled (1 piece ≈ 15 g)	100	3.5 oz	262	<.1	.081
Lobster, cooked, moist heat	145	1 c	142	.3	.413
Mackerel:					
Baked/broiled, average	100	3.5 oz	262	.2	.580
Baked/broiled, Spanish	100	3.5 oz	158	.5	.810
Canned, Jack, tall can	361	1 can	563	.2	1.10
Mussels, Blue,					
meat only, steamed	85	3 oz	147	.2	.330
Northern pike, baked/broiled	100	3.5 oz	113	.4	.400
Ocean Perch:					
Baked or broiled	100	3.5 oz	121	.3	.370
Breaded, fried	85	3 oz	185	.1	.248
OYSTERS:					
Raw, Eastern	248	1 c	170	.3	.456
Raw, Pacific	248	1 c	200	.5	.900
Breaded, fried, medium	88	6 ea	173	.2	.260
Simmered, Eastern	100	3.5 oz	137	.3	.400
Pollock, baked or broiled	100	3.5 oz	99	.3	.343
Roe, raw, mixed species	28	1 oz	39	1.1	.425
SALMON:					
Baked or broiled	85	3 oz	183	.5	.935
Sockeye, cooked	100	3.5 oz	216	.5	1.10
Chinook, smoked	85	3 oz	99	.7	.740
Atlantic, No. 1/4 can	220	1 can	281	1.1	3.00
Pink, drained, #1 can	454	1 can	631	.5	2.90
Sockeye, drained, #1 can	369	1 can	566	.3	1.94
Sardines, canned, drained:					
Atlantic, 2 sardines ≈ 24g	92	1 can	192	.3	.591
Pacific, 1 sardine ≈ 38g	100	3.5 oz	178	.4	.730
Seatrout (Steelhead), cooked	100	3.5 oz	131	1.7	2.25
Shark, batter fried	85	3 oz	194	.2	.462
SHRIMP:					
Boiled, 2 large ≈ 11g	100	3.5 oz	99	.3	.300
Fried, 2 large ≈ 15g	90	12 ea	218	.1	.139
Canned, drained	128	1 c	154	.2	.310
Canned with liquid	100	3.5 oz	102	.2	.210
Snapper, baked/broiled	100	3.5 oz	128	.2	.237
SOLE (Flounder):					
Baked/broiled	85	3 oz	99	.5	.500
Breaded, fried	100	3.5 oz	188	.2	.366
Swordfish, baked/broiled	100	3.5 oz	155	.3	.476
Trout, baked/broiled	85	3 oz	129	.9	1.17
TUNA, light, canned:					
Oil pack, drained	171	1 can	339	.2	.612
Water pack, drained	165	1 can	216	.3	.612

Meats
Beef, Pork, Ham, Rabbit, Frog legs, Venison and Veal

	Wt (g)	Svg	Cal	Panto (mg) 100 Cal	Svg
BEEF:					
Chuck blade, pot roasted:					
Lean & fat	85	3 oz	325	.1	.252
Lean only	85	3 oz	230	.1	.299
Ground beef, cooked average:					
Extra Lean, 17% fat, raw	85	3 oz	215	.1	.247
Lean, 20.7% fat, raw	85	3 oz	231	.1	.275
Regular, 26.6% fat, raw	85	3 oz	250	.1	.286
Frozen patty, broiled, 23% fat	85	3 oz	240	.1	.323
Rib, choice, oven roasted:					
Lean & fat	85	3 oz	324	.1	.298
Lean only	85	3 oz	204	.2	.377
Round steak, broiled, choice:					
Lean & fat	85	3 oz	233	.1	.309
Lean only	85	3 oz	165	.2	.345
Round tip, roasted:					
Lean & fat	85	3 oz	213	.2	.366
Lean only	85	3 oz	162	.2	.400
Sirloin steak, broiled:					
Lean & fat	85	3 oz	238	.1	.281
Lean only	85	3 oz	172	.2	.311
T-bone steak, broiled:					
Lean & fat	85	3 oz	276	.1	.247
Lean only	85	3 oz	182	.2	.283
Dried beef, cured	28	1 oz	47	.6	.303
Corned beef, canned	85	3 oz	213	.2	.340
Beef heart	85	3 oz	140	.5	.740
Beef liver	85	3 oz	184	2.7	5.03
FROG LEGS, raw meat only	100	3.5 oz	173	.5	.370
HAM (see cured Pork, Bacon, Sausages and Lunchmeats group, and Turkey ham)					
LAMB:					
Arm chop, braised:					
Lean & fat	70	1 ea	244	.18	.430
Lean only	55	1 ea	152	.22	.340
Loin chop, broiled:					
Lean & fat	64	1 ea	201	.2	.410
Lean only	46	1 ea	100	.3	.310
Cutlet, lean, ckd average	85	3 oz	175	.3	.587
Leg of lamb, roasted:					
Lean & fat	85	3 oz	219	.3	.580
Lean only	85	3 oz	162	.4	.600
PORK:					
Bacon, cooked:					
Regular	19	3 pce	109	.2	.200
Canadian style	47	2 pce	86	.3	.242
Breakfast strips	34	3 pce	156	.2	.313
Chop, center loin:					
Braised, lean & fat	75	1 ea	266	.2	.467
Braised, lean only	61	1 ea	166	.3	.451
Broiled, lean & fat	82	1 ea	284	.2	.512
Broiled, lean only	72	1 ea	166	.3	.498
Pan-fried, lean & fat	89	1 ea	334	.3	.899
Pan-fried, lean only	67	1 ea	178	.3	.531
Roasted, lean & fat	88	1 ea	268	.2	.681
Roasted, lean only	72	1 ea	180	.3	.489

	Wt (g)	Svg	Cal	Panto (mg) 100 Cal	Panto (mg) Svg
PORK, continued:					
Chop, center rib:					
Braised, lean & fat	67	1 ea	246	.1	.367
Braised, lean only	53	1 ea	147	.2	.350
Broiled, lean & fat	77	1 ea	264	.2	.484
Broiled, lean only	63	1 ea	162	.3	.470
Pan fried, lean & fat	88	1 ea	343	.2	.525
Pan fried, lean only	62	1 ea	160	.3	.492
Roasted, lean & fat	79	1 ea	252	.2	.536
Roasted, lean only	66	1 ea	162	.2	.393
Pork roast, leg, lean	85	3 oz	187	.3	.570
Pork roast, loin / rib:					
Lean & fat	85	3 oz	265	.2	.618
Lean only	85	3 oz	206	.3	.524
Shoulder, braised, yield from 6.8 oz raw w/bone:					
Lean & fat	85	3 oz	293	.2	.472
Lean only	67	2.4 oz	166	.3	.450
Spareribs, cooked from 1 lb raw:	177	6.25 oz	703	.2	1.33
Pork heart	145	1 c	214	1.7	3.58
Pork liver	85	3 oz	141	2.9	4.06
RABBIT, roasted meat	85	3 oz	175	.4	.620
VEAL (calf)					
Cutlet, lean, ckd average	85	3 oz	166	.7	1.13
Rib roast, lean	85	3 oz	151	.8	1.17
Liver, pan-fried	85	3 oz	134	3.1	4.12

Meats: Poultry
Chicken, Turkey, Duck, Goose, etc.

	Wt (g)	Svg	Cal	Panto 100 Cal	Panto Svg
CHICKEN:					
All types:					
Fried	140	1 c	307	.5	1.63
Roasted	140	1 c	266	.6	1.55
Stewed	140	1 c	248	.4	1.04
Dark meat only:					
Fried	85	3 oz	203	.5	1.07
Roasted	85	3 oz	174	.6	1.03
Stewed	85	3 oz	163	.5	.759
Light meat only:					
Fried	85	3 oz	163	.5	.874
Roasted	85	3 oz	147	.6	.826
Stewed	85	3 oz	135	.4	.487
Breast*, meat & skin:					
Batter-fried	140	1 ea	364	.3	1.15
Flour-fried	98	1 ea	218	.5	.981
Roasted	98	1 ea	193	.5	.917
Stewed	110	1 ea	202	.3	.602
Breast*, meat only:					
Fried	86	1 ea	161	.6	.894
Roasted	86	1 ea	142	.6	.830
Stewed	95	1 ea	144	.4	.544
* 2 pieces per bird					
Drumstick, meat & skin:					
Batter-fried	72	1 ea	193	.4	.725
Flour-fried	49	1 ea	120	.5	.595
Roasted	52	1 ea	112	.6	.630
Stewed	57	1 ea	116	.4	.489
Drumstick, meat only:					
Fried	42	1 ea	82	.7	.554
Roasted	44	1 ea	76	.8	.574
Stewed	46	1 ea	78	.6	.442
Thigh, meat & skin:					
Batter-fried	86	1 ea	238	.4	.845
Flour-fried	62	1 ea	162	.5	.735
Roasted	62	1 ea	153	.4	.687
Stewed	68	1 ea	158	.3	.528

	Wt (g)	Svg	Cal	Panto 100 Cal	Panto Svg
CHICKEN, continued:					
Thigh, meat only:					
Fried	52	1 ea	113	.6	.668
Roasted	52	1 ea	109	.6	.616
Stewed	55	1 ea	107	.4	.481
Wing, meat & skin:					
Batter-fried	49	1 ea	159	.2	.348
Roasted	34	1 ea	99	.3	.305
Wing, meat only, roasted	21	1 ea	43	.5	.210
Chicken liver, simmered	20	1 ea	30	3.6	1.08
Chicken heart	3.3	1 ea	6	1.5	.087
Chicken gizzard	22	1 ea	34	.5	.158
DUCK, domestic, roasted:					
Meat & skin	85	3 oz	286	.3	.932
Meat only	85	3 oz	171	.7	1.28
TURKEY:					
Ground turkey, cooked	100	3.5 oz	229	.4	.814
Roasted:					
All types	85	3 oz	202	.55	1.12
Dark meat	85	3 oz	223	.68	1.53
Light meat	85	3 oz	186	.43	.806
Turkey patty, breaded, fried	64	1 ea	181	.1	.260
Turkey liver, cooked	75	1 ea	127	3.5	4.47

Meats: Sausages and Lunchmeats

	Wt (g)	Svg	Cal	Panto 100 Cal	Panto Svg
Barbecue loaf, pork and beef	23	1 pce	40	.9	.360
Braunschweiger sausage	18	1 pce	65	.9	.608
Chicken roll, light meat	57	2 pce	90	.6	.550
FRANKFURTERS (hotdogs):					
Beef or Pork, 8/package	57	1 ea	184	.11	.195
Chicken, 10/package	45	1 ea	115	.13	.154
Turkey, 10/package	45	1 ea	102	.15	.154
HAM lunchmeat:					
Extra lean	57	2 oz	75	.4	.264
Regular	57	2 oz	103	.2	.253
Ham & cheese roll (loaf)	57	2 oz	147	.2	.297
Keilbasa	26	1 pce	81	.3	.210
Liverwurst, pork	18	1 pce	59	.9	.532
Luxury loaf	57	2 oz	80	.4	.292
Olive loaf	57	2 oz	133	.3	.437
Pastrami, turkey	57	2 oz	73	.9	.650
Pepperoni, small slice	22	4 pce	109	.4	.411
Pickle & pimento loaf	57	2 oz	149	.3	.450
Salami:					
Beef salami	23	1 pce	58	.4	.229
Beef & pork, dry	20	2 pce	85	.3	.220
Pork and beef	57	2 oz	143	.3	.480
Smoked link sausage:					
Beef/pork	68	1 ea	229	.1	.300
Pork	68	1 ea	265	.2	.530
Turkey ham	57	2 oz	73	.9	.650
Turkey loaf, breast meat	43	2 pce	46	.5	.251
Turkey roll, light or dark	57	2 oz	84	.3	.290

Mixed Dishes & Fast Foods

	Wt (g)	Svg	Cal	Panto 100 Cal	Panto Svg
Beef & vegetable stew:					
Recipe	245	1 c	220	.1	.298
Canned	245	1 c	194	.1	.288
Beef pot pie, fr/frozen	234	1 ea	426	.1	.285
Beef, macaroni, tomato sauce, recipe	226	1 c	189	.3	.531

Values are for edible portion of foods

Pantothenic Acid   mg

	Wt (g)	Svg	Cal	Panto (mg) 100 Cal	Panto (mg) Svg
Mixed Foods & Fast Foods, continued:					
Burrito:					
Bean burrito	174	1 ea	322	.1	.313
Beef burrito	177	1 ea	463	.1	.590
Beef and bean	175	1 ea	390	.2	.627
Deluxe combination	198	1 ea	424	.1	.569
Cheese souffle, recipe	112	1 c	221	.5	1.11
Chicken & noodles, recipe	240	1 c	365	.2	.550
Chicken a la king, recipe	245	1 c	470	.2	.883
Chicken chow mein, recipe	250	1 c	255	.3	.690
Chicken pot pie, from frozen	230	1 ea	430	.3	1.10
Chicken salad w/celery	78	1/2 c	266	.2	.485
Chili with beans, canned	255	1 c	286	1.3	3.62
Corn pudding	250	1 c	271	.2	.615
Corned beef hash, canned	220	1 c	382	.2	.640
Egg salad	183	1 c	438	.6	2.66
ENCHILADA:					
Beef enchilada	120	1 ea	292	.1	.304
Cheese enchilada	120	1 ea	330	.1	.308
Chicken enchilada	120	1 ea	269	.2	.505
French toast, recipe	65	1 pce	123	.4	.434
LASAGNA, recipe:					
With meat	245	1 pce	398	.2	.621
Without meat	218	1 pce	316	.1	.411
Macaroni & cheese:					
Recipe	200	1 c	430	.1	.392
Canned	240	1 c	230	.1	.290
Manicotti, frozen entree	225	1 ea	271	.2	.460
Meat loaf:					
Beef only	87	1 pce	193	.2	.430
Beef & 1/3 pork	87	1 pce	212	.2	.491
Moussaka, lamb & eggplant	250	1 c	250	.4	1.00
PIZZA, cheese:					
Regular crust, 1/8 of 15"	120	1 pce	290	<.1	.158
Thick crust, 1/2 of 10"	208	1 pce	519	.1	.413
Potato salad w/mayo & eggs	250	1 c	358	.4	1.34
Quiche Lorraine, 1/8 Pie	176	1 pce	600	.1	.840
Ravioli, beef, canned	226	1 c	220	.3	.596
SANDWICHES, fast food items:					
Cheeseburger, 3 oz meat	112	1 ea	300	.1	.324
Cheeseburger, 4 oz meat	194	1 ea	524	.1	.640
Chicken patty sandwich	157	1 ea	436	.2	.783
English muffin, egg, cheese and bacon	138	1 ea	360	.2	.730
Fish sandwich:					
Large, without cheese	170	1 ea	470	.1	.470
Regular, with cheese	140	1 ea	420	.1	.476
Hamburger, 3 oz meat	98	1 ea	245	.1	.284
Hamburger, 4 oz meat	174	1 ea	445	.1	.531
Hotdog (frankfurter) w/bun	85	1 ea	260	.1	.372
SANDWICHES (on part whole wheat bread, except when stated as rye):					
Avocado, cheese, tomato, & sprouts	195	1 ea	432	.3	1.11
Bacon, lettuce & tomato	135	1 ea	327	.2	.547
Chicken salad sandwich	100	1 ea	294	.2	.490
Egg salad sandwich	111	1 ea	319	.3	.927
Grilled cheese	117	1 ea	393	.1	.519
Ham & American cheese	151	1 ea	363	.2	.652
Ham & swiss on rye	145	1 ea	350	.2	.621
Ham sandwich	122	1 ea	256	.2	.513
Peanut butter & jam	100	1 ea	341	.1	.460
Roast beef	122	1 ea	280	.2	.586
Tuna salad sandwich	116	1 ea	303	.1	.422
Turkey	122	1 ea	271	.2	.597

	Wt (g)	Svg	Cal	Panto (mg) 100 Cal	Panto (mg) Svg
SPAGHETTI, pasta & tomato sauce w/cheese:					
Homemade	250	1 c	260	.2	.448
Canned	250	1 c	190	.2	.448
Spinach souffle	136	1 c	218	.4	.914
Taco:					
Beef	78	1 ea	207	.1	.231
Chicken	78	1 ea	172	.3	.532
Tostada:					
W/refried beans	157	1 ea	212	.2	.413
W/beans & beef	192	1 ea	332	.2	.606
W/beans & chicken	157	1 ea	249	.3	.795
Tuna noodle casserole, recipe	202	1 c	251	.2	.473
Tuna salad	205	1 c	383	.2	.642
Turkey pot pie, f/frozen	233	1 ea	416	.3	1.10

Nuts & Seeds

	Wt (g)	Svg	Cal	Panto (mg) 100 Cal	Panto (mg) Svg
ALMONDS, dried, whole	142	1 c	837	.1	.669
Almonds, dry roasted	138	1 c	810	<.1	.351
Brazil nuts, dry	140	1 c	919	<.1	.330
Cashews:					
Dry roasted	137	1 c	787	.2	1.67
Oil roasted	130	1 c	748	.2	1.55
Coconut:					
Fresh, grated	80	1 c	283	.1	.240
Sweetened, shredded, pkg	93	1 c	466	.2	.754
Dried, unsweetened	78	1 c	515	.1	.624
Filberts (hazelnuts) whole	135	1 c	853	.2	1.55
Macadamias, dried	134	1 c	940	.1	1.23
MIXED NUTS with peanuts (cashews, peanuts, brazil nuts, filberts, almonds & pecans):					
Dry roasted	137	1 c	814	.2	1.65
Oil roasted	142	1 c	876		1.77
MIXED NUTS, without peanuts (cashews, almonds, brazil nuts, pecans & filberts):					
Oil roasted	144	1 c	886	.2	1.39
PEANUTS, roasted average	144	1 c	837	.2	2
Peanut butter	258	1 c	1517	.2	2.37
Pecans, dried, chopped	119	1 c	794	.3	2.03
Pine nuts, dried	28	1 oz	154	.2	.256
Pistachios (no shells) dried	128	1 c	739	.1	.900
Sesame seeds, dried:					
Kernels (hulled)	150	1 c	882	.1	1.02
Whole	144	1 c	825	.1	.720
Sesame flour:					
High fat	28	1 oz	149	.6	.832
Lowfat	28	1 oz	95	.8	.780
Soybeans, roasted	172	1 c	810	.1	.779
SUNFLOWER seed kernels:					
Dry kernels	144	1 c	821	.2	2.02
Oil roasted	135	1 c	830	.2	1.77
Walnuts, chopped:					
Black	125	1 c	759	.1	.789
English	120	1 c	770	.1	.757

77

ESHA Research

	Wt (g)	Svg	Cal	Panto (mg) 100 Cal	Svg

Soups & Sauces

SAUCES, also see Other:

	Wt (g)	Svg	Cal	100 Cal	Svg
Cheese sauce:					
Homemade	101	1/2 c	216	.2	.327
From mix with milk	279	1 c	305	.3	.766
Spaghetti sauce:					
Homemade	220	1 c	179	.3	.542
Canned	249	1 c	272	.4	.980
White sauce:					
Homemade	250	1 c	395	.2	.807
From mix with milk	264	1 c	240	.3	.700

SOUPS: All soups are canned unless otherwise stated. For soups prep w/milk, assume whole milk. RTS = Ready To Serve.

	Wt (g)	Svg	Cal	100 Cal	Svg
Beef broth (bouillon) w/water	240	1 c	16	3.3	.528
Black bean soup	247	1 c	116	.2	.198
Celery, cream of, w/milk	248	1 c	165	.2	.390
Chicken, cream of, w/milk	248	1 c	191	.2	.400
Chili beef, w/water	250	1 c	169	.2	.350
Clam chowder:					
New England style	248	1 c	163	.2	.400
Manhattan style	244	1 c	78	.16	.122
Lentil and ham, RTS	248	1 c	140	.2	.347
Mushroom, cream of:					
Prep w/milk	248	1 c	205	.3	.683
Prep fr/dry	253	1 c	96	.3	.243
Oyster stew, w/milk	245	1 c	134	.3	.390
Potato, cream of, w/milk	248	1 c	148	.6	.838
Tomato vegetable, from dry	253	1 c	55	.3	.144
Vegetable beef	244	1 c	79	.4	.337
Vegetarian vegetable	241	1 c	70	.5	.337

Vegetables & Legumes

	Wt (g)	Svg	Cal	100 Cal	Svg
Alfalfa seed sprouts	33	1 c	10	1.9	.186
Artichoke, globe, cooked					
(300g with refuse)	120	1 ea	60	.7	.410
Artichoke hearts:					
Cooked from frozen	240	9 oz	108	.4	.480
Marinated	170	6 oz	168	.2	.360
Asparagus, pieces:					
Fresh, uncooked	67	1/2 c	15	22	.330
Ckd from fresh	90	1/2 c	23	1.3	.290
Ckd from frozen	180	1 c	50	.6	.284
BEANS (see also Garbanzo,					
Lentils, Soybeans):					
Baked beans (dry white beans					
with spices and sauce):					
Canned, plain/vegetarian	254	1 c	235	.1	.244
Canned w/franks	257	1 c	356	.1	.357
Canned w/pork	253	1 c	268	.1	.253
Canned w/pork, sweet sce	253	1 c	282	.1	.261
Canned w/pork, tomato sce	253	1 c	247	.5	1.34
Black beans, cooked	172	1 c	227	.2	.416
Broadbeans:					
Ckd f/dry	170	1 c	186	.1	.267
Canned	256	1 c	183	.2	.305
Great northern, ckd f/dry	177	1 c	210	.2	.471
Green beans (snap):					
Fresh, uncooked	110	1 c	34	.6	.200
Cooked from fresh	125	1 c	44	.5	.240
Canned with liquid	240	1 c	36	.7	.254

BEANS, continued:

	Wt (g)	Svg	Cal	100 Cal	Svg
Kidney beans:					
Ckd f/dry	177	1 c	225	.2	.389
Canned	256	1 c	208	.2	.374
Lima beans:					
Ckd from fresh	170	1 c	208	.2	.437
Ckd from frozen	170	1 c	170	.2	.278
Ckd from dry	188	1 c	217	.4	.793
Canned with liquid	241	1 c	191	.3	.624
Navy beans, cooked f/dry	182	1 c	259	.2	.464
Pinto beans:					
Ckd f/dry	171	1 c	235	.2	.487
Canned	240	1 c	186	.2	.326
Refried beans, canned	253	1 c	270	.2	.606
White, small, cooked f/dry	179	1 c	253	.2	.449
Yellow wax: see green beans.					
Bean sprouts (mung beans):					
Fresh sprouts	104	1 c	31	1.3	.396
Ckd from fresh, stir fried	124	1 c	62	1.0	.601
Ckd from fresh, boiled	124	1 c	26	1.2	.301
Canned, drained	125	1 c	16	1.8	.280
Beet greens, ckd fr/fresh	144	1 c	40	1.2	.474
BROCCOLI, chopped:					
Fresh, uncooked	88	1 c	24	2.0	.470
Ckd from fresh	156	1 c	44	1.8	.792
Ckd from frozen	184	1 c	51	1.0	.504
Brussels sprouts, ckd:					
From fresh	156	1 c	60	.9	.560
From frozen	155	1 c	65	.8	.530
CABBAGE:					
Common, shredded	35	1/2 c	8	.6	.049
Common, cooked	75	1/2 c	16	.3	.047
Red cabbage, shredded	70	1 c	19	1.2	.227
Red cabbage, cooked	75	1/2 c	16	1.0	.165
CARROTS:					
Whole, 7-1/2" x 1-1/8"	72	1 ea	31	.46	.142
Raw, shredded	55	1/2 c	24	.45	.108
Cooked from fresh, slices	78	1/2 c	35	.7	.237
Cooked from frozen	73	1/2 c	26	.45	.118
Canned, drained, slices	73	1/2 c	17	.58	.099
Carrot juice	123	1/2 c	49	.6	.280
Celery, diced:					
Fresh, raw (7.5" stalk = 40g)	60	1/2 c	10	1.1	.112
Cooked from fresh	150	1 c	27	1.1	.292
Chard, Swiss:					
Raw, chopped	36	1 c	7	.9	.062
Cooked	175	1 c	35	.8	.285
Collards:					
Ckd fr/fresh	128	1 c	35	.19	.067
Ckd fr/frozen	170	1 c	63	.31	.196
CORN:					
Fresh, uncooked	77	1/2 c	66	.9	.585
Ckd fr/fresh	82	1/2 c	89	.8	.720
Ckd fr/frozen	82	1/2 c	67	.27	.178
Canned, drained	82	1/2 c	66	.3	.206
Canned, w/liquid	128	1/2 c	79	.8	.668
Canned, vacuum pack	210	1 c	166	.9	1.42
Corn, cream style, canned	128	1/2 c	93	.2	.230
Cucumber, whole, 8" x 2"	301	1 ea	39	1.9	.752
Eggplant, cooked cubes	160	1 c	43	.2	.120
Endive, fresh, chopped	25	1/2 c	4	5.6	.225
Escarole (curly endive)	50	1 c	8.5	5.3	.450
Garbanzo beans (chickpeas):					
Ckd from dry	164	1 c	269	.17	.469
Canned	240	1 c	285	.25	.718

Vegetables & Legumes, continued:

	Wt (g)	Svg	Cal	Panto (mg) 100 Cal	Panto (mg) Svg
Jerusalem artichoke, fresh	150	1 c	114	.36	.405
Kale, chopped:					
Raw, fresh	67	1 c	33	.18	.061
Ckd from fresh	130	1 c	41	.16	.064
Ckd from frozen	130	1 c	39	.18	.069
Kohlrabi:					
Fresh slices	140	1 c	38	.6	.231
Ckd fr/fresh	165	1 c	48	.4	.168
Lentils, cooked from dry	198	1 c	231	.5	1.26
Lentils, sprouted:					
Fresh sprouts	77	1 c	81	.5	.445
Stir fried	100	3.5 oz	101	.6	.571
LETTUCE, chopped:					
Butterhead	56	1 c	7	2.5	.176
Iceberg	56	1 c	7	.4	.026
Looseleaf/Romaine	56	1 c	10	1.1	.112
Romaine	56	1 c	9	1.2	.112
Mushrooms:					
Fresh slices	35	1/2 c	9	8.8	.770
Ckd from fresh	78	1/2 c	21	8.0	1.69
Canned, drained	78	1/2 c	19	6.0	1.13
Mustard greens:					
Fresh, chopped	56	1 c	14	.84	.118
Ckd from fresh	70	1/2 c	11	.76	.084
Ckd from frozen	75	1/2 c	14	.09	.012
Okra:					
Ckd fr/fresh, pods	85	8 pods	27	.67	.181
Ckd fr/frozen, slices	92	1/2 c	34	.7	.221
Onions:					
Fresh, chopped	160	1 c	61	.28	.170
Cooked from fresh	105	1/2 c	46	.26	.119
Onion rings, breaded, f/frozen	70	7 rings	285	.06	.161
Parsley, chopped	30	1/2 c	10	.9	.09
Parsnips:					
Fresh slices	133	1 c	100	8.0	7.98
Ckd from fresh, slices	156	1 c	125	.7	.918
PEAS:					
Black-eyed peas:					
Ckd from frozen	170	1 c	224	.2	.362
Canned	240	1 c	184	.2	.456
Ckd from dry	171	1 c	198	.4	.703
Green peas:					
Fresh, uncooked	78	1/2 c	63	.13	.081
Ckd from fresh	160	1 c	134	.2	.245
Ckd from frozen	80	1/2 c	63	.18	.114
Green peas, edible-pods:					
Fresh, uncooked	145	1 c	61	1.8	1.09
Ckd from fresh	160	1 c	67	1.6	1.08
Ckd from frozen	80	1/2 c	42	1.3	.540
Split peas, cooked f/dry	196	1 c	231	.5	1.17
Peas, mature, sprouted:					
Fresh sprouts	120	1 c	154	.8	1.24
Cooked fr/fresh	100	3.5 oz	118	.6	.683
Peas & carrots:					
Ckd from frozen	80	1/2 c	38	.34	.130
Canned with liquid	128	1/2 c	48	.32	.154
PEPPERS, HOT:					
Red or green chili peppers	68	1/2 c	17	2.7	.462
Jalapeno, chopped, canned	68	1/2 c	17	4.3	.734
PEPPERS, SWEET:					
Raw, chopped	50	1/2 c	14	.29	.040
Cooked, chopped	68	1/2 c	19	.39	.054

	Wt (g)	Svg	Cal	Panto (mg) 100 Cal	Panto (mg) Svg
POTATOES:					
Baked:					
Flesh and skin	202	1 ea	220	.5	1.12
Flesh only	156	1 ea	145	.6	.866
Potato skin	58	1 ea	115	.2	.256
Boiled. flesh only:					
Cooked in skin	136	1 ea	119	.6	.707
Cooked without skin	135	1 ea	116	.6	.687
Canned, 1" diam.	70	2 ea	42	.6	.248
French fries, fr/frozen:					
Fried in oil	50	10 strips	158	.2	.328
Oven heated	50	10 strips	111	.3	.329
Hash brown, fr/frozen	156	1 c	340	.2	.696
Mashed potatoes prepared:					
With milk	210	1 c	162	.6	1.00
Instant	220	1 c	239	.4	.846
Potato puff (tater tots)	62	1/2 c	138	.3	.408
Potatoes au gratin:					
Recipe	245	1 c	322	.3	.948
From dry mix	245	1 c	228	.3	.586
Scalloped potatoes:					
Recipe	245	1 c	210	.6	1.26
From dry mix	245	1 c	228	.4	.801
Potato pancakes	76	1 ea	237	.3	.713
Pumpkin:					
Cooked from fresh	245	1 c	50	1.5	.726
Canned	123	1/2 c	42	1.2	.490
Radish seeds, sprouted	38	1 c	16	1.8	.279
Radishes, raw, red	45	10 ea	7	.6	.040
Rutabaga:					
Fresh cubes	140	1 c	51	.4	.224
Cooked cubes	85	1/2 c	29	.4	.116
Sauerkraut, canned, with liquid	236	1 c	44	.5	.219
Soybeans, ckd fr/dry	172	1 c	298	.1	.308
Soybeans, mature, sprouted:					
Fresh	35	1/2 c	45	.7	.325
Steamed	94	1 c	76	.9	.698
Stir fried	100	1 c	125	.9	1.20

SOYBEAN PRODUCTS: see tofu this section; miso & tempeh in Other; roasted soybeans in Nuts & Seeds; soy milk in Dairy; soy flour in Grains.

	Wt (g)	Svg	Cal	Panto (mg) 100 Cal	Panto (mg) Svg
Spinach:					
Fresh, chopped	56	1 c	12	.3	.036
Cooked from fresh	180	1 c	41	.6	.261
Cooked from frozen, leaf	95	1/2 c	27	.3	.079
Canned, drained	107	1/2 c	25	.2	.050
Spirulina, (seaweed) dried	28	1 oz	82	1.2	.987
SQUASH, SUMMER varieties:					
Crookneck, fresh slices	130	1 c	26	.5	.133
Crookneck, cooked	180	1 c	36	.7	.247
Zucchini, fresh slices	130	1 c	19	.6	.108
Zucchini, cooked	90	1/2 c	14	.7	.103
SQUASH, WINTER, mashed:					
Acorn /Danish, baked	245	1 c	137	.9	1.23
Acorn /Danish, boiled	245	1 c	83	.9	.742
Butternut, baked	245	1 c	99	.9	.880
Butternut, ckd f/frozen	240	1 c	94	.4	.370
Hubbard, baked	240	1 c	120	.9	1.07
Hubbard, boiled	236	1 c	70	1.0	.701
Spaghetti, baked or boiled	155	1 c	45	1.2	.550
Succotash:					
Ckd f/frozen	85	1/2 c	79	.1	.081
Canned w/liquid	128	1/2 c	81	.5	.396

ESHA Research

	Wt (g)	Svg	Cal	Panto (mg) 100 Cal	Svg
Vegetables & Legumes, continued:					
Sweet potatoes:					
Baked w/skin, flesh only	114	1 ea	118	.6	.736
Boiled w/o skin	151	1 ea	160	.5	.803
Canned, mashed	128	1/2 c	129	.5	.665
Vacuum packed, mashed	255	1 c	233	.6	1.33
Candied, recipe, 2.5" x 2" pce	105	1 pce	144	.3	.450
Tofu:					
Firm	126	1/2 c	183	.1	.168
Regular	124	1/2 c	94	.1	.084
TOMATOES:					
Fresh, whole, 2.6" diam.	123	1 ea	26	1.2	.304
Fresh, chopped	180	1 c	38	1.2	.445
Ckd from fresh	240	1 c	65	1.1	.708
Canned, whole	240	1 c	47	.9	.401
Tomato juice, canned	244	1 c	42	1.5	.610
Tomato paste, canned	262	1 c	220	.9	1.97
Tomato puree, canned	250	1 c	102	1.1	1.10
Tomato sauce, canned	245	1 c	74	1.0	.757
Turnips:					
Fresh cubes	130	1 c	35	.7	.260
Cooked from fresh	78	1/2 c	14	.8	.111
Turnip greens:					
Cooked from fresh	144	1 c	29	1.4	.395
Cooked from frozen	82	1/2 c	30	.23	.069
Vegetable juice cocktail	242	1 c	46	1.2	.536
VEGETABLES, MIXED combinations, cooked fr/frozen:					
Broccoli with water chestnuts	95	1/2 c	33	.7	.220
Mixed vegetables (corn, peas, carrots, green beans, limas):					
Cooked from frozen	182	1 c	107	.3	.275
Canned, drained	163	1 c	77	.3	.233
Peas, carrots & onions	91	1/2 c	54	.4	.220
Peas, cauliflower w/cream sce	95	1/2 c	118	.2	.250
Peas, onions in cheese sce	142	1/2 c	165	.3	.430
Peas, pasta & corn w/cream sce	95	1/2 c	132	.2	.300
Peas, pasta, mushroom & cream sauce	95	1/2 c	129	.3	.430
Water cress, chopped	17	1/2 c	2	2.7	.053
Yams, orange: see Sweet potatoes.					
Yams, white, cooked cubes	136	1 c	158	.3	.423
Zucchini: see Squash, summer.					

Other

Beverages, Cooking ingredients, Condiments, Flavorings, Spices, Sweeteners, & Miscellaneous

	Wt (g)	Svg	Cal	Panto (mg) 100 Cal	Svg
BEVERAGES: also see Milk beverages, Fruit juices & Vegetable juices.					
Beer, 12 fl oz = 1.5 c	356	1.5 c	146	.1	.206
Light beer, 12 fl oz = 1.5 c	354	1.5 c	100	.1	.127
Cranberry juice cocktail	190	3/4 c	108	.1	.106
Pineapple & orange drink	250	1 c	125	.1	.143
CANDY & CANDY BARS:					
Chocolate coated almonds	165	1 c	935	.1	.674
Chocolate coated peanuts	170	1 c	954	.3	2.49
M&M's plain candies	48	1 pkg	237	.1	.289
M&M's peanut candies	47	1 pkg	240	.1	.284
Milk chocolate w/peanuts	28	1 oz	155	.3	.500
Honey	339	1 c	1030	.1	.678
Hummous, humous	246	1 c	420	.2	.708
Miso (soybean product)	138	1/2 c	284	.1	.356
Molasses, blackstrap	40	2 T	85	.2	.200

	Wt (g)	Svg	Cal	Panto (mg) 100 Cal	Svg
Salsa:					
Picante, Tostitos	85	1 oz	40	.8	.300
Recipe	108	1/2 c	46	.6	.286
Tempeh (soybean product)	83	1/2 c	165	.2	.295
Yeast:					
Brewer's	8	1 T	25	2.7	.676
Dry active	30	4 T	80	4.1	3.30

Vitamin C   (Ascorbic Acid)

Vitamin C is a very reactive vitamin. It is one of the most versatile in the body, and the most unstable in food. It is convertible in different forms, and as a water soluble compound is excreted rapidly when excessive amounts are taken.

It is important in maintaining and forming the protein collagen, an essential part of the connective tissue which binds the body's cells together. Bones and teeth continually need vitamin C to repair their connective tissues; cuts and burns cannot heal without well-formed collagen; and it keeps capillaries and other blood vessels strong.

Vitamin C is also involved in the metabolism of several amino acids (proteins) and the precursor hormones epinephrine and norepinephrine. It is needed for the synthesis of thyroxin, which regulates metabolism. Extreme stress can increase the need for vitamin C as well.

As an anti-oxidant, it is a bodyguard to other substances and becomes oxidized to protect other items from a similar fate. Because of this property it is sometimes added to food products to protect them from oxidation (turning brown) as well as to improve their nutritional value. It is now known that vitamin C eaten with iron can double or triple the absorption of iron from foods.

A severe deficiency can result in scurvy, which is how the vitamin was discovered and named. Ascorbic acid was derived from the anti-scorbutic (anti-scurvy) factor.

Vitamin Losses

Vitamin C is more readily destroyed than the other vitamins. It is stable in growing plants, but when they are cut or bruised, an enzyme is activated that destroys the vitamin. Blanching the vegetables inactivates the enzyme. It is also easily destroyed by exposure to the air, heat, iron and copper pans, and will leach easily into cooking water. It is most stable in acid fruits (they don't have the destroying enzyme). Citrus fruit juices stored in the refrigerator in a sealed container and small airspace will retain most of their vitamin C.

Recommended Dietary Allowances for Adults

USA RDA: Adults, 60 mg*   (Pregnant, 70 mg; Lactating/nursing, 95 mg)
Canadian RNI: Men, 60 mg; Women, 45mg (last two trimesters of pregnancy, 65mg; Lactating/nursing, 75 mg)

*The 1989 RDA subcommittee recommends that regular cigarette smokers ingest at least 100 mg of vitamin C daily, to compensate for the affect the cigarettes have on lowering blood serum levels of this vitamin.

We probably don't know all the things vitamin C does in or for the body, thus the recommended amounts vary all over the scientific map. But the official recommendations relate to *basic* body functions and needs, and not to pharmaceutical use of the Vitamin.

Food Sources

The primary sources are fresh vegetables and fruits, especially citrus fruits.

	Wt (g)	Svg	Cal	Vit C (mg) 100 Cal	Vit C (mg) Svg

Baked Goods

Grains are not a source of vitamin C. It is present in baked goods only because of added ingredients.

	Wt (g)	Svg	Cal	100 Cal	Svg
Apple crisp, 3" x 3"	78	1 pce	146	2.5	3.7
Cherry crisp, 3" x 3"	138	1 pce	157	2.0	3.2
Muffin, bran, recipe	45	1 ea	125	2.4	3.0
Peach crisp, 3" x 3"	139	1 pce	166	2.9	4.8
PIE, piece = 1/6 of 9" pie.					
Blueberry	158	1 pce	380	1.6	6.0
Lemon meringue	140	1 pce	355	1.2	4.2
Peach pie	158	1 pce	405	1.2	4.7
Poptart-type pastry, fortified	54	1 ea	210	1.9	4.0

Beverages

(Also see Fruit juices, Vegetable juices & Milk beverages.)
Vitamin C is fortified in some drinks, check label.

	Wt (g)	Svg	Cal	100 Cal	Svg
Alcoholic:					
Bloody mary, 5 fl oz	148	1 ea	116	18	20
Screwdriver, 7 fl oz	213	1 ea	174	38	66
Tequila sunrise, 5.5 fl oz	172	1 ea	189	18	33
Non-alcoholic:					
Clam & tomato juice	166	2/3 c	77	9	7
Fruit punch drink:					
From frozen	247	1 c	113	96	108
Canned	253	1 c	118	64	75
Grape drink	250	1 c	112	76	85
Kool-aid, sugar & C added	240	1 c	100	6	6
Lemonade, fr/frozen	248	1 c	100	10	10
Limeade, fr/frozen	247	1 c	102	7	7
Pineapple-grapefruit drink	250	1 c	117	98	115
Pineapple-orange drink	250	1 c	125	45	56

Dairy & Dairy Products

	Wt (g)	Svg	Cal	100 Cal	Svg
CHEESE: There is no Vitamin C in cheese.					
Kefir, beverage	233	1 c	160	4	6.0
MILK (cow):					
Skim	245	1 c	86	2.8	2.4
Lowfat 1%	244	1 c	102	2.3	2.4
Lowfat 2%	244	1 c	121	1.9	2.3
Whole (3.3% fat)	244	1 c	150	1.5	2.3
Buttermilk (< 1% fat)	245	1 c	99	2.4	2.4
Canned:					
Skim, evaporated	255	1 c	200	1.5	3.0
Whole, evaporated	252	1 c	340	1.4	4.7
Sweet, condensed	306	1 c	982	.8	8.0
Dry, instant nonfat	68	1 c	244	1.6	3.8
Dry buttermilk	120	1 c	464	1.5	6.8
MILK (other):					
Goat milk	244	1 c	168	2	3.2
Human breast milk, mature	246	1 c	171	7	12.3
Soy milk	240	1 c	79	0	0
MILK BEVERAGES and mixes:					
Instant Breakfast, dry mix	37	1 env	130	20.8	27
Shakes (1.25 cup = 10 fl oz):					
Chocolate	283	1.25 c	360	.4	1.3
Strawberry	283	1.25 c	319	.7	2.1
Vanilla	283	1.25 c	314	.7	2.2
Sherbet (2% fat)	193	1 c	270	1.4	3.9

Fruits & Fruit Juices

	Wt (g)	Svg	Cal	100 Cal	Svg
Acerola juice, fresh	242	1 c	51	7592	3872
APPLE, 2.75" diam, with or without peel	138	1 ea	80	10	7.8
Apple juice, canned, bottled	248	1 c	116	2	2.3
Applesauce, unsweetened	244	1 c	106	3	2.9
APRICOTS:					
Fresh, pitted	106	3 ea	51	21	11
Canned halves:					
Juice pack	84	3 ea	40	10	4
Heavy syrup	85	3 ea	70	27	2.64
Dried, halves	130	1 c	310	1	3
Apricot nectar, canned:					
Regular	251	1 c	141	1.4	2
Vitamin C added	251	1 c	141	97	136
Avocado, whole:					
California	173	1 ea	305	4.5	14
Florida	304	1 ea	340	7	24
Banana, 8.75",175g w/peel	114	1 ea	105	10	10
Blackberries:					
Fresh berries	144	1 c	74	41	30.2
Frozen, unthawed	151	1 c	97	5	5
Canned	256	1 c	236	5	12
Blackberry juice, fresh	250	1 c	93	27	25
Black currant juice	250	1 c	138	294	405
Blueberries:					
Fresh berries	145	1 c	82	24	20
Frozen, unthawed	155	1 c	78	5	4
Canned	256	1 c	225	1	3
Boysenberries:					
Frozen, unthawed	132	1 c	66	6	4
Canned	256	1 c	225	7	16
Cherries, sour:					
Frozen	155	1 c	72	3.6	2.6
Canned	244	1 c	90	5.7	5.1
CHERRIES, SWEET:					
Fresh, pitted	68	10 ea	49	10	5
Frozen	259	1 c	232	6	13
Canned with liquid	257	1 c	213	4	9
Cranberries, whole	95	1 c	46	28	13
Cranberry apple juice	253	1 c	169	48	81
Cranberry juice cocktail	253	1 c	145	62	90
Cranberry orange relish	275	1 c	490	10	50
Cranberry sauce, canned	277	1 c	419	1.3	5.5
Currants:					
Fresh, black	112	1 c	71	286	203
Fresh, red or white	112	1 c	63	73	46
Dried (Zante)	144	1 c	407	1.6	7
Fruit cocktail, canned:					
Juice pack	248	1 c	115	6	7
Heavy syrup	255	1 c	185	3	5
Gooseberries:					
Fresh berries	150	1 c	67	62	42
Canned with liquid	252	1 c	185	14	25
GRAPEFRUIT:					
(half= 241g with refuse)					
Pink or red half	123	1 ea	37	127	47
White half	118	1 ea	39	101	39
Canned, sections	254	1 c	152	36	54
Grapefruit juice:					
Prepared from frozen	247	1 c	102	82	83
Canned, unsweetened	247	1 c	93	77	72
Grapes, fresh:					
Thompson, seedless	50	10 ea	35	15	5
Tokay or Emperor	57	10 ea	40	16	6

Values are for edible portion of foods

Fruits & Fruit Juices, continued:	Wt (g)	Svg	Cal	Vit C (mg) 100 Cal	Svg
Grape juice:					
Bottled, canned	253	1 c	155	.1	0.2
From frozen, Vitamin C added	250	1 c	128	47	60
Kiwi fruit	76	1 ea	46	162	74.5
Lemon peel	6	1 T	3	257	7.7
Lemon juice:					
Fresh	15.2	1 T	3.8	184	7
Bottled	15	1 T	5	76	4
Lime juice:					
Fresh	15	1 T	4	113	4.5
Bottled	246	1 c	50	32	16
Loganberries:					
Fresh	150	1 c	105	50	52
Frozen, unthawed	147	1 c	80	28	23
Mandarin oranges, canned	252	1 c	155	32	50
Mango, fresh slices	165	1 c	108	42	46
MELON, also see Watermelon:					
Cantaloupe, cubes	160	1 c	57	118	68
Casaba, cubes	170	1 c	45	60	27
Honeydew, cubes	170	1 c	60	70	42
Frozen, melon balls, mixed	173	1 c	55	20	11
Nectarine (1 med = 1 c slices)	136	1 ea	67	11	7
ORANGE:					
180g whole, 2-5/8" diam.	131	1 ea	60	116	70
Orange juice:					
Fresh juice	248	1 c	111	112	124
Prepared from frozen	249	1 c	110	88	97
Canned, unsweetened	249	1 c	105	82	86
Orange grapefruit juice, cnd	247	1 c	105	69	72
Orange peel, grated	6	1 T	5	164	8
Papaya, 454g w/refuse	304	1 ea	117	161	188
Papaya nectar, canned	250	1 c	142	5	8
PEACHES:					
Fresh, whole, 2.5" diam.	87	1 ea	37	15	6
Frozen slices, thawed (Vitamin C added)	250	1 c	235	100	236
Canned half, juice pack	77	1 ea	34	8	3
Canned half, heavy syrup	81	1 ea	60	4	2
Peach nectar, canned	249	1 c	134	10	13
PEARS:					
Fresh, Bartlett	166	1 ea	98	7	7
Canned, juice pack	248	1 c	123	3	4
Canned, heavy syrup	255	1 c	188	1.5	3
PINEAPPLE:					
Fresh chunks	155	1 c	76	31	24
Canned:					
Slice, heavy syrup	58	1 ea	45	10	4
Pieces, juice pack	250	1 c	150	16	24
Pieces, heavy syrup	255	1 c	199	10	120
Pineapple juice:					
Prepared f/frozen	250	1 c	129	23	30
Canned, unsweetened	250	1 c	140	19	27
Plantain slices:					
Fresh	148	1 c	181	15	27
Cooked	154	1 c	179	9	17
Plums:					
Medium, 2-1/8" dm.	66	1 ea	36	18	6
Canned, whole, jce pack	95	3 ea	55	5	3
Prunes, dried, pitted	84	10 ea	201	1.4	3
Prune juice, bottled	256	1 c	181	6	11
Raisins, dark, unpacked	145	1 c	435	1	5

	Wt (g)	Svg	Cal	Vit C (mg) 100 Cal	Svg
Raspberries:					
Fresh berries	123	1 c	60	51	31
Frozen, thawed	250	1 c	255	16	41
Canned with liquid	256	1 c	234	10	22
Raspberry juice, fresh	240	1 c	98	37	36
Rhubarb:					
Fresh dices	122	1 c	26	38	10
Cooked with sugar	240	1 c	279	3	8
STRAWBERRIES:					
Fresh berries	149	1 c	45	188	85
Frozen, unthawed	149	1 c	52	118	61
Frzn, thawed, sweetened	255	1 c	245	43	106
Tangerine	84	1 ea	37	70	26
Tangerine juice:					
From frozen	241	1 c	110	53	58
Canned, sweetened	249	1 c	125	44	55
Watermelon cubes	160	1 c	50	31	15

Grains & Grain Products
Dry grains are not a source of vitamin C.

CEREALS, COLD (Ready to eat)
Vitamin C may be added as fortification, check label.

	Wt (g)	Svg	Cal	100 Cal	Svg
Wheat berries, sprouted, fresh	106	1 c	236	1	3

Meats: Fish & Shellfish

CLAMS:	Wt (g)	Svg	Cal	100 Cal	Svg
Canned, drained	160	1 c	236	2	6
Minced w/liquid, sm. can	183	1 ea	145	2	3
Breaded, fried, small	188	20 ea	379	1	4
Steamed, meat only	90	20 ea	133	3	3.5
CRAB, meat only:					
Blue crab, cooked	135	1 c	138	1	2
Dungeness, cooked	101	3/4 c	85	2	2
Crayfish, ckd, moist heat	85	3 oz	97	3	3
Mackerel					
Baked/broiled, Atlantic	100	3.5 oz	262	1.5	4
Canned, Jack,1 tall can	361	1 can	563	.6	3
OYSTERS:					
Raw, Eastern	248	1 c	170	14	24
Raw, Pacific	248	1 c	200	12	24
Simmered, Eastern	100	3.5 oz	137	5	7
Bread fried, medium	88	6 ea	173	4	7
Pike, Northern, baked/broiled	100	3.5 oz	113	3	4
Roe, raw, mixed species	28	1 oz	39	10	4
Salmon, cnd, Atlantic	220	1 can	281	1	4
Squid, flour-fried	85	3 oz	149	2	3.5
Trout, baked/broiled	85	3 oz	129	2	3

Meats
Beef, Lamb, Pork, and Veal

	Wt (g)	Svg	Cal	100 Cal	Svg
BEEF liver, fried	85	3 oz	184	11	19
LAMB liver, pan-fried	85	3 oz	202	5	11
PORK, CURED:					
Bacon, cooked:					
Regular	19	3 pce	109	6	6.4
Canadian style	47	2 pce	86	12	10
Breakfast strips	34	3 pce	156	10	15

83

	Wt (g)	Svg	Cal	Vit C (mg) 100 Cal	Svg
PORK, CURED, HAM:					
Canned, roasted, average	85	3 oz	142	14	20
VEAL liver, pan fried	85	3 oz	208	9	18

Meats: Sausages & Lunchmeats

	Wt (g)	Svg	Cal	Vit C (mg) 100 Cal	Svg
Beef lunchmeat:					
Loaf or roll	28	1 oz	87	5	4
Thin sliced	28	1 oz	50	8	4
Beerwurst (beer salami):					
Beef salami	23	1 pce	75	4	3
Pork salami	23	1 pce	55	12	7
BOLOGNA:					
Beef bologna	23	1 pce	72	6	4
Beef and pork	28	1 oz	89	7	6
Cured pork	23	1 pce	57	14	8
Braunsweiger	57	2 oz	205	3	5.4
Brotwurst link	70	1 ea	226	9	20
Cheesefurter (cheese smoki)	43	1 ea	141	6	8
Dutch brand loaf	28	1 oz	68	7	5
FRANKFURTER (hotdog):					
Beef, 8/pkg	57	1 ea	184	8	14
Beef & pork, 8/pkg	57	1 ea	183	8	15
HAM:					
Chopped, canned	21	1 pce	50	16	8
Lunchmeat:					
Thin sliced	28	1 oz	37	19	7
Extra lean	57	2 oz	75	20	15
Regular	57	2 oz	103	15	16
Minced	21	1 pce	55	11	6
Ham & cheese roll/loaf	57	2 oz	147	10	14
Ham salad spread	240	1 c	518	3	14
Knockwurst link	68	1 ea	209	9	18
Luxury loaf	57	2 oz	80	15	12
Peppered loaf	28	1 oz	42	17	7
Pickle & pimento loaf	57	2 oz	149	5	8
Salami:					
Beef and pork, dry	20	2 pce	85	7	6
Pork and beef	57	2 oz	143	5	7
Smoked link sausage, beef and pork	68	1 ea	229	6	13
Summer sausage	23	1 pce	80	6	5

Mixed Dishes & Fast Foods

	Wt (g)	Svg	Cal	Vit C (mg) 100 Cal	Svg
Beef & vegetable stew:					
Recipe	245	1 c	220	8	17
Canned	245	1 c	194	4	7
Beef, macaroni & tomato sauce, recipe	226	1 c	189	8	16
BURRITO:					
Bean burrito	174	1 ea	322	2	5
Beef burrito	177	1 ea	463	.6	3
Beef and bean	175	1 ea	390	1	5
Deluxe combination	198	1 ea	424	1.6	7
Chicken a la king, recipe	245	1 c	470	3	12
Chicken chow mein, cnd	250	1 c	95	14	13
Chicken pot pie, 1/3 recipe	232	1 pce	545	1	5
Chili w/beans, canned	255	1 c	286	1.5	4
Chop suey w/beef & pork	250	1 c	300	11	33
Cole slaw	120	1 c	84	47	39
Corn pudding	250	1 c	271	3	7
Corned beef hash, canned	220	1 c	382	2	8

	Wt (g)	Svg	Cal	Vit C (mg) 100 Cal	Svg
Enchilada:					
Beef enchilida	120	1 ea	292	3	9
Cheese	120	1 ea	330	2	8
Chicken	120	1 ea	269	3	9
Enchirito	207	1 ea	441	3	15
Fried rice	180	1 c	286	1	3
LASAGNA:					
Recipe with meat	245	1 pce	398	2	7
Recipe without meat	218	1 pce	316	2	7
Frozen entree	205	1 pce	275	2	6
Macaroni & cheese, frozen	200	1 c	254	1	2
Macaroni salad, no cheese	141	1 c	371	1	3
Manicotti, frzn entree	225	1 ea	271	1.5	4
Moussaka, lamb & eggplant	250	1 c	250	3	7
PIZZA, cheese:					
Regular crust 1/8 of 15"	120	1 pce	290	.7	2
Thick crust 1/2 of 10"	208	1 pce	519	.5	3
Potato salad, w/ mayo & eggs	250	1 c	358	7	25
Quiche Lorraine, 1/4 pie	242	1 pce	825	.02	.137
Ravioli, beef, canned	226	1 c	220	6	13
SANDWICHES, fast food:					
Cheeseburger, 3 oz meat	112	1 ea	300	.3	1
Cheeseburger, 4 oz meat	194	1 ea	524	.6	3
Chicken patty	157	1 ea	436	.8	4
Fish sandwich with cheese	140	1 ea	420	.7	3
Hamburger, 3 oz meat	98	1 ea	245	.4	1
Hamburger, 4 oz meat	174	1 ea	445	.4	2
Hotdog (frankfurter w/bun)	85	1 ea	260	5	12
Roast beef & bun	150	1 ea	345	.6	2
SANDWICHES on part whole wheat bread:					
Avocado, cheese, sprouts & tomato	195	1 ea	432	2	11
Bacon, lettuce & tomato	135	1 ea	327	4	13
Ham sandwich	122	1 ea	256	6	14
SPAGHETTI, pasta & tomato sauce w/cheese:					
Homemade	250	1 c	260	5	13
Canned	250	1 c	190	5	10
TOSTADAS:					
Beans and beef	192	1 ea	332	1.8	6
Beans and chicken	157	1 ea	249	1.4	3
Refried beans	157	1 ea	212	2.8	6

Nuts & Seeds

	Wt (g)	Svg	Cal	Vit C (mg) 100 Cal	Svg
Chestnuts, roasted	143	1 c	350	11	37
Coconut, raw, grated	80	1 c	283	.9	3
Coconut water, raw	240	1 c	46	1	6
Pecans, dried, chopped	119	1 c	794	.3	2
Walnuts, English, chopped	120	1 c	770	.5	4

Soups, Sauces & Gravies

	Wt (g)	Svg	Cal	Vit C (mg) 100 Cal	Svg
Gravy, chicken f/mix	260	1 c	85	3.5	3
SAUCES:					
Au Jus, canned	238	1 c	38	6	2
Spaghetti sauce:					
Homemade	220	1 c	179	11	20
Canned	249	1 c	272	10	28

Left column

	Wt (g)	Svg	Cal	Vit C (mg) 100 Cal	Vit C (mg) Svg
SOUPS: All soups are canned unless otherwise stated. RTS = Ready To Serve. For soups prepared with milk, assume whole milk.					
Beef, chunky, RTS	240	1 c	171	4	7
Chicken, chunky, RTS	240	1 c	167	3	6
Chicken gumbo, regular	244	1 c	56	9	5
Chicken noodle soup	241	1 c	75	—	trace
Lentil and ham, RTS	248	1 c	140	3	4
Minestrone, chunky, RTS	240	1 c	127	4	5
Mushroom, cream of:					
Condensed	251	1 c	257	.9	2
From dry	260	1 c	105	4	4
Prepared with milk	248	1 c	205	1	2
Onion soup	248	1 c	100	2	2
Split pea with ham, RTS	240	1 c	184	4	7
Tomato, cream of:					
Prepared with milk	248	1 c	160	43	68
Prepared with water	244	1 c	86	77	67
Prepared from dry	265	1 c	102	5	5
Tomato rice	247	1 c	120	12	15
Tomato vegetable, f/pkt	189	3/4 c	41	12	5
Turkey, chunky, RTS	236	1 c	136	11	14
Vegetable beef	244	1 c	79	3	2
Vegetable, Chunky, RTS	240	1 c	122	5	6

Vegetables & Legumes

	Wt (g)	Svg	Cal	Vit C (mg) 100 Cal	Vit C (mg) Svg
Alfalfa sprouts	33	1 c	10	27	3
Amaranth:					
Fresh, chopped	28	1 c	7	166	12
Boiled	132	1 c	28	194	54
Artichoke, globe, cooked	120	1 ea	60	20	12
Artichoke, hearts:					
Cooked from frozen	240	9 oz	108	11	12
Marinated	170	6 oz	168	31	52
Asparagus pieces:					
Fresh pieces	67	1/2 c	15	147	22
Cooked from fresh	90	1/2 c	23	109	25
Cooked from frozen	180	1 c	50	87	44
Canned, drained	121	1/2 c	16	113	18
BEANS (see also Garbanzo, Lentils, Soybeans):					
Baked beans (dry White beans w/spices & sauce):					
Home prepared	253	1 c	382	.7	3
Canned w/franks	257	1 c	366	2	6
Canned w/pork	253	1 c	268	2	5
Canned, pork & sweet sce	253	1 c	282	3	8
Canned, pork & tomato sce	253	1 c	247	3	8
Green (snap) beans:					
Fresh, uncooked	110	1 c	34	53	18
Ckd fr/resh	125	1 c	44	28	12
Ckd fr/frozen	135	1 c	36	31	11
Canned, drained	135	1 c	26	25	6
Canned w/liquid	240	1 c	36	27	10
Kidney beans, all types:					
Cooked fr/dry	177	1 c	225	.9	2
Canned w/liquid	256	1 c	208	1.4	3
Lima beans:					
Ckd from fresh	170	1 c	208	8	17
Ckd fr/frozen, large	85	1/2 c	85	13	11
Ckd fr/frozen, baby	90	1/2 c	94	6	5
Canned, drained	170	1 c	164	6	10
Canned w/liquid	241	1 c	191	11	21
Pinto beans:					
Canned	240	1 c	186	1	1.7
Ckd fr/dry	171	1 c	235	1.5	3.5
Refried beans, canned	253	1 c	270	6	15

Right column

	Wt (g)	Svg	Cal	Vit C (mg) 100 Cal	Vit C (mg) Svg
BEANS, continued:					
Winged, ckd fr/fresh	62	1 c	23	27	6
Yellow wax: see green beans.					
Bean sprouts (Mung beans):					
Fresh sprouts	104	1 c	31	44	14
Ckd, stir fried	124	1 c	62	32	20
Ckd, boiled, drained	124	1 c	26	54	14
Beets:					
Cooked from fresh	100	2 ea	31	18	6
Canned, diced, drained	85	1/2 c	27	13	4
Canned w/liquid	123	1/2 c	36	13	5
Pickled, slices	114	1/2 c	74	4	3
Beet greens, cooked	144	1 c	40	90	36
BROCCOLI, chopped:					
Fresh, uncooked	88	1 c	24	342	82
Ckd from fresh	156	1 c	44	264	116
Ckd from frozen:	184	1 c	51	145	74
W/cheese sauce	142	1/2 c	166	21	35
W/hollandaise sauce	95	1/2 c	105	39	41
Brussels sprouts:					
Ckd from fresh	156	1 c	60	161	97
Ckd from frozen	155	1 c	65	109	71
CABBAGE, fresh:					
Common, shredded	70	1 c	16	206	33
Common, cooked	150	1 c	32	114	36
Bok choy, shredded	70	1 c	9	350	32
Bok choy, cooked	170	1 c	20	221	44
Pe-tsai, shredded	76	1 c	11	186	21
Pe-tsai, cooked	119	1 c	16	118	19
Red, shredded	70	1 c	19	210	40
Red, cooked	75	1/2 c	16	161	26
Savoy, shredded	70	1 c	20	109	22
Savoy, cooked	145	1 c	35	71	25
CARROTS:					
Fresh (7.5" x 1-1/8")	72	1 ea	31	22	7
Fresh, grated	55	1/2 c	24	21	5
Ckd fr/resh, slices	78	1/2 c	35	6	2
Ckd fr/frozen, slices	73	1/2 c	26	8	2
Canned, drained	73	1/2 c	17	12	2
Canned with liquid	123	1/2 c	28	12	3
Carrot juice	123	1/2 c	49	21	11
Cauliflower:					
Fresh pieces	50	1/2 c	12	298	36
Ckd fr/fresh	62	1/2 c	15	229	34
Ckd fr/frozen	180	1 c	34	166	56
Celery, fresh:					
Outer stalk (7.5" = 40g)	40	1 ea	6	47	2.8
Chopped	60	1/2 c	10	42	4.2
Chard, Swiss:					
Fresh, chopped	36	1 c	7	158	11
Ckd from fresh	175	1 c	35	90	32
Collards:					
Fresh, chopped	36	1 c	11	73	8
Ckd from fresh	128	1 c	35	44	15.5
Ckd from frozen	170	1 c	63	71	45
CORN:					
Fresh kernels	77	1/2 c	66	8	5
Ckd from fresh	82	1/2 c	89	6	5
Ckd from frozen	82	1/2 c	67	3	2
Canned, drained	82	1/2 c	66	5	3
Canned with liquid	128	1/2 c	79	11	8.5
Canned, vacuum pack	210	1 c	166	10	17
Corn, cream style, canned	128	1/2 c	93	6	6
Dandelion greens:					
Fresh, chopped	55	1 c	25	77	19
Cooked	105	1 c	35	54	19

	Wt (g)	Svg	Cal	Vit C (mg) 100 Cal	Svg
Vegetables & Legumes, continued:					
Dock, (sorrel) greens:					
Fresh chopped	133	1 c	29	220	64
Cooked	100	1 c	20	132	26
Eggplant, cooked	160	1 c	45	5	2
Garbanzo beans, ckd fr/dry	164	1 c	269	1	2
Garden cress:					
Fresh, chopped	25	1/2 c	8	216	17
Cooked from fresh	135	1 c	31	100	31
Jerusalem artichoke, fresh	150	1 c	114	5	6
Jicama, fresh slices	100	3.5 oz	20	69	14
Kale, fresh, chopped:	67	1 c	33	244	80
Ckd from fresh	130	1 c	42	128	53
Ckd from frozen	130	1 c	39	84	33
Kohlrabi:					
Fresh slices	140	1 c	38	228	87
Ckd f/fresh	165	1 c	48	186	89
Lambquarters:					
Fresh, chopped	56	1 c	24	187	45
Cooked, chopped	90	1/2 c	29	115	67
Leeks, chopped:					
Fresh	104	1 c	63	20	13
Cooked from fresh	52	1/2 c	24	8	2
Lentils, cooked from dry	198	1 c	231	1.3	3
Lentils, sprouted:					
Fresh sprouts	77	1 c	81	16	13
Stir-fried	100	3.5 oz	101	13	13
LETTUCE, chopped:					
Butterhead	56	1 c	7.3	62	4.5
Iceberg	56	1 c	7.3	30	2.2
Loose leaf	56	1 c	10	100	10
Romaine	56	1 c	9	150	13
Lotus root slices:					
Fresh slices	81	10 ea	45	78	36
Cooked	89	10 ea	59	42	24
Mushrooms:					
Sliced (1 average = 18g)	35	1/2 c	9	14	1
Cooked from fresh	78	1/2 c	21	15	3
Mustard greens:					
Fresh, chopped	56	1 c	15	269	39
Ckd fr/fresh	140	1 c	21	169	35
Ckd fr/frozen	150	1 c	29	73	20
Okra:					
Pods, ckd fr/fresh	85	8 ea	27	51	14
Slices, ckd fr/rozen	92	1/2 c	34	33	11
ONIONS:					
Fresh, chopped	160	1 c	61	17	10.2
Cooked from fresh	105	1/2 c	46	13	6
Dehydrated flakes	14	1/4 c	45	23	10.5
Onions, Spring, chopped	50	1/2 c	16	59	9.4
Parsley:					
Fresh, chopped	30	1/2 c	10	270	27
Freeze dried	1.4	1/4 c	4	53	2
Parsnips:					
Fresh slices	133	1 c	100	23	23
Cooked from fresh	156	1 c	125	16	20
PEAS:					
Black-eyed peas:					
Cooked from fresh	165	1 c	160	2.3	3.6
Cooked from frozen	170	1 c	224	2.0	4.5
Canned	240	1 c	184	3.5	6.5
Green peas:					
Fresh, uncooked	145	1 c	118	50	58
Cooked from fresh	160	1 c	134	17	23
Cooked from frozen	80	1/2 c	63	13	8
Canned, drained	85	1/2 c	59	14	8
Canned with liquid	124	1/2 c	61	22	14

	Wt (g)	Svg	Cal	Vit C (mg) 100 Cal	Svg
PEAS, continued:					
Green edible-pod peas:					
Fresh, uncooked	145	1 c	61	143	87
Cooked from fresh	160	1 c	67	114	77
Cooked from frozen	80	1/2 c	42	42	18
Peas & carrots:					
Cooked from frozen	80	1/2 c	38	17	7
Canned with liquid	128	1/2 c	48	18	8
Peas, sprouted, mature:					
Fresh peas	120	1 c	154	8.1	12.5
Cooked	100	3.5 oz	118	5.6	6.6
PEPPER, SWEET, green:					
Fresh, chopped	50	1/2 c	14	319	44.7
Ckd from fresh	68	1/2 c	19	266	50.6
PEPPER, SWEET, red:					
Fresh, chopped	50	1/2 c	14	679	95
Ckd from fresh	68	1/2 c	19	611	116
PEPPER, HOT (data is for green, red varieties have more):					
Chili pepper, green:					
Fresh, chopped, peeled	75	1/2 c	30	607	182
Canned w/liquid	68	1/2 c	17	272	46
Jalapeno, canned pod	68	1/2 c	17	52	9
Pimento, canned	57	2 oz	13	369	48
Poi, two finger	240	1 c	269	4	10
POTATOES ("C" content declines with storage time):					
Baked in oven:					
Flesh and skin	202	1 ea	220	12	26
Flesh only	156	1 ea	145	14	20
Potato skin	58	1 ea	115	7	8
Baked in microwave:					
Flesh and skin	202	1 ea	212	14	31
Flesh only	156	1 ea	156	15	24
Boiled:					
Peeled after	136	1 ea	119	15	18
Peeled before	135	1 ea	116	9	10
Canned, 1" diam.	70	2 ea	42	9	4
Cooked f/frozen, small	70	1 ea	46	14	7
French fries, frozen:					
Cooked in oil	50	10 ea	158	3	5
Oven heated	50	10 ea	111	5	6
Hash browns, f/frozen	156	1 c	340	3	10
Mashed:					
Prepared w/milk	210	1 c	162	9	14
Prepared fr/nstant	215	1 c	239	10	25
Potato puffs (tater tots)					
heated from frozen	62	1/2 c	138	3	4
Potato dishes:					
Au gratin, recipe	245	1 c	322	8	24
Au gratin, from mix	245	1 c	228	3	8
Scalloped, recipe	245	1 c	210	12	26
Scalloped, from mix	245	1 c	228	4	8
Potato chips	28	14 ea	148	8	12
Potato flour	179	1 c	628	5	34
Pumpkin, mashed:					
Cooked from fresh	245	1 c	50	23	12
Canned	123	1/2 c	42	12	5
Purslane, boiled	115	1 c	21	58	12
Radishes, red	45	10 ea	7	147	10
Radish seeds, sprouted	38	1 c	16	71	11
Rutabaga, cubes:					
Fresh cubes	140	1 c	51	69	35
Cooked from fresh	85	1/2 c	29	64	19
Sauerkraut, canned, w/liquid	236	1 c	44	79	35

	Wt (g)	Svg	Cal	Vit C (mg) 100 Cal	Vit C (mg) Svg
Seaweed:					
Lavar, fresh	28	1 oz	10	111	11
Spirulina, dried	28	1 oz	82	4	3
Soybeans, sprouted:					
Fresh beans	35	1/2 c	45	12	5
Steamed	94	1 c	76	10	7.8
SOYBEAN PRODUCTS: see tofu this section, miso, natto, tempeh in Other; and roasted soybeans in Nuts and Seeds.					
Spinach:					
Fresh, chopped	56	1 c	12	128	16
Cooked from fresh	180	1 c	41	98	40
Cooked from frozen	190	1 c	53	44	23
Canned, drained	214	1 c	50	62	31
SQUASH, SUMMER, slices:					
Crookneck, fresh slices	130	1 c	24	45	11
Crookneck, ckd	180	1 c	36	28	10
Scallop, fresh slices	130	1 c	24	98	23
Scallop, ckd	90	1/2 c	14	69	10
Zucchini, fresh slices	130	1 c	19	62	12
Zucchini, ckd	180	1 c	29	29	8
SQUASH, WINTER, mashed:					
Acorn, baked	245	1 c	137	19	26
Acorn, boiled	245	1 c	83	19	16
Butternut, baked	245	1 c	99	37	37
Butternut, boiled	240	1 c	94	9	8
Butternut, ckd fr/rzn	240	1 c	94	9	8
Hubbard, baked	240	1 c	120	19	23
Hubbard, boiled	236	1 c	70	22	15
Spaghetti, baked/boiled	155	1 c	45	12	6
Succotash:					
Ckd fr/fresh	192	1 c	222	7	16
Ckd fr/frozen	170	1 c	158	6	10
Sweet potato, 5" x 2":					
Baked with skin	114	1 ea	118	24	28
Boiled, flesh only	151	1 ea	160	16	26
Candied, recipe	105	1 pce	144	5	7
Canned, mashed:					
Regular	128	1/2 c	129	5	6.5
Vacuum pack	128	1/2 c	117	29	34
TOMATOES:					
Fresh, 2.6" diam.	123	1 ea	26	85	22
Fresh, chopped	180	1 c	38	84	32
Cooked from fresh	240	1 c	65	85	55
Canned, whole	240	1 c	47	77	36
Tomato juice, canned	244	1 c	42	108	45
Tomato paste, canned	262	1 c	220	51	111
Tomato puree, canned	250	1 c	102	87	88
Tomato sauce, canned	245	1 c	74	43	32
Turnips:					
Fresh cubes	130	1 c	35	78	27
Cooked from fresh	78	1/2 c	14	64	9
Turnip greens:					
Cooked from fresh	144	1 c	29	136	40
Cooked from frozen	82	1/2 c	24	75	18
Vegetable juice cocktail	242	1 c	46	146	67
VEGETABLE COMBINATIONS, cooked f/ frozen:					
Broccoli, carrots and pasta	95	2/3 c	88	30	26
Broccoli, carrots & water chestnuts	91	2/3 c	32	122	39
Broccoli, cauliflower & red peppers	95	2/3 c	25	240	60
Broccoli & water chestnuts	95	1/2 c	33	167	55
Cantonese stir fry	95	1/2 c	53	83	44

	Wt (g)	Svg	Cal	Vit C (mg) 100 Cal	Vit C (mg) Svg
VEGETABLES, MIXED, continued:					
Chinese stir fry	95	1/2 c	31	58	18
Green beans, spaetzle	95	1/2 c	108	4	4
Japanese vegetables	95	1/2 c	29	90	26
Mixed vegetables (corn, peas, carrots, green beans, lima beans):					
Cooked fr/frozen	182	1 c	107	5	6
Canned, drained	163	1 c	77	11	8
Peas, carrots, onions	91	1/2 c	54	22	12
Peas, carrots, onions, pasta	95	1/2 c	122	7	8
Peas, cauliflower, cream sce	95	1/2 c	118	16	19
Peas & cheese sauce	142	1/2 c	165	9	15
Peas & mushrooms	95	1/2 c	73	27	20
Peas & onions	95	1/2 c	71	27	19
Peas, onions, carrots, & butter sce	71	2/3 c	100	13	13
Peas, potatoes, cream sauce	76	1/2 c	140	9	13
Peas, rice, mushrooms	66	2/3 c	108	9	10
Peas, pasta, corn, cream sce	95	1/2 c	132	10	13
Peas, pasta, mushrooms & cream sauce	95	1/2 c	129	11	14
Spinach & water chestnuts	95	1/2 c	29	62	18
Water chestnuts, fresh	62	1/2 c	66	4	2.5
Watercress, fresh	17	1/2 c	2	365	7
Yam, orange: see Sweet potato.					
Yam, white, cooked cubes	136	1 c	158	10	16.5
Zucchini: see Squash.					

Other

Cooking Ingredients, condiments, flavorings, spices, sweeteners, etc.

	Wt (g)	Svg	Cal	Vit C (mg) 100 Cal	Vit C (mg) Svg
Baking chocolate, unsweetened	28	1 oz	145	0	0
Barbecue sauce	250	1 c	160	8	13
Catsup	17	1 T	18	14	2.6
Chili sauce:					
Tomato based	273	1 c	284	16	44
Hot red pepper	31	2 T	6	154	9
Dill pickle	65	1 ea	5	80	4
Honey	339	1 c	1030	.3	3
Hummous, Humous	246	1 c	420	5	19
Jelly	18	1 T	49	1.4	.7
Marmalade	20	1 T	52	4	2
Pickle slices, fresh pack	30	4 pce	20	10	2
Salsa, recipe	108	1/2 c	46	49	39
SPICES:					
Cayenne, red pepper	5.3	1 T	17	24	4
Chili powder	7.5	1 T	24	20	5
Chives, freeze dried	.8	1/4 c	2	265	5.3
Cloves, ground	6.6	1 T	21	25	5.3
Coriander leaf, dried	1.8	1 T	5	204	10
Garlic cloves	12	4 ea	18	21	4
Oregano, ground	4.5	1 T	14	21	3
Paprika	6.9	1 T	20	25	5

ESHA Research

Vitamin D

Vitamin D regulates the metabolism of calcium and phosphorus, and promotes the growth of strong bones. It is essential for increasing the absorption of calcium into our bodies.

This is an unusual nutrient in that most of the body's intake is not from food, but rather from the action of sunlight (ultraviolet radiation) forming the nutrient in the skin. Given enough sun, a person doesn't need to consume any vitamin D. The few foods that do contain vitamin D have been fortified.

Darker skinned people take longer to synthesize vitamin D from the sun, which may explain why such people in northern, smoggy cities are more prone to vitamin D deficiency. Rickets is the name of the disease caused by vitamin D deficiency, and it is most dramatically seen in growing children. . . with bowed legs and misformed ribs.

Fortifying milk with vitamin D has largely solved the problems of any deficiency. It should be noted that the safe range for vitamin D is fairly small, and too much taken in the form of supplements can be toxic to the system. This cannot happen from "D" formed from sunlight, because the body controls the amount formed.

Vitamin D is a fat-soluble vitamin, and it appears that the body can store enough during warm weather to last through the winter if necessary. This ability to store the vitamin is also why too much from supplements can be toxic.

Vitamin Losses

Vitamin D is relatively stable in foods and will survive storing, processing, and cooking fairly well. Some drugs will interfere with vitamin D metabolism—cortisone, barbituates, anti-cholesterol drugs, and some anti-convulsants.

Recommended Dietary Allowances for Adults

USA RDA: 5 mcg per day (200 IU); (Pregnant or Nursing: 10 mcg per day).
   Children and young adults (19-24 years), 10 mcg per day.
Canadian RNI: 2.5 mcg per day (5.0 mcg for pregnancy or lactation).

Read above about Vitamin D from sunlight.

Food Sources

Because so few foods contain vitamin D, it was not included in the first versions of The Food Processor® nutrition systems. Except for eggs and liver, and some fish, the other food items will have vitamin D because of fortification of milk, and therefore foods prepared with milk. Cheese and most commercial yogurts are made with milk which is *not* fortified with vitamin D, however. A few food sources have been listed here for your information.

	Wt (g)	Svg	Vitamin D IU / Svg

Dairy & Dairy Products

CHEESE & Cheese Products:
(made with Vitamin D fortified milk only)

	Wt (g)	Svg	IU / Svg
American cheese food	28	1 oz	8
Cheddar	28	1 oz	3
Cheese Whiz	28	1 oz	1
Cottage (all)	226	1 c	5
Cream cheese	28	1 oz	2
Swiss	28	1 oz	28

Cream, dairy:

Sweet, heavy, fluid	15	1 T	15
Sweet, light, fluid	15	1 T	8
Sour cream	15	1 T	1

MILK:
Fluid (cow):

All milks, fortified	244	1 c	100
Canned, evaporated	126	1/2 c	110
Goat milk	244	1 c	5
Human milk	246	1 c	1-24

MILK DESSERTS:

Pudding, prepared	130	1/2 c	50

Eggs

Vitamin D is in the yolk: 1 yolk = 16.4 g

Cooked, whole egg	50	1 ea	23
Raw, whole egg	50	1 ea	27

Fats & Spreads

Butter	4.7	1 tsp	1.4
Cod Liver Oil	4.7	1 tsp	400
Margarine	4.7	1 tsp	15.0
Mayonnaise	13.8	1 T	.055

Fish & Shellfish

Eel, smoked	28	1 oz	1814
Herring, raw	28	1 oz	255
Oysters, raw, medium	57	4 ea	2.9
Salmon, Atlantic, canned	28	1 oz	142
Sardines, canned	28	1 oz	85
Shrimp, canned	28	1 oz	30

Grains & Grain Products

Cold cereals (ready to serve) are often fortified
with vitamin D. Brands vary; check label.

Most fortified brands	28	1 oz	49
Stuffing, from mix, Stove Top	≈106	1/2 c	9-13

Meats

Bacon, cooked:

	Wt (g)	Svg	IU / Svg
Regular, medium	6.3	1 pce	3
Canadian	28	1 oz	10
Ham	85	3 oz	21.3

Liver, cooked:

Calf, fried	85	3 oz	11.9
Chicken, simmered	20	1 ea	10.0

Lunchmeats & Sausages:

Bologna, beef	23	1 pce	7.0
Braunschweiger	28	1 oz	9-12
Frankfurter, 10/pkg	45	1 ea	11-16
Pork sausag, cooked	27	1 patty	23.0
Salami	23	1 pce	8-10
Summer sausage	23	1 pce	11-15
Turkey breast meat	28	1 oz	5.3

Other

Any recipe or combination food with egg or vitamin D
fortified milk will contain the proportional amount
of vitamin D.

ESHA Research

Vitamin E

Vitamin E acts as an anti-oxidant to prevent cell-membrane damage. In other words, it is a scavenger for destructive substances that would react against the unsaturated fats in the body. It also acts to detoxify radicals (sort of a good word for destructive substances) that occur elsewhere. It aids in the growth of good tissue and the healing of tissue from surgery and burns. We actually know more about vitamin E deficiencies in animals than we do in humans, and the deficiencies may not always translate to humans. However, if vitamin E in the bloodstream falls too low, the red blood cells tend to break open and cause a form of anemia. There is also evidence that environmental conditions like air pollution may increase human need for this vitamin.

Vitamin E is present in several different forms called tocopherols. The most active is α-tocopherol, and the current RDA (1980) is for "the amount of vitamin E activity equivalent to that of 10 milligrams of D-α-Tocopherol." Other forms are considered less active*, but they still make a contribution, and recent studies show evidence that various tocopherols may need to be present together -- they are synergistic to each other. The list below shows Alpha, Alpha equivalents and Total tocopherol information for a selection of oils. The listings in the tables are for Total Vitamin E.

D-Alpha-Tocopherol is the natural form of the vitamin. The manufactured versions often contain certain amounts of "mirror image" versions of the vitamin, called L-Alpha-Tocopherol, and we don't know if the body uses that form or ignores it. It generally has 74% of the activity of the natural Vitamin. Natural or manufactured, high doses of Vitamin E may interefere with blood thinning (anti-blood clotting) drugs.

Vitamin Losses

Vitamin E is present in good quantity in whole grains, but is readily destroyed by processing. Milling of grains takes out a high percentage, and bleaching of flour takes the rest out. Refining of vegetable oil reduces about 25% of its vitamin E content. If protected from light, the losses are fairly small during the first 4 to 6 weeks of storage.

There is also evidence that freezing reduces the amount of vitamin E, and so does long storage. Although vitamin E can be destroyed by high heat processing (such as frying), it is stable at regular cooking and boiling temperatures. Mineral oil will also absorb vitamin E and carry it out of the body.

Recommended Dietary Allowances for Adults (mg of D-alpha-tocopherol equivalents*)
USA RDA: Men, 10 mg; Women, 8 mg (Pregnant, 10 mg; Lactating, 11 mg)
Canadian RNI: Men, 10 mg for age 19-24 and 9 mg for age 25-49. Women are recommended at 7 mg for age 19-24 and 6 mg for age 25-49 (an additional 2 mg for pregnancy, and an additional 3 mg for lactation).

Food Sources

Fresh or lightly processed foods are the best sources of this vitamin. A continual diet of processed convenience foods over a long period of time may contribute to deficiency.

It is found in good quantity in grains, before processing, and in foods that are high in unsaturated fats, such as unrefined vegetable oils. There is more vitamin E in mature plants and in dark vegetables than in root vegetables and pale-colored fruits.

Amounts per 100 g of oil (1T = 13.6g)

	Total Vit E (mg)	Alpha Equivalents	Alpha only
Almond oil	40	39	39
Barley oil	150	49	26
Canola oil	66	23	19
Corn oil	104	33	26
Cottonseed oil	65	38	35
Olive oil	13	12	12
Palm oil	26	8	6
Peanut oil	13-25	9-13	9-12
Rice Bran oil	51	36	36
Safflower oil	38	34	34
Sesame oil	29	1.8	1.4
Soybean oil	104	17	10
Sunflower oil	63	62	62
Wheat Germ oil	255	182	149

* Where the activity level of D-α-tocopherol = 1.0; β- tocopherol = .5; γ - tocopherol = 0.1; α - tocotrienol = .3; and δ - tocopherol = activity level of 0.01.

	Wt (g)	Svg	Cal	Vit E (mg) 100 Cal	Svg

Baked Goods
Breads, Cakes, Cookies, Crackers, Muffins, Pancakes, Pastries, Pies, Rolls, and some desserts.

Item	Wt (g)	Svg	Cal	100 Cal	Svg
Apple crisp, 3" x 3"	78	1 pce	146	.7	1.08
Bagel, 3.5" diam	68	1 ea	180	1.0	1.80
Biscuit:					
Homemade	28	1 ea	100	1.3	1.30
From mix	28	1 ea	94	.8	.78
From refrig dough	20	1 ea	65	.8	.50
BREADS:					
Banana nut, 1/2" slice	50	1 pce	161	.2	.39
Boston brown, canned	45	1 pce	95	.4	.37
Cornbread, muffin	45	1 ea	145	1.4	2.0
Cracked wheat	25	1 pce	65	.3	.23
Mixed grain bread	25	1 pce	65	.5	.31
Pita pocket bread, 6.5" dm.	60	1 ea	165	.3	.43
Pumpernickel, 5" x 4" x 3/8"	32	1 pce	80	.5	.37
Raisin bread	25	1 pce	68	.3	.20
Rye, light, 5" x 3.5" x 7/16"	25	1 pce	65	<.1	.05
Wheat (blend of white & whole wheat flour)	25	1 pce	65	.1	.08
White bread	28	1 pce	75	<.1	.02
Whole wheat bread	28	1 pce	70	.4	.27
Brownies:					
Homemade, w/nuts	20	1 ea	95	.5	.50
From frozen, w/frosting	25	1 ea	103	.4	40
Commercial, nuts & frosting	25	1 ea	100	.4	.44
CAKES, piece = 1/16th cake unless stated otherwise.					
Boston cream pie, 1/8	120	1 pce	260	.8	2.0
Carrot, cream cheese frosting, 2.5" x 3"	112	1 pce	406	2.4	9.7
Cheesecake, 1/12	92	1 pce	278	.9	2.5
Chocolate cake, choc frosting	68	1 pce	248	.6	1.6
Coffee cake fr/mix, 2.4" x 2.8"	72	1 pce	230	.9	2.16
Fruitcake, dark	43	1 pce	165	.4	.60
Pound cake, 1/2" slice	30	1 pce	115	.2	.27
Snack cake, creamed filled:					
Chocolate (like Ding-Dongs)	28	1 ea	105	.3	.30
Sponge cake (like Twinkies)	42	1 ea	155	.5	.70
Sponge cake, 1/12 cake	66	1 pce	194	.2	.40
White cake:					
Chocolate frosting	77	1 pce	291	.2	.60
White/coconut frosting	70	1 pce	265	.1	.27
Yellow, chocolate frosting:					
From mix	69	1 pce	240	.9	2.07
Commercial	69	1 pce	245	.8	2.0
Cherry cobbler, 3" x 3"	129	1 pce	199	.5	.96
Cherry crisp, 3" x 3"	138	1 pce	157	.5	.79
Chips: see corn & tortilla this section; potato chips under Vegetables.					
COOKIES:					
Butter cookies	25	5 ea	115	.7	.75
Chocolate chip:					
Recipe	40	4 ea	185	.6	1.2
Commercial	42	4 ea	180	.9	1.7
Fr/refrigerated dough	48	4 ea	225	.8	1.8
Oatmeal raisin	52	4 ea	245	1.0	2.4
Peanut butter, recipe	48	4 ea	245	2.5	6.1
Vanilla wafers	40	10 ea	185	.6	1.1
Sandwich cookies, all	40	4 ea	195	.8	1.65
Shortbread cookies	32	4 ea	155	.6	.93
Snickerdoodle	20	1 ea	110	.4	.45
Sugar, f/refrig dough	48	4 ea	235	.6	1.4
Corn chips	28	1 oz	155	1.2	1.8
Cracker bread, Armenian	28	4 pce	117	.3	.33
Cream puff, custard filling	110	1 ea	280	.2	.63

Item	Wt (g)	Svg	Cal	100 Cal	Svg
Croissant, 4.5" x 4" x 2"	57	1 ea	235	.2	.50
Danish pastry, plain	57	1 ea	220	1.1	2.31
Danish pastry with fruit	65	1 ea	235	1.1	2.63
Doughnut:					
Cake type, medium	50	1 ea	210	1.0	2.02
Yeast raised, plain	60	1 ea	235	1.0	2.43
Yeast raised, jelly filled	65	1 ea	226	.8	1.70
Eclair w/custard filling & chocolate icing	94	1 ea	262	.2	.65
English muffin, plain or sourdough	57	1 ea	130	.6	.80
MUFFINS, from mix:					
Blueberry	45	1 ea	140	1.0	1.35
Bran (wheat)	45	1 ea	140	1.5	2.14
Cornmeal	45	1 ea	145	1.4	2.0
Pancakes:					
Plain, 4" dm, recipe	27	1 ea	60	2.2	1.32
Plain, 4" dm, fr/mix	27	1 ea	60	.5	.30
Buckwheat, 4" fr/mix	27	1 ea	55	2.2	1.22
Whole wheat, 5" fr/mix	52	1 ea	94	1.4	1.3
Patty shell (puff pastry)	71	1 ea	250	1.0	2.5
Peach cobbler, 3" x 3"	130	1 pce	130	2.0	2.6
Peach crisp, 3" x 3"	139	1 pce	166	1.8	2.97
PIES: piece = 1/6 of 9" pie					
Apple pie	158	1 pce	405	.3	1.08
Banana cream	198	1 pce	319	.4	1.40
Blueberry	158	1 pce	380	.2	.88
Cherry	158	1 pce	410	.2	.85
Coconut custard	165	1 pce	384	.2	.73
Custard	152	1 pce	293	.2	.56
Lemon meringue	140	1 pce	355	.3	.97
Mincemeat	160	1 pce	395	.6	2.45
Peach pie	158	1 pce	405	.2	.90
Pecan pie	138	1 pce	583	.2	1.02
Pumpkin	200	1 pce	367	.3	1.23
Popover	51	1 ea	96	.3	.32
Poptart-type pastry	54	1 ea	210	5.0	10.5
Pretzels, thin twists	60	10 ea	240	.2	.46
ROLLS:					
Dinner roll, avg, white	35	1 ea	120	.1	.16
Rye, dark	28	1 ea	79	.3	.27
Rye, light	28	1 ea	76	.4	32
Submarine roll (hoagie)	135	1 ea	400	<.1	.22
Whole wheat roll	35	1 ea	88	.6	.53
Stove Top stuffing, prepared	108	1/2 c	176	.3	.50
Taco shell	13.6	1 ea	59	.4	.25
TORTILLAS:					
Corn, 6" diam, fried	30	1 ea	87	1.7	1.5
Flour, 8" diam	35	1 ea	105	.6	.66
Flour, 10.5" diam	57	1 ea	168	.6	1.06
Tortilla chips, all	28	1 oz	139	.8	1.05
Waffles, from mix, 7" diam	75	1 ea	205	1.1	2.2
Waffles, frozen, 4" diam	35	1 ea	98	1.0	1.0

Dairy & Dairy Products

Item	Wt (g)	Svg	Cal	100 Cal	Svg
CHEESE (1.5" cube ≈ 1 oz):					
Average all hard cheeses	28	1 oz	varies	.1-.2	.179
Cottage cheese:					
Lowfat 1%	226	1 c	164	.9	1.45
Lowfat 2%	226	1 c	205	.7	1.45
Creamed, large curd	225	1 c	235	.6	1.35
Creamed, small curd	210	1 c	215	.6	1.34
Creamed, with fruit	226	1 c	279	.5	1.45
Dry curd	145	1 c	123	.8	.93
Ricotta, w/part skim	246	1 c	340	.5	1.57

ESHA Research

	Wt (g)	Svg	Cal	Vit E (mg) 100 Cal	Vit E (mg) Svg
CREAM, SWEET, liquid:					
Half and half	242	1 c	315	.1	.27
Light whipping cream	239	1 c	699	.2	1.51
Heavy whipping cream	238	1 c	821	.2	1.50
CREAM, SWEET, whipped:					
Heavy cream	119	1 c	410	.2	.75
Pressurized	60	1 c	154	.2	.25
CREAM, SOUR, dairy	230	1 c	493	.3	1.5
Cream, sour, Imitation, non-dairy	230	1 c	479	0	0
MILK (cow):					
Skim	282	1 c	216	3.5	7.5
Whole	281	1 c	280	2.8	7.7
Canned:					
Skim, evaporated	255	1 c	200	.1	.23
Whole, evap	252	1 c	340	.1	.45
Sweet, condensed	306	1 c	982	.1	.65
Milk, human, mature	246	1 c	171	1.4	2.44
MILK BEVERAGES and mixes:					
Chocolate flavored, to be mixed w/water:					
Powder (includes dry milk)	28	1 oz	100	.2	.22
Drink, prepared	206	3/4 c	100	.2	.22
Chocolate flavored, to be mixed w/milk:					
Powder	21.6	3/4 oz	75	.4	.27
Drink, prepared	266	1 c	226	.1	.22
Milkshakes (1.25 c = 10 fl oz):					
Chocolate	283	1.25 c	360	.2	.70
Strawberry or vanilla	283	1.25 c	317	.2	.70
MILK DESSERTS:					
Custard, baked	265	1 c	305	.3	.80
Ice cream, vanilla:					
Regular	133	1 c	269	.2	.49
Rich	148	1 c	349	.1	.52
Soft serve	173	1 c	377	.2	.61
Ice milk, vanilla:					
Regular	131	1 c	184	.2	.46
Soft serve	175	1 c	223	.3	.61
Pudding, average, Instant or cooked	258	1 c	295	.1	.25
Yogurt, frozen, avg	174	1 c	220	<.1	.09

Eggs

	Wt (g)	Svg	Cal	Vit E (mg) 100 Cal	Vit E (mg) Svg
Chicken egg, cooked or raw:					
Whole (large)	50	1 ea	77.5	.6	.42
Yolk only	16.6	1 ea	63	.7	.43

Fats, Oils & Salad Dressings

	Wt (g)	Svg	Cal	Vit E (mg) 100 Cal	Vit E (mg) Svg
Bacon fat	14	1 T	126	.8	1.00
Beef fat drippings	12.8	1 T	115	.3	.38
Butter	14.2	1 T	100	.2	.22
Butter oil, ghee	205	1 c	1795	.3	5.80
Chicken fat:					
Melted	12.8	1 T	115	.3	.30
Separable, raw	102	1/2 c	647	.4	2.80
Lard	205	1 c	1849	.1	2.67
Margarine:					
Regular, 80% fat	14	1 T	100	1.6	1.56
Spread, 60% fat	14	1 T	75	1.6	1.17
Soft, 40% fat (imitation)	14.2	1 T	50	1.5	.75
Mayonnaise	13.8	1 T	100	1.0	1.02

	Wt (g)	Svg	Cal	Vit E (mg) 100 Cal	Vit E (mg) Svg
OILS:					
Almond oil	218	1 c	1927	4.4	87
Apricot kernel	218	1 c	1927	.5	8.7
Barley oil	218	1 c	1927	17	327
Canola oil	218	1 c	1927	2.6-7.4	50-143
Cocoa butter	218	1 c	1927	.2	3.9
Coconut oil	218	1 c	1927	.2	3.6
Cod liver	14	1 T	126	2.4	3.07
Corn oil	218	1 c	1927	9.4-11.7	181-226
Cottonseed	218	1 c	1927	4.0-7.4	77-142
Olive oil	216	1 c	1909	1.4	28.0
Palm kernel	218	1 c	1927	.7	13.5
Palm oil	218	1 c	1927	2.2-3.0	42-57
Peanut	216	1 c	1909	1.5-2.8	28-54
Rice bran oil	218	1 c	1927	5.8	111
Safflower	218	1 c	1927	4.3-5.3	83-103
Sesame	218	1 c	1925	3.3	63.0
Soybean	218	1 c	1927	7.3-11.8	141-227
Soybean/cottonseed	218	1 c	1927	6.5	125.0
Sunflower	218	1 c	1927	7.1-7.3	137-141
Wheat germ oil	13.6	1 T	120	28.8	34.6
SALAD DRESSING (low cal dressings vary. See label):					
1000 island	16	1 T	60	8.3	5.0
1000 island, low cal	15	1 T	24	3.3	.8
Blue cheese	15	1 T	77	1.5	1.1
Cooked type	16	1 T	25	10.0	2.5
French	16	1 T	85	5.2	4.4
French, low cal	16	1 T	24	3.4	.8
Italian dressing	14.7	1 T	75	5.8	4.4
Italian, low calorie	15	1 T	5	6.0	.3
Mayonnaise type	15	1 T	58	1.7	1.0
Mayonnaise, low cal	15	1 T	35	1.7	.6
Ranch dressing	14.9	1 T	54	4.6	2.5
Russian	15.3	1 T	76	5.0	3.8
Vinegar & oil	16	1 T	70	4.3	3.0

Fruits & Fruit Juices

	Wt (g)	Svg	Cal	Vit E (mg) 100 Cal	Vit E (mg) Svg
APPLE, 2.75" diam:					
With peel	138	1 ea	80	.8	.66
Without peel	128	1 ea	72	.4	.28
Applesauce, unsweetened	244	1 c	106	.4	.47
APRICOTS:					
Fresh, pitted	106	3 ea	51	1.9	.94
Canned, halves:					
Juice pack	84	3 ea	40	1.9	.75
Heavy syrup	85	3 ea	70	1.1	.76
Dried halves	35	10 ea	83	2.6	2.18
Apricot nectar, canned	251	1 c	141	.5	.70
Avocado, whole:					
California	173	1 ea	305	.8	2.32
Florida	304	1 ea	340	1.2	4.07
Banana, 8-3/4" 176g whole	114	1 ea	105	.3	.27
Blackberries:					
Fresh berries	144	1 c	74	1.2	.9
Frozen, unthawed	151	1 c	97	1.3	1.3
Canned	256	1 c	236	.9	2.2
Blueberries:					
Fresh berries	145	1 c	82	3.3	2.72
Frozen, unthawed	155	1 c	78	4.0	3.1
Canned	256	1 c	225	1.9	4.2
Boysenberries:					
Frozen, unthawed	132	1 c	66	.5	.34
Canned	256	1 c	225	.1	.30
Cantaloupe: see Melon.					

Fruits & Fruit Juices, continued:

	Wt (g)	Svg	Cal	Vit E (mg) 100 Cal	Vit E (mg) Svg
CHERRIES, SWEET:					
Fresh, pitted	68	10 ea	49	1.2	.61
Frozen, thawed	259	1 c	232	.1	.32
Canned with liquid	257	1 c	213	.1	.26
Cranberries, fresh	95	1 c	46	2.1	.95
Cranberry juice cocktail	253	1 c	145	.2	.30
Cranberry sauce, canned	277	1 c	419	.1	.45
Currants, fresh, all	112	1 c	67	1.7	1.12
Fruit cocktail, canned:					
Juice pack	248	1 c	115	1.9	2.17
Heavy syrup	255	1 c	185	1.2	2.17
Gooseberries:					
Fresh berries	150	1 c	67	.8	.56
Canned with liquid	252	1 c	185	.5	.93
GRAPEFRUIT					
(half = 241g w/refuse):					
Fresh half, pink/red	123	1 ea	37	.8	.31
Fresh half, white	118	1 ea	39	.7	.27
Canned sections	254	1 c	152	.5	.80
Grapefruit juice:					
Fresh juice	247	1 c	96	.5	.45
Prepared f/frozen	247	1 c	102	.4	.40
Canned, unsweetened	247	1 c	93	.3	.32
Grapes:					
Thompson seedless	50	10 ea	35	.9	.33
Tokay or Emperor	57	10 ea	40	.9	.37
Canned, heavy syrup	256	1 c	187	.2	.29
Lemon juice, fresh, frozen					
or bottled	244	1 c	55	1.0	.54
Loganberries, fresh/frzn	147	1 c	80	.4	.32
Mandarin oranges, cnd	252	1 c	155	.8	1.26
Mango, fresh (300g whole)	207	1 med	135	1.7	2.32
Melon: also see Watermelon.					
Cantaloupe, cubes	160	1 c	57	.5	.29
Frozen melon balls, mixed	173	1 c	55	.4	.20
ORANGE, avg, 2-5/8" dm,					
180g w/peel	131	1 ea	60	.5	.31
Orange juice:					
Fresh squeezed	248	1 c	111	.7	.79
Chilled or f/frozen	249	1 c	110	.4	.48
Canned, unsweetened	249	1 c	105	.4	.37
Orange grapefruit jce, cnd	247	1 c	105	.2	.22
PEACHES:					
Fresh, whole, 2.5"	87	1 ea	37	2.4	.9
Frozen slices, thawed	250	1 c	235	2.0	4.8
Canned, juice pack	77	1 half	34	4.0	1.4
Canned, heavy syrup	81	1 half	60	2.3	1.4
Dried, halves	13	1 half	31	1.6	.5
Peach nectar, canned	249	1 c	134	.2	.25
PEARS:					
Bartlett (180g whole)	166	1 ea	98	.8	.83
Canned, juice pack	77	1 half	38	.8	.31
Canned, heavy syrup	79	1 half	59	.5	.31
Dried halves	17	1 half	46	.6	.27
PLUMS:					
Fresh, 2-1/8" dm.	66	1 ea	36	1.6	.57
Canned, juice pack	95	3 ea	55	.4	.20
Canned, heavy syrup	110	3 ea	98	.8	.78
Prunes, dried, pitted	84	10 ea	201	1.0	2.1
Prune juice, bottled	256	1 c	181	.3	.52
Raisins, dark, unpacked	145	1 c	435	.1	.42
Raspberries:					
Fresh berries	123	1 c	60	.6	.37
Canned, with liquid	256	1 c	234	.2	.45

	Wt (g)	Svg	Cal	Vit E (mg) 100 Cal	Vit E (mg) Svg
Rhubarb:					
Fresh dices	122	1 c	26	.9	.24
Cooked w/sugar	240	1 c	279	.2	.48
STRAWBERRIES:					
Fresh berries	149	1 c	45	.7	.30
Frozen	149	1 c	52	.6	.32
Frzn, thawed, swtnd	255	1 c	245	.2	.54
Tangerine	84	1 ea	37	.7	.27
Tangerine juice:					
From frozen	241	1 c	110	.4	.43
Canned	249	1 c	125	.7	.87
Watermelon pieces	160	1 c	50	.3	.16

Grains & Grain Products

Cereals, Grains, Flours, Noodles, Pastas, Popcorn

	Wt (g)	Svg	Cal	Vit E (mg) 100 Cal	Vit E (mg) Svg
Barley, cooked, pearled	157	1 c	193	.2	.43
BRAN: see Oat, Rice, Wheat.					
Buckwheat flour:					
Dark	98	1 c	338	2.3	7.75
Light	98	1 c	340	1.3	4.50
Bulgar wheat, cooked	182	1 c	151	.3	.48
CEREALS, COLD (Ready to eat)					
Brands vary. See label.					
CEREALS, HOT, cooked:					
Corn grits, cooked	242	1 c	145	<.1	.048
Cream of Wheat	244	1 c	140	.2	.31
Farina	233	1 c	116	.3	.30
Oatmeal, cooked:					
From rolled oats	234	1 c	145	.3	.40
Instant, fortified, f/pkt:					
Plain	177	3/4 c	104	.6	.67
Flavored, average	164	3/4 c	160	.4	.67
Ralston	253	1 c	134	1.0	1.40
Roman Meal	181	3/4 c	111	.6	.72
Wheatena	243	1 c	135	1.1	1.43
Whole wheat cereal	242	1 c	151	.5	.68
Corn grits: see Cereal, Hot.					
Corn flour:					
Regular	117	1 c	422	.3	1.23
Masa Harina, enriched	114	1 c	416	.6	2.34
Cornmeal:					
Degermed, cooked	240	1 c	120	.8	1.01
Unbolted, whole, dry	122	1 c	442	1.5	6.60
FLOUR: see specific grain,					
nut, or vegetable					
Millet, cooked	120	1/2 c	143	.2	.29
Oat bran (1 T ≈ 6 g)	94	3.5 oz	132	4.9	6.50
Oats, rolled, dry	81	1 c	311	.4	1.22
PASTA: See Macaroni, Noodles, Spaghetti.					
RICE, cooked:					
Brown rice	195	1 c	217	.6	1.40
White, converted	200	1 c	186	.3	.47
White, regular	205	1 c	264	.2	.46
White, instant	165	1 c	162	.1	.22
Wild rice, cooked	164	1 c	166	1.9	3.2
Rice bran	83	1 c	262	5.7	14.9
Rice flour	158	1 c	578	1.0	6.0
Rye flour:					
Dark	128	1 c	415	.9	3.53
Medium	102	1 c	374	.5	1.98
Light	102	1 c	361	.3	.95

ESHA Research

Values are for edible portion of foods

	Wt (g)	Svg	Cal	Vit E (mg) 100 Cal	Svg
Grains & Grain Products, continued:					
Soy flour:					
Full fat	85	1 c	370	1.9	7.15
Low fat	88	1 c	326	1.0	3.40
Spaghetti noodles:					
White, cooked firm	140	1 c	197	<.1	.039
Whole wheat, cooked	140	1 c	174	.7	1.16
WHEAT:					
Bran, dry	30	1/2 c	65	1.5	.97
FLOUR, unbleached (bleached flour has no Vitamin E):					
All purpose, white, unsifted	125	1 c	455	.1	.313
Cake, sifted	96	1 c	348	.4	1.29
Whole wheat flour	120	1 c	407	.8	3.11
Wheat germ, raw	100	1 c	360	4.4	15.8
Wheat germ, toasted	113	1 c	432	7.2	31.1
Wheat, rolled:					
Cooked	240	1 c	142	1.5	2.1
Dry	85	1 c	289	1.2	3.50
Whole grain wheat (wheat berries) cooked	50	1/3 c	28	1.5	.42
Whole grain wheat, sprouted	108	1 c	214	1.2	2.50

Meats: Fish & Shellfish

	Wt (g)	Svg	Cal	Vit E (mg) 100 Cal	Svg
Bass, baked/broiled	100	3.5 oz	140	.7	1.0
Bluefish:					
Baked/broiled	100	3.5 oz	159	.6	1.0
Fried in crumbs	100	3.5 oz	205	1.0	2.0
Carp, baked/broiled	100	3.5 oz	162	.5	.73
Catfish, cornmeal fried	100	3.5 oz	229	.4	1.0
COD:					
Baked/broiled	100	3.5 oz	105	1.2	1.3
Batter fried	100	3.5 oz	199	.8	1.6
Smoked	100	3.5 oz	79	.3	.23
CRAB:					
Blue crab meat, ckd	135	1 c	138	.9	1.22
Blue crab meat, cnd	135	1 c	133	.6	.82
Dungeness, ckd meat	101	3/4 c	85	1.4	1.22
Crayfish, ckd moist heat	85	3 oz	97	1.0	.95
Fish cakes, fried	100	3.5 oz	172	.6	1.0
Fish sticks fr/frozen	57	2 ea	155	.1	.18
Haddock:					
Baked/broiled/steamed	85	3 oz	95	.4	.37
Breaded, fried	85	3 oz	175	.6	1.02
Smoked	100	3.5 oz	116	.4	.44
HALIBUT:					
Baked/broiled	85	3.5 oz	119	.4	.50
Smoked	100	3.5 oz	224	.4	1.0
Herring:					
Baked/broiled	100	3.5 oz	203	.6	1.2
Cnd in oil, w/liquid	100	3.5 oz	208	.4	.88
Smoked or kippered	100	3.5 oz	217	.4	.80
Pickled (1 piece ≈ 15g)	100	3.5 oz	262	.3	.77
Lobster meat, cooked	145	1 c	142	1.1	1.5
Mackerel:					
Baked/broiled, Atlantic	100	3.5 oz	262	.6	1.52
Canned,1 tall can	361	1 can	563	.2	1.0
Mussels, Blue, meat only, steamed	85	3 oz	147	.6	.9
OYSTERS:					
Raw, Eastern	248	1 c	170	1.2	2.04
Raw, Pacific	248	1 c	200	1.0	2.04
Breaded, fried, med.	88	6 ea	173	.5	.9
Simmered, Eastern	100	3.5 oz	137	.8	1.1

	Wt (g)	Svg	Cal	Vit E (mg) 100 Cal	Svg
Perch, Ocean:					
Baked/broiled	100	3.5 oz	121	1.1	1.3
Breaded, fried	85	3 oz	185	1.1	2.0
Pike, Northern, bkd/broiled	100	3.5 oz	113	.2	.23
Pollock, baked/broiled	100	3.5 oz	100	.5	.47
Roe, raw, mixed species	28	1 oz	39	2.6	1.0
SALMON, cooked:					
Baked/broiled, average	85	3 oz	183	.7	1.34
Smoked, Chinook	85	3 oz	99	1.0	1.02
Canned, drained:					
Atlantic, small can	220	1 can	281	1.0	2.67
Pink, #1 can	454	1 can	631	.8	4.8
Sockeye, #1 can	369	1 can	566	1.0	5.5
Sardines, canned:					
Atlantic, 2 sardines = 24g	92	1 can	192	.1	.28
Pacific, 1 sardine = 38g	100	3.5 oz	178	.2	.30
Scallops:					
Breaded, fried	93	6 ea	200	2.8	5.58
Steamed	100	3.5 oz	113	1.1	1.2
Seatrout (Steelhead) cooked	100	3.5 oz	131	.2	.23
Shark, batter fried	85	3 oz	194	1.1	2.1
SHRIMP:					
Breaded, fried, large	90	12 ea	218	2.4	5.3
Boiled, large	44	8 ea	44	1.5	.7
Canned w/liquid	100	3.5 oz	102	1.1	1.1
Canned, drained	128	1 c	154	.7	1.0
Snapper, baked/broiled	100	3.5 oz	128	.6	.83
SOLE (also Flounder):					
Baked or broiled	85	3 oz	99	1.6	1.6
Batter fried	85	3 oz	250	.7	1.71
Breaded, fried	100	3.5 oz	188	.9	1.61
Steamed	100	3.5 oz	92	1.8	1.61
Squid, fried in flour	85	3 oz	149	.7	1.1
Swordfish, baked/broiled	100	3.5 oz	155	.4	.67
TUNA, canned, drained, light, No.1/2 can:					
Canned in oil	171	1 can	339	.1	.46
Water packed	165	1 can	216	.2	.45

Meats
Beef, Ham, Pork, Lamb, Rabbit, Veal

	Wt (g)	Svg	Cal	Vit E (mg) 100 Cal	Svg
BEEF (Vitamin E is present in meat in low amounts as alpha tocopherol):					
Cooked beef, average	85	3 oz	220-320	.04-.05	.12
Beef liver, fried	85	3 oz	184	.8	1.38
Separable fat, cooked	28	1 oz	193	.2	.39
HAM: see Pork, cured; Turkey ham; and Sausages and Lunchmeats.					
LAMB:					
Chop, avg, cooked:					
Lean and fat	85	3 oz	250	<.1	.12
Lean only	85	3 oz	175	.1	.16
Shoulder roast:					
Lean and fat	85	3 oz	235	<.1	.12
Lean only	85	3 oz	173	.1	.15
PORK:					
Bacon, cooked:					
Regular	19	3 pces	109	.1	.11
Canadian style	47	2 pce	86	.3	.28
Chops, avg of broiled, pan-fried, roasted:					
Lean and fat	89	1 ea	268-368	.1-.2	.53
Lean only	62	1 ea	175	.2	.37

Values are for edible portion of foods

Vitamin E (Tocopherols)　mg

	Wt (g)	Svg	Cal	Vit E (mg) 100 Cal	Svg
PORK, continued:					
Chops, average, braised:					
Lean and fat	75	1 ea	250	.2	.45
Lean only	61	1 ea	166	.2	.37
Pork roast, average:					
Lean and fat	85	1 pce	265	.2	.51
Lean only	85	1 pce	206	.2	.40
Shoulder, braised (yield fr/6.8 oz raw w/bone):					
Lean and fat	85	3 oz	293	.1	.32
Lean only	67	2.4 oz	166	.2	.26
Spareribs, ckd yield fr/ 1 lb raw	177	6.25 oz	703	.1	.67
PORK CURED, HAM, roasted:					
Lean & fat	85	3 oz	207	.6	1.21
Lean only	85	3 oz	133	.3	.40
Canned, average	85	3 oz	142	.3	.45
RABBIT, cooked	85	3 oz	175	.1	.21
VEAL:					
Cutlet, lean, cooked	85	3 oz	166	.2	.36
Rib, roasted	85	3 oz	151	.2	.31

Meats: Poultry
Chicken, Turkey, Duck, Goose

	Wt (g)	Svg	Cal	Vit E (mg) 100 Cal	Svg
CHICKEN:					
All types of meat:					
Fried	140	1 c	307	.3	.80
Roasted	140	1 c	266	.3	.77
Stewed	140	1 c	248	.3	.70
Canned, boned w/broth	142	5 oz	235	.2	.57
Dark meat only:					
Fried	85	3 oz	203	.2	.47
Roasted	85	3 oz	174	.3	.47
Stewed	85	3 oz	163	.3	.43
Light meat only:					
Fried	85	3 oz	163	.3	.47
Roasted	85	3 oz	147	.3	.47
Stewed	85	3 oz	135	.3	.47
Breast*, meat & skin:					
Batter-fried	140	1 ea	364	.3	1.1
Flour-fried	98	1 ea	218	.3	.62
Stewed	110	1 ea	202	.2	.50
Roasted	98	1 ea	193	.3	.62
Breast*, meat only:					
Fried	86	1 ea	161	.2	.33
Roasted	86	1 ea	142	.2	.31
Stewed	95	1 ea	144	.2	.25
* 2 pieces per bird					
Drumstick, meat & skin:					
Batter-fried	72	1 ea	193	.2	.34
Flour-fried	49	1 ea	120	.2	.27
Roasted	52	1 ea	112	.3	.29
Stewed	57	1 ea	116	.2	.27
Drumstick, meat only:					
Fried	42	1 ea	82	.3	.25
Roasted	44	1 ea	76	.3	.24
Stewed	46	1 ea	78	.3	.23
Thigh, meat & skin:					
Batter-fried	86	1 ea	238	.2	.47
Flour-fried	62	1 ea	162	.2	.34
Roasted	62	1 ea	153	.2	.34
Stewed	68	1 ea	158	.2	.29
Thigh, meat only:					
Fried	52	1 ea	113	.3	.29
Roasted	52	1 ea	109	.3	.29
Stewed	55	1 ea	107	.2	.24
Liver, simmered	20	1 ea	30	1.0	.29

	Wt (g)	Svg	Cal	Vit E (mg) 100 Cal	Svg
DUCK, domestic, roasted:					
Meat & skin	85	3 oz	286	.4	1.1
Meat only	85	3 oz	171	.9	1.5
GOOSE, domestic, roasted:					
Meat & skin	85	3 oz	259	.8	2.2
Meat only	85	3 oz	202	1.2	2.4
TURKEY:					
Ground turkey, cooked	100	3.5 oz	229	.2	.35
Roasted:					
All types	85	3 oz	145	.3	.38
Dark meat	85	3 oz	159	.3	.54
Light meat	85	3 oz	133	<.1	.05
Patty, breaded, fried	64	1 ea	181	.4	.64
Frozen slices with gravy	142	5 oz	95	.3	.33

Meats: Sausages and Lunchmeats

	Wt (g)	Svg	Cal	Vit E (mg) 100 Cal	Svg
Brotwurst, link	70	1 ea	226	.2	.35
Cheesefurter (cheese smoki)	43	1 ea	141	.2	.25
Chicken roll, light meat	57	2 oz	90	.3	.31
Corned beef loaf, jellied	28	1 oz	46	.5	.22
Dutch brand loaf	28	1 oz	68	.6	.40
FRANKFURTER (hotdog):					
Beef/ beef & pork, 8/pkg	57	1 ea	184	.13	.24
Chicken frank, 10/pkg	45	1 ea	115	.15	.18
Turkey frank, 10/pkg	45	1 ea	102	<.1	.01
HAM lunchmeat (also see Pork, cured):					
Chopped, packaged	42	2 pce	98	.2	.21
Regular	57	2 oz	103	.3	.28
Italian sausage link, cooked	67	1 ea	216	.1	.21
Knockwurst link	68	1 ea	209	.2	.39
Luxury loaf	57	2 oz	80	.6	.51
Olive loaf	57	2 oz	133	.4	.50
Peppered loaf	28	1 oz	42	1.2	.50
Pickle and pimento loaf	57	2 oz	149	.3	.50
Salami:					
Pork and beef	57	2 oz	143	.3	.39
Turkey	57	2 oz	111	.3	.38
Smoked link sausage:					
Beef and pork	68	1 ea	229	.1	.22
Pork link	68	1 ea	265	.1	.22
Turkey ham	57	2 oz	73	.5	.36
Turkey pastrami	57	2 oz	74	.5	.38

Mixed Dishes & Fast Foods

	Wt (g)	Svg	Cal	Vit E (mg) 100 Cal	Svg
Beef & vegetable stew:					
Recipe	245	1 c	220	1.0	2.15
Canned	245	1 c	194	.5	.93
Beef, macaroni, tomato sauce casserole fr/recipe	226	1 c	189	.4	.77
BURRITOS:					
Bean burrito	174	1 ea	322	.5	1.62
Beef burrito	177	1 ea	463	.4	1.84
Beef & bean	175	1 ea	390	.4	1.67
Deluxe combination	198	1 ea	424	.4	1.90
Cheese soufflé, recipe	112	1 c	221	1.4	3.00
Chicken a la king, recipe	245	1 c	470	.2	1.12
Chicken chow mein:					
Recipe	250	1 c	255	3.5	8.96
Canned	250	1 c	95	.9	.90
Chicken pot pie, fr/frzn	230	1 ea	430	.9	4.00
Chicken salad w/celery	78	1/2 c	266	4.1	11.00

95

ESHA Research

Values are for edible portion of foods

Mixed Dishes & Fast Foods, continued:

	Wt (g)	Svg	Cal	Vit E (mg) 100 Cal	Svg
Chili with beans, canned	255	1 c	286	.5	1.40
Cole slaw	120	1 c	84	11.9	10.00
Corn dog	111	1 ea	330	.4	1.24
Corn fritter, recipe	45	1 ea	116	3.2	3.67
Corn pudding	250	1 c	271	.3	.70
Egg salad	183	1 c	438	2.8	12.10
Enchiladas:					
Beef enchilada	120	1 ea	292	.8	2.19
Cheese enchilada	120	1 ea	330	.7	2.19
Chicken enchilada	120	1 ea	269	.8	2.16
French toast, recipe	65	1 pce	123	.4	.46
LASAGNA:					
Recipe w/meat	245	1 pce	398	.3	1.21
Recipe w/o meat	218	1 pce	316	.3	1.03
Macaroni & cheese:					
Recipe	200	1 c	430	.8	3.52
Canned	240	1 c	230	.1	.32
Manicotti, frozen entree	225	1 ea	271	.8	2.10
Meat loaf	87	1 pce	193	.2	.36
Moussaka, lamb & eggplant	250	1 c	250	.4	.99
PIZZA, cheese:					
Regular crust 1/8 of 15"	120	1 pce	290	.1	.36
Thick crust 1/2 of 10"	208	1 pce	519	.2	.82
Quiche Lorraine, 1/8 pie	176	1 pce	600	.9	5.50
Pies, fried, commercial:					
Apple pie	85	1 ea	255	.3	.64
Cherry	85	1 ea	250	.3	.64
Ravioli, beef, canned	226	1 c	220	.3	.67
SANDWICHES, fast food:					
Cheeseburger, 3 oz meat	112	1 ea	300	.3	.78
Cheeseburger, 4 oz meat	194	1 ea	524	.2	.89
Chicken patty sandwich	157	1 ea	436	.1	.54
English muffin w/egg, cheese & bacon	138	1 ea	360	.4	1.27
Fish sandwich:					
Regular, w/cheese	140	1 ea	420	.5	2.22
Large, w/o cheese	170	1 ea	470	.5	2.20
Hamburger, 3 oz meat	98	1 ea	245	.2	.59
Hamburger, 4 oz meat	174	1 ea	445	.3	1.33
Hotdog (frankfurter w/ bun)	85	1 ea	260	.1	.35
SANDWICHES (on part whole wheat bread, except where stated as on rye):					
Avocado & cheese w/sprouts, tomato	195	1 ea	432	.6	2.46
Bacon, lettuce, tomato	135	1 ea	327	.5	1.61
Chicken salad sandwich	100	1 ea	294	2.4	6.96
Corned beef & swiss on rye	147	1 ea	429	.8	3.42
Egg salad sandwich	111	1 ea	319	1.4	4.31
Grilled cheese	117	1 ea	393	.4	1.44
Ham sandwich	122	1 ea	256	.5	1.37
Ham & Am. cheese	151	1 ea	363	.4	1.55
Ham on rye	116	1 ea	242	.6	1.38
Ham salad sandwich	125	1 ea	339	1.2	4.04
Patty melt, ground beef on rye	177	1 ea	567	.4	2.50
Peanut butter & jam	100	1 ea	341	.6	2.12
Reuben sandwich, grilled	233	1 ea	480	.7	3.51
Roast beef sandwich	122	1 ea	280	.5	1.27
Tuna salad sandwich	116	1 ea	303	3.0	9.14
Turkey sandwich	122	1 ea	271	1.0	2.75
Turkey ham sandwich	122	1 ea	253	1.2	2.93
SPAGHETTI (pasta & tomato sce):					
Homemade	220	1 c	179	1.8	3.30
Canned	249	1 c	272	1.4	3.70
Spinach soufflé	136	1 c	218	.5	1.01

	Wt (g)	Svg	Cal	Vit E (mg) 100 Cal	Svg
Taco, beef	78	1 ea	207	.3	.62
Taco, chicken	78	1 ea	172	.3	.59
TOSTADAS:					
Beans and beef	192	1 ea	332	.5	1.79
Beans and chicken	157	1 ea	249	.6	1.51
Refried beans	157	1 ea	212	.7	1.49
Tuna salad	205	1 c	383	5.8	2.2
Turkey pot pie, frozen	233	1 ea	416	.7	3.0
Waldorf salad	142	1 c	424	3.7	15.5

Nuts & Seeds

	Wt (g)	Svg	Cal	Vit E (mg) 100 Cal	Svg
ALMONDS:					
Dried, whole	142	1 c	837	3.6	30.2
Oil roasted	157	1 c	970	3.3	32.0
Almond butter	16	1 T	101	3.8	3.8
Brazil nuts, dried (≈ 7)	28	1 oz	186	1.2	2.2
Cashews, dry or oil roasted	133	1 c	768	1.3	10.3
Coconut, raw, grated	80	1 c	283	.2	.6
Filberts, hazelnuts, whole	135	1 c	853	3.8	32.3
Macadamias:					
Dried	134	1 c	940	2.3	22
Oil roasted	134	1 c	962	2.0	19
MIXED NUTS, w/ peanuts (cashews, peanuts, brazil nuts, filberts, almonds, pecans):					
Dry roasted	137	1 c	814	1.4	11.0
Oil roasted	142	1 c	876	1.3	11.0
MIXED NUTS, w/o peanuts (cashews, almonds, brazil nuts, pecans, filberts):					
oil roasted	144	1 c	886	.8	7.5
PEANUTS:					
Dried	146	1 c	827	1.8	14.6
Oil roasted	144	1 c	837	1.2	9.94
Peanut butter	32	2 T	188	1.1	2.05
Pecans, dried, chopped	119	1 c	794	.5	3.7
Pine nuts, dried, average	28	1 oz	154	1.1	1.7
Pistachios, dried, shelled	128	1 c	739	.9	6.7
Pumpkin/squash seeds:					
Dried kernels	138	1 c	747	2.0	15.0
Roasted kernels	227	1 c	1185	1.4	16.0
Whole seeds, roasted	64	1 c	285	.9	2.5
Sesame seeds:					
Dried, hulled	38	1/4 c	221	.4	.86
Dried, whole	36	1/4 c	206	.4	.82
Soybeans, roasted	86	1/2 c	405	1.3	5.13
SUNFLOWER seeds:					
Dried seeds	144	1 c	821	3.5	29.1
Oil roasted	135	1 c	830	5.2	43.1
Sunflower seed butter	16	1 T	93	5.1	4.7
Walnuts, black chopped	125	1 c	759	.4	3.2
Walnuts, English, chopped	120	1 c	770	.5	4.0

Soups, Sauces & Gravies

	Wt (g)	Svg	Cal	Vit E (mg) 100 Cal	Svg
Beef gravy:					
Recipe	135	1/2 c	151	.2	.37
Canned	233	1 c	124	.5	.64
Brown gravy fr/dry mix	258	1 c	75	.3	.26
Chicken gravy:					
Recipe	130	1/2 c	163	.3	.53
Canned	238	1 c	189	.5	.97
Chili beef soup	250	1 c	169	.8	1.40

	Wt (g)	Svg	Cal	Vit E (mg) 100 Cal	Svg
Soups, Sauces & Gravies, continued:					
Cheese sauce:					
Recipe	101	1/2 c	216	.5	1.1
From mix with milk	279	1 c	305	.1	.2
Hollandaise sauce, recipe	160	1 c	867	1.3	10.9
White sauce, recipe	250	1 c	395	.9	3.5

Vegetables & Legumes

	Wt (g)	Svg	Cal	Vit E (mg) 100 Cal	Svg
Artichoke:					
Globe (300g whole)	120	1 ea	60	<.1	.02
Hearts, marinated in oil	170	6 oz	168	30	50
Asparagus, pieces:					
Fresh, uncooked	67	1/2 c	15	8.9	1.33
Cooked from fresh	90	1/2 c	23	10.9	2.45
Cooked from frozen	180	1 c	50	4.8	2.44
Canned pieces, drained	121	1/2 c	16	6.9	1.1
BEANS: see also Garbanzo, Lentils, Soybeans.					
Baked beans (dry White beans with spices and sauce):					
Canned w/frankfurters	257	1 c	366	.4	1.53
Canned w/pork, sweet sce	253	1 c	282	.5	1.53
Canned w/pork, tomato sce	253	1 c	247	.6	1.53
Black beans, ckd fr/dry	172	1 c	227	.5	1.03
Great northern, ckd fr/dry	177	1 c	210	.9	1.95
Kidney beans, all types:					
Ckd fr/ dry	177	1 c	225	.5	1.15
Canned with liquid	256	1 c	208	.5	1.0
Lima beans:					
Cooked from fresh	170	1 c	208	1.3	2.7
Ckd fr/ frozen, baby	90	1/2 c	94	1.7	1.57
Ckd fr/frozen, large	85	1/2 c	85	1.7	1.47
Ckd fr/ dry	188	1 c	217	2.9	6.2
Canned, drained	170	1 c	164	1.8	3.0
Canned w/liquid	241	1 c	191	1.6	3.14
Navy, ckd fr/dry	182	1 c	259	.8	2.05
Pinto, ckd fr/dry	171	1 c	235	.4	.90
Refried beans, canned	253	1 c	270	.5	1.45
White beans, ckd from dry	179	1 c	253	.8	2.0
Bean sprouts (Mung beans) fresh	104	1 c	31	.7	.21
Beet greens:					
Fresh, pieces	19	1/2 c	4	7.1	.28
Cooked from fresh, drained	144	1 c	40	6.0	2.39
BROCCOLI, chopped:					
Fresh, chopped	88	1 c	24	2.5	.60
Ckd from fresh	156	1 c	44	4.1	1.80
Ckd from frozen	184	1 c	51	3.9	1.97
Brussels sprouts:					
Ckd from fresh	156	1 c	60	2.2	1.33
Ckd from frozen	155	1 c	65	1.5	1.0
CABBAGE:					
Common, fresh	70	1 c	16	.3	.041
Common, cooked	150	1 c	32	.1	.043
Bok choy, raw	70	1 c	9	1.0	.091
Bok choy, cooked	170	1 c	20	.6	.114
Pe-Tsai, raw	76	1 c	11	.9	.099
Pe-Tsai, cooked	119	1 c	16	1.0	.155
Red cabbage, raw	70	1 c	19	.74	.14
Red cabbage, cooked	75	1/2 c	16	.75	.12
Savoy, raw,	70	1 c	20	.7	.14
Savoy, cooked	145	1 c	35	.8	.29

	Wt (g)	Svg	Cal	Vit E (mg) 100 Cal	Svg
CARROTS:					
Fresh, 7.5"x1-1/8" dm.	72	1 ea	31	0.9	.29
Fresh, grated	55	1/2 c	24	.9	.22
Ckd fr/fresh slices	78	1/2 c	35	2.0	.71
Ckd fr/frozen slices	73	1/2 c	26	2.6	.66
Canned, drained	73	1/2 c	17	3.9	.66
Canned with liquid	123	1/2 c	28	2.5	.70
Cauliflower:					
Fresh pieces	50	1/2 c	12	1.1	.13
Cooked fr/fresh	62	1/2 c	15	.7	.10
Cooked from frozen	180	1 c	34	.6	.20
Celeriac, celery root, cooked	100	3.5 oz	25	2.0	.50
Celery:					
Fresh stalk (7.5" = 40g)	40	1 ea	6.4	3.4	.22
Cooked, diced	150	1 c	27	.7	.20
Chard, Swiss:					
Fresh, chopped	36	1 c	7	7.9	.54
Cooked	175	1 c	35	8.6	3.0
CORN:					
Fresh kernels	77	1/2 c	66	.8	.51
Cooked from fresh	82	1/2 c	89	.5	.40
Cooked from frozen	82	1/2 c	67	.3	.22
Canned, vacuum pack	210	1 c	166	.7	1.2
Corn, cream style, canned	128	1/2 c	93	.7	.64
Cucumber, whole, 8" x 2+"	301	1 ea	39	1.2	.45
Dandelion greens:					
Fresh, chopped	55	1 c	25	5.5	1.38
Cooked from fresh	105	1 c	35	6.7	2.35
Escarole, chopped	50	1 c	8.5	2.4	.20
Garbanzo beans, ckd from dry	164	1 c	269	.7	1.92
Jerusalem artichoke, fresh	150	1 c	114	.2	.27
Kale, chopped:					
Fresh, chopped	67	1 c	33	16.2	5.36
Ckd fr/fresh	130	1 c	42	17.8	7.41
Ckd fr/frozen	130	1 c	39	15.7	6.13
Leeks, chopped, cooked	52	1/2 c	24	1.9	.45
Lentils, ckd fr/dry	198	1 c	231	.5	1.16
Lettuce, chopped:					
Butterhead	56	1 c	7.3	7.9	.58
Iceberg	56	1 c	7.3	.8	.06
Loose leaf	56	1 c	10.1	2.2	.22
Romaine	56	1 c	9.0	2.5	.22
Mustard greens:					
Fresh, chopped	56	1 c	15	7.7	1.13
Ckd fr/fresh	140	1 c	21	14.3	3.0
Ckd fr/frozen	150	1 c	29	9.8	2.8
Onions, chopped:					
Fresh, chopped	160	1 c	61	.8	.50
Ckd from fresh	105	1/2 c	46	.7	.31
Parsley:					
Fresh, chopped	30	1/2 c	10	5.0	.50
Freeze dried	1.4	1/4 c	4	5.0	.20
Parsnips, sliced, ckd	156	1 c	125	1.0	1.2
PEAS:					
Green peas:					
Fresh, uncooked	145	1 c	118	1.7	2.03
Ckd from fresh	160	1 c	134	2.5	3.4
Ckd from frozen	80	1/2 c	63	.6	.40
Canned, drained	85	1/2 c	59	1.3	.76
Canned with liquid	124	1/2 c	61	1.3	.77
Green peas, edible-pods:					
Fresh, uncooked	145	1 c	61	6.4	3.9
Ckd from fresh	160	1 c	67	7.0	4.7
Ckd from frozen	80	1/2 c	42	5.5	2.3
Split peas, ckd from dry	196	1 c	231	.7	1.55
Peas & carrots:					
Ckd fr/frozen	80	1/2 c	38	2.4	.90
Canned w/liquid	128	1/2 c	48	1.5	.73

ESHA Research

	Wt (g)	Svg	Cal	Vit E (mg) 100 Cal	Vit E (mg) Svg
Vegetables & Legumes, continued:					
PEPPERS, SWEET:					
Fresh, chopped	50	1/2 c	14	2.4	.34
Cooked from fresh, chopped	68	1/2 c	19	1.3	.244
PEPPERS, HOT:					
Fresh, chopped	75	1/2 c	30	1.7	.50
Canned, drained	68	1/2 c	17	1.4	.24
Jalapeno, canned, chopped	68	1/2 c	17	2.7	.45
POTATOES:					
Baked potato	202	1 ea	220	<.1	.06
Boiled potato	135	1 ea	118	<.1	.06
Cottage fried	50	10 strips	109	.3	.29
French fries, from frozen:					
Oven heated	50	10 strips	111	.3	.29
Fried in oil	50	10 strips	158	.2	.29
Mashed, prepared with milk	210	1 c	162	.1	.21
Potatoes au gratin, recipe	245	1 c	322	.2	.65
Potato chips, plain	28	14 ea	148	1.3	1.87
Pumpkin:					
Cooked from fresh	245	1 c	50	3.3	1.63
Canned	123	1/2 c	42	2.6	1.1
Seaweed (kelp), fresh	28	1 oz	12.2	2.0	.25
Soybeans, cooked from dry	172	1 c	298	4.0	11.8
SOYBEAN PRODUCTS: see tofu in this section; and roasted soybeans in Nuts and Seeds.					
Spinach:					
Fresh, chopped	56	1 c	12	12.2	1.5
Cooked from fresh	180	1 c	41	9.8	4.0
Cooked from frozen, leaf	190	1 c	53	6.6	3.5
Canned, drained	214	1 c	50	1.0	.50
SQUASH, SUMMER varieties:					
Crookneck, fresh slices	130	1 c	24	1.7	.40
Crookneck, cooked slices	180	1 c	36	.7	.25
Scallop, cooked slices	90	1/2 c	14	1.8	.25
Scallop, fresh slices	130	1 c	24	1.3	.30
Zucchini, fresh slices	130	1 c	19	2.7	.51
Zucchini, cooked slices	180	1 c	29	.9	.25
SQUASH, WINTER varieties:					
Acorn (Danish) baked	245	1 c	137	1.2	1.61
Acorn (Danish) boiled	245	1 c	83	1.4	1.20
Butternut, baked	245	1 c	99	1.6	1.61
Butternut, cooked f/ frozen	240	1 c	94	1.4	1.35
Succotash:					
Cooked from fresh	192	1 c	222	1.0	2.15
Cooked from frozen	170	1 c	158	1.1	1.74
Sweet potato:					
Baked, peeled after	114	1 ea	118	4.4	5.2
Boiled, peeled	151	1 ea	160	3.8	6.0
Candied, recipe	105	1 pce	144	.5	.8
Canned, mashed	128	1/2 c	129	3.5	4.5
Canned, vacuum pack	255	1 c	233	3.9	9.0
Tofu (soybean curd):	124	1/2 c	94	4.5	4.2
TOMATOES:					
Fresh, whole	123	1 ea	26	3.3	.86
Fresh, chopped	180	1 c	38	3.3	1.26
Cooked from fresh	240	1 c	65	2.5	1.6
Canned, whole	240	1 c	47	1.9	.89
Tomato juice	244	1 c	42	1.3	.54
Tomato products, canned:					
Tomato paste	262	1 c	220	1.6	3.52
Tomato puree	250	1 c	102	.6	.60
Tomato sauce	245	1 c	74	1.6	1.2

	Wt (g)	Svg	Cal	Vit E (mg) 100 Cal	Vit E (mg) Svg
Turnip greens:					
Cooked from fresh	144	1 c	29	8.5	2.46
Cooked from frozen	82	1/2 c	24	5.8	1.4
VEGETABLES, MIXED:					
Broccoli, cauliflower & red pepper	95	2/3 c	25	2.4	.61
Mixed vegetables (corn, peas, carrots, green beans, lima beans):					
Cooked from frozen	182	1 c	107	.7	.76
Canned, drained	163	1 c	77	.9	.68
Vegetable juice cocktail	242	1 c	46	2.6	1.2
Yam, orange: see Sweet potatoes.					

Other
Cooking Ingredients, Condiments, Flavorings, Spices, Sweeteners, Misc.

	Wt (g)	Svg	Cal	Vit E (mg) 100 Cal	Vit E (mg) Svg
CANDY & CANDY BARS:					
Almonds, sugar coated	28	7 ea	146	2.7	3.9
Chocolate covered:					
Almonds	165	1 c	935	3.1	28.7
Coconut	28	1 oz	133	.4	.5
Peanuts	170	1 c	954	1.7	16.5
Raisins	187	1 c	733	.5	3.5
Divinity with nuts	20	1 pce	80	.7	.6
Fudge, chocolate with nuts	28	1 oz	114	.6	.7
Milk chocolate:					
Plain	28	1 oz	145	1.0	1.40
With almonds	28	1 oz	150	1.8	2.75
With peanuts	28	1 oz	155	1.3	2.05
With rice cereal	28	1 oz	140	.8	1.12
REESE's peanut butter cup	45	2 ea	240	1.8	4.41
Catsup	245	1 c	255	.3	.82
Chili sauce:					
Tomato based	273	1 c	284	.3	.82
Hot, red pepper	31	2 T	6	4.2	.25
CHOCOLATE:					
Baking, unsweetened	28	1 oz	145	1.2	1.68
Bittersweet	28	1 oz	141	1.2	1.68
Choc. chips, semi-sweet	170	1 c	860	1.2	10.20
Dark, sweet	28	1 oz	150	.3	.45
Hot fudge topping	38	2 T	129	.3	.38
Syrup, thin type	38	2 T	85	.4	.38
Cocoa powder	86	1 c	224	.9	1.94
Mustard, prepared	5	1 tsp	3.9	5.5	.22
Olives, pitted:					
Green	39	10 ea	45	1.4	.62
Ripe, black	45	10 ea	52	2.3	1.20
Paprika	6.9	1 T	20	2.8	.56
Poppyseed	8.8	1 T	47	2.1	.97
Salsa, homemade	108	1/2 c	46	2.3	1.79
Yeast:					
Brewer's	8	1 T	25	1.6	.40
Dry active	30	4 T	80	1.9	1.50

Vitamin K

This vitamin is necessary for the normal clotting of blood. It is partially synthesized by bacteria in the intestine, so that people taking antibiotics, especially sulfa drugs, may become deficient because these drugs destroy the intestinal bacteria.

Recommended Dietary Allowances for Adults

USA RDA: Men, 80 mcg per day; Women, 65 mcg per day. This amount is based on 1 mcg/kg of body weight. These RDA's were first established in the 1989 recommendations, and replace the previous recommended range of 70-140 mcg/day.
Canadian RNI: States that there is evidence that a basic need may range from 0.03 to 1.5 mcg per kilogram of body weight per day. Greatest need may be during the time immediately after birth.

Food Sources

Many foods contain good amounts of vitamin K, especially the green leafy vegetables, milk, and members of the cabbage family. Soybean oil and egg yolk also contain this vitamin.

	Wt (g)	Serving	Vitamin K mcg / svg		Wt (g)	Serving	Vitamin K mcg / svg
Dairy & Eggs				**Vegetables**, continued:			
				Cabbage, raw, shredded	70	1 c	104
Egg yolk, large egg	16.6	1 ea	24.4	Carrot, raw whole	72	1 ea	9
Milk, cow, fluid	244	1 c	10	Cauliflower, raw, chopped	100	1 c	96
Milk, dry instant	23	1/3 c	2.3	Cucumber, raw slices, 1/8" thick	28	7 pces	1.4
Milk, human	246	1 c	5-9	Garbanzo beans, dry	100	1/2 c	132
Milk formulas: consult label				Garbanzo beans, sprouted	100	3.5 oz	48
				Green beans:			
Fruits				Fresh, uncooked	110	1 c	28
				Cooked from frozen	135	1 c	43
Apple w/skin, 2.75" diam	138	1 ea	4	Lentils, dry	96	1/2 c	214
Orange, 180g w/peel	131	1 ea	6.6	Lentils, sprouted, fresh	38	1/2 c	15
Strawberries, fresh	149	1 c	21	Lettuce, chopped	56	1 c	63
				Mushrooms, fresh, sliced	35	1/2 c	2.8
				Peas, mature, dry	99	1/2 c	80
Grains & Grain Products				Peas, sprouted, raw	120	1 c	34
				Potato, baked	202	1 ea	8
Oats, rolled, dry	81	1 c	63	Seaweed:			
Wheat:				Dulse/rockweed, dried	28	1 oz	482
Bran	9	1/4 c	7.5	Seagrass	28	1 oz	70
Flour, whole wheat	120	1 c	36	Sealettuce	28	1 oz	19
Wheat germ	19	1/4 c	7.3	Soybeans, mature, dry	93	1/2 c	177
				Spinach:			
Meats				Raw, chopped	56	1/2 c	149
				Frozen	90	1/2 c	124
				Tomato, raw, 2.6"diam			
Ground beef, raw	100	3.5 oz	4	Red, ripe	123	1 ea	28
Liver, beef, raw	85	3 oz	88	Green	123	1 ea	58
Liver, chicken, raw	20	1 ea	16	Turnip greens, raw, chopped	56	1 c	364
Liver, calf (veal)	85	3 oz	23	Watercress, chopped	17	1/2 c	9.7
Vegetables & Legumes				**Other**			
Asparagus:				Coffee, dry, rounded tsp	1.8	1+ tsp	0.7
Raw spears	58	4 ea	39	Fats & Oils			
Cooked from frozen	180	1 c	49	Corn oil	13.6	1 T	6.8
Bean sprouts, fresh (Mung)	104	1 c	34	Palm oil	13.6	1 T	1.1
Beets, raw, 2" diam	163	2 ea	8	Soybean oil	13.6	1 T	68.0
Broccoli:				Honey	21	1 T	5.3
Raw, chopped	88	1 c	176	Tea, green, dry	0.7	1 tsp	5
Frozen	184	1 c	125				

ESHA Research

Calcium

About 99 percent of all the body's calcium is in the bones and teeth. It makes them strong. But, it is not in there permanently. These deposits are constantly being moved back into the bloodstream and redeposited back into the bones, according to the body's needs and the current balance of other nutrients.

The remaining one percent of calcium plays important roles in the fluid of the blood and the body. Calcium is necessary for the transmission of nerve impulses, the ability of the blood to clot, and for the proper balance of muscle contractions, including the rhythm of the heart muscle. It is also involved in the function of several enzymes, in the synthesis of acetylcholine and the absorption of vitamin B-12.

Calcium is regulated by hormones in the body, which is rather unique, and requires a hormone made with vitamin D. Hormones regulate the comings and goings of calcium between the bones and the fluids. Calcium works with magnesium in tooth enamel and is a strong partner with phosphorus in our bone structure. Our bones are basically calcium phosphate and are, by weight at least, about 40 percent calcium and 45 percent phosphorus. Other dietary items that are essential for, or complementary to, the proper utilization of calcium include copper, zinc, manganese, flourine, silicon and boron.

Nutrient Losses

Vitamin D is required for the absorption and use of calcium, so a deficiency of sunlight or of Vitamin D fortified food will reduce the calcium available to the body. Large amounts of dietary protein promote the urinary excretion of calcium, so that a high protein diet (usually twice what the body needs) can contribute to a loss of calcium in the system. Coincidentally, the general American and Canadian diets are very high in protein. Dietary sodium and caffeine also increase urinary excretion of calcium.

Recommended Dietary Allowances for Adults

The greatest need for calcium is in the growing years. Since experimental evidence indicates that bone density in the elderly is related to peak bone density achieved during the period of bone formation, special attention should be given to intakes throughout childhood to age 25 years.

USA RDA: Adults age 25 and over 800 mg.; young adults 19-24, 1200mg (pregnant or nursing: 1200 mg)
Canadian RNI: Men, 800 mg; Women, 700 mg (1300 mg for pregnancy or lactation/nursing).
World Health Organization (WHO): 400-500 mg per day for adults.

Canada and the United States recommend twice the amount of the World Health Organization, partly based on the greater protein intake in our diets. Because of the prevalence of Osteoporosis in our American society, there is pressure to increase the recommended amount for women over 50 years of age. However, there are many causes of this bone thinning disease, and lack of dietary calcium is only one factor. Major factors include lack of exercise (bones are stronger if they get continual use), lack of other needed nutrients such as sunlight or vitamin D, diets excessive in protein, sodium and caffeine, and changes in hormones (occurring primarily in women after menopause or after a hysterectomy).

Food Sources

The best sources are milk, cheese, related dairy products, dark green leafy vegetables, legumes (dried beans and peas), lime-processed tortillas, soft bones of canned fish, and tips of poultry leg bones.

Toxicity

Supplementation to a level much above the current RDA is not recommended. An excessive calcium intake can inhibit the absorption of essential minerals such as iron and zinc. Supplementation to 2,500 mg of Calcium per day can place susceptible males at high risk of kidney stone formation.

	Wt (g)	Svg	Cal	Calcium (mg) 100 Cal	Svg

Baked Goods

Breads, Cakes, Cookies, Crackers, Muffins, Pancakes, Pastries, Pies, Rolls, and some desserts.

	Wt (g)	Svg	Cal	100 Cal	Svg
Bagel, plain or egg, 3.5" diam.	68	1 ea	180	11	20
Biscuit, from mix	28	1 ea	94	63	59
BREADS:					
Boston brown, cnd, 1/2" slice	45	1 pce	95	43	41
Cornbread, muffin:					
Recipe	45	1 ea	145	46	66
From mix	45	1 ea	145	21	30
Cracked wheat bread	25	1 pce	65	25	16
French, 5" x 2.5" x 1"	35	1 pce	100	39	39
Mixed grain	25	1 pce	65	42	27
Oatmeal bread	25	1 pce	65	23	15
Pita pocket bread, 6.5" diam.	60	1 ea	165	30	49
Pumpernickel, 5" x 4" x 3/8"	32	1 pce	80	29	23
Raisin bread	25	1 pce	68	37	25
Rye, light, 5" x 3.5" x 7/16"	25	1 pce	65	31	20
Wheat bread (white & whole wheat flour)	25	1 pce	65	49	32
White bread	28	1 pce	75	47	35
Whole wheat bread	28	1 pce	70	29	20
CAKES, piece = 1/16th cake unless stated otherwise.					
Angel food, 1/12 tube	53	1 pce	125	35	44
Carrot, cream cheese frosting	112	1 pce	406	7	27
Chocolate, choc frosting	69	1 pce	235	17	41
Cheesecake:					
Recipe, 1/12	92	1 pce	278	19	52
Fr/mix, 1/8	103	1 pce	300	60	181
Coffee cake, mix, 2.4" x 2.8"	72	1 pce	230	19	44
Fruitcake, dark, 2/3" arc	43	1 pce	165	25	41
Gingerbread, mix, 3" x 3"	63	1 pce	174	33	57
Pound cake, mix, 1/2" slice	30	1 pce	115	17	20
White cake:					
Choc frosting	77	1 pce	291	13	38
Coconut or white frost	70	1 pce	265	13	34
Yellow cake choc. frosting	69	1 pce	235	27	63
Cheese puffs (Cheetos)	28	1 oz	158	11	18
Cherry crisp, 3" x 3" piece	138	1 pce	157	18	29
Chips: see corn this section; potato chip under Vegetable.					
COOKIES:					
Butter cookies	25	5 ea	115	28	32
Choc. Chip-avg all	43	4 ea	197	7	14
Fig bars	56	4 ea	210	19	40
Sandwich type-all	40	4 ea	195	6	12
Sugar cookies, fr/ refrigerated dough	48	4 ea	235	21	50
Corn chips	28	1 oz	155	23	35
CRACKERS:					
Armenian cracker bread	28	4 pce	117	18	21
Cheese crackers	10	10 ea	50	22	11
Cheese, w/peanut butter fill	30	4 ea	150	18	26
Round, like Ritz	9	3 ea	45	20	9
Rye wafers, whole grain	14	2 ea	55	13	7
Sesame crackers	12	4 ea	60	33	20
Wheat cracker, thin	8	4 ea	35	9	3
Cream puff, custard filling	110	1 ea	280	23	64
Croissant, 4.5" x 4" x 2"	57	1 ea	235	9	20
Danish pastry:					
Plain	57	1 ea	220	27	60
With fruit	65	1 ea	235	7	17
Doughnuts:					
Cake type, medium	50	1 ea	210	11	23
Yeast, jelly-filled	65	1 ea	226	12	28

	Wt (g)	Svg	Cal	Calcium (mg) 100 Cal	Svg
Eclair, custard filled, choc icing	94	1 ea	262	24	62
English muffin, enriched:					
Plain muffin	57	1 ea	140	69	96
Sourdough	56	1 ea	129	87	112
With raisins	56	1 ea	146	53	78
MUFFINS:					
Blueberry, recipe	45	1 ea	135	40	54
Blueberry, fr/mix	45	1 ea	140	11	15
Bran, wheat, recipe	45	1 ea	125	48	60
Bran, wheat, fr/mix	45	1 ea	140	19	27
Cornmeal, recipe	45	1 ea	145	46	66
Cornmeal, fr/mix	45	1 ea	145	21	30
Pancakes, 4" diam:					
Buckwheat, fr/mix	27	1 ea	55	107	59
Plain, recipe	27	1 ea	60	45	27
Plain, fr/mix	27	1 ea	60	60	36
Whole wheat, 5" dm, f/mix	52	1 ea	94	55	52
Peach crisp, 3" x 3"	139	1 pce	166	14	24
PIES: piece= 1/16 of 9" pie.					
Banana cream	198	1 pce	319	30	100
Chocolate cream	175	1 pce	311	51	160
Coconut cream	172	1 pce	343	43	146
Coconut custard	165	1 pce	384	38	146
Custard pie	152	1 pce	293	42	124
Pumpkin	200	1 pce	367	58	212
Rhubarb	160	1 pce	405	25	102
PopTart type pastry, fortified	54	1 ea	210	50	104
Pretzels, thin twists	60	10 ea	240	7	16
ROLLS:					
Cinnamon bun, small	50	1 ea	158	17	27
Dinner roll, 2.5" x 2"	28	1 ea	85	39	33
Hamburger bun	45	1 ea	129	47	61
Hard roll, white	50	1 ea	155	15	24
Hotdog bun	40	1 ea	115	47	54
Rye roll, dark	28	1 ea	79	36	28
Rye roll, light	28	1 ea	76	29	22
Submarine roll (Hoagie)	135	1 ea	400	25	100
Whole wheat roll	35	1 ea	88	28	25
Stove Top Stuffing	108	1/2 c	176	23	41
Taco shell	13.6	1 ea	59	44	26
Tortillas:					
Corn, enr, 6" diam., fried	30	1 ea	87	48	42
Flour, 10.5" diam.	57	1 ea	168	20	34
Flour, 8" diam.	35	1 ea	105	20	21
Tortilla chips:					
Plain	28	1 oz	139	59	82
Doritos, taco flavor	28	1 oz	140	32	45
Waffles:					
Homemade, 7" diam.	75	1 ea	245	63	154
From mix, 7" diam.	75	1 ea	205	87	179
From frozen, 4" diam.	35	1 ea	98	30	29

Dairy & Dairy Products

	Wt (g)	Svg	Cal	Calcium (mg) 100 Cal	Svg
CHEESE (1.5" cube ≈ 1 oz):					
American, processed	28	1 oz	106	164	174
American cheese food, cold pack	28	1 oz	94	154	145
American cheese food, jar	28	1 oz	93	175	163
American cheese spread	28	1 oz	82	194	159
Blue cheese	28	1 oz	100	150	150
Brick cheese	28	1 oz	105	182	191
Brie cheese	28	1 oz	95	55	52
Camembert	28	1 oz	85	129	110
Caraway	28	1 oz	107	179	191
Cheddar cheese	28	1 oz	114	179	204
Cheshire	28	1 oz	110	165	182

ESHA Research

	Wt (g)	Svg	Cal	Calcium (mg) 100 Cal	Svg		Wt (g)	Svg	Cal	Calcium (mg) 100 Cal	Svg
Dairy & Dairy Products, Cheese, continued:						**MILK BEVERAGES** & mixes:					
Colby cheese	28	1 oz	112	173	194	**Chocolate flavored, to be mixed w/water:**					
Cottage cheese:						Powder (includes dry milk)	28	1 oz	100	89	89
Lowfat 1%	226	1 c	164	84	138	Drink, prepared	206	3/4 c	100	89	89
Lowfat 2%	226	1 c	205	76	155	**Chocolate flavored, to be mixed w/milk:**					
Creamed	225	1 c	225	58	130	Powder	21.6	3/4 oz	75	11	8
Creamed, w/fruit	226	1 c	279	39	108	Drink, w/whole milk	266	1 c	226	133	300
Dry curd	145	1 c	123	37	46	Instant Breakfast, dry powder	37	1 env	130	8	10
Cream cheese (1 T = 15g)	28	1 oz	99	23	23	Milkshakes (10 fl oz = 1.25 c):					
Edam cheese	28	1 oz	101	205	207	Chocolate	283	1.25 c	360	89	319
Feta cheese	28	1 oz	75	187	140	Strawberry	283	1.25 c	319	100	320
Fontina	28	1 oz	110	142	156	Vanilla	283	1.25 c	314	110	344
Gjetost	28	1 oz	132	86	113						
Gorgonzola	28	1 oz	111	134	149	**MILK DESSERTS:**					
Gouda cheese	28	1 oz	101	196	198	Custard, baked from recipe	265	1 c	305	97	297
Gruyere	28	1 oz	117	245	287	Ice cream bars:					
Liederkranz	28	1 oz	87	126	110	Creamsicle	66	1 ea	103	45	46
Limburger	28	1 oz	93	152	141	Drumstick	60	1 ea	186	36	67
Monterey jack	28	1 oz	106	200	212	Fudgesicle	73	1 ea	91	142	129
Mozzarella, part skim, low moisture	28	1 oz	80	259	207	Ice cream, vanilla:					
Muenster	28	1 oz	104	195	203	Regular	133	1 c	269	65	176
Neufchatel	28	1 oz	74	28	21	Rich	148	1 c	349	43	151
Parmesan, grated (1 T = 5g)	28	1 oz	129	302	390	Soft serve	173	1 c	377	63	236
Pimento, processed	28	1 oz	106	164	174	Ice milk, vanilla:					
Provolone	28	1 oz	100	214	214	Hard	131	1 c	184	96	176
Ricotta, w/part skim milk	246	1 c	340	197	669	Soft serve	175	1 c	223	123	274
Romano	28	1 oz	128	236	302	**PUDDINGS**, prepared:					
Roquefort	28	1 oz	105	179	188	(5 oz can ≈ 1/2 + cup)					
Swiss cheese	28	1 oz	107	254	272	Assorted flavors, low calorie	130	1/2 ea	69	239	165
Swiss cheese food	28	1 oz	92	223	205	Chocolate:					
Swiss, processed	28	1 oz	95	231	219	Ckd or instant-mix	260	1 c	305	90	276
						Canned	142	1 can	205	36	74
CREAM, SWEET, fluid:						Coconut, from instant	149	1/2 c	184	80	148
Coffee or table cream	15	1 T	30	47	14	Lemon, from instant	149	1/2 c	178	83	147
Half and half	15	1 T	20	80	16	Rice pudding:					
Light whipping cream	239	1 c	699	24	166	Ckd from mix	132	1/2 c	155	86	133
Heavy whipping cream	238	1 c	821	19	154	From instant	149	1/2 c	175	86	150
						Tapioca pudding:					
CREAM, SWEET, whipped:						Ckd from mix	130	1/2 c	145	90	131
Heavy cream	119	1 c	410	19	77	Canned	142	1 can	160	74	119
Pressurized	60	1 c	154	40	61	Vanilla:					
						Ckd or instant fr/mix	130	1/2 c	148	89	130
CREAM, SOUR:						Canned	142	1 can	220	36	79
Cultured, dairy	14	1 T	30	54	16	Pudding pops:					
Half and half	15	1 T	20	80	16	Average other flavors	57	1 ea	94	81	76
Cream, sour, Imitation, non-dairy	14	1 T	30	1	.4	Chocolate or fudge	57	1 ea	99	88	87
						Sherbet (2% fat)	193	1 c	270	38	103
CREAM SUBSTITUTES (non-dairy):						**Yogurt, frozen**, average	174	1 c	220	109	240
Coffee whitener, liquid	15	1 T	20	7	1.4						
Coffee whitener, powdered	6	1 T	32	4	1.3	**YOGURT:**					
Dessert topping (e.g.Coolwhip)	75	1 c	239	2	5	Lowfat, plain	227	1 c	144	288	415
Kefir beverage	233	1 c	160	219	350	Lowfat, fruit	227	1 c	231	149	345
						Lowfat, coffee or vanilla	227	1 c	193	201	388
MILK (cow):						Nonfat	227	1 c	127	356	452
Buttermilk (< 1% fat)	245	1 c	99	288	285	Whole	227	1 c	138	199	275
Lowfat 1%	244	1 c	102	294	300	Yogurt cheese, recipe	208	1 c	222	322	715
Lowfat 2%	244	1 c	121	245	297						
Skim milk	245	1 c	86	351	302						
Whole milk (3.3% fat)	244	1 c	150	194	291						
Canned:						**Eggs**					
Skim, evaporated	255	1 c	200	369	738						
Whole, evap	252	1 c	340	193	657	Eggs, chicken, cooked or raw:					
Sweetened cond	306	1 c	982	88	868	Whole egg	50	1 ea	77.5	32	25
Dry Instant nonfat	68	1 c	244	343	837	White	33.4	1 ea	17	12	2
Dry buttermilk	120	1 c	464	306	1421	Yolk	16.6	1 ea	59	39	23
Milk (other):											
Goat milk	244	1 c	168	194	326						
Human breast milk (mature)	246	1 c	171	46	79						
Soy milk	240	1 c	79	13	10						

Fruits & Fruit Juices

	Wt (g)	Svg	Cal	Calcium (mg) 100 Cal	Svg
Acerola juice, fresh	242	1 c	51	47	24
APPLE, 2.75" diam:					
With peel	138	1 ea	80	13	10
Without peel	128	1 ea	72	7	5
Apple juice, bottled	248	1 c	116	15	17
Applesauce, unsweetened	244	1 c	106	7	7
APRICOTS:					
Fresh, whole, pitted	106	3 ea	51	29	15
Cnd halves, juice pack	84	3 ea	40	25	10
Cnd halves, heavy syrup	85	3 ea	70	10	7
Dried halves	35	10 ea	83	19	16
Apricot nectar, canned	251	1 c	141	13	18
Avocados, whole:					
California	173	1 ea	305	6	19
Florida	304	1 ea	340	10	33
Bananas, 8.75", 176 g whole	114	1 ea	105	7	7
Blackberries:					
Fresh berries	144	1 c	74	62	46
Frozen, unthawed	151	1 c	97	45	44
Canned	256	1 c	236	23	54
Blackberry juice, fresh	250	1 c	93	32	30
Blueberries:					
Fresh berries	145	1 c	82	11	9
Frozen, unsweetened	155	1 c	78	15	12
Frozen, sweetened, thawed	230	1 c	185	7	13
Canned	256	1 c	225	6	14
Boysenberries:					
Frozen, unthawed	132	1 c	66	55	36
Canned	256	1 c	225	20	46
Cantaloupe/Casaba: see Melons.					
Cherries, sour:					
Frozen, unthawed	155	1 c	72	28	20
Canned	244	1 c	90	30	27
CHERRIES, SWEET:					
Fresh, whole, pitted	68	10 ea	49	20	10
Frozen, thawed measure	259	1 c	232	13	31
Canned, with liquid	257	1 c	213	11	23
Cranberry apple juice	253	1 c	169	11	18
Cranberry juice cocktail	253	1 c	145	6	8
Cranberry sauce, canned	277	1 c	419	3	11
Currants:					
Fresh, Black	112	1 c	71	86	61
Fresh, Red or white	112	1 c	63	59	37
Dried, Zante	144	1 c	407	30	124
Dates, whole, pitted	83	10 ea	228	12	27
Elderberries, fresh	145	1 c	105	52	55
Figs:					
Fresh, medium	50	1 ea	37	49	18
Dried	19	1 ea	48	56	27
Fruit cocktail:					
Juice pack	248	1 c	115	17	20
Heavy syrup	255	1 c	185	9	16
Gooseberries, cnd, w/liquid	252	1 c	185	22	40
Grape juice, bottled	253	1 c	155	14	22
GRAPEFRUIT:					
Half, 241g w/rind	120	1 ea	38	35-36	13-14
Canned sections	254	1 c	152	24	36
Grapefruit juice, fr/frzn	247	1 c	102	20	20
Honeydew: see Melons.					
Kiwi fruit	76	1 ea	46	43	20
Lemon juice:					
Fresh	244	1 c	60	30	18
Bottled	244	1 c	52	50	26
Lime juice:					
Fresh	246	1 c	65	34	22
Bottled	246	1 c	50	60	30
Loganberries	145	1 c	90	43	39
Loganberry juice, fresh	246	1 c	100	16	16
Mandarin oranges, canned	252	1 c	155	12	18
Mango, fresh, slices	165	1 c	108	16	17
Melons: see also Watermelon.					
Cantaloupe cubes	160	1 c	57	32	18
Casaba cubes	170	1 c	45	20	9
Honeydew cubes	170	1 c	60	17	10
Mixed melon balls, frozen	173	1 c	55	31	17
Nectarine, medium	136	1 ea	67	9	6
ORANGE, 2-5/8", 180 g whole	131	1 ea	60	87	52
Orange juice:					
Fresh juice	248	1 c	111	24	27
Chilled	249	1 c	110	23	25
Prepared from frozen	249	1 c	110	20	22
Canned, unsweetened	249	1 c	105	19	20
Orange grapefruit juice, canned	247	1 c	105	19	20
Papaya nectar, canned	250	1 c	142	17	24
Papaya, 454 g with refuse	304	1 ea	117	62	72
PEACHES:					
Fresh slices, peeled	170	1 c	73	12	9
Canned, juice pack	77	1 half	34	15	5
Canned, heavy syrup	81	1 half	60	5	3
Peach nectar, canned	249	1 c	134	10	13
PEARS:					
Fresh, Bartlett, 180g w/refuse	166	1 ea	98	19	19
Canned, juice pack	77	1 half	38	18	7
Canned, heavy syrup	79	1 half	59	7	4
Dried, halves	175	10 ea	459	13	59
Pear nectar, canned	250	1 c	149	7	11
PINEAPPLE:					
Frozen, sweetened	245	1 c	208	11	22
Canned pieces, juice pack	250	1 c	150	23	35
Pineapple juice:					
Canned, unsweetened	250	1 c	140	31	43
From frozen concentrate	250	1 c	129	22	28
Plums:					
Fresh, medium, 2-1/8" diam.	66	1 ea	36	8	3
Canned, juice pack	95	3 ea	55	17	10
Prickly pear fruit, fresh	103	1 ea	42	138	58
Prunes, dried, pitted	84	10 ea	201	21	43
Prune juice, bottled	256	1 c	181	17	31
Raisins, dark, unpacked	145	1 c	435	16	71
Raspberries:					
Fresh berries	123	1 c	60	45	27
Frozen, unsweetened	250	1 c	255	15	38
Canned, with liquid	256	1 c	234	12	27
Raspberry juice, fresh	240	1 c	98	59	58
Rhubarb:					
Fresh, diced	122	1 c	26	654	170
Ckd with sugar	240	1 c	279	125	348
STRAWBERRIES:					
Fresh berries	149	1 c	45	47	21
Frozen, unthawed	149	1 c	52	44	23
Tangerine	84	1 ea	37	32	12
Tangerine juice:					
Canned	249	1 c	125	36	45
From frozen	241	1 c	110	16	18
Watermelon, diced pieces	160	1 c	50	26	13

ESHA Research

Values are for edible portion of foods

	Wt (g)	Svg	Cal	Calcium (mg) 100 Cal	Calcium (mg) Svg
Grains & Grain Products					
Cereals, Grains, Flours, Noodles, Pasta, Popcorn					
Amaranth grain, dry	195	1 c	729	41	298
Barley, cooked:					
Pearled	157	1 c	193	9	17
Whole	200	1 c	200	10	19
BRAN: see Oat, Rice, or Wheat.					
Buckwheat flour, dark	98	1 c	338	9	32
Buckwheat flour, light	98	1 c	340	3	11
Bulgar wheat, cooked	182	1 c	151	12	18
CEREALS, COLD (Ready to eat):					
Some cereals may be fortified					
w/calcium. Check label.					
CEREALS, HOT, cooked:					
Corn grits, cooked	242	1 c	145	0	0
Cream of Rice	244	1 c	126	6	8
Cream of Wheat	244	1 c	140	39	54
Maypo cereal	180	3/4 c	128	73	94
Oatmeal, cooked rolled oats:					
Regular, Quick, or Instant	234	1 c	145	13	18.7
Oatmeal, fortified instant:					
Plain	177	3/4 c	104	157	163
With bran and raisins	195	7/8 c	158	109	173
Other flavors averaged	164	3/4 c	160	105	168
Roman Meal	181	3/4 c	111	20	22
Whole wheat cereal	242	1 c	151	11	17
Corn flour:					
Regular	117	1 c	422	2	8.19
Masa Harina, enriched	114	1 c	416	39	161
Cornmeal, dry:					
Degermed, enriched	138	1 c	505	1	7
Nearly whole, bolted	122	1 c	441	5	21
FLOUR: see specific grain, nut, or vegetable.					
MACARONI, cooked:					
Enriched	140	1 c	197	5	9.8
Whole wheat	140	1 c	174	12	21
Vegetable, enriched	134	1 c	172	8.5	14.7
Noodles:					
Chow mein noodles, dry	45	1 c	237	6	14
Egg noodles, cooked	160	1 c	213	9	19
Spinach noodles, cooked	140	1 c	182	23	42
Oat bran (1T ≈ 6g)	94	3.5 oz	132	41	54.5
Oats, rolled:					
Dry	80	1 c	311	14	42
Cooked: see Cereals, hot.					
PASTA: see noodles, macaroni, spaghetti.					
Quinoa grain, dry	170	1 c	635	16	102
RICE, cooked:					
Brown rice	195	1 c	217	9	19.5
White, regular	205	1 c	264	9	22.6
White, converted	200	1 c	186	18	33
White instant	165	1 c	162	8	13.2
Wild rice	164	1 c	166	3	5
Rice bran	83	1 c	262	6.6	47.3
Rice polish	158	1 c	578	3	15.8
Rye flour:					
Light	102	1 c	361	6	21.4
Dark	128	1 c	415	17	71.7
Soy flour:					
Full fat, average	85	1 c	370	45	168
Lowfat	88	1 c	326	51	165

	Wt (g)	Svg	Cal	Calcium (mg) 100 Cal	Calcium (mg) Svg
Spaghetti, noodles, cooked:					
Enriched	140	1 c	197	5	10
Whole wheat spaghetti	140	1 c	174	12	21
WHEAT:					
Bran	30	1/2 c	65	34	22
FLOUR:					
All purpose, white, unsifted	125	1 c	455	4	19
Cake, sifted	96	1 c	348	4	13.4
Gluten	140	1 c	529	11	56
Self-rising	125	1 c	442	96	423
Semolina	167	1 c	601	5	28.4
Whole wheat flour	120	1 c	407	10	40.8
Wheat germ, raw	100	1 c	360	11	39
Wheat germ, toasted	113	1 c	432	12	51
Wheat, rolled:					
Cooked	240	1 c	142	12	17
Dry	85	1 c	289	11	31
Wheat, sprouted	108	1 c	214	14	30.2
Meats: Fish & Shellfish					
Anchovies (includes bones) cnd	45	11 ea	95	109	104
Bass, baked/broiled	100	3.5 oz	125	69	86
Bluefish, baked/broiled	100	3.5 oz	159	6	9
Carp, baked/broiled	100	3.5 oz	162	32	52
Catfish, fried in cornmeal	100	3.5 oz	229	19	44
Clams:					
Canned, drained	160	1 c	236	63	148
Breaded, fried, small	188	20 ea	379	31	119
Steamed, meat only	90	20 ea	133	62	83
Clam nectar, canned	240	1 c	6	517	31
Cod:					
Baked/broiled	100	3.5 oz	105	13	14
Fried in batter	100	3.5 oz	199	40	80
CRAB meat cooked:					
Alaska King, leg	134	1 ea	129	62	80
Blue crab, ckd, unpkd	135	1 c	138	101	140
Blue crab, cnd	135	1 c	133	103	137
Dungeness	101	3/4 c	85	54	46
Crab, imitation fr/surimi	85	3 oz	87	13	11
Crayfish, ckd, moist heat	85	3 oz	97	27	26
Eel:					
Baked/broiled	100	3.5 oz	236	11	26
Smoked	100	3.5 oz	330	29	95
Fish cakes, fried, fr/frozen	100	3.5 oz	213	33	70
Fish sticks, frozen, heated	57	2 ea	155	7	11
Gefiltefish, sweet, commercial	42	1 pce	35	29	10
HADDOCK:					
Baked/broiled	85	3 oz	95	38	36
Breaded, fried	85	3 oz	175	19	34
Smoked	100	3.5 oz	116	42	49
Halibut, baked/broiled	85	3 oz	119	43	51
Herring:					
Baked or broiled	100	3.5 oz	203	36	74
Canned in oil	100	3.5 oz	208	71	147
Smoked or kippered	100	3.5 oz	217	39	84
Pickled	30	2 pce	79	29	23
Lobster, meat only, ckd	145	1 c	142	62	88
Mackerel:					
Baked/broiled, Atlantic	100	3.5 oz	262	6	15
Canned, Jack	361	1 can	563	155	870
Mullet, baked/broiled	85	3 oz	127	20	26
Ocean perch:					
Baked/broiled	100	3.5 oz	121	113	137
Breaded, fried	85	3 oz	185	50	92

	Wt (g)	Svg	Cal	Calcium (mg) 100 Cal	Svg
OYSTERS:					
Raw, Eastern	248	1 c	170	65	111
Raw, Pacific	248	1 c	200	10-75	20-150
Breaded, fried, Eastern	88	6 ea	173	31	54
Simmered, Eastern	100	3.5 oz	137	65	89
Perch, baked/broiled	92	2 ea	108	87	94
Pike, Northern, baked/broiled	100	3.5 oz	113	65	73
Pollock, baked/broiled:					
Mixed species	100	3.5 oz	99	69	68
Walleye	100	3.5 oz	113	5	6
Pompano, baked/broiled	100	3.5 oz	211	20	43
Rockfish, baked/broiled	100	3.5 oz	121	10	12
SALMON:					
Baked or broiled-avg	85	3 oz	183	3	6
Smoked, Chinook	85	3 oz	99	9	9
Canned, drained (calcium reduced if de-boned):					
Atlantic, small can	220	1 can	281	9	24
Chum, w/bones, #1 tall	369	1 can	521	177	920
Pink, w/bones, #1	454	1 can	631	154	969
Sockeye, w/bones, #1	369	1 can	566	156	883
Sardines, canned, drained: (calcium reduced if deboned)					
Atlantic, 2 sardines = 24 g	92	1 can	192	183	351
Pacific, 1 sardine = 38g	100	3.5 oz	176	136	240
Sea Bass, baked/broiled	100	3.5 oz	124	10	13
Seatrout (Steelhead), ckd	100	3.5 oz	131	15	20
Scallops:					
Breaded, fried	93	6 ea	200	20	39
Steamed	100	3.5 oz	113	27	30
Shark, batter-fried	85	3 oz	194	22	42
SHRIMP:					
Boiled, 2 large ≈ 11 g	100	3.5 oz	99	39	39
Fried, breaded, 2 large ≈ 15 g	90	12 ea	218	28	60
Canned, drained	128	1 c	154	49	75
Canned, w/liquid	100	3.5 oz	102	58	59
Shrimp, imitation fr/surimi	85	3 oz	86	19	16
Smelt, Rainbow, ckd	85	3 oz	106	61	65
Snapper, baked/broiled	100	3.5 oz	128	31	40
SOLE (Flounder):					
Baked, broiled, poached	85	3 oz	99	16	16
Breaded, fried	100	3.5 oz	188	21	40
Fried in batter	85	3 oz	250	20	50
Squid, fried in flour	85	3 oz	149	22	33
Surimi, processed walleye pollock (also see Imitation crab, scallops, shrimp, lobster)	85	3 oz	84	9	7
Swordfish, baked/broiled	100	3.5 oz	155	4	6
Trout, baked/broiled, Rainbow	85	3 oz	129	57	73
Tuna, canned, drained:					
Light tuna, canned in oil	171	1 can	339	7	23
Light tuna, water pack	165	1 can	216	9	20

Meats

Beef, Pork, Ham, Rabbit, Frog legs, Venison, Veal

	Wt (g)	Svg	Cal	Calcium (mg) 100 Cal	Svg
BEEF:					
Chuck blade, pot roasted:					
Lean and fat	85	3 oz	325	3	11
Lean only	85	3 oz	230	5	11
Ground beef, cooked avg:					
Extra lean	85	3 oz	215	3	6
Lean	85	3 oz	231	4	8
Regular	85	3 oz	250	3	8
Frozen patty	85	3 oz	240	4	9

	Wt (g)	Svg	Cal	Calcium (mg) 100 Cal	Svg
Rib, choice, oven roasted:					
Lean and fat	85	3 oz	324	3	10
Lean only	85	3 oz	204	5	10
Round steak, broiled, choice:					
Lean and fat	85	3 oz	233	3	6
Lean only	85	3 oz	165	3	5
Steak, average:					
Lean and fat	85	3 oz	238-276	3	7-8
Lean only	85	3 oz	177	4	6-7
Corned beef, canned	85	3 oz	213	8	17
FROG LEGS, flour-fried	144	6 ea	418	7	28
HAM: See Pork, cured; Turkey Ham; and Sausages & Lunchmeats.					
PORK:					
Pork chop, avg loin/rib, ckd:					
Lean & fat	81	1 ea	281	2	4-9
Lean only	65	1 ea	165	3	3-8
Pork roast, average:					
Lean & fat	85	3 oz	265	3	6.8
Lean only	85	3 oz	206	4	8.0
Shoulder, braised:					
Lean & fat	85	3 oz	293	2	6
Lean part only	67	2.4 oz	166	3	5
Spareribs	177	6.25 oz	703	12	83
HAM — cured pork:					
Roasted, lean and fat	85	3 oz	207	3	6
Roasted, lean only	85	3 oz	133	5	6
Canned, roasted	85	3 oz	140	4	6
Pickled pigs feet	28	1 oz	58	15	9
LAMB:					
Chop, loin, broiled:					
Lean & fat	64	1 ea	201	7	13
Lean only	46	1 ea	100	9	9
Cutlet, lean, ckd avg	85	3 oz	175	7	13
Leg of lamb, roasted:					
Lean & fat	85	3 oz	219	4	9
Lean only	85	3 oz	162	4	7
RABBIT, roasted	85	3 oz	175	7	13
VEAL:					
Cutlet, lean, ckd avg	85	3 oz	166	12	20
Rib roast	85	3 oz	151	7	10
Liver, pan-fried	83	3 oz	208	5	10
VENISON (deer) roasted	85	3 oz	134	4	6

Meat: Poultry

	Wt (g)	Svg	Cal	Calcium (mg) 100 Cal	Svg
CHICKEN:					
All types, cooked:					
Fried	140	1 c	307	8	24
Roasted	140	1 c	266	8	21
Stewed	140	1 c	248	8	20
Canned, boned, w/broth	142	5 oz	235	9	20
Dark meat only, cooked:					
Fried	85	3 oz	203	7	15
Roasted	85	3 oz	174	7	13
Stewed	85	3 oz	163	7	12
Light meat only, cooked:					
Fried	85	3 oz	163	8	13
Roasted	85	3 oz	147	9	13
Stewed	85	3 oz	135	8	11
Breast, roasted, 2 pieces per bird:					
Meat & skin	98	1 ea	193	7	14
Meat only	86	1 ea	142	9	13

Values are for edible portion of foods

	Wt (g)	Svg	Cal	Calcium (mg) 100 Cal	Svg
Poultry, continued:					
CHICKEN, continued:					
Thigh, roasted:					
Meat & skin	62	1 ea	153	5	8
Meat only	52	1 ea	109	7	8
TURKEY:					
Ground turkey, ckd	100	3.5 oz	229	11	25
Roasted:					
All types	85	3 oz	145	14	21
Dark meat	85	3 oz	159	17	27
Light meat	85	3 oz	133	12	16
Frozen, slices w/gravy	142	5 oz	95	21	20
Patty, breaded & fried	64	1 ea	181	5	9

Meat: Sausages & Lunchmeats

	Wt (g)	Svg	Cal	Calcium (mg) 100 Cal	Svg
Barbecue loaf, pork & beef	23	1 pce	40	33	13
Bologna, turkey	28	1 oz	56	42	23
Brotwurst link	70	1 ea	226	15	34
Cheesefurter (cheese smoki)	43	1 ea	141	18	25
Chicken roll, light meat	57	2 oz	90	27	24
Dutch brand loaf	28	1 oz	68	35	24
FRANKFURTER (hot dog):					
Beef, 8/pkg	57	1 ea	184	4	7
Beef and pork, 8/pkg	57	1 ea	183	3	6
Chicken, 10/pkg	45	1 ea	115	37	43
Turkey, 10/pkg	45	1 ea	102	57	58
Ham & cheese roll or loaf	57	2 oz	147	22	33
Luxury loaf	57	2 oz	80	26	21
Olive loaf	57	2 oz	133	47	62
Peppered loaf	28	1 oz	42	36	15
Pickle & pimento loaf	57	2 oz	149	36	54
Pork sausage, ckd patty	27	1 patty	100	9	9
Turkey ham,1 piece = 1 oz	57	2 pce	73	7	5
Turkey roll:					
Light & dark meat	57	2 pce	84	21	18
Light meat only	57	2 pce	83	28	23

Mixed Dishes & Fast Foods

	Wt (g)	Svg	Cal	Calcium (mg) 100 Cal	Svg
Beef & vegetable stew:					
Recipe	245	1 c	220	13	29
Frozen entree	200	7 oz	208	13	26
Beef, macaroni & tomato					
sauce, recipe	226	1 c	189	16	30
BURRITO:					
Bean burrito	174	1 ea	322	56	181
Beef burrito	177	1 ea	463	32	148
Beef & bean burrito	175	1 ea	390	42	165
Deluxe combination	198	1 ea	424	43	183
Cheese soufflé	112	1 c	221	97	214
Chicken a la king, recipe	245	1 c	470	27	127
Chicken chow mein:					
From recipe	250	1 c	255	23	58
Canned	250	1 c	95	47	45
Chicken pot pie, fr/frozen	230	1 ea	430	7	30
Chicken salad w/celery	78	1/2 c	266	6	16
Chili w/beans, canned	255	1 c	286	42	119
Chop suey w/beef & pork	250	1 c	300	20	60
Cole slaw	120	1 c	84	64	54
Corn dog	111	1 ea	330	10	34
Corn pudding	250	1 c	271	37	100
Corned beef hash, canned	220	1 c	382	8	29
Egg salad	183	1 c	438	21	94

	Wt (g)	Svg	Cal	Calcium (mg) 100 Cal	Svg
ENCHILADA:					
Beef enchilada	120	1 ea	292	87	255
Cheese enchilada	120	1 ea	330	138	457
Chicken enchilada	120	1 ea	269	95	256
French toast, recipe	65	1 pce	123	64	79
LASAGNA, recipe:					
With meat	245	1 pce	398	116	460
Without meat	218	1 pce	316	145	457
Macaroni & cheese:					
Recipe	200	1 c	430	84	362
From frozen	200	1 c	254	71	181
Manicotti, frozen entree	225	1 ea	271	101	274
Moussaka, lamb & eggplant	250	1 c	250	52	129
Pizza, cheese:					
Regular crust 1/ 8 of 15"	120	1 pce	290	76	220
Thick crust 1/2 of 10"	208	1 pce	519	69	359
Potato salad, w/mayo & eggs	250	1 c	358	13	48
Quiche Lorraine, 1/8 pie	176	1 pce	600	35	211
SANDWICHES, Fast food:					
Cheeseburger, 3 oz patty	112	1 ea	300	45	135
Cheeseburger, 4 oz patty	194	1 ea	524	45	236
Chicken patty sandwich	157	1 ea	436	10	44
English muffin with egg,					
cheese, bacon	138	1 ea	360	55	197
Fish sandwich:					
Regular with cheese	140	1 ea	420	31	132
Large, without cheese	170	1 ea	470	13	61
Hamburger, 3 oz patty	98	1 ea	245	23	56
Hamburger, 4 oz patty	174	1 ea	445	17	75
Hotdog (frankfurter) on bun	85	1 ea	260	23	59
Roast beef on bun	150	1 ea	345	17	60
SANDWICHES (on part whole wheat					
bread, unless specified as on rye):					
Avocado, cheese,					
tomato & sprouts	195	1 ea	432	69	299
Bacon, lettuce & tomato	135	1 ea	327	24	80
Chicken salad	99.7	1 ea	294	27	79
Corned beef & swiss, on rye	147	1 ea	429	77	331
Egg salad sandwich	111	1 ea	319	30	95
Grilled cheese	117	1 ea	393	108	424
Ham sandwich	122	1 ea	256	31	79
Ham sandwich on rye	116	1 ea	242	20	49
Ham & American cheese	151	1 ea	363	71	256
Ham & swiss on rye	145	1 ea	350	93	325
Peanut butter & jam	100	1 ea	341	24	82
Reuben, grilled, on rye	233	1 ea	480	75	358
Tuna salad sandwich	116	1 ea	303	26	80
Turkey sandwich	122	1 ea	271	28	76
Spinach soufflé	136	1 c	218	106	230
SPAGHETTI (pasta &					
tomato sauce w/cheese):					
Homemade	250	1 c	260	31	80
Canned	250	1 c	190	21	40
Taco:					
Beef	78	1 ea	207	41	85
Chicken	78	1 ea	172	51	87
Tostada:					
Bean tostada	157	1 ea	212	83	177
Beef tostada	192	1 ea	332	56	186
Chicken tostada	157	1 ea	249	65	162
Tuna noodle casserole	202	1 c	251	15	37
Tuna salad	205	1 c	383	9	35
Turkey pot pie, fr/frozen	233	1 ea	416	15	64
Waldorf salad	142	1 c	424	10	44

	Wt (g)	Svg	Cal	Calcium (mg) 100 Cal	Svg
Nuts & Seeds					
ALMONDS, dried, whole	142	1 c	837	45	378
Almond butter	16	1 T	101	43	43
Brazil nuts, dried	140	1 c	919	27	246
Cashews, dry or oil roasted	130	1 c	748	7	53
Filberts (Hazelnuts), whole	135	1 c	853	30	253
Macadamias, oil roasted	134	1 c	962	6	60
MIXED NUTS:					
(cashews, peanuts, brazil nuts, filberts, almonds & pecans)					
Oil roasted	142	1 c	876	17	153
PEANUTS, oil roasted	144	1 c	837	15	126
Peanut butter, smooth	32	2 T	188	6	11
Pecans, dried, chopped	119	1 c	794	5	43
Pistachios, dried, no shells	128	1 c	739	23	173
Pumpkin/squash seeds:					
Kernels, roasted	227	1 c	1185	8	98
Whole, roasted	64	1 c	285	12	35
Sesame seed kernels, dried	38	1/4 c	221	22	49
Sesame flour:					
High fat	28	1 oz	149	30	45
Lowfat	28	1 oz	95	44	42
Partially defatted	28	1 oz	109	39	43
Soybeans, roasted	86	1/2 c	405	5	119
Sunflower seed kernels:					
Dried	144	1 c	821	20	167
Oil roasted	135	1 c	830	9	76
Walnuts, chopped:					
Black	125	1 c	759	10	72
English	120	1 c	770	15	113
Soups, Sauces & Gravies					
GRAVIES:					
Beef gravy:					
Homemade	135	1/2 c	151	6	9
Canned	233	1 c	124	11	14
Brown gravy, from dry mix	258	1 c	75	89	67
Chicken gravy:					
Homemade	130	1/2 c	163	4	6
Canned	238	1 c	189	25	48
From dry packet	260	1 c	85	46	39
SAUCES (also see Other):					
Cheese sauce:					
Recipe	101	1/2 c	216	130	281
Fr/mix w/milk	279	1 c	305	187	569
Hollandaise sauce, recipe	160	1 c	867	16	138
Spaghetti sauce, plain:					
Homemade	220	1 c	179	29	52
Canned	249	1 c	272	26	70
White sauce:					
Recipe	250	1 c	395	74	292
Fr/mix, with milk	264	1 c	240	177	425
SOUPS: Canned unless otherwise stated. RTS = Ready To Serve. For soups with milk, assume whole milk (usually the source of calcium).					
Bean w/ham, chunky, RTS	243	1 c	231	34	79
Beef broth (bouillon):					
Canned	240	1 c	16	94	15
From dry	244	1 c	20	49	9.8
Beef, chunky, RTS	240	1 c	171	18	31
Black bean soup	247	1 c	116	39	45
Celery, cream of, w/milk	248	1 c	165	113	186

	Wt (g)	Svg	Cal	Calcium (mg) 100 Cal	Svg
SOUPS, continued:					
Cheese soup:					
Prep w/milk	251	1 c	230	125	288
Prep w/water	247	1 c	155	92	142
Chicken, chunky, RTS	251	1 c	178	13	24
Chicken, cream of:					
Prep w/milk	248	1 c	191	94	180
Prep w/water	244	1 c	115	30	34
Prep from dry	261	1 c	107	71	76
Chicken broth (bouillon):					
Prep w/water	244	1 c	39	23	9
From dry	244	1 c	21	71	15
Chicken noodle:					
Prep w/water	241	1 c	75	23	17
Chunky, RTS	240	1 c	114	21	24
W/meatballs	248	1 c	99	30	30
Prep fr/mix, w/water	252	1 c	53	60	32
Chicken vegetable, chunky, RTS	240	1 c	167	15	25
Chili beef soup	250	1 c	169	25	43
Clam Chowder:					
New England style	248	1 c	163	115	187
Tomato based:					
Manhatten	244	1 c	78	34	26.8
Manhatten, chunky, RTS	240	1 c	133	50	67
Lentil and ham, RTS	248	1 c	140	30	42
Minestrone, chunky, RTS	240	1 c	127	48	61
Mushroom, cream of:					
Prep w/milk	248	1 c	205	87	178
Condensed, undiluted	251	1 c	257	25	64
Prep fr/dry	253	1 c	96	70	67
Onion soup:					
Canned	241	1 c	57	46	26
Prep fr/dry packet	184	3/4 c	20	45	9
Oyster stew, w/milk	245	1 c	134	125	167
Potato, cream of, w/milk	248	1 c	148	112	166
Split pea soup:					
Canned	253	1 c	189	12	22
Prep from dry	255	1 c	133	17	22
Split pea & ham, chunky, RTS	240	1 c	184	18	33
Tomato:					
Prep w/milk	248	1 c	160	99	159
Prep w/water	244	1 c	86	15	13
Prep fr/dry	265	1 c	102	53	54
Tomato beef noodle	244	1 c	140	13	18
Turkey, chunky, RTS	236	1 c	136	37	50
Turkey noodle	244	1 c	69	17	12
Turkey vegetable soup	241	1 c	74	23	17
Vegetable, chunky, RTS	240	1 c	122	46	56
Vegetable beef	244	1 c	79	22	17
Vegetarian soup	241	1 c	70	30	21
Vegetables & Legumes					
Amaranth leaves:					
Chopped	28	1 c	7	824	61
Boiled	132	1 c	28	986	276
Artichoke, globe, cooked	120	1 ea	60	90	54
Artichoke hearts:					
Frozen, cooked- pkg	240	9 oz	108	46	50
Marinated-jar	170	6 oz	168	23	39
Asparagus, pieces:					
Fresh, uncooked	67	1/2 c	15	100	15
Ckd from fresh	90	1/2 c	22.5	98	22
Ckd from frozen	180	1 c	50	82	4
Canned	121	1/2 c	16	138	22

ESHA Research

Calcium (Ca) mg

Values are for edible portion of foods

	Wt (g)	Svg	Cal	Calcium (mg) 100 Cal	Svg
Vegetables & Legumes, continued:					
BEANS: see also Garbanzo, Lentils, Soybeans.					
Baked beans (dry white beans w/spices & sauces):					
Home prepared	253	1 c	382	41	155
Canned, plain or vegetarian	254	1 c	235	54	128
Canned w/franks	257	1 c	356	34	123
Canned w/pork	253	1 c	268	50	133
Canned w/pork, sweet sce	253	1 c	282	55	155
Canned w/pork, tomato sce	253	1 c	247	57	141
Black beans, cooked	172	1 c	227	21	47
Broadbeans:					
Ckd f/fresh veg	100	3.5 oz	56	32	18
Ckd f/dry legumes	170	1 c	186	33	62
Great northern, ckd f/dry	177	1 c	210	58	121
Green beans (snap):					
Fresh, uncooked	110	1 c	34	121	41
Ckd from fresh	125	1 c	44	132	58
Ckd from frozen	135	1 c	36	169	61
Canned, drained	135	1 c	26	138	36
Canned, with liquid	240	1 c	36	160	58
Hyacinth beans:					
Ckd from fresh (veg)	87	1 c	43	84	36
Ckd from dry (legumes)	194	1 c	228	34	77
Kidney beans, all types:					
Ckd fr/dry	177	1 c	225	22	50
Canned, w/liquid	256	1 c	208	33	69
Lima beans:					
Ckd fr/fresh	170	1 c	208	26	54
Ckd f/frozen, large	170	1 c	170	22	38
Ckd f/frozen, small/baby	90	1/2 c	94	27	25
Ckd f/dry legume	188	1 c	217	15	32
Canned, drained	170	1 c	164	29	48
Canned, w/liquid	241	1 c	191	26	50
Navy beans, ckd f/dry	182	1 c	259	49	128
Pinto beans:					
Canned	240	1 c	186	48	89
Ckd f/dry	171	1 c	235	35	82
Refried beans, canned	253	1 c	270	44	118
White, ckd f/dry	179	1 c	253	52	131
Winged, ckd f/fresh	62	1 c	23	165	38
Yardlong, ckd f/fresh	104	1 c	49	94	46
Yellow wax: See green beans.					
Bean sprouts (Mung beans):					
Fresh sprouts	104	1 c	31.2	45	14
Cked from fresh	124	1 c	62	26	16
Canned, drained	125	1 c	16	113	18
Beet greens, ckd, drained	144	1 c	40	413	165
Beets:					
Cooked, whole	100	2 ea	31	35	11
Canned, diced, drained	85	1/2 c	27	48	13
BROCCOLI, chopped:					
Fresh, uncooked	88	1 c	24	175	42
Ckd from fresh	156	1 c	44	164	72
Ckd from frozen	184	1 c	51	184	94
W/cheese sauce	142	1/2 c	166	72	119
W/hollandaise sauce	95	1/2 c	105	23	24
Brussels sprouts:					
Ckd from fresh	156	1 c	60	93	56
Ckd from frozen	155	1 c	65	58	38
CABBAGE:					
Common, raw	70	1 c	16	200	32
Common, ckd	150	1 c	32	156	50
Bok choy, raw	70	1 c	9	822	74
Bok choy, ckd	170	1 c	20	790	158
Pe-tsai, raw	76	1 c	11	536	59
Pe-tsai, ckd	119	1 c	16	238	38

	Wt (g)	Svg	Cal	Calcium (mg) 100 Cal	Svg
CABBAGE, continued:					
Red cabbage, raw	70	1 c	19	189	36
Red, ckd	75	1/2 c	16	175	28
Savoy, raw	70	1 c	20	125	25
Savoy, ckd	145	1 c	35	126	44
CARROTS:					
Fresh, 7.5" x 1-1/8" dm	72	1 ea	31	61	19
Ckd from fresh, sliced	78	1/2 c	35	69	24
Cooked from frozen	73	1/2 c	26	81	21
Canned, drained	73	1/2 c	17	112	19
Canned, with liquid	123	1/2 c	28	111	31
Carrot juice	123	1/2 c	49	59	29
Cauliflower:					
Fresh pieces	50	1/2 c	12	117	14
Ckd from fresh	62	1/2 c	15	113	17
Ckd from frozen	180	1 c	34	91	31
Celeriac (celery root), ckd	100	3.5 oz	25	104	26
Celery:					
Outer stalk, 7.5" long	40	1 ea	6.4	250	16
Cooked, diced	150	1 c	27	233	63
Chard, Swiss:					
Chopped, fresh	36	1 c	7	269	18
Cooked	175	1 c	35	291	102
Collards:					
Fresh, chopped	36	1 c	11	95	10.4
Ckd from fresh	128	1 c	35	84	29.4
Ckd from frozen	170	1 c	61	89	54
CORN:					
Cooked from frozen	82	1/2 c	67	2	1.6
Canned, drained	82	1/2 c	66	6	4
Cucumber, whole, 8" x 2+"	301	1 ea	39	108	42
Dandelion greens:					
Fresh	55	1 c	25	412	103
Cooked	105	1 c	35	420	147
Dock (sorrel greens):					
Fresh	133	1 c	29	203	59
Cooked	100	1 c	20	190	38
Eggplant, cooked	160	1 c	45	21	10
Endive, fresh, chopped	25	1/2 c	4	325	13
Escarole or curly endive	50	1 c	8.5	306	26
Garbanzo beans (chickpeas):					
Ckd from dry	164	1 c	269	30	80
Canned	240	1 c	285	27	78
Garden cress, fresh	25	1/2 c	8	250	20
Garden cress, cooked	135	1 c	31	265	82
Kale, chopped:					
Fresh	67	1 c	33	273	90
Ckd from fresh	130	1 c	42	226	94
Ckd from frozen	130	1 c	39	459	179
Kohlrabi:					
Fresh slices	140	1 c	38	89	34
Ckd from fresh	165	1 c	48	85	41
Lambquarters, chopped:					
Fresh	56	1 c	24	721	173
Cooked	180	1 c	58	800	464
Leeks, chopped:					
Fresh	104	1 c	63	97	61
Cooked	52	1/2 c	24	67	16
Lentils, ckd fr/dry	198	1 c	231	16	37
Lentils, sprouted:					
Fresh	77	1 c	81	23	19
Stir fried	100	3.5 oz	101	14	14
LETTUCE, chopped, fresh:					
Butterhead	56	1 c	7.3	257	19
Iceberg	56	1 c	7.3	146	11
Loose leaf	56	1 c	10	376	38
Romaine	56	1 c	9	225	20
Lotus root, slices, cooked	89	10 ea	59	39	23

ESHA Research

108

Values are for edible portion of foods

Calcium (Ca) mg

Vegetables & Legumes, continued:

Food	Wt (g)	Svg	Cal	Calcium 100 Cal	Calcium Svg
Mustard greens:					
Fresh, chopped	56	1 c	15	397	58
Ckd from fresh	140	1 c	21	495	104
Ckd from frozen	150	1 c	29	533	152
Okra:					
Ckd f/fresh, pods	85	8 ea	27	197	54
Ckd f/frozen, slices	92	1/2 c	34	259	88
Onions:					
Fresh, chopped	160	1 c	61	52	32
Ckd f/fresh, chopped	105	1/2 c	46	50	23
Ckd f/frozen, chopped	105	1/2 c	30	57	17
Dry, dehydrated flakes	14	1/4 c	45	80	36
Parsley, fresh, chopped	30	1/2 c	10	390	39
Parsnips:					
Fresh slices	133	1 c	100	47	47
Cooked	156	1 c	125	46	58
PEAS:					
Black-eyed peas:					
Ckd from fresh	165	1 c	160	132	211
Ckd from frozen	170	1 c	224	18	40
Ckd from dry	171	1 c	198	21	42
Canned	240	1 c	184	26	48
Green peas:					
Fresh, uncooked	145	1 c	118	31	36
Ckd from fresh	160	1 c	134	33	44
Ckd from frozen	80	1/2 c	63	30	19
Canned, drained	85	1/2 c	59	29	17
Canned, with liquid	124	1/2 c	61	36	22
Green peas, edible-pods:					
Fresh, uncooked	72	1/2 c	30	102	31
Frozen, unprepared	72	1/2 c	30	120	36
Ckd from fresh	160	1 c	67	100	67
Ckd from frozen	80	1/2 c	42	114	48
Split peas, cooked f/dry	196	1 c	231	11	26
Peas & carrots:					
Ckd from frozen	80	1/2 c	38	47	18
Canned, with liquid	128	1/2 c	48	60	29
Peas, mature, sprouted:					
Fresh	120	1 c	154	28	43
Cooked	100	3.5 oz	118	22	26
PEPPERS, hot, green or red:					
Chili pepper, fresh, chpd	75	1/2 c	30	43	13
Chili pepper, canned	68	1/2 c	17	29	5
Jalapeno peppers, canned	68	1/2 c	17	106	18
PEPPERS, SWEET, raw, chopped	50	1/2 c	14	32	4.5
Poi, two finger	240	1 c	269	14	37
POTATOES:					
Baked, flesh and skin	202	1 ea	220	10	21
Baked, potato skins	58	1 ea	96	24	23
Boiled, flesh only	136	1 ea	118	8	9
Dried, dehydrated flakes	45	1 c	159	43	69
French fried potatoes	50	10 strips	148	6	9
Hash browns, from frozen	156	1 c	340	7	24
Mashed:					
Prep w/milk	210	1 c	162	34	55
Prep fr/Instant	215	1 c	239	38	92
Potatoes au gratin, recipe	245	1 c	322	91	292
Potatoes, scalloped, recipe	245	1 c	210	67	140
Potato chips	28	14 ea	148	5	7
Pumpkin, mashed:					
Ckd from fresh	245	1 c	50	74	37
Canned	123	1/2 c	42	76	32
Radish, Daikon	44	1/2 c	8	150	12
Radishes, red	45	10 ea	7	129	9
Radish seeds, sprouted	38	1 c	15.6	124	19
Rutabaga, fresh cubes	140	1 c	51	127	65
Ckd from fresh	85	1/2 c	29	124	36

Food	Wt (g)	Svg	Cal	Calcium 100 Cal	Calcium Svg
Salsify, cooked slices	135	1 c	92	70	64
Sauerkraut, cnd, w/liquid	236	1 c	44	164	72
Seaweed:					
Fresh, Agar	28	1 oz	7.4	207	15
Fresh, Kelp	28	1 oz	12	390	48
Fresh, Lavar	28	1 oz	10	200	20
Fresh, Wakame	28	1 oz	13	332	42
Dried, Agar	28	1 oz	87	203	177
Dried, Spirulina	28	1 oz	82	41	34
Soybeans:					
Fresh, green	256	1 c	376	134	504
Ckd from fresh	180	1 c	255	102	261
Ckd fr/dry legume	172	1 c	298	59	175
Soybeans, mature, sprouted:					
Fresh	35	1/2 c	45	53	24
Steamed from fresh	94	1 c	76	72	55

SOYBEAN PRODUCTS: See tofu this section; miso, natto & tempeh in Other; roasted soybeans in Nuts & Seeds; soy milk in Dairy; soy flour in Grains.

Food	Wt (g)	Svg	Cal	Calcium 100 Cal	Calcium Svg
Spinach:					
Fresh, chopped	56	1 c	12.3	450	55.4
Ckd from fresh	180	1 c	41	595	244
Ckd from frozen	190	1 c	53	521	277
Canned, drained	214	1 c	50	542	271
SQUASH, SUMMER varieties:					
Crookneck, fresh slices	130	1 c	24	117	28
Crookneck, ckd	180	1 c	36	133	48
Scallop, fresh slices	130	1 c	24	104	25
Scallop, ckd	90	1/2 c	14	100	14
Zucchini, fresh slices	130	1 c	19	105	20
Zucchini, ckd	180	1 c	29	79	23
SQUASH, WINTER, mashed:					
Acorn (Danish) baked	245	1 c	137	79	108
Acorn (Danish) boiled	245	1 c	83	78	65
Butternut, baked	245	1 c	99	102	100
Butternut, ckd f/frozen	240	1 c	94	49	46
Hubbard, baked	240	1 c	120	34	41
Hubbard, boiled	236	1 c	70	33	23
Spaghetti, baked or boiled	155	1 c	45	73	33
Succotash:					
Ckd from fresh	192	1 c	222	14	32
Ckd from frozen	170	1 c	158	16	25
Sweet potatoes:					
Baked in skin, then peeled	114	1 ea	118	27	32
Boiled, flesh only	151	1 ea	160	20	32
Cnd, regular, mashed	128	1/2 c	129	30	38
Cnd, vacuum packed	255	1 c	233	24	56
Candied sweet potatoes	105	1 pce	144	19	27
Tofu, raw:					
Firm	126	1/2 c	183	141	258
Regular	124	1/2 c	94	138	130
TOMATOES:					
Fresh, whole	123	1 ea	26	24	6.2
Fresh, chopped	180	1 c	38	24	9
Ckd from fresh	240	1 c	65	22	14.4
Canned, whole (calcium added as firming agent)	240	1 c	47	134	63
Tomato juice, cnd	244	1 c	41.5	53	22
Tomato paste, cnd	262	1 c	220	42	92
Tomato puree, cnd	250	1 c	102	36	37
Tomato sauce, cnd	245	1 c	74	46	34
Turnips, cubed:					
Fresh	130	1 c	35	111	39
Ckd fr/ fresh	78	1/2 c	14	129	18
Turnip greens, ckd fr/fresh	144	1 c	29	683	198
Turnip greens, ckd fr/frozen	82	1/2 c	24	521	125

109

ESHA Research

	Wt (g)	Svg	Cal	Calcium (mg) 100 Cal	Calcium (mg) Svg
VEGETABLE COMBINATIONS, ckd from frozen:					
Broccoli, carrots & pasta	95	2/3 c	88	34	30
Broccoli, cauliflower, red pepper	95	2/3 c	25	116	29
Chinese stir fry vegetables	95	1/2 c	31	206	64
Green beans & spaetzle, Bavarian	95	1/2 c	108	38	41
Mixed vegetables (corn, peas, carrots, green beans, lima beans):					
Cooked from frozen	182	1 c	107	43	46
Canned, drained	163	1 c	77	57	44
Peas, carrots and onions	91	1/2 c	54	48	26
Peas, carrots, onions & butter sauce	71	2/3 c	100	27	27
Peas, carrots, onions & pasta	95	1/2 c	122	59	72
Peas, cauliflower, cream sce	95	1/2 c	118	35	41
Peas, onions, cheese sce	142	1/2 c	165	55	90
Peas, potatoes, cream sce	76	1/2 c	140	31	43
Peas, pasta, corn, cream sce	95	1/2 c	132	34	45
Peas, pasta & mushrooms, cream sauce	95	1/2 c	129	28	36
Vegetable juice cocktail	242	1 c	46	58	27
Water chestnuts, canned slices	70	1/2 c	35	9	3
Watercress, fresh	17	1/2 c	2	1000	20
Yam, orange: see Sweet Potato.					
Yam, white, cubes, cooked	136	1 c	158	25	40
Zucchini: see Squash, summer.					

Other

Cooking Ingredients, condiments, flavorings, other beverages, spices, sweets, misc.

	Wt (g)	Svg	Cal	Calcium (mg) 100 Cal	Calcium (mg) Svg
Beer (12 fl oz = 1.5 c):					
Regular	356	1.5 c	146	12	18
Light	354	1.5 c	100	18	18
BAKING POWDER:					
Mono calcium phosphate monohydrate	3	1 tsp	5	1160	58
Calcium sulfate	2.9	1 tsp	5	3660	183
Straight phosphate	3.8	1 tsp	5	4780	239
Low sodium	4.3	1 tsp	5	4140	207
BEVERAGES, carbonated (12 fl oz = 1.5 c):					
Cola, regular	370	1.5 c	151	6	9
Cola, diet	355	1.5 c	2	600	12
Club soda	355	1.5 c	0	17	17
Cream soda	371	1.5 c	191	10	19
Diet soda, assorted, avg	355	1.5 c	2	700	14
Fruit flavored	372	1.5 c	170	9	15
Ginger ale	366	1.5 c	124	10	12
Butterscotch topping	50	3 T	156	15	24
CANDY AND CANDY BARS:					
Almonds, sugar coated	28	7 ea	146	27	40
Caramel (butterscotch/choc)	28	1 oz	115	37	42
Chocolate kisses	28	6 pce	154	34	53
Chocolate covered almonds	165	1 c	935	30	278
Chocolate covered peanuts	170	1 c	954	21	197
Chocolate covered raisins	187	1 c	733	11	81
Fudge, chocolate	28	1 oz	115	19	22
Fudge, vanilla	28	1 oz	118	25	30
KIT KAT	43	1 ea	210	31	65
KRACKLE	34	1 ea	179	34	60
M & M's plain candies	48	1 pkg	237	33	79
M & M's peanut candies	47	1 pkg	240	25	59
Malted milk balls	28	14 pce	135	47	63
MARS bar	50	1 ea	240	35	85
MILKY WAY bar	60	1 ea	260	33	86
Milk chocolate:					
Plain	28	1 oz	145	34	50
With peanuts	28	1 oz	155	21	32
With almonds	28	1 oz	150	41	61
With rice cereal	28	1 oz	140	34	48
MR. GOODBAR	47	1 ea	250	26	65
REESE'S Peanut Butter Cup	45	2 ea	240	14	35
SNICKERS bar, 2.2 oz	61	1 bar	290	24	70
Caramel topping	50	3 T	155	18	28
Carob flour	103	1 c	185	194	359
Catsup	245	1 c	255	21	47
Chili sauce, tomato based	273	1 c	284	19	55
CHOCOLATE:					
Baking, unsweetened	28	1 oz	145	15	22
Bittersweet	28	1 oz	141	9	13
Choc. chips, semi-sweet	170	1 c	860	6	51
Hot fudge topping	300	1 c	1020	29	300
Syrup , thin	300	1 c	680	7	45
Clam & tomato juice, can	166	.7 c	77	27	21
Cocoa powder	86	1 c	224	55	124
Gatorade	230	1 c	39	59	23
Honey	339	1 c	1030	2	17
Horseradish, prepared	15	1 T	6	150	9
Hummous (humous)	246	1 c	420	30	124
Miso (soybean product)	275	1 c	565	32	183
Molasses:					
Blackstrap	40	2 T	85	322	274
Light	20	1 T	43	77	33
Mustard, prepared	125	1/2 c	94	112	105
Natto (soybean product)	100	3.5 oz	167	62	103
Olives, pitted:					
Green	39	10 ea	45	53	24
Ripe, large black	45	10 ea	52	77	40
Perrier water, 6.5 fl. oz	192	1 ea	0	26	26
Pickle, dill	65	1 ea	12	49	5.9
Ranch style salad dressing	14	1 T	54	28	15
Salsa, recipe	108	1/2 c	46	39	18
Sugar:					
Brown	220	1 c	820	23	187
White, granulated	200	1 c	770	<1	3
SPICES:					
Basil, dried	4.5	1 T	11	864	95
Cinnamon	2.3	1 tsp	6	475	28
Cumin seed	6	1 T	22	255	56
Dill weed, dried	3.1	1 T	8	625	50
Garlic cloves, whole	12	4 ea	18	121	22
Garlic powder	8.4	1 T	28	25	7
Oregano, ground	1.5	1 tsp	4.7	504	24
Poppyseed	8.8	1 T	47	270	127
Tarragon, ground	4.8	1 T	14	393	55
Thyme, ground	4.3	1 T	12	675	81
SYRUPS:					
Corn syrup, dark	328	1 c	944	16	151
Corn syrup, light	328	1 c	912	16	146
Maple syrup	20	1 T	50	66	33
Pancake	84	1/4 c	244	1	2
Sorghum	20	1 T	52	58	30
Tempeh (soybean product)	83	1/2 c	165	47	77
Water (varies by location)	237	1 c	0	5	5
Wine:					
Red	118	1/2 c	85	11	9
Rosé	118	1/2 c	84	12	10
White, dry	119	1/2 c	79	14	11
Yeast:					
Brewer's	8	1 T	25	68	17
Dry active (1 pkt = 7 g)	30	4 T	80	30	24

Copper

Copper is essential in some key roles as part of several enzymes and as a catalyst. It is involved in the making of red blood cells, absorption and transportation of iron, healing of wounds, making and maintaining the sheath around nerve fibers, bone formation, synthesis of RNA, and making of collagen. It also plays a part in controlling the oxidation of vitamin C and in the metabolism of fatty acids.

Severe copper deficiency is rare in human beings, and in normal circumstances, deficiency from diet is not known. Deficiencies can occur however where there is impaired copper utilization (extremely rare inherited disease); if a person or baby is on supplemental feeding only over an extended period of time; or if a person consumes supplements of zinc, over a period of time.

Data on copper deficiencies in *animals* indicate the following abnormalities: anemia, skeletel defects, demyelination and degeneration of of the nervous system, reproductive failure, myocardial degeneration, decreased arterial elasticity, and defects in pigmentation and structure of hair or wool of the animal.

Nutrient Losses

Milling and processing , such as wheat to flour, and peanuts to peanut butter, can result in significant losses of copper. Copper can leach into cooking water and, depending on the food item, 15 percent to 30 percent can be lost. If the cooking water is used, then most of the nutrient is retained.

High zinc intake (that can result from taking supplements) is known to cause a copper deficiency, as does the medication B-penicillamine.

Estimated Safe and Adequate Amounts for Adults

USA: 1.5 to 3.0 mg per day. Because there is less information upon which to base allowances, a range of 1.5 mg to 3.0 mg is given by the Food and Nutrition Board as an estimated safe and adequate daily intake.

Canada: 1.0 to 2.0 mg per day. This amount is not an official recommendation, but is the amount per day that will cover nutritional requirements according to "Recommended Nutrient Intakes for Canadians."

Food Sources

The copper content of legumes is quite high and can be much greater than the amount found in meats. Seafood and shellfish, organ meats (liver, heart, etc.), whole grains, nuts and seeds, and vegetables all have good amounts of copper. Drinking water can be a good (but variable) source of copper, particularly in locations where copper pipe is used in plumbing or the mineral content of the water is high. As a result, some canned goods often have higher copper content than normally expected. Some copper-containing fungicides sprayed on agricultural products may also end up inadvertantly contributing copper to the diet.

ESHA Research

Baked Goods

	Wt (g)	Svg	Cal	Copper (mg) 100 Cal	Svg
Bagels, plain or egg, 3.5" dm.	68	1 ea	180	.1	.115
Biscuit:					
From mix	28	1 ea	94	<.1	.031
Homemade	28	1 ea	100	<.1	.020
From refrig dough	20	1 ea	65	<.1	.020
BREADS:					
Banana nut, 1/2" slice	50	1 pce	161	.1	.089
Boston brown bread, 1/2"	45	1 pce	95	.2	.200
Cornbread muffin, avg	45	1 ea	145	<.1	.024
Cracked wheat bread	25	1 pce	65	.1	.067
French, 5" x 2.5" x 1"	35	1 pce	100	.1	.051
Mixed grain	25	1 pce	65	.1	.071
Oatmeal bread	25	1 pce	65	.1	.055
Pita pocket bread, 6.5" dm	60	1 ea	165	.1	.108
Pumpernickel, 5" x 4" x 3/8"	32	1 pce	80	.1	.087
Raisin bread	25	1 pce	68	.1	.043
Rye bread, light, 5" x 3.5"	25	1 pce	65	.1	.048
Wheat bread (white and whole wheat flour)	28	1 pce	72	.1	.067
White bread	28	1 pce	75	.1	.039
Whole wheat bread	28	1 pce	70	.1	.096
CAKES, pce = 1/16th cake unless stated:					
Carrot cake, cream cheese frosting, 2.5" x 3"	112	1 pce	406	<.1	.127
Chocolate, choc frosting	69	1 pce	235	<.1	.090
Fruitcake, dark, 2/3" arc	43	1 pce	165	<.1	.072
Sponge cake, 1/12 cake	66	1 pce	194	<.1	.049
White cake:					
W/chocolate frosting	77	1 pce	291	.1	.149
W/coconut frosting	70	1 pce	270	<.1	.050
W/white frosting	71	1 pce	260	<.1	.030
Yellow, chocolate frosting	69	1 pce	240	<.1	.060
Cheese puffs (Cheetos)	28	1 oz	158	<.1	.040
Cherry crisp, 3" x 3"	138	1 pce	157	.1	.140
Chips: See corn & tortilla chips in this section; see potato chips under vegetables.					
COOKIES:					
Animal cookies (box ≈ 1 oz)	28	27 ea	120	<.1	.024
Chocolate chip:					
Homemade	40	4 ea	185	.1	.129
Refrigerator dough	48	4 ea	225	<.1	.053
Commercial	42	4 ea	180	<.1	.080
Fig bars	56	4 ea	210	.1	.160
Oatmeal raisin	52	4 ea	245	<.1	.061
Peanut butter, recipe	48	4 ea	245	<.1	.075
Sandwich type, all	40	4 ea	195	<.1	.030
Shortbread, commercial	32	4 ea	155	<.1	.020
Sugar, f/chilled dough	48	4 ea	235	<.1	.063
Vanilla wafer cookies	40	10 ea	185	<.1	.030
Corn chips	28	1 oz	155	<.1	.075
CRACKERS:					
Armenian cracker bread	28	4 pce	117	.1	.080
Cheese crackers	10	10 ea	50	.1	.030
Cheese, w/peanut butter	30	4 ea	150	<.1	.030
Graham crackers	14	2 ea	60	.1	.030
Round, like Ritz	9	3 ea	45	<.1	.020
Rye wafers, whole grain	14	2 ea	55	.1	.040
Saltines	12	4 ea	50	<.1	.020
Sesame crackers	12	4 ea	60	.1	.030
Soda crackers	17	6 ea	75	.1	.043
Whole wheat crackers	8	2 ea	35	.2	.066
Cream puff, custard filled	110	1 ea	280	<.1	.072
Croissant, 4.5" x 4" x 2"	57	1 ea	235	<.1	.057

	Wt (g)	Svg	Cal	Copper (mg) 100 Cal	Svg
Danish pastry:					
Plain pastry	57	1 ea	220	<.1	.068
With fruit	65	1 ea	235	<.1	.078
Doughnut:					
Cake type, medium	50	1 ea	210	<.1	.065
Yeast raised, plain	60	1 ea	235	<.1	.084
Eclair, custard filled/choc iced	94	1 ea	262	<.1	.086
English muffin, avg	53	1 ea	135	.1	.176
MUFFINS:					
Blueberry, average	45	1 ea	138	<.1	.036
Bran (wheat) average	45	1 ea	132	<.1	.042
Cornmeal muffin	45	1 ea	145	<.1	.024
Pancakes, from mix:					
Buckwheat, 4" dm	27	1 ea	55	.2	.092
Plain, regular, 4" dm	27	1 ea	60	<.1	.016
Whole wheat, 5" dm	52	1 ea	94	.1	.071
PIES: pce = 1/6 of 9" pie.					
Apple pie	158	1 pce	405	<.1	.100
Banana cream, commercial	198	1 pce	319	<.1	.149
Cherry pie	158	1 pce	410	<.1	.154
Chocolate cream	175	1 pce	311	<.1	.101
Coconut cream	172	1 pce	343	<.1	.108
Peach pie	158	1 pce	405	<.1	.122
Pecan pie	138	1 pce	583	<.1	.287
Pumpkin pie	200	1 pce	367	.1	.221
Poptart type toaster pastry	54	1 ea	210	<.1	.081
Pretzel, dutch twist	16	1 ea	65	<.1	.023
Pretzel, thin twists	60	10 ea	240	<.1	.088
ROLLS:					
Cinnamon bun, small	50	1 ea	158	.1	.080
Dinner roll, 2.5" x 2"	28	1 ea	85	<.1	.023
Hamburger bun	45	1 ea	129	.1	.074
Hard roll, white	50	1 ea	155	.1	.080
Hotdog bun	40	1 ea	115	.1	.066
Rye roll, dark	28	1 ea	79	.1	.060
Rye roll, light	28	1 ea	76	<.1	.034
Submarine (hoagie) roll	135	1 ea	400	<.1	.135
Whole wheat roll	35	1 ea	88	.1	.116
Stuffing, prepared:					
From dry bread crumbs	140	1 c	500	.1	.300
Stove Top Stuffing	108	1/2 c	176	<.1	.083
Tortilla chips:					
Plain	28	1 oz	139	.1	.110
Doritos, flavored	28	1 oz	140	.1	.080
Tortillas:					
Corn, 6" dm, fried	30	1 ea	87	<.1	.005
Flour, 10.5" dm	57	1 ea	168	<.1	.074
Flour, 8" dm	35	1 ea	105	<.1	.046
Waffles:					
Homemade, 7" dm	75	1 ea	245	<.1	.053
From mix, 7" dm	75	1 ea	205	<.1	.062
Frozen, 4" dm	35	1 ea	98	<.1	.021

Dairy & Dairy Products

	Wt (g)	Svg	Cal	Copper (mg) 100 Cal	Svg
CHEESE (1.5" cube ≈ 1 oz):					
American cheese	28	1 oz	106	<.1	.008
Brie cheese	28	1 oz	95	.1	.059
Caraway	28	1 oz	107	<.1	.048
Cheddar cheese, hard	28	1 oz	114	<.1	.009
Cottage cheese:					
Lowfat 1%	226	1 c	164	<.1	.045
Lowfat 2%	226	1 c	205	<.1	.063
Creamed	222	1 c	225	<.1	.059
Creamed, with fruit	226	1 c	279	<.1	.045
Dry curd	145	1 c	123	<.1	.029

Dairy & Dairy Products, continued:

	Wt (g)	Svg	Cal	100 Cal	Svg
CHEESE, continued:					
Gjetost	28	1 oz	132	.1	.110
Gouda	28	1 oz	101	.1	.059
Gruyere	28	1 oz	117	.1	.059
Neufchatel	28	1 oz	74	.1	.042
Parmesan, grated, 1 T = 5g	28	1 oz	129	<.1	.011
Ricotta, made w/part skim	246	1 c	340	<.1	.084
Romano, grated, 1 T = 5g	28	1 oz	128	.1	.069
Swiss cheese	28	1 oz	107	<.1	.009
CREAM, SWEET, fluid:					
Coffee or table	240	1 c	469	.1	.264
Half and half	242	1 c	315	.1	.174
Heavy whipping cream	238	1 c	821	<.1	.264
Light whipping cream	239	1 c	699	<.1	.264
CREAM, SWEET whipped:					
Heavy cream, unsweetened	119	1 c	410	<.1	.132
Pressurized	60	1 c	154	<.1	.056
CREAM, SOUR, dairy	230	1 c	493	.1	.253
Cream, sour, Imitation, non-dairy	230	1 c	479	<.1	.110
Kefir beverage	233	1 c	160	.1	.140
MILK (cow):					
Skim milk	245	1 c	86	.1	.049
Lowfat 1%	244	1 c	102	.1	.061
Lowfat 2%	244	1 c	121	.1	.073
Whole (3.3% fat)	244	1 c	150	<.1	.054
Buttermilk	245	1 c	99	<.1	.024
Canned:					
Evaporated, skim	255	1 c	200	.1	.188
Evaporated, whole	252	1 c	340	.1	.328
Sweetened, condensed	306	1 c	982	.1	.673
Dry, nonfat, instant	68	1 c	244	.1	.326
Dry, buttermilk	120	1 c	464	<.1	.110
Milk (other):					
Goat milk	244	1 c	168	<.1	.081
Soy milk	240	1 c	79	.4	.288
MILK BEVERAGES & MIXES:					
Chocolate flavored to be mixed with water:					
Dry powder-includes dry milk	28	1 oz	100	.1	.108
Drink	206	3/4 c	100	.1	.108
Chocolate flavored to be mixed with milk:					
Dry powder	22	3/4 oz	75	.2	.152
Drink	266	1 c	226	.1	.176
Instant Breakfast, dry, fortified	37	1 env	130	.4	.500
Milkshakes, 10 fl oz = 1.25 c:					
Chocolate	283	1.25 c	360	.1	.184
Strawberry	283	1.25 c	319	<.1	.062
Vanilla	283	1.25 c	314	<.1	.144
MILK DESSERTS:					
Custard, baked recipe	265	1 c	305	<.1	.106
Ice cream, vanilla:					
Regular	133	1 c	269	<.1	.027
Rich	148	1 c	349	<.1	.030
Soft serve	173	1 c	377	<.1	.035
Ice milk, hard	131	1 c	184	<.1	.079
Ice milk, soft serve	175	1 c	223	<.1	.105
PUDDINGS, prepared:					
Canned (5oz can ≈ 1/2 c):					
Chocolate	142	1 can	205	<.1	.065
Tapioca	142	1 can	160	<.1	.065
Vanilla	142	1 can	220	<.1	.065
Chocolate, ckd/inst-mix	260	1 c	305	<.1	.133
Coconut, f/instant	149	1/2 c	184	<.1	.039
Tapioca, ckd from mix	130	1/2 c	145	<.1	.041

	Wt (g)	Svg	Cal	100 Cal	Svg
PUDDING, continued:					
Vanilla, ckd/inst-mix	130	1/2 c	148	<.1	.041
Pudding pops, chocolate/fudge	57	1 ea	99	.1	.059
Sherbet	193	1 c	270	<.1	.100
Yogurt, frozen, average	174	1 c	220	<.1	.018
YOGURT:					
Lowfat, plain	227	1 c	144	.1	.090
Lowfat, coffee or vanilla	227	1 c	193	<.1	.090
Lowfat, with fruit	227	1 c	231	<.1	.090
Nonfat	227	1 c	127	.1	.066
Whole	227	1 c	138	.1	.114
Yogurt cheese, recipe	208	1 c	222	.1	.180

Eggs

	Wt (g)	Svg	Cal	100 Cal	Svg
CHICKEN, raw or cooked:					
Whole egg	50	1 ea	77.5	<.1	.007
White only	33.4	1 ea	17	<.1	.002
Yolk only	16.6	1 ea	59	<.1	.004

Fats, Oils, Salad Dressings

	Wt (g)	Svg	Cal	100 Cal	Svg
Butter	14	1 T	102	<.1	.004
Margarine:					
Regular, 80% fat	14	1 T	102	<.1	.024
Spread, 60% fat	14	1 T	76	<.1	.018
Soft, 40% fat (imitation)	14	1 T	49	<.1	.006
Mayonnaise	14	1 T	100	<.1	.038
OILS:					
Cod liver oil	14	1 T	126	.1	.095
Corn oil	218	1 c	1927	<.1	.020
Olive oil	216	1 c	1909	<.1	.151
Peanut oil	216	1 c	1909	<.1	.022
Safflower oil	218	1 c	1927	<.1	.283
Sesame oil	218	1 c	1925	<.1	.283
Soybean oil	218	1 c	1927	<.1	.04
Salad dressing:					
Ceasar's salad dressing	92	1/2 c	408	<.1	.058
Ranch type	119	1/2 c	435	<.1	.169
Tartar sauce	14	1 T	74	<.1	.038

Fruits & Fruit Juices

	Wt (g)	Svg	Cal	100 Cal	Svg
Apple, 2.75" diam:					
With skin	138	1 ea	80	.1	.057
Without peel	128	1 ea	72	.1	.040
Apple juice, cnd/bottled	248	1 c	116	<.1	.055
Applesauce, unsweetened	244	1 c	106	.1	.063
APRICOTS:					
Fresh, pitted	106	3 ea	51	.2	.094
Canned halves, juice pack	84	3 ea	40	.1	.045
Canned halves, heavy syrup	85	3 ea	70	.1	.066
Dried halves	35	10 ea	83	.2	.150
Apricot nectar, canned	251	1 c	141	.1	.183
Avocado, whole:					
California (227g w/refuse)	173	1 ea	305	.2	.460
Florida (454g w/refuse)	304	1 ea	340	.2	.763
Banana, 8.75", 176g w/ peel	114	1 ea	105	.1	.119
Blackberries:					
Fresh berries	144	1 c	74	.3	.202
Frozen, unthawed	151	1 c	97	.2	.181
Canned	256	1 c	236	.1	.340

ESHA Research

Values are for edible portion of foods

	Wt (g)	Svg	Cal	Copper (mg) 100 Cal	Svg
Fruits & Fruit Juices, continued:					
Blueberries:					
Fresh berries	145	1 c	82	.1	.088
Frozen, unthawed	155	1 c	78	.1	.051
Canned	256	1 c	225	.1	.136
Boysenberries:					
Frozen, unthawed	132	1 c	66	.2	.106
Canned	256	1 c	225	.1	.179
Breadfruit, fresh	220	1 c	227	.1	.270
Cantaloupe / Casaba: see Melon.					
Cherries, sour:					
Frozen	155	1 c	72	.2	.140
Canned	244	1 c	90	.2	.171
CHERRIES, SWEET:					
Fresh, pitted, whole	68	10 ea	49	.1	.065
Frozen, thawed	259	1 c	232	<.1	.062
Canned w/liquid	257	1 c	213	.1	.294
Cranberries, fresh, whole	95	1 c	46	.1	.055
Cranberry apple juice	253	1 c	169	<.1	.020
Cranberry juice cocktail	253	1 c	145	<.1	.033
Cranberry sauce, canned	277	1 c	419	<.1	.055
Currants:					
Black, fresh	112	1 c	71	.1	.096
Red or white, fresh	112	1 c	63	.2	.120
Dried, Zante	144	1 c	407	.2	.674
Dates, whole, pitted	83	10 ea	228	.1	.239
Figs, whole:					
Fresh, medium	50	1 ea	37	.1	.035
Canned, heavy syrup	85	3 ea	75	.1	.090
Dried figs	19	1 ea	48	.1	.059
Fruit cocktail, canned:					
Juice pack	248	1 c	115	.1	.154
Heavy syrup	255	1 c	185	.1	.176
Gooseberries:					
Fresh	150	1 c	67	.2	.105
Canned, with liquid	252	1 c	185	.3	.547
Grapes, fresh:					
Thompson seedless	50	10 ea	35	.2	.085
Tokay (Emperor)	57	10 ea	40	.2	.096
Grape juice:					
Bottled or canned	253	1 c	155	<.1	.071
From frozen	250	1 c	128	<.1	.033
GRAPEFRUIT (half = 241g w/rind):					
Fresh half, pink/red	123	1 ea	37	.1	.054
Fresh half, white	118	1 ea	39	.2	.059
Canned, sections	254	1 c	152	.1	.168
Grapefruit juice:					
Fresh juice	247	1 c	96	.1	.082
From frozen	247	1 c	102	.1	.103
Canned, unsweetened	247	1 c	93	.1	.094
Canned, sweetened	250	1 c	115	.1	.120
Guava, fresh	90	1 ea	45	.2	.093
Honeydew: see Melon.					
Kiwi fruit	76	1 ea	46	.1	.036
Lemon juice:					
Fresh	244	1 c	60	.1	.071
Frozen or bottled	244	1 c	53	.1	.082
Lime juice:					
Fresh	246	1 c	65	.1	.074
Bottled	246	1 c	50	.1	.074
Loganberries:					
Fresh berries	150	1 c	105	.2	.210
Frozen, unthawed	147	1 c	80	.2	.172
Lychees, canned	100	3.5 oz	68	.2	.110
Mandarin oranges, canned	252	1 c	155	.1	.125
Mango, fresh slices	165	1 c	108	.2	.182
Melon: see also Watermelon.					
Cantaloupe cubes	160	1 c	57	.1	.067
Casaba cubes	170	1 c	45	.1	.050

	Wt (g)	Svg	Cal	Copper (mg) 100 Cal	Svg
Melon, continued:					
Honeydew cubes	170	1 c	60	.1	.070
Frozen melon balls, mixed	173	1 c	55	.2	.104
Nectarine (1 med = 1 c slices)	136	1 med	67	.1	.099
ORANGE:					
Average 2-5/8" 180g w/ peel	131	1 ea	60	.1	.059
California navel, 206g w/peel	140	1 ea	65	.1	.078
Orange juice:					
Fresh	248	1 c	111	.1	.109
Chilled	249	1 c	110	.1	.100
From frozen	249	1 c	110	.1	.110
Canned, unsweetened	249	1 c	105	.1	.142
Orange grapefruit jce, canned	247	1 c	105	.2	.188
Papaya, 454g w/refuse	304	1 ea	117	<.1	.049
Papaya nectar, canned	250	1 c	142	<.1	.033
PEACHES:					
Fresh, 2.5" diam.	87	1 ea	37	.2	.059
Frozen, thawed slices	250	1 c	235	<.1	.060
Canned half, juice pack	77	1 ea	34	.1	.039
Canned half, heavy syrup	81	1 ea	60	.1	.041
Peach nectar, canned	249	1 c	134	.1	.172
PEARS:					
Fresh, Bartlett, 181g w/refuse	166	1 ea	98	.2	.188
Canned half, juice pack	77	1 ea	38	.1	.041
Canned half, heavy syrup	79	1 ea	59	.1	.039
Pear nectar, canned	250	1 c	149	.1	.168
PINEAPPLE:					
Fresh chunks	155	1 c	76	.2	.171
Canned slice, juice pack	58	1 ea	35	.1	.050
Canned slice, heavy syrup	58	1 ea	45	.1	.059
Pineapple juice, canned, unswt	250	1 c	140	.2	.225
Plantain:					
Fresh slices	148	1 c	181	.1	.200
Cooked	154	1 c	179	.1	.150
Plums:					
Fresh, 2-1/8" dm	66	1 ea	36	.1	.028
Canned, juice pack	252	1 c	146	.1	.136
Canned, heavy syrup	258	1 c	230	<.1	.095
Pomegranate juice, fresh	100	2/5 c	44	.2	.070
Prunes, dried, pitted	84	10 ea	201	.2	.361
Prune juice, bottled	256	1 c	181	.1	.174
Raisins, dark, packed meas.	165	1 c	494	.1	.510
Raspberries:					
Fresh	123	1 c	60	.2	.091
Frozen, thawed	250	1 c	255	.1	.263
Rhubarb:					
Fresh, diced	122	1 c	26	.1	.026
Cooked with sugar	240	1 c	279	<.1	.065
STRAWBERRIES:					
Fresh berries	149	1 c	45	.2	.073
Frozen, unthawed	149	1 c	52	.1	.073
Frozen, thawed, sweetened	255	1 c	245	<.1	.051
Tangerine	84	1 ea	37	.1	.024
Tangerine juice	241	1 c	110	.1	.060
Watermelon pieces	160	1 c	50	.1	.051

Grains & Grain Products
Cereals, Grains, Flours, Noodles, Pasta, Popcorn

	Wt (g)	Svg	Cal	Copper (mg) 100 Cal	Svg
Amaranth grain, dry	195	1 c	729	.2	1.52
Barley:					
Pearled, cooked	157	1 c	193	.1	.165
Whole, cooked	200	1 c	200	.2	.470
Bran: see Oat, Rice, Wheat.					

Copper (Cu) mg

Grains & Grain Products, continued:	Wt (g)	Svg	Cal	100 Cal	Svg
Buckwheat flour:					
Dark	98	1 c	338	.2	.686
Light	98	1 c	340	.2	.545
Bulgar wheat, cooked	182	1 c	151	.1	.137
CEREALS, COLD (Ready to eat)					
Copper content varies & some cereals are fortified, check label.					
CEREALS, HOT, cooked:					
Corn grits, enriched	242	1 c	145	<.1	.029
Cream of rice	244	1 c	126	.1	.083
Cream of wheat	244	1 c	140	<.1	.068
Farina	233	1 c	116	<.1	.026
Malt-O-Meal	240	1 c	122	<.1	.026
Maypo	180	3/4 c	128	.1	.119
Oatmeal (reg, quick, inst)	234	1 c	145	.1	.129
Oatmeal, fortified, instant:					
Plain, packet	177	3/4 c	104	.1	.119
Flavors, averaged	155	3/4 c	135-175	.1	.121
W/bran & raisin	195	7/8 c	158	.2	.284
Ralston cereal	253	1 c	134	.1	.200
Roman Meal	181	3/4 c	111	.2	.241
Wheatena cereal	243	1 c	135	.1	.126
Whole wheat cereal	242	1 c	151	.1	.201
Corn flour, Masa Harina, enr.	114	1 c	416	<.1	.193
Cornmeal, dry:					
Degermed, enriched	138	1 c	505	<.1	.12
Nearly whole, bolted	122	1 c	441	.1	.235
FLOUR: see specific grain, nut, or vegetable.					
MACARONI, cooked:					
Enriched	140	1 c	197	.1	.137
Vegetable, enriched	134	1 c	172	.1	.123
Whole wheat	140	1 c	174	.1	.234
Millet, cooked	120	1/2 c	143	.2	.193
Noodles, cooked:					
Egg noodles	160	1 c	213	.1	.138
Spinach noodles	140	1 c	182	.2	.287
Oat bran, 1T ≈ 6 g	94	3.5 oz	132	.2	.378
Oats, rolled:					
Dry	81	1 c	311	.1	.278
Cooked: see Oatmeal (Cereals, hot).					
Popcorn, popped in oil	11	1 c	55	.1	.033
Quinoa, dry	170	1 c	635	.22	1.39
RICE, cooked:					
Brown rice	195	1 c	217	.1	.195
White, regular	205	1 c	264	<.1	.129
White, converted	200	1 c	186	.1	.165
White, instant	165	1 c	162	.1	.107
Wild rice	164	1 c	166	.1	.198
Rice bran	83	1 c	262	.2	.604
Rye flour:					
Dark	128	1 c	415	.2	.96
Medium	115	1 c	374	.1	.293
Light	102	1 c	361	.1	.255
Soy flour, stirred:					
Low fat	44	1/2 c	163	1.4	2.2
Defatted	50	1/2 c	164	1.0	2.0
Full fat, raw	42	1/2 c	182	.7	1.2
Spaghetti noodles, cooked:					
Enriched	140	1 c	197	.1	.137
Whole wheat spaghetti	140	1 c	174	.1	.234
WHEAT:					
Bran	30	1/2 c	65	.5	.3
Flours, unbleached:					
All-purpose, white, unsifted	125	1 c	455	<.1	.18
Cake flour, sifted	96	1 c	348	<.1	.134

Wheat Flours, continued:	Wt (g)	Svg	Cal	100 Cal	Svg
Self-rising flour	125	1 c	442	<.1	.14
Semolina flour	167	1 c	601	.1	.316
Whole wheat flour	120	1 c	407	.1	.458
Wheat germ:					
Raw	100	1 c	360	.2	.796
Toasted	113	1 c	432	.2	.701
Wheat, rolled, cooked	240	1 c	142	.1	.2
Wheat, rolled, dry	85	1 c	289	.2	.45
Whole grain wheat (wheat berries) cooked	50	1/3 c	28	.1	.04
Whole wheat, sprouted	108	1 c	214	.1	.282

Meats: Fish & Shellfish

	Wt (g)	Svg	Cal	100 Cal	Svg
Anchovies, canned in oil	45	11 ea	95	.2	.153
Bass, baked/broiled	100	3.5 oz	125	.1	.100
Bluefish:					
Baked/broiled	100	3.5 oz	159	<.1	.070
Fried in crumbs	100	3.5 oz	205	<.1	.060
Carp, baked/broiled	100	3.5 oz	162	<.1	.073
Catfish, fried in cornmeal	100	3.5 oz	229	.1	.210
Clams, meat only:					
Canned, drained	160	1 c	236	.5	1.10
Canned w/liquid, minced	183	1 can	145	18.7	27.1
Breaded, fried, small	188	20 ea	379	.2	.669
Steamed, whole, meat only	100	3.5 oz	148	.5	.688
Clam nectar, canned	240	1 c	6	15.6	.934
Cod:					
Baked/broiled	100	3.5 oz	105	<.1	.036
Smoked	100	3.5 oz	79	.2	.170
Dried, salted, Atlantic	28	1 oz	81	.1	.049
CRAB:					
Alaska King (leg) ckd	134	1 ea	129	1.2	1.58
Blue crab, cooked	135	1 c	138	.6	.871
Blue crab, canned	135	1 c	133	.8	1.03
Dungeness crab, cooked	101	3/4 c	85	.8	.680
Crab, imitation f/surimi	85	3 oz	87	<.1	.030
Crab cakes from recipe	60	1 ea	93	.4	.366
Crayfish, ckd meat, moist heat	85	3 oz	97	.5	.475
Eel, smoked	100	3.5 oz	330	<.1	.060
Fish cakes, fried f/frozen	100	3.5 oz	213	.1	.135
Fish sticks, f/frozen	57	2 ea	155	<.1	.058
Gefiltefish, sweet, commercial	42	1 pce	35	.2	.082
Grouper, cooked	100	3.5 oz	118	<.1	.045
Haddock, baked/broiled	85	3 oz	95	<.1	.028
HALIBUT:					
Baked/broiled	85	35 oz	119	<.1	.030
Smoked	100	3.5 oz	224	<.1	.027
Herring:					
Baked/broiled	100	3.5 oz	203	.1	.118
Canned with liquid	100	3.5 oz	208	.1	.239
Pickled (1 pce ≈ 15g)	100	3.5 oz	262	<.1	.105
Smoked, kippered	100	3.5 oz	217	.1	.135
Lobster meat, cooked	145	1 c	142	2.0	2.81
Mackerel:					
Baked/broiled	100	3.5 oz	210	<.1	.080
Canned, Jack, #1 tall can	361	1 can	563	.1	.531
Mullet, baked/broiled	85	3 oz	127	.1	.120
Octopus, raw	100	3.5 oz	82	.5	.435
OYSTERS:					
Raw, Eastern	248	1 c	170	6.5	11.1
Raw, Pacific	248	1 c	200	2.0	3.91
Breaded, fried, Eastern	88	6 med	173	2.2	3.78
Simmered, Eastern	100	3.5 oz	137	6.5	8.92

ESHA Research

	Wt (g)	Svg	Cal	Copper (mg) 100 Cal	Svg
Fish & Shellfish, continued:					
Ocean perch:					
Baked/broiled	100	3.5 oz	121	<.1	.033
Breaded, fried	85	3 oz	185	<.1	.022
Pike, Northern, baked/broiled	100	3.5 oz	113	.1	.065
Pollock, baked/broiled	100	3.5 oz	99	.1	.057
Pompano, baked/broiled	100	3.5 oz	211	<.1	.078
Rockfish, baked/broiled	100	3.5 oz	121	<.1	.037
SALMON, cooked:					
Baked/broiled	85	3 oz	183	.1	.255
Smoked, Chinook	85	3 oz	99	.2	.196
Cnd, Atlantic, small can	220	1 can	281	.1	.396
Cnd, Pink, drained, #1 can	454	1 can	631	.1	.463
Cnd, Sockeye, drained, #1 can	369	1 can	566	.1	.310
Sardines, canned:					
Atlantic (2 = 24g)	92	1 can	192	.1	.171
Pacific (1 = 38g)	100	3.5 oz	178	.2	.272
Scallops:					
Breaded, fried	93	6 ea	200	<.1	.073
Steamed	100	3.5 oz	113	.1	.065
Sea bass, baked/broiled	100	3.5 oz	124	<.1	.024
Seatrout (Steelhead), cooked	100	3.5 oz	131	<.1	.035
Shark, batter-fried	85	3 oz	194	<.1	.036
SHRIMP:					
Boiled (2 lrg ≈ 11g)	100	3.5 oz	99	.2	.193
Breaded, fried (2 lrg ≈15g)	90	12 ea	218	.1	.247
Canned, drained	128	1 c	154	.2	.384
Canned w/liquid	100	3.5 oz	102	.2	.170
Snapper, baked/broiled	100	3.5 oz	128	<.1	.046
SOLE (Flounder) baked/broiled	85	3 oz	99	<.1	.022
Squid, flour-fried	85	3 oz	149	1.2	1.80
Sturgeon, cooked	85	3 oz	115	<.1	.045
Sturgeon, smoked	85	3 oz	147	<.1	.064
Surimi, processed walleye					
(Alaska) pollock, see imitation crab.					
Swordfish, baked/broiled	100	3.5 oz	155	.1	.162
Trout, baked/broiled	85	3 oz	129	.1	.120
TUNA:					
Ckd fr/fresh, Bluefin	85	3 oz	157	.1	.094
Canned in oil, Light tuna	171	1 can	339	<.1	.121
Canned in water, Light tuna	165	1 can	216	<.1	.018

Meats

Beef, Pork, Ham, Lamb, Rabbit, Veal, Venison

	Wt (g)	Svg	Cal	Copper (mg) 100 Cal	Svg
BEEF:					
Chuck roast, pot roasted:					
Lean & fat (5.4 oz raw)	85	3 oz	325	<.1	.101
Lean only	85	3 oz	230	.1	.126
GROUND BEEF, cooked avg:					
Extra lean (17% fat raw)	85	3 oz	215	<.1	.066
Lean (21% fat raw)	85	3 oz	231	<.1	.061
Regular (26.6% fat raw)	85	3 oz	250	<.1	.066
Frzn patty, broiled (23% fat)	85	3 oz	240	<.1	.055
Rib, choice, roasted:					
Lean & fat (5 oz raw)	85	3 oz	324	<.1	.081
Lean only	85	3 oz	204	<.1	.083
Round steak, broiled:					
Lean & fat (4.5 oz raw)	85	3 oz	233	<.1	.082
Lean only	85	3 oz	165	.1	.091
Round tip, roasted:					
Lean & fat	85	3 oz	213	<.1	.103
Lean only	85	3 oz	162	.1	.106

	Wt (g)	Svg	Cal	Copper (mg) 100 Cal	Svg
BEEF, continued:					
Sirloin steak, broiled:					
Lean & fat	85	3 oz	238	<.1	.088
Lean only	85	3 oz	172	.1	.091
T-bone steak, broiled:					
Lean & fat	85	3 oz	276	<.1	.111
Lean only	85	3 oz	182	.1	.122
Variety meats:					
Brains, pan fried	85	3 oz	167	.1	.187
Heart, simmered	85	3 oz	140	.4	.629
Kidney, cooked	140	1 ea	201	.5	.952
Liver, fried	85	3 oz	184	1.3	2.40
Corned beef, canned	85	3 oz	213	<.1	.054
Dried beef, cured (6-7 pces)	28	1 oz	47	.1	.045
HAM: see Pork, Cured; Turkey ham and					
Sausages & Lunchmeats.					
LAMB:					
Arm chop, braised, raw ≈ 5.6 oz w/bone:					
Lean & fat	70	1 ea	244	<.1	.098
Lean only	55	1 ea	152	.1	.084
Loin chop, broiled, raw ≈ 4.5 oz w/bone:					
Lean & fat	64	1 ea	201	<.1	.083
Lean only	46	1 ea	100	.1	.068
Cutlet, lean, cooked avg	85	3 oz	175	.1	.109
Leg of lamb, roasted:					
Lean & fat	85	3 oz	219	<.1	.098
Lean only	85	3 oz	162	.1	.102
Shoulder roast:					
Lean & fat	85	3 oz	235	<.1	.092
Lean only	85	3 oz	173	.1	.096
Lamb heart, braised	85	3 oz	158	.3	.518
Lamb liver, pan-fried	85	3 oz	202	4.1	8.36
PORK:					
Bacon, cooked:					
Regular	19	3 pce	109	<.1	.032
Canadian style	47	2 pce	86	<.1	.025
Breakfast strips	34	3 pce	156	<.1	.052
Chop, average, loin/rib cuts					
Braised:					
Lean & fat	71	1 ea	256	<.1	.066
Lean only	57	1 ea	156	<.1	.058
Broiled/fried/roasted:					
Lean & fat	85	1 ea	289	<.1	.064
Lean only	67	1 ea	168	<.1	.055
Pork roast, leg:					
Lean & fat	85	3 oz	250	<.1	.085
Lean only	85	3 oz	187	<.1	.092
Pork roast, avg loin & rib:					
Lean & fat	85	3 oz	265	<.1	.064
Lean only	85	3 oz	206	<.1	.068
Shoulder, braised, from 6.8 oz raw w/bone:					
Lean and fat	85	3 oz	293	<.1	.117
Lean only	67	2.4 oz	166	.1	.108
Spareribs, f/1 lb raw	177	6.25 oz	703	<.1	.251
Pork heart, cooked	145	1 c	214	.3	.737
Pork liver, pan-fried	85	3 oz	141	.4	.539
PORK, CURED—HAM, roasted:					
Lean & fat	85	3 oz	207	<.1	.071
Lean only	85	3 oz	133	.1	.074
Canned	85	3 oz	140	<.1	.066
RABBIT, roasted	85	3 oz	175	.1	.126
VEAL:					
Cutlet, lean, ckd average	85	3 oz	166	.1	.102
Rib roast	85	3 oz	151	.1	.090
Heart, braised	85	3 oz	158	.2	.367
Liver, pan-fried	85	3 oz	208	.4	8.40
VENISON (deer) roasted	85	3 oz	134	.2	.255

Copper (Cu) mg

	Wt (g)	Svg	Cal	Copper (mg) 100 Cal	Copper (mg) Svg

Meats: Poultry

CHICKEN:

All types of meat:

Fried	140	1 c	307	<.1	.105
Roasted	140	1 c	266	<.1	.094
Stewed	140	1 c	248	<.1	.085
Canned, boned, w/broth	142	5 oz	235	<.1	.058

Dark meat:

Fried	85	3 oz	203	<.1	.076
Roasted	85	3 oz	174	<.1	.068
Stewed	85	3 oz	163	<.1	.064

Light meat:

Fried	85	3 oz	163	<.1	.046
Roasted	85	3 oz	147	<.1	.042
Stewed	85	3 oz	135	<.1	.038

Breast*, meat & skin (145g raw ≈ 181g w/bone):

Batter-fried	140	1 ea	364	<.1	.084
Flour-fried	98	1 ea	218	<.1	.056
Roasted	98	1 ea	193	<.1	.049
Stewed	110	1 ea	202	<.1	.048

Breast, meat (118g raw):

Pan-fried	86	1 ea	161	<.1	.046
Roasted	86	1 ea	142	<.1	.042
Stewed	95	1 ea	144	<.1	.041

*2 pieces per bird

Drumstick, meat & skin (73g raw ≈ 110g w/bone):

Batter-fried	72	1 ea	193	<.1	.055
Flour-fried	49	1 ea	120	<.1	.039
Roasted	52	1 ea	112	<.1	.040
Stewed	57	1 ea	116	<.1	.040

Drumstick, meat (62g raw):

Pan-fried	42	1 ea	82	<.1	.034
Roasted	44	1 ea	76	<.1	.035
Stewed	46	1 ea	78	<.1	.035

Thigh, meat & skin (94g raw ≈ 120g w/bone):

Batter-fried	86	1 ea	238	<.1	.071
Flour-fried	62	1 ea	162	<.1	.055
Roasted	62	1 ea	153	<.1	.048
Stewed	68	1 ea	158	<.1	.048

Thigh, meat (69g raw):

Pan-fried	52	1 ea	113	<.1	.047
Roasted	52	1 ea	109	<.1	.042
Stewed	55	1 ea	107	<.1	.042
Chicken liver	20	1 ea	30	.2	.074

TURKEY:

Ground turkey, cooked	100	3.5 oz	229	<.1	.090

Roasted:

All types	140	1 c	238	.1	.132
Dark meat	140	1 c	262	.1	.224
White meat	140	1 c	195	.1	.120
Frozen slices w/gravy	142	5 oz	95	<.1	.029
Gizzard, cooked	67	1 ea	109	.1	.116
Heart, cooked	16	1 ea	28	.4	.100
Liver, cooked	75	1 ea	127	.3	.420
Hickory smoked breast meat	28	1 oz	35	.2	.060
Turkey patty, breaded, cooked	64	1 ea	181	<.1	.035

Meats: Sausages & Lunchmeats

Beef lunchmeat, loaf/roll	28	1 oz	87	<.1	.030
Braunschweiger	18	1 pce	65	.1	.043
Brotwurst, link	70	1 ea	226	<.1	.050
Chicken roll	57	2 oz	90	<.1	.020

FRANKFURTER (hotdog):

Beef, 8/pkg	57	1 ea	184	<.1	.030
Beef and pork, 8/pkg	57	1 ea	183	<.1	.050
Chicken, 10/pkg	45	1 ea	115	<.1	.020
Turkey, 10/pkg	45	1 ea	102	<.1	.022

HAM lunchmeats:

Chopped, packaged	42	2 pce	98	<.1	.021
Extra lean	57	2 oz	75	.1	.042
Regular	57	2 oz	103	.1	.056
Patty, cooked	60	1 ea	203	<.1	.060
Ham & cheese loaf	57	2 oz	147	<.1	.043
Ham salad spread	240	1 c	518	<.1	.174
Italian sausage link, cooked	67	1 ea	216	<.1	.054
Keilbasa	26	1 pce	81	<.1	.030
Knockwurst, link	68	1 ea	209	<.1	.040
Liverwurst, pork	18	1 pce	59	.1	.046
Luxury loaf	57	2 oz	80	.1	.057
Pickle & pimento loaf	57	2 oz	149	<.1	.071
Pork sausage patty, cooked	27	1 pce	100	<.1	.038

Turkey lunchmeats:

Ham	57	2 oz	73	.1	.068
Pastrami	57	2 oz	74	.1	.045
Loaf, breast meat	42.5	2 pce	46	.1	.023
Roll, light & dark meat	57	2 pce	84	<.1	.040
Roll, light meat only	57	2 pce	83	<.1	.020
Summer sausage	28	1 oz	50	.1	.040

Mixed Dishes & Fast Foods

Beef & vegetable stew:

Recipe	245	1 c	220	.1	.186
Canned	245	1 c	194	.1	.149
Beef, macaroni & tomato sauce, recipe	226	1 c	189	.2	.330
Beef pot pie, f/frozen	234	1 ea	426	<.1	.106

BURRITO:

Bean burrito	174	1 ea	322	.1	.315
Beef burrito	177	1 ea	463	<.1	.198
Beef and bean	175	1 ea	390	.1	.255
Deluxe combination	198	1 ea	424	.1	.279
Cheese souffle, recipe	112	1 c	221	<.1	.086
Chicken a la king, recipe	245	1 c	470	.1	.283
Chicken & noodles, recipe	240	1 c	365	<.1	.174

Chicken chow mein:

Recipe	250	1 c	255	.2	.400
Canned	250	1 c	95	.2	.200
Chicken pot pie, f/frzn	230	1 ea	430	<.1	.130
Chicken salad w/celery	78	1/2 c	266	<.1	.104
Chili with beans, canned	255	1 c	286	.1	.395
Corn dog	111	1 ea	330	<.1	.090
Corn fritter, recipe	45	1 ea	116	<.1	.055
Corn pudding	250	1 c	271	<.1	.108
Corned beef hash, canned	220	1 c	382	.1	.316
Egg salad	183	1 c	438	.1	.240

ENCHILADA:

Beef enchilada	120	1 ea	292	<.1	.085
Cheese	120	1 ea	330	<.1	.059
Chicken	120	1 ea	269	<.1	.069
French toast, recipe	65	1 pce	123	.1	.063

LASAGNA, recipe:

With meat	245	1 pce	398	.1	.461
Without meat	218	1 pce	316	.1	.423

Macaroni & cheese:

Recipe	200	1 c	430	<.1	.094
Canned	240	1 c	230	<.1	.080
Macaroni salad, w/o cheese	141	1 c	371	.1	.208

Mixed Foods & Fast Foods, continued:	Wt (g)	Svg	Cal	Copper (mg) 100 Cal	Svg
Manicotti:					
W/meat & tomato sce	233	1 ea	320	<.1	.074
Frozen entree	225	1 ea	271	.1	.230
Meat loaf, beef only	87	1 pce	193	.1	.110
Moussaka (lamb & eggplant)	250	1 c	250	.1	.348
PIZZA, cheese:					
Regular crust, 1/8 of 15"	120	1 pce	290	.1	.243
Thick crust, 1/2 of 10"	208	1 pce	519	.1	.615
Potato salad w/mayo & egg	250	1 c	358	.1	.295
Quiche Lorraine, 1/8	176	1 pce	600	<.1	.165
Ravioli, beef, canned	226	1 c	220	.1	.228
SANDWICHES, fast food:					
Cheeseburger, 3 oz beef	112	1 ea	300	<.1	.105
Cheeseburger, 4 oz beef	194	1 ea	524	<.1	.155
Chicken patty sandwich	157	1 ea	436	<.1	.101
English muffin, egg,					
cheese & bacon	138	1 ea	360	<.1	.121
Fish sandwich:					
Large, w/o cheese	170	1 ea	470	<.1	.136
Regular, w/cheese	140	1 ea	420	<.1	.098
Hamburger, 3 oz beef	98	1 ea	245	<.1	.095
Hamburger, 4 oz beef	174	1 ea	445	<.1	.157
Hotdog (frankfurter & bun)	85	1 ea	260	<.1	.106
SANDWICHES: on part whole wheat bread, unless stated as on rye.					
Avocado, cheese,					
sprouts & tomato	195	1 ea	432	.1	.379
Bacon, lettuce, tomato	135	1 ea	327	.1	.221
Chicken salad sandwich	100	1 ea	294	.1	.193
Corned beef & swiss on rye	147	1 ea	429	.1	.241
Egg salad sandwich	111	1 ea	319	.1	.219
Grilled cheese	117	1 ea	393	<.1	.158
Ham sandwich	122	1 ea	256	.1	.208
Ham & cheese	151	1 ea	363	.1	.216
Ham on rye bread	116	1 ea	242	.1	.134
Ham & swiss on rye	145	1 ea	350	<.1	.143
Ham salad sandwich	125	1 ea	339	.1	.203
Patty melt on rye	177	1 ea	567	<.1	.207
Peanut butter & jam	100	1 ea	341	.1	.338
Reuben, grilled	233	1 ea	480	.1	.335
Roast beef	122	1 ea	280	.1	.213
Tuna salad sandwich	116	1 ea	303	.1	.241
Turkey sandwich	122	1 ea	271	.1	.190
Turkey ham	122	1 ea	253	.1	.235
Turkey ham & cheese	151	1 ea	361	.1	.243
SPAGHETTI (pasta, tomato sauce & cheese):					
Homemade	250	1 c	260	.2	.400
Canned	250	1 c	190	.1	.200
Spinach souffle	136	1 c	218	.1	.120
Taco, beef	78	1 ea	207	<.1	.062
Taco, chicken	78	1 ea	172	<.1	.037
Tostada, with:					
Beans & beef	192	1 ea	332	.1	.244
Beans & chicken	157	1 ea	249	.1	.186
Refried beans	157	1 ea	212	.1	.273
Tuna noodle casserole, recipe	202	1 c	251	.1	.225
Tuna salad	205	1 c	383	.1	.297
Turkey pot pie, f/frozen	233	1 ea	416	.1	.289

Nuts & Seeds

	Wt (g)	Svg	Cal	Copper (mg) 100 Cal	Svg
ALMONDS, dried, whole	142	1 c	837	.2	1.34
Almond butter	250	1 c	1583	.1	2.25
Brazil nuts, dry	140	1 c	919	.3	2.48
Cashews:					
Dry roasted	137	1 c	787	.4	3.04
Oil roasted	130	1 c	748	.4	2.82
Chestnuts, roasted	143	1 c	350	.2	.725
Coconut:					
Raw, grated	80	1 c	283	.1	.348
Dried, unsweetened	78	1 c	515	.1	.621
Flaked, sweetened, packaged	74	1 c	351	.1	.223
Coconut milk, canned	226	1 c	445	.1	.504
Filberts (hazelnuts), whole	135	1 c	853	.2	2.04
Macadamias, oil roasted	134	1 c	962	<.1	.402
MIXED NUTS w/peanuts (almonds, brazil nuts, cashews, filberts, peanuts & pecans):					
Dry roasted	137	1 c	814	.2	1.75
Oil roasted	142	1 c	876	.3	2.36
MIXED NUTS w/o peanuts (cashews, almonds, brazil nuts, pecans & filberts):					
Oil roasted	144	1 c	886	.3	2.59
PEANUTS, oil roasted	144	1 c	837	.2	1.87
Peanut butter	32	2 T	188	.1	.178
Peanut flour, defatted	60	1 c	196	.6	1.08
Pecans, dried, chopped	119	1 c	794	.2	1.41
Pine nuts, dried:					
Pignola	28	1 oz	146	.2	.291
Pinyon	28	1 oz	161	.2	.294
Pistachio nuts, dried, shelled	128	1 c	739	.2	1.52
Pumpkin/squash seeds:					
Kernels, roasted	227	1 c	1185	.3	3.14
Whole, roasted	64	1 c	285	.2	.442
Sesame seeds, dried:					
Kernels	150	1 c	882	.7	5.80
Whole	144	1 c	825	.7	5.88
Soybeans, roasted	86	1/2 c	405	.2	.712
Sunflower seed kernels:					
Dried	144	1 c	821	.3	2.52
Oil roasted	135	1 c	830	.3	2.44
Tahini (sesame butter)	15	1 T	91	.3	.241
Walnuts, chopped:					
Black	125	1 c	759	.2	1.28
English	120	1 c	770	.2	1.65

Soups, Sauces & Gravies

	Wt (g)	Svg	Cal	Copper (mg) 100 Cal	Svg
GRAVY:					
Beef gravy:					
Recipe	135	1/2 c	151	<.1	.105
Canned	233	1 c	124	.2	.233
Chicken gravy:					
Recipe	130	1/2 c	163	<.1	.076
Canned	238	1 c	189	.1	.238
SAUCES (also see Other):					
Hollandaise, recipe	160	1 c	867	<.1	.289
Spaghetti sauce, plain:					
Recipe	220	1 c	179	.5	.950
Canned	249	1 c	272	.1	.284
White sauce, recipe	250	1 c	395	<.1	.051

	Wt (g)	Svg	Cal	Copper (mg) 100 Cal	Svg
SOUPS: soups are prepared from canned unless otherwise stated. For soups prepared w/milk, assume whole milk. RTS = Ready To Serve.					
Bean and 'frank'	250	1 c	187	.2	.395
Bean with bacon	253	1 c	173	.2	.402
Beef broth:					
Prepared with water	240	1 c	16	1.2	.185
From cube	241	1 c	8	11.6	.927
From dry	244	1 c	19	.1	.015
Beef noodle	244	1 c	84	.2	.139
Black bean	247	1 c	116	.3	.385
Celery, cream of:					
Prepared w/water	244	1 c	90	.2	.142
Prepared w/milk	248	1 c	165	.1	.154
Chicken soup, chunky, RTS	251	1 c	178	.1	.251
Chicken broth:					
Prepared w/water	244	1 c	39	.3	.124
From cube	243	1 c	13	7.2	.935
From dry	244	1 c	21	4.5	.939
Chicken, cream of:					
Prepared w/milk	248	1 c	191	.1	.139
Condensed, undiluted	251	1 c	233	.1	.251
Chicken noodle:					
Prepared w/water	241	1 c	75	.3	.195
From dry, pkt	188	3/4 c	40	.1	.026
Chicken rice	241	1 c	60	.2	.118
Chicken vegetable:					
Chunky, RTS	240	1 c	167	.1	.123
From dry	251	1 c	49	.1	.025
Chili beef soup	250	1 c	169	.2	.396
Clam chowder, New England	248	1 c	163	.1	.139
Clam chowder, Manhattan:					
Prepared w/water	244	1 c	78	.2	.132
Chunky, RTS	240	1 c	133	.2	.240
Consomme with gelatin	241	1 c	29	.8	.246
Minestrone	241	1 c	80	.2	.123
Mushroom, cream of:					
Prepared w/milk	248	1 c	205	.1	.139
Prepared f/dry	253	1 c	96	<.1	.030
Condensed, undiluted	251	1 c	257	.1	.251
Mushroom & beef stock	244	1 c	85	.3	.251
Onion soup	241	1 c	57	.2	.123
Oyster stew	245	1 c	134	1.2	1.60
Pea soup, prep w/milk	254	1 c	239	.2	.391
Potato, cream of	248	1 c	148	.2	.263
Split pea:					
Prepared w/water	253	1 c	189	.2	.369
From dry	255	1 c	133	.1	.192
Tomato soup:					
Prepared w/milk	248	1 c	160	.2	.268
From dry	265	1 c	102	.1	.093
Tomato beef noodle	244	1 c	140	.1	.124
Tomato vegetable, f/dry	253	1 c	55	.1	.033
Turkey, chunky, RTS	236	1 c	136	.4	.532
Turkey noodle	244	1 c	69	.2	.124
Vegetable, chunky, RTS	240	1 c	122	.2	.240
Vegetable beef:					
Canned, average	242	1 c	80	.2	.147
From dry	253	1 c	53	.1	.030
Vegetarian vegetable soup	241	1 c	70	.3	.220

Vegetables & Legumes

	Wt (g)	Svg	Cal	Copper (mg) 100 Cal	Svg
Alfalfa sprouts	33	1 c	10	.5	.052
Amaranth leaves:					
Fresh, chopped	28	1 c	7	.6	.046
Boiled	132	1 c	28	.6	.175
Artichoke, globe, cooked (300g w/refuse)	120	1 ea	60	.5	.28

	Wt (g)	Svg	Cal	Copper (mg) 100 Cal	Svg
Artichoke hearts:					
Cooked f/frozen pkg	240	9 oz	108	.1	.146
Marinated (jar)	170	6 oz	168	.1	.144
Asparagus pieces:					
Fresh pieces	67	1/2 c	15	.7	.103
Cooked from fresh	90	1/2 c	23	.4	.090
Cooked from frozen	180	1 c	50	.6	.308
Canned, drained	121	1/2 c	16	.7	.116
Canned with liquid	122	1/2 c	17	.8	.131
BEANS: also see Garbanzo, Lentils, Soybeans.					
Baked beans (dry white beans w/spices & sauce):					
Home prepared	253	1 c	382	.11	.402
Canned, plain/vegetarian	254	1 c	235	.22	.523
Canned w/franks	257	1 c	366	.1	.547
Canned w/pork	263	1 c	268	.20	.544
Canned w/pork, sweet sce	253	1 c	282	.1	.253
Canned w/pork, tomato sce	253	1 c	247	.3	.643
Black beans ckd from dry	172	1 c	227	.16	.359
Broadbeans, mature:					
Cooked from dry	170	1 c	186	.24	.440
Canned	256	1 c	183	.15	.279
Great northern:					
Cooked from dry	177	1 c	210	.21	.437
Canned	262	1 c	300	.14	.419
Green (snap) beans:					
Fresh, uncooked	110	1 c	34	.2	.076
Ckd from fresh	125	1 c	44	.3	.129
Ckd from frozen	135	1 c	36	.3	.090
Canned, drained	135	1 c	26	.2	.052
Canned with liquid	240	1 c	36	.5	.168
Hyacinth, ckd from dry	194	1 c	228	.3	.662
Kidney beans, all types:					
Cooked from dry	177	1 c	225	.2	.428
Canned with liquid	256	1 c	208	.2	.384
Lima beans:					
Ckd from fresh	170	1 c	208	.2	.519
Ckd from frozen (large)	85	1/2 c	85	.1	.047
Ckd from frozen (baby)	90	1/2 c	94	.2	.177
Ckd from dry legume	188	1 c	217	.2	.442
Canned, drained	170	1 c	164	.2	.360
Canned with liquid	241	1 c	191	.2	.434
Navy, ckd from dry	182	1 c	259	.21	.537
Pinto beans:					
Canned	240	1 c	186	.18	.336
Cooked from dry	171	1 c	235	.19	.439
Refried, canned	253	1 c	270	.4	1.04
White, ckd from dry	179	1 c	253	.1	.267
Winged, ckd from dry	172	1 c	252	.5	1.33
Yardlong, ckd from dry	171	1 c	202	.2	.385
Yellow wax: see green beans.					
Bean sprouts (Mung beans):					
Fresh sprouts	104	1 c	31	.5	.171
Stir fried	124	1 c	62	.5	.319
Boiled, drained	124	1 c	26	.6	.151
Canned, drained	125	1 c	16	.8	.120
Beets:					
Ckd from fresh, whole	100	2 ea	31	.2	.057
Canned, drained, diced	85	1/2 c	27	.2	.050
Canned with liquid, diced	123	1/2 c	36	.3	.119
Pickled slices	114	1/2 c	74	.2	.132
Beet greens, cooked	144	1 c	40	.9	.361
BROCCOLI, chopped:					
Fresh chopped	88	1 c	24	.2	.040
Cooked from fresh	156	1 c	44	.2	.068
Cooked from frozen	184	1 c	51	.2	.079
w/cheese sauce	142	1/2 c	166	<.1	.030
w/hollandaise sauce	95	1/2 c	105	<.1	.027

119

Values are for edible portion of foods

Vegetables & Legumes, continued:	Wt (g)	Svg	Cal	Copper (mg) 100 Cal	Svg
Brussels sprouts:					
Ckd from fresh (7-8 ≈ 1 c)	156	1 c	60	.2	.130
Ckd from frozen	155	1 c	65	.2	.109
CABBAGES:					
Common, shredded	70	1 c	16	.1	.016
Common, cooked	150	1 c	32	.1	.042
Bok-choy, shredded	70	1 c	9	.4	.038
Bok-choy, cooked	170	1 c	20	.3	.057
Pe-tsai, shredded	76	1 c	12	.2	.027
Pe-tsai, cooked	119	1 c	16	.2	.035
Red cabbage, shredded	70	1 c	19	.4	.068
Red cabbage, cooked	75	1/2 c	16	.3	.052
Savoy cabbage, shredded	70	1 c	20	.2	.049
Savoy cabbage, cooked	145	1 c	35	.3	.106
CARROTS:					
Fresh, 7.5" x 1-1/8" whole	72	1 ea	31	.1	.034
Fresh, grated	55	1/2 c	24	.1	.026
Cooked from fresh, sliced	78	1/2 c	35	.3	.105
Cooked from frozen	73	1/2 c	26	.2	.053
Canned, drained	73	1/2 c	17	.4	.076
Canned with liquid	123	1/2 c	28	.5	.127
Carrot juice	123	1/2 c	49	.1	.057
Cauliflower:					
Fresh, uncooked	50	1/2 c	12	.1	.016
Cooked from fresh	62	1/2 c	15	.4	.056
Cooked from frozen	180	1 c	34	.1	.043
Celeriac (celery root), ckd	100	3.5 oz	25	.5	.130
Celery: (7.5" stalk = 40 g)					
Fresh, chopped	60	1/2 c	10	1.2	.12
Cooked, diced	150	1 c	27	.2	.054
Chard, Swiss:					
Fresh, chopped	36	1 c	7	5.7	.392
Cooked	175	1 c	35	.5	.172
Collards:					
Fresh, chopped	36	1 c	11	.1	.014
Cooked from fresh	128	1 c	35	.1	.041
Cooked from frozen	170	1 c	63	.2	.094
CORN:					
Fresh kernels, uncooked	77	1/2 c	66	.1	.042
Ckd from fresh	82	1/2 c	89	<.1	.043
Ckd from frozen	82	1/2 c	67	<.1	.027
Canned, drained	82	1/2 c	66	.1	.048
Canned w/liquid	128	1/2 c	79	.1	.072
Canned, vacuum pack	210	1 c	166	.6	1.00
Canned, cream style corn	128	1/2 c	93	.1	.052
Cucumber, whole, 8" x 2+" dm	301	1 ea	39	.3	.120
Dandelion greens:					
Fresh, chopped	55	1 c	25	.3	.082
Cooked	105	1 c	35	.3	.116
Eggplant, cooked	160	1 c	45	.4	.173
Endive, fresh, chopped	25	1/2 c	4	.6	.025
Escarole (curly endive), chopped	50	1 c	9	.6	.050
Garbanzo beans (chickpeas):					
Cooked from dry	164	1 c	269	.2	.577
Canned	240	1 c	285	.15	.418
Jerusalem artichoke, fresh	150	1 c	114	.2	.189
Jicama	100	3.5 oz	20	.2	.040
Kale:					
Fresh, chopped	67	1 c	33	.6	.194
Cooked from fresh	130	1 c	42	.5	.203
Cooked from frozen	130	1 c	39	.2	.061
Kohlrabi:					
Fresh slices	140	1 c	38	.5	.196
Cooked from fresh	165	1 c	48	.4	.196
Leeks, fresh, chopped	104	1 c	63	.3	.169
Leeks, cooked	52	1/2 c	24	.2	.045

	Wt (g)	Svg	Cal	Copper (mg) 100 Cal	Svg
Lentils, cooked from dry	198	1 c	231	.2	.497
Lentils, sprouted:					
Fresh, uncooked	77	1 c	81	.3	.271
Stir fried	100	3.5 oz	101	.3	.337
LETTUCE, chopped:					
Butterhead	56	1 c	7.3	.2	.013
Iceberg	56	1 c	7.3	.2	.016
Looseleaf	56	1 c	10	.3	.028
Romaine	56	1 c	9	.2	.015
Mushrooms:					
Fresh, sliced (1≈ 18g)	35	1/2 c	9	2.0	.172
Cooked from fresh	78	1/2 c	21	1.9	.393
Canned, drained	78	1/2 c	19	1.0	.183
Mustard greens:					
Fresh chopped	56	1 c	15	.5	.066
Cooked from fresh	140	1 c	21	.6	.120
Cooked from frozen	150	1 c	29	.3	.087
Okra, cooked:					
Pods, ckd from fresh	85	8 ea	27	.3	.073
Slices, ckd from frozen	92	1/2 c	34	.3	.089
Onion:					
Fresh, chopped	160	1 c	61	.2	.096
Ckd from fresh, chopped	105	1/2 c	46	.2	.07
Ckd from frozen, chopped	105	1/2 c	30	<.1	.020
Dehydrated flakes	14	1/4 c	45	.1	.058
Parsley:					
Fresh, chopped	30	1/2 c	10	.2	.017
Freeze dried	2	1/4 c	4	.2	.006
Parsnips, ckd from fresh	156	1 c	125	.2	.216
PEAS:					
Black-eyed peas:					
Cooked from fresh	165	1 c	160	.1	.219
Cooked from frozen	170	1 c	224	.1	.313
Cooked from dry	167	1 c	562	.3	1.41
Canned	240	1 c	184	.2	.281
Green peas:					
Fresh, uncooked	145	1 c	118	.2	.255
Ckd from fresh	160	1 c	134	.2	.277
Ckd from frozen	80	1/2 c	63	.2	.111
Canned, drained	85	1/2 c	59	.1	.070
Canned with liquid	124	1/2 c	61	.2	.134
Green peas, edible-pods:					
Fresh	145	1 c	61	.2	.120
Cooked from fresh	160	1 c	67	.2	.123
Cooked from frozen	80	1/2 c	42	.1	.049
Split peas, cooked from dry	196	1 c	231	.2	.355
Peas, sprouted, fresh	120	1 c	154	.2	.326
Peas and carrots:					
Cooked from frozen	80	1/2 c	38	.2	.061
Canned, with liquid	128	1/2 c	48	.3	.132
PEPPER, HOT chili, green or red:					
Fresh, chopped	75	1/2 c	30	.4	.131
Canned peppers	68	1/2 c	17	.4	.075
Canned, Jalapeño, chopped	68	1/2 c	17	.6	.095
PEPPER, SWEET, green or red:					
Fresh, chopped	50	1/2 c	14	.2	.033
Cooked from fresh, chopped	68	1/2 c	19	.2	.044
Pimento, canned	57	2 oz	13	.2	.028
Poi, two finger	240	1 c	269	<.1	.092
POTATOES:					
Baked in oven:					
Flesh and skin	202	1 ea	220	.3	.616
Flesh only	156	1 ea	145	.2	.335
Potato skin	58	1 ea	115	.4	.474
Boiled w/skin, flesh only	136	1 ea	119	.2	.256
Boiled without skin	135	1 ea	116	.2	.225
Canned, 1" dm.	70	2 ea	42	.1	.040

	Wt (g)	Svg	Cal	Copper (mg) 100 Cal	Svg
Vegetables & Legumes, continued:					
POTATOES, continued:					
French fried f/frzn:					
Fried in oil	50	10 ea	158	<.1	.069
Oven heated	50	10 ea	111	.1	.082
Hash browns, fr/frozen	156	1 c	340	.1	.237
Mashed potatoes:					
Prepared with milk	210	1 c	162	.2	.294
From instant	215	1 c	239	<.1	.097
Potato puffs (tator tots)	62	.5 c	138	<.1	.037
Potato dishes, prepared:					
Au gratin, recipe	245	1 c	322	.1	.392
Au gratin, fr/mix	245	1 c	228	<.1	.113
Scalloped, recipe	245	1 c	210	.2	.399
Scalloped, fr/mix	245	1 c	228	.1	.120
Potato chips (1 oz ≈14 chips)	28	14 ea	148	<.1	.057
Potato pancakes	76	1 ea	237	.1	.273
Pumpkin, mashed:					
Cooked from fresh	245	1 c	50	.4	.194
Canned	123	1/2 c	42	.3	.131
Radish, Daikon, chopped	44	1/2 c	8	.8	.061
Radish, red	45	10 ea	7	.3	.018
Radish seeds, sprouted	38	1 c	16	.3	.046
Rutabaga:					
Fresh cubes	140	1 c	51	.1	.056
Cooked	85	1/2 c	29	.1	.031
Salsify, cooked slices	135	1 c	92	.2	.162
Sauerkraut, canned w/liquid	236	1 c	44	.5	.227
Seaweed, fresh:					
Irish moss	28	1 oz	14	.3	.042
Kelp	28	1 oz	12	.3	.037
Lavar	28	1 oz	10	.8	.075
Wakame	28	1 oz	13	.6	.081
Soybeans, cooked from dry	172	1 c	298	.2	.700
Soybeans, sprouted:					
Fresh, uncooked	35	1/2 c	45	.3	.150
Steamed	94	1/2 c	76	.4	.310
Stir fried	100	1 c	125	.4	.527
SOYBEAN PRODUCTS: See tofu, this section; miso, in Other; roasted soybeans in Nuts & Seeds; soy milk in Dairy; soy flour in Grains.					
Spinach:					
Fresh, chopped	56	1 c	12	.6	.073
Cooked from fresh, chopped	180	1 c	41	.8	.313
Cooked from frozen	190	1 c	53	.5	.268
Canned, drained	214	1 c	50	.8	.385
SQUASH, SUMMER varieties:					
Crookneck, fresh slices	130	1 c	24	.6	.133
Crookneck, cooked	180	1 c	36	.5	.185
Scallop, fresh slices	130	1 c	24	.6	.133
Scallop, cooked	180	1 c	28	.5	.150
Zucchini, fresh slices	130	1 c	19	.4	.074
Zucchini, cooked	180	1 c	29	.5	.155
SQUASH, WINTER varieties:					
Acorn / Danish, baked	245	1 c	137	.2	.210
Acorn /Danish, boiled	245	1 c	83	.2	.127
Butternut, baked	245	1 c	99	.2	.159
Butternut boiled	240	1 c	94	.1	.086
Butternut, ckd f/frozen	240	1 c	94	.1	.086
Hubbard , baked	240	1 c	120	.1	.108
Hubbard, boiled	236	1 c	70	.2	.111
Spaghetti, baked/boiled	155	1 c	45	.1	.054
Succotash:					
Cooked from fresh	192	1 c	222	.2	.344
Cooked from frozen	170	1 c	158	.1	.102

	Wt (g)	Svg	Cal	Copper (mg) 100 Cal	Svg
Sweet potato:					
Baked in skin	114	1 ea	118	.2	.237
Boiled, peeled	151	1 ea	160	.2	.243
Canned, mashed	128	1/2 c	129	.3	.355
Canned, vacuum pack	255	1 c	233	.2	.354
Candied	105	1 pce	144	.1	.107
Tofu, firm	126	1/2 c	183	.3	.476
Tofu, regular	124	1/2 c	94	.3	.239
TOMATOES					
Fresh, whole	123	1 ea	26	.4	.091
Fresh, chopped	180	1 c	38	.4	.133
Cooked from fresh	240	1 c	65	.3	.223
Canned, whole	240	1 c	47	.6	.264
Tomato juice, canned	244	1 c	42	.6	.246
Tomato paste, canned	262	1 c	220	.7	1.55
Tomato puree, canned	250	1 c	102	.4	.408
Tomato pauce, canned	245	1 c	74	.6	.480
Turnips:					
Fresh cubes	130	1 c	35	.2	.078
Cooked from fresh	78	1/2 c	14	.2	.031
Turnip greens:					
Cooked from fresh	144	1 c	29	1.3	.364
Cooked from frozen	82	1/2 c	24	.5	.123
Vegetable juice cocktail	242	1 c	46	1.1	.484
VEGETABLES, MIXED ckd fr/frozen:					
Broccoli, carrots, pasta	95	2/3 c	88	<.1	.040
Broccoli, carrots, water chestnuts	91	2/3 c	32	.1	.037
Broccoli, cauliflower, red pepper	95	2/3 c	25	.1	.030
Cantonese stir fry	95	1/2 c	53	.1	.042
Chinese stir fry	95	1/2 c	31	.1	.040
Green beans & spaetzle, Bavarian style	95	.5 c	108	<.1	.046
Japanese style	95	1/2 c	29	.3	.077
Mixed vegetables (corn, peas, carrots, green beans, limas):					
Cooked from frozen	182	1 c	107	.1	.151
Canned, drained	163	1 c	77	.2	.119
Peas, mixed with:					
Carrots, onions	91	1/2 c	54	.2	.082
Carrots, onions, pasta	95	1/2 c	122	.1	.069
Cauliflower, cream sce	95	1/2 c	118	<.1	.045
Corn, pasta, cream sce	95	1/2 c	132	.1	.070
Mushrooms	95	1/2 c	73	.1	.098
Mushrooms, pasta, crm sce	95	.5 c	129	.1	.140
Onions	95	1/2 c	71	.1	.083
Onions in cheese sce	142	1/2 c	165	.1	.100
Rice & mushrooms	66	2/3 c	108	<.1	.040
Water chestnuts, canned slices	70	1/2 c	35	.2	.070
Yam, orange: see Sweet potato.					
Yam, white, cubed, cooked	136	1 c	158	.1	.207
Zucchini: see Squash, summer.					

Other

Cooking Ingredients, condiments, flavorings, other beverages, spices, sweeteners.

	Wt (g)	Svg	Cal	Copper (mg) 100 Cal	Copper (mg) Svg
Apple butter	35	2 T	66	.2	.127
BEVERAGES:					
Beer (12 fl oz = 1.5 c):					
Regular	356	1.5 c	146	<.1	.032
Light	354	1.5 c	100	.1	.085
Carbonated (12 fl oz = 1.5 c):					
Cola, regular	370	1.5 c	151	<.1	.041
Cream soda	371	1.5 c	191	<.1	.030
Fruit flavored soda	372	1.5 c	170	<.1	.061
Ginger ale	366	1.5 c	124	.1	.066
Grape soda	372	1.5 c	161	.1	.082
Lemon lime soda	368	1.5 c	149	<.1	.044
Orange soda	372	1.5 c	177	<.1	.056
"Pepper" type soda	368	1.5 c	151	<.1	.022
Root beer	370	1.5 c	152	<.1	.026
Clam nectar, canned	240	1 c	6	16	.934
Coffee, brewed	240	1 c	1-2	1.0	.016
Fruit punch:					
From frozen conc.	247	1 c	113	.1	.074
Canned	253	1 c	118	.1	.129
Grape drink	250	1 c	112	<.1	.030
Lemonade, f/frozen	248	1 c	100	<.1	.005
Limeade, f/frozen	247	1 c	102	<.1	.028
Perrier Water	240	1 c	0	0	0
Pineapple grapefruit drink	250	1 c	117	.1	.113
Pineapple orange drink	250	1 c	125	.1	.103
Tea, brewed:					
Regular	240	1 c	2	1.2	.024
Herbal	178	3/4 c	1	2.7	.027
Water, municipal: varies by location and type of plumbing. Water from copper pipes can have more copper in it.					
Wine:					
Dessert, dry	118	1/2 c	149	<.1	.053
Red	118	1/2 c	85	<.1	.024
White	118	1/2 c	80	<.1	.015
CANDY and CANDY BARS:					
Almonds, sugar coated	28	7 ea	146	.1	.126
Chocolate covered:					
Almonds	165	1 c	935	.1	1.37
Coconut	28	1 oz	133	.1	.130
Peanuts	170	1 c	954	.1	1.29
Raisins	187	1 c	733	.2	1.12
Chocolate kisses	28	6 pce	154	.1	.110
Divinity with nuts	20	1 pce	80	<.1	.032
Krackle candy bar	34	1 ea	179	.1	.140
M&Ms Plain choc candies	48	1 pkg	237	.1	.150
M&Ms Peanut candies	47	1 pkg	240	.1	.170
Mars bar	50	1 ea	240	.1	.130
Milk chocolate:					
Plain	28	1 oz	145	.1	.111
With almonds	28	1 oz	150	.1	.150
With peanuts	28	1 oz	155	.1	.200
With rice cereal	28	1 oz	140	.1	.089
Milky Way bar	60	1 ea	260	<.1	.090
Mr. Goodbar	47	1 ea	250	.1	.240
Reese's peanut butter cups	45	2 ea	240	.1	.257
Snickers, 2.2 oz bar	61	1 ea	290	<.1	.140
Carob flour	103	1 c	185	.3	.588
Catsup	17	1 T	18	.2	.04
Chili sauce, tom base	273	1 c	284	.4	1.10

	Wt (g)	Svg	Cal	Copper (mg) 100 Cal	Copper (mg) Svg
CHOCOLATE:					
Baking, unsweetened	28	1 oz	145	2.6	.567
Bittersweet	28	1 oz	141	.2	.250
Chocolate chips, semi-sweet	170	1 c	860	.3	2.17
Dark, sweet	28	1 oz	150	<.1	.057
Hot fudge topping	38	2 T	129	.1	.113
Syrup, thin	38	2 T	85	.2	.188
Cocoa powder	86	1 c	224	1.4	3.07
Enchilada dip, Fritos	28	1 oz	35	.3	.110
Falafel (2.25" patty = 17g)	17	1 ea	57	.1	.044
Gelatin, dry, plain, env	7	1 env	25	.1	.031
Gelatin salad, dessert	120	1/2 c	70	.1	.075
Honey	339	1 c	1030	<.1	.110
Horseradish, prepared	15	1 T	6	.4	.021
Hummous (Humous)	246	1 c	420	.1	.561
Jalapeno bean dip, Fritos	28	1 oz	33	.3	.090
Miso (soybean product)	138	1/2 c	284	.2	.603
Molasses:					
Blackstrap	40	2 T	85	.7	.568
Light	20	1 T	43	.7	.284
Mustard, prepared	5	1 tsp	3.9	.5	.021
Olives, pitted:					
Green	39	10 ea	45	.3	.133
Black, large ripe	45	10 ea	52	.2	.113
Pickles:					
Dill, medium	65	1 ea	12	.4	.051
Sweet, medium	35	1 ea	41	.1	.037
Fresh pack, slices	30	4 pce	20	.2	.03
Pickle relish, sweet	15	1 T	20	.3	.055
Salsa, recipe	105	1/2 c	45	.2	.079
Soy sauce:					
Regular, f/soy & wheat, Shoyu	18	1 T	9	.2	.021
Tamari, from soy	18	1 T	11	.2	.024
F/hydrolyzed veg protein	18	1 T	7	.9	.060
SPICES:					
Basil, dried	4.5	1 T	11	.5	.059
Black pepper	2.1	1 tsp	5.4	.4	.024
Caraway seed	6.7	1 T	22	.4	.085
Garlic, cloves	12	4 ea	18	.2	.031
Garlic, powder	2.8	1 tsp	9	.2	.021
Ginger, ground	5.4	1 T	19	.1	.026
Nutmeg, ground	7	1 T	37	.2	.070
Onion powder	2.1	1 tsp	5	1.3	.067
Poppyseed	8.8	1 T	47	.4	.189
Sugar:					
Brown	220	1 c	820	.1	.770
White, granulated	200	1 c	770	<.1	.030
White, powdered, sifted	100	1 c	385	<.1	.020
Syrup:					
Corn syrup, dark	328	1 c	944	.1	.110
Corn syrup, light	328	1 c	912	<.1	.072
Maple syrup	20	1 T	50	.2	.090
Tempeh (soybean product)	166	1 c	331	.3	1.1
Tobasco sauce	15	1 T	1.6	2.1	.035
Vinegar, cider	240	1 c	29	.5	.156
Yeast:					
Active dry	30	4 T	80	1.9	1.50
Brewer's yeast	8	1 T	25	1.0	.261

Iodine

Iodine is an important constituent of the thyroid hormones, which play a lot of roles in the body, including regulating metabolic rate, body temperature, growth, reproduction, making blood cells, muscle function, and nerve function.

Iodine deficiency diseases include goiter, where the thyroid glands in the neck become greatly enlarged, and cretinism, a form of mental retardation. These diseases were occurring with high incidence in the Central Plains states, where the soil is low in iodine. Plant sources of iodine vary widely with the iodine content of the soil they grow in.

To handle this problem, iodine was added to something everyone would eat—salt, and iodized salt has greatly reduced the prevalence of these diseases. But now there is some evidence that there may be an over abundance of iodine in the American diet. This may, in part, be from the amount of salt (iodized salt) in processed foods. It may also be from bread, where commercial dough conditioners include iodine in the form of iodate. Interestingly enough, very high intakes of iodine can depress thyroid activity.

Recommended Dietary Allowances for Adults
USA RDA: 150 mcg per day (Pregnant, 175 mcg; nursing, 200 mcg)
Canadian RNI: 160 mcg per day (Pregnant, 185 mcg; nursing, 210 mcg)

Food Sources

In the seashore areas, seafoods, water, and even the ocean mist are important sources of iodine. Further inland, the plant and animal products vary depending on fertilizing and feeding practices. The most reliable consistent source is iodized salt. Dairy products contain iodine because of feed additives and disinfectants; breads can contain iodine from the iodates in dough conditioners. There may be much more iodine in foods than we know, but research literature does not discuss many studies or analyses. As a result, there are only a few known food sources that we can list here. It can be added to foods (fortification), so check labels.

	Wt (g)	Svg	Iodine (mcg) Svg		Wt (g)	Svg	Iodine (mcg) Svg
Dairy, Eggs & Related Products				**Meats & Seafood**			
CHEESE:				Beef, ground	85	3 oz	7.7
Cheddar	28	1 oz	5-23	Cod	85	3 oz	87
Cottage cheese, creamed	105	1/2 c	18-71	Haddock	85	3 oz	104-145
Cottage cheese, low fat	105	1/2 c	26-71	Perch	85	3 oz	18
Cream cheese	28	1 oz	9-11	Pork chop	85	3 oz	4.3
Feta	28	1 oz	1.4-2.6	Shrimp	85	3 oz	21-37
Mozzarella	28	1 oz	23-44				
Cheese spread (Velveeta)	28	1 oz	17				
EGG, large, whole	50	1 ea	18-26	**Vegetables, Nuts & Legumes**			
MILK							
Buttermilk, cultured	245	1 c	69-127	Almonds, whole, dry	28	1 oz	4
Whole milk, fluid	245	1 c	51-140	Beans, dried, average	100	1/2 c	7
Canned, evaporated	127	1/2 c	52-54	Broccoli:			
Milk beverage mix, dry powder:				Fresh chopped	88	1 c	4.2
Milk chocolate	28	1 oz	11	Cooked from frozen	92	1/2 c	4.6
Rich chocolate	28	1 oz	2	Potato, medium	150	1 ea	7.5
				Spinach, cooked from frozen	95	1/2 c	5.1
Baked Goods & Grain Products				Spinach, canned	107	1/2 c	8.6
				Other			
Bread, commercial, made with iodate dough conditioner, made with:				Light Salt (Morton)	6	1 tsp	400
Continuous mix process	28	1 pce	142	Salt, iodized	6	1 tsp	400
Conventional process	28	1 pce	35	Salt, regular	6	1 tsp	<100

ESHA Research

Iron

The greatest portion of iron in our bodies is in the hemoglobin of blood cells; it is what makes our blood red. Hemoglobin carries oxygen from the lungs to all the cells in the body, and it needs iron to work. At the receiving point of the transported oxygen is myogloblin, which also contains iron. It stores the oxygen in the muscle for use in muscle contractions. Iron also plays a role in transporting carbon dioxide from the cells back to the lungs, and is part of several enzymes and important proteins, and is involved in energy metabolism.

Iron deficiency anemia is a condition of dietary iron deficiency, but iron deficiency can also be caused indirectly by a deficiency in other nutrients. For example, deficiencies in vitamin B-6, vitamin E, Folacin, vitamin B-12, vitamin A, vitamin C, or copper can affect the hemoglobin level in the body, and can indirectly create an iron deficiency.

Iron deficiency in children can show up as psychological disturbances such as a short attention span, apathy, irratability, hyperactivity, and reduced ability to learn, before overt anemia is evident. A severe, longstanding deficiency can result in reduced intelligence.

Iron comes in heme and nonheme forms, and they are absorbed by different mechanisms. Iron in animal meat averages about 40% heme iron, which is highly absorbable. The remaining 60% of the iron in meats, and 100% of the iron in vegetables is present in nonheme compounds. Nonheme iron is more difficult for the body to absorb. When it is eaten with foods containing vitamin C, absorption is increased. Also, some iron is absorbed into foods cooked in iron vessels. However, there are also other factors that affect iron uptake (see Nutrient Losses). Because of the inhibiting and enhancing factors, nonheme iron absorption may vary up to ten fold, depending on the content of the *entire* diet.

Nutrient Losses

Iron absorption can be lowered by the presence of antacids, phosphate salts, calcium phosphate, phytates, bran, and tannic acid/polyphenols in tea. Studies have also shown that drinking coffee before a meal can reduce iron absorption as well. Absorption of iron can be increased by eating food sources of vitamin C and iron together.

Recommended Dietary Allowances for Adults

USA RDA: Men, 10 mg; Women, 15 mg (pregnant, 30 mg; nursing, 30 mg); after menopause, 10 mg per day.
Canadian RNI: Men, 8 mg; Women, 14 mg (pregnant or nursing: 20 mg); after menopause, 7 mg per day.

The body's absorption of iron is influenced by various complex factors, as discussed above. Iron in meat is absorbed more easily than iron from plant sources. In addition, absorption of iron into the body also depends on the individual's iron status. Average absorption of dietary iron is relatively low when body stores are high, and may be increased when stores are low. Therefore an iron intake that is below the recommended amounts may not necessarily predict an iron deficiency.

The recommendations take such differences into account and reflect the different needs between men and women. Even though the body has mechanisms to store and recycle most of its iron, minor losses occur, and major losses occur through bleeding from a wound, internal hemorrhaging, or menstruation. People who consume no meat, fish or poultry (good sources of iron) as well as very little vitamin C, may require higher amounts of iron.

Toxicity

There are approximately 2,000 cases of iron poisoning each year in the U.S., mostly among children who ingest iron supplements formulated for adults. Also, some people are genetically at risk from iron overload or hemochromatosis.

Food Sources

Good sources of iron are lean meats, fish, poultry, organ meats (liver, kidney, and heart), legumes (dried beans and peas), nuts and seeds, whole grains, dark molasses, and green leafy vegetables.

Values are for edible portion of foods

Iron (Fe) mg

Baked Goods

Breads, Cakes, Cookies, Crackers, Muffins, Pancakes, Pastries, Pies, Rolls, and some desserts. Recipes here use flour enriched with iron, according to U.S. standards.

	Wt (g)	Svg	Cal	Iron (mg) 100 Cal	Iron (mg) Svg
Apple crisp, 3" x 3"	78	1 pce	146	.5	.765
Bagel, 3.5" dm, plain/egg	68	1 ea	180	1.2	2.10
Biscuits:					
Homemade	28	1 ea	100	.7	.700
From refrig dough	20	1 ea	65	.7	.474
From mix	28	1 ea	94	.6	.581
BREADS:					
Banana nut, 1/2" slice	50	1 pce	161	.5	.830
Boston brown, cnd, 1/2"	45	1 pce	95	.9	.900
Cornbread muffin f/mix	45	1 ea	145	.9	1.30
Cracked wheat	25	1 pce	65	1.0	.666
French, 5" x 2.5"	35	1 pce	100	1.1	1.08
Mixed grain	25	1 pce	65	1.2	.800
Oatmeal bread	25	1 pce	65	1.1	.700
Pita pocket, enr. 6.5" dm.	60	1 ea	165	.9	1.45
Pumpernickel 5" x 4" x 3/8"	32	1 pce	80	1.1	.877
Raisin bread	25	1 pce	68	1.1	.775
Rye, light, 5" x 3.5" x 7/16"	25	1 pce	65	1.0	.680
Vienna bread	25	1 pce	70	1.1	.770
Wheat (white and whole					
wheat flour)	28	1 pce	72	1.4	.974
White bread	28	1 pce	75	1.1	.796
Whole wheat bread	35	1 pce	86	1.4	1.19
Bread crumbs:					
Dry, grated	100	1 c	390	1.1	4.10
Soft	45	1 c	120	1.1	1.28
Bread cubes, soft	30	1 c	80	1.1	.852
Bread dressing: see Stuffing.					
Bread pudding w/raisins	165	1 c	349	.5	1.77
Bread sticks, 4" x 1/2" dm.	100	10 ea	384	.2	.900
Brownies:					
Homemade, w/nuts	20	1 ea	95	.4	.400
Frozen, frosted	25	1 ea	103	.4	.400
Commercial, frosted w/nuts	25	1 ea	100	.6	.605
CAKES, Cupcakes ≈ 42g; piece = 1/16th cake unless otherwise noted:					
Angel food, 1/12 tube cake	53	1 ea	126	.2	.228
Carrot, cream cheese					
frosting, 2.5"x3"	112	1 pce	406	.3	1.20
Cheesecake, recipe, 1/12	92	1 pce	278	.2	.441
Chocolate, chocolate frosting	69	1 pce	235	.6	1.40
Coffee, f/mix, 2.4" x 2.8"	72	1 pce	230	.5	1.22
Gingerbread: 3" x 3"					
Homemade	110	1 pce	351	.8	2.74
From mix	63	1 pce	174	.7	1.20
Pound cake, 1/2" slice	30	1 pce	115	.4	.500
Sheet cake, 3" x 3"					
Plain	86	1 pce	315	.4	1.30
White frosting	121	1 pce	445	.3	1.20
Snack cake, filled:					
Spongecake, like Twinkies	42	1 ea	155	.4	.600
Chocolate, like Ding-Dongs	28	1 ea	105	1.0	1.00
Sponge cake, 1/12	66	1 pce	194	.6	1.22
White:					
Chocolate frosting	77	1 pce	291	.5	1.40
Coconut or white frosting	70	1 pce	265	.4	1.06
Yellow, chocolate frosting:					
From mix	69	1 pce	235	.4	.965
Commercial	69	1 pce	245	.5	1.24
Cherry crisp, 3" x 3"	138	1 pce	157	1.4	2.16

Chips: see corn & tortilla this section; potato chips under vegetables.

	Wt (g)	Svg	Cal	Iron (mg) 100 Cal	Iron (mg) Svg
COOKIES:					
Animal cookies	28	27 ea	120	.8	.920
Chocolate chip:					
Homemade	40	4 ea	185	.5	1.00
Commercial	42	4 ea	180	.4	.800
From refrigerator dough	48	4 ea	225	.5	1.04
Fig bars	56	4 ea	210	.6	1.36
Lady fingers	44	4 ea	158	.5	.800
Oatmeal raisin	52	4 ea	245	.4	1.10
Peanut butter, homemade	48	4 ea	245	.4	1.10
Sandwich, all types	40	4 ea	195	.7	1.40
Shortbread, homemade	28	2 ea	145	.4	.550
Shortbread, commercial	32	4 ea	155	.5	.800
Snickerdoodle	20	1 ea	110	.6	.622
Sugar, f/refrig dough	48	4 ea	235	.4	.900
Vanilla wafers	40	10 ea	185	.4	.800
Corn chips	28	1 oz	155	.3	.500
CRACKERS:					
Armenian cracker bread	28	4 pce	117	.4	.450
Cheese crackers	10	10 ea	50	.7	.350
Cheese, pnut butter filled	30	4 ea	150	.8	1.20
Graham crackers	14	2 ea	60	.6	.367
Round (like Ritz)	9	3 ea	45	.7	.300
Rye wafers	14	2 ea	55	.9	.500
Saltines	12	4 ea	50	1.0	.500
Sesame	12	4 ea	60	.7	.400
Cream puff, custard filling	110	1 ea	280	.4	1.07
Crepe (without filling)	27	1 ea	47	1.0	.449
Croissant, 4.5" x 4" x 2"	57	1 ea	235	.9	2.10
Danish pastry:					
Plain pastry	57	1 ea	220	.5	1.10
Pastry with fruit	65	1 ea	235	.6	1.30
Doughnuts:					
Cake, medium	50	1 ea	210	.4	.800
Yeast raised, plain	60	1 ea	235	.6	1.40
Yeast raised, jelly filled	65	1 ea	226	.4	.800
Eclair, custard filling,					
chocolate icing	94	1 ea	262	.3	.856
English muffin:					
Plain muffin	57	1 ea	140	1.2	1.70
Sourdough	56	1 ea	129	1.1	1.40
With raisins	56	1 ea	146	1.2	1.68
MUFFINS: also see English muffins.					
Blueberry, recipe/mix	45	1 ea	138	.7	.900
Bran, wheat:					
From recipe	45	1 ea	125	1.1	1.40
From mix	45	1 ea	140	1.2	1.70
Cornbread:					
From recipe	45	1 ea	145	.6	.900
From mix					
Cornmeal muffin f/mix	45	1 ea	145	.9	1.30
Pancakes:					
Buckwheat, fr/mix, 4" dm.	27	1 ea	55	.7	.400
Regular, recipe, 4" dm	27	1 ea	60	.8	.500
Regular, fr/ mix, 4" dm.	27	1 ea	60	1.2	.700
Whole wheat, 5" dm.	52	1 ea	94	.9	.817
Peach crisp, 3" x 3"	139	1 pce	166	.6	.997
PIES: pce is 1/6 of a 9"pie.					
Apple	158	1 pce	405	.4	1.67
Blueberry	158	1 pce	380	.6	2.10
Banana cream, recipe	198	1 pce	319	.3	1.09
Cherry	158	1 pce	410	.8	3.17
Chocolate cream	175	1 pce	311	.3	1.08
Coconut cream	172	1 pce	343	.3	1.20
Coconut custard	165	1 pce	384	.5	1.77
Custard pie	152	1 pce	293	.5	1.44
Lemon meringue	140	1 pce	355	.4	1.40
Mincemeat	160	1 pce	395	.5	2.03

125

ESHA Research

	Wt (g)	Svg	Cal	Iron (mg) 100 Cal	Svg
Baked Goods, continued:					
PIES, continued:					
Peach pie	158	1 pce	405	.5	1.90
Pecan pie	138	1 pce	583	.3	1.85
Pumpkin	200	1 pce	367	.7	2.63
Popovers	51	1 ea	96	.9	.858
PopTart-type pastry, fortified	54	1 ea	210	1.0	2.16
Pretzel, dutch twist	16	1 ea	65	.5	.315
Pretzel, thin twists	60	10 ea	240	.5	1.20
ROLLS:					
Cinnamon, small	50	1 ea	158	.6	.900
Dinner roll, 2.5" x 2"	28	1 ea	85	1.0	.810
Hamburger bun	45	1 ea	129	1.0	1.34
Hard roll, white	50	1 ea	155	.9	1.40
Hotdog bun	40	1 ea	115	1.0	1.19
Rye roll, light	28	1 ea	76	1.0	.762
Rye roll, dark	28	1 ea	79	1.0	.809
Submarine (hoagie)	135	1 ea	400	1.0	3.80
Whole wheat roll	35	1 ea	88	1.3	1.17
Stove Top stuffing	108	1/2 c	176	.7	1.17
Tortillas:					
Corn, enr, fried, 6'	30	1 ea	87	.7	.600
Flour, 10.5" diam.	57	1 ea	168	.5	.880
Flour, 8" diam.	35	1 ea	105	.5	.549
Tortilla chips:					
Plain chips	28	1 oz	139	.7	1.00
Doritos, Nacho flavor	28	1 oz	139	.3	.400
Doritos, Taco flavor	28	1 oz	140	.5	.700
Waffles:					
Homemade, 7" diam.	75	1 ea	245	.6	1.50
From mix, 7" diam.	75	1 ea	205	.6	1.20
From frozen, 4" diam.	35	1 ea	98	1.7	1.70

Dairy & Dairy Products

	Wt (g)	Svg	Cal	Iron (mg) 100 Cal	Svg
CHEESE (1.5" cube ≈ 1 oz):					
American cheese	28	1 oz	106	.10	.110
American cheese food	28	1 oz	94	.25	.240
American cheese spread	28	1 oz	82	.11	.091
Blue cheese	28	1 oz	100	.09	.090
Brie cheese	28	1 oz	95	.15	.140
Cheddar, hard	28	1 oz	114	.17	.197
Cottage cheese:					
Lowfat 1%	226	1 c	164	.2	.320
Lowfat 2%	226	1 c	205	.2	.360
Creamed	218	1 c	225	.1	.275
Dry curd	145	1 c	123	.3	.330
Cream cheese (1 T = 15g)	28	1 oz	99	.3	.338
Feta cheese	28	1 oz	75	.24	.181
Gjetost cheese	28	1 oz	132	.10	.130
Parmesan, grated (1 T = 5g)	28	1 oz	129	.21	.27
Monterey Jack	28	1 oz	106	.19	.201
Mozzerella, part skim, low moisture	28	1 oz	80	.10	.076
Ricotta, part skim	246	1 c	340	.3	1.09
Ricotta, whole milk	246	1 c	428	.2	.940
Swiss cheese	28	1 oz	107	<.1	.050
Swiss processed cheese	28	1 oz	95	.18	.170
CREAM, SWEET, fluid:					
Coffee or table	240	1 c	469	<.1	.100
Half & half	214	1 c	315	.1	.170
Light whipping cream	239	1 c	699	<.1	.070
Heavy whipping cream	238	1 c	821	<.1	.070
CREAM SWEET, whipped:					
Whipped (heavy) unsweetened	119	1 c	410	<.1	.035
Pressurized	60	1 c	154	<.1	.030

	Wt (g)	Svg	Cal	Iron (mg) 100 Cal	Svg
CREAM, SOUR, dairy:					
Cultured, dairy	14	1 T	31	<.1	.009
Half & half, dairy (12% fat)	15	1 T	20	.1	.010
Cream, sour, Imitation, non-dairy	14	1 T	30	<.1	.01
CREAM SUBSTITUTES (non-dairy):					
Coffee whitener:					
Liquid or frozen	120	1/2 c	163	<.1	.040
Powdered	94	1 c	514	.2	1.08
Dessert Toppings, non-dairy:					
Frozen (like Coolwhip)	75	1 c	239	<.1	.090
Pressurized					
Kefir beverage	233	1 c	160	.3	.500
MILK (cow):					
Skim milk	245	1 c	86	.1	.100
Lowfat 1%	244	1 c	102	.1	.120
Lowfat 2%	244	1 c	121	.1	.120
Whole (3.3% fat)	244	1 c	150	.1	.120
Canned, skim, evap	255	1 c	200	.4	.700
Canned, whole, evap	252	1 c	340	.1	.480
Dry, nonfat, instant	68	1 c	244	.1	.210
Milk (other):					
Goat	244	1 c	168	.1	.120
Human breast milk	246	1 c	171	.04	.070
Soy milk	240	1 c	79	1.7	1.38
MILK BEVERAGES & mixes:					
Carob flavored mix, dry	12	3 T	45	1.2	.550
Chocolate milk, commercial					
Lowfat 1% - whole	250	1 c	160-210	.3-4	.600
Chocolate flavored, to be mixed with water:					
Powder (with dry milk)	28	1 oz	100	.3	.286
Drink, prepared	206	3/4 c	100	.3	.286
Chocolate flavored, to be mixed with milk:					
Powder	21.6	3/4 oz	75	.9	.680
Drink w/whole milk	266	1 c	226	.4	.800
Cocoa, hot, w/whole milk	250	1 c	218	.4	.780
Eggnog, commercial	254	1 c	342	.1	.510
Instant Breakfast, dry	37	1 env	130	6.1	7.90
Malted milk, w/whole milk:					
Chocolate flavor	265	1 c	229	.3	.600
Natural flavor	265	1 c	237	.1	.270
Milkshake (10 fl oz = 1.25 c):					
Chocolate	283	1.25 c	360	.2	.880
Strawberry	283	1.25 c	319	.1	.300
Vanilla	283	1.25 c	314	.1	.260
MILK DESSERTS:					
Custard, baked	265	1 c	305	.4	1.1
Ice cream, vanilla:					
Regular	133	1 c	269	<.1	.12
Rich	148	1 c	349	<.1	.10
Soft serve	173	1 c	377	.1	.43
Ice milk, regular	131	1 c	184	.15	.28
Ice milk, soft serve	175	1 c	223	.14	.31
PUDDINGS, prepared:					
(5 oz can ≈ 1/2 cup)					
Chocolate, ckd from mix	260	1 c	300	.1	.40
Chocolate, from instant	260	1 c	310	.2	.60
Chocolate, canned	142	1 can	205	.6	1.2
Rice pudding, cooked or instant-mix	141	1/2 c	165	.3	.52
Tapioca, ckd from mix	130	1/2 c	145	.1	.10
Tapioca, canned	142	1 can	160	.2	.30
Vanilla, ckd from mix	130	1/2 c	145	.1	.10
Vanilla, from instant	130	1/2 c	150	.1	.10
Vanilla, canned	142	1 can	220	.1	.20

	Wt (g)	Svg	Cal	Iron (mg) 100 Cal	Iron (mg) Svg
Dairy, continued:					
MILK DESSERTS, continued:					
Pudding pops, chocolate	57	1 ea	99	.4	.36
Sherbet (2% fat)	193	1 c	270	.1	.31
Yogurt, frozen, avg	174	1 c	220	.05	.104
YOGURT:					
Coffee/vanilla flavor	227	1 c	193	.08	.159
Lowfat, plain (1.5% fat)	227	1 c	144	.13	.180
Lowfat, fruit (1.15% fat)	227	1 c	231	.07	.160
Nonfat, plain	227	1 c	127	.16	.204
Whole milk yogurt	227	1 c	138	.08	.114

Eggs

	Wt (g)	Svg	Cal	Iron (mg) 100 Cal	Iron (mg) Svg
Chicken egg, raw or ckd:					
Whole egg	50	1 ea	77.5	.9	.720
White only	33.4	1 ea	17	.1	.010
Yolk only	16.6	1 ea	59	1.0	.59

Fruits & Fruit Juices

	Wt (g)	Svg	Cal	Iron (mg) 100 Cal	Iron (mg) Svg
Acerola juice, fresh	242	1 c	51	2.4	1.21
Apple, 2.75" diam:					
With peel	138	1 ea	80	.3	.250
Peeled	72	1 ea	72	.2	.160
Applesauce, unsweetened	244	1 c	106	.3	.290
Applesauce, sweetened	255	1 c	195	.5	1.00
Apple juice:					
From frozen	239	1 c	111	.5	.610
Canned/bottled	248	1 c	116	.8	.920
APRICOTS:					
Fresh, pitted	106	3 ea	51	1.1	.580
Canned, juice pack	84	3 ea	40	.6	.252
Canned, heavy syrup	85	3 ea	70	.4	.260
Dried halves	35	10 ea	83	2.0	1.65
Apricot nectar, canned	251	1 c	141	.7	.960
Avocados, whole:					
California	173	1 ea	305	.7	2.04
Florida	304	1 ea	340	.5	1.60
Bananas, 8.75" (176g whole)	114	1 ea	105	.3	.353
Blackberries:					
Fresh berries	144	1 c	74	1.1	.800
Frozen, unthawed	151	1 c	94	.1	.044
Canned	256	1 c	236	.7	1.66
Blackberry juice, fresh	250	1 c	93	2.4	2.25
Blueberries:					
Fresh berries	145	1 c	82	.3	.24
Frozen, sweet, thawed	230	1 c	185	.5	.900
Canned	256	1 c	225	.4	.840
Boysenberries:					
Frozen	132	1 c	66	1.7	1.12
Canned	256	1 c	225	.5	1.10
Cantaloupe: see Melon.					
Cherries, sour:					
Frozen	155	1 c	72	1.1	.820
Canned	244	1 c	90	3.7	3.34
CHERRIES, SWEET:					
Fresh, pitted (10 = 68g)	45	1 c	104	.5	.565
Frozen	259	1 c	232	.4	.907
Canned with liquid	257	1 c	213	.4	.910
Cranberries, whole, fresh	95	1 c	46	.4	.19
Cranberry juice cocktail	253	1 c	145	.3	.380
Cranberry sauce, canned	277	1 c	419	.1	.610

	Wt (g)	Svg	Cal	Iron (mg) 100 Cal	Iron (mg) Svg
Currants:					
Fresh, Black	112	1 c	71	2.2	1.53
Fresh, Red or white	112	1 c	63	2.1	1.34
Dried, Zante	144	1 c	407	1.1	4.65
Dates, whole, pitted	83	10 ea	228	.4	1.00
Elderberries, fresh	145	1 c	105	2.2	2.32
Figs:					
Fresh, medium	50	1 ea	37	.5	.18
Dried	19	1 ea	48	.9	.42
Fruit cocktail, canned:					
Juice pack	248	1 c	115	.5	.530
Heavy syrup	255	1 c	185	.4	.730
Gooseberries:					
Fresh	150	1 c	67	.7	.470
Canned	252	1 c	185	.5	.840
Grapes, Thompson seedless	160	1 c	114	.4	.410
Grape juice:					
From frozen	216	3/4 c	111	.7	.778
Bottled/canned	253	1 c	155	.4	.607
GRAPEFRUIT					
(Half=241g w/refuse):					
Fresh, pink or red	123	1 half	37	.4	.148
Fresh, white	118	1 half	39	..2	.070
Canned sections	254	1 c	152	.7	1.02
Grapefruit juice:					
Fresh juice	247	1 c	96	.5	.490
From frozen	247	1 c	102	.3	.340
Canned, sweetened	250	1 c	115	.8	.900
Kiwi fruit, fresh	76	1 ea	46	.7	.300
Lemon juice, bottled	244	1 c	52	.6	.310
Lime juice, bottled	246	1 c	50	1.2	.600
Loganberries:					
Fresh berries	100	2/3 c	70	.9	.640
Frozen, unthawed	147	1 c	80	1.2	.940
Mandarin oranges, canned	252	1 c	155	.6	.900
Mango, fresh slices	165	1 c	108	1.9	2.10
Melon (also see Watermelon):					
Cantaloupe, cubes	160	1 c	57	.6	.336
Casaba, cubes	170	1 c	45	1.5	.680
Honedew, cubes	170	1 c	60	.2	.120
Frozen mixed melon balls	176	1 c	55	1.3	.700
ORANGE, 2-5/8" dm, avg	131	1 ea	60	.2	.136
Orange juice:					
Fresh juice	248	1 c	111	.5	.500
Prepared f/frozen	249	1 c	110	.25	.274
Chilled	249	1 c	110	.4	.423
Canned, unsweetened	249	1 c	105	1.0	1.10
Orange grapefruit jce, canned	247	1 c	105	1.0	1.10
Papaya, (454g w/refuse)	304	1 ea	117	.3	.300
Papaya nectar, canned	250	1 c	142	.6	.300
Passion fruit juice:					
Purple	247	1 c	126	.5	.590
Yellow	247	1 c	149	.6	.890
PEACHES:					
Fresh, 2.5" dm. pitted, peeled	87	1 ea	37	.26	.096
Canned, heavy syrup	81	1 half	60	.4	.220
Canned, juice pack	77	1 half	34	.6	.210
Peach nectar, canned	249	1 c	134	.4	.470
PEARS:					
Bartlett, 180g whole	166	1 ea	98	.4	.415
Canned, juice pack	77	1 ea	38	.6	.223
Canned, heavy syrup	79	1 c	59	.3	.170
Pear nectar, canned	250	1 c	149	.4	.647

ESHA Research

Fruits & Fruit Juices, continued:

	Wt (g)	Svg	Cal	Iron 100 Cal	Iron Svg
PINEAPPLE:					
Fresh chunks	155	1 c	76	.8	.574
Canned (chunks, tidbits, crushed):					
Juice pack	250	1 c	150	.5	.700
Heavy syrup	255	1 c	199	.5	.969
Pineapple juice, canned, unswt	250	1 c	140	.5	.700
Plantain slices, fresh or ckd	151	1 c	180	.5	.890
PLUMS:					
Fresh, 2-1/8" diam.	66	1 ea	36	.2	.070
Canned, juice pack	95	3 ea	55	.6	.323
Canned, heavy syrup	110	3 ea	98	.9	.924
Prunes, dried, pitted	84	10 ea	201	1.0	2.08
Prune juice, bottled	256	1 c	181	1.7	3.02
Raisins:					
Dark, unpacked	145	1 c	435	.7	3.02
Golden, unpacked	145	1 c	437	.6	2.60
Raspberries:					
Fresh berries	123	1 c	60	1.2	.701
Frozen, thawed	250	1 c	255	.6	1.62
Canned with liquid	256	1 c	234	.5	1.08
Raspberry juice, fresh	240	1 c	98	20.4	20.0
Rhubarb:					
Fresh dices	122	1 c	26	1.0	.268
Ckd with sugar	240	1 c	279	0.2	.504
STRAWBERRIES:					
Fresh berries	149	1 c	45	1.3	.566
Frozen	149	1 c	52	2.2	1.12
Frozen, sweet, thawed	255	1 c	245	.6	1.50
Tangerines, whole	84	1 ea	37	.2	.084
Tangerine juice, canned, sweetened	249	1 c	125	.4	.500
Watermelon cubes	160	1 c	50	.5	.272

Grains & Grain Products

Cereals, Flour, Grains, Noodles, Pasta, Popcorn

	Wt (g)	Svg	Cal	Iron 100 Cal	Iron Svg
Amaranth grain, dry	195	1 c	729	2	14.8
Barley, cooked:					
Pearled	157	1 c	193	1.1	2.09
Whole	200	1 c	200	.8	1.55
Bran: see Oat, Rice, Wheat.					
Buckwheat flour, dark	98	1 c	338	.7	2.50
Buckwheat flour, light	98	1 c	340	.3	1.00
Bulgar wheat, cooked	182	1 c	151	1.1	1.75
CEREALS, COLD (Ready to eat). Iron is fortified in cereals. Brands vary. Check label.					
CEREALS, HOT (Cooked):					
Corn grits, yellow or white:					
Enriched	242	1 c	145	1.1	1.55
Unenriched	242	1 c	145	.3	.484
Cream of Rice	244	1 c	126	.3	.400
Cream of Wheat	244	1 c	140	7.8	10.9
Farina, enriched	233	1 c	116	1.0	1.17
Farina, unenriched	233	1 c	116	.04	.047
Malt-O-Meal	240	1 c	122	7.9	9.60
Oatmeal, cooked rolled oats (reg, quick, ins)	234	1 c	145	1.1	1.59
Oatmeal, instant, fortified, f/packet:					
Plain	177	3/4 c	104	6.1	6.32
With bran & raisins	195	7/8 c	158	4.8	7.61
Average of flavors	155	3/4 c	160	4.0	6.41
Ralston	253	1 c	134	1.2	1.64
Roman Meal	181	3/4 c	111	1.4	1.59
Wheatena	243	1 c	135	1.0	1.36
Whole wheat cereal	242	1 c	150	1.0	1.50

	Wt (g)	Svg	Cal	Iron 100 Cal	Iron Svg
Corn flour:					
Regular	117	1 c	422	.7	2.79
Masa Harina, enriched flour	114	1 c	416	2.0	8.22
Cornmeal:					
Degermed, enr, cooked	240	1 c	120	1.2	1.48
Degermed, enr, dry	138	1 c	505	1.1	5.7
Bolted, nearly whole	122	1 c	441	1.0	4.2
FLOUR: see specific grain, nut, or vegetable.					
MACARONI, cooked:					
Enriched	140	1 c	197	1.0	1.96
Whole wheat	140	1 c	174	.9	1.48
Vegetable, enriched	134	1 c	172	.4	.657
Millet, cooked	120	1/2 c	143	.5	.756
Noodles:					
Chow mein, dry	45	1 c	237	.9	2.13
Egg, cooked, enriched	160	1 c	213	1.2	2.54
Spinach, cooked	140	1 c	182	.8	1.46
Oat bran	94	1 c	132	3.8	5.08
Oats, rolled:					
Dry	81	1 c	311	1.1	3.41
Cooked: see Cereals, hot.					
PASTA: see macaroni, noodles, spaghetti.					
Quinoa grain, dry	170	1 c	635	2.5	15.7
RICE, cooked:					
Brown rice	195	1 c	217	.4	.819
White, regular	205	1 c	264	.9	2.26
White, converted	175	1 c	200	1.0	1.98
White, instant	165	1 c	162	.6	1.04
Wild rice	164	1 c	166	.6	.984
Rice bran	83	1 c	262	5.9	15.4
Rice polish	158	1 c	578	.1	.553
Rye flour:					
Dark	128	1 c	415	2.0	8.26
Light	102	1 c	361	.5	1.84
Soy flour, stirred:					
Defatted	100	1 c	327	2.8	9.24
Low fat	88	1 c	326	1.6	5.27
Full fat, raw/roasted	85	1 c	371	1.4	5.18
Spaghetti noodles, ckd:					
Enriched	140	1 c	197	1.0	1.96
Whole wheat spaghetti	140	1 c	174	.8	1.48
WHEAT:					
Bran	30	1/2 c	65	4.9	3.17
FLOUR (iron is added to white, all-purpose flours in U.S.A.):					
All-purpose, white, unsifted	125	1 c	455	1.3	5.8
Cake flour, sifted	96	1 c	348	2.0	7.03
Self rising	125	1 c	442	1.3	5.84
Semolina flour	167	1 c	601	1.2	7.28
Whole wheat flour	120	1 c	407	1.1	4.66
Wheat germ:					
Raw	100	1 c	360	1.7	6.26
Toasted	113	1 c	432	2.4	10.3
Wheat, rolled, cooked	240	1 c	142	1.1	1.5
Wheat, rolled, dry	85	1 c	289	1.0	2.9
Whole grain wheat, (wheat berries) cooked	50	1/3 c	28	1.1	.294
Whole wheat, sprouted	108	1 c	214	1.1	2.31

Meats: Fish & Shellfish

	Wt (g)	Svg	Cal	Iron (mg) 100 Cal	Iron (mg) Svg
Bass, freshwater, baked or broiled	100	3.5 oz	125	1.3	.610
Bluefish:					
Baked/broiled	100	3.5 oz	159	.4	.615
Fried in crumbs	100	3.5 oz	205	.3	.533
Carp, baked/broiled	100	3.5 oz	162	1.0	1.59
Catfish, cornmeal fried	100	3.5 oz	229	.6	1.43
Caviar, 1T = 16g	16	1 T	40	4.7	1.89
CLAMS, meat only:					
Canned, drained	160	1 c	236	18.9	44.7
Canned w/liquid, minced, small can	183	1 can	145	19.0	27.5
Breaded, fried, small	188	20 ea	379	6.9	26.2
COD:					
Baked/broiled	100	3.5 oz	105	.5	.490
Batter-fried	100	3.5 oz	199	.3	.500
Canned w/liquid, 11oz can	312	1 can	327	.5	1.52
CRAB:					
Alaska King leg, ckd	134	1 leg	129	.8	1.01
Blue, unpacked meat:					
Cooked from fresh	135	1 c	138	.9	1.22
Canned	135	1 c	133	.8	1.13
Dungeness, ckd f/fresh	101	3/4 c	85	.4	.370
Crab, imitation f/surimi	85	3 oz	87	.4	.330
Crab cakes, recipe	60	1 ea	93	.7	.650
Crayfish, ckd, moist heat	85	3 oz	97	2.8	2.67
Eel:					
Baked/broiled	100	3.5 oz	236	.3	.640
Smoked	100	3.5 oz	330	.2	.700
Fish cakes, fried f/frzn	100	3.5 oz	213	.5	1.00
Fish sticks, heated f/frzn	57	2 ea	155	.3	.420
Haddock:					
Baked/broiled	85	3 oz	95	1.2	1.14
Breaded, fried	85	3 oz	175	.7	1.15
Smoked	100	3.5 oz	116	1.2	1.40
HALIBUT:					
Baked/broiled	85	35 oz	119	.8	.910
Smoked	100	3.5 oz	224	.4	.840
Herring					
Baked/broiled	100	3.5 oz	203	.7	1.41
Canned in oil	100	3.5 oz	208	1.5	3.10
Pickled	100	6.67 pce	262	.5	1.22
Smoked/kippered	100	3.5 oz	217	.7	1.51
Lobster meat, cooked	145	1 c	142	.4	.570
Mackerel:					
Baked/broiled, Atlantic	100	3.5 oz	262	.6	1.57
Canned, Jack, #1 tall can	361	1 can	563	1.3	7.36
Mullet, avg, baked/broiled	85	3 oz	127	.9	1.20
Mussels, Blue, steamed meat	85	3 oz	147	3.9	5.71
Ocean perch:					
Baked/broiled	100	3.5 oz	121	1.0	1.18
Breaded, fried	85	3 oz	185	.6	1.20
Octopus, raw	100	3.5 oz	82	6.5	5.30
OYSTERS:					
Raw Eastern	248	1 c	170	9.8	16.6
Raw Pacific	248	1 c	200	6.4	12.7
Batter/breaded, fried, med	88	6 ea	173	3.5	6.12
Simmered, Eastern	100	3.5 oz	137	9.8	13.4
Perch, freshwater, baked or broiled fillet	92	2 ea	108	1.0	1.06
Pike, Northern, baked/broiled	100	3.5 oz	113	.6	.710
Pollock, baked/broiled	100	3.5 oz	99	.5	.525
Rockfish, baked/broiled	100	3.5 oz	121	.4	.530
Roe, raw, mixed species	28	1 oz	39	1.1	.439

	Wt (g)	Svg	Cal	Iron (mg) 100 Cal	Iron (mg) Svg
SALMON:					
Average, baked/broiled	85	3 oz	183	.3	.470
Chinook, smoked	85	3 oz	99	.7	.720
Coho, steamed/poached	100	3.5 oz	185	.5	.890
Canned, drained:					
Atlantic, small can	220	1 can	281	.6	1.58
Pink, # 1 can	454	1 can	631	.6	3.83
Sockeye, #1 can	369	1 can	566	.7	3.90
SARDINES, canned, drained:					
Atlantic, 2 = 24g	92	1 can	192	1.4	2.68
Pacific, 1 = 38g	10	3.5 oz	178	1.3	2.30
Scallops:					
Breaded or fried	93	6 ea	200	.4	.763
Steamed	100	3.5 oz	113	.3	.354
Sea bass, baked/broiled	100	3.5 oz	124	.3	.370
Seatrout (Steelhead), ckd	100	3.5 oz	131	.2	.311
Shad, baked w/bacon	100	3.5 oz	201	.3	.600
Shark, batter-fried	85	3 oz	194	.5	.940
SHRIMP:					
Breaded, fried (2 large = 5g	90	12 ea	218	.5	1.13
Boiled, 2 large	100	3.5 oz	99	3.1	3.09
Canned, drained	128	1 c	154	2.3	3.50
Canned with liquid	100	3.5 oz	102	1.7	1.70
Shrimp, imitation f/surimi	85	3 oz	86	.6	.510
Smelt, Rainbow, cooked	85	3 oz	106	.9	.980
SOLE (also Flounder):					
Batter-fried	85	3 oz	250	.3	.717
Breaded, fried	100	3.5 oz	188	.2	.450
Squid, flour-fried	85	3 oz	149	.6	.860
Surimi, processed walleye (Alaska) pollock: also see imitation crab & shrimp.	85	3 oz	84	.3	.220
Swordfish, baked/broiled	100	3.5 oz	155	.7	1.04
Trout, baked/broiled	85	3 oz	129	1.6	2.07
TUNA:					
Ckd f/fresh, Bluefin	85	3 oz	157	.7	1.11
Canned, drained, No. 1/2 can:					
Light, canned in oil	171	1 can	339	.7	2.38
Light water pack	165	1 can	216	1.1	2.38

Meats
Beef, Pork, Ham, Rabbit, Frog legs, Venison, and Veal

	Wt (g)	Svg	Cal	Iron (mg) 100 Cal	Iron (mg) Svg
BEEF:					
Breakfast strips, cured beef, cooked	34	3 strips	153	.7	1.07
Chuck blade, pot roasted:					
Lean and fat (5.4 oz raw)	85	3 oz	325	.8	2.52
Lean only	85	3 oz	230	1.4	3.13
GROUND BEEF, cooked avg:					
Extra lean (17% fat, raw)	85	3 oz	215	.9	2.00
Lean (20.7% fat, raw)	85	3 oz	231	.8	1.82
Regular (26.6% fat, raw)	85	3 oz	250	.8	2.08
Frozen patty, broiled (23%)	85	3 oz	240	.7	1.79
Rib, choice, oven roasted:					
Lean & fat (5 oz raw)	85	3 oz	324	.6	1.80
Lean only	85	3 oz	204	1.1	2.22
Round steak, broiled, choice:					
Lean & fat (4.5 oz raw)	85	3 oz	233	.9	2.05
Lean only	85	3 oz	165	1.4	2.28
Round tip, roasted:					
Lean & fat	85	3 oz	213	1.1	2.30
Lean only	85	3 oz	162	1.5	2.50

ESHA Research

	Wt (g)	Svg	Cal	Iron (mg) 100 Cal	Iron (mg) Svg
Meats, continued:					
BEEF, continued:					
Sirloin steak, broiled, all grades					
(11.3 oz raw = 8.2 oz cooked w/o bone;					
6.9 oz lean only):					
Lean & fat	85	3 oz	238	.7	1.60
Lean only	85	3 oz	172	1.2	2.10
T-bone steak, choice, broiled					
(16 oz raw = 9.7 oz cooked w/o bone;					
7.4 oz lean only):					
Lean & fat	85	3 oz	276	.8	2.16
Lean only	85	3 oz	182	1.4	2.5
Beef variety meats:					
Heart, simmered	85	3 oz	148	4.3	6.38
Liver, pan-fried	85	3 oz	184	2.9	5.34
Dried beef, cured (6-7 pieces)	28	1 oz	47	2.7	1.28
FROG LEGS, flour-fried	144	6 ea	418	.5	2.00
HAM: see Pork, cured; Sausages & Lunchmeats; and Turkey ham.					
LAMB:					
Arm chop, braised (5.6 oz raw w/bone):					
Lean & fat (2.5 oz ckd)	70	1 ea	244	.7	1.68
Lean only (1.9 oz ckd)	55	1 ea	152	1.0	1.48
Loin chop, broiled (4.2 oz raw w/bone):					
Lean & fat (2.3 oz ckd)	64	1 ea	201	.6	1.16
Lean only (1.6 oz ckd)	46	1 ea	100	.9	.930
Cutlet, lean, ckd, avg	85	3 oz	175	1.0	1.74
Leg of lamb, roasted:					
Lean and fat	85	3 oz	219	.8	1.69
Lean only	85	3 oz	162	1.1	1.81
Shoulder roast:					
Lean and fat	85	3 oz	235	.7	1.67
Lean only	85	3 oz	173	1.0	1.81
PORK:					
Bacon, cooked:					
Regular	19	3 pce	109	.3	.323
Canadian style	47	2 pce	86	.4	.380
Breakfast strips,	34	3 pce	156	.4	.670
Blade chop, braised, cut 3 per lb, = 151g					
(5.3 oz) raw w/bone; ≈ 110g (3.9 oz)					
w/o bone. Cooked values follow:					
Lean & fat	67	1 ea	275	.3	.871
Lean only	50	1 ea	156	.5	.810
Center loin chop, cut 3 per lb, = 151g					
(5.3 oz) raw w/ bone; ≈124g (4.4 oz)					
w/o bone. Cooked values follow:					
Braised, lean & fat	75	1 ea	266	.2	.620
Braised, lean only	61	1 ea	166	.3	.580
Broiled, lean & fat	82	1 ea	284	.3	.710
Broiled, lean only	72	1 ea	166	.4	.662
Pan-fried, lean & fat	89	1 ea	334	.4	.748
Pan-fried, lean only	67	1 ea	178	.4	.670
Roasted, lean & fat	88	1 ea	268	.3	.870
Roasted, lean only	72	1 ea	180	.4	.820
Center rib chop, cut 3 per lb. ≈ 151g					
(5.3 oz) raw w/bone; ≈ 112g (3.9 oz)					
w/o bone. Cooked values follow:					
Braised, lean & fat	67	1 ea	246	.3	.710
Braised, lean only	53	1 ea	147	.5	.670
Broiled, lean & fat	77	1 ea	264	.2	.560
Broiled, lean only	63	1 ea	162	.3	.510
Pan-fried, lean & fat	88	1 ea	343	.2	.580
Pan-fried, lean only	62	1 ea	160	.3	.490
Roasted, lean & fat	79	1 ea	252	.3	.710
Roasted, lean only	66	1 ea	162	.4	.660
Pork roast, avg loin & rib:					
Lean & fat	85	1 pce	265	.3	.800
Lean only	85	1 pce	206	.4	.890

	Wt (g)	Svg	Cal	Iron (mg) 100 Cal	Iron (mg) Svg
Shoulder, braised (yield from					
6.8 oz raw w/bone):					
Lean & fat	85	3 oz	293	.5	1.40
Lean only	67	2.4 oz	166	.8	1.31
Spareribs, ckd f/1 lb raw	177	6.25 oz	703	.5	3.27
Heart, braised	145	1 c	214	3.9	8.45
Liver, braised	85	3 oz	141	10.8	15.2
PORK, CURED-- HAM (also see bacon under Pork):					
Roasted, lean and fat	85	3 oz	207	.4	.740
Roasted, lean only	85	3 oz	133	.6	.800
Roasted, canned	85	3 oz	140	.7	.911
RABBIT, roasted	85	3 oz	175	.9	1.51
VEAL (calf):					
Cutlet, ckd average	85	3 oz	166	.6	.99
Rib, roasted	85	3 oz	151	.5	.82
Liver, pan-fried	85	3 oz	208	2.1	4.45
VENISON (deer) roasted	85	3 oz	134	2.8	3.80

Meats: Poultry

	Wt (g)	Svg	Cal	Iron (mg) 100 Cal	Iron (mg) Svg
CHICKEN: A 3 lb chicken ≈ 1.45 lbs					
raw meat; ≈ 1.1 lbs cooked.					
All types:					
Fried	140	1 c	307	.6	1.89
Roasted	140	1 c	266	.6	1.69
Stewed	140	1 c	248	.7	1.63
Canned, boned w/broth	142	5 oz	235	.9	2.20
Dark meat only:					
Fried	85	3 oz	203	.6	1.27
Roasted	85	3 oz	174	.7	1.13
Stewed	85	3 oz	163	.7	1.15
Light meat only:					
Fried	85	3 oz	163	.6	.965
Roasted	85	3 oz	147	.6	.905
Stewed	85	3 oz	135	.6	.795
Breast*, meat & skin (181g raw with bone):					
Batter-fried	140	1 ea	364	.5	1.75
Roasted	98	1 ea	193	.5	1.04
Breast*, meat only (118g raw):					
Fried	86	1 ea	161	.6	.980
Roasted	86	1 ea	142	.6	.894
*2 pieces per bird					
Drumstick, meat & skin					
(110g raw with bone):					
Batter-fried	72	1 ea	193	.5	.970
Flour-fried	49	1 ea	120	.6	.660
Drumstick, meat only (62g raw):					
Fried	42	1 ea	82	.7	.550
Roasted	44	1 ea	76	1	.760
Thigh, meat & skin (120g raw with bone):					
Batter-fried	86	1 ea	238	.5	1.24
Flour-fried	62	1 ea	162	.6	.930
Roasted	62	1 ea	153	.5	.830
Thigh, meat only (69g raw):					
Fried	52	1 ea	113	.7	.760
Roasted	52	1 ea	109	.6	.680
Liver, simmered	22	1 ea	34	5	1.70
DUCK, domestic:					
Meat & skin, roasted	85	3 oz	286	.8	2.3
Meat only, roasted	85	3 oz	171	1.3	2.3
GOOSE, domestic:					
Meat & skin, roasted	85	3 oz	259	.9	2.4
Meat only, roasted	85	3 oz	202	1.2	2.4

TURKEY:

	Wt (g)	Svg	Cal	Iron (mg) 100 Cal	Svg
Ground, cooked	100	3.5 oz	229	.8	1.93
Roasted:					
All types	140	3 oz	238	.6	1.51
Dark meat	85	3 oz	159	1.3	1.99
Light meat	85	3 oz	133	.9	1.14
Frozen slices w/gravy	142	5 oz	95	1.4	1.32
Patty, breaded, fried	64	1 ea	181	.8	1.41
Turkey gizzard, ckd	67	1 ea	109	3.3	3.64
Turkey heart, ckd	16	1 ea	28	3.9	1.10
Turkey liver, ckd	75	1 ea	127	4.6	5.85

Meats: Sausages & Lunchmeats

	Wt (g)	Svg	Cal	Iron (mg) 100 Cal	Svg
Beef lunchmeat, loaf	28	1 oz	87	.8	.660
Beef lunchmeat, thin sliced	28	1 oz	50	1.5	.760
Beerwurst (beer salami), beef	23	1 pce	75	.4	.310
BOLOGNA:					
Beef bologna	23	1 pce	72	.4	.320
Beef and pork	28	1 oz	89	.5	.430
Turkey	28	1 oz	56	.8	.433
Braunschweiger	57	2 oz	205	2.6	5.32
Chicken roll	57	2 oz	90	.6	.550
Corned beef loaf, jellied	28	1 oz	46	1.3	.580
FRANKFURTER (hotdog):					
Beef, 8/package	57	1 ea	184	.4	.760
Beef and pork, 8/pkg	57	1 ea	183	.4	.660
Chicken, 10/package	45	1 ea	115	.8	.900
Turkey, 10/package	45	1 ea	102	.8	.770
Ham & cheese loaf	57	2 pce	147	.4	.520
HAM:					
Patty, cooked	60	1 ea	203	.5	.960
Lunchmeat, extra lean	57	2 oz	75	.6	.431
Lunchmeat, regular	57	2 oz	103	.5	.561
Chopped, packaged	42	2 pce	98	.4	.399
Hotdog: see Frankfurter.					
Italian sausage link, cooked	67	1 ea	216	.5	1.01
Keilbasa sausage	26	1 pce	81	.5	.380
Liverwurst, pork	18	1 pce	59	1.9	1.15
Luncheon sausage, beef & pork	23	1 pce	60	.6	.330
Luxury loaf	57	2 oz	80	.7	.590
Pastrami:					
Beef, cured	57	2 oz	198	.3	.540
Turkey, cured	57	2 oz	74	1.1	.810
Peppered loaf	28	1 pce	42	.7	.300
Pepperoni, small slice	22	4 pce	109	.3	.308
Pickle & pimento loaf	57	2 oz	149	.4	.580
Polish sausage	28	1 oz	92	.4	.410
Pork sausage patty, ckd	27	1 pce	100	.3	.340
SALAMI:					
Beef salami	23	1 pce	58	.8	.460
Dry, beef and pork	20	2 pce	85	.4	.300
Turkey	57	2 oz	111	.8	.930
Smoked link sausage:					
Beef and pork	68	1 ea	229	.4	.990
Pork link	68	1 ea	265	.3	.790
Turkey	28	1 oz	55	.7	.410
Summer sausage	23	1 oz	80	.6	.470
Turkey breakfast sausage	28	1 oz	65	.8	.520
Turkey ham, 1 pce =1 oz	57	2 oz	73	2.1	1.56
Turkey roll:					
Light & dark meat	57	2 oz	84	.9	.760
Light meat only	57	2 oz	83	.9	.720
Turkey summer sausage	28	1 oz	50	1.1	.530

Mixed Dishes & Fast Foods

	Wt (g)	Svg	Cal	Iron (mg) 100 Cal	Svg
Beef & vegetable stew, recipe	245	1 c	220	1.3	2.90
Beef, macaroni, tomato sauce, recipe	226	1 c	189	1.3	2.39
Beef pot pie f/frzn	234	1 ea	426	.8	3.60
Canned	245	1 c	194	1.6	3.18
BURRITO:					
Bean burrito	174	1 ea	322	.8	2.53
Beef burrito	177	1 ea	463	.6	2.87
Beef and bean	175	1 ea	390	.7	2.70
Deluxe combination	198	1 ea	424	.6	2.75
Cheese soufflé, recipe	112	1 c	221	.6	1.35
Chicken & noodles, recipe	240	1 c	365	.6	2.35
Chicken a la king, recipe	245	1 c	470	.5	2.50
Chicken chow mein:					
Recipe	250	1 c	255	1.0	2.50
Canned	250	1 c	95	1.4	1.30
Chicken egg roll	100	1 ea	242	.4	1.00
Chicken pot pie f/frzn	230	1 ea	430	.7	3.10
Chicken salad w/celery	78	3/4 c	266	.2	.657
Chili w/beans, canned	255	1 c	286	3.1	8.75
Chop suey, beef & pork, recipe	250	1 c	300	1.6	4.80
Cole slaw	120	1 c	84	.8	.701
Corn dog	111	1 ea	330	.6	1.94
Corn pudding	250	1 c	271	.5	1.40
Corned beef hash, canned	220	1 c	382	1.2	4.40
ENCHILADA:					
Beef enchilada	120	1 ea	292	.6	1.77
Cheese	120	1 ea	330	.4	1.37
Chicken	120	1 ea	269	.6	1.52
French toast, recipe	65	1 pce	123	.9	1.08
LASAGNA, recipe:					
Without meat	218	1 pce	316	.8	2.38
With meat	245	1 pce	398	.8	3.08
Macaroni & cheese:					
Recipe	200	1 c	430	.4	1.80
Canned	240	1 c	230	.4	1.00
Macaroni salad, no cheese	141	1 c	371	.3	1.14
Manicotti, frozen entree	225	1 ea	271	.8	2.20
Meat loaf:					
Beef & 1/3rd pork	87	1 pce	212	.7	1.39
Beef only	87	1 pce	193	1.0	1.90
Moussaka, lamb & eggplant	250	1 c	250	1.1	2.75
Pies, fried, commercial:					
Apple	85	1 ea	255	.4	.935
Cherry	85	1 ea	250	.3	.700
PIZZA, cheese:					
Regular crust 1/8 of 15"	120	1 pce	290	.6	1.6
Thick crust 1/2 of 10"	208	1/2 ea	519	.7	3.48
Potato salad, w/mayo & eggs	250	1 c	358	.5	1.63
Quiche Lorraine, 1/8	176	1 pce	600	.2	1.40
Ravioli, beef, canned	226	1 c	220	1.0	2.12
SANDWICHES (Fast food):					
Cheeseburger, 3 oz patty	112	1 ea	300	.8	2.30
Cheeseburger, 4 oz patty	194	1 ea	524	.8	4.45
Chicken patty sandwich	157	1 ea	436	.4	1.87
English muffin, egg, cheese and bacon	138	1 ea	360	.9	3.10
Fish sandwich:					
Large, no cheese	170	1 ea	470	.5	2.23
Regular, w/cheese	140	1 ea	420	.4	1.85
Hamburger, 3 oz patty	98	1 ea	245	.9	2.20
Hamburger, 4 oz patty	174	1 ea	445	1.1	4.84
Hotdog (frankfurter & bun)	85	1 ea	260	.7	1.71
Roast beef & bun	150	1 ea	345	.9	3.22

	Wt (g)	Svg	Cal	Iron (mg) 100 Cal	Svg		Wt (g)	Svg	Cal	Iron (mg) 100 Cal	Svg
Mixed Dishes & Fast Foods, continued:						**MIXED NUTS with peanuts**					
SANDWICHES on part whole wheat unless						(cashews, peanuts, brazil nuts, filberts,					
stated as on rye:						almonds, pecans):					
Avocado, cheese, tomato,						Dry roasted	137	1 c	814	.6	5.07
sprouts	195	1 ea	432	.7	3.09	Oil roasted	142	1 c	876	.5	4.56
Bacon, lettuce & tomato	135	1 ea	327	.8	2.57	**MIXED NUTS, without peanuts**					
Chicken salad sandwich	100	1 ea	294	.8	2.30	(cashews, almonds, brazil nuts,					
Corned beef & swiss on rye	147	1 ea	429	.9	3.98	pecans, filberts):					
Egg salad sandwich	111	1 ea	319	.9	2.85	Oil roasted	144	1 c	886	.4	3.70
Grilled cheese	117	1 ea	393	.6	2.17						
Ham sandwich	122	1 ea	256	1.0	2.52	**PEANUTS:**					
Ham on rye	116	1 ea	242	.8	1.94	Dry roasted	146	1 c	855	.4	3.30
Ham & cheese	151	1 ea	363	.7	2.64	Oil roasted	144	1 c	837	.3	2.63
Ham & swiss on rye	145	1 ea	350	.6	1.99	Peanut butter:					
Patty melt, on rye	177	1 ea	567	.6	3.33	Chunky	258	1 c	1525	.31	4.67
Peanut butter & jam	100	1 ea	341	.8	2.59	Smooth	258	1 c	1517	.28	4.30
Reuben on rye	233	1 ea	480	1.1	5.20	Pecans, dried, chopped	119	1 c	794	.3	2.53
Tuna salad sandwich	116	1 ea	303	.9	2.63	Pine nuts, dried:					
Turkey sandwich	122	1 ea	271	.8	2.23	Pignola	28	1 oz	146	1.8	2.61
Turkey ham	122	1 ea	253	1.4	3.63	Pinyon	28	1 oz	161	.5	.870
Turkey ham & cheese	151	1 ea	361	1.0	3.74	Pistachios, dried, shelled	128	1 c	739	1.2	8.68
						Pumpkin/squash seeds:					
SPAGHETTI, pasta & tomato						Roasted kernels	227	1 c	1185	2.9	33.9
sauce w/cheese:						Roasted, whole	64	1 c	285	.7	2.12
Homemade	250	1 c	260	.9	2.30	Sesame seeds:					
Canned	250	1 c	190	1.5	2.80	Kernels, dried	150	1 c	882	1.3	11.7
Spinach soufflé	136	1 c	218	.6	1.34	Whole, roasted	28	1 oz	161	2.6	4.19
Stuffed cabbage rolls	228	8 oz	218	.8	1.72	Soybeans, roasted	86	1/2 c	405	.8	3.4
Stuffed green pepper	172	1 ea	217	1.1	2.32	Sunflower seed kernels:					
Taco, beef	78	1 ea	207	.6	1.29	Dried seeds	144	1 c	821	1.2	9.75
Taco, chicken	78	1 ea	172	.5	.899	Oil roasted	135	1 c	830	1.1	9.05
Tostada:						Tahini/sesame butter:					
Refried beans	157	1 ea	212	.9	1.93	Fr/unroasted kernels	15	1 T	91	1.0	.953
Beans & beef	192	1 ea	332	.7	2.16	Fr/rstd & tstd kernels	15	1 T	89	1.5	1.34
Beans & chicken	157	1 ea	249	.7	1.69	Walnuts, chopped:					
Tuna noodle casserole	202	1 c	251	.8	1.94	Black	125	1 c	759	.5	3.84
Tuna salad	205	1 c	383	.5	2.04	English	120	1 c	770	.4	2.93
Turkey pot pie	233	1 ea	416	.5	2.10						

Nuts & Seeds

Soups, Sauces, Gravies

	Wt (g)	Svg	Cal	Iron (mg) 100 Cal	Svg
ALMONDS:					
Dry roasted	138	1 c	810	.6	5.25
Whole, dried	142	1 c	837	.6	5.20
Almond butter	16	1 T	101	.6	.592
Brazil nuts, dry	140	1 c	919	.5	4.76
Cashews:					
Oil roasted	130	1 c	748	.7	5.33
Dry roasted	137	1 c	787	1.0	8.22
Chestnuts, roasted	143	1 c	350	.4	1.30
COCONUT:					
Raw, grated	80	1 c	283	.7	1.94
Dried, unsweetened	78	1 c	515	.5	2.59
Flaked, sweet, packaged	74	1 c	351	.4	1.33
Coconut cream, raw	240	1 c	792	.7	5.47
Coconut milk:					
Raw	240	1 c	552	.7	3.94
From frozen	240	1 c	485	.4	1.95
Canned	226	1 c	445	1.7	7.46
Coconut water, raw	240	1 c	45.6	1.5	.696
Filberts (hazelnuts), chopped	115	1 c	727	.5	3.76
Macadamias:					
Dried	134	1 c	940	.3	3.23
Oil roasted	134	1 c	962	.3	2.41

	Wt (g)	Svg	Cal	Iron (mg) 100 Cal	Svg
GRAVIES:					
Beef gravy, recipe	135	1/2 c	151	.3	.455
Beef, canned	233	1 c	124	1.3	1.63
Chicken, recipe	130	1/2 c	163	.3	.500
Chicken, f/dry mix	260	1 c	85	.4	.300
Mushroom, canned	238	1 c	120	1.3	1.60
Turkey gravy, canned	238	1 c	122	1.4	1.67
SAUCES (also see Other):					
Au Jus, canned	238	1 c	38	3.8	1.43
Cheese sauce:					
Recipe	101	1/2 c	216	.2	.399
From mix, w/milk	279	1 c	305	.1	.300
Hollandaise, recipe	160	1 c	867	.4	3.86
Hollandaise, mix w/water	259	1 c	240	.4	.900
Spaghetti sce, recipe	220	1 c	179	1.6	2.95
White sauce, medium:					
Home recipe	250	1 c	395	.2	.900
From mix w/milk	264	1 c	240	.1	.300
SOUPS: soups are prep. from canned unless					
otherwise stated. RTS = Ready To Serve.					
For soups prep. w/milk, assume whole milk.					
Bean & 'frank' w/water	250	1 c	187	1.3	2.35
Bean w/bacon	253	1 c	173	1.2	2.05
Bean w/ham, chunky, RTS	243	1 c	231	1.4	3.23
Beef bouillon, canned	240	1 c	16	2.6	.410

Iron (Fe) mg

Food	Wt (g)	Svg	Cal	Iron 100 Cal	Iron Svg
Soups, continued:					
Beef noodle:					
Regular	250	1 c	187	1.3	2.35
Chunky, RTS	240	1 c	171	1.4	2.33
From dry mix	251	1 c	41	.8	.330
Black bean soup	247	1 c	116	1.9	2.16
Celery, cream of, w/milk	248	1 c	165	.4	.690
Cheese soup, w/milk	251	1 c	230	.4	.810
Chicken broth, canned	244	1 c	39	1.3	.510
Chicken bouillon, f/dry	244	1 c	21	.4	.08
Chicken, cream of, w/milk	248	1 c	191	.4	.670
Chicken noodle soup:					
Regular	241	1 c	75	1.0	.780
Chunky, RTS	240	1 c	114	1.3	1.44
From dry	252	1 c	53	.9	.50
Chicken rice	241	1 c	60	1.3	.750
Chicken, chunky, RTS	251	1 c	178	1.1	1.87
Chicken vegetable	241	1 c	74	1.2	.870
Chicken vegetable, f/dry	251	1 c	49	1.2	.590
Chicken & vegetable	240	1 c	167	.9	1.47
Chili beef soup	250	1 c	169	1.4	2.40
Clam chowder, New England	248	1 c	163	.9	1.48
Clam chowder, Manhattan:					
Regular	244	1 c	78	2.1	1.64
Chunky, RTS	240	1 c	133	2.0	2.64
Consomme w/gelatin	241	1 c	29	1.8	.530
Mushroom, cream of, w/milk	248	1 c	205	.3	.590
Potato, cream of, w/water	244	1 c	73	.7	.480
Green pea, prep. w/water	250	1 c	85	2.4	2.00
Lentil & ham, RTS	248	1 c	140	1.9	2.64
Minestrone:					
Chunky, RTS	240	1 c	127	1.4	1.77
Prep. w/water	241	1 c	80	1.2	.920
Onion, prep. w/water	241	1 c	57	1.2	.670
Oyster stew, prep. w/milk	245	1 c	134	.8	1.04
Split pea:					
Canned	253	1 c	189	1.2	2.28
Prep. from dry	255	1 c	133	.8	1.01
Tomato, cream of:					
Prep. with milk	248	1 c	160	1.1	1.82
Prep. with water	244	1 c	86	2.0	1.76
Prep. from dry mix	265	1 c	102	.4	.420
Tomato beef noodle	244	1 c	140	.8	1.12
Tomato rice	247	1 c	120	.7	.790
Tomato vegetable, f/mix	253	1 c	55	1.1	.630
Turkey noodle	244	1 c	69	1.4	.940
Turkey vegetable	241	1 c	74	1.0	.760
Vegetable beef	244	1 c	79	1.4	1.11
Vegetable beef f/mix	253	1 c	53	1.6	.850
Vegetarian vegetable	241	1 c	70	1.5	1.08

Vegetables & Legumes

Food	Wt (g)	Svg	Cal	Iron 100 Cal	Iron Svg
Alfalfa sprouts	33	1 c	10	3.2	.317
Amaranth leaf:					
Fresh chopped	14	1 ea	3.6	9.0	.325
Cooked, boiled	132	1 c	28	10.6	2.98
Artichoke:					
Globe, ckd from fresh	120	1 ea	60	2.6	1.55
Hearts, marinated	170	6 oz jar	168	1.0	1.62
Asparagus, pieces:					
Fresh pieces	67	1/2 c	15	3.0	.456
Cooked from fresh	90	1/2 c	22	2.6	.594
Cooked from frozen	180	1 c	50	2.3	1.15
Canned, drained	121	1/2 c	16	6.3	1.00
Canned with liquid	122	1/2 c	17	4.2	.708
Bamboo shoots, slices, canned	131	1 c	25	1.7	.420

Food	Wt (g)	Svg	Cal	Iron 100 Cal	Iron Svg
BEANS: also see garbanzo, lentils, soybeans, yambeans.					
Baked beans (dry white beans w/spices & sauce):					
Home prepared	253	1 c	382	1.3	5.04
Canned, plain/vegetarian	254	1 c	235	.3	.740
Canned with franks	257	1 c	366	1.2	4.45
Canned with pork	253	1 c	268	1.6	4.31
Canned, pork & sweet sce	253	1 c	282	1.5	4.20
Canned, pork & tomato sce	253	1 c	247	3.4	8.30
Black beans, cooked	172	1 c	227	1.6	3.60
Broadbeans, cooked:					
From fresh veg	100	3.5 oz	56	2.7	1.5
From dry beans, legumes	170	1 c	186	1.4	2.54
Great northern, ckd from dry	177	1 c	210	1.8	3.77
Great northern, canned	262	1 c	300	1.4	4.11
Green beans (snap):					
Fresh, uncooked	110	1 c	34	4.1	1.41
Cooked from fresh	125	1 c	44	3.6	1.60
Cooked from frozen	135	1 c	36	3.1	1.11
Canned, drained	135	1 c	26	4.7	1.22
Canned with liquid	240	1 c	36	5.9	2.11
Hyacinth, cooked:					
From fresh veg	87	1 c	43	1.5	.66
From dry legume	194	1 c	228	3.9	8.88
Kidney beans, all types:					
Canned, solids & liquid	256	1 c	208	1.5	3.14
Cooked from dry	177	1 c	225	2.3	5.20
Lima beans, vegetable:					
Ckd from fresh	170	1 c	208	2.0	4.17
Ckd from frozen:					
Large type	85	1/2 c	85	1.4	1.16
Small (baby)	90	1/2 c	94	1.9	1.76
Ckd from dry	188	1 c	217	2.1	4.50
Canned, drained	170	1 c	164	1.8	2.92
Canned w/liquid, large	241	1 c	191	2.3	4.35
Navy, cooked from dry	182	1 c	259	1.7	4.51
Pinto:					
Cooked from dry	171	1 c	235	1.9	4.47
Canned	240	1 c	186	2.1	3.85
Refried beans, canned	253	1 c	270	1.7	4.47
White, ckd from dry					
Small	179	1 c	253	2.0	5.10
Regular	179	1 c	249	2.7	6.61
Winged beans:					
Ckd from fresh veg	31	1/2 c	12	2.9	.34
Ckd from dry legume	172	1 c	252	3.0	7.45
Yardlong, mature legumes:					
Cooked from fresh veg	104	1 c	49	2.0	1.02
Cooked from dry legume	171	1 c	202	2.2	4.51
Yellow wax: see green beans.					
Bean sprouts (Mung beans):					
Fresh, uncooked	104	1 c	31	3.0	.946
Ckd, stir fried	124	1 c	62	3.9	2.40
Ckd, boiled, drained	124	1 c	26	3.1	.810
Canned, drained	125	1 c	16	3.4	.540
Beets:					
Ckd from fresh	100	2 ea	31	2.0	.620
Cnd, dices, drained	85	1/2 c	27	5.7	1.55
Pickled slices	114	1/2 c	74	6	.467
Beet greens:					
Fresh pieces	19	1/2 c	4	16.0	63
Ckd f/fresh, drained	144	1 c	40	6.9	2.74
Borage:					
Fresh	44	1/2 c	9	16.1	1.45
Cooked	100	3.5 oz	25	14.6	3.64
BROCCOLI, chopped:					
Fresh, uncooked	88	1 c	24	3.3	.78
Ckd from fresh	156	1 c	44	3.0	1.31
Ckd from frozen	184	1 c	51	2.2	1.12

ESHA Research

	Wt (g)	Svg	Cal	Iron (mg) 100 Cal	Iron (mg) Svg		Wt (g)	Svg	Cal	Iron (mg) 100 Cal	Iron (mg) Svg
Vegetables & Legumes, continued:						Leeks:					
BROCCOLI, ckd from frozen, continued:						Chopped, fresh	104	1 c	63	3.4	2.18
W/cheese sauce	142	1/2 c	166	.3	.55	Chopped, cooked	52	1/2 c	24	2.4	.570
W/hollandaise sauce	95	1/2 c	105	.6	.590	Lentils, dry	192	1 c	649	2.7	17.3
Brussels sprouts:						Lentils, sprouted:					
Ckd from fresh	156	1 c	60	3.1	1.88	Fresh	77	1 c	81	3.0	2.47
Ckd from frozen	155	1 c	65	1.8	1.15	Stir fried	100	3.5 oz	101	3.1	3.10
CABBAGES:						**LETTUCE,** fresh:					
Common, shredded	70	1 c	16	2.5	.400	Butterhead	56	1 c	7	2.3	.168
Common, cooked	150	1 c	32	1.8	.585	Iceberg	56	1 c	7	3.8	.280
Bok choy, shredded	70	1 c	9	6.2	.560	Loose leaf, chopped	56	1 c	10	7.8	.784
Bok choy, cooked	170	1 c	20	8.9	1.77	Romaine, chopped	56	1 c	9	6.9	.616
Red, shredded	70	1 c	19	1.8	.350	Lotus root:					
Red, cooked	75	1/2 c	16	1.7	.27	Fresh slices	81	10 ea	45	2.1	.940
Pe-Tsai, shredded	76	1 c	11	2.1	.23	Cooked slices	89	10 ea	59	1.4	.801
Pe-tsai, cooked	119	1 c	16	2.3	.360	Mushroom (1 avg=18g):					
Savoy, cooked	145	1 c	35	1.6	.550	Fresh, sliced	35	1/2 c	9	5.0	.434
						Cooked from fresh	78	1/2 c	21	6.5	1.36
CARROTS:						Canned, drained	78	1/2 c	19	3.3	.616
Fresh (7-1/2" x 1-1/8")	72	1 ea	31	1.2	.360	Mustard greens:					
Ckd from fresh, slices	78	1/2 c	35	1.4	.484	Fresh greens	56	1 c	15	5.6	.818
Ckd from frozen, slices	73	1/2 c	26	1.3	.350	Ckd from fresh	140	1 c	21	7.4	1.56
Canned, drained	73	1/2 c	17	2.7	.467	Ckd from frozen	150	1 c	29	5.9	1.68
Canned with liquid	123	1/2 c	28	2.7	.750	Okra, cooked:					
Carrot juice	123	1/2 c	49	1.2	.564	From fresh, pods	85	8 ea	27	1.4	.383
Cauliflower:						From frozen, slices	92	1/2 c	34	1.8	.617
Fresh, uncooked	50	1/2 c	12	2.4	.290	Onion, chopped:					
Ckd from fresh	62	1/2 c	15	1.7	.260	Fresh	160	1 c	61	.6	.352
Ckd from frozen	180	1 c	34	2.2	.738	Cooked f/fresh	105	1/2 c	46	.5	.252
Chard, Swiss:						Spring, all	50	1/2 c	16	4.6	.74
Fresh	36	1 c	7	9.5	.648	Onion rings, f/frozen	20	2 ea	81	.4	.340
Cooked	175	1 c	35	11	3.96	Parsley:					
Collards:						Fresh, chopped	30	1/2 c	10	18.6	1.86
Fresh, chopped	36	1 c	11	.6	.068	Freeze dried	1.4	1/4 c	4	18.8	.750
Ckd from fresh	128	1 c	35	.6	.205	Parsnips:					
Ckd from frozen	170	1 c	63	3	1.90	Fresh slices	133	1 c	100	.8	.785
						Cooked slices	156	1 c	125	.7	.900
CORN, kernels:											
Fresh, uncooked	77	1/2 c	66	.6	.400	**PEAS:**					
Ckd from fresh	82	1/2 c	89	.6	.500	Black-eyed peas:					
Ckd from frozen	82	1/2 c	67	.7	.5	Cooked from fresh	165	1 c	160	1.2	1.85
Canned, vacuum pack	210	1 c	166	.5	.882	Cooked from frozen	170	1 c	224	1.6	3.60
Corn, cream style, canned	128	1/2 c	93	.5	.486	Cooked from dry	171	1 c	198	2.2	4.29
Cucumber, whole, 8" x 2+"	301	1 ea	39	2.2	.840	Canned	240	1 c	184	1.3	2.34
Dandelion greens:						Dry beans	167	1 c	562	2.5	13.8
Fresh	55	1 c	25	6.8	1.71	Green peas:					
Cooked	105	1 c	35	5.4	1.89	Fresh, uncooked	145	1 c	118	1.8	2.13
Dock (sorrel) greens:						Cooked from fresh	160	1 c	134	1.8	2.47
Fresh greens	133	1 c	29	11.0	3.19	Cooked from frozen	80	1/2 c	63	2.0	1.25
Cooked greens	100	1 c	20	10.4	2.08	Canned, drained	85	1/2 c	59	1.4	.810
Eggplant, cooked cubes	160	1 c	44.8	1.3	.560	Green peas, edible-pods:					
Garbanzo beans:						Fresh	145	1 c	61	4.9	3.01
Cooked f/ry	164	1 c	269	1.8	4.74	Cooked from fresh	160	1 c	67	4.7	3.15
Canned	240	1 c	285	1.1	3.23	Cooked from frozen	80	1/2 c	42	3.8	1.58
Garden cress:						Split peas, ckd from dry	196	1 c	231	1.1	2.52
Fresh	25	1/2 c	8	4.1	.330	Peas & carrots f/frozen	80	1/2 c	38	2.0	.750
Cooked	135	1 c	31	3.5	1.08	Peas, sprouted, mature:					
Jerusalem artichoke, fresh slices	150	1 c	114	4.5	5.10	Fresh	120	1 c	154	1.8	2.71
Jicama, fresh	100	3.5 oz	20	2.5	.500	Cooked	100	3.5 oz	118	1.4	1.67
Jute (potherb):											
Kale, chopped:						**PEPPERS, HOT** chili, green/red:					
Ckd from fresh	130	1 c	42	2.8	1.17	Fresh, chopped	75	1/2 c	30	3.0	.900
Ckd from frozen	130	1 c	39	3.1	1.22	Canned	68	1/2 c	17	2.0	.340
Kohlrabi:						Jalapeno, canned	68	1/2 c	17	11.2	1.90
Fresh slices	140	1 c	38	1.5	.560						
Ckd from fresh	165	1 c	48	1.4	.660	**PEPPERS, SWEET,** green/red					
Lambquarters, chopped:						Fresh, chopped	50	1/2 c	14	1.6	.23
Fresh, chopped	56	1 c	24	2.8	.672	Ckd from fresh	68	1/2 c	19	1.6	.313
Cooked	180	1 c	58	2.2	1.26	Pimento, canned	57	2 oz	13	7.4	.958
						Poi, two finger	240	1 c	269	.8	2.11

	Wt (g)	Svg	Cal	Iron (mg) 100 Cal	Svg		Wt (g)	Svg	Cal	Iron (mg) 100 Cal	Svg
Vegetables & Legumes, continued:						**SQUASH, WINTER:**					
POTATOES:						Acorn / Danish, baked	245	1 c	137	1.7	2.28
Baked in oven:						Acorn / Danish, boiled	245	1 c	83	1.7	1.37
Flesh & skin	202	1 ea	220	1.3	2.75	Butternut, baked	245	1 c	99	1.5	1.47
Flesh only	156	1 ea	145	.4	.550	Butternut, ckd from frzn	240	1 c	94	1.5	1.40
Potato skins	58	1 ea	115	1.9	2.20	Hubbard, baked	240	1 c	120	.9	1.13
Baked in microwave:						Hubbard, boiled	236	1 c	70	1.0	.670
Flesh only	156	1 ea	156	.4	.640	Spaghetti baked/boiled	155	1 c	45	1.2	.520
Potato skin	58	1 ea	77	4.5	3.44	Succotash:					
Boiled, flesh only	135	1 ea	118	.4	.420	Ckd from fresh	192	1 c	222	1.3	2.93
Canned, 1" diam.	70	2 ea	42	2.1	.880	Ckd from frozen	170	1 c	158	1.0	1.51
Ckd from frozen, small	70	1 ea	46	1.3	.588	Sweet potatoes:					
Cottage fried f/frozen	50	10 ea	109	.7	.750	Baked with peel	114	1 ea	118	.4	.520
French fried, animal/veg oil:						Boiled, peeled first	151	1 ea	160	.5	.800
From fresh	50	10 ea	137	.5	.700	Canned, mashed	128	1/2 c	129	1.3	1.70
From frozen	50	10 ea	158	.2	.380	Vacuum pack, mashed	255	1 c	233	1.0	2.27
Oven heated	50	10 ea	111	.6	.670	Candied sweet potatoes	105	1 pce	144	.8	1.20
Hash browns, f/frozen	156	1 c	340	.7	2.36	Tofu (soybean curd):					
Mashed, prepared w/milk	210	1 c	162	.4	.570	Firm, raw	126	1/2 c	183	7.2	13.2
Potato puffs (tator tots)						Regular, raw	124	1/2 c	94	7.1	6.65
heated fr/frozen	62	1/2 c	138	.7	.970						
Potato dishes:						**TOMATO:**					
Au gratin, recipe	245	1 c	322	.5	1.56	Fresh, whole	123	1 ea	26	2.1	.554
Au gratin, from mix	245	1 c	228	.3	.784	Fresh, chopped	180	1 c	38	2.1	.81
Scalloped, recipe	245	1 c	210	.7	1.41	Cooked from fresh	240	1 c	65	2.1	1.34
Scalloped, from mix	245	1 c	228	.4	.931	Canned, whole	240	1 c	47	3.1	1.45
Potato chips	28	14 ea	148	.2	.340	Tomato products, canned:					
Potato flour	179	1 c	628	4.9	30.8	Tomato juice	244	1 c	42	3.4	1.41
Potato pancakes	76	1 ea	237	.5	1.21	Tomato paste	262	1 c	220	3.6	7.84
						Tomato puree	250	1 c	102	2.3	2.32
PUMPKIN, mashed:						Tomato sauce	245	1 c	74	2.5	1.88
Cooked from fresh	245	1 c	50	2.8	1.40	Turnips:					
Canned	123	1/2 c	42	4.1	1.71	Fresh cubes	130	1 c	35	1.1	.390
Purslane:						Ckd from fresh	78	1/2 c	14	1.2	.170
Fresh	43	1 c	7	12.3	.860	Turnip greens:					
Boiled	115	1 c	21	4.2	.890	Cooked from fresh	144	1 c	29	4.0	1.15
Rutabaga:						Cooked from frozen	82	1/2 c	24	6.6	1.59
Fresh cubes	140	1 c	51	1.4	.730						
Cooked cubes	85	1/2 c	29	1.4	.400	**VEGETABLE COMBINATIONS,** cooked from frozen:					
Sauerkraut, cnd w/liquid	236	1 c	44	7.9	3.47	Broccoli, carrots &					
Seaweed:						water chestnuts	91	2/3 c	32	1.9	.600
Agar, fresh	28	1 oz	7	7.1	.527	Broccoli, cauliflower					
Irish moss	28	1 oz	14	18.1	2.52	& red pepper	95	2/3 c	25	2.4	.600
Kelp, fresh	28	1 oz	12	6.6	.808	Cantonese stir fry	95	1/2 c	53	.6	.330
Lavar, fresh	28	1 oz	90	5.2	.510	Chinese stir fry	95	1/2 c	31	3.0	.920
Wakame, fresh	28	1 oz	113	4.8	.618	Japanese style	95	1/2 c	29	2.7	.780
Spirulina, dried	28	1 oz	82	9.8	8.08	**Mixed vegetables** (corn, peas,					
Soybeans:						carrots, green beans & lima beans):					
Cooked from fresh	90	1/2 c	127	1.8	2.25	Cooked fr/frozen	182	1 c	107	1.4	1.49
Cooked from dry	172	1 c	298	3.0	8.84	Canned, drained	163	1 c	77	2.2	1.71
Soybeans, sprouted:						Peas, carrots, onion	91	1/2 c	54	1.9	1.03
Fresh sprouts	35	1/2 c	45	1.6	.740	Peas, carrots, onion, pasta	95	1/2 c	122	.7	.900
Steamed	94	1 c	76	1.6	1.23	Peas, cauliflower, cream sce	95	1/2 c	118	.6	.660
						Peas & mushrooms	95	1/2 c	73	1.3	.930
SOYBEAN PRODUCTS: See tofu this section;						Peas & onions	95	1/2 c	71	1.2	.850
miso, natto, tempeh in Other; roasted soybeans						Peas & onions, carrots, btr sce	71	2/3 c	100	.9	.900
in Nuts and Seeds; soy milk in Dairy; soy flour in Grains.						Peas & onions, cheese sauce	142	1/2 c	165	.7	1.20
Spinach:						Peas, pasta, corn, crm sauce	95	1/2 c	132	.8	1.07
Fresh, chopped	56	1 c	12	12.4	1.52	Peas, pasta, mushroom, crm sce	95	1/2 c	129	1.0	1.26
Cooked from fresh	180	1 c	41	15.7	6.42	Peas, potatoes, cream sauce	76	1/2 c	140	.3	.390
Cooked from frozen	190	1 c	53	5.4	2.89	Peas, rice, mushrooms	66	2/3 c	108	.8	.840
Canned, drained	214	1 c	50	9.8	4.92	Spinach & water chestnuts	95	1/2 c	29	6.0	1.75
						Vegetable juice cocktail	242	1 c	46	2.2	1.02
SQUASH, SUMMER:						Water chestnuts, cnd slices	70	1/2 c	35	1.7	.610
Crookneck, fresh slices	130	1 c	24	2.6	620	Yams:					
Crookneck, cooked	180	1 c	36	1.8	.640	Hawaii mountain, steamed	145	1 c	119	.5	.624
Scallop, fresh slices	130	1 c	24	2.2	.520	Orange: see Sweet Potato.					
Scallop, cooked	180	1 c	28	2.1	.580	White, cooked cubes	136	1 c	158	.4	.707
Zucchini, fresh slices	130	1 c	19	2.9	.550	Zucchini: see Squash.					
Zucchini, cooked	180	1 c	29	2.2	.630						

135

Other
Cooking Ingredients, condiments, salad dressings,
other beverages, spices, sweets, etc.

	Wt (g)	Svg	Cal	Iron (mg) 100 Cal	Svg
Barbecue sauce	250	1 c	160	1.3	2.00
BEVERAGES: other than Dairy & fruit & vegetable juices					
Beer (12 fl oz = 1.5 c):					
Regular	356	1.5 c	146	.07	.107
Light	354	1.5 c	≈100	.14	.142
Clam & tomato juice	166	2/3 c	77	1.3	1.00
Coffee, brewed	240	1 c	2	6.0	.120
Cola (avg of all)	370	1.5 c	151	.1	.130
Diet cola	355	1.5 c	2	5.0	.110
Fruit punch, canned	253	1 c	118	.4	.517
Ginger ale	366	1.5 c	124	.5	.660
Grape drink	250	1 c	112	.4	.412
Grape soda, carbonated	372	1.5 c	161	.2	.310
Lemonade, f/frozen	248	1 c	100	.4	.410
Pineapple grapefruit drink	250	1 c	117	.7	.770
Pineapple orange drink	250	1 c	125	.5	.670
Wine:					
Red	118	1/2 c	85	.6	.508
Rosé	118	1/2 c	84	.5	.448
White, medium	118	1/2 c	80	.4	.354
CANDY & CANDY BARS:					
Almond Joy	28	1 oz	151	.5	.78
Caramel, plain/choc	28	1 oz	115	.3	.40
Chocolate covered:					
Almonds	165	1 c	935	.7	6.34
Coconut	28	1 oz	133	.5	.615
Peanuts	170	1 c	954	.4	4.13
Raisins	187	1 c	733	.6	4.37
Milk chocolate, plain	28	1 oz	145	.3	.40
Milk chocolate, w/almonds	28	1 oz	150	.4	.56
Milk chocolate, w/peanuts	28	1 oz	155	.4	.68
Mr. Goodbar	47	1 ea	250	.4	.94
REESE's peanut butter cup	45	2 ea	240	.3	.68
SNICKERS, 2.2 oz bar	61	1 bar	290	.2	.49
Sugar coated almonds	28	7 ea	146	.5	.78
Carob flour	103	1 c	185	1.6	3.03
Catsup	17	1 T	18	.7	.12
Chili sauce, tomato base	17	1 T	18	.8	.137
CHOCOLATE:					
Baking, unsweetened	28	1 oz	145	1.3	1.9
Bittersweet	28	1 oz	141	.7	1.0
Choc. chips, semi-sweet	170	1 c	860	.7	5.8
Hot fudge topping	38	2 T	129	.4	.50
Syrup, thin type	38	2 T	85	.9	.75
Cocoa powder	86	1 c	224	5.2	11.7
Dill pickle	65	1 ea	11.7	2.9	.345
Garlic cloves	9	3 ea	13.4	1.14	.153
Granola bar	28	1 ea	127	.7	.91
Honey	339	1 c	1030	.2	1.7
Hummous (Humous)	246	1 c	420	.9	3.87
Miso (soybean product)	275	1 c	565	1.3	7.52
Molasses:					
Blackstrap	40	2 T	85	11.9	10.1
Light	20	1 T	43	2.5	1.0
Mustard, prepared	125	1/2 c	94	2.7	2.50
Natto (soybean product)	88	1/2 c	187	4.0	7.57
Olives:					
Green, pitted	39	10 ea	45	1.3	.600
Ripe, large pitted	45	10 ea	52	2.9	1.49
Pickle relish	15	1 T	20	.6	.125
Pickle slices, fresh pack	30	4 pce	20	2.7	.547

	Wt (g)	Svg	Cal	Iron (mg) 100 Cal	Svg
SALAD DRESSING:					
1000 island	250	1 c	943	.2	1.50
1000 island, low cal	245	1 c	389	.4	1.50
Blue cheese	245	1 c	1235	<.1	.490
French	250	1 c	1321	.1	1.00
French, low cal	260	1 c	386	.3	1.00
Italian	235	1 c	1098	<.1	.500
Italian, low calorie	240	1 c	130	.4	.500
Mayo type, low cal	240	1 c	556	.1	.500
Ranch	119	1/2 c	435	.1	.313
Russian	245	1 c	1210	.1	1.50
Salsa:					
Picante by Tostitos	85	6 T	40	1.2	.480
Recipe	108	1/2 c	46	1.1	.86
Soy sauce:					
Regular (wheat & soy) shoyu	18	1 T	9	4.0	.360
Tamari (soy)	18	1 T	11	3.9	.430
Fr/hydrolyzed veg protein	18	1 T	7	3.9	.270
SPICES:					
Allspice	6	1 T	16	2.7	.424
Basil, dried	4.5	1 T	11	17.2	1.89
Caraway seed	6.7	1 T	22	5.0	1.09
Cardamom	5.8	1 T	18	4.5	.810
Celery seed	2.0	1 tsp	8.5	10.6	.900
Chili powder	2.6	1 tsp	8	4.6	.370
Cilantro: see Coriander.					
Cinnamon	2.3	1 tsp	5.9	14.6	.864
Coriander, fresh	4	1/4 c	1	9.8	.08
Coriander leaf, dried	1.8	1 T	5	15.2	.760
Coriander seed	5	1 T	15	5.4	.816
Cumin seed	6	1 T	22	18.1	3.98
Curry powder	2	1 tsp	6	9.9	.592
Dill weed, dried	3.1	1 T	8	18.8	1.50
Fenugreek seed	11.1	1 T	36	10.3	3.72
Garlic powder	8.4	1 T	28	.8	.231
Ginger, ground	5.4	1 T	19	3.3	.620
Mace, ground	5.3	1 T	25	3.0	.740
Marjoram, dried	1.7	1 T	5	28.2	1.41
Oregano, ground	1.5	1 tsp	4.7	14.0	.660
Paprika	2.1	1 tsp	6.1	8.2	.500
Pepper:					
Black	2.1	1 tsp	5.4	11.3	.606
Cayenne (red)	5.3	1 T	17	2.4	.413
White	7.1	1 T	21	4.9	1.02
Poultry seasoning	3.7	1 T	11	11.9	1.31
Poppyseed	8.8	1 T	47	1.8	.827
Pumpkin pie spice	5.6	1 T	19	5.8	1.10
Rosemary, dried	3.3	1 T	11	8.8	.965
Sage, ground	2	1 T	6	9.4	.562
Tarragon, ground	4.8	1 T	14	11.1	1.55
Thyme, ground	4.3	1 T	12	44.2	5.31
Turmeric, ground	6.8	1 T	24	11.8	2.82
Sugar:					
Brown, packed measure	220	1 c	820	.6	4.80
White, granulated	200	1 c	770	<.1	.10
Syrups (other):					
Corn, dark	328	1 c	944	.3	3.3
Sorghum	20	1 T	52	4.6	2.4
Treacle, black	20	1 T	53	3.4	1.8
Tempeh (soybean product)	83	1/2 c	165	1.1	1.88
Teriyaki sauce, f/mix	18	1 T	15	2.1	.310
Tobasco sauce	15	1 T	1.6	6.0	.100
Vinegar, cider	240	1 c	29	4.8	1.40
Yeast:					
Active dry (1pkt=7g)	30	4 T	80	7.4	5.90
Brewer's	8	1 T	25	5.6	1.39

Magnesium

Magnesium is found throughout the body and has a wide variety of functions. Over half of the body's magnesium is stored in the bones, and there appears to be a relationship between calcium and magnesium. For example, magnesium promotes resistance to tooth decay by holding calcium in tooth enamel.

About 40% of Magnesium is in the muscles and soft tissues. Magnesium helps relax muscles after contraction and plays a role in the conduction of nerve impulses. It is an important enzyme activator (more than 300 enzymes) and is essential to most of the enzymes that use thiamin (vitamin B1), riboflavin (vitamin B2), pyridoxine (vitamin B6), vitamin C and vitamin E. As a result, a magnesium deficiency may result in symptoms that resemble the deficiencies of these other vitamins. Because it is so necessary to the function of several enzymes, magnesium is important for protein metabolism as well.

Deficiencies are not common, but can occur when a person is taking diuretics, as a result of alcohol abuse, suffering severe vomiting, or from continual diarrhea. Muscular weakness and spasms, behavioral disturbances, and growth problems in youth are some of the effects of magnesium deficiency.

Nutrient Losses

As a mineral, there are no losses due to heat or light. However, some magnesium is lost in processed foods, and a small amount is leached into cooking water. More than 80% of magnesium is lost in whole grains when they are milled into flour and this mineral is not added back through "enrichment."

Recommended Dietary Allowances for Adults

USA RDA: Men, 350 mg; Women, 280 mg; (Pregnant, 320 mg; Nursing, 355 mg)
Canadian RNI: Men, 250 mg; Women, 200 mg (Pregnant, 215-225 mg; Nursing, 280 mg)

Food Sources

Magnesium tends to follow protein and phosphorus in foods. Good sources include whole grains (NOT processed), legumes, nuts and seeds, chocolate, green vegetables, some seafood and poultry, avocados, bananas and some berries. Most fish, meat and milk are relatvely poor sources.

ESHA Research

Magnesium (Mg) mg

Values are for edible portion of foods

Baked Goods

Breads, Cakes, Cookies, Crackers, Muffins, Pancakes, Pastries, Pies, Rolls, and some desserts.

	Wt (g)	Svg	Cal	Magnes 100 Cal	Magnes Svg
Apple crisp, 3" x 3"	78	1 pce	146	8	12
Bagel, 3.5" diam:					
Plain bagel	68	1 ea	180	8	15
Egg bagel	68	1 ea	180	10	18
BREADS:					
Banana nut, 1/2"	50	1 pce	161	9	14
Boston brown, canned	45	1 pce	95	42	40
Cornbread muffin	45	1 ea	145	8	11
Cracked wheat	25	1 pce	65	18	12
French 5" x 2.5" pce	35	1 pce	100	7	3
Mixed grain	25	1 pce	65	19	12
Oatmeal bread	25	1 pce	65	13	8.5
Pumpernickel	32	1 pce	80	28	22
Raisin	25	1 pce	68	9	6.2
Rye bread, light	25	1 pce	65	12	8
Wheat (white & whole wheat flour)	28	1 pce	72	18	13
White bread	28	1 pce	75	8	6
Whole wheat bread	28	1 pce	70	37	26
Brownies, w/nuts, commercial	25	1 ea	100	14	14

CAKES (cupcake ≈ 42g); piece = 1/16th cake unless stated otherwise.

	Wt (g)	Svg	Cal	100 Cal	Svg
Angel food, 1/12 tube	53	1 pce	125	3	4.3
Boston cream, 1/8	120	1 pce	260	4	11
Carrot, cream cheese frosting	112	1 pce	405	3	11.5
Cheesecake fr/mix, 1/8	103	1 pce	300	7	22
Chocolate, choc frosting	69	1 pce	235	9	22
Fruitcake, dark, 2/3" arc	43	1 pce	165	6	11
Gingerbread, 3" x 3":					
Recipe	110	1 pce	351	5	17
From mix	63	1 pce	174	3	4.5
Pound cake, 1/2"	30	1 pce	115	3	3
Sheet cake, 3" x 3":					
Plain	86	1 pce	315	4	12
White frosting	121	1 pce	445	3	12
White cake:					
Chocolate frosting	77	1 pce	291	2	4.5
Coconut frosting	70	1 pce	270	3	7
White frosting	71	1 pce	260	1	4
Yellow, chocolate frosting	69	1 pce	240	2	4.5
Cherry crisp, 3" x 3"	138	1 pce	157	10	16

Chips: see corn & tortilla this section; potato chips under Vegetables.

COOKIES:	Wt (g)	Svg	Cal	100 Cal	Svg
Chocolate chip:					
Homemade	40	4 ea	185	8	14
Fr/refrigerator dough	48	4 ea	225	4	10
Commercial	42	4 ea	180	6	10
Fig bars	56	4 ea	210	7	15
Oatmeal raisin	52	4 ea	245	11	26
Peanut butter, homemade	48	4 ea	245	8	19
Sandwich cookie, all	40	4 ea	195	8	15
Corn chips	28	1 oz	155	14	21
CRACKERS:					
Armenian cracker bread	28	4 pce	117	35	41
Graham crackers	14	2 ea	60	10	6
Round, like Ritz	9	3 ea	45	4	2
Rye wafers, whole grain	14	2 ea	55	29	16
Sesame	12	4 ea	60	28	17
Danish pastry:					
Plain	57	1 ea	220	5	11
With fruit	65	1 ea	235	5	13

Doughnut:	Wt (g)	Svg	Cal	100 Cal	Svg
Cake, medium	50	1 ea	210	5	11.5
Yeast, plain	60	1 ea	235	6	13
Yeast, jelly filled	65	1 ea	226	7	16
Eclair, custard filling, chocolate icing	94	1 ea	262	5	12
English muffin, plain or sourdough	57	1 ea	135	8	11
MUFFINS, prepared:					
Blueberry, recipe	45	1 ea	135	8	11
Blueberry, mix	45	1 ea	140	5	6.5
Bran, wheat, recipe	45	1 ea	125	27	34
Bran, wheat, mix	45	1 ea	140	20	28
Cornmeal, mix or recipe	45	1 ea	145	8	11
Pancakes:					
Buckwheat fr/mix, 4" diam	27	1 ea	55	33	18
Plain, 4" diam.	27	1 ea	60	11	7
Whole wheat, 5" diam	52	1 ea	94	24	22
Peach crisp, 3" x 3"	139	1 pce	166	11	18
PIES: piece = 1/6 of 9" pie.					
Banana cream, recipe	198	1 pce	319	10	31
Chocolate cream	175	1 pce	311	9	29
Coconut cream	172	1 pce	343	8	26
Coconut custard	165	1 pce	384	8	32
Custard pie	152	1 pce	293	6	18
Pecan	138	1 pce	583	5	32
Lemon meringue	140	1 pce	355	2.5	9
Mincemeat	160	1 pce	395	5	19
Pumpkin	200	1 pce	367	11	40
Pretzels, thin twists	60	10 ea	240	6	15
ROLLS:					
Cinnamon, small	50	1 ea	158	9	14
Hard roll, white	50	1 ea	155	9	14
Rye, dark	28	1 ea	79	17	13
Submarine (hoagie)	135	1 ea	400	8	31
Whole wheat roll	35	1 ea	88	36	32
Stuffing, Stove Top	108	1/2 c	176	7	13
Tortillas:					
Corn, enr 6" diam, fried	30	1 ea	87	23	20.0
Flour, 10.5" diam.	57	1 ea	168	11	18.5
Flour, 8" diam.	35	1 ea	105	11	11.5
Tortilla chips:					
Plain chips	28	1 oz	139	16	22
Doritos, Nacho flavor	28	1 oz	139	9	13
Doritos, Taco flavor	28	1 oz	140	19	27
Waffles, 7" dm., fr/mix	75	1 ea	205	7	14

Dairy & Dairy Products

CHEESE [1.5" cube ≈ 1oz]:	Wt (g)	Svg	Cal	100 Cal	Svg
American processed cheese	28	1 oz	106	6	6.5
Blue cheese	28	1 oz	100	7	6.8
Cheddar, shredded	28	1 oz	114	7	7.7
Cottage cheese:					
Lowfat 1%	226	1 c	164	7	12
Lowfat 2%	226	1 c	205	7	14
Creamed, large curd	225	1 c	235	5	11
Creamed, small curd	210	1 c	215	5	11
Parmesan, grated (1 T = 5g)	28	1 oz	111	11	12
Ricotta, made w/whole milk	246	1 c	428	7	28
Romano, grated (1T = 5g)	28	1 oz	128	11	14
Swiss cheese	28	1 oz	107	9	10
CREAM, SWEET, fluid:					
Coffee or table cream	240	1 c	469	4	21
Half & half	242	1 c	315	8	25

	Wt (g)	Svg	Cal	Magnes (mg) 100 Cal	Magnes (mg) Svg
Dairy, continued:					
CREAM, SWEET, fluid, continued:					
Light whipping cream	239	1 c	699	2	17
Heavy whipping cream	238	1 c	821	2	17
CREAM, SWEET, WHIPPED:					
Heavy cream, unsweetened	119	1 c	410	2	8.5
Pressurized	60	1 c	154	4	6
CREAM, SOUR, cultured, dairy	230	1 c	493	5	26
Cream substitutes, non-dairy:					
Coffee whitener, liquid	120	1/2 c	163	.2	.3
Coffee whitener, powder	94	1 c	514	.8	4
Dessert Toppings, non-dairy:					
Frozen (like Coolwhip)	75	1 c	239	.4	1
Pressurized, non-dairy	70	1 c	185	.5	1
Kefir beverage	233	1 c	160	18	28
MILK (cow):					
Skim milk	245	1 c	86	33	28
Lowfat 1%	244	1 c	102	33	34
Lowfat 2%	244	1 c	121	27	33
Whole	244	1 c	150	22	33
Buttermilk	245	1 c	99	26	26
Canned, skim, evap	255	1 c	200	34	68
Canned, whole, evap	252	1 c	340	18	60
Dry, nonfat, instant	68	1 c	244	33	80
Milk (other):					
Goat	244	1 c	168	20	34
Human breast milk (mature)	246	1 c	171	5	8
Soybean	240	1 c	79	57	45
MILK BEVERAGES & mixes:					
Chocolate flavored, to be mixed with water:					
Powder (w/dry milk)	28	1 oz	100	23	23
Drink, prepared w/water	206	3/4 c	100	23	23
Chocolate flavored, to be mixed with milk:					
Powder	22	3/4 oz	75	28	21
Drink prepared	266	1 c	226	24	54
Eggnog, commercial	254	1 c	342	14	47
Instant Breakfast, dry, fortified	37	1 env	130	62	80
Malted milk, with whole milk:					
Chocolate flavor	265	1 c	229	21	47
Natural flavor	265	1 c	237	22	52
Milkshakes (10 fl oz = 1.25 c):					
Chocolate	283	1.25 c	360	13	47
Strawberry	283	1.25 c	319	11	36
Vanilla	283	1.25 c	314	11	35
MILK DESSERTS:					
Custard, baked	133	1/2 c	153	12	19
Custard, prep. from mix	143	1/2 c	161	14	23
Ice cream, vanilla:					
Regular	133	1 c	269	7	18
Rich	148	1 c	349	5	16
Soft serve	173	1 c	377	7	25
Ice cream bar, Fudgesicle	73	1 ea	91	15	14
Ice milk, regular	131	1 c	184	10	19
Ice milk, soft serve	175	1 c	223	13	29
PUDDINGS, prepared					
(5 oz can ≈ 1/2 c):					
Assorted low calorie	130	1/2 ea	69	25	17
Chocolate, instant or cooked-mix	130	1/2 c	153	16	24
Chocolate, canned	142	1 can	205	12	24
Coconut, fr/instant	149	1/2 c	184	11	20
Lemon, fr/instant	149	1/2 c	178	10	17
Rice, fr/instant	149	1/2 c	175	11	20
Tapioca, fr/instant	130	1/2 c	145	11	16
MILK DESSERTS, continued:					
Puddings, continued:					
Tapioca, canned	142	1 can	160	15	24
Vanilla, fr/instant	130	1/2 c	150	11	16
Vanilla, canned	142	1 can	220	11	24
Pudding pop, chocolate	57	1 ea	99	15	15
Sherbet	193	1 c	270	6	15
Yogurt, frozen (average)	87	1/2 c	110	10	11
YOGURT:					
Lowfat, plain	227	1 c	144	28	40
Lowfat, w/fruit	227	1 c	231	13	31
Lowfat, coffee or vanilla	227	1 c	193	19	36
Nonfat	227	1 c	127	34	43
Whole	227	1 c	138	20	27
Yogurt cheese, recipe	208	1 c	222	27	60

Eggs

	Wt (g)	Svg	Cal	Magnes (mg) 100 Cal	Magnes (mg) Svg
Eggs, chicken:					
Whole, raw or cooked	50	1 ea	77.5	6	5
White only	33.4	1 ea	17	24	4
Yolk only	16.6	1 ea	59	3	1.5

Fruits & Fruit Juices

	Wt (g)	Svg	Cal	Magnes (mg) 100 Cal	Magnes (mg) Svg
Acerola juice, fresh	242	1 c	51	57	29
APPLE 2.75" diam:					
With peel	138	1 ea	80	8	6
Without peel	128	1 ea	72	6	4
Apple juice f/rozen	239	1 c	111	11	12
APRICOTS:					
Fresh halves	155	1 c	74	16	12
Canned, juice pack	84	3 ea	40	20	8
Canned, heavy syrup	85	3 ea	70	9	6
Dried halves	35	10 ea	83	19	16
Apricot nectar, canned	251	1 c	141	9	13
Avocado, whole:					
California	173	1 ea	305	23	70
Florida	304	1 ea	340	31	104
Banana, 8.75" (176g w/peel)	114	1 ea	105	31	32
Blackberries:					
Fresh berries	144	1 c	74	39	29
Frozen, unthawed	151	1 c	97	34	33
Canned	256	1 c	236	19	44
Boysenberries:					
Frozen, unthawed	132	1 c	66	32	21
Canned	256	1 c	225	12	28
Cantaloupe: see Melon.					
Cherries, sour:					
Frozen, unthawed	155	1 c	72	18	13
Canned	244	1 c	90	17	15
CHERRIES, SWEET:					
Fresh pitted (10 = 68g)	145	1 c	104	15	16
Frozen, thawed	259	1 c	232	11	26
Canned with liquid	257	1 c	213	10	22
Currants:					
Fresh, Black	112	1 c	71	38	27
Fresh, Red or white	112	1 c	63	22	14
Dried, Zante	144	1 c	407	14	59
Dates, whole, pitted	8	1 ea	23	13	3
Fig:					
Fresh, medium	50	1 ea	37	22	8
Dried	19	1 ea	48	23	11
Fruit cocktail, canned:					
Juice pack	248	1 c	115	15	17
Heavy syrup	255	1 c	185	8	14

ESHA Research

Magnesium (Mg) mg   Values are for edible portion of foods

	Wt (g)	Svg	Cal	Magnes (mg) 100 Cal	Svg
Fruits & Fruit Juices, continued:					
Gooseberries:					
Fresh berries	150	1 c	67	22	15
Canned w/liquid	252	1 c	185	8	15
Grapes, Thompson seedless (10 =50g)	160	1 c	114	9	10
Grape juice:					
Bottled/canned	253	1 c	155	15	24
Prep. from frozen	250	1 c	128	9	11
GRAPEFRUIT half = 241g w/rind:					
Pink or red half	123	1 ea	37	27	10
White half	118	1 ea	39	28	11
Canned, sections	254	1 c	152	16	25
Grapefruit juice:					
Fresh	247	1 c	96	31	30
Prepared from frozen	247	1 c	102	25	26
Canned, unsweetened	247	1 c	93	26	24
Canned, sweetened	250	1 c	115	21	24
Kiwi fruit	76	1 ea	46	50	23
Lemon juice:					
Fresh	244	1 c	60	27	16
Bottled	244	1 c	52	42	22
Lime juice:					
Fresh	246	1 c	65	22	14
Bottled	246	1 c	50	32	16
Loganberries:					
Fresh berries	100	2/3 c	70	31	22
Frozen	147	1 c	80	40	32
Mandarin oranges, canned	252	1 c	155	14	22
Mango, fresh slices	165	1 c	108	14	15
MELONS: also see Watermelon.					
Cantaloupe cubes	160	1 c	57	32	18
Casaba cubes	170	1 c	45	31	14
Honeydew cubes	170	1 c	60	20	12
Frozen melon balls, mixed	173	1 c	55	44	24
Nectarines (1 med = 1 c slices)	136	1 med	67	16	11
ORANGE, 2-5/8" diam. (180g with peel)	131	1 ea	60	22	13
Orange juice:					
Fresh juice	248	1 c	111	24	27
Chilled f/fresh	249	1 c	110	25	28
Prep. from frozen	249	1 c	110	22	24
Canned, unsweetened	249	1 c	105	26	27
Orange grapefruit juice, cnd	247	1 c	105	23	24
Papaya (454g with refuse)	304	1 ea	117	26	31
PEACHES:					
Fresh peeled slices	170	1 c	73	15	11
Frozen, thawed	250	1 c	235	5	12
Canned slices (1 half = 80g):					
Juice pack	248	1 c	109	17	18
Heavy syrup	256	1 c	190	7	13
Dried halves	130	10 ea	311	17	54
Peach nectar, canned	249	1 c	134	8	11
PEARS:					
Fresh, Bartlett	166	1 ea	98	9	9
Fresh, D'Anjou	200	1 ea	120	9	11
Canned, juice pack	77	1 ea	38	14	5.4
Canned, heavy syrup	79	1 ea	59	6	3
Persimmon, Japanese, fresh	168	1 ea	118	13	15
PINEAPPLE:					
Fresh chunks	155	1 c	76	28	21
Frozen, sweetened	245	1 c	208	12	25
Canned pieces:					
Juice pack	250	1 c	150	23	35
Heavy syrup	255	1 c	199	20	40
Pineapple juice:					
From frozen	250	1 c	129	18	23
Canned, unsweetened	250	1 c	140	24	34
Plantain slices:					
Fresh	148	1 c	181	30	55
Cooked	154	1 c	179	27	49
PLUMS:					
Fresh, medium 2-1/8" dm.	66	1 ea	36	11	4
Canned, juice pack	95	3 ea	55	14	8
Canned, heavy syrup	110	3 ea	98	6	5.5
Prickly pear fruit, fresh	103	1 ea	42	210	88
Prunes, dried, pitted	84	10 ea	201	19	38
Prune juice, bottled	256	1 c	181	20	36
Raisins, dark, unpacked	145	1 c	435	11	48
Raspberries:					
Fresh berries	123	1 c	60	37	22
Frozen, thawed, unsweetened	250	1 c	255	13	32
Canned with liquid	256	1 c	234	13	31
Rhubarb:					
Fresh, diced	122	1 c	26	54	14
Cooked with sugar	240	1 c	279	11	30
STRAWBERRIES:					
Fresh berries	149	1 c	45	36	16
Frozen, unsweetened	149	1 c	52	31	16
Tangerine, fresh, whole	84	1 ea	37	27	10
Tangerine juice, canned	249	1 c	125	16	20
Watermelon, fresh pieces	160	1 c	50	34	17

Grains & Grain Products
Cereals, Flours, Grains,Noodles, Pasta, Popcorn

	Wt (g)	Svg	Cal	Magnes (mg) 100 Cal	Svg
Amaranth grain, dry	195	1 c	729	71	519
Barley, cooked:					
Pearled	157	1 c	193	18	34.5
Whole	200	1 c	200	17	33
Bran: see Oat, Rice, Wheat.					
Buckwheat, whole grain, dry	175	1 c	586	57	335
Buckwheat flour:					
Dark	98	1 c	338	40	135
Light	98	1 c	340	14	47
Bulgar wheat, cooked	182	1 c	151	34	58
CEREALS, COLD (Ready to eat):					
Cereals may be fortified with magnesium. Amounts vary. Check the label.					
CEREALS, HOT (Cooked):					
Cereals may be fortified with magnesium. Check the label.					
Corn grits, enriched	242	1 c	145	7	9.68
Cream of Wheat	244	1 c	140	9	12
Farina, enriched	233	1 c	116	4	4.66
Malt-O-Meal	240	1 c	122	11	14
Maypo	180	3/4 c	128	30	38
Oatmeal, fr/rolled oats, unenriched	234	1 c	145	39	56
Oatmeal, from packet, instant, fortified:					
Plain	177	3/4 c	104	49	51
With bran & raisins	195	7/8 c	158	36	57
Assorted flavors	156	3/4 c	159	21	51
Ralston	253	1 c	134	44	59
Roman Meal	181	3/4 c	111	74	82
Wheatena	243	1 c	135	36	49
Whole wheat cereal	242	1 c	150	35	53
Corn flour:					
Regular	117	1 c	422	26	109
Masa Harina flour, enr.	114	1 c	416	30	125

ESHA Research   140

Grains & Grain Products, continued:

Food	Wt (g)	Svg	Cal	Magnes 100 Cal	Magnes Svg
Cornmeal:					
Enr., degermed, dry	138	1 c	505	11	55
Nearly whole, bolted	122	1 c	441	35	154
FLOUR: see specific grain, nut, or vegetable.					
MACARONI, cooked:					
Enriched	140	1 c	197	13	25.2
Whole wheat	140	1 c	174	24	42
Vegetable, enriched	134	1 c	172	15	25.5
Millet, cooked	120	1/2 c	143	37	53
Noodles, cooked:					
Egg noodles, enriched	160	1 c	213	14	30.4
Spinach	140	1 c	182	48	87
Oat bran, 1T ≈ 6 g	94	1 c	132	91	120
Oatmeal: see Cereal, hot.					
Oats, rolled, dry	81	1 c	311	39	120
PASTA: see Macaroni, Noodles, Spaghetti.					
Popcorn, popped in oil	11	1 c	55	45	25
Quinoa grain, dry	170	1 c	635	56	357
RICE, cooked:					
Brown	195	1 c	217	39	84
White, regular	205	1 c	264	10	27
White, converted	175	1 c	200	11	21
White, instant	165	1 c	162	5	8.3
Wild rice	164	1/2 c	166	32	52.5
Rice bran	83	1 c	262	25	64.8
Rice flour	158	1 c	578	10	55.3
Rye flour:					
Dark	128	1 c	415	76	317
Light	102	1 c	361	20	71.4
Soy flour, stirred:					
Low fat	88	1 c	370	62	202
Full fat, raw	85	1 c	368	99	364
Spaghetti noodles, cooked:					
Enriched	140	1 c	197	13	25
Whole wheat spaghetti	140	1 c	174	24	42
WHEAT:					
Bran, dry	30	1/2 c	65	282	183
FLOUR, unbleached:					
All-purpose, white, unsifted	125	1 c	455	6	27.5
Cake, sifted	96	1 c	348	4	15.4
Self-rising	125	1 c	442	5	23.8
Semolina	167	1 c	601	13	78.5
Whole wheat flour	120	1 c	407	41	166
Wheat germ:					
Raw	100	1 c	360	81	239
Toasted	113	1 c	432	84	362
Wheat, rolled:					
Cooked	240	1 c	142	41	58
Dry	85	1 c	289	45	130
Wheat, sprouted	108	1 c	214	41	88.6
Whole grain wheat (wheat berries) cooked	50	1/3 c	28	42	12

Meats: Fish & Shellfish

Food	Wt (g)	Svg	Cal	Magnes 100 Cal	Magnes Svg
Bass, Freshwater, bkd/broiled	100	3.5 oz	125	26	32
Bluefish:					
Baked or broiled	100	3.5 oz	159	26	42
Fried in crumbs	100	3.5 oz	205	18	37
Carp, baked/broiled	100	3.5 oz	162	23	38
Catfish, cornmeal fried	100	3.5 oz	229	13	30
CLAMS, meat only:					
Breaded, fried, small	188	20 ea	379	7	27
Steamed	100	3.5 oz	148	12	18
Canned, drained	160	1 c	236	13	30
Canned w/liquid, minced	183	1 ea	145	226	327
Clam nectar, canned	240	1 c	6	433	26
COD:					
Baked/broiled/poached	100	3.5 oz	105	40	42
Fried in batter	100	3.5 oz	199	18	36
Smoked	100	3.5 oz	79	32	25
CRAB, meat only:					
Blue crab, ckd	135	1 c	138	33	45
Blue crab, cnd	135	1 c	133	39	52
Dungeness crab, cooked	101	3/4 c	85	54	46
Crab, Imitation from surimi	85	3 oz	87	40	35
Crab cakes, from recipe	60	1 ea	93	22	20
Crayfish, cooked, moist heat	85	3 oz	97	28	27
Eel:					
Baked or broiled	100	3.5 oz	236	21	50
Smoked	100	3.5 oz	330	15	50
Fish cakes, fried f/frzn	100	3.5 oz	213	8	18
Fish sticks, heated f/frzn	57	2 ea	155	9	14
Haddock:					
Baked/broiled/poached	85	3 oz	95	45	43
Breaded, fried	85	3 oz	175	15	26
Smoked	100	3.5 oz	116	47	54
HALIBUT:					
Baked/broiled	85	35 oz	119	76	91
Smoked	100	3.5 oz	224	37	83
Herring:					
Baked or broiled	100	3.5 oz	203	20	41
Canned in oil	100	3.5 oz	208	24	49
Smoked or kippered	100	3.5 oz	217	21	46
Lobster meat, cooked	145	1 c	142	36	51
Mackerel:					
Baked/broiled, Atlantic	100	3.5 oz	262	37	97
Baked/broiled, Spanish	100	3.5 oz	158	24	38
Canned, Jack, #1 tall can	361	1 can	563	24	133
Mullet, baked/broiled	85	3 oz	127	22	28
Ocean Perch, baked/broiled	85	3 oz	103	32	33
Oysters:					
Breaded, fried, Eastern	88	6 ea	173	29	51
Simmered, Eastern	100	3.5 oz	137	80	109
Perch, freshwater, bkd/broiled	85	3 oz	99	33	33
Pollock, baked/broiled	100	3.5 oz	99	77	76
Pompano, baked/broiled	100	3.5 oz	211	15	31
Rockfish, baked/broiled	100	3.5 oz	121	28	34
SALMON:					
Average, baked/broiled	85	3 oz	183	14	26
Chinook, smoked	85	3 oz	99	15	15
Sockeye, baked/broiled	100	3.5 oz	216	14	31
Canned:					
Atlantic, small can	220	1 can	281	19	54
Pink, drained, #1 can	454	1 can	631	24	152
Sockeye, drained, #1 can	369	1 ea	566	19	107
Sardines, canned, drained:					
Atlantic, 2 sardines ≈ 24g	92	1 can	192	19	36
Pacific, 1 sardine ≈ 38g	100	3.5 oz	178	19	34
Scallops:					
Breaded, fried	93	6 ea	200	28	55
Steamed	100	3.5 oz	113	60	68
Sea bass, baked/broiled	100	3.5 oz	124	43	53
Seatrout (steelhead), cooked	100	3.5 oz	131	27	36
Shad, baked w/bacon	100	3.5 oz	201	16	33
Shark, batter-fried	85	3 oz	194	19	37

	Wt (g)	Svg	Cal	Magnes (mg) 100 Cal	Svg
Fish & Shellfish, continued:					
SHRIMP:					
Boiled, 2 large = 11g	100	3.5 oz	99	34	34
Canned, drained	128	1 c	154	34	53
Canned with liquid	100	3.5 oz	102	59	60
Smelt, Rainbow, cooked	85	3 oz	106	31	33
Snapper, baked/broiled	100	3.5 oz	128	29	37
SOLE (Flounder):					
Baked/broiled	85	3 oz	99	51	50
Batter-fried	85	3 oz	250	13	33
Breaded, fried	100	3.5 oz	188	23	43
Steamed or poached	100	3.5 oz	92	52	48
Squid, flour-fried	85	3 oz	149	22	33
Surimi, processed walleye (Alaska) pollock: see imitation crab.					
Swordfish, baked/broiled	100	3.5 oz	155	22	34
Trout, baked/broiled	85	3 oz	129	26	33
TUNA, Light, canned, drained (No. 1/2 can):					
Canned in oil	171	1 ea	339	16	53
Water pack	165	1 ea	216	23	49

Meats

Beef, Pork, Ham, Rabbit, Frog leg, Venison &Veal

	Wt (g)	Svg	Cal	Magnes (mg) 100 Cal	Svg
BEEF:					
Breakfast strips, cured beef, cooked	34	3 ea	153	6	9
Chuck blade, pot roasted:					
Lean & fat (5.4 oz raw)	85	3 oz	325	5	15
Lean only	85	3 oz	230	8	19
GROUND BEEF, avg cooked:					
Extra lean, 17% fat, raw	85	3 oz	215	8	17
Lean, 20.7% fat, raw	85	3 oz	231	7	16
Regular, 26.6% fat, raw	85	3 oz	250	6	16
Frozen patty, 23% fat, raw	85	3 oz	240	7	17
Rib, choice, oven roasted:					
Lean & fat	85	3 oz	324	5	16
Lean only	85	3 oz	204	10	21
Round steak, broiled:					
Lean & fat	85	3 oz	233	9	21
Lean only	85	3 oz	165	15	24
Round tip, roasted:					
Lean & fat	85	3 oz	213	10	21
Lean only	85	3 oz	162	14	23
Sirloin steak, broiled:					
Lean & fat	85	3 oz	238	8	20
Lean only	85	3 oz	172	13	23
T-bone steak, broiled:					
Lean & fat	85	3 oz	276	7	20
Lean only	85	3 oz	182	14	25
Corned beef, canned	85	3 oz	213	6	12
Heart, simmered	85	3 oz	140	16	22
Liver, pan-fried	85	3 oz	184	11	20
HAM: see Pork, cured; Turkey ham; Sausages and Lunchmeats.					
LAMB:					
Arm chop, braised:					
Lean & fat	70	1 ea	244	7	18
Lean only	55	1 ea	152	11	16
Loin chop, broiled:					
Lean & fat	64	1 ea	201	7	15
Lean only	46	1 ea	100	13	13

	Wt (g)	Svg	Cal	Magnes (mg) 100 Cal	Svg
LAMB, continued:					
Cutlet, ckd average, lean	85	3 oz	175	13	22
Leg of lamb, roasted:					
Lean & fat	85	3 oz	219	9	20
Lean only	85	3 oz	162	14	22
Rib roast:					
Lean & fat	85	3 oz	305	6	17
Lean only	85	3 oz	197	10	20
Shoulder roast:					
Lean & fat	85	3 oz	235	6	15
Lean only	85	3 oz	173	11	19
PORK:					
Bacon, cooked:					
Regular	19	3 ea	109	5	5
Canadian style	47	2 ea	86	12	10
Breakfast strips	34	3 ea	156	6	9
Blade chop:					
Braised, lean & fat	67	1 ea	275	4	11
Braised, lean only	50	1 ea	156	6	10
Pan-fried, lean & fat	89	1 ea	368	5	17
Pan-fried, lean only	62	1 ea	175	9	15
Center loin chop:					
Braised, lean & fat	75	1 ea	266	5	13
Braised, lean only	61	1 ea	166	8	14
Broiled, lean & fat	82	1 ea	284	7	19
Broiled, lean only	72	1 ea	166	13	21
Pan-fried, lean & fat	89	1 ea	333	7	23
Pan-fried, lean only	67	1 ea	178	12	22
Roasted, lean & fat	88	1 ea	268	6	17
Roasted, lean only	72	1 ea	180	9	16
Center rib chop:					
Braised, lean & fat	67	1 ea	246	5	13
Braised, lean only	53	1 ea	147	8	12
Broiled, lean & fat	77	1 ea	264	7	19
Broiled, lean only	63	1 ea	162	12	19
Pan-fried, lean & fat	88	1 ea	343	6	19
Pan-fried, lean only	62	1 ea	160	11	17
Roasted, lean & fat	79	1 ea	252	6	15
Roasted, lean only	68	1 ea	162	9	14
Pork roast:					
Leg, lean & fat	85	3 oz	250	7	18
Leg, lean only	85	3 oz	187	11	21
Rib/loin, lean & fat	85	3 oz	265	6	16
Rib/loin, lean only	85	3 oz	206	9	18
Shoulder, braised:					
Lean & fat	85	3 oz	293	5	16
Lean only	67	2.4 oz	166	9	15
Spareribs, cooked	177	6.25 oz	703	6	43
Heart, cooked	145	1 c	214	16	35
Liver, cooked	85	3 oz	141	8	12
PORK, CURED, HAM:					
Roasted, lean & fat	85	3 oz	207	13	27
Roasted, lean only	85	3 oz	133	23	31
Canned, heated	85	3 oz	140	19	27
Lean ham, unheated	140	1 c	206	12	25
RABBIT, roasted	85	3 oz	175	8	14
VEAL (calf):					
Cutlet, ckd average	85	3 oz	166	14	24
Rib, roasted	85	3 oz	151	13	20
Liver, pan-fried	85	3 oz	208	11	22
VENISON (deer) roasted	85	3 oz	134	15	20

Meats: Poultry

	Wt (g)	Svg	Cal	Magnes (mg) 100 Cal	Svg
CHICKEN:					
All types:					
Fried	140	1 c	307	12	38
Roasted	140	1 c	266	13	35
Stewed	140	1 c	248	12	29
Canned, boned w/broth	142	5 oz	235	7	17
Dark meat only:					
Fried	85	3 oz	203	10	21
Roasted	85	3 oz	174	12	20
Stewed	85	3 oz	163	10	17
Light meat only:					
Fried	85	3 oz	163	15	25
Roasted	85	3 oz	147	16	23
Stewed	85	3 oz	135	14	19
Breast*, meat & skin:					
Batter-fried	140	1 ea	364	9	34
Flour-fried	98	1 ea	218	13	29
Roasted	98	1 ea	193	14	27
Stewed	110	1 ea	202	12	24
Breast*, meat only:					
Fried	86	1 ea	161	17	27
Roasted	86	1 ea	142	18	25
Stewed	95	1 ea	144	15	22
*2 pieces per bird					
Drumstick, meat & skin:					
Batter-fried	72	1 ea	193	7	14
Flour-fried	49	1 ea	120	9	11
Roasted	52	1 ea	112	11	12
Stewed	57	1 ea	116	9	11
Drumstick, meat only:					
Fried	42	1 ea	82	12	10
Roasted	44	1 ea	76	14	11
Stewed	46	1 ea	78	13	10
Thigh, meat & skin:					
Batter-fried	86	1 ea	238	8	18
Flour-fried	62	1 ea	162	9	15
Roasted	62	1 ea	153	9	14
Stewed	68	1 ea	158	8	13
Thigh, meat only:					
Fried	52	1 ea	113	12	14
Stewed	55	1 ea	107	11	12
Wing, meat, roasted	21	1 ea	43	16	7
Liver, simmered	140	7 ea	219	13	29
DUCK domestic, roasted:					
Meat & skin	85	3 oz	286	5	14
Meat only	85	3 oz	171	10	17
GOOSE domestic, roasted:					
Meat & skin	85	3 oz	259	7	19
Meat only	85	3 oz	202	11	21
TURKEY:					
Ground turkey, ckd	100	3.5 oz	229	10	24
Roasted:					
All types	85	3 oz	144	16	22
Dark meat	85	3 oz	159	13	21
White meat	85	3 oz	133	18	24
Frozen slices w/gravy	142	5 oz	95	12	11
Gizzard	67	1 ea	109	11	13
Liver	75	1 ea	127	9	11
Turkey patty, breaded, fried	64	1 ea	181	7	12

Meats: Sausages and Lunchmeats

	Wt (g)	Svg	Cal	Magnes (mg) 100 Cal	Svg
Brotwurst, link	70	1 ea	226	5	11
Chicken roll, light meat	57	2 oz	90	11	10
Frankfurters:					
Beef/Beef & Pork, 8/pkg	57	1 ea	184	4	7
Turkey/Chicken, 10/pkg	45	1 ea	102-115	7.5	8
HAM lunchmeat, regular	57	2 oz	103	10	11
Ham & cheese roll/loaf	57	2 oz	147	6	9
Ham salad spread	240	1 c	518	4	23
Italian link, cooked	67	1 ea	216	6	12
Luxury loaf	57	2 oz	80	14	11
Olive loaf	57	2 oz	133	8	11
Pastrami:					
Beef cured	57	2 oz	198	5	10
Turkey cured	57	2 oz	74	14	10
Pickle & pimento loaf	57	2 oz	149	7	10
Smoked link sausage, pork	68	1 ea	265	5	13
Turkey ham	57	2 oz	73	16	12
Turkey roll, light & dark	57	2 pce	84	12	10

Mixed Dishes & Fast Foods

	Wt (g)	Svg	Cal	Magnes (mg) 100 Cal	Svg
Beef & vegetable stew:					
Recipe	245	1 c	220	18	40
Canned	245	1 c	194	20	39
Beef, macaroni, tomato sauce, recipe	226	1 c	189	20	37
BURRITOS:					
Bean burrito	174	1 ea	322	24	76
Beef burrito	177	1 ea	463	10	47
Beef and bean	175	1 ea	390	16	61
Deluxe combination	198	1 ea	424	15	64
Cheese soufflé, recipe	112	1 c	221	9	19
Chicken & noodles, recipe	240	1 c	365	10	37
Chicken a la king, recipe	245	1 c	470	4	20
Chicken chow mein:					
Recipe	250	1 c	255	11	28
Canned	250	1 c	95	15	14
Chicken pot pie, f/frozen	230	1 ea	430	7	30
Chicken salad w/celery	78	1/2 c	266	4	11
Chili with beans, canned	255	1 c	286	40	115
Chop suey, beef & pork	250	1 c	300	11	32
Cole slaw	120	1 c	84	14	12
Corn dog	111	1 ea	330	7	22
Corn fritter, recipe	45	1 ea	116	12	14
Corn pudding	250	1 c	271	14	38
Egg salad	183	1 c	438	5	24
ENCHILADAS:					
Beef enchilada	120	1 ea	292	12	36
Cheese enchilada	120	1 ea	330	11	37
Chicken enchilada	120	1 ea	269	13	36
French toast, recipe	65	1 pce	123	10	12
LASAGNA:					
Recipe w/meat	245	1 pce	398	10	41
Recipe w/o meat	218	1 pce	316	11	35
Macaroni & cheese:					
Recipe	200	1 c	430	9	37
Canned	240	1 c	230	13	31
Macaroni salad, no cheese	141	1 c	371	6	23
Manicotti, frozen entree	225	1 ea	271	17	46
Meat loaf	85	3 oz	198	9	18
Moussaka, lamb & eggplant	250	1 c	250	18	44

	Wt (g)	Svg	Cal	Magnes (mg) 100 Cal	Svg
Mixed Dishes & Fast Foods, continued:					
PIZZA, cheese:					
Regular crust 1/8 of 15"	120	1 pce	290	27	78
Thick crust 1/2 of 10"	208	1 pce	519	12	64
Potato salad w/mayo & eggs	250	1 c	358	11	39
Quiche Lorraine, 1/8	176	1 pce	600	4	23
Ravioli, beef, canned	226	1 c	220	19	41
SANDWICHES (Fast Food):					
Cheeseburger, 3 oz patty	112	1 ea	300	7	22
Cheeseburger, 4 oz patty	194	1 ea	524	8	43
Chicken patty sandwich	157	1 ea	436	7	30
English muffin w/ egg, cheese, & bacon	138	1 ea	360	8	28
Fish sandwich:					
Regular with cheese	140	1 ea	420	7	29
Large, no cheese	170	1 ea	470	7	34
Frankfurter (hotdog) w/bun	85	1 ea	260	5	13
Hamburger, 3 oz patty	98	1 ea	245	8	19
Hamburger, 4 oz patty	174	1 ea	445	9	38
SANDWICHES on part whole wheat bread, unless stated as on rye:					
Avocado, cheese, tomato & sprouts	195	1 ea	432	15	66
Bacon, lettuce, tomato	135	1 ea	327	11	36
Chicken salad sandwich	100	1 ea	294	11	32
Corned beef & swiss on rye	147	1 ea	429	7	32
Egg salad sandwich	111	1 ea	319	10	31
Grilled cheese	117	1 ea	393	10	39
Ham sandwich	122	1 ea	256	15	39
Ham on rye bread	116	1 ea	242	10	25
Ham & cheese	151	1 ea	363	12	45
Ham & swiss, on rye	145	1 ea	350	10	35
Ham salad sandwich	125	1 ea	339	9	32
Patty melt, ground beef on rye	177	1 ea	567	7	40
Peanut butter & jam	100	1 ea	341	20	69
Reuben, grilled on rye	233	1 ea	480	9	44
Roast beef sandwich	122	1 ea	280	13	37
Turkey sandwich	122	1 ea	271	14	37
Turkey ham	122	1 ea	253	16	40
Turkey ham on rye	116	1 ea	239	11	27
Turkey ham & cheese	151	1 ea	361	13	47
Turkey ham & cheese on rye	145	1 ea	347	10	33
SPAGHETTI (pasta & tomato sauce with cheese):					
Homemade	250	1 c	260	10	26
Canned	250	1 c	190	11	21
SPAGHETTI (pasta & tomato sauce with meat):					
Homemade	248	1 c	330	12	40
Canned	250	1 c	260	8	20
Spinach soufflé	136	1 c	218	17	37
Taco:					
Beef	78	1 ea	207	11	23
Chicken	78	1 ea	172	13	23
TOSTADA:					
Beans only	157	1 ea	212	29	62
Beans and beef	192	1 ea	332	16	52
Beans and chicken	157	1 ea	249	19	48
Tuna noodle casserole, recipe	202	1 c	251	12	31
Tuna salad, fr/recipe	205	1 c	383	10	40
Turkey pot pie, fr/frozen	233	1 ea	416	6	25
Waldorf salad	142	1 c	424	10	41

Nuts & Seeds

	Wt (g)	Svg	Cal	Magnes (mg) 100 Cal	Svg
ALMONDS, dried	142	1 c	837	50	420
Brazil nuts, dry	140	1 c	919	34	315
Cashews, dry roasted	137	1 c	787	45	356
COCONUT:					
Raw, grated	80	1 c	283	9	26
Dried, unsweetened	78	1 c	515	14	70
Flaked, packaged	74	1 c	351	10	36
Coconut milk, fresh	240	1 c	552	16	89
Coconut water, raw	240	1 c	45.6	132	60
Filberts (hazelnuts) chopped	115	1 c	727	45	328
Macadamias, dried	134	1 c	940	16	155
MIXED NUTS with peanuts (cashews, peanuts, brazil nuts, filberts, almonds & pecans):					
Dry roasted	137	1 c	814	38	308
Oil roasted	142	1 c	876	38	334
MIXED NUTS, without peanuts (cashews, almonds, brazil nuts, pecans, filberts):					
Oil roasted	144	1 c	886	41	361
PEANUTS:					
Dried	146	1 c	855	31	263
Oil roasted	145	1 c	837	32	266
Peanut butter	32	2 T	190	26	50
Peanut flour, defatted	60	1 c	196	113	222
Pecans, dried, chopped	119	1 c	794	19	152
Pine nuts, dried	28	1 oz	153	44	67
Pistachios, dried, no shells	28	1 oz	164	27	45
Pumpkin/squash seeds:					
Dry kernels	138	1 c	747	99	738
Oil roasted kernels	227	1 c	1185	102	1212
Whole, roasted	64	1 c	285	59	168
Sesame seeds:					
Dry kernels	38	1/4 c	221	59	130
Roasted, whole	28	1 oz	161	63	101
Soybeans, roasted	86	1/2 c	405	31	125
Sunflower seed kernels:					
Dry kernels	36	1/4 c	205	62	128
Oil roasted	135	1 c	830	21	171
Sunflower seed butter	16	1 T	93	64	59
Tahini, sesame butter	15	1 T	91	58	14-53
Walnuts, chopped:					
Black	125	1 c	759	33	253
English	120	1 c	770	26	203

Soups, Sauces & Gravy

	Wt (g)	Svg	Cal	Magnes (mg) 100 Cal	Svg
Gravy, brown, fr/dry mix	258	1 c	74.8	14	10.3
SAUCES (also see "Other"):					
Cheese sauce:					
Recipe	101	1/2 c	216	8	17
From mix w/milk	279	1 c	305	10	32
Hollandaise, recipe	160	1 c	867	2	16
Spaghetti sauce, canned	249	1 c	272	22	60
White/cream sauce:					
Homemade	250	1 c	395	8	30
From mix with milk	264	1 c	240	13	30

SOUPS

All soups are canned unless otherwise stated. RTS = Ready To Serve. For soups prepared with milk, assume whole milk.

	Wt (g)	Svg	Cal	Magnes (mg) 100 Cal	Svg
Bean & 'frank'	250	1 c	187	26	49
Bean with bacon	253	1 c	173	25	44
Cheese soup, w/milk	251	1 c	230	9	20
Celery, cream of, w/milk	248	1 c	165	13	22
Chicken, cream of, w/milk	248	1 c	191	9	18
Chicken vegetable, fr/dry	251	1 c	49	43	21
Chili beef soup	250	1 c	169	18	30
Clam chowder:					
New England style	248	1 c	163	14	23
Manhatten style	244	1 c	78	16	12.2
Mushroom, cream of, cond.	248	1 c	205	10	20
Oyster stew, w/milk	245	1 c	134	16	21
Split pea soup:					
With or without ham	253	1 c	189	25	48
Prep. fr/dry	255	1 c	133	35	46
Tomato soup:					
Prep. w/milk	248	1 c	160	14	23
Prep. fr/dry	265	1 c	102	15	15
Tomato vegetable, fr/dry	253	1 c	55	36	20

Vegetables & Legumes

	Wt (g)	Svg	Cal	Magnes (mg) 100 Cal	Svg
Amaranth leaves:					
Fresh, chopped	28	1 c	7.4	211	16
Cooked	132	1 c	28	261	73
Artichoke, whole:					
Globe, ckd from fresh	120	1 ea	60	120	72
Artichoke hearts:					
Ckd f/frozen pkg	240	9 oz	108	69	74
Marinated-jar	170	6 oz	168	29	48
Asparagus:					
Fresh pieces	67	1/2 c	15	80	12
Ckd from fresh	90	1/2 c	23	76	17
Ckd from frozen	180	1 c	50	46	23
Canned, drained	121	1/2 c	16	75	12
Canned with liquid	122	1/2 c	17	65	11
BEANS: also see garbanzo, lentils, soybeans.					
Baked beans (dry white beans w/spices & sauce):					
Home prepared	253	1 c	382	29	110
Canned, plain/vegetarian	254	1 c	235	35	82
Canned w/franks	257	1 c	366	19	71
Canned w/pork	253	1 c	268	32	85
Canned, pork & sweet sce	253	1 c	282	31	87
Canned, pork & tomato sce	253	1 c	247	36	88
Black beans, cooked	172	1 c	227	53	121
Broadbeans, cooked:					
From fresh veg	109	1 c	79	52	41
From dry legumes	170	1 c	186	39	73
Great northern, ckd fr/dry	177	1 c	210	42	88
Green beans (snap):					
Fresh, uncooked	110	1 c	34	79	27
Ckd from fresh	125	1 c	44	73	32
Ckd from frozen	135	1 c	36	81	29
Canned, drained	135	1 c	26	69	18
Canned with liquid	240	1 c	36	87	31
Hyacinth beans, cooked:					
Fr/fresh vegetables	109	1 c	37	105	39
Fr/dry legumes	194	1 c	228	70	159
Kidney beans:					
Canned w/liquid	256	1 c	208	38	79
Cooked fr/dry	177	1 c	225	36	80

	Wt (g)	Svg	Cal	Magnes (mg) 100 Cal	Svg
BEANS, continued:					
Lima beans:					
Ckd from fresh	170	1 c	208	61	126
Ckd from frozen, large	85	1/2 c	85	34	29
Ckd from frozen, baby	90	1/2 c	94	53	50
Ckd from dry	188	1 c	217	38	82
Canned, drained	170	1 c	164	43	70
Canned with liquid	241	1 c	191	49	94
Navy beans, ckd fr/dry	182	1 c	259	41	107
Pinto beans:					
Ckd from dry	171	1 c	235	40	95
Canned	240	1 c	186	34	64
Refried beans	253	1 c	270	37	99
White beans, ckd fr/dry	179	1 c	253	48	122
Winged beans, cooked:					
Fr/fresh veg	62	1 c	23	78	1
Fr/dry legumes	172	1 c	252	37	94
Yardlong beans, ckd fr/dry	171	1 c	202	83	167
Yellow wax: see green beans.					
Bean sprouts (mung beans):					
Fresh, uncooked	104	1 c	31	71	22
Boiled f/fresh, drained	124	1 c	26	69	18
Stir fried f/fresh	124	1 c	62	61	38
Canned, drained	125	1 c	16	69	11
Beets:					
Canned dices	85	1/2 c	27	48	13
Pickled slices	114	1/2 c	74	23	17
Beet greens:					
Fresh, chopped	19	1/2 c	4	350	14
Cooked f/fresh, drained	144	1 c	40	242	97
Borage:					
Fresh, chopped	44	1/2 c	9	256	23
Cooked from fresh	100	3.5 oz	25	228	57
BROCCOLI, chopped:					
Fresh, uncooked	88	1 c	24	92	22
Ckd from fresh	156	1 c	44	85	37.4
Ckd from frozen	184	1 c	51	73	37
W/ cheese sauce	142	1/2 c	166	11	18
W/ hollandaise sauce	95	1/2 c	105	10	10
Brussels sprouts:					
Ckd from fresh	156	1 c	60	53	32
Ckd from frozen	155	1 c	65	57	37
CABBAGE:					
Common, fresh	70	1 c	16	63	10
Common, ckd	150	1 c	32	70	23
Bok choy, fresh	70	1 c	9	144	13
Bok choy, ckd	170	1/2 c	20	90	18
Pe-tsai, fresh	76	1 c	11	91	10
Pe-tsai, ckd	119	1 c	16	75	12
Red, fresh	70	1 c	19	58	11
Red, ckd	75	1/2 c	16	50	8
Savoy, fresh	70	1 c	20	100	20
Savoy, ckd	145	1 c	35	97	34
CARROTS:					
Fresh (7-1/2" x 1-1/8")	72	1 ea	31	35	11
Ckd f/fresh slices	78	1/2 c	35	29	10
Canned w/liquid	123	1/2 c	28	39	11
Carrot juice	123	1/2 c	49	35	17
Cauliflower:					
Fresh	50	1/2 c	12	58	7
Ckd fr/fresh	62	1/2 c	15	47	7
Ckd fr/frozen	90	1/2 c	17	47	8
Celery:					
Fresh (7-1/2" stalk = 40g)	40	1 stalk	6.4	69	4.4
Cooked, diced	150	1 c	27	67	18
Chard, Swiss:					
Fresh, chopped	36	1 c	6.8	427	29
Cooked	175	1 c	35	429	150

ESHA Research

Values are for edible portion of foods

	Wt (g)	Svg	Cal	Magnes 100 Cal	Magnes Svg
Vegetables & Legumes, continued:					
Collards:					
Ckd from fresh	128	1 c	35	26	9
Ckd from frozen	170	1 c	63	85	52
CORN, kernels:					
Ckd from fresh	82	1/2 c	89	29	26
Ckd from frozen	82	1/2 c	67	22	15
Canned, drained	82	1/2 c	66	23	15
Canned with liquid	128	1/2 c	79	25	20
Corn, creamed, canned	128	1/2 c	93	23	22
Cucumber (8" x 2+")	301	1 ea	39	85	33
Dandelion greens:					
Fresh	55	1 c	25	80	20
Cooked	105	1 c	35	74	26
Dock (sorrel) greens:					
Fresh chopped	133	1 c	29	472	137
Cooked from fresh	100	1 c	20	445	89
Eggplant, cooked	160	1 c	45	47	21
Garbanzo beans (chickpeas):					
Cooked from dry	164	1 c	269	29	78
Canned	240	1 c	285	25	70
Jerusalem artichoke, fresh	150	1 c	114	23	26
Jicama, sliced	100	3.5 oz	20	50	10
Kale, chopped:					
Fresh, chopped	67	1 c	33	70	23
Ckd from fresh	130	1 c	42	55	23
Ckd from frozen	130	1 c	39	59	23
Kohlrabi:					
Fresh slices	140	1 c	38	71	27
Cooked	104	1 c	63	46	29
Lentils, cooked from dry	198	1 c	231	31	71
Lentils, sprouted:					
Fresh, uncooked	77	1 c	81	35	28
Stir fried	100	3.5 oz	101	35	35
LETTUCE, chopped:					
Butterhead	56	1 c	7.3	85	6.2
Iceberg	56	1 c	7.3	11	5.0
Looseleaf	56	1 c	10	11	1.1
Romaine	56	1 c	9	38	3.4
Mushrooms:					
Fresh slices	35	1/2 c	9	40	3.5
Ckd from fresh	78	1/2 c	21	45	9.4
Mustard greens:					
Fresh greens	56	1 c	15	123	18
Ckd from fresh	140	1 c	21	100	21
Ckd from frozen	150	1 c	28	68	20
Okra, cooked:					
Fr/fresh, pods	85	8 ea	27	176	48
Fr/frozen, slices	92	1/2 c	34	138	47
Onion, chopped:					
Fresh	160	1 c	61	26	16
Ckd from fresh	105	1/2 c	46	26	12
Spring, fresh	50	1/2 c	16	63	10
Parsley, fresh, chopped	30	1/2 c	10	130	13
Parsnips, ckd slices	156	1 c	125	37	46
PEAS:					
Black-eyed peas:					
Ckd from fresh	165	1 c	160	54	85.8
Ckd from dry	171	1 c	198	46	91
Canned	240	1 c	184	36	66
Green peas:					
Fresh, uncooked	145	1 c	118	41	48
Ckd from fresh	160	1 c	134	47	63
Ckd from frozen	80	1/2 c	63	37	23
Canned, drained	85	1/2 c	59	25	15
Green peas, edible-pods:					
Fresh, uncooked	145	1 c	61	57	35
Ckd from fresh	160	1 c	67	63	42

	Wt (g)	Svg	Cal	Magnes 100 Cal	Magnes Svg
PEAS, continued:					
Green peas, edible pods, continued:					
Ckd from frozen	80	1/2 c	42	52	22
Split peas, ckd fr/dry	196	1 c	231	31	71
Peas & carrots f/frozen	80	1/2 c	38	34	13
Peas, mature, sprouted:					
Fresh sprouts	120	1 c	154	44	67
Cooked from fresh	100	3.5 oz	118	35	41
PEPPER, HOT,					
green or red, chopped	75	1/2 c	30	63	19
PEPPER, SWEET,					
green or red chopped, cooked	68	1 ea	19	37	7
Poi, two finger	240	1 c	269	22	58
POTATOES:					
Baked, 4.75" x 2.3":					
Flesh & skin	202	1 ea	220	25	55
Flesh only	156	1 ea	145	27	39
Potato skins	58	1 ea	115	22	25
Boiled, flesh only:					
Peeled before cooking	135	1 ea	116	22	26
Peeled after	136	1 ea	119	25	30
French fried f/frozen:					
Fried in oil	50	10 ea	158	11	17
Oven heated	50	10 ea	111	10	11
Hash brown, f/frozen	156	1 c	340	8	26.5
Mashed, with milk	210	1 c	162	24	39
Canned, 1" diam.	70	2 ea	42	24	10
Puffs (tator tots) frzn, heated	62	1/2 c	138	9	12
Potato dishes, prepared:					
Au gratin, recipe	245	1 c	322	15	48
Au gratin, dry mix	245	1 c	228	44	100
Scalloped, recipe	245	1 c	210	22	46
Scalloped, dry mix	245	1 c	228	15	34
Potato chips	28	14 ea	148	11	17
Potato pancakes	76	1 ea	237	10	24
PUMPKIN, mashed:					
Cooked from fresh	245	1 c	50	44	22
Canned	123	1/2 c	42	67	28
Purslane:					
Fresh	43	1 c	7	414	29
Ckd from fresh	115	1 c	21	367	77
Rutabaga:					
Fresh cubes	140	1 c	51	63	32
Cooked from fresh	85	1/2 c	29	62	18
Salsify, cooked slices	135	1 c	92	26	24
Sauerkraut, cnd w/liquid	236	1 c	44	70	31
Seaweed:					
Agar, fresh	28	1 oz	7	257	19
Kelp, fresh	28	1 oz	12	281	34
Wakame, fresh	28	1 oz	13	237	30
Spirulina, dried	28	1 oz	82	67	55
Soybeans, ckd from dry	172	1 c	298	50	148
Soybeans, sprouted:					
Fresh sprouts	35	1/2 c	45	56	25
Steamed	94	1 c	76	75	57
Stir fried	100	3.5 oz	125	77	96

SOYBEAN PRODUCTS: See tofu this section, miso, natto, tempeh in Other; roasted soybeans in Nuts & Seeds; soy flour in Grains; and soy milk in Dairy.

	Wt (g)	Svg	Cal	Magnes 100 Cal	Magnes Svg
Spinach:					
Fresh, chopped	56	1 c	12	367	44
Ckd from fresh, chopped	180	1 c	41	383	157
Ckd from frozen, leaf	190	1 c	53	247	131
Canned, drained	214	1 c	50	324	162

	Wt (g)	Svg	Cal	Magnes (mg) 100 Cal	Svg

Vegetables & Legumes, continued:

SQUASH, SUMMER, sliced:

	Wt (g)	Svg	Cal	100 Cal	Svg
Crookneck, fresh	130	1 c	24	113	27
Crookneck, cooked	180	1 c	36	122	44
Scallop, fresh	130	1 c	24	125	30
Scallop, cooked	90	1/2 c	14	121	17
Zucchini, fresh	130	1 c	19	147	28
Zucchini, cooked	180	1 c	29	138	40

SQUASH, WINTER, mashed:

	Wt (g)	Svg	Cal	100 Cal	Svg
Acorn/Danish, baked	245	1 c	137	76	104
Acorn/Danish, boiled	245	1 c	83	76	63
Butternut, baked	245	1 c	99	71	71
Butternut, ckd f/frozen	240	1 c	94	23	22
Hubbard, baked	240	1 c	120	44	53
Hubbard, boiled	236	1 c	70	46	32
Spaghetti, baked/boiled	155	1 c	45	38	17
Succotash:					
Ckd from fresh	192	1 c	222	46	102
Ckd from frozen	170	1 c	158	25	39
Sweet potato:					
Baked in skin	114	1 ea	118	19	23
Boiled, peeled first	151	1 ea	160	9	15
Canned, mashed:					
Regular	128	1/2 c	129	23	30
Vacuum pack	255	1 c	233	24	57
Candied sweet potatoes	105	1 pce	144	8	12
Tofu (soybean curd):					
Firm, raw	126	1/2 c	183	64	118
Regular, raw	124	1/2 c	94	135	127

TOMATO:

	Wt (g)	Svg	Cal	100 Cal	Svg
Fresh, whole, 2.6" dm	123	1 ea	26	54	14
Ckd from fresh	240	1 c	65	52	34
Canned, whole	240	1 c	47	62	29
Tomato products, canned:					
Tomato juice	244	1 c	42	65	27
Tomato paste	262	1 c	220	61	134
Tomato puree	250	1 c	102	59	60
Tomato sauce	245	1 c	74	62	46
Turnip cubes:					
Fresh cubes	130	1 c	35	40	14
Ckd from fresh	78	1/2 c	14	43	6
Turnip greens:					
Ckd from fresh	144	1 c	29	110	32
Ckd from frozen	82	1/2 c	24	88	21

VEGETABLE COMBINATIONS,
cooked from frozen:

	Wt (g)	Svg	Cal	100 Cal	Svg
Broccoli, carrots, pasta	95	2/3 c	88	15	13
Broccoli, carrots					
& water chestnuts	91	2/3 c	32	41	13
Broccoli, cauliflower					
& red pepper	95	2/3 c	25	56	14
Cantonese stir fry	95	1/2 c	53	21	11
Chinese stir fry	95	1/2 c	31	77	24
Green beans & spaetzle	95	1/2 c	108	14	15
Japanese style	95	1/2 c	29	38	11
Mixed vegetables (corn, peas, carrots,					
green beans, lima beans):					
Ckd from frozen	182	1 c	107	37	40
Canned, drained	95	1/2 c	128	13	17
Peas, carrots, onions	91	1/2 c	54	31	17
Peas, carrots, onion, pasta	95	1/2 c	122	15	18
Peas, cauliflower, cream sce	95	1/2 c	118	14	17
Peas & mushrooms	95	1/2 c	73	26	19
Peas & onions	95	1/2 c	71	25	18
Peas, onions, cheese sauce	142	1/2 c	165	16	2
Peas, pasta, corn, cream sce	95	1/2 c	132	17	23
Peas, pasta, mushrooms	95	1/2 c	129	17	22

VEGETABLE COMBINATIONS, continued:

	Wt (g)	Svg	Cal	100 Cal	Svg
Peas, potatoes, cream sce	76	1/2 c	140	10	14
Vegetable juice cocktail	242	1 c	46	58	27
Yams, orange: see Sweet Potato.					
Yams, white, cubed, cooked	136	1 c	158	16	26
Zucchini: see Squash, summer.					

Other

Cooking Ingredients, condiments, flavorings,
other beverages, spices, sweets

	Wt (g)	Svg	Cal	100 Cal	Svg
Barbecue sauce	16	1 T	10	9	.9
BEVERAGES, other than fruit juices,					
veg juices or milk beverages:					
Alcoholic:					
Beer, 12 fl oz = 1.5 c:					
Regular	356	1.5 c	146	16	23
Light	354	1.5 c	100	18	18
Bloody mary, 5 fl oz	148	1 ea	116	9	11
Screwdriver, 7 fl oz	213	1 ea	174	10	17
Tequila sunrise, 5.5 oz	172	1 ea	189	6	12
Wine;					
Red	118	1/2 c	85	18	15.0
Rosé	118	1/2 c	84	14	11.4
White, medium	118	1/2 c	80	16	12.6
Non-alcoholic:					
Coffee, brewed	240	1 c	2	675	13.5
Pineapple grapefruit drink	250	1 c	117	13	15
Pineapple orange drink	250	1 c	125	11	14
Tea, brewed	240	1 c	2	68	1.4

CANDY and CANDY BARS:

	Wt (g)	Svg	Cal	100 Cal	Svg
Almonds, sugar coated	28	7 ea	146	31	46
Chocolate covered:					
Almonds	165	1 c	935	39	364
Coconut candy	28	1 oz	133	12	16
Mints	28	1 oz	116	14	16
Peanuts	170	1 c	954	30	282
Raisins	187	1 c	733	16	119
Chocolate kisses	28	6 pce	154	12	18
Fudge, chocolate	28	1 oz	115	12	14
Fudge, chocolate w/nuts	28	1 oz	114	13	15
Kit Kat bar	43	1 ea	210	9	19
Krackle bar	34	1 ea	179	11	20
M &M's plain candies	48	1 pkg	237	13	30
M &M's peanut candies	47	1 pkg	240	16	38
MARS bar	50	1 ea	240	15	37
MILKY WAY bar	60	1 ea	260	8	22
Milk chocolate, plain:	28	1 oz	145	11	16
With almonds	28	1 oz	150	22	33
With peanuts	28	1 oz	155	23	35
With rice cereal	28	1 oz	140	9	13
MR. GOODBAR	47	1 ea	250	18	45
REESE's peanut butter cup	45	2 ea	240	20	47
SNICKERS bar (2.2 oz)	61	1 bar	290	13	39
Carob flour	103	1 c	185	30	56
Catsup	15	1 T	16	23	3.75
Chili sauce, tom. base	17	1 T	18	11	1.9

CHOCOLATE:

	Wt (g)	Svg	Cal	100 Cal	Svg
Baking chocolate, unswt	28	1 oz	145	57	82
Bittersweet	28	1 oz	141	20	28
Choc. chips, semi sweet	170	1 c	860	27	230
Dark, sweet	28	1 oz	150	21	32
Hot fudge topping	38	2 T	129	14	18
Syrup, thin	38	2 T	85	31	26

ESHA Research

	Wt (g)	Svg	Cal	Magnes (mg) 100 Cal	Svg		Wt (g)	Svg	Cal	Magnes (mg) 100 Cal	Svg
Other, continued:											
Cocoa powder	86	1 c	224	186	417						
Falafel (2.25" patty)	17	1 ea	57	25	14						
Hummous (Humous)	246	1 c	420	17	71						
Miso (soybean product)	138	1/2 c	284	20	58						
Molasses, blackstrap	40	2 T	85	121	103						
Mustard, prepared	125	1/2 c	94	64	60						
Natto (soybean product)	88	1/2 c	187	54	101						
Olives, pitted:											
Black, large	45	10 ea	52	3	1.8						
Green	39	10 ea	45	19	8.6						
SPICES:											
Basil, dried	4.5	1 T	11	167	18						
Black pepper	6.4	1 T	16	74	12						
Caraway seed	6.7	1 T	22	77	17						
Chili powder	7.5	1 T	24	56	13.5						
Coriander leaf, dried	1.8	1 T	5	240	12						
Coriander seed	5	1 T	15	113	17						
Cumin seed	6	1 T	22	100	22						
Curry powder	6.3	1 T	18	89	16						
Dill weed, dried	3.1	1 T	8	163	13						
Fenugreek seed	11.1	1 T	36	58	21						
Ginger, ground	5.4	1 T	19	53	10						
Nutmeg, ground	7	1 T	37	35	13						
Oregano, ground	4.5	1 T	14	86	12						
Paprika	6.9	1 T	20	64	13						
Poppyseed	8.8	1 T	47	62	29						
Tarragon, ground	4.8	1 T	14	121	17						
Turmeric, ground	6.8	1 T	24	54	13						
Sugar, brown	220	1 c	820	16	135						
Tempeh (soybean product)	83	1/2 c	165	35	58						
Teriyaki sauce	18	1 T	15	73	11						
Yeast:											
Brewer's	8	1 T	25	74	18.5						
Dry active, pkg	7	1 pkg	20	80	16						

Phosphorus

About 80 percent to 90 percent of the phosphorus in the body is found in the bones and in the teeth, combined with calcium phosphate. In addition to bone formation, phosphorus has a wide variety of functions.

As a part of phosphoric acid, it is found in all cells; and as a part of DNA and RNA, it is necessary for new cell formation and growth. Many enzymes and the B vitamins rely on the presence of phosphorus to function. As a result, many key functions in the metabolism of foods—carbohydrates, protein and fat—require phosphorus. It also helps transport some fats in the bloodstream. It helps move nutrients into and out of cells, and it is used in many hormones.

Nutrient Losses

Losses will occur in processing and cooking, primarily from leaching into cooking water. However, phosphorus is readily available and deficiencies are unknown.

Phosphorus, as a part of phytic acid, is a part of fiber, which can reduce the absorption of other minerals as well as the phosphorus in the phytic acid. However, the yeast-leavening of bread can inactivate this inhibitory effect of phytic acid.

Recommended Dietary Allowances for Adults

USA RDA: Adults, 800 mg; Young adults 18-24, 1200 mg; (pregnant or lactating: 1200 mg)
Canadian RNI: Men 800 mg; Women 700 mg (pregnant or lactating, 1300 mg)

Because of its relationship with calcium, the recommended daily amounts of these two minerals in the diet is about the same.

Food sources

Phosphorus is present in nearly all foods. Good sources are protein-rich foods (meats, poultry, fish, milk, and milk products), eggs, grains, and legumes.

ESHA Research

	Wt (g)	Svg	Cal	Phos (mg) 100 Cal	Phos (mg) Svg
Baked Goods					
Bagel, 3.5" diam.	68	1 ea	180	34	61
Biscuits:					
Homemade	28	1 ea	100	36	36
From mix	28	1 ea	94.4	137	129
From refrig dough	20	1 ea	65.2	120	78
BREADS:					
Banana nut bread, 1/2" slc	50	1 pce	161	30	48
Boston brown bread, cnd	45	1 pce	95	76	72
Cornbread, muffin	45	1 ea	145	88	128
Cracked wheat	25	1 pce	65	49	32
French bread, 5" x 2.5"	35	1 pce	100	30	30
Italian bread, 4.75" x 4" x 1/2"	30	1 pce	83	28	23
Mixed grain	25	1 pce	65	85	55
Oatmeal bread	25	1 pce	65	48	31
Pita pocket bread, 6.5" dm	60	1 ea	165	36	60
Pumpernickel, 5" x 4" x 3/8"	32	1 pce	80	89	71
Raisin bread	25	1 pce	68	32	22
Rye, light, 5" x 3.5" x 7/16"	25	1 pce	65	55	36
Wheat bread (white and whole wheat flour)	28	1 pce	72	72	52
White bread	28	1 pce	75	40	30
Whole wheat bread	28	1 pce	70	106	74
Breadstick, 4" x 1/2" diam.	100	10 ea	384	26	99
Brownies, with frosting	25	1 ea	100	31	31
CAKE, piece = 1/16th cake unless stated otherwise:					
Angel food, 1/12 tube cake	53	1 pce	125	73	91
Boston cream, 1/8 cake	120	1 pce	260	27	70
Carrot cake, cream cheese frosting, 2.5" x 3"	112	1 pce	406	16	65
Cheesecake:					
Recipe, 1/12	92	1 pce	278	29	81
From mix, 1/8	103	1 pce	300	64	193
Chocolate, choc frosting	69	1 pce	235	31	72
Coffee cake, f/mix, 2.4" x 2.8"	72	1 pce	230	54	125
Dark fruitcake 2/3" arc	43	1 pce	165	30	50
Gingerbread, 3" x 3"	63	1 pce	174	36	63
Pound cake, f/mix, 1/2" slice	30	1 pce	115	24	28
Sheet cake, plain, 3" x 3"	86	1 pce	315	28	88
Sheet cake, white frosting	121	1 pce	445	20	91
Snack cake, cream-filled:					
Choc (like Ding-dongs)	28	1 ea	105	25	26
Sponge (like Twinkies)	42	1 ea	155	28	44
Sponge cake, 1/12 tube	66	1 pce	194	36	70
White cake, choc frosting	77	1 pce	291	42	122
White cake, coconut frosting	70	1 pce	270	39	106
Yellow cake, choc frosting	69	1 pce	240	51	122
Cheese puffs (Cheetos)	28	1 oz	158	18	29
Chips: see corn & tortilla this section; potato chips under Vegetables.					
COOKIES:					
Chocolate chip:					
Homemade	40	4 ea	185	18	34
Commercial	42	4 ea	180	23	41
Refrigerator dough	48	4 ea	225	15	34
Fig bars	56	4 ea	210	16	34
Lady fingers	44	4 ea	158	46	72
Oatmeal raisin cookies	52	4 ea	245	24	58
Peanut butter, recipe	48	4 ea	245	24	60
Sandwich type, all	40	4 ea	195	21	40
Shortbread, commercial	32	4 ea	155	25	39
Sugar, f/refrig dough	48	4 ea	235	39	91
Vanilla wafers	40	10 ea	185	19	36
Corn chips	28	1 oz	155	34	52

	Wt (g)	Svg	Cal	Phos (mg) 100 Cal	Phos (mg) Svg
CRACKERS:					
Cheese, w/peanut butter	30	4 ea	150	63	94
Graham crackers	14	2 ea	60	33	20
Round, like Ritz	9	3 ea	45	40	18
Rye wafers, whole grain	14	2 ea	55	80	44
Saltines	12	4 ea	50	24	12
Wheat, thin, small	8	4 ea	35	43	15
Croissant, 4.5" x 4" x 2"	57	1 ea	235	27	64
Cream puff w/custard filling	110	1 ea	280	37	104
Danish pastry:					
Plain pastry	57	1 ea	220	26	58
With fruit	65	1 ea	235	34	80
Doughnut, cake type	50	1 ea	210	53	111
Doughnut, yeast raised	60	1 ea	235	23	55
Eclair, custard fill, choc icing	94	1 ea	262	35	92
English muffins, all	57	1 ea	135	50	67
MUFFINS: see English muffin.					
Blueberry, recipe	45	1 ea	135	34	46
Blueberry, fr/mix	45	1 ea	140	64	90
Bran, wheat, recipe	45	1 ea	125	100	125
Bran, wheat, fr/mix	45	1 ea	140	130	182
Cornbread, recipe	45	1 ea	145	41	59
Pancakes, 4" diam:					
Buckwheat, fr/mix	27	1 ea	55	165	91
Plain, fr/mix	27	1 ea	60	118	71
Plain, recipe	27	1 ea	60	63	38
Whole wheat, 5" dm, fr/mix	52	1 ea	94	101	95
PIES, piece = 1/6 of 9" pie:					
Apple pie	158	1 pce	405	12	50
Banana cream	198	1 pce	319	39	125
Blueberry	158	1 pce	380	12	46
Cherry	158	1 pce	410	14	58
Chocolate cream	175	1 pce	311	47	147
Coconut cream	172	1 pce	343	41	140
Coconut custard	165	1 pce	384	51	195
Cream pie, commercial	152	1 pce	455	34	154
Custard	152	1 pce	293	50	147
Lemon meringue	140	1 pce	355	19	69
Mincemeat	160	1 pce	395	18	70
Peach	158	1 pce	405	14	55
Pecan	138	1 pce	583	22	130
Pumpkin	200	1 pce	367	57	211
Poptart-type pastry, fortified	54	1 ea	210	50	104
Pretzels, thin twists	60	10 ea	240	23	55
ROLLS:					
Cinnamon bun, small	50	1 ea	158	32	51
Dinner roll, 2.5" x 2"	28	1 ea	85	52	44
Hamburger bun	45	1 ea	129	38	50
Hotdog bun	40	1 ea	115	38	44
Hard roll, white	50	1 ea	155	30	46
Rye roll, dark	28	1 ea	79	61	48
Rye roll, light	28	1 ea	76	53	40
Submarine roll (hoagie)	135	1 ea	400	29	115
Whole wheat roll	35	1 ea	88	102	90
Stuffing:					
From dry bread cubes	140	1 c	500	27	136
Stove Top	108	1/2 c	176	35	62
Taco shell	14	1 ea	59	56	33
Tortillas:					
Corn, 6" diam, fried	30	1 ea	87	46	40
Flour, 10.5" diam.	57	1 ea	168	56	94
Flour, 8" diam.	35	1 ea	105	56	59
Tortilla chips, plain	28	1 oz	139	53	74
Tortilla chips, Doritos, flavored	28	1 oz	140	68	95
Waffles:					
Homemade, 7" diam.	75	1 ea	245	55	135
From mix, 7" diam.	75	1 ea	205	125	257
Frozen, 4" diam.	35	1 ea	98	137	134

	Wt (g)	Svg	Cal	Phos (mg) 100 Cal	Phos (mg) Svg

Dairy & Dairy Products

CHEESE (1.5" cube ≈ 1 oz):

	Wt (g)	Svg	Cal	100 Cal	Svg
American, processed	28	1 oz	106	199	211
American cheese food, cold pack	28	1 oz	94	126	118
American cheese food, jar	28	1 oz	93	140	130
American cheese spread	28	1 oz	82	245	201
Blue cheese	28	1 oz	100	110	110
Brick cheese	28	1 oz	105	122	128
Brie cheese	28	1 oz	95	56	53
Camembert	28	1 oz	85	115	98
Caraway	28	1 oz	107	130	139
Cheddar cheese	28	1 oz	114	128	146
Cheshire	28	1 oz	110	119	131
Colby cheese	28	1 oz	112	115	129
Cottage cheese:					
Lowfat 1%	226	1 c	164	184	302
Lowfat 2%	226	1 c	205	166	340
Creamed, large curd	225	1 c	235	126	297
Creamed, small curd	210	1 c	215	129	277
Creamed, with fruit	226	1 c	279	85	236
Dry curd	145	1 c	123	123	151
Cream cheese (1 T = 15g)	28	1 oz	99	30	30
Edam	28	1 oz	101	150	152
Feta	28	1 oz	75	128	96
Fontina	28	1 oz	110	155	170
Gjetost	28	1 oz	132	95	126
Gorgonzola	28	1 oz	111	109	121
Gouda	28	1 oz	101	153	155
Gruyere	28	1 oz	117	147	172
Liederkranz	28	1 oz	87	115	100
Limburger	28	1 oz	93	119	111
Monterey jack	28	1 oz	106	119	126
Mozzarella:					
Part skim, low moisture	28	1 oz	80	186	149
Whole milk, low moisture	28	1 oz	90	130	117
Muenster	28	1 oz	104	128	133
Neufchatel	28	1 oz	74	53	39
Parmesan , grated (1 T = 5g)	5	1 T	23	174	40
Pimento processed cheese	28	1 oz	106	199	211
Port du salut	28	1 oz	100	102	102
Provolone	28	1 oz	100	141	141
Ricotta, made with:					
Part skim milk	246	1 c	340	132	449
Whole milk	246	1 c	428	91	389
Romano, grated	28	6 T	128	195	250
Roquefort	28	1 oz	105	115	121
Swiss cheese	28	1 oz	107	160	171
Swiss, processed	28	1 oz	95	227	216
Swiss cheese food	28	1 oz	92	162	149
Tilsit cheese	28	1 oz	96	148	142

CREAM, SWEET, fluid:

	Wt (g)	Svg	Cal	100 Cal	Svg
Coffee or table	240	1 c	469	41	192
Half and half	242	1 c	315	73	230
Light whipping cream	239	1 c	699	21	146
Heavy whipping cream	238	1 c	821	18	149

CREAM, SWEET, whipped:

	Wt (g)	Svg	Cal	100 Cal	Svg
Heavy cream, unsweetened	119	1 c	410	18	75
Pressurized	60	1 c	154	35	54

CREAM, SOUR, dairy:

	Wt (g)	Svg	Cal	100 Cal	Svg
Cultured, dairy	230	1 c	493	40	195
Half and half, dairy	15	1 T	20	70	14
Cream, sour, Imitation (non-dairy)	230	1 c	479	21	102

CREAM SUBSTITUTES (non-dairy):

	Wt (g)	Svg	Cal	100 Cal	Svg
Coffee whitener, liquid	15	1 T	20	47	10
Coffee whitener, powder	5.8	1 T	32	77	25
Dessert Toppings:					
Frozen (like Coolwhip)	75	1 c	239	3	6
Dessert powder, dry	43	1.5 oz	245	13	31
Pressurized, nondairy	70	1 c	185	7	13
Kefir beverage	233	1 c	160	199	319

MILK (cow):

	Wt (g)	Svg	Cal	100 Cal	Svg
Fresh, fluid:					
Skim	245	1 c	86	287	247
Lowfat 1%	244	1 c	102	230	235
Lowfat 2%	244	1 c	121	192	232
Whole (3.3% fat)	244	1 c	150	152	228
Buttermilk	245	1 c	99	221	219
Canned:					
Skim, evaporated	255	1 c	200	249	497
Whole, evaporated	252	1 c	340	150	510
Sweetened condensed	306	1 c	982	79	775
Dry, nonfat, instant	68	1 c	244	275	670
Dry, buttermilk	120	1 c	464	241	1119
Milk (other):					
Goat milk	244	1 c	168	161	270
Human breast milk	246	1 c	171	20	34
Soy milk	240	1 c	79	148	117

MILK BEVERAGES & MIXES:

	Wt (g)	Svg	Cal	100 Cal	Svg
Carob flavor mix, w/milk	256	1 c	195	117	228
Chocolate flavored to be mixed with water:					
Powder (includes dry milk)	28	1 oz	100	88	88
Drink	206	3/4 c	100	88	88
Chocolate flavored to be mixed with milk:					
Powder	21.6	3/4 oz	75	37	28
Drink	266	1 c	226	113	256
Eggnog, commercial	254	1 c	342	81	278
Instant breakfast, dry	37	1 env	130	11	15
Malted milk, w/whole milk:					
Chocolate flavor	265	1 c	229	116	265
Natural flavor	265	1 c	237	128	303
Milkshakes, 10 fl oz = 1.25 c:					
Chocolate	283	1.25 c	360	80	288
Strawberry	283	1.25 c	319	89	283
Vanilla	283	1.25 c	314	92	289

MILK DESSERTS:

	Wt (g)	Svg	Cal	100 Cal	Svg
Custard, baked	265	1 c	305	102	310
Custard, from mix	143	1/2 c	161	108	174
Ice cream bars:					
Creamsicle	66	1 ea	103	36	37
Drumstick	60	1 ea	186	32	59
Fudgesicle	73	1 ea	91	109	99
Ice cream, vanilla:					
Regular	133	1 c	269	50	134
Rich	148	1 c	349	33	115
Soft serve	173	1 c	377	53	199
Ice milk, regular	131	1 c	184	70	129
Ice milk, soft serve, 3% fat	175	1 c	223	91	202
PUDDINGS, prepared (5 oz can ≈ 1/2 c):					
Assorted, low calorie	130	1/2 c	69	199	137
Chocolate:					
Cooked from mix	260	1 c	300	80	240
Instant from mix	260	1 c	310	212	658
Canned, 5 oz	142	1 can	205	57	117
Coconut, instant	149	1/2 c	184	159	292
Lemon, instant	149	1/2 c	178	183	326
Rice, cooked or instant	141	1/2 c	165	71	118

ESHA Research

	Wt (g)	Svg	Cal	Phos 100 Cal	Phos Svg
Dairy & Dairy Products, continued:					
MILK DESSERTS, Pudding, continued:					
Tapioca:					
Prep. from mix	130	1/2 c	145	71	103
Canned, 5 oz	142	1 can	160	71	113
Vanilla:					
Cooked from mix	130	1/2 c	145	70	102
Instant from mix	130	1/2 c	150	182	273
Canned, 5 oz	142	1 can	220	43	94
Pudding pops:					
Banana	57	1 ea	94	67	63
Butterscotch	57	1 ea	94	67	63
Chocolate	57	1 ea	99	76	75
Chocolate fudge	57	1 ea	99	80	79
Vanilla	57	1 ea	93	68	63
Sherbet (2% fat)	193	1 c	270	27	74
Yogurt, frozen, average	174	1 c	220	891	1960
YOGURT:					
Lowfat, plain	227	1 c	144	226	326
Lowfat, with fruit	227	1 c	231	141	325
Lowfat, coffee or vanilla	227	1 c	193	159	306
Nonfat milk	227	1 c	127	279	354
Whole milk	227	1 c	138	157	216
Yogurt cheese, recipe	208	1 c	222	243	540

Eggs

	Wt (g)	Svg	Cal	Phos 100 Cal	Phos Svg
Egg, chicken, raw/cooked:					
Whole	50	1 ea	77.5	112	86
White only	33.4	1 ea	17	24	4
Yolk only	16.6	1 ea	59	137	81
Egg substitutes (can vary, check label):					
Frozen	60	1/4 c	96	45	43
Liquid	251	1 c	211	144	304
Powder	10	.35 oz	44	107	47

Fruits & Fruit Juices

	Wt (g)	Svg	Cal	Phos 100 Cal	Phos Svg
Apple, 2.75" diam, with or without peel	138	1 ea	80	13	10
Applesauce	244	1 c	106	17	18
APRICOTS:					
Fresh, pitted	106	3 ea	51	41	21
Canned, juice pack	84	3 ea	40	42	17
Canned, heavy syrup	85	3 ea	70	15	10
Dried halves	35	10 ea	83	49	41
Avocado, whole:					
California, 227g w/refuse	173	1 ea	305	24	73
Florida, 454g w/refuse	304	1 ea	340	35	119
Banana, 8.75" long, 176g w/peel	114	1 ea	105	21	22
Black currant juice	250	1 c	138	29	40
Blackberries:					
Fresh berries	144	1 c	74	41	30
Frozen, unthawed	151	1 c	97	47	46
Canned	256	1 c	236	15	36
Blackberry juice, fresh	250	1 c	93	32	30
Blueberries:					
Fresh berries	145	1 c	82	18	15
Canned	256	1 c	225	12	26
Boysenberries, frozen	132	1 c	66	55	36
Breadfruit, fresh	220	1 c	227	29	66
Cantaloupe: see Melon.					
CHERRIES, SWEET:					
Fresh, whole, pitted	145	1 c	104	27	28
Frozen, thawed	259	1 c	232	18	41
Canned with liquid	257	1 c	213	22	46

	Wt (g)	Svg	Cal	Phos 100 Cal	Phos Svg
Currants:					
Black, fresh	112	1 c	71	93	66
Red or white, fresh	112	1 c	63	78	49
Dried (Zante)	144	1 c	407	44	180
Dates, whole, pitted	83	10 ea	228	14	33
Figs, dried	187	10 ea	477	27	128
Fruit cocktail, canned:					
Juice pack	248	1 c	115	30	34
Heavy syrup	255	1 c	185	15	28
Grapes:					
Fresh, Thompson seedless	50	10 ea	35	20	7
Canned, heavy syrup	256	1 c	187	24	44
Grape juice, bottled or canned	253	1 c	155	17	27
GRAPEFRUIT, half = 241 w/peel:					
Half, pink/red	123	1 ea	37	29.7	11
Half, white	118	1 ea	39	23.1	9
Canned sections	254	1 c	152	16.4	25
Grapefruit juice:					
Fresh juice	247	1 c	96	39	37
Prepared fr/frozen	247	1 c	102	33	34
Canned, unsweetened	247	1 c	93	29	27
Kiwi fruit, fresh	76	1 ea	46	65	30
Loganberries, fresh	100	2/3 c	70	37	26
Melon: also see Watermelon.					
Cantaloupe, cubes	160	1 c	57	47	27
Casaba, cubes	170	1 c	45	27	12
Honeydew, cubes	170	1 c	60	28	17
Frozen, mixed melon balls	173	1 c	55	40	22
ORANGE, 2-5/8" dm,180g w/peel	131	1 ea	60	30	18
Orange juice:					
Fresh juice	248	1 c	111	38	42
Chilled	249	1 c	110	25	27
Canned, unsweetened	249	1 c	105	33	35
Prepared from frozen	249	1 c	110	36	40
Orange grapefruit juice, canned	247	1 c	105	33	35
PEACHES:					
Fresh slices, peeled	170	1 c	73	27	20
Frozen slices, thawed	250	1 c	235	12	28
Canned half, juice	77	1 ea	34	39	13
Canned half, heavy syrup	81	1 ea	60	15	9
Dried halves	130	10 ea	311	50	155
PEARS:					
Fresh, Bartlett, 181g w/refuse	166	1 ea	98	18	18
Canned half, juice pack	77	1 ea	38	24	9
Canned half, heavy syrup	79	1 ea	59	8	5
PINEAPPLE:					
Fresh pieces	155	1 c	76	14	11
Canned pieces, juice pack	250	1 c	150	10	15
PLUMS:					
Fresh, med, 2-1/8" diam.	66	1 ea	36	19	7
Canned, juice pack	95	3 ea	55	27	15
Canned, heavy syrup	110	3 ea	98	14	14
Prunes, dried	84	10 ea	201	33	66
Prune juice, bottled	256	1 c	181	35	64
Raisins, dark, unpacked	145	1 c	435	32	140
Raspberries:					
Fresh berries	123	1 c	60	25	15
Frozen, thawed	250	1 c	255	17	43
Raspberry juice, fresh	240	1 c	98	29	28
STRAWBERRIES:					
Fresh berries	149	1 c	45	62	28
Frozen, sweetened, thawed	255	1 c	245	13	33
Tangerine	84	1 ea	37	22	8
Watermelon, 1" x 10" diam	482	1 pce	152	27	41

	Wt (g)	Svg	Cal	Phos (mg) 100 Cal	Svg		Wt (g)	Svg	Cal	Phos (mg) 100 Cal	Svg

Grains & Grain Products
Cereals, Flours, Grains, Noodles, Pasta, Popcorn

	Wt (g)	Svg	Cal	100 Cal	Svg
Amaranth grain, dry	195	1 c	729	122	887
Barley, cooked:					
Pearled	157	1 c	193	44	85
Whole barley	200	1 c	200	85	170
Bran: see Oat, Rice, Wheat.					
Buckwheat flour, dark	98	1 c	338	88	298
Buckwheat flour, light	98	1 c	340	25	86
Bulgar wheat, cooked	182	1 c	151	48	73
CEREALS, COLD (Ready to eat)					
check label for values.					
CEREALS, HOT (cooked):					
Corn grits, cooked	242	1 c	145	20	29
Cream of rice	244	1 c	126	33	42
Cream of wheat	244	1 c	140	31	43
Farina, cooked	233	1 c	116	24	28
Maypo	180	3/4 c	128	145	186
Oatmeal, rolled oats					
Regular, Quick, Instant	234	1 c	145	123	178
Oatmeal, instant, from packet:					
Plain	177	3/4 c	104	128	133
Flavors, averaged	164	3/4 c	160	93	148
Ralston	253	1 c	134	110	148
Roman Meal, cooked	181	3/4 c	111	146	162
Wheatena	243	1 c	135	108	146
Whole wheat cereal, cooked	242	1 c	150	112	168
Corn flour:					
Regular	117	1 c	422	75	318
Masa Harina, enriched	114	1 c	416	61	254
Corn grits (hominy): see Cereals.					
Cornmeal, dry:					
Degermed, enriched	138	1 c	505	23	116
Nearly whole, bolted	122	1 c	441	62	272
FLOUR: see specific grain,					
nut, or vegetable.					
Macaroni, cooked:					
Enriched	140	1 c	197	39	76
Vegetable, enriched	134	1 c	172	39	67
Whole wheat	140	1 c	174	72	125
Millet, cooked	120	1/2 c	143	84	120
Noodles:					
Chow mein noodles, dry	45	1 c	237	30	72
Egg noodles, cooked, enriched	160	1 c	213	52	110
Spinach noodles, cooked	140	1 c	182	83	151
Oat bran, 1 T ≈ 6 g	94	1 c	132	523	690
Oats, rolled, ckd: see Cereals, hot.					
Oats, rolled, dry	81	1 c	311	123	384
PASTA: see Macaroni, Noodles, Spaghetti.					
Popcorn, cooked in oil	11	1 c	55	56	31
Quinoa grain, dry	170	1 c	635	110	697
RICE, cooked:					
Brown rice	195	1 c	217	75	162
White, converted	175	1 c	200	48	96.4
White, regular	205	1 c	264	28	73.5
White, instant	165	1 c	162	14	23.1
Wild rice	164	1 c	166	81	135
Rice bran	83	1 c	262	531	1392
Rice polish	158	1 c	578	27	155
Rye flour:					
Dark	128	1 c	415	195	809
Light	102	1 c	361	55	198

	Wt (g)	Svg	Cal	100 Cal	Svg
Soy flour, stirred:					
Defatted	100	1 c	327	206	674
Low-fat	88	1 c	326	160	522
Full fat, raw	85	1 c	368	114	420
Spaghetti noodles, cooked:					
Enriched	140	1 c	197	39	76
Whole wheat spaghetti	140	1 c	174	72	125
WHEAT:					
Wheat bran	30	1/2 c	65	468	304
FLOUR:					
All purpose, white, unsifted	125	1 c	455	30	135
Cake flour, sifted	96	1 c	348	23	81.6
Gluten flour	140	1 c	529	37	196
Self rising flour	125	1 c	442	168	744
Semolina flour	167	1 c	601	38	227
Whole wheat flour	120	1 c	407	102	415
Wheat germ:					
Raw	100	1 c	360	234	842
Toasted	113	1 c	432	300	1295
Wheat, rolled:					
Cooked	240	1 c	142	92	130
Dry	85	1 c	289	93	270
Whole grain wheat					
(wheatberries) cooked	50	1/3 c	28	93	26
Whole grain wheat, sprouted	108	1 c	214	101	216

Meats: Fish & Shellfish

	Wt (g)	Svg	Cal	100 Cal	Svg
Abalone, fried	85	3 oz	161	99	160
Anchovies, canned in oil	45	11 ea	95	119	113
Bass, baked/broiled	100	3.5 oz	125	173	216
Bluefish:					
Baked/broiled	100	3.5 oz	159	182	290
Fried in crumbs	100	3.5 oz	205	139	285
Carp, baked/broiled	100	3.5 oz	162	328	531
Catfish, fried in cornmeal	100	3.5 oz	229	94	216
Caviar (1 T ≈ 16g)	100	3.5 oz	252	141	355
Clams:					
Breaded, fried, small	188	20 ea	379	93	353
Steamed, meat only	90	20 ea	133	229	304
Canned, drained	160	1 c	236	229	540
Canned, w/liquid, small can	183	1 can	145	420	609
COD:					
Baked or broiled	100	3.5 oz	105	131	138
Batter fried	100	3.5 oz	199	101	200
Smoked	100	3.5 oz	79	241	190
CRAB, cooked meat:					
Alaska King crab leg	134	1 ea	129	291	376
Blue crab, cooked	135	1 c	138	201	278
Blue crab, canned	135	1 c	133	264	351
Dungeness, cooked	101	3/4 c	85	216	184
Crab, imitation from surimi	85	3 oz	87	133	116
Crab cakes, from recipe	60	1 ea	93	138	128
Crayfish, cooked, moist heat	85	3 oz	97	289	280
Eel, baked/broiled	100	3.5 oz	236	117	277
Eel, smoked	100	3.5 oz	330	64	211
Fish cakes, from frozen	100	3.5 oz	213	52	110
Fish sticks, heated fr/frozen	57	2 ea	155	66	103
Gefiltefish, sweet	42	1 pce	35	89	31
HADDOCK:					
Baked / broiled	85	3 oz	95	216	205
Breaded, fried	85	3 oz	175	105	183
Smoked	100	3.5 oz	116	216	251
Halibut, baked / broiled	85	3.5 oz	119	203	242

ESHA Research

	Wt (g)	Svg	Cal	Phos (mg) 100 Cal	Svg		Wt (g)	Svg	Cal	Phos (mg) 100 Cal	Svg
Fish & Shellfish, continued:						**BEEF**, continued:					
Herring:						**Ground beef**, cooked average:					
Baked / broiled	100	3.5 oz	203	149	303	Extra lean/Lean	85	3 oz	223	57	126
Canned in oil, w/liquid	100	3.5 oz	208	143	297	Regular	85	3 oz	250	54	135
Smoked / kippered	100	3.5 oz	217	150	325	**Rib, choice**, oven roasted:					
Pickled	100	3.5 oz	262	34	89	Lean and fat (5 oz raw)	85	3 oz	324	44	144
Lobster, meat only, cooked	145	1 c	142	189	268	Lean only	85	3 oz	204	89	181
Mackerel:						**Round steak**, broiled:					
Baked/broiled, Atlantic	100	3.5 oz	262	106	278	Lean and fat (4.5 oz raw)	85	3 oz	233	77	179
Canned, Jack, 1 tall can	361	1 can	563	193	1087	Lean only	85	3 oz	165	122	201
Mullet, baked / broiled	85	3 oz	127	163	207	**Round tip**, roasted:					
Ocean perch:						Lean and fat	85	3 oz	213	88	188
Baked / broiled	100	3.5 oz	121	229	277	Lean only	85	3 oz	162	127	205
Breaded, fried	85	3 oz	185	103	191	**Sirloin steak**, broiled, 11.3 oz raw steak =					
						8.2 oz lean & fat cooked; 6.9 oz lean only:					
OYSTERS:						Lean and fat	85	3 oz	238	71	168
Raw, Eastern	248	1 c	170	202	344	Lean only	85	3 oz	172	108	185
Raw, Pacific	248	1 c	200	201	402	**T-bone steak**, broiled, 16 oz raw = 9.7 oz lean &					
Breaded, fried, Eastern	88	6 med	173	81	140	fat cooked; 7.4 oz lean:					
Simmered, Eastern	100	3.5 oz	137	203	278	Lean and fat	85	3 oz	276	54	150
Perch, baked / broiled	85	3 oz	100	219	218	Lean only	85	3 oz	182	97	177
Pollock, baked / broiled	100	3.5 oz	99	255	252	Dried beef, cured, 6-7 pces	28	1 oz	47	104	49
Rockfish, baked / broiled	100	3.5 oz	121	188	228	Corned beef, canned	85	3 oz	213	44	94
						Variety meats:					
SALMON, cooked:						Brains, pan fried	85	3 oz	167	196	328
Broiled / baked	85	3 oz	183	128	234	Heart, simmered	85	3 oz	140	152	213
Smoked, Chinook	85	3 oz	99	140	139	Kidney, cooked	140	1 ea	201	213	428
Canned, Atlantic, small can	220	1 can	281	141	396	Liver, fried	85	3 oz	184	213	392
Pink, drained, #1 can	454	1 can	631	190	1202	Tongue, cooked	85	3 oz	241	50	121
Sardines, canned, drained:						**HAM**: see Pork, cured;					
Atlantic, 2 sardines = 24g	92	1 can	192	235	451	Lunchmeats group & Turkey ham.					
Pacific, 1 sardine = 38g	100	3.5 oz	178	206	366						
Scallops:						**LAMB**:					
Breaded, fried	93	6 ea	200	110	219	**Arm chop**, braised, raw = 5.6 oz w/ bone:					
Steamed	100	3.5 oz	113	236	267	Lean and fat	70	1 ea	244	59	145
Sea Bass, baked / broiled	100	3.5 oz	124	200	248	Lean part	55	1 ea	152	84	127
Seatrout, Steelhead, cooked	100	3.5 oz	131	220	288	**Loin chop**, broiled, raw = 4.2 oz w/bone:					
Shad, baked with bacon	100	3.5 oz	201	156	313	Lean and fat	64	1 ea	201	62	125
Shark, batter-fried	85	3 oz	194	85	165	Lean only	46	1 ea	100	105	105
						Cutlet, lean, cooked average	85	3 oz	175	102	179
SHRIMP:						**Leg of lamb**, roasted:					
Boiled, 2 large ≈ 11g	100	3.5 oz	99	138	137	Lean and fat	85	3 oz	219	74	162
Broiled, fried, 2 large ≈ 15g	90	12 ea	218	90	196	Lean only	85	3 oz	162	108	175
Canned with liquid	100	3.5 oz	102	149	152	**Shoulder roast**:					
Canned, drained	128	1 c	154	194	299	Lean and fat	85	3 oz	235	66	156
Snapper, baked / broiled	100	3.5 oz	128	157	201	Lean only	85	3 oz	173	98	170
SOLE (Flounder):						**PORK**:					
Baked / broiled	85	3 oz	99	248	246	Bacon:					
Batter-fried	85	3 oz	250	64	161	Regular, cooked	19	3 pce	109	59	64
Breaded, fried	100	3.5 oz	188	112	210	Canadian style	47	2 pce	86	160	138
Squid, fried in flour	85	3 oz	149	143	213	Breakfast strips, cooked	34	3 pce	156	58	90
Swordfish, broiled / baked	100	3.5 oz	155	217	337	**Center loin chop**:					
Trout, baked / broiled	85	3 oz	129	211	272	Braised, lean & fat	75	1 ea	266	61	161
						Braised, lean only	61	1 ea	166	92	153
TUNA, drained, No. 1/2 can:						Broiled, lean & fat	82	1 ea	284	65	184
Light, canned in oil	171	1 can	339	157	532	Broiled, lean only	72	1 ea	166	106	17
Light, canned in water	165	1 can	216	142	306	Pan-fried, lean & fat	89	1 ea	333	57	191
						Pan-fried, lean only	67	1 ea	178	100	178
						Roasted, lean & fat	88	1 ea	268	65	173
Meats						Roasted, lean only	72	1 ea	180	91	164
Beef, Pork, Ham, etc.						**Center rib chop**:					
						Braised, lean & fat	67	1 ea	246	57	141
BEEF:						Braised, lean only	53	1 ea	147	90	133
Breakfast strips,						Broiled, lean & fat	77	1 ea	264	66	175
cured beef, cooked	34	3 strips	153	52	80	Broiled, lean only	63	1 ea	162	104	168
Chuck blade, pot roasted:						Pan-fried, lean & fat	88	1 ea	343	53	183
						Pan-fried, lean only	62	1 ea	160	105	168
Lean and fat (5.4 oz raw)	85	3 oz	325	50	162	Roasted, lean & fat	79	1 ea	252	70	177
Lean only	85	3 oz	230	87	200	Roasted, lean only	66	1 ea	162	104	169

	Wt (g)	Svg	Cal	Phos (mg) 100 Cal	Phos (mg) Svg
PORK, continued:					
Pork roast, leg:					
Lean and fat	85	3 oz	250	84	210
Lean only	85	3 oz	187	128	239
Pork roast, rib / loin:					
Lean and fat	85	3 oz	265	68	179
Lean only	85	3 oz	206	98	201
Shoulder, braised, yield from 6.8 oz raw w/bone:					
Lean and fat	85	3 oz	293	55	162
Lean only	67	2.4 oz	166	91	151
Spareribs from 1 lb raw w/bone	177	6.25 oz	703	66	462
PORK, CURED - HAM (also see Bacon under Pork):					
Roasted, lean and fat	140	1 c	341	88	300
Roasted, lean only	140	1 c	219	145	318
Canned, roasted	85	3 oz	140	134	188
RABBIT, roasted	85	3 oz	131	134	175
VEAL (calf):					
Cutlet, lean, cooked avg	85	3 oz	166	128	213
Rib, roasted	85	3 oz	151	117	176
Liver, pan-fried	85	3 oz	208	179	373
VENISON (deer) roasted	85	3 oz	134	143	192

Meats: Poultry

CHICKEN: A 3 lb chicken ≈ 1.45 lbs raw meat, ≈ 1.1 lbs cooked.

	Wt (g)	Svg	Cal	Phos (mg) 100 Cal	Phos (mg) Svg
All meats:					
Fried	140	1 c	307	93	287
Roasted	140	1 c	266	103	273
Stewed	140	1 c	248	85	210
Canned, boned with broth	142	5 oz	235	67	158
Dark meat only:					
Fried	85	3 oz	203	78	158
Roasted	85	3 oz	174	87	152
Stewed	85	3 oz	163	75	122
Light meat only:					
Fried	85	3 oz	163	121	196
Roasted	85	3 oz	147	125	183
Stewed	85	3 oz	135	100	135
Breast*, meat & skin (145g raw ≈181g w/bone):					
Batter-fried	140	1 ea	364	71	258
Flour-fried	98	1 ea	218	105	228
Roasted	98	1 ea	193	109	210
Stewed	110	1 ea	202	85	172
Breast* meat = 118g raw:					
Fried	86	1 ea	161	132	212
Roasted	86	1 ea	142	138	196
* Two pieces per bird					
Drumstick, meat & skin (73g raw ≈110g w/bone):					
Batter-fried	72	1 ea	193	55	106
Flour-fried	49	1 ea	120	72	86
Roasted	52	1 ea	112	81	91
Stewed	57	1 ea	116	69	80
Drumstick, meat = 62g raw:					
Fried	42	1 ea	82	95	78
Roasted	44	1 ea	76	107	81
Thigh, meat and skin (94g raw ≈120g w/bone):					
Batter-fried	86	1 ea	238	56	134
Flour-fried	62	1 ea	162	72	116
Roasted	62	1 ea	153	71	108
Stewed	68	1 ea	158	59	94
Thigh, meat = 69g raw:					
Fried	52	1 ea	113	91	103
Roasted	52	1 ea	109	87	95
Stewed	55	1 ea	107	77	82

	Wt (g)	Svg	Cal	Phos (mg) 100 Cal	Phos (mg) Svg
CHICKEN, continued:					
Wing meat & skin (49g raw ≈ 90g w/bone):					
Batter-fried	49	1 ea	159	37	59
Flour-fried	32	1 ea	103	47	48
Roasted	34	1 ea	99	51	51
Stewed	40	1 ea	100	48	48
Wing meat = 29g raw:					
Fried	20	1 ea	42	79	33
Roasted	21	1 ea	43	81	35
Stewed	24	1 ea	43	74	32
DUCK, domestic, roasted:					
Meat and skin	85	3 oz	286	46	132
Meat only	85	3 oz	171	101	173
GOOSE, domestic, roasted:					
Meat and skin	85	3 oz	259	89	230
Meat only	85	3 oz	202	130	263
TURKEY:					
Breast meat, barbecued	28	1 oz	40	185	74
Breast meat, hickory smoked	28	1 oz	35	226	79
Ground turkey, cooked	100	3.5 oz	229	86	196
Roasted:					
All types	140	1 c	238	125	298
Dark meat	85	3 oz	159	109	174
White meat	85	3 oz	133	140	186
Frozen slices w/gravy	142	5 oz	95	121	115
Turkey patty, breaded, fried	64	1 ea	181	96	173

Meats: Sausages & Lunchmeats

	Wt (g)	Svg	Cal	Phos (mg) 100 Cal	Phos (mg) Svg
Barbecue loaf, pork/beef	23	1 pce	40	75	30
Beef lunchmeat, loaf/roll	28	1 oz	87	39	34
Beef, thin sliced	28	1 oz	50	96	48
Berliner sausage	23	1 pce	53	57	30
Bologna:					
Beef and pork	28	1 oz	89	29	26
Cured pork	23	1 pce	57	58	33
Turkey (1 pce = 1 ounce)	28	1 oz	56	66	37
Braunschweiger sausage	57	2 oz	205	47	96
Brotwurst, link	70	1 ea	226	42	94
Cheesefurter (cheese smoki)	43	1 ea	141	54	76
Chicken roll, light meat	57	2 oz	90	99	89
Dutch brand loaf	28	1 oz	68	68	46
FRANKFURTER (hotdog):					
Beef, 8 /pkg	57	1 ea	184	26	47
Beef and pork, 8 /pkg	57	1 ea	183	27	49
Chicken, 10 /package	45	1 ea	115	42	48
Turkey, 10 /package	45	1 ea	102	81	83
HAM:					
Chopped ham	42	2 pce	98	59	58
Ham, extra lean	57	2 oz	75	165	124
Ham, regular	57	2 oz	103	136	140
Minced	21	1 pce	55	60	33
Ham patty, cooked	60	1 ea	203	30	60
Ham & cheese roll/loaf	57	2 oz	147	97	143
Ham salad spread	240	1 c	518	55	286
Hotdog: see Frankfurter					
Italian sausage, cooked	67	1 link	216	53	114
Keilbasa sausage	26	1 pce	81	47	38
Knockwurst sausage	68	1 link	209	32	67
Liverwurst, pork	18	1 pce	59	69	41
Luncheon sausage, beef & pork	23	1 pce	60	47	28
Luxury loaf	57	2 oz	80	131	105
Olive loaf	57	2 oz	133	54	72

ESHA Research

	Wt (g)	Svg	Cal	Phos (mg) 100 Cal	Phos (mg) Svg
Meats: Sausages & Lunchmeats, continued:					
Pastrami:					
Cured beef	57	2 oz	198	43	85
Cured turkey	57	2 oz	74	192	142
Peppered loaf	28	1 oz	42	114	48
Pepperoni, small slice	22	4 pce	109	24	26
Pickle & pimento loaf	57	2 oz	149	53	79
Polish sausage	28	1 oz	92	42	39
Pork sausage patty, cooked	27	1 pce	100	50	50
Salami, pork and beef	57	2 oz	143	46	65
Salami, turkey	57	2 oz	111	66	73
Smoked link sausage:					
Beef and pork	68	1 ea	229	32	73
Pork	68	1 ea	265	42	110
Turkey	28	1 oz	55	67	37
Turkey:					
Breakfast sausage	28	1 oz	65	80	52
Ham	57	2 oz	73	189	138
Loaf, breast meat	43	2 pce	46	211	97
Roll, light and dark	57	2 pce	84	113	95
Roll, light meat	57	2 pce	83	125	104
Summer sausage	28	1 oz	50	136	68

Mixed Dishes & Fast Foods

	Wt (g)	Svg	Cal	Phos (mg) 100 Cal	Phos (mg) Svg
Beef & vegetable stew:					
Recipe	245	1 c	220	84	184
Canned	245	1 c	194	29	56
Beef, macaroni, & tomato sauce, recipe	226	1 c	189	62	118
Beef pot pie, from frozen	234	1 ea	426	28	121
BURRITO:					
Bean	174	1 ea	322	75	243
Beef	177	1 ea	463	66	306
Beef and bean	175	1 ea	390	70	274
Deluxe combination	198	1 ea	424	68	289
Cheese soufflé, recipe	112	1 c	221	99	219
Chicken & noodles, recipe	240	1 c	365	68	247
Chicken a la king, recipe	245	1 c	470	76	358
Chicken chow mein:					
Recipe	250	1 c	255	115	293
Canned	250	1 c	95	89	85
Chicken pot pie, fr/frozen	230	1 ea	430	41	177
Chicken salad w/celery	78	1/2 c	266	30	80
Chili w/beans, canned	255	1 c	286	137	393
Chop suey w/beef & pork	250	1 c	300	83	248
Cole slaw	120	1 c	84	45	38
Corn dog	111	1 ea	330	92	303
Corn fritter, recipe	45	1 ea	116	44	51
Corn pudding	250	1 c	271	53	143
Corned beef hash, canned	220	1 c	382	38	147
Egg salad	183	1 c	438	64	282
Enchilada:					
Beef	120	1 ea	292	54	159
Cheese	120	1 ea	330	79	260
Chicken	120	1 ea	269	63	170
French toast, recipe	65	1 pce	123	67	82
LASAGNA:					
With meat	245	1 pce	398	99	393
Without meat	218	1 pce	316	109	345
Macaroni & cheese:					
Recipe	200	1 c	430	75	322
Canned	240	1 c	230	79	182
Macaroni salad, w/o cheese	141	1 c	371	14	50
Manicotti, frozen entree	225	1 ea	271	97	264
Meat loaf, average	85	3 oz	198	62	123
Moussaka, lamb & eggplant	250	1 c	250	98	245

	Wt (g)	Svg	Cal	Phos (mg) 100 Cal	Phos (mg) Svg
Pies, fried pastry, commercial:					
Apple, fried pastry	85	1 ea	255	13	34
Cherry, fried pastry	85	1 ea	250	16	41
PIZZA, cheese:					
Regular crust, 1/8 of 15"	120	1 pce	290	74	216
Thick crust, 1/2 of 10"	208	1 pce	519	67	346
Potato salad w/ mayo & eggs	250	1 c	358	36	130
Quiche Lorraine, 1/8 pie	176	1 pce	600	46	276
Ravioli, beef, canned	226	1 c	220	49	108
SANDWICHES, Fast Food items:					
Cheeseburger, 3 oz meat	112	1 ea	300	58	174
Cheeseburger, 4 oz meat	194	1 ea	524	61	320
Chicken patty sandwich	157	1 ea	436	40	173
English muffin w/egg, cheese & bacon	138	1 ea	360	81	290
Fish sandwich:					
Regular, with cheese	140	1 ea	420	53	223
Large, without cheese	170	1 ea	470	52	246
Hamburger, 3 oz meat	98	1 ea	245	44	107
Hamburger, 4 oz meat	174	1 ea	445	51	225
Hotdog (frankfurter) & bun	85	1 ea	260	32	83
SANDWICHES, on part whole wheat bread, except when stated as rye:					
Avocado, cheese, tomato & sprouts	195	1 ea	432	63	274
Bacon, lettuce & tomato	135	1 ea	327	55	181
Chicken salad sandwich	100	1 ea	294	49	144
Corned beef & swiss on rye	147	1 ea	429	72	310
Grilled cheese	117	1 ea	393	135	531
Ham sandwich	122	1 ea	256	91	234
Ham on rye	116	1 ea	242	84	203
Ham and cheese	151	1 ea	363	123	447
Ham and swiss	145	1 ea	350	107	376
Patty melt, ground beef on rye	177	1 ea	567	75	423
Peanut butter & jam	100	1 ea	341	57	195
Reuben sandwich, grilled	233	1 ea	480	68	328
Roast beef	122	1 ea	280	71	200
Tuna salad sandwich	116	1 ea	303	58	176
Turkey sandwich	122	1 ea	271	87	235
Turkey ham	122	1 ea	253	97	245
Turkey ham on rye	116	1 ea	239	90	214
SPAGHETTI (pasta, & tomato sce w/cheese):					
Homemade	250	1 c	260	52	135
Canned	250	1 c	190	46	88
SPAGHETTI, & tomato sce w/meat:					
Homemade	248	1 c	330	72	236
Canned	250	1 c	260	43	113
Spinach soufflé	136	1 c	218	106	231
Taco, beef	78	1 ea	207	68	141
Taco, chicken	78	1 ea	172	91	156
TOSTADA:					
With beans & beef	192	1 ea	332	74	247
With beans & chicken	157	1 ea	249	97	242
With refried beans	157	1 ea	212	92	195
Tuna noodle casserole, recipe	202	1 c	251	73	182
Tuna salad	205	1 c	383	95	365
Turkey pot pie, frozen	233	1 ea	416	33	137
Waldorf salad	142	1 c	424	21	88

	Wt (g)	Svg	Cal	Phos (mg) 100 Cal	Phos (mg) Svg		Wt (g)	Svg	Cal	Phos (mg) 100 Cal	Phos (mg) Svg
Nuts & Seeds						Spaghetti sauce, plain:					
						Recipe	220	1 c	179	48	86
						Canned	249	1 c	272	33	90
Almonds, dried, whole	142	1 c	837	88	738	Spaghetti sauce with meat:					
Brazil nuts, dry	140	1 c	919	91	840	Recipe	248	1 c	297	46	136
Cashew butter	16	1 T	94	78	73	Canned	206	.8 c	220	48	106
Cashew nuts, oil roasted	130	1 c	748	74	554	White sauce:					
Chestnuts, roasted	143	1 c	350	44	153	Recipe, medium	250	1 c	395	60	238
Coconut:						From mix w/milk	264	1 c	240	107	256
Raw, grated	80	1 c	283	32	90						
Dried unsweetened	78	1 c	515	31	161	**SOUPS**: Prepared fr/canned unless otherwise					
Flaked, sweet, packaged	74	1 c	351	21	74	stated. RTS = Ready To Serve. For soups					
Coconut cream, raw	240	1 c	792	37	293	prep. w/milk, assume whole milk.					
Filberts (hazelnuts) whole	135	1 c	853	49	421	Bean with bacon soup	253	1 c	173	76	132
Macadamias, oil roasted	134	1 c	962	28	268	Beef broth /bouillon:					
						Prepared from dry	244	1 c	19	137	26
MIXED NUTS w/peanuts						From canned	240	1 c	16	194	31
(cashews, peanuts, brazil nuts,						From dry	244	1 c	19.5	125	24.4
filberts, almonds, pecans):						Beef noodle soup:					
Dry roasted	137	1 c	814	73	596	Canned	244	1 c	84	55	46
Oil roasted	142	1 c	876	75	659	Canned, chunky, RTS	240	1 c	171	70	120
MIXED NUTS, w/o peanuts						From dry mix	251	1 c	41	98	40
(cashews, almonds, brazil nuts,						Celery, cream of, w/milk	248	1 c	165	92	151
pecans, filberts):						Chicken broth, from canned	244	1 c	39	187	73
Oil roasted	144	1 c	886	73	646	Chicken, cream of:					
						Prepared with milk	248	1 c	191	80	152
PEANUTS:						From dry mix	261	1 c	107	90	96
Dry roasted	146	1 c	855	62	523	**Chicken noodle**:					
Oil roasted	144	1 c	837	89	744	Canned	241	1 c	75	48	36
Peanut butter	32	2 T	188	55	103	Canned, Chunky, RTS	251	1 c	178	63	113
Peanut flour, defatted	60	1 c	196	233	456	Prepared from dry	252	1 c	53	62	33
Pecans, dried, chopped	119	1 c	794	44	346	Chicken vegetable:					
Pine nuts, pignola, dried	28	1 oz	146	99	144	Canned	241	1 c	74	55	41
Pistachios, w/o shells, dried	128	1 c	739	87	644	Canned, chunky, RTS	240	1 c	167	25	41
Pumpkin/squash seeds:						Prepared from dry	251	1 c	49	67	33
Kernels, dried	138	1 c	747	217	1620	Chili beef	250	1 c	169	88	148
Kernels, roasted	227	1 c	1185	219	2600	Clam chowder:					
Whole, roasted	64	1 c	285	21	59	New England	248	1 c	163	96	157
Sesame seeds:						Manhatten, RTS	240	1 c	133	63	84
Kernels, dried	150	1 c	882	132	1164	Lentil and ham, RTS	248	1 c	140	131	184
Whole, dried	36	1/4 c	206	110	227	Minestrone	241	1 c	80	70	56
Sesame flour:						Mushroom, cream of:					
High fat	28	1 oz	149	154	229	Prepared from dry	253	1 c	96	80	77
Lowfat	28	1 oz	95	226	215	Condensed, undiluted	251	1 c	257	33	84
Soybeans, roasted	86	1/2 c	405	77	312	Potato, cream of, w/milk	248	1 c	148	108	160
Sunflower seed kernels:						Split pea soup:					
Dry roasted	128	1 c	745	198	1478	With or without ham	253	1 c	189	113	213
Oil roasted	135	1 c	830	185	1538	Prepared f/dry mix	255	1 c	133	101	134
Tahini, sesame butter	15	1 T	91	131	119	**Tomato**, cream of:					
Walnuts, Black, chopped	125	1 c	759	76	580	Prepared with milk	248	1 c	160	93	148
Walnuts, English, chopped	120	1 c	770	49	380	Prepared f/dry	265	1 c	102	65	66
						Tomato rice	247	1 c	120	28	33
						Tomato vegetable, f/dry	253	1 c	55	53	29
Soups, Sauces & Gravies						Turkey noodle:					
						From canned	244	1 c	69	70	48
						Chunky, RTS	236	1 c	136	172	234
GRAVIES:						Turkey vegetable	241	1 c	74	54	40
Beef gravy, canned	233	1 c	124	56	70	Vegetable beef:					
Brown gravy, from dry mix	258	1 c	74.8	59	44	Prepared w/water	244	1 c	79	52	41
Chicken gravy:						Prepared fr/dry	253	1 c	53	70	37
Recipe	130	1/2 c	163	26	43	Chunky, RTS	240	1 c	122	59	72
From dry mix	260	1 c	85	55	47	Vegetable, cream of, f/ dry	260	1 c	105	51	54
Canned	238	1 c	189	37	69	Vegetable w/beef broth	241	1 c	81	48	39
Mushroom gravy, canned	238	1 c	120	30	36	Vegetarian vegetable	241	1 c	70	50	35
SAUCES (also see Other):											
Cheese sce, fr/mix, w/milk	279	1 c	305	144	438						
Hollandaise sauce:											
Recipe	160	1 c	867	42	366						
From mix w/water	259	1 c	240	53	127						

Vegetables & Legumes

	Wt (g)	Svg	Cal	Phos 100 Cal	Phos Svg
Amaranth leaves:					
Fresh, chopped	28	1 c	7	200	14
Boiled	132	1 c	28	339	95
Artichoke, globe, cooked					
300 g with refuse	120	1 ea	60	172	103
Artichoke hearts:					
Cooked fr/frzn-pkg	240	9 oz	108	135	146
Marinated-jar	170	6 oz	168	61	102
Asparagus, pieces:					
Fresh pieces	67	1/2 c	15	233	35
Cooked from fresh	90	1/2 c	23	244	55
Cooked from frozen	180	1 c	50	196	99
Canned, drained	121	1/2 c	16	394	63
Canned, with liquid	122	1/2 c	17	271	46
Bamboo shoots, sliced, canned	131	1 c	25	132	33
BEANS: see also Garbanzo,					
Lentils, Soybeans.					
Baked beans (dry white					
beans w/spices & sauce):					
Home prepared	253	1 c	382	72	275
Canned, plain/vegetarian	254	1 c	235	112	264
Canned, w/franks	257	1 c	366	73	267
Canned, w/pork	253	1 c	268	102	274
Canned, w/sweet sce	253	1 c	282	94	266
Canned, w/tomato sce	253	1 c	247	120	297
Black, cooked from dry	172	1 c	227	106	241
Broadbeans:					
Cooked f/fresh veg	100	3.5 oz	56	130	73
Cooked f/dry legume	170	1 c	186	114	212
Canned legumes	256	1 c	183	110	202
Great northern f/ dry	177	1 c	210	140	293
Green (snap) beans:					
Fresh, uncooked	110	1 c	34	124	42
Cooked from fresh	125	1 c	44	109	48
Cooked from frozen	135	1 c	36	92	33
Canned, drained	135	1 c	26	100	26
Canned w/liquid	240	1 c	36	128	46
Hyacinth beans:					
Cooked f/fresh (veg)	87	1 c	43	84	36
Cooked f/dry (legume)	194	1 c	228	102	233
Kidney beans:					
Cooked f/dry	177	1 c	225	112	252
Canned, with liquid	256	1 c	208	129	269
Lima beans:					
Cooked f/fresh	170	1 c	208	106	221
Cooked f/frozen, large	85	1/2 c	85	64	54
Cooked f/frozen, baby	90	1/2 c	94	107	101
Cooked f/dry, average	185	1 c	223	98	220
Canned, drained	170	1 c	164	73	120
Canned, &liquid	241	1 c	191	93	178
Navy, ckd fr/dry	182	1 c	259	110	285
Pinto:					
Ckd fr/dry	171	1 c	235	116	273
Canned	240	1 c	186	118	220
Refried beans, canned	253	1 c	270	79	214
White, ckd from dry	179	1 c	253	119	302
Winged beans:					
Fresh slices (veg) pod = 16g	44	1 c	22	73	16
Ckd f/fresh, slices	62	1 c	23	70	16
Ckd f/dry legume	172	1 c	252	105	264
Yardlong beans:					
Fresh slices (veg) pod = 12g	91	1 c	43	126	54
Ckd f/fresh, slices	104	1 c	49	120	59
Ckd f/dry legume	171	1 c	202	153	309
Yellow wax: see green beans.					

	Wt (g)	Svg	Cal	Phos 100 Cal	Phos Svg
Bean sprouts (Mung beans):					
Fresh sprouts	104	1 c	31	179	56
Ckd f/fresh stir fried	124	1 c	62	113	70
Ckd f/fresh, boiled	124	1 c	26	131	34
Canned, drained	125	1 c	16	250	40
Beets:					
Whole, 2"dm, ckd f/fresh	100	2 ea	31	100	31
Canned, diced	85	1/2 c	27	56	15
Beet greens, ckd f/fresh	144	1 c	40	145	58
BROCCOLI, chopped:					
Fresh, uncooked	88	1 c	24	242	58
Ckd f/fresh	156	1 c	44	209	92
Ckd f/frozen	184	1 c	51	198	101
W/cheese sauce	142	1/2 c	166	66	109
W/hollandaise sauce	95	1/2 c	105	50	52
Brussels sprouts:					
Ckd f/fresh	156	1 c	60	145	87
Ckd f/frozen	155	1 c	65	129	84
CABBAGE:					
Bok choy, fresh, shredded	70	1 c	9	289	26
Bok choy, cooked	170	1 c	20	245	49
Common, fresh, shredded	70	1 c	16	100	16
Common, cooked	150	1 c	32	119	38
Pe-tsai, fresh, shredded	76	1 c	12	183	22
Pe-tsai, cooked	119	1 c	16	288	46
Red, fresh, shredded	70	1 c	19	153	29
Red, cooked	150	1 c	32	131	42
Savoy, fresh, shredded	70	1 c	20	145	29
Savoy, cooked	145	1 c	35	137	48
CARROTS:					
Fresh, whole (7.5" x 1-1/8")	72	1 ea	31	103	32
Cooked from fresh, slices	78	1/2 c	35	69	24
Cooked from frozen	73	1/2 c	26	92	24
Carrot juice	123	1/2 c	49	104	51
Cauliflower:					
Fresh pieces	50	1/2 c	12	192	23
Ckd from fresh	124	1 c	30	147	44
Ckd from frozen	180	1 c	34	126	43
Celery:					
Fresh (7.5" stalk = 40g)	40	1 stalk	6	167	10
Ckd, diced	150	1 c	27	141	38
Chard, swiss:					
Fresh, chopped	36	1 c	7	243	17
Ckd from fresh	175	1 c	35	166	58
Collards:					
Ckd f/fresh	128	1 c	35	29	10.2
Ckd f/frozen	170	1 c	63	73	46
CORN:					
Fresh kernels, uncooked	77	1/2 c	66	105	69
Ckd f/frozen, kernels	82	1/2 c	67	58	39
Canned, drained	82	1/2 c	66	98	65
Canned with liquid	128	1/2 c	79	82	65
Canned, Vacuum pack	210	1 c	166	81	134
Cream style, canned	128	1/2 c	93	70	65
Cucumber, whole, 8" x 2+"	301	1 ea	39	131	51
Dandelion greens:					
Fresh greens	55	1 c	25	144	36
Cooked	105	1 c	35	126	44
Eggplant, cooked	160	1 c	45	78	35
Garbanzo beans, chickpeas:					
Canned	240	1 c	285	76	216
Cooked from dry	164	1 c	269	102	275
Jerusalem artichoke, fresh	150	1 c	114	103	117
Kale, chopped:					
Fresh, chopped	67	1 c	33	115	38
Ckd from fresh	130	1 c	42	87	36
Ckd from frozen	130	1 c	39	92	36

Vegetables & Legumes, continued:

	Wt (g)	Svg	Cal	Phos (mg) 100 Cal	Phos (mg) Svg
Kohlrabi:					
Fresh, slices	140	1 c	38	168	64
Ckd from fresh	165	1 c	48	154	74
Leeks, chopped, fresh	104	1 c	63	57	36
Lentils, ckd from dry	198	1 c	231	154	356
Lentils, sprouted:					
Fresh sprouts	77	1 c	81	164	133
Stir fried	100	3.5 oz	101	151	153
LETTUCE, fresh, chopped:					
Butterhead	56	1 c	7	178	13
Iceberg	56	1 c	7	157	11
Looseleaf	56	1 c	10	140	14
Romaine	56	1 c	9	278	25
Lotus root, sliced, cooked	89	10 ea	59	118	69
Mushrooms:					
Fresh, sliced (1 avg = 18g)	35	1/2 c	9	416	36
Cooked from fresh	78	1/2 c	21	324	68
Canned, drained	78	1/2 c	19	278	52
Mustard greens:					
Fresh, chopped	56	1 c	15	160	24
Cooked from fresh	140	1 c	21	271	57
Cooked from frozen	150	1 c	29	126	36
Okra:					
Pods ckd f/fresh	85	8 ea	27	176	48
Slices, ckd f/frozen	92	1/2 c	34	124	42
Onions, fresh, chopped	160	1 c	61	87	53
Onions, dehydrated flakes	14	1/4 c	45	93	42
Parsnips:					
Fresh slices	133	1 c	100	94	94
Ckd fr/fresh	156	1 c	125	86	108
PEAS:					
Black-eyed peas:					
Ckd from fresh	165	1 c	160	53	84
Ckd from frozen	170	1 c	224	93	208
Ckd from dry	171	1 c	198	134	266
Canned	240	1 c	184	91	167
Green peas:					
Fresh, uncooked	145	1 c	118	133	157
Ckd from fresh	160	1 c	134	140	187
Ckd from frozen	80	1/2 c	63	114	72
Canned, drained	85	1/2 c	59	97	57
Canned, with liquid	124	1/2 c	61	108	66
Green peas, edible-pods:					
Fresh, uncooked	145	1 c	61	126	77
Ckd from fresh	160	1 c	67	133	89
Ckd from frozen	80	1/2 c	42	110	46
Split peas, cooked from dry	196	1 c	231	84	195
Peas, mature, sprouted:					
Fresh sprouts	120	1 c	154	129	198
Ckd fr/fresh, boiled	100	3.5 oz	118	20	24
Peas & carrots:					
Ckd from frozen	80	1/2 c	38	103	39
Canned w/liquid	128	1/2 c	48	121	58
PEPPERS, HOT chili, raw, chopped	75	1/2 c	30	113	34
PEPPERS, SWEET, all, fresh:					
Chopped, (pod≈74 g)	50	1/2 c	14	68	9.5
Cooked, chopped	68	1/2 c	19	63	12
Poi, two finger	240	1 c	269	35	94
POTATOES:					
Baked:					
Flesh and skin	202	1 ea	220	52	115
Flesh only	156	1 ea	145	54	78
Potato skin	58	1 ea	115	51	59
Boiled, 2.5 diam, flesh only:					
Cooked in skin	136	1 ea	119	50	60
Boiled without skin	135	1 ea	116	47	54

POTATOES, continued:

	Wt (g)	Svg	Cal	Phos (mg) 100 Cal	Phos (mg) Svg
Cottage fried, from frozen	50	10 ea	109	30	33
Dehydrated potato flakes	200	1 c	722	63	457
French fries, from frozen:					
Oven heated	50	10 ea	111	39	43
Fried in oil	50	10 ea	158	30	47
Hash browns f/frozen	156	1 c	340	33	112
Mashed potatoes:					
Prepared with milk	210	1 c	162	62	100
Prepared fr/instant	215	1 c	239	45	108
Potato puffs (tater tots)	62	1/2 c	138	22	30
Potato dishes, prepared:					
Au gratin, recipe	245	1 c	322	86	277
Au gratin, from mix	245	1 c	228	102	233
Scalloped, recipe	245	1 c	210	73	154
Scalloped, from mix	245	1 c	228	60	137
Potato chips (1 oz ≈ 14 chips)	28	1 oz	148	29	43
Potato flour	179	1 c	628	51	319
Potato pancakes	76	1 ea	237	33	78
Pumpkin, mashed:					
Cooked from fresh	245	1 c	50	148	74
Canned	123	1/2 c	42	102	43
Purslane, boiled	115	1 c	21	205	43
Radish seeds, sprouted	38	1 c	16	276	43
Rutabaga:					
Fresh cubes	140	1 c	51	159	81
Ckd from fresh	85	1/2 c	29	145	42
Salsify, cooked, slices	135	1 c	92	82	75
Sauerkraut, cnd w/liquid	236	1 c	44	105	46
Seaweed:					
Irish moss, fresh	28	1 oz	14	320	45
Spirulina, dried	28	1 oz	82	41	33
Soybeans:					
Ckd f/fresh veg	180	1 c	255	111	284
Ckd f/dry mature legume	172	1 c	298	141	421
Soybeans, mature, sprouted:					
Fresh (10 sprouts = 10g)	35	1/2 c	45	127	57
Steamed	94	1/2 c	76	167	127
Stir fried	100	3.5 oz	125	173	216

SOYBEAN PRODUCTS: see Tofu this section; miso, natto & tempeh in Other; roasted soybeans in Nuts & Seeds; soy milk in Dairy and soy flour in Grains.

	Wt (g)	Svg	Cal	Phos (mg) 100 Cal	Phos (mg) Svg
Spinach:					
Fresh, chopped	56	1 c	12	223	27
Cooked from fresh	180	1 c	41	244	100
Cooked from frozen, leaf	190	1 c	53	171	91
Canned, drained	214	1 c	50	188	94
SQUASH, SUMMER, slices:					
Crookneck, fresh	130	1 c	24	175	42
Crookneck, cooked	180	1 c	36	192	69
Scallop, fresh	130	1 c	24	196	47
Scallop, cooked	180	1 c	36	192	69
Zucchini, fresh	130	1 c	19	221	42
Zucchini, cooked	180	1 c	29	248	72
SQUASH, WINTER, mashed:					
Acorn/Danish, baked	245	1 c	137	81	111
Acorn/Danish, boiled	245	1 c	83	81	67
Butternut, baked	245	1 c	99	67	66
Butternut, ckd fr/frozen	240	1 c	94	36	34
Hubbard, baked	240	1 c	120	46	55
Hubbard, boiled	236	1 c	70	47	33
Succotash:					
Ckd from fresh	192	1 c	222	101	224
Ckd from frozen	170	1 c	158	75	119
Sweet potato:					
Baked with peel	114	1 ea	118	53	63
Boiled without peel	151	1 ea	160	26	41
Candied potatoes	105	1 pce	144	19	27

ESHA Research

	Wt (g)	Svg	Cal	Phos (mg) 100 Cal	Phos (mg) Svg
Vegetables & Legumes, continued:					
Sweet potato, continued:					
Canned, mashed	128	1/2 c	129	52	67
Canned, vacuum pack	255	1 c	233	54	125
Taro chips	23	10 ea	110	27	30
Tofu, firm	126	1/2 c	183	131	239
Tofu, regular	124	1/2 c	94	128	120
TOMATOES:					
Fresh, whole	123	1 ea	26	115	30
Fresh, chopped	180	1 c	38	113	43
Ckd from fresh	240	1 c	65	114	74
Canned, whole	240	1 c	47	98	46
Tomato juice, canned	244	1 c	42	112	46
Tomato paste, canned	262	1 c	220	94	207
Tomato puree, canned	250	1 c	102	97	99
Tomato sauce, canned	245	1 c	74	105	78
Turnips:					
Fresh, cubes	130	1 c	35	100	35
Ckd from fresh	78	1/2 c	14	107	15
Turnip greens:					
Ckd from fresh	144	1 c	29	141	41
Ckd from frozen	82	1/2 c	24	113	27
Vegetable juice cocktail	242	1 c	46	89	41
VEGETABLES, MIXED:					
Broccoli, cauliflower, red pepper	95	2/3 c	25	176	44
Broccoli & water chestnuts	95	1/2 c	33	155	51
Cantonese stir fry vegetables	95	1/2 c	53	85	45
Chinese stir fry vegetables	95	1/2 c	31	119	37
Green beans & spaetzle, Bavarian style	95	1/2 c	108	32	35
Japanese style vegetables	95	1/2 c	29	117	34
Mixed vegetables (corn, peas, carrots, green beans, limas):					
Cooked	182	1 c	107	87	93
Canned, drained	163	1 c	77	88	68
Mexicana style vegetables	95	1/2 c	125	46	58
Peas, carrots & onions	91	1/2 c	54	104	56
Peas, carrots, onions, pasta	95	1/2 c	122	71	87
Peas, cauliflower in cream sce	95	1/2 c	118	54	64
Peas, mushrooms	95	1/2 c	73	96	70
Peas, onions	95	1/2 c	71	83	59
Peas, onions & cheeese sce	142	1/2 c	165	74	122
Peas, onions, carrots, butter sce	71	2/3 c	100	63	63
Peas, potatoes in cream sce	76	1/2 c	140	44	62
Peas, rice & mushrooms	66	2/3 c	108	36	39
Peas, pasta, corn, cream sce	95	1/2 c	132	67	88
Peas, pasta, mushrooms, & cream sce	95	1/2 c	129	67	87
Water chestnuts, canned, slices	70	1/2 c	35	40	14
Winged bean parts (see Beans):					
Leaves, fresh	100	1 c	74	85	63
Tuber, fresh	100	1 c	159	28	45
Yam, orange: see Sweet potatoes.					
Yam, white, cooked cubes	136	1 c	158	47	75
Zucchini: see Squash, summer.					

Other

Condiments, fats, flavorings, other beverages, sweets, spices, etc.

	Wt (g)	Svg	Cal	Phos (mg) 100 Cal	Phos (mg) Svg
Baking powder:					
#1 with monocalcium phosphate monohydrate	3	1 tsp	5	1740	87
#2 with monocalcium phosphate monohydrate calcium sulfate	2.9	1 tsp	5	900	45
#3 straight phosphate	3.8	1 tsp	5	7180	359
#4 low sodium	4.3	1 tsp	5	6280	314
Barbecue sauce	250	1 c	160	31	50
Beer (12 fl oz = 1.5 c):					
Regular	356	1.5 c	146	30	44
Light	354	1.5 c	100	43	43
Beverages, carbonated (12 fl oz = 1.5 c):					
Cola beverage, regular	370	1.5 c	151	30	46
Diet cola, all	355	1.5 c	2	1500	30
Diet soda, assorted	355	1.5 c	2	1900	38
Pepper type soda	368	1.5 c	151	27	41
Butter	113	1/2 c	813	3	26
CANDY:					
Almonds, sugar coated	28	7 ea	146	59	86
Caramel, plain or chocolate	28	1 oz	115	30	35
Chocolate kisses	28	6 pce	154	56	87
Chocolate coated:					
Almonds	165	1 c	935	67	627
Coconut	28	1 oz	133	22	29
Peanuts	170	1 c	954	53	507
Raisins	187	1 c	733	27	197
M&M's plain candies	48	1 pkg	237	27	65
M&M's peanut candies	47	1 pkg	240	27	64
Malted milk balls	28	14 pce	135	64	86
Milk chocolate:					
Plain	28	1 oz	145	42	61
With almonds	28	1 oz	150	51	77
With peanuts	28	1 oz	155	56	87
With rice cereal	28	1 oz	140	41	57
Reese's peanut butter cup	45	2 ea	240	36	87
Snickers, 2.2 oz	62	1 bar	290	26	75
CHOCOLATE:					
Baking, unsweetened	28	1 oz	145	75	109
Bittersweet	28	1 oz	141	43	60
Chips, semi-sweet	170	1 c	860	21	178
Dark, sweet	28	1 oz	150	27	41
Hot fudge topping	38	2 T	129	47	60
Syrup, thin type	38	2 T	85	57	49
Corn syrup:					
Dark	328	1 c	944	6	52
Light	328	1 c	912	4	39
Carob flour	103	1 c	185	44	81
Catsup	245	1 c	255	38	96
Chili sauce, tomato based	273	1 c	284	50	142
Cocoa powder	86	1 c	224	276	618
Enchilada dip, Fritos	28	1 oz	35	83	29
Falafel, 2.25" patty	17	1 patty	57	58	33
Granola bar	28	1 ea	127	62	79
Hummous, Humous	246	1 c	420	65	275
Margarine:					
Regular, 80% fat	113	1/2 c	812	3	26
Soft spread, 60% fat	227	1 c	1225	3	37
Mayonnaise	220	1 c	1577	4	62
Miso (soybean product)	138	1/2 c	284	74	211
Molasses, blackstrap	40	2 T	85	40	34
Mustard, prepared	125	1/2 c	94	98	92
Natto, soybean product	88	1/2 c	187	82	153
Pickle relish	245	1 c	320	11	34
SALAD DRESSINGS:					
1000 island:					
Regular dressing	250	1 c	943	4	42
Low calorie	245	1 c	389	11	42
Blue cheese dressing	245	1 c	1235	14	176
Ceasar's salad dressing	92	1/2 c	408	32	129
French, regular	250	1 c	1321	1	16
French, low calorie	260	1 c	386	20	78
Mayonnaise type:					
Regular	235	1 c	916	7	61
Low calorie	240	1 c	556	13	70

Values are for edible portion of foods

Phosphorus (P) mg

	Wt (g)	Svg	Cal	Phos (mg) 100 Cal	Phos (mg) Svg
Other, continued:					
SALAD DRESSING, continued:					
Ranch style dressing	119	1/2 c	435	23	100
Russian salad dressing	245	1 c	1210	8	91
Salsa, Picante by Tostitos	85	6 T	40	150	60
Salsa, recipe	108	1/2 c	46	51	41
Soy sauce:					
Regular, wheat+soy, shoyu	18	1 T	9	222	20
Tamari (soy)	18	1 T	11	209	23
Fr/hydrolyzed veg protein	18	1 T	7	243	17
SPICES and Seasonings:					
Caraway seed	7	1 T	22	173	38
Cumin seed	6	1 T	22	136	30
Fenugreek seed	11.1	1 T	36	92	33
Garlic powder	8.4	1 T	28	125	35
Poppyseed	8.8	1 T	47	160	75
Salt substitute by Morton	6	1 T	0	28	28
Sugar, brown	220	1 c	820	7	56
Tempeh (soybean product)	83	1/2 c	165	104	171
Teriyaki sauce	18	1 T	15	187	28
Tobasco sauce	15	1 T	1.6	165	2.7
Yeast:					
Brewer's yeast	8	1 T	25	560	140
Dry active	30	4 T	80	484	387

161

ESHA Research

Potassium

Potassium plays a vital role in maintaining fluid balance in the body and does this in a roughly balanced way with sodium. When there is significant water loss from the body, sodium is lost and potassium is pulled out of the cells and excreted. Because a potassium deficiency affects the brain cells fairly early, the victim is unable to perceive that he or she needs water. Nerve and muscle cells are especially rich in potassium, and they need it to function well. It is also believed that potassium plays a catalytic role in the metabolism of carbohydrates and protein.

Symptoms of deficiency include weakness, nausea, listlessness, apprehension, drowsiness, irrational behavior and anorexia. Severe deficiency may result in irregular heartbeats that can be fatal. Loss of potassium can result from prolonged vomiting, chronic diarrhea, laxative abuse, or from the use of diuretics. Such loss can affect the heart muscle and heart beat. Only minimal mounts of potassium are lost from sweat.

Estimated Requirements for Adults

USA: Adults: Minimum requirement = 2000 mg, with a recommendation to eat more fruits and vegetables, which would raise intake of adults to about 3,500 mg per day -- an amount which may reduce the incidence of hypertension.
Canada: 30 mg of Potassium per kilogram of body weight, per day.

Generally, we should consume about the same amounts of potassium and sodium, since there is a balanced function between the two minerals in the body. Since there has been a concern about high salt intake (sodium chloride), there has been an increased use of salt substitutes, which are potassium-based instead. It will be interesting to see the developments in the future, since too much potassium may create as much of an imbalance in the body as too much sodium.

Nutrient Losses

Potassium is processed out of whole grains when they are milled into flour. When foods are cut into pieces and boiled, some potassium will dissolve into the cooking water. If the water is retained and used, most of the potassium will be recovered.

Food Sources

Potassium is widely distributed in foods and should be easily obtained in the diet. Good sources include fruits, vegetables, whole grains, most meats, and dairy products. The less fat in the food, the more potassium. Drinking water is a poor source of potassium.

Baked Goods
Breads, Cakes, Cookies, Crackers, Pancakes, Pastries, Pies, Rolls, some desserts.

	Wt (g)	Svg	Cal	100 Cal	Svg
Apple crisp, 3" x 3"	78	1 pce	146	77	112
Bagel, plain/egg, 3.5" dm.	68	1 ea	180	36	65
Biscuits:					
Homemade	28	1 ea	100	32	32
From mix	28	1 ea	94	60	57
From refrig dough	20	1 ea	65	28	18
BREADS:					
Banana nut, 1/2" slice	50	1 pce	161	65	104
Boston brown, cnd, 1/2"	45	1 pce	95	138	131
Cornbread muffin fr/mix	45	1 ea	145	21	31
Cracked wheat	25	1 pce	65	52	34
French, 5" x 2.5"	35	1 pce	100	32	32
Mixed grain	25	1 pce	65	86	56
Oatmeal bread	25	1 pce	65	60	39
Pita pocket, 6.5" dm.	60	1 ea	165	43	71
Pumpernickel, 5" x 4" x 3/8"	32	1 pce	80	176	141
Raisin bread	25	1 pce	68	87	59
Rye, light, 5" x 3.5" x 7/16"	25	1 pce	65	78	51
Wheat (white & whole wheat flour)	25	1 pce	65	54	35
White bread	25	1 pce	65	43	28
Whole wheat bread	35	1 pce	86	72	62
Brownies with nuts	25	1 ea	100	50	50
CAKES, cupcakes = 42 g; Piece = 1/16 cake unless stated otherwise:					
Angel food, 1/12 tube cake	53	1 pce	125	57	71
Boston cream pie, 1/8	120	1 pce	260	15	40
Carrot, cream cheese frosting, 2.5" x 3"	112	1 pce	406	29	116
Cheesecake:					
Recipe, 1/12	92	1 pce	278	32	90
From mix, 1/8	103	1 pce	300	80	241
Chocolate, choc frosting	69	1 pce	235	38	90
Coffeecake from mix, 2.4"x 2.8" piece	72	1 pce	230	34	78
Dark fruitcake, 2/3" arc	43	1 pce	165	118	194
Gingerbread, 1/9 of 8" sq.					
Home recipe	110	1 pce	351	61	214
From mix	63	1 pce	174	99	173
Pound cake, 1/2"	30	1 pce	115	23	27
Sheet cake, 3" x 3" piece:					
Plain	86	1 pce	315	22	68
White frosting	121	1 pce	445	17	74
Snack cake, cream filled:					
Chocolate (like Ding Dongs)	28	1 ea	105	32	34
Sponge cake (like Twinkies)	42	1 ea	155	24	37
Sponge cake, 1/12	66	1 pce	194	26	50
White cake:					
Chocolate frosting	77	1 pce	291	35	102
Coconut frosting	70	1 pce	270	27	73
White frosting	71	1 pce	260	20	52
Yellow, chocolate frosting:					
Commercial	69	1 pce	245	50	123
From mix	69	1 pce	235	32	75
Cherry crisp, 3" x 3"	138	1 pce	157	104	163
Chips: see corn & tortilla this section; potato chips under vegetables.					
COOKIES:					
Animal cookies	28	27 ea	120	22	26
Chocolate chip:					
Homemade	40	4 ea	185	44	82

COOKIES, Chocolate Chip, continued:	Wt (g)	Svg	Cal	100 Cal	Svg
From refrigerator dough	48	4 ea	225	28	62
Commercial	42	4 ea	180	31	56
Fig bars	56	4 ea	210	77	162
Lady fingers	44	4 ea	158	33	52
Oatmeal raisin	52	4 ea	245	37	90
Peanut butter, recipe	48	4 ea	245	45	110
Sandwich type, all	40	4 ea	195	34	66
Shortbread, commercial	32	4 ea	155	25	38
Sugar cookies fr/refrig dough	48	4 ea	235	14	33
Vanilla wafers	40	10 ea	185	27	50
Corn chips	28	1 oz	155	34	52
CRACKERS:					
Armenian cracker bread	28	4 pce	117	66	77
Cheese crackers	10	10 ea	50	34	17
Cheese, pnut butter filled	30	4 ea	150	43	64
Graham crackers	14	2 ea	60	60	36
Round (like Ritz)	9	3 ea	45	27	12
Rye wafers, whole grain	14	2 ea	55	118	65
Whole wheat crackers	8	2 ea	35	89	31
Cream puff, custard filled	110	1 ea	280	30	85
Crepe, without filling	26.8	1 ea	46.6	72	34
Croissant, 4.5" x 4" x 2"	57	1 ea	235	29	68
Danish pastry:					
Plain	57	1 ea	220	24	53
With fruit	65	1 ea	235	24	57
Doughnut:					
Cake type, medium	50	1 ea	210	28	58
Yeast raised, plain	60	1 ea	235	27	64
Eclair, custard filled, chocolate icing	94	1 ea	262	34	90
English muffin, plain/sourdough	57	1 ea	135	245	331
MUFFINS: also see English muffin.					
Blueberry, recipe	45	1 ea	135	35	47
Blueberry, from mix	45	1 ea	140	39	54
Bran, wheat, recipe	45	1 ea	125	79	99
Bran, wheat, from mix	45	1 ea	140	36	50
Cornbread, recipe	45	1 ea	145	21	57
Cornbread, from mix	45	1 ea	145	21	31
Pancakes:					
Buckwheat, 4" dm., f/mix	27	1 ea	55	120	66
Plain, 4" dm, f/recipe	27	1 ea	60	55	33
Plain, 4" dm, f/mix	27	1 ea	60	72	43
Whole wheat, 5" dm.	52	1 ea	94	109	102
Patty shell, puff pastry	71	1 ea	250	15	38
Peach crisp, 3" x 3"	139	1 pce	166	119	197
PIES: piece is 1/6th of 9" pie.					
Apple pie	158	1 pce	405	25	100
Banana cream, commercial	152	1 pce	319	97	308
Blueberry pie	158	1 pce	380	33	126
Cherry pie	158	1 pce	410	37	153
Chocolate cream	175	1 pce	311	71	222
Coconut cream	172	1 pce	343	65	222
Coconut custard	165	1 pce	384	75	289
Cream pie, commercial	152	1 pce	455	29	133
Custard pie	152	1 pce	293	59	173
Lemon meringue	140	1 pce	355	20	70.0
Mincemeat pie	160	1 pce	395	88	349
Peach pie	158	1 pce	405	58	235
Pecan pie	138	1 pce	583	22	130
Pumpkin pie	200	1 pce	367	109	400
Strawberry chiffon, recipe	162	1 pce	372	41	151
Popovers	51	1 ea	96	72	69
PopTart type pastry	54	1 ea	210	43	91
Pretzels, thin twists	60	10 ea	240	25	61

163

	Wt (g)	Svg	Cal	Potassium (mg) 100 Cal	Svg		Wt (g)	Svg	Cal	Potassium (mg) 100 Cal	Svg
Baked Goods, continued:						**CHEESE**, continued:					
ROLLS:						Cheddar cheese	28	1 oz	114	25	28
Cinnamon bun, small	50	1 ea	158	39	62	Cheshire	28	1 oz	110	25	27
Dinner rolls, 2.5" x 2"	28	1 ea	85	42	36	Colby cheese	28	1 oz	112	32	36
Hamburger bun	45	1 ea	129	49	63	**Cottage cheese:**					
Hard roll, white	50	1 ea	155	32	49	Lowfat 1%	226	1 c	164	118	193
Hotdog bun	40	1 ea	115	49	56	Lowfat 2%	226	1 c	205	106	217
Rye roll, dark	28	1 ea	79	103	82	Creamed, large curd	225	1 c	235	81	190
Rye roll, light	28	1 ea	76	75	57	Creamed, small curd	210	1 c	215	82	177
Submarine (hoagie) roll	135	1 ea	400	32	128	Creamed, with fruit	226	1 c	279	54	151
Whole wheat roll	35	1 ea	88	69	61	Dry curd	145	1 c	123	38	47
Stuffings, from enr. bread:						Cream cheese (1T = 15g)	28	1 oz	99	34	34
From dry	140	1 c	500	25	126	Edam cheese	28	1 oz	101	52	53
Stove Top Stuffing	108	1/2 c	176	59	103	Fontina	28	1 oz	110	23	25
Taco shell, 13.6g	13.6	1 ea	59	42	25	Gjetost	28	1 oz	132	30	40
Tortillas:						Gorgonzola	28	1 oz	111	23	26
Corn, fried, 6" dm.	30	1 ea	87	49	43	Gouda	28	1 oz	101	34	34
Flour, 10.5" dm	56.8	1 ea	168	33	56	Liederkranz	28	1 oz	87	78	68
Flour, 8" dm	35.4	1 ea	105	33	35	Limburger	28	1 oz	93	39	36
Tortilla chips:						Mozzarella, skim, low moisture	28	1 oz	80	34	27
Plain chips	28	1 oz	139	22	30	Muenster	28	1 oz	104	37	38
Doritos, nacho	28	1 oz	139	78	109	Neufchatel	28	1 oz	74	45	33
Doritos, taco	28	1 oz	140	51	72	Parmesan, grated (1 oz = 5.6T)	28	1 oz	129	23	30
Waffles:						Pimento processed	28	1 oz	106	43	46
Homemade, 7" dm.	75	1 ea	245	53	129	Provolone	28	1 oz	100	39	39
From frozen, 7" dm.	35	1 ea	98	74	73	Ricotta cheese, part skim milk	246	1 c	340	90	307
Prep from mix, 4" dm.	75	1 ea	205	71	146	Ricotta cheese, whole milk	246	1 c	428	60	257
						Romano, grated (1 T = 5g)	28	1 oz	128	26	33
						Roquefort	28	1 oz	105	25	26
						Swiss cheese	28	1 oz	107	29	31
						Swiss, processed	28	1 oz	95	64	61
						Swiss cheese food	28	1 oz	92	88	81

Beverages

Also see Dairy; and juices in Fruits & Vegetable sections.

	Wt (g)	Svg	Cal	Potassium (mg) 100 Cal	Svg		Wt (g)	Svg	Cal	Potassium (mg) 100 Cal	Svg
Alcoholic:						**CREAM, SWEET, fluid:**					
Beer (12 fl oz = 1.5 cup)	356	1.5 c	146	61	89	Coffee or table	240	1 c	469	62	292
Beer, light	354	1.5 c	100	64	64	Half and half	242	1 c	315	100	314
Bloody mary, 5 fl oz	148	1 ea	116	186	216	Light whipping cream	239	1 c	699	33	231
Champagne	119	1/2 c	91	104	95	Heavy whipping cream	238	1 c	821	22	179
Pina colada, 4.5 fl oz	141	1 ea	262	38	100	**CREAM, SWEET, whipped:**					
Screwdriver, 7 fl oz	213	1 ea	174	187	325	Heavy cream, unsweetened	119	1 c	410	22	90
Tequila sunrise, 5.5 fl oz	172	1 ea	189	94	178	Pressurized	60	1 c	154	57	88
Whiskey sour, 3 fl oz	90	1 ea	123	39	48						
Wine:						**CREAM, SOUR**, dairy:					
Red	118	1/2 c	85	154	131	Cultured, dairy	230	1 c	493	67	331
Rosé	118	1/2 c	84	139	117	Half & Half, dairy	15	1 T	20	95	19
White wine, medium	118	1/2 c	80	118	94	Cream, sour, Imitation, non-dairy	230	1 c	479	77	369
Non-alcoholic:											
Clam & tomato juice	166	2/3 c	77	194	149	**CREAM SUBSTITUTES, non-dairy:**					
Coffee, brewed	240	1 c	2	4800	128	Coffee whitener, liquid	120	1/2 c	163	140	229
Coffee, prep f/instant	240	1 c	2	4340	87	Coffee whitener, powder	94	1 c	514	148	763
Fruit punch, f/frozen	247	1 c	113	27	31	Dessert Toppings, (non-dairy):					
Fruit punch, canned	253	1 c	118	54	64	Frozen (like Coolwhip)	75	1 c	239	6	14
Lemonade, f/frozen	248	1 c	100	38	38	Dessert powder, dry mix	43	1.5 oz	245	29	71
Limeade, f/frozen	247	1 c	102	31	32	Pressurized	70	1 c	185	7	13
Pineapple grapefruit drink	250	1 c	117	132	154	Kefir, beverage	233	1 c	160	128	205
Pineapple orange drink	250	1 c	125	93	116						
Tea, brewed	240	1 c	2	4450	89	**MILK (cow):**					
						Skim	245	1 c	86	472	406
						Lowfat 1%	244	1 c	102	374	381

Dairy & Dairy Products

	Wt (g)	Svg	Cal	Potassium (mg) 100 Cal	Svg		Wt (g)	Svg	Cal	Potassium (mg) 100 Cal	Svg
						Lowfat 2%	244	1 c	121	312	377
						Whole (3.3% fat)	244	1 c	150	247	370
						Buttermilk (<1% fat)	245	1 c	99	375	371
CHEESE (1.5" cube ≈ 1 oz):						Canned, skim, evap	255	1 c	200	423	845
American processed	28	1 oz	106	43	46	Canned, whole	252	1 c	340	225	764
American cheese food, cold	28	1 oz	94	111	104	Dry, instant nonfat, envelope	91	1 ea	326	476	1552
American cheese food, jar	28	1 oz	93	85	79	Dried, buttermilk	120	1 c	464	412	1910
American cheese spread	28	1 oz	82	84	69						
Blue cheese	28	1 oz	100	73	73	**Milk (other):**					
Brick cheese	28	1 oz	105	36	38	Goat milk	244	1 c	168	297	499
Brie cheese	28	1 oz	95	45	43	Human breast milk	246	1 c	171	74	126
Camembert	28	1 oz	85	62	53	Soy milk	240	1 c	79	428	338
Caraway	28	1 oz	107	26	28						

	Wt (g)	Svg	Cal	Potassium (mg) 100 Cal	Svg		Wt (g)	Svg	Cal	Potassium (mg) 100 Cal	Svg
MILK BEVERAGES & MIXES:											
Carob flavor, with milk	256	1 c	195	190	370	**Fruits & Fruits Juices**					
Chocolate milk, commercial											
Lowfat 1%	250	1 c	160	266	425	Acerola juice, fresh	242	1 c	51	461	235
Lowfat 2%	250	1 c	180	234	422	**APPLE, 2.75" dm:**					
Whole (3.3% fat)	250	1 c	210	199	417	With peel	138	1 ea	80	199	159
Chocolate flavor mix						Without peel	128	1 ea	72	200	144
to be mixed w/water:						Applesauce, unsweetened	244	1 c	106	173	183
Powder (w/dry milk)	28	1 oz	100	223	223	Apple juice, canned/bottled	248	1 c	116	254	295
Drink, prep. w/water	206	3/4 c	100	223	223						
Chocolate flavor mix						**APRICOTS:**					
to be mixed w/milk:						Fresh, pitted	106	3 ea	51	614	313
Powder	21.6	3/4 oz.	75	171	128	Canned, juice pack	84	3 ea	40	348	139
Drink, prep. w/wh milk	266	1 c	226	220	498	Canned, heavy syrup	85	3 ea	70	170	119
Eggnog, commercial	254	1 c	342	123	420	Dried halves	35	10 ea	83	581	482
Malted milk, w/whole milk:						Apricot nectar, canned	251	1 c	141	203	286
Chocolate flavor	265	1 c	229	218	499	Avocado, whole:					
Natural flavor	265	1 c	237	223	529	California	173	1 ea	305	360	1097
Milkshakes, 10 fl oz = 1.25 cups:						Florida	304	1 ea	340	436	1484
Chocolate	283	1.25 c	360	158	567	Banana, 8.75" (176g w/peel)	114	1 ea	105	430	451
Strawberry	283	1.25 c	319	162	516	Blackberries:					
Vanilla	283	1.25 c	314	157	492	Fresh berries	144	1 c	74	381	282
						Frozen, unthawed	151	1 c	97	218	211
MILK DESSERTS:						Canned	256	1 c	236	108	254
Custard, baked f/recipe	265	1 c	305	127	387	Blackberry juice, fresh	250	1 c	93	457	425
Custard, from mix	143	1/2 c	161	158	254	Blueberries:					
Ice cream, vanilla						Fresh berries	145	1 c	82	157	129
Regular	133	1 c	269	96	257	Frozen, unthawed	155	1 c	78	106	83
Rich	148	1 c	349	63	221	Canned	256	1 c	225	45	102
Soft serve	173	1 c	377	90	338	Blueberry juice, fresh	250	1 c	135	206	278
Ice milk, regular	131	1 c	184	144	265	Boysenberries:					
Ice milk, soft serve, 3% fat	175	1 c	223	185	412	Frozen, unthawed	132	1 c	66	277	183
PUDDINGS, prepared (5 oz can≈.55c):						Canned	256	1 c	225	102	230
Assorted, low cal	130	1/2 ea	69	339	234	Cantaloupe: see Melon.					
Chocolate, mix-ckd/inst, avg	260	1 c	305	120	366	Cherries, sour:					
Chocolate, canned, 5 oz size	142	1 can	205	124	254	Frozen	155	1 c	72	267	192
Coconut, from instant	149	1/2 c	184	114	210	Canned	244	1 c	90	267	240
Lemon, from instant	149	1/2 c	178	108	192						
Rice, mix-ckd/inst, avg	141	1/2 c	165	107	176	**CHERRIES, SWEET:**					
Tapioca, ckd from mix	130	1/2 c	145	115	167	Fresh, pitted (10 = 68g)	145	1 c	104	313	325
Tapioca, canned, 5 oz	142	1 can	160	133	212	Frozen, sweetened, thawed	259	1 c	232	222	514
Vanilla, from instant	130	1/2 c	150	109	164	Canned w/liquid	257	1 c	213	175	373
Vanilla, canned, 5 oz	142	1 can	220	70	155	Cranberries, whole	95	1 c	46	146	67
Pudding pops:						Cranberry juice cocktail	253	1 c	145	31	45
Average	57	1 ea	94	88	83	Cranberry apple juice	253	1 c	169	40	68
Chocolate/choc. fudge	57	1 ea	99	126	125	Cranberry sauce, canned	277	1 c	419	17	72
Sherbet	193	1 c	270	73	198	Currants:					
Yogurt, frozen, average	174	1 c	220	135	296	Black, fresh	112	1 c	71	508	361
						Red or white, fresh	112	1 c	63	489	308
YOGURT:						Zante, dried	144	1 c	407	316	1285
Lowfat, plain	227	1 c	144	369	531	Dates, whole, pitted	83	10 ea	228	237	541
Lowfat with fruit	227	1 c	231	191	442	Elderberries, fresh	145	1 c	105	387	406
Lowfat, coffee/vanilla	227	1 c	193	258	497	Figs, fresh, medium	50	1 ea	37	314	116
Nonfat	227	1 c	127	456	579	Figs, dried	19	1 ea	48	279	133
Whole	227	1 c	138	255	352	Fruit cocktail, canned:					
Yogurt cheese, recipe	208	1 c	222	300	666	Juice pack	248	1 c	115	204	235
						Heavy syrup	255	1 c	185	121	224
						Gooseberries:					
Eggs						Fresh berries	150	1 c	67	443	297
						Canned w/liquid	252	1 c	185	105	194
Egg, chicken, raw/cooked:											
Whole large egg	50	1 ea	77.5	81	63	**GRAPES:**					
White only	33.4	1 ea	17	300	48	Thompson, seedless (10 ≈50g)	160	1 c	114	260	296
Yolk only	16.6	1 ea	59	27	16	Fresh, Tokay or Emperor	57	10 ea	40	263	105
Egg substitutes vary by brand.						Canned, heavy syrup	256	1 c	187	141	264
Check label:						Grape juice:					
Frozen	60	1/4 c	96	133	128	Prepared from frozen	250	1 c	128	41	53
Liquid	251	1 c	211	392	828	Canned/bottled	253	1 c	155	215	334
Powder	10	.35 oz	44	168	74						

	Wt (g)	Svg	Cal	Potassium (mg) 100 Cal	Svg
Fruits & Fruit Juices, continued:					
GRAPEFRUIT, half = 241g w/refuse					
Fresh half, pink/red	123	1 ea	37	427	158
Fresh half, white	118	1 ea	39	449	175
Canned sections	254	1 c	152	216	328
Grapefruit juice:					
Fresh juice	247	1 c	96	417	400
Prepared from frozen	247	1 c	102	330	337
Canned, unsweetened	247	1 c	93	406	378
Canned, sweetened	250	1 c	115	352	405
Guava, fresh	90	1 ea	45	569	256
Honeydew: see Melon.					
Kiwi fruit	76	1 ea	46	548	252
Lemon juice:					
Fresh	244	1 c	60	505	303
Bottled	244	1 c	52	477	248
Lime juice:					
Fresh	246	1 c	65	412	268
Bottled	246	1 c	50	370	185
Loganberries:					
Fresh	100	2/3 c	70	207	145
Frozen, unthawed	147	1 c	80	266	213
Mandarin oranges, canned	252	1 c	155	127	197
Mango, fresh slices	165	1 c	108	238	257
Melon: also Watermelon.					
Cantaloupe cubes	160	1 c	57	867	494
Casaba cubes	170	1 c	45	793	357
Honeydew cubes	170	1 c	60	768	461
Frozen, mixed melon balls	173	1 c	55	880	484
Nectarine (1 med = 1 c slices)	136	1 med	67	430	288
Oheloberries, fresh	140	1 c	39	138	54
ORANGE, 2-5/8" dm, 180g whole	131	1 ea	60	395	237
Orange juice:					
Fresh juice	248	1 c	111	447	496
Chilled f/fresh	249	1 c	110	430	473
Prep. from frozen	249	1 c	110	431	474
Canned, unsweetened	249	1 c	105	415	436
Orange grapefruit jce, cnd	247	1 c	105	371	390
Papaya (454g whole)	304	1 ea	117	667	780
Passion fruit	18	1 ea	18	350	63
Passion fruit juice:					
Purple	247	1 c	126	272	343
Yellow	247	1 c	149	461	687
PEACHES:					
Fresh, 2.5" diam.	87	1 ea	37	462	171
Cnd half, juice pack	77	1 ea	34	289	98
Cnd half, heavy syrup	81	1 ea	60	125	75
Peach nectar, canned	249	1 c	134	75	101
PEARS:					
Fresh, Bartlett (181g whole)	166	1 ea	98	212	208
Cnd half, juice pack	77	1 ea	38	195	74
Cnd half, heavy syrup	79	1 ea	59	86	51
Pear nectar, canned	250	1 c	149	22	33
Persimmon, fresh	168	1 ea	118	229	270
PINEAPPLE:					
Fresh chunks	155	1 c	76	230	175
Canned pieces (slice≈ 58g):					
Heavy syrup	255	1 c	199	133	265
Juice pack	250	1 c	150	203	305
Pineapple juice, cnd, unswtnd	250	1 c	140	241	338
Plantain, fresh slices	148	1 c	181	408	739
Plantain, cooked	154	1 c	179	400	716
PLUMS:					
Fresh, pitted, 2-1/8" dm.	66	1 ea	36	317	114
Canned, juice pack	95	3 ea	55	267	147
Canned, heavy syrup	110	3 ea	98	102	100

	Wt (g)	Svg	Cal	Potassium (mg) 100 Cal	Svg
Pomegranate (275g w/refuse)	154	1 ea	104	384	399
Pomegranate juice, fresh	125	1/2 c	55	455	250
Prunes, dried, pitted	84	10 ea	201	311	626
Prune juice, bottled	256	1 c	181	391	707
Raisins, dark, unpacked	145	1 c	435	250	1089
Raspberries:					
Fresh berries	123	1 c	60	312	187
Frozen, thawed	250	1 c	255	112	285
Canned with liquid	256	1 c	234	103	241
Rhubarb, fresh dices	122	1 c	26	1350	351
Rhubarb, cooked with sugar	240	1 c	279	82	230
STRAWBERRIES:					
Fresh berries	149	1 c	45	549	247
Frozen, unsweetened	149	1 c	52	423	220
Frozen, thawed, sweetened	255	1 c	245	102	250
Tangerine	84	1 ea	37	357	132
Tangerine juice, f/frozen	241	1 c	110	247	272
Watermelon pieces	160	1 c	50	372	186

Grains & Grain Products
Cereals, Flour, Grains, Pasta, Noodles, Popcorn

	Wt (g)	Svg	Cal	Potassium (mg) 100 Cal	Svg
Amaranth grain	195	1 c	729	98	714
Barley, cooked:					
Pearled	157	1 c	193	76	146
Whole	200	1 c	200	85	170
BRAN: see oat, rice, wheat.					
Buckwheat flour, dark	98	1 c	338	145	490
Buckwheat flour, light	98	1 c	340	92	314
Bulgar wheat, cooked	182	1 c	151	82	124
CEREALS, COLD (Ready to eat): Amount & fortification varies by brand - check label.					
CEREALS, HOT, cooked:					
Corn grits	242	1 c	145	37	53
Cream of Rice	244	1 c	126	39	49
Cream of Wheat	244	1 c	140	33	46
Malt-O-Meal	240	1 c	122	25	31
Farina	233	1 c	116	26	30
Maypo	180	3/4 c	128	123	158
Oatmeal (rolled oats)					
Regular or instant	234	1 c	145	90	131
Oatmeal, fortified instant, fr/pkt:					
Plain	177	3/4 c	104	95	99
With bran & raisin	195	3/4 c	158	149	236
Others averaged	164	3/4 c	160	86	137
Ralston cereal	253	1 c	134	114	153
Roman Meal	181	3/4 c	111	205	227
Whole wheat cereal	242	1 c	150	114	171
Corn flour:					
Regular	117	1 c	422	87	369
Masa Harina, enriched	114	1 c	416	82	340
Cornmeal, dry:					
Regular, degermed	138	1 c	505	44	224
Nearly whole, bolted	122	1 c	441	69	303
FLOUR: see specific grain, nut, vegetable.					
Macaroni, cooked:					
Regular, enriched	140	1 c	197	22	43
Vegetable, enriched	134	1 c	172	24	41.5
Whole wheat	140	1 c	174	36	62
Millet, cooked	120	1/2 c	143	52	74
NOODLES:					
Chow mein, dry	45	1 c	237	23	54
Egg noodles, ckd, enriched	160	1 c	213	21	45
Spinach noodles, ckd	140	1 c	182	45	81

	Wt (g)	Svg	Cal	Potassium (mg) 100 Cal	Svg		Wt (g)	Svg	Cal	Potassium (mg) 100 Cal	Svg
Grains & Grain Products, continued:						**CRAB** meat, unpacked measure:					
Oat bran (1T≈ 6g)	94	1 c	132	403	532	Alaska king leg, ckd	134	1 ea	129	271	350
Oats, rolled:						Blue crab, ckd	135	1 c	138	317	437
Dry oats	81	1 c	311	91	284	Blue crab, canned	135	1 c	133	380	505
Cooked (Oatmeal): see Cereals, hot.						Dungeness, cooked	101	3/4 c	85	422	359
						Crab, imitation f/surimi	85	3 oz	87	89	77
PASTA: see macaroni, noodles, spaghetti.						Crab cakes, homemade	60	1 ea	93	210	195
Popcorn, popped in oil	11	1 c	55	35	19	Crayfish, ckd, moist heat	85	3 oz	97	307	298
Quinoa grain, dry	170	1 c	635	198	1258	Eel, baked/broiled	100	3.5 oz	236	148	349
						Eel, smoked	100	3.5 oz	330	72	239
RICE, cooked:						Fish cakes:					
Brown rice	195	1 c	217	39	84	Home-fried	100	3.5 oz	172	192	330
White rice, regular	205	1 c	264	30	80	Fried f/frozen	100	3.5 oz	213	122	260
White rice, converted	175	1 c	200	200	65	Fish sticks, f/frozen	57	2 ea	155	96	149
Wild rice	164	1 c	166	100	166						
Rice bran	83	1 c	262	471	1233	**HADDOCK:**					
Rice polish	158	1 c	578	21	120	Baked/broiled	85	3 oz	95	357	339
Rye flour:						Breaded, fried	85	3 oz	175	154	270
Dark	128	1 c	415	225	934	Smoked	100	3.5 oz	116	358	415
Light	102	1 c	361	66	238	Halibut, baked/broiled	85	3 oz	119	412	490
Soy flour, stirred:						Herring:					
Low fat flour	44	1/2 c	163	694	1131	Baked/broiled	100	3.5 oz	203	206	419
Defatted	50	1/2 c	164	727	1192	Canned w/liquid	100	3.5 oz	208	231	480
Full fat, raw	42	1/2 c	182	587	1069	Smoked, kippered	100	3.5 oz	217	206	447
Spaghetti noodles, cooked:						Pickled (1 piece ≈ 15 g)	100	3.5 oz	262	26	69
Enriched	140	1 c	197	22	43	Lobster meat, cooked	145	1 c	142	359	510
Whole wheat spaghetti	140	1 c	174	36	62	Mackerel:					
Tapioca, dry	152	1 c	518	3	17	Baked/broiled, Pacific	100	3.5 oz	262	153	401
						Baked/broiled, Spanish	100	3.5 oz	158	351	554
WHEAT:						Canned, Jack, #1 can	361	1 can	563	124	700
Wheat bran	30	1/2 c	65	546	355	Mullet, fried	85	3 oz	127	306	389
FLOURS, unbleached:						Ocean perch:					
All-purpose, white, unsifted	125	1 c	455	29	134	Baked/broiled	100	3.5 oz	121	289	350
Cake, sifted	96	1 c	348	29	101	Breaded, fried	85	3 oz	185	130	241
Gluten	140	1 c	529	16	84						
Self-rising	125	1 c	442	35	155	**OYSTERS:**					
Semolina	167	1 c	601	52	311	Raw, Eastern	248	1 c	170	334	568
Whole wheat	120	1 c	407	119	486	Raw, Pacific	248	1 c	200	209	417
Wheat germ:						Fried, med, Eastern	88	6 ea	173	124	215
Raw	100	1 c	360	248	892	Simmered, Eastern	100	3.5 oz	137	334	458
Toasted	113	1 c	432	249	1070	Perch (freshwater)					
Wheat, rolled, cooked	240	1 c	142	142	165	Baked/broiled	92	2 ea	108	293	316
Wheat, rolled, dry	85	1 c	289	112	323	Pollock, baked/broiled	100	3.5 oz	99	404	400
Whole grain wheat (wheat						Rockfish, baked/broiled	100	3.5 oz	121	430	520
berries), cooked	50	1/3 c	28	118	33	Roe, raw, mixed species	28	1 oz	39	182	71
Wheat, sprouted	108	1 c	214	86	183						
						SALMON:					
						Average, baked/broiled	85	3 oz	183	174	319
## Meats: Fish & Shellfish						Chinook, smoked	85	3 oz.	99	151	149
						Coho, steamed/poached	100	3.5 oz	185	289	534
						Canned, Atlantic, small can	220	1 can	281	345	970
Anchovies, canned						Canned, Pink, drained, #1can	454	1 can	631	235	1482
in oil, drained	45	11 ea	95	258	245	**Sardines,** canned, drained:					
Bass, baked/broiled	100	3.5 oz	125	308	385	Atlantic, 2 sardines = 24g	92	1 can	192	190	365
Bluefish:						Canned, 1 sardine = 38g	100	3.5 oz	178	192	341
Baked/broiled	100	3.5 oz	159	300	477						
Fried in crumbs	100	3.5 oz	205	201	413	**SCALLOPS:**					
Carp, baked/broiled	100	3.5 oz	162	264	427	Breaded, fried	93	6 ea	200	155	310
Catfish, fried in cornmeal	100	3.5 oz	229	148	340	Steamed	100	3.5 oz	113	348	393
Caviar (1T = 16g)	100	3.5 oz	252	71	180	Imitation (surimi) fried	85	3 oz	84	105	88
						Sea bass, baked/broiled	100	3.5 oz	124	265	328
CLAMS:						Seatrout (Steelhead), cooked	100	3.5 oz	131	300	393
Canned, drained	160	1 c	236	426	1005	Shad, baked w/bacon	100	3.5 oz	201	188	377
Canned w/liquid, minced	183	1 can	145	35	51	Shark, batter-fried	85	3 oz	194	68	132
Breaded, fried small	188	20 ea	379	161	612						
						SHRIMP:					
COD:						Boiled, 2 large ≈ 11g	100	3.5 oz	99	184	182
Baked/broiled	100	3.5 oz	105	232	244	Breaded, fried (2 lrg ≈ 15g)	90	12 ea	218	93	203
Batter-fried	100	3.5 oz	199	186	370	Canned w/liquid	100	3.5 oz	102	176	180
Canned w/liquid, 11 oz can	312	1 can	327	504	1647	Canned meat	128	1 c	154	175	269
						Shrimp, Imitation, f/surimi	85	3 oz	86	88	76

	Wt (g)	Svg	Cal	Potassium (mg) 100 Cal	Svg		Wt (g)	Svg	Cal	Potassium (mg) 100 Cal	Svg
Fish & Shellfish, continued:						**LAMB:**					
Smelt, Rainbow, cooked	85	3 oz	106	298	316	**Arm chop**, braised					
Snapper, baked/broiled	100	3.5 oz	128	408	522	(raw = 5.6 oz w/ bone):					
						Lean & fat	70	1 ea	244	88	216
SOLE (Flounder):						Lean only)	55	1 ea	152	122	185
Baked/broiled	85	3 oz	99	295	292	**Loin chop**, broiled					
Batter-fried	85	3 oz	250	76	191	(raw= 4.2 oz w/bone):					
Breaded, fried	100	3.5 oz	188	133	250	Lean & fat	64	1 ea	201	104	209
Squid, flour-fried	85	3 oz	149	159	237	Lean only	46	1 ea	100	175	175
Sturgeon, cooked	85	3 oz	115	269	309	**Cutlet**, lean, cooked avg.	85	3 oz	175	167	293
Surimi, processed walleye						**Leg of lamb**, roasted:					
(Alaska) pollock: also see						Lean & fat	85	3 oz	219	121	266
imitation crab, scallops, shrimp	100	3.5 oz	99	113	112	Lean only	85	3 oz	162	177	287
Swordfish, baked/broiled	100	3.5 oz	155	238	369	**Rib roast:**					
Trout, baked/broiled	85	3 oz	129	418	539	Lean & fat	85	3 oz	305	76	231
						Lean only	85	3 oz	197	136	268
TUNA:						**Shoulder roast:**					
Baked/broiled, Bluefin	85	3 oz	157	175	275	Lean & fat	85	3 oz	235	91	214
Canned, drained, #1/2 can:						Lean only	85	3 oz	173	130	225
Canned in oil	171	1 can	339	104	354						
Water pack	165	1 can	216	240	518	**PORK:**					
						Bacon, cooked:					
						Regular	19	3 pce	109	84	92
Meats						Canadian style	47	2 pce	86	210	181
Beef, Pork, Ham, Rabbit, Frog legs, Venison and Veal						Breakfast strips	34	3 pce	156	101	158
						Blade chop: Cut 3 per lb = 151g					
						(5.3 oz) raw w/bone; 110g (3.9 oz)					
BEEF:						w/o bone. Cooked values follow.					
Breakfast strips	34	3 pce	153	92	140	Pan-fried, lean & fat	89	1 ea	368	72	264
Chuck blade, pot roasted, all:						Pan-fried, lean only	62	1 ea	175	139	244
Lean & fat (5.4 oz raw)	85	3 oz	325	58	190	**Center loin chop:** Cut 3 per lb = 151g					
Lean only	85	3 oz	230	97	223	(5.3 oz) raw w/bone; 124g (4.4 oz)					
GROUND BEEF, average						w/o bone. Cooked values follow.					
of broiled & pan-fried:						Braised, lean & fat	75	1 ea	266	89	238
Extra lean (17% fat, raw)	85	3 oz	215	124	266	Braised, lean only	61	1 ea	166	137	227
Lean (20.7% fat, raw)	85	3 oz	231	110	255	Broiled, lean & fat	82	1 ea	284	110	312
Regular (26.6% fat, raw)	85	3 oz	250	101	252	Broiled, lean only	72	1 ea	166	182	302
Frzn patty, broiled (23% fat, raw)	85	3 oz	240	104	250	Pan-fried, lean & fat	89	1 ea	334	97	323
Rib, choice, oven roasted:						Pan-fried, lean only	67	1 ea	178	171	305
Lean & fat (5 oz raw)	85	3 oz	324	77	250	Roasted, lean & fat	88	1 ea	268	100	284
Lean only	85	3 oz	204	157	320	Roasted, lean only	72	1 ea	180	150	271
Round steak, choice, broiled:						**Center rib chop:** Cut 3 per lb = 151g					
Lean & fat	85	3 oz	233	133	311	(5.3 oz) raw w/bone; 112g (3.9 oz)					
Lean only	85	3 oz	165	213	352	w/o bone. Cooked values follow.					
Round tip, all, roasted:						Braised, lean & fat	67	1 ea	246	113	277
Lean & fat	85	3 oz	213	141	300	Braised, lean only	53	1 ea	147	181	266
Lean only	85	3 oz	162	202	328	Broiled, lean & fat	77	1 ea	264	108	285
Sirloin steak, all, broiled						Broiled, lean only	63	1 ea	162	171	276
(11.3 oz raw = 8.2 oz cooked,						Pan-fried, lean & fat	88	1 ea	343	90	309
lean & fat; 6.9 oz lean only):						Pan-fried, lean only	62	1 ea	160	180	288
Lean & fat	85	3 oz	238	126	299	Roasted, lean & fat	79	1 ea	252	115	291
Lean only	85	3 oz	172	195	336	Roasted, lean only	66	1 ea	162	172	279
T-bone steak, choice, broiled						Pork roast, leg:					
(16 oz raw = 9.7 oz cooked,						Lean & fat	85	3 oz	250	112	280
lean & fat; 7.4 oz lean only):						Lean only	85	3 oz	187	170	317
Lean & fat	85	3 oz	276	104	288	Pork roast, average loin & rib:					
Lean only	85	3 oz	182	190	346	Lean & fat	85	3 oz	265	111	294
Variety meats:						Lean only	85	3 oz	206	162	333
Brains, pan fried	85	3 oz	167	180	301	Shoulder, braised (yield from					
Heart, simmered	85	3 oz	140	141	198	6.8 oz raw meat w/bone:					
Kidney, cooked	140	1 ea	201	124	250	Lean & fat	85	3 oz	293	98	286
Liver, fried	85	3 oz	184	168	309	Lean only	67	2.4 oz	166	163	271
Tongue, cooked	85	3 oz	241	63	153	Spareribs, cooked f/1 lb raw	177	6.25 oz	703	81	566
Beef, dried, cured (6 - 7 pces)	28	1 oz	47	268	126	Variety meats:					
Beef, corned, canned	85	3 oz	213	54	116	Heart, braised	145	1 c	214	140	299
						Liver, braised	85	3 oz	141	91	128
FROG LEGS, flour fried	144	6 ea	418	55	231						
						CURED PORK - HAM: also see Lunchmeats, Turkey ham.					
HAM: see Pork, cured						Roasted, lean & fat	85	3 oz	207	117	243
(also see Lunchmeats, Turkey ham)						Roasted, lean only	85	3 oz	133	202	269
						Canned, roasted	85	3 oz	140	213	298

Meats: Poultry

	Wt (g)	Svg	Cal	100 Cal	Svg
RABBIT, roasted	85	3 oz	175	146	255
VEAL (calf):					
Cutlet, braised/broiled	100	3.5 oz	181	159	288
Rib, roasted	85	3 oz	151	175	264
Liver, pan-fried	85	3 oz	208	178	372
VENISON (deer) roasted	85	3 oz	134	213	285
CHICKEN: A 3 lb chicken					
≈ 1.45 lbs raw; ≈ 1.1 lbs cooked.					
All types of meat:					
Fried	140	1 c	307	117	360
Roasted	140	1 c	266	128	340
Stewed	140	1 c	248	102	252
Canned, boned w/broth	142	5 oz	235	83	196
Dark meat:					
Fried	85	3 oz	203	106	215
Roasted	85	3 oz	174	117	204
Stewed	85	3 oz	163	94	154
Light meat:					
Fried	85	3 oz	163	137	223
Roasted	85	3 oz	147	143	209
Stewed	85	3 oz	135	113	153
Breast*, meat & skin =145g raw					
(181g raw w/bone):					
Batter-fried	140	1 ea	364	77	282
Flour-fried	98	1 ea	218	116	253
Roasted	98	1 ea	193	124	240
Stewed	110	1 ea	202	97	195
Breast*, meat only = 118g raw:					
Fried	86	1 ea	161	147	237
Roasted	86	1 ea	142	155	220
Stewed	95	1 ea	144	124	178
*2 pieces per bird					
Drumstick, meat & skin = 73g raw					
(110g raw w/bone):					
Batter-fried	72	1 ea	193	69	134
Flour-fried	49	1 ea	120	93	112
Roasted	52	1 ea	112	106	119
Stewed	57	1 ea	116	91	105
Drumstick, meat only = 62g raw:					
Fried	42	1 ea	82	128	105
Roasted	44	1 ea	76	142	108
Stewed	46	1 ea	78	118	92
Thigh, meat & skin= 94g raw					
(120g raw w/bone):					
Batter-fried	86	1 ea	238	69	165
Flour-fried	62	1 ea	162	91	147
Roasted	62	1 ea	153	90	137
Stewed	68	1 ea	158	73	115
Thigh, meat only = 69g raw:					
Fried	52	1 ea	113	119	134
Roasted	52	1 ea	109	114	124
Stewed	55	1 ea	107	94	101
Wing, meat & skin = 49g raw					
(90g raw w/bone):					
Batter-fried	49	1 ea	159	43	68
Flour-fried	32	1 ea	103	55	57
Roasted	102	3 ea	297	63	186
Stewed	40	1 ea	100	56	56
Wing, meat only = 29g raw:					
Fried	20	1 ea	42	100	42
Roasted	21	1 ea	43	102	44
Chicken liver, simmered	20	1 ea	30	93	28
Chicken gizzard, simmered	22	1 ea	34	115	39

	Wt (g)	Svg	Cal	100 Cal	Svg
DUCK, domestic, roasted:					
Meat & skin	85	3 oz	286	61	174
Meat only	85	3 oz	171	125	214
GOOSE, domestic, roasted:					
Meat & skin	85	3 oz	259	108	280
Meat only	85	3 oz	202	163	330
TURKEY:					
Breast meat, seasoned:					
Barbecued	28	1 oz	40	143	57
Hickory smoked	28	1 oz	35	169	59
Ground turkey, cooked	100	3.5 oz	229	118	270
Roasted:					
All types	140	1 c	238	176	418
Dark meat	85	3 oz	159	155	246
Light meat	85	3 oz	133	195	259
Frozen slices w/gravy	142	5 oz	95	92	87
Patty, breaded, fried	64	1 ea	181	97	176
Variety meats, cooked:					
Gizzard	67	1 ea	109	129	141
Heart	16	1 ea	28	104	29
Liver	75	1 ea	127	115	146

Meats: Sausages & Lunchmeats

	Wt (g)	Svg	Cal	100 Cal	Svg
Barbecue loaf, pork/beef	23	1 pce	40	190	76
Beef, thin sliced	28	1 oz	50	244	122
Beerwurst (beer salami), pork	23	1 pce	55	105	58
Berliner sausage	23	1 pce	53	123	65
BOLOGNA:					
Beef	23	1 pce	72	50	36
Beef and pork	28	1 oz	89	57	51
Cured pork	23	1 pce	57	114	65
Turkey	28	1 oz	56	100	56
Brotwurst, link	70	1 ea	226	87	197
Chicken roll, light meat	57	2 oz	90	143	129
Dutch brand loaf	28	1 oz	68	157	107
FRANKFURTER (hotdog):					
Beef, 8/pkg	57	1 ea	184	49	90
Beef and pork, 8/pkg	57	1 ea	183	52	95
Chicken, 10/pkg	45	1 ea	115	33	38
Turkey, 10/pkg	45	1 ea	102	86	88
Ham & cheese roll/loaf	57	2 oz	147	113	166
Ham salad spread	240	1 c	518	69	359
HAM lunchmeat:					
Extra lean	57	2 oz	75	264	198
Regular	57	2 oz	103	183	188
Thin sliced, 3 slices≈1 oz	28	1 oz	37	268	99
Chopped ham, packaged	42	2 pce	98	121	119
Minced ham	21	1 pce	55	118	65
Hot dog: see frankfurter					
Italian sausage link, ckd	67	1 ea	216	94	204
Knockwurst link	68	1 ea	209	65	136
Luncheon meat, canned	21	1 pce	70	64	45
Luncheon sausage, beef/pork	23	1 pce	60	93	56
Luxury loaf	57	2 oz	80	266	213
Olive loaf	57	2 oz	133	127	169
Pastrami:					
Cured beef	57	2 oz	198	65	129
Cured turkey	57	2 oz	74	209	155
Pepperoni sausage, sm. slice	22	4 pce	109	70	76
Pickle & pimento loaf	57	2 oz	149	130	193
Polish sausage	28	1 oz	92	73	67
Pork sausage, cooked:					
Link	13	1 ea	48	98	47
Patty	27	1 pce	100	97	97

	Wt (g)	Svg	Cal	Potassium (mg) 100 Cal	Svg		Wt (g)	Svg	Cal	Potassium (mg) 100 Cal	Svg
Meats: Sausages & Lunchmeats, continued:						Onion rings, cooked:					
Salami:						From fresh	115	1 c	39	456	178
Beef salami	23	1 pce	58	90	52	From frozen	20	2 ea	81	32	26
Beef and pork, dry	20	2 pce	85	89	76	Pies, fried pastry, commer.					
Pork and beef	57	2 oz	143	78	112	Apple pie	85	1 ea	255	16	42
Turkey	57	2 oz	111	113	125	Cherry pie	85	1 ea	250	24	61
Sausage, brown & serve links	13	1 ea	50	50	25						
Smoked link sausage:						**PIZZA**, cheese:					
Beef and pork	68	1 ea	229	56	129	Regular crust, 1/8 of 15"	120	1 pce	290	163	474
Pork	68	1 ea	265	86	228	Thick crust, 1/2 of 10"	208	1 pce	519	71	367
Turkey	28	1 oz	55	107	59	Potato salad w/mayo & eggs	250	1 c	358	177	635
Summer sausage	23	1 pce	80	66	53	Quiche Lorraine, 1/8 pie	176	1 pce	600	47	283
TURKEY lunchmeats (other):						Ravioli, beef, canned ≈ 16/cup	226	1 c	220	251	553
Breakfast sausage	28	1 oz	65	117	76						
Ham	57	2 oz	73	223	163	**SANDWICHES**, fast food:					
Loaf, breast meat	43	2 pce	46	257	118	Cheeseburger, 3 oz beef	112	1 ea	300	73	219
Roll, light & dark	57	2 pce	84	182	153	Cheeseburger, 4 oz beef	194	1 ea	524	78	407
Roll, light meat	57	2 pce	83	171	142	Chicken patty sandwich	157	1 ea	436	44	194
Summer sausage	28	1 oz	50	130	65	English muffin, egg, cheese,					
						and bacon	138	1 ea	360	56	201
						Fish sandwich:					
Mixed Dishes & Fast Foods						Large, w/o cheese	170	1 ea	470	80	375
						Regular, w/cheese	140	1 ea	420	65	274
Beef & vegetable stew:						Hamburger, 3 oz beef	98	1 ea	245	82	202
Recipe	245	1 c	220	279	613	Hamburger, 4 oz beef	174	1 ea	445	91	404
Canned	245	1 c	194	215	417	Frankfurter (hotdog) w/bun	85	1 ea	260	43	113
Beef, macaroni, tomato						Roast beef w/bun	150	1 ea	345	98	338
sauce, recipe	226	1 c	189	297	562						
Beef pot pie, homemade	210	1 pce	515	65	334	**SANDWICHES** on part whole wheat bread,					
						unless stated as rye:					
BURRITO:						Avocado, cheese,					
Bean burrito	174	1 ea	322	133	427	tomato, sprouts	195	1 ea	432	130	562
Beef burrito	177	1 ea	463	78	363	Bacon, lettuce, & tomato	135	1 ea	327	82	269
Beef & bean burrito	175	1 ea	390	99	388	Chicken salad sandwich	100	1 ea	294	52	152
Deluxe combination	198	1 ea	424	102	433	Corned beef & swiss on rye	147	1 ea	429	41	174
Cheese souffle, recipe	112	1 c	221	77	170	Egg salad sandwich	111	1 ea	319	42	135
Chicken a la king, recipe	245	1 c	470	86	404	Grilled cheese	117	1 ea	393	44	174
Chicken & noodles, recipe	240	1 c	365	58	211	Ham & cheese	151	1 ea	363	92	334
Chicken chow mein:						Ham & swiss on rye	145	1 ea	350	98	342
Homemade	250	1 c	255	185	473	Ham on rye	116	1 ea	242	129	311
Canned	250	1 c	95	440	418	Ham sandwich	122	1 ea	256	112	287
Chicken curry, homemade	337	1.5 c	305	134	410	Patty melt, on rye	177	1 ea	567	72	410
Chicken pot pie, recipe, 1/3	232	1 pce	545	63	343	Peanut butter & jam	100	1 ea	341	77	262
Chicken salad w/celery	78	1/2 c	266	52	137	Reuben, grilled	233	1 ea	480	65	313
Chili w/beans, canned	255	1 c	286	326	932	Roast beef sandwich	122	1 ea	280	112	314
Chop suey, beef/pork	250	1 c	300	142	425	Tuna salad sandwich	116	1 ea	303	71	215
Cole slaw	120	1 c	84	260	218	Turkey sandwich	122	1 ea	271	88	239
Corn dog	111	1 ea	330	50	164	Turkey ham	122	1 ea	253	98	249
Corn fritter, recipe	45	1 ea	116	83	96	Turkey ham & cheese	151	1 ea	361	82	296
Corn pudding	250	1 c	271	148	402	Turkey ham on rye	116	1 ea	239	114	273
Corned beef hash, canned	220	1 c	382	115	440	Turkey ham & cheese on rye	145	1 ea	347	92	319
Egg salad	183	1 c	438	48	211						
Enchilada:						**SPAGHETTI**, pasta &					
Beef enchilada	120	1 ea	292	66	193	tomato sauce w/ cheese:					
Cheese enchilada	120	1 ea	330	41	135	Homemade	250	1 c	260	157	408
Chicken enchilada	120	1 ea	269	65	176	Canned	250	1 c	190	159	303
French toast, homemade	65	1 pce	123	78	96	**SPAGHETTI**, pasta &					
						tomato sauce w/ meat:					
LASAGNA, recipe:						Homemade	248	1 c	330	202	665
With meat	245	1 pce	398	127	507	Canned	250	1 c	260	94	245
Without meat	218	1 pce	316	134	424	Spinach souffle	136	1 c	218	93	202
Macaroni & cheese:						Taco, beef	78	1 ea	207	88	183
Homemade	200	1 c	430	56	240	Taco, chicken	78	1 ea	172	92	158
Canned	240	1 c	230	60	139	Tostada:					
Macaroni salad, w/o cheese	141	1 c	371	44	162	Beans & beef	192	1 ea	332	133	442
Manicotti, frozen entree	225	1 ea	271	128	347	Beans & chicken	157	1 ea	249	144	358
Meat loaf:						Refried beans	157	1 ea	212	199	422
Beef and 1/3 pork	87	1 pce	212	112	238	Tuna noodle casserole, recipe	202	1 c	251	89	224
Beef only	87	1 pce	193	118	227	Tuna salad	205	1 c	383	139	531
Moussaka (lamb & eggplant)	250	1 c	250	278	695	Turkey pot pie, frozen	233	1 ea	416	33	138
						Waldorf salad	142	1 c	424	66	279

Nuts & Seeds

	Wt (g)	Svg	Cal	100 Cal	Svg
ALMONDS dried, whole	142	1 c	837	124	1034
Brazil nuts, dry	140	1 c	919	91	840
Cashews:					
Dry roasted	137	1 c	787	98	774
Oil roasted	130	1 c	748	92	689
Cashew butter	16	1 T	94	93	87
Chestnuts, roasted	143	1 c	350	242	846
Coconut:					
Raw, grated	80	1 c	283	101	285
Dried, unsweetened	78	1 c	515	82	423
Flaked, sweetened, cnd/pkgd	76	1 c	346	70	242
Coconut cream, raw	240	1 c	792	98	780
Coconut milk, canned	226	1 c	445	112	497
Coconut water, raw	240	1 c	46	1316	600
Filberts (hazelnuts), whole	135	1 c	853	70	601
Macadamias, oil roasted	134	1 c	962	46	441
MIXED NUTS w/peanuts					
(almonds, brazil nuts, cashews,					
filberts, peanuts & pecans):					
Dry roasted	137	1 c	814	100	817
Oil roasted	142	1 c	876	94	825
MIXED NUTS, w/o peanuts					
(cashews, almonds, brazil					
nuts, pecans & filberts):					
Oil roasted	144	1 c	886	88	783
PEANUTS:					
Dry roasted	146	1 c	855	112	960
Oil roasted	144	1 c	837	117	982
Peanut butter	16	1 T	95	121	115
Pecans, dried, chopped	119	1 c	794	59	466
Pine nuts, dried,					
pignola/pinyon	28	1 oz	154	113	174
Pistachios, dried, shelled	128	1 c	739	189	1399
Poppyseed	8.8	1 T	47	132	62
Pumpkin seed:					
Dried kernels	28	1 oz	154	149	229
Roasted kernels	227	1 c	1185	154	1830
Whole, roasted	64	1 c	285	206	588
Sesame seeds:					
Whole seed, dried	144	1 c	825	82	674
Kernels, dried	150	1 c	882	69	611
Sesame flour:					
Low fat	28	1 oz	95	119	113
High fat	28	1 oz	149	81	120
Soybeans, roasted	86	1/2 c	405	312	1264
Sunflower seed kernels:					
Dried seeds	144	1 c	821	121	992
Oil roasted	135	1 c	830	79	652
Tahini (sesame butter)	15	1 T	89	74	66
Walnuts, chopped:					
Black	125	1 c	759	86	655
English	120	1 c	770	78	602

Soups, Sauces, Gravies

	Wt (g)	Svg	Cal	100 Cal	Svg
GRAVY:					
Beef gravy, homemade	135	1/2 c	151	48	73
Beef gravy, canned	233	1 c	124	152	189
Brown gravy from dry mix	258	1 c	74.8	76	57
Chicken gravy:					
Homemade	130	1/2 c	163	69	112
From dry mix	260	1 c	85	73	62
Canned	238	1 c	189	138	260
Mushroom gravy, canned	238	1 c	120	210	252

	Wt (g)	Svg	Cal	100 Cal	Svg
SAUCES: also see Other.					
Cheese sauce:					
Recipe	101	1/2 c	216	59	127
Mix with milk	279	1 c	305	181	552
Hollandaise:					
Recipe	160	1 c	867	13	115
Mix with water	259	1 c	240	52	124
Spaghetti sauce, plain:					
Homemade	220	1 c	179	511	915
Canned	249	1 c	272	352	957
Spaghetti sauce w/meat:					
Homemade	248	1 c	297	207	615
Canned	206	7/8 c	220	202	444
White sauce:					
Recipe, medium	250	1 c	395	96	381
Mix with milk	264	1 c	240	185	444
SOUPS: soups are prepared from canned					
unless otherwise stated. RTS = Ready To					
Serve. For soup prep. w/milk, assume					
whole milk.					
Bean w/bacon	253	1 c	173	233	403
Beef bouillon	240	1 c	16	813	130
Beef broth f/dry	244	1 c	19	193	36.6
Beef noodle	244	1 c	84	119	100
Black bean soup	247	1 c	116	235	273
Celery, cream of, w/milk	248	1 c	165	187	309
Cheese soup w/milk	251	1 c	230	148	340
Chicken bouillon	244	1 c	39	538	210
Chicken broth f/dry	244	1 c	21	119	25
Chicken, chunky, RTS	251	1 c	178	99	176
Chicken, cream of, w/milk	248	1 c	191	143	273
Chicken noodle	241	1 c	75	73	55
Chicken rice w/water	241	1 c	60	168	101
Chicken vegetable	241	1 c	74	208	154
Chicken vegetable f/dry	251	1 c	49	139	68
Chicken veg, chunky, RTS	240	1 c	167	92	154
Chili beef	250	1 c	169	311	525
Clam chowder:					
New England style	248	1 c	163	184	300
Manhatten style	244	1 c	78	241	188
Gazpacho soup, RTS	248	1 c	140	254	356
Lentil & ham, RTS	248	1 c	140	254	356
Minestrone soup	241	1 c	80	390	312
Mushroom, cream of:					
Prepared w/milk	248	1 c	205	132	270
From dry mix	253	1 c	96	207	199
Mushroom in beef stock	244	1 c	85	186	158
Onion soup	241	1 c	57	121	69
Onion soup, f/dry pkt	184	3/4 c	20	240	48
Oyster stew w/milk	245	1 c	134	175	235
Potato, cream of, w/milk	248	1 c	148	218	323
Split pea	253	1 c	189	211	399
Split pea f/dry	255	1 c	133	179	238
Tomato soup:					
Prep with milk	248	1 c	160	281	450
Prep with water	244	1 c	86	306	263
Prep from dry	265	1 c	102	289	295
Tomato beef noodle	244	1 c	140	158	221
Tomato rice soup	247	1 c	120	275	330
Tomato vegetable f/dry	253	1 c	55	187	103
Turkey soup, chunky, RTS	236	1 c	136	599	814
Turkey noodle	244	1 c	69	109	75
Turkey vegetable	241	1 c	74	236	175
Vegetable, chunky, RTS	240	1 c	122	325	396
Vegetable beef	244	1 c	79	219	173
Vegetarian vegetable	241	1 c	70	299	209

ESHA Research

	Wt (g)	Svg	Cal	Potassium (mg) 100 Cal	Svg

Vegetables & Legumes

	Wt (g)	Svg	Cal	100 Cal	Svg
Alfalfa sprouts	33	1 c	10	260	26
Amaranth leaves:					
Fresh chopped	28	1 c	7	2471	173
Boiled from fresh	132	1 c	28	3021	846
Artichoke, globe, ckd, 300g whole	120	1 ea	60	708	425
Artichoke hearts, marinated-jar	170	6 oz	168	261	438
ASPARAGUS pieces:					
Fresh pieces	67	1/2 c	15	1347	202
Cooked from fresh	90	1/2 c	23	1240	279
Cooked from frozen	180	1 c	50	778	392
Canned, drained	121	1/2 c	16	1250	200
Canned with liquid	122	1/2 c	17	1094	186
Bamboo shoots, sliced:					
Cooked f/fresh	120	1 c	15	4267	640
Canned	131	1 c	25	420	105
BEANS: Also see garbanzo, lentils, soybeans.					
Baked beans (dry white beans w/spices & sauce):					
Home prepared	253	1 c	382	237	907
Canned, plain/vegetarian	254	1 c	235	320	752
Canned with franks	257	1 c	366	165	604
Canned with pork	253	1 c	268	291	781
Canned w/pork, sweet sce	253	1 c	282	239	673
Canned w/pork, tomato sce	253	1 c	247	307	759
Black beans, ckd f/dry	172	1 c	227	269	611
Broadbeans:					
Cooked f/fresh veg.	100	3.5 oz	56	345	193
Cooked f/dry legume	170	1 c	186	245	456
Canned legume	256	1 c	183	339	620
Great Northern beans, cooked f/dry	177	1 c	210	330	692
Green (snap) beans:					
Fresh, uncooked	110	1 c	34	676	230
Cooked from fresh	125	1 c	44	848	373
Cooked from frozen	135	1 c	36	419	151
Canned, drained	135	1 c	26	565	147
Canned with liquid	240	1 c	36	653	235
Hyacinth beans, cooked f/dry	194	1 c	228	286	653
Kidney beans, all types:					
Cooked from dry	177	1 c	225	317	713
Canned	256	1 c	208	316	658
Lima beans:					
Cooked from fresh	170	1 c	208	466	969
Ckd from frozen, baby	90	1/2 c	94	394	370
Ckd from frozen, large	85	1/2 c	85	408	347
Ckd from dry	188	1 c	217	440	955
Canned, drained	170	1 c	164	230	378
Canned with liquid	241	1 c	191	278	531
Navy beans:					
Cooked from dry	182	1 c	259	258	669
Canned	262	1 c	296	255	755
Pinto beans:					
Canned	240	1 c	186	389	723
Cooked from dry	171	1 c	235	340	800
Refried beans, canned	253	1 c	270	368	994
White beans, ckd f/dry	179	1 c	253	327	828
Winged beans:					
Ckd f/fresh veg slices	62	1 c	23	465	170
Ckd f/dry legume	172	1 c	252	191	481
Yambean (veg) cooked slices	100	1 c	46	393	181
Yardlong beans:					
Ckd f/fresh veg, slices	104	1 c	49	616	302
Ckd f/dry legume	171	1 c	202	267	539
Yellow wax : see green beans.					

	Wt (g)	Svg	Cal	100 Cal	Svg
Bean sprouts, (Mung beans):					
Fresh sprouts	104	1 c	31	494	154
Boiled, drained	124	1 c	26	481	125
Stir fried	124	1 c	62	323	200
Canned, drained	125	1 c	16	213	34
Beets:					
Ckd from fresh, diced	85	1/2 c	26	1023	266
Canned, diced	85	1/2 c	27	467	126
Pickled, slices	114	1/2 c	74	228	169
Beet greens, ckd, drained	144	1 c	40	3270	1308
Black-eyed peas: see Peas.					
BROCCOLI, chopped:					
Fresh, raw	88	1 c	24	1192	286
Ckd from fresh	156	1 c	44	552	456
Ckd from frozen	184	1 c	51	649	331
W/cheese sauce	142	1/2 c	166	120	199
W/hollandaise sauce	95	1/2 c	105	173	182
Brussels sprouts:					
Cooked from fresh	156	1 c	60	818	491
Cooked from frozen	155	1 c	65	775	504
CABBAGE:					
Common, shredded	70	1 c	16	1075	172
Common, cooked	150	1 c	32	963	308
Bok choy, shredded	70	1 c	9	1956	176
Bok choy, cooked	170	1 c	20	3150	630
Pe-tsai, shredded	76	1 c	11	1645	181
Pe-tsai, cooked	119	1 c	16	1675	268
Red cabbage, shredded	70	1 c	19	758	144
Red cabbage, cooked	75	1/2 c	16	656	105
Savoy, shredded	70	1 c	20	805	161
Savoy, cooked	145	1 c	35	763	267
CARROTS:					
Fresh 7.5" x 1-1/8" dm.	72	1 ea	31	752	233
Fresh, grated	55	1/2 c	24	742	178
Cooked slices:					
From fresh	78	1/2 c	35	506	177
From frozen	73	1/2 c	26	442	115
Canned with liquid	123	1/2 c	28	761	213
Canned, drained	73	1/2 c	17	771	131
Carrot juice	123	1/2 c	49	731	358
Cauliflower:					
Fresh, uncooked	50	1/2 c	12	1483	178
Cooked from fresh	62	1/2 c	15	1333	200
Cooked from frozen	180	1 c	34	735	250
Celery (7.5" stalk=40g=6.4 Cal):					
Fresh, chopped	60	1/2 c	10	1720	172
Ckd from fresh, diced	150	1 c	27	1578	426
Chard, Swiss:					
Fresh, chopped	36	1 c	7	1988	136
Cooked from fresh	175	1 c	35	2746	961
Chayote, fresh	132	1 c	32	619	198
Chayote, boiled, drained	160	1 c	38	726	276
Collards:					
Fresh, chopped	36	1 c	11	553	60.8
Cooked from fresh	128	1 c	35	480	168
Cooked from frozen	170	1 c	63	487	307
CORN, kernels:					
Fresh, uncooked	77	1/2 c	66	315	208
Ckd from fresh	82	1/2 c	89	229	204
Ckd from frozen	82	1/2 c	67	170	114
Canned, vacuum pack	210	1 c	166	235	390
Canned, drained	82	1/2 c	66	191	126
Canned with liquid	128	1/2 c	79	248	196
Corn, cream style, canned	128	1/2 c	93	185	172
Cucumber slices w/peel	28	7 pce	4	1135	42

	Wt (g)	Svg	Cal	100 Cal	Svg
Vegetables & Legumes, continued:					
Dandelion greens:					
Fresh, chopped	55	1 c	25	872	218
Cooked from fresh	105	1 c	35	697	244
Dock (sorrel) greens:					
Fresh pieces	133	1 c	29	1790	519
Cooked from fresh	100	1 c	20	1605	321
Eggplant, ckd from fresh	160	1 c	45	886	397
Endive, fresh, chopped	25	1/2 c	4	1975	79
Escarole (curly endive) chopped	50	1 c	9	1847	157
Garbanzo beans:					
Cooked from dry	164	1 c	269	177	477
Canned	241	1 c	285	72	206
Garden cress:					
Fresh, chopped	25	1/2 c	8.0	1900	152
Cooked from fresh	135	1 c	31	1539	477
Jerusalem artichoke, fresh slices	150	1 c	114	565	644
Jicama, fresh	120	1 c	49	429	210
Kale, chopped:					
Fresh, raw	67	1 c	33	906	299
Cooked from fresh	130	1 c	42	712	296
Cooked from frozen	130	1 c	39	1069	417
Kohlrabi:					
Fresh slices	140	1 c	38	1289	490
Cooked	165	1 c	48	1169	561
Leeks, chopped, ckd f/fresh	52	1/2 c	24	188	45
Lentils:					
Cooked from dry	198	1 c	231	316	731
Fresh sprouts	77	1 c	81	306	248
Stir fried sprouts	100	3.5 oz	101	281	284
LETTUCE, chopped:					
Butterhead	56	1 c	7.3	1973	144
Iceberg	56	1 c	7.3	1212	88.5
Looseleaf	56	1 c	10	1465	148
Romaine	56	1 c	9	1800	162
Lima beans: see Beans.					
Lotus root, ckd f/fresh	89	10 ea	59	550	323
Mushrooms:					
Fresh slices (1 avg = 18g)	35	1/2 c	9	1486	130
Cooked from fresh	78	1/2 c	21	1324	278
Canned, drained	78	1/2 c	19	540	101
Mustard greens:					
Fresh, chopped	56	1 c	15	1356	198
Cooked from fresh	140	1 c	21	1348	283
Cooked from frozen	150	1 c	29	733	209
Okra pods, ckd f/fresh	85	8 ea	27	1007	274
Okra slices, ckd f/frzn	92	1/2 c	34	632	215
Onions, globe, chopped:					
Fresh, raw	160	1 c	61	411	251
Cooked from fresh	105	1/2 c	46	378	174
Cooked from frozen	105	1/2 c	30	380	114
Onions, Spring, all, chopped	50	1/2 c	16	864	138
Onion rings:					
Cooked from fresh	115	1 c	39	456	178
Cooked from frozen	20	2 ea	81	32	26
Onion flakes, dehydrated	14	1/4 c	45	504	227
Parsley:					
Fresh, chopped	30	1/2 c	10	1610	161
Freeze dried	1.4	1/4 c	4	2200	88
Parsnips:					
Fresh slices	133	1 c	100	499	499
Cooked from fresh	156	1 c	125	458	573
PEAS:					
Black-eyed peas (cowpeas):					
Cooked from fresh	165	1 c	160	431	690
Cooked from frozen	170	1 c	224	384	860
Canned	240	1 c	184	224	413
Cooked from dry	171	1 c	198	240	476
PEAS, continued:					
Green peas:					
Fresh, uncooked	145	1 c	118	300	354
Cooked from fresh	160	1 c	134	324	434
Cooked from frozen	80	1/2 c	63	213	134
Canned with liquid	124	1/2 c	61	177	108
Canned, drained	85	1/2 c	59	249	147
Green peas, edible-pods:					
Fresh, uncooked	145	1 c	61	475	290
Cooked from fresh	160	1 c	67	572	383
Cooked from frozen	80	1/2 c	42	412	173
Split peas, ckd f/dry	196	1 c	231	307	710
Peas, mature, sprouted:					
Fresh sprouts	120	1 c	154	297	457
Cooked	100	3.5 oz	118	227	268
Peas & carrots:					
From frozen	80	1/2 c	38	334	127
Canned, with liquid	128	1/2 c	48	267	128
PEPPERS, HOT, chili, green/red:					
Fresh pod (1=45g)	45	1 ea	18	850	153
Fresh, chopped	75	1/2 c	30	850	255
Canned with liquid	68	1/2 c	17	841	143
Jalapeño, chopped, cnd	68	1/2 c	17	541	92
PEPPERS, SWEET, green or red:					
Fresh pod (1=74g)	74	1 ea	20	655	131
Fresh, chopped	50	1/2 c	14	636	89
Ckd f/fresh, chopped	68	1/2 c	19	595	113
Poi, two finger	240	1 c	269	163	439
POTATOES:					
Baked in oven, 4.75" x 2.3":					
Flesh and skin	202	1 ea	220	384	844
Flesh only	156	1 ea	145	421	610
Potato skin	58	1 ea	115	289	332
Boiled w/o skin, flesh only, 2.5"	135	1 ea	116	382	443
Boiled w/skin, flesh only	136	1 ea	119	433	515
Canned, 1" diam.	70	2 ea	42	381	160
Cottage fried, f/frzn	50	10 pces	109	220	240
French fries:					
Oven heated	50	10 ea	111	206	229
Fried in oil	50	10 ea	158	232	366
Hash brown, ckd f/frzn	156	1 c	340	200	680
Mashed potatoes:					
Prep w/milk	210	1 c	162	388	628
Prep w/milk/marg	210	1 c	222	273	607
From instant	215	1 c	239	179	428
Dehydrated flakes	45	1 c	159	307	488
Potato puffs (tater tot)					
heated from frozen	62	1/2 c	138	171	236
Potato dishes, prepared:					
Au gratin, recipe	245	1 c	322	301	970
Au gratin, mix	245	1 c	228	236	537
Scalloped, recipe	245	1 c	210	441	926
Scalloped, mix	245	1 c	228	218	497
Potato chips (14 chips≈1 oz)	28	1 oz	148	249	369
Potato flour	179	1 c	628	453	2843
Potato pancakes	76	1 ea	237	227	538
Pumpkin, mashed:					
Cooked from fresh	245	1 c	50	1128	564
Canned (can be a					
mixture with squash)	123	1/2 c	42	600	252
Radishes:					
Daikon	44	1/2 c	8	1250	100
Red	45	10 ea	7	1486	104
Radish seeds, sprouted	38	1 c	16	212	33
Rutabaga:					
Fresh cubes	140	1 c	51	924	471
Cooked	85	1/2 c	29	841	244
Salsify, cooked	135	1 c	92	414	381

Vegetables & Legumes, continued:

	Wt (g)	Svg	Cal	Potassium (mg) 100 Cal	Svg
Sauerkraut, cnd w/liquid	236	1 c	44	911	401
Seaweed:					
Agar, fresh	28	1 oz	7	865	64
Agar, dried	28	1 oz	87	470	386
Kelp, fresh	28	1 oz	12	207	25
Lavar, fresh	28	1 oz	10	1020	101
Spirulina, dried	28	1 oz	82	470	386
Soybeans, ckd f/dry	172	1 c	298	297	886
Soybeans, sprouted:					
Fresh sprouts	35	1/2 c	45	376	169
Steamed	94	1/2 c	76	439	334
Stir fried	100	1 c	125	454	567

SOYBEAN PRODUCTS: see tofu this section; miso, natto, tempeh in Other; roasted soybeans in Nuts & Seeds; soy milk in Dairy; soy flour in Grains.

	Wt (g)	Svg	Cal	Potassium (mg) 100 Cal	Svg
Spinach:					
Fresh, chopped	56	1 c	12	2537	312
Cooked from fresh	180	1 c	41	2044	838
Cooked from frozen	190	1 c	53	1064	566
Canned, drained	214	1 c	50	1480	740

SQUASH, SUMMER, sliced:

	Wt (g)	Svg	Cal	100 Cal	Svg
Crookneck, fresh	130	1 c	24	1150	276
Crookneck, ckd f/fresh	180	1 c	36	961	346
Scallop, fresh	130	1 c	24	983	236
Scallop, ckd f/fresh	90	1/2 c	14	900	126
Zucchini, fresh	130	1 c	19	1695	322
Zucchini, ckd f/fresh	180	1 c	29	1569	455

SQUASH, WINTER, mashed:

	Wt (g)	Svg	Cal	100 Cal	Svg
Acorn (Danish), baked	245	1 c	137	782	1071
Acorn (Danish), boiled	245	1 c	83	777	645
Butternut, baked	245	1 c	99	704	697
Butternut, ckd f/frozen	240	1 c	94	339	319
Hubbard, baked	240	1 c	120	716	859
Hubbard, boiled	236	1 c	70	720	504
Spaghetti, baked/boiled	155	1 c	45	404	182
Succotash:					
Cooked from fresh	192	1 c	222	355	787
Cooked from frozen	170	1 c	158	285	451
Sweet potatoes:					
Baked in skin, flesh only 5" x 2"	114	1 ea	118	336	397
Boiled, peeled 5" x 2"	151	1 ea	160	174	278
Canned, mashed, regular	128	1/2 c	129	208	268
Canned, mashed, vacuum pack	255	1 c	233	342	796
Candied, recipe, 2.5" x 2" piece	105	1 pce	144	138	198
Taro, fresh slices	104	1 c	112	549	615
Taro, ckd f/fresh slices	132	1 c	187	341	638
Taro chips	23	10 ea	110	172	189
Tofu, firm	126	1/2 c	183	163	298
Tofu, regular	124	1/2 c	94	160	150

TOMATOES:

	Wt (g)	Svg	Cal	100 Cal	Svg
Fresh, whole (2-3/5" dm)	123	1 ea	26	1050	273
Fresh, chopped	180	1 c	38	1053	400
Cooked from fresh	240	1 c	65	1031	670
Canned, whole	240	1 c	47	1126	529
Tomato products, canned:					
Tomato juice	244	1 c	42	1294	537
Tomato paste	262	1 c	220	1110	2442
Tomato puree	250	1 c	102	1030	1051
Tomato sauce	245	1 c	74	1227	908
Turnips, cubes:					
Fresh, raw	130	1 c	35	709	248
Cooked f/fresh	78	1/2 c	14	757	106
Turnip greens:					
Cooked from fresh	144	1 c	29	1010	293
Cooked from frozen	82	1/2 c	24	767	184

	Wt (g)	Svg	Cal	Potassium (mg) 100 Cal	Svg
Vegetable juice cocktail	242	1 c	46	1015	467

VEGETABLE COMBINATIONS, prepared from frozen:

	Wt (g)	Svg	Cal	100 Cal	Svg
Broccoli, carrots, pasta	95	2/3 c	88	209	184
Broccoli, carrots & water chestnuts	91	2/3 c	32	772	247
Broccoli, cauliflower, red peppers	95	2/3 c	25	812	203
Broccoli, water chestnuts	95	1/2 c	33	761	251
Cantonese stir fry	95	1/2 c	53	257	136
Chinese stir fry	95	1/2 c	31	477	148
Green beans & spaetzle (Bavarian)	95	1/2 c	108	74	80
Japanese style	95	1/2 c	29	400	116
Mexicana style	95	1/2 c	125	140	175
Mixed Vegetables (corn, peas, carrots, green beans, lima beans):					
Cooked from frozen	182	1 c	107	288	308
Canned, drained	163	1 c	77	616	474
New England style	95	1/2 c	128	126	161
Peas, carrots, onions	91	1/2 c	54	293	158
Peas, carrots, onions, pasta	95	1/2 c	122	124	151
Peas, cauliflower, cream sce	95	1/2 c	118	178	210
Peas & mushrooms	95	1/2 c	73	241	176
Peas & onions	95	1/2 c	71	242	172
Peas, onions, carrots, btr sce	71	2/3 c	100	119	119
Peas, onions, cheese sce	142	1/2 c	165	130	215
Peas, potatoes, cream sce	76	1/2 c	140	169	23.0
Peas, pasta, corn, cream sce	95	1/2 c	132	127	167
Peas, pasta, mushrooms & cream sauce	95	1/2 c	129	125	161
Water chestnuts, cnd slices	70	1/2 c	35	234	82
Watercress, fresh chopped	17	1/2 c	2	2800	56
Winged bean leaves, fresh	100	1 c	74	238	176
Yams:					
Hawaii mountain, steamed	145	1 c	119	603	718
Orange: see Sweet potato.					
White, cooked cubes	136	1 c	158	325	513
Zucchini: see Squash, summer.					

Other

Cooking ingredients, condiments, fats, flavorings, spices, sweets, etc.

	Wt (g)	Svg	Cal	100 Cal	Svg
Apple butter	35	2 T	66	135	89
Butter	113	1/2 c	813	4	29
Butterscotch topping	50	3 T	156	22	34
Baking powder, low sodium	4.3	1 tsp	7	6728	471
Barbecue sauce	250	1 c	160	272	435

CANDY and CANDY BARS:

	Wt (g)	Svg	Cal	100 Cal	Svg
Almonds, sugar coated	28	7 ea	146	96	132
Caramel, plain/chocolate	28	1 oz	115	47	54
Chocolate coated:					
Almonds	165	1 c	935	108	1011
Coconut candy	28	1 oz	133	56	75
Mints	28	1 oz	116	22	26
Peanuts	170	1 c	954	90	857
Raisins	187	1 c	733	157	1153
Chocolate kisses	28	6 pce	154	75	115
English toffee	32	1 ea	220	23	50
Fudge:					
Chocolate w/nuts	28	1 oz	114	40	46
Chocolate w/o nuts	28	1 oz	115	37	42
Vanilla w/nuts	28	1 oz	122	31	38
Vanilla w/o nuts	28	1 oz	118	31	36
Kit Kat bar	43	1 ea	210	61	129
Krackle bar	34	1 ea	179	65	116
M&M's plain candies (≈1.7 oz)	48	1 pkg	237	72	171
M&M's peanut candies	47	1 pkg	240	68	162

	Wt (g)	Svg	Cal	Potassium (mg) 100 Cal	Potassium (mg) Svg
Other, continued:					
CANDY, continued:					
Malted milk balls	28	14 pce	135	84	113
Mars bar	50	1 ea	240	73	176
Milk chocolate, plain	28	1 oz	145	66	96
Milk chocolate, with almonds	28	1 oz	150	83	125
Milk chocolate, with peanuts	28	1 oz	155	100	155
Milk chocolate, with rice cereal	28	1 oz	140	71	100
Milky Way	60	1 ea	260	64	167
Mr. Goodbar	47	1 ea	250	85	212
Reese's peanut butter cup	45	2 ea	240	70	168
Snickers bar, 2.2 oz	61	1 ea	290	72	209
Carob flour	103	1 c	185	461	852
Catsup	15	1 T	16	513	82
Chili sauce:					
Tomato based	273	1 c	284	356	1010
Hot, red pepper	31	2 T	6	767	46
CHOCOLATE:					
Baking chocolate, unsweetened	28	1 oz	145	162	235
Bittersweet	28	1 oz	141	91	129
Choc.chips, semi-sweet	170	1 c	860	69	593
Dark, sweet	28	1 oz	150	57	86
Hot fudge topping	38	2 T	129	64	82
Syrup, thin	38	2 T	85	100	85
Cocoa powder	86	1 c	224	446	1000
Enchilada dip, Fritos	28	1 oz	35	331	116
Falafel (2-1/4" patty)	17	1 ea	57	174	99
Ginger root, sliced, raw	11	5 pce	8	575	46
Gelatin salad, dessert	120	1/2 c	70	130	91
Honey	339	1 c	1030	17	173
Horseradish, prepared	15	1 T	6	733	44
Hummous (Humous)	246	1 c	420	102	427
Jalapeno bean dip, Fritos	28	1 oz	33	233	77
Margarine:					
80% fat	113	1/2 c	812	6	48
Spread, 60% fat	113	1/2 c	610	6	34
Soft 40% fat	113	1 c	392	7	29
Marmalade	20	1 T	52	24	12
Mayonnaise	220	1 c	1577	5	78
Miso (soybean product)	138	1/2 c	284	80	226
Molasses:					
Blackstrap	40	2 T	85	1378	1171
Light	20	1 T	43	426	183
Mustard, prepared	125	1/2 c	94	172	162
Natto (soybean product)	88	1/2 c	187	674	1276
Pectin	14	1 T	2	1650	33
Pickles:					
Dill pickle	65	1 ea	11.7	1701	199
Sweet pickle, medium	35	1 ea	41	27	11.2
Fresh pack, slices	30	4 pce	20	300	20
Pickle relish, sweet	15	1 T	20	150	30
SALAD DRESSINGS:					
1000 island	250	1 c	943	30	282
1000 island, low cal	245	1 c	389	71	277
Blue cheese	245	1 c	1235	8	96
Ceasar's salad dressing	92	1/2 c	408	13	55
French dressing	250	1 c	1321	2	31
French dressing, low cal	260	1 c	386	12	47
Italian dressing	235	1 c	1098	3	35
Italian dressing, low cal	240	1 c	130	28	36
Ranch style	119	1/2 c	435	36	158
Russian dressing	123	1/2 c	605	32	192
Salsa:					
Picante by Tostitos	85	6 T	40	428	171
Recipe	108	1/2 c	46	443	204

	Wt (g)	Svg	Cal	Potassium (mg) 100 Cal	Potassium (mg) Svg
Salt substitutes vary, check label:					
Lite Salt (Morton)	6	1 tsp	0	high	1500
Salt Substitute (Morton)	5.5	1 tsp	0	high	2800
Seasoned Salt Substitute (Morton)	4.8	1 tsp	2.4	high	2100
Soy sauce:					
Regular (wheat & soy) shoyu	18	1 T	9	356	32
Tamari (soy)	18	1 T	11	345	38
Fr/hydrolized veg protein	18	1 T	7	386	27
Sugar:					
Brown	220	1 c	820	92	757
White granulated	200	1 c	770	1	7
Syrups (other):					
Corn syrup, light	328	1 c	912	6	54
Maple syrup	20	1 T	50	52	26
Sorghum syrup	20	1 T	52	231	120
SPICES:					
Allspice	6	1 T	16	394	63
Basil, dried	4.5	1 T	11	1400	154
Caraway seed	6.7	1 T	22	414	91
Cardamom	5.8	1 T	18	361	65
Celery seed	2.0	1 tsp	8.5	353	30
Chili powder	2.6	1 tsp	8	625	50
Cinnamon	6.8	1 T	17.7	192	34
Cloves, ground	6.6	1 T	21	348	73
Coriander/Cilantro:					
Fresh leaves	4	1/4 c	1	2200	22
Dried leaf	.6	1 tsp	1.7	1617	27
Coriander seed	5	1 T	15	420	63
Cream of tartar	10	1 T	7	5157	361
Cumin seed	6	1 T	22	486	107
Curry powder	2	1 tsp	6	517	31
Dill weed, dried	3.1	1 T	8	1375	110
Fenugreek seed	11.1	1 T	36	236	85
Garlic, cloves	12	4 ea	17.9	268	48
Garlic, powder	2.8	1 tsp	9.3	333	31
Ginger, ground	5.4	1 T	19	358	68
Mace, ground	5.3	1 T	25	100	25
Marjoram, dried	1.7	1 T	5	520	26
Onion powder	6.5	1 T	15	400	60
Oregano, ground	1.5	1 tsp	4.7	532	25
Paprika	2.1	1 tsp	6.1	803	49
Pepper:					
Black	2.1	1 tsp	5.4	485	26
Cayenne (red)	5.3	1 T	17	629	107
Poppyseed	8.8	1 T	47	132	62
Poultry seasoning	3.7	1 T	11	227	25
Pumpkin pie spice	5.6	1 T	19	195	37
Rosemary, dried	3.3	1 T	11	300	33
Tarragon, ground	4.8	1 T	14	1036	145
Thyme, ground	4.3	1 T	12	292	35
Turmeric, ground	6.8	1 T	24	717	172
Tempeh (soybean product)	166	1 c	331	184	609
Teriyaki sauce, f/dry mix	18	1 T	15	273	41
Tobasco sauce	15	1 T	1.65	12	0.2
Vinegar, cider	240	1 c	29	828	240
Yeast:					
Brewer's yeast	8	1 T	25	608	152
Dry active, packaged	7	1 pkg	20	700	140

ESHA Research

Selenium

Selenium is an important part of an enzyme which acts as an anti-oxident, and works with vitamin E in regulating free radicals (destructive ions). It is also known that a selenium deficiency can affect the heart, and a severe deficiency can cause heart failure.

There are a lot of questions about selenium to be clarified, especially relating to the indication that it may offer some protection from heavy metals, such as mercury. Selenium requirements could also conceivably be increased because of exposure to various environmental pollutants, such as mercury, cadmium, or arsenic, which either tie up with selenium in an inactive form, or selenium may promote their excretion.

Recent studies indicate that pregnant women tend to conserve slightly more dietary selenium than nonpregnant women. At this time, however, there is not a clear understanding of why.

Selenium has been found to have a desirable affect on growth and development of certain farm animals. In 1979 the FDA accepted selenium as an additive to the feed of most species because of its nutritional importance.

Toxicity

It is important to know that high doses of selenium are toxic, but the amount to cause such poisoning in humans is not known with certainty. Symptoms include damage to the nervous system, skin lesions, thickening but more fragile nails, loss of hair and nails, nausea, abdominal pain, diarrhea, fatique and irritability.

Like several other nutrients, intake of selenium in moderate amounts is absolutely necessary—in big doses it is dangerous. Some selenium compounds smell like garlic and, interestingly enough, garlic tends to be higher in selenium than other vegetables. One known case of selenium toxicity was diagnosed because the patient had garlic breath — but never consumed garlic in his diet.

Nutrient Losses

Food loses most of its selenium during milling, such as from grain to flour, and it is not replaced by enrichment. There are differences in stability during cooking because selenium occurs in different forms. As a general rule, selenium follows the protein in foods, and little is lost when cooked.

Some molecules are leached into the cooking water (from asparagus and mushrooms), and some forms of selenium are volatile, which means a small portion can go up with the steam when cooking.

Recommended Dietary Allowances for Adults

USA RDA: Adult men, 70 mcg; Adult women, 55 mcg (pregnant, 65 mcg; nursing, 75 mcg per day). These amounts for the reference man and woman relate to the recommended selenium allowance of 0.87 mcg /kg of body weight/day.
Canada: There is no official recommendation at this time, however their documentation states there is evidence that 25-70 mcg a day may cover possible requirements and maintain plasma levels in adults.

Food Sources

Because selenium tends to follow the protein in food, it is found primarily in seafoods, meats, eggs, whole grains, and legumes. Brazil nuts are an unusually high source of selenium. Higher fat meats, sausages, and cream have lower selenium levels, and fruits and vegetables contain only small amounts.

The selenium present in the soil where the food is grown, or where the animal grazes, will determine the amount of selenium in the food grown or raised in the area. Many animal feeds now contain selenium to compensate for such deficiencies, so the selenium content of meats and poultry may be higher than soil samples would suggest.

Baked Goods

	Wt (g)	Svg	Cal	Selenium (mcg) 100 Cal	Svg
Apple crisp, 3" x 3"	78	1 pce	146	2	2.6
Bagel, 3.5" dm, plain	68	1 ea	180	12	21
Bagel, 3.5" dm, egg	68	1 ea	180	14	26
Biscuit:					
Recipe	28	1 ea	100	8	8
From mix	28	1 ea	94.4	8	8
From refrig dough	20	1 ea	65.2	5	3.4
BREADS:					
Banana nut, 1/2" slice	50	1 pce	161	4	6
Boston brown, cnd, 1/2"	45	1 pce	95	16	15
Cornbread muffin	45	1 ea	145	3	4.5
Cracked wheat	25	1 pce	65	25	16
French, 5" x 2.5" x 1"	35	1 pce	100	8	8
Italian, 4.75" x 4" x 1/2"	30	1 pce	83	9	7
Mixed grain	25	1 pce	65	17	11
Oatmeal bread	25	1 pce	65	12	8
Pita pocket, 6.5" dm.	60	1 ea	165	11	18
Pumpernickel, 5" x 4" x 3/8"	32	1 pce	80	14	11.5
Raisin bread	25	1 pce	68	11	7.5
Rye, light, 5" x 3.5" x 7/16"	25	1 pce	65	14	9
Vienna bread	25	1 pce	70	11	7.5
Wheat (white & whole wheat flour)	28	1 pce	72	15	11
White bread	28	1 pce	75	10	8
Whole wheat bread	28	1 pce	70	22	16
Brownies w/frosting	25	1 ea	100	3	3
CAKES: pce = 1/16th cake unless stated otherwise. (Cupcake ≈ 42 g)					
Angel food, 1/12 tube cake	53	1 pce	125	3	3
Boston cream pie, 1/8	120	1 pce	260	1	3
Carrot cake, cream cheese frosting, 2.5" x 3"	112	1 pce	406	1	8.5
Cheesecake, 1/12	92	1 pce	278	2	6
Chocolate, chocolate frosting	69	1 pce	235	1	2
Coffee cake, f/mix , 2.4" x 2.8"	72	1 pce	230	1	2.5
Gingerbread, 3" x 3":					
Recipe	110	1 pce	351	4	14
From mix	63	1 pce	174	2	3
Pound cake, f/mix, 1/2"	30	1 pce	115	3	3
Sheet cake, 3" x 3" pce:					
Plain	86	1 pce	315	2	7
White frosting	121	1 pce	445	2	7
Snack cake, cream filled:					
Chocolate, like Ding-Dongs	28	1 ea	105	1	1
Spongecake, like Twinkies	42	1 ea	155	1	1
Sponge cake, 1/12 tube cake	66	1 pce	194	6	11
White cake:					
Chocolate frosting	77	1 pce	291	1	3
Coconut frosting	70	1 pce	270	1	4
White frosting	71	1 pce	260	1	3
Yellow, chocolate frosting	69	1 pce	240	1	3
Cherry crisp, 3" x 3"	138	1 pce	157	2	2
Cheese puffs/Cheetos	28	1 oz	158	.4	.7
Chips: see corn & tortalla this section; potato chips under Vegetable.					
COOKIES:					
Animal cookie	28	27 ea	120	1.6	2
Butter cookie	25	5 ea	115	1.5	2
Shortbread, commercial	32	4 ea	155	1.2	2
Lady fingers	44	4 ea	158	2.4	4
Oatmeal raisin	52	4 ea	245	1.9	5
Peanut butter, recipe	48	4 ea	245	1.2	3
Sandwich cookies, all	40	4 ea	195	.2	0.4

	Wt (g)	Svg	Cal	Selenium (mcg) 100 Cal	Svg
COOKIES, continued:					
Snickerdoodle	20	1 ea	110	2.9	3
Sugar, f/chilled dough	48	4 ea	235	.7	2
Vanilla wafers	40	10 ea	185	1.6	3
Corn chips	28	1 oz	155	1.6	3
CRACKERS:					
Armenian cracker bread	28	4 pce	117	17	20
Cheese crackers	10	10 ea	50	2	1
Cheese, peanut butter filling	30	4 ea	150	2	2
Graham cracker	14	2 ea	60	3	1.5
Round, like Ritz	9	3 ea	45	2	1
Rye wafers, whole grain	14	2 ea	55	9	5
Saltine	12	4 ea	50	2	1
Sesame	12	4 ea	60	1	0.7
Soda	17	6 ea	75	1	1
Wheat, thin	8	4 ea	35	1	0.5
Whole wheat crackers	8	2 ea	35	1	0.4
Cream puff, custard filled	110	1 ea	280	3	10
Crepe, plain	27	1 ea	47	9	4
Croissant, 4.5" x 4" x 2"	57	1 ea	235	2	5
Croutons, dry bread cubes	30	1 c	111	11	12
Danish pastry:					
Plain pastry	57	1 ea	220	4	9
With fruit	65	1 ea	235	4	10
Doughnut:					
Cake type, medium	50	1 ea	210	2	5
Yeast raised, plain	60	1 ea	235	2	6
Eclair, custard filled, choc iced	94	1 ea	262	3	7
English muffin, plain/sourdough	57	1 ea	135	9	11.5
MUFFINS, average:					
Blueberry	45	1 ea	138	5	7
Bran (wheat)	45	1 ea	133	8	11
Cornbread	45	1 ea	145	3	5
Pancakes:					
Plain, 4", recipe	27	1 ea	60	4	2.4
Plain, 4", mix	27	1 ea	60	4	2.4
Buckwheat, 4", from mix	27	1 ea	55	4	2.4
Whole wheat, 5"	52	1 ea	94	14	13
Patty shell (puff pastry)	71	1 ea	250	4	10
Peach crisp, 3" x 3"	139	1 pce	166	2	3
PIES: pce is 1/6 of 9" pie.					
Apple	158	1 pce	405	2	9
Banana cream pie	198	1 pce	319	2	6
Blueberry	158	1 pce	380	2	8
Cherry	158	1 pce	410	2	8
Chocolate cream	175	1 pce	311	3	9
Coconut cream	172	1 pce	343	2	8
Coconut custard	165	1 pce	384	3	12
Custard pie	152	1 pce	293	4	12
Lemon meringue	140	1 pce	355	3	9
Mincemeat	160	1 pce	395	2	8
Peach	158	1 pce	405	2	9
Pecan	138	1 pce	583	3	19
Pumpkin	200	1 pce	367	3	10
Strawberry chiffon, recipe	162	1 pce	372	2	5
Pretzels:					
Dutch twist	16	1 ea	65	8	5
Thin sticks	3	10 ea	10	10	1
Thin twists	60	10 ea	240	8	20
ROLLS:					
Cinnamon, small	50	1 ea	158	6	10
Dinner roll, 2.5" x 2"	28	1 ea	85	9	7
Hamburger bun	45	1 ea	129	10	13
Hard roll, white	50	1 ea	155	9	14
Hotdog bun	40	1 ea	115	10	11
Rye roll, dark	28	1 ea	79	12	10

ESHA Research

Selenium (Se) mcg

	Wt (g)	Svg	Cal	Selenium (mcg) 100 Cal	Svg
Baked Goods, continued:					
ROLLS, continued:					
Rye roll, light	28	1 ea	76	12	9
Submarine (hoagie)	135	1 ea	400	8	32
Whole wheat roll	35	1 ea	88	26	23
Stuffing, prep w/enr bread:					
From dry	140	1 c	500	2	9
Stove Top stuffing	108	1/2 c	176	7	12
Taco shell	13.6	1 ea	59	2	1
Tortilla chips:					
Plain	28	1 oz	139	2	3
Doritos, nacho/taco	28	1 oz	139	2	3
Tortillas:					
Corn, 6" fried	30	1 ea	87	3	3
Flour, 10.5", large	57	1 ea	168	6	10
Flour, 8"	35	1 ea	105	7	7
Waffles:					
From mix, 7"	75	1 ea	205	7	14
Frozen, 4"	35	1 ea	98	6	6

Dairy & Dairy Products

	Wt (g)	Svg	Cal	100 Cal	Svg
CHEESE (1.5" cube ≈ 1 oz):					
American, processed	28	1 oz	106	2	3
American cheese food	28	1 oz	94	4	4
American cheese spread	28	1 oz	82	4	3
Average — hard cheeses	28	1 oz	85-105	1-3	1-3
Cheddar cheese	28	1 oz	114	4	4
Cottage cheese:					
Lowfat 1%	226	1 c	164	8	14
Lowfat 2%	226	1 c	205	7	14
Creamed, large curd	225	1 c	235	6	14
Creamed, small curd	210	1 c	215	6	13
Creamed, with fruit	226	1 c	279	5	14
Dry curd	145	1 c	123	10	12
Ricotta, w/ part skim	246	1 c	340	1.5	5
Ricotta, w/ whole milk	246	1 c	428	1	5
Swiss cheese, hard	28	1 oz	107	4	4
Swiss, processed	28	1 oz	95	4	4
Swiss cheese food	28	1 oz	92	4	3
Kefir beverage	233	1 c	160	3	4
MILK (cow):					
Skim	245	1 c	86	8	7
Lowfat 1%	244	1 c	102	7	7
Lowfat 2%	244	1 c	121	5	6
Whole (3.3% fat)	244	1 c	150	2	3
Buttermilk (< 1% fat)	245	1 c	99	3	3
Canned:					
Skim, evap	255	1 c	200	2	3
Whole, evap (7.56% fat)	252	1 c	340	1	3
Dry, nonfat	68	1 c	244	5	12
Milk (other):					
Goat	244	1 c	168	1	2
Human breast milk	246	1 c	171	3	4
Sheep	245	1 c	264	1	2
MILK BEVERAGES & mixes:					
Chocolate milk, commercial:					
Lowfat 1%	250	1 c	160	5	8
Lowfat 2%	250	1 c	180	3	6
Whole (3.3% fat)	250	1 c	210	1.5	3
Chocolate flavor, to be mixed with water:					
Powder (includes dry milk)	28	1 oz	100	2	2
Drink, prep w/water	206	3/4 c	100	2	2

	Wt (g)	Svg	Cal	100 Cal	Svg
BEVERAGES & MIXES, continued:					
Chocolate flavor, to be mixed with milk:					
Powder	21.6	3/4 oz	75	1	1
Drink, prep w/milk	266	1 c	226	1	3
Malted milk, prep w/whole milk:					
Chocolate flavor	265	1 c	229	1	3
Natural flavor	265	1 c	237	1	3
MILK DESSERTS:					
Custard, baked	265	1 c	305	6	18
Ice cream, vanilla:					
Regular	133	1 c	269	4	9
Rich	148	1 c	349	2	8
Soft serve	173	1 c	377	3	13
Ice milk, vanilla:					
Regular	131	1 c	184	3	5
Soft serve	175	1 c	223	6	13
PUDDINGS, prepared, 5 oz can ≈ 1/2 c:					
Chocolate, ckd/instant mix	260	1 c	305	1	4
Chocolate, 5 oz can	142	1 can	205	1	3
Rice, ckd f/mix	132	1/2 c	155	1	2
Tapioca, recipe	165	1 c	220	4	10
Tapioca, ckd f/mix	130	1/2 c	145	1	2
Tapioca, 5 oz can	142	1 can	160	2	3
Vanilla, recipe	255	1 c	285	2	6
Vanilla, ckd/instant mix	130	1/2 c	148	1	2
Vanilla, 5 oz can	142	1 can	220	1	3
Sherbet	193	1 c	270	2	4
Yogurt, frozen, average	174	1 c	220	2	5
YOGURT, commercial:					
Lowfat, plain	227	1 c	144	4	6
Lowfat, fruit	227	1 c	231	1	3
Lowfat, coffee/vanilla	227	1 c	193	3	5
Nonfat	227	1 c	127	6	8
Whole	227	1 c	138	2	2
Yogurt cheese, recipe	208	1 c	222	5	11

Eggs

	Wt (g)	Svg	Cal	100 Cal	Svg
Egg, chicken, ckd/raw:					
Whole egg	50	1 ea	77.5	16	12.4
White only	33.4	1 ea	17	.3	4.4
Yolk only	16.6	1 ea	59	13.5	8

Fruits & Fruit Juices

	Wt (g)	Svg	Cal	100 Cal	Svg
APPLE w/peel, 2.75" dm.	138	1 ea	80	.7	.55
Applesauce, unsweetened	244	1 c	106	1	1
Apple juice, canned/bottled	248	1 c	116	<1	.4
APRICOTS, fresh, pitted	106	3 ea	51	3	1.4
Banana, 8.75", 176g whole	114	1 ea	105	1	1
Blackberries, fresh	144	1 c	74	1	.7
Cherries, sour, canned	244	1 c	90	2	2
CHERRIES, SWEET, fresh, pitted	68	10 ea	49	2	1
Cranberry apple juice	253	1 c	169	.2	.4
Dates, whole, pitted	83	10 ea	228	.7	2
Figs, fresh, medium	50	1 ea	37	3	1
Figs, dried	187	10 ea	477	2	11
GRAPEFRUIT (half = 241g w/refuse):					
Pink or red half	123	1 ea	37	3	1
White half	118	1 ea	39	3	1
Grapefruit juice, cnd, unswtnd	247	1 c	93	.5	.5

Fruits & Fruit Juices, continued:	Wt (g)	Svg	Cal	Selenium (mcg) 100 Cal	Svg
Grapes, fresh:					
Thompson seedless	50	10 ea	35	14	5
Tokay/Emperor	57	10 ea	40	14	6
Slip skin w/seeds	92	1 c	58	15	9
Grapes, cnd, hvy syrup	256	1 c	187	4	7
Grape juice:					
From frozen	250	1 c	128	2	3
Cnd/bottled	253	1 c	155	2	3
Mandarin oranges, canned	252	1 c	155	1	2
Melons: also see Watermelon.					
Cantaloupe, cubes	160	1 c	57	1.4	1
Casaba, cubes	170	1 c	45	1.9	1
Honeydew, cubes	170	1 c	60	1.4	1
Frozen melon balls, mixed	173	1 c	55	1.3	1
Nectarines (med = 1 c slices)	136	1 ea	67	1.8	1
ORANGE, 2-5/8" dm., 180g whole	131	1 ea	60	2.0	1
Orange juice, f/frozen	249	1 c	110	<1	.5
Orange grapefruit jce, cnd	247	1 c	105	2.1	2
Peaches, fresh, 2.5"	87	1 ea	37	3	1
Pears, fresh, Bartlett	166	1 ea	98	1	1
Pineapple:					
Fresh chunks	155	1 c	76	1	1
Canned pces, juice pack	250	1 c	150	1	1.6
Pineapple juice, cnd, unswtnd	250	1 c	140	.6	1
Plums, med, 2-1/8" dm.	66	1 ea	36	1	.5
Prune juice, bottled	256	1 c	181	.7	1
Prunes, dried, pitted	84	10 ea	201	1	2
Raisins, dark, unpacked	145	1 c	435	2	9
Strawberries, fresh	149	1 c	45	3	1
Tangerine, fresh	84	1 ea	37	2	.6
Watermelon pieces	160	1 c	50	1	.6

Grains & Grain Products
Cereals, Grains, Flours, Noodles, Pasta, Popcorn

	Wt (g)	Svg	Cal	100 Cal	Svg
Barley, cooked:					
Pearled	157	1 c	193	11	22
Whole	200	1 c	200	14	27
Bran: see Rice, Wheat.					
Buckwheat flour, dark	98	1 c	338	3	9
Buckwheat flour, light	98	1 c	340	2	6
Bulgar wheat, cooked	182	1 c	151	10	15
CEREALS, COLD (Ready To Eat):					
100% Bran	66	1 c	178	4	6
100% Natural	28	1/4 c	135	2	3
40% Bran flakes	43	1 c	139	6	8
All-Bran	28	1/3 c	70	5	3
Cheerios	23	1 c	89	10	9
Corn Flakes	28	1.25 c	110	2	2
Cracklin' Oat Bran	60	1 c	229	2	5
Crispy Wheat 'N Raisins	43	1 c	150	2	3
Nutri-Grain, corn	42	1 c	160	8	12
Nutri-Grain, wheat	44	1 c	158	9	14
Raisin Bran	53	1 c	164	2	4
Rice Krispies, Kelloggs	29	1 c	112	6	6
Frosted Mini-Wheats	31	4 ea	111	2	3
Granola, homemade	122	1 c	595	4	23
Granola, Nature Valley	113	1 c	503	2	10
Grape Nuts	57	1/2 c	202	3	5
Grape Nuts flakes	28	7/8 c	102	18	18
Honey & Nut Corn flakes	28	3/4 c	113	2	2
Honey Nut Cheerios	33	1 c	125	12	15
Shredded Wheat, sm. biscuits	32	3/4 c	115	1	2
Special K	32	1.5 c	125	16	20
Super Golden Crisp	33	1 c	123	21	26
Total cereal, wheat w/calcium	33	1 c	122	3	4

CEREALS, COLD, continued:	Wt (g)	Svg	Cal	Selenium (mcg) 100 Cal	Svg
Wheat Chex	46	1 c	169	2	4
Wheat & Raisins Chex	54	1 c	185	2	4
Wheaties	29	1 c	101	4	4.4
CEREAL, HOT (Cooked):					
Corn grits, enriched	242	1 c	145	5	7.5
Cream of Rice	244	1 c	126	7	9
Cream of Wheat	244	1 c	140	19	27
Farina	233	1 c	116	22	26
Malt-O-Meal	240	1 c	122	7	8
Maypo	180	3/4 c	128	7	9
Oatmeal, from rolled oats					
(Reg, Quick or Instant)	234	1 c	145	14	20
Oatmeal, fortified, from packet:					
Plain	177	3/4 c	104	14	15
Other flavors, avg	164	3/4 c	160	9	14
Ralston	253	1 c	134	15	20
Roman Meal	181	3/4 c	111	6	7
Wheatena	243	1 c	135	20	27
Whole wheat cereal	242	1 c	150	13	19
Corn, flour:					
Regular	117	1 c	422	.7	3
Masa Harina, enr.	114	1 c	416	2	9
Cornmeal, dry:					
Degermed, enriched	138	1 c	505	2	9
Bolted, nearly whole	122	1 c	441	2	10
FLOUR: see specific grain, nut, vegetable.					
Macaroni noodles, cooked:					
Regular, enriched	140	1 c	197	19	38
Vegetable, enriched	134	1 c	172	20	35
Whole wheat	140	1 c	174	24	41
Noodles:					
Chow mein noodles, dry	45	1 c	237	8	19
Egg noodles, cooked, enriched	160	1 c	213	20	43
Spinach noodles, cooked	140	1 c	182	15	28
Oats, rolled, dry	81	1 c	311	7	22
PASTA: see Macaroni, Noodles, Spaghetti.					
Popcorn, oil popped, salted	11	1 c	55	2	1
RICE, cooked:					
Brown	195	1 c	217	12	26
White, regular	205	1 c	264	7	19
White, converted	175	1 c	200	9	17
White, instant	165	1 c	162	7	12
Wild rice	164	1 c	166	3	4.6
Rice bran	83	1 c	262	12	32
Rice polish	158	1 c	578	6	32
Rye flour:					
Dark	128	1 c	415	11	45
Light	102	1 c	361	6	22
Soy flour:					
Full fat, raw	85	1 c	371	2	6
Low fat	88	1 c	326	2	8
Spaghetti noodles, cooked:					
Enriched	140	1 c	197	21	41
Whole wheat spaghetti	140	1 c	174	16	28
WHEAT:					
Bran	30	1/2 c	65	28	18
Flours:					
All purpose, white, unsifted	125	1 c	455	8	38
Cake flour, sifted	96	1 c	348	7	23
Self-rising	125	1 c	442	9	38
Semolina	167	1 c	601	3	20
Whole wheat	120	1 c	407	24	96
Wheat germ, raw	100	1 c	360	20	71
Wheat germ, toasted	113	1 c	432	26	114

ESHA Research

	Wt (g)	Svg	Cal	Selenium (mcg) 100 Cal	Svg		Wt (g)	Svg	Cal	Selenium (mcg) 100 Cal	Svg
Grains & Grain Products, continued:						**SALMON,** cooked:					
Wheat, rolled, cooked	240	1 c	142	14	20	Baked/broiled, mixed	85	3 oz	183	38	70
Wheat, rolled, dry	85	1 c	289	5	15	Chinook, smoked	85	3 oz	99	47	46
Whole grain wheat						Salmon, canned:					
(wheat berries) cooked	50	1/3 c	28	19	5.2	Atlantic, small can	220	1 can	281	59	165
Wheat, sprouted	108	1 c	214	16	34	Pink meat w/bone, # 1 can	454	1 can	631	54	341
						Sockeye, drained, #1 can	369	1 can	566	49	277
						Sardines, canned, drained:					
Meats: Fish & Shellfish						Atlantic, 2 ea = 24g	92	1 can	192	25	48
						Pacific, 1 ea = 38g	100	3.5 oz	178	32	57
						Scallop:					
Anchovies, cnd, drained	45	11 ea	95	26	25	Breaded, fried	93	6 ea	200	37	73
Bass, freshwater, bkd/brld	100	3.5 oz	125	34	43	Steamed	100	3.5 oz	113	72	82
Bluefish:						Seatrout (Steelhead), cooked	100	3.5 oz	131	84	110
Baked/broiled	100	3.5 oz	159	31	50	Shad, baked w/bacon	100	3.5 oz	201	10	20
Fried in crumbs	100	3.5 oz	205	22	44						
Carp, baked/broiled	100	3.5 oz	162	98	159	**SHRIMP:**					
Catfish, fried in cornmeal	100	3.5 oz	229	17	38	Boiled, 2 large ≈ 11g	100	3.5 oz	99	65	64
Caviar (1 T = 16g)	100	3.5 oz	252	56	140	Breaded, fried, 2 lrg≈15g	90	12 ea	218	24	53
						Canned, drained	128	1 c	154	53	82
CLAMS:						Canned with liquid	100	3.5 oz	102	54	55
Canned, drained	160	1 c	236	35	83	Smelt, Rainbow, cooked	85	3 oz	106	34	36
Minced w/liquid, small can	183	1 can	145	202	293	Snapper, baked/broiled	100	3.5 oz	128	135	173
Breaded, fried, small	188	20 ea	379	16	60	Sole (Flounder):					
Steamed meat	90	20 ea	133	39	52	Baked/broiled	85	3 oz	99	44	44
Clam nectar, canned	240	1 c	6	38	2.30	Batter-fried	85	3 oz	250	11	28
COD:						Breaded, fried	100	3.5 oz	188	20	37
Baked/broiled	100	3.5 oz	105	51	53	Squid, flour-fried	85	3 oz	149	46	68
Batter-fried	100	3.5 oz	199	23	45	Swordfish, baked/broiled	100	3.5 oz	155	22	34
Canned w/liquid, 11 oz	312	1 can	327	34	110	Trout, baked/broiled	85	3 oz	129	47	60
Smoked	100	3.5 oz	79	53	42						
CRAB:						**TUNA:**					
Blue crab:						Cooked f/fresh, Bluefin	85	3 oz	157	50	78
Cooked meat	135	1 c	138	43	59	Canned, drained (No. 1/2 can):					
Canned meat	135	1 c	133	38	50	Light, oil pack	171	1 can	339	36	123
Dungeness crab, cooked	101	3/4 c	85	69	59	Light, water pack	165	1 can	216	55	119
Crayfish, cooked, moist heat	85	3 oz	97	88	85						
Eel, baked/broiled	100	3.5 oz	236	11	25						
Eel, smoked	100	3.5 oz	330	6	20	**Meats**					
Fish cakes, fried f/frzn	100	3.5 oz	213	7	15	Beef, Pork, Ham, Rabbit, Venison and Veal					
Fish sticks, from frzn	57	2 ea	155	5	8						
						BEEF:					
HADDOCK:						**Chuck blade**, pot roasted, all:					
Baked/broiled	85	3 oz	95	43	41	Lean & fat	85	3 oz	325	7	22
Breaded, fried	85	3 oz	175	26	45	Lean only	85	3 oz	230	12	28
Smoked	100	3.5 oz	116	54	63	**Ground beef**, ckd, avg.	85	3 oz	238	9	22
Halibut:						**Rib**, oven roasted:					
Baked/broiled	85	3.5 oz	119	111	132	Lean & fat	85	3 oz	324	6	20
Smoked	100	3.5 oz	224	61	136	Lean only	85	3 oz	204	12	25
Herring:						**Round steak**, broiled:					
Baked/broiled	100	3.5 oz	203	39	80	Lean & fat	85	3 oz	233	11	25
Canned in oil, w/liquid	100	3.5 oz	208	26	54	Lean only	85	3 oz	165	17	28
Smoked/kippered	100	3.5 oz	217	65	14	**Variety** meats:					
Pickled (1 pce≈15g)	100	3.5 oz	262	16	42	Heart, simmered	85	3 oz	140	13	18
Lobster meat, cooked	145	1 c	142	79	112	Kidney, cooked	85	3 oz	122	77	94
Mackerel:						Liver, pan-fried	85	3 oz	184	26	48
Baked/broiled, Atlantic	100	3.5 oz	262	16	42	Tongue, cooked	85	3 oz	241	8	18
Canned, Jack, tall can	361	1 can	563	14	80	Dried beef, cured	28	1 oz	47	19	9
Mullet, baked/broiled	85	3 oz	127	43	55	Corned beef, canned	85	3 oz	213	8	16
Octopus, raw	100	3.5 oz	82	92	75						
						HAM: see Pork, cured					
OYSTERS:						**LAMB:**					
Raw, Eastern	248	1 c	170	83	141	Cutlet, lean, cooked average	85	3 oz	175	8	15
Raw, Pacific	248	1 c	200	71	141						
Breaded, fried, Eastern	88	6 ea	173	25	44	**PORK:**					
Simmered, Eastern	100	3.5 oz	137	41	56	Bacon, cooked:					
Perch, baked/broiled	85	3 oz	185	14	25	Regular	19	3 pce	109	4	4
Pollock:						Canadian style	47	2 pce	86	19	16
Baked/broiled	100	3.5 oz	99	222	220	Breakfast strips	34	3 pce	156	5	7
Walleye, baked/broiled	100	3.5 oz	113	194	220						

PORK, continued:	Wt (g)	Svg	Cal	Selenium (mcg) 100 Cal	Svg
Center rib chop:					
Broiled, lean & fat	77	1 ea	264	10	27
Broiled, lean only	63	1 ea	162	14	22
Pan-fried, lean & fat	88	1 ea	343	9	31
Pan-fried, lean only	62	1 ea	160	14	22
Roasted, lean & fat	79	1 ea	252	11	28
Roasted, lean only	66	1 ea	162	14	23
Center loin chop:					
Broiled, lean & fat	82	1 ea	284	10	29
Broiled, lean only	72	1 ea	166	14	23
Roasted, lean & fat	88	1 ea	268	11	29
Roasted, lean only	72	1 ea	180	13	24
Pork roast, leg:					
Lean and fat	85	3 oz	250	12	30
Lean only	85	3 oz	187	16	30
Pork roast, loin/rib:					
Lean and fat	85	3 oz	265	6	17
Lean only	85	3 oz	206	9	18
Shoulder, braised:					
Lean and fat	85	3 oz	293	10	30
Lean only	67	2.4 oz	166	14	23
Spareribs, ckd from 1 lb raw					
with bone	177	6.25 oz	703	9	62
Pork heart	85	3 oz	130	35	46
Pork liver	85	3 oz	141	17	24
PORK, CURED, HAM (Also see Bacon under Pork):					
Ham, roasted:					
Lean & fat	85	3 oz	207	12	25
Lean only	85	3 oz	133	22	30
Canned, average	85	3 oz	142	15	21
RABBIT, cooked	85	3 oz	175	.6	1
VEAL (calf)					
Cutlet, braised/broiled	100	3.5 oz	181	7	13
Liver, simmered	85	3 oz	222	22	48
Rib roast	100	3.5 oz	160	8	13
VENISON (deer) cooked	85	3 oz	134	31	42

Meats: Poultry

CHICKEN: A 3 lb chicken ≈ 1.45 lbs raw meat; 1.1 lbs cooked.

	Wt (g)	Svg	Cal	Selenium (mcg) 100 Cal	Svg
All types of meat:					
Fried	140	1 c	307	9	29
Roasted	140	1 c	266	11	29
Stewed	140	1 c	248	9	23
Canned, boned w/broth	142	5 oz	235	9	22
Dark meat:					
Fried	85	3 oz	203	9	17.8
Roasted	85	3 oz	174	10	17.8
Stewed	85	3 oz	163	9	14.2
Light meat:					
Fried	85	3 oz	163	10	17
Roasted	85	3 oz	147	12	17
Stewed	85	3 oz	135	10	13
Breast,* meat & skin (145g raw ≈ 181g raw w/bone):					
Batter-fried	140	1 ea	364	8	28
Flour-fried	98	1 ea	218	13	28
Roasted	98	1 ea	193	10	20
Stewed	110	1 ea	202	4	8.5
Breast,* meat only (118g raw):					
Fried	86	1 ea	161	11	17
Roasted	86	1 ea	142	12	17
Stewed	95	1 ea	144	6	8.4
*2 pieces per bird					

CHICKEN, continued:	Wt (g)	Svg	Cal	Selenium (mcg) 100 Cal	Svg
Drumstick, meat & skin (73g raw; 110g w/bone):					
Batter-fried	72	1 ea	193	8	15
Flour-fried	49	1 ea	120	12	15
Roasted	52	1 ea	112	10	11
Stewed	57	1 ea	116	4	5
Drumstick, meat only (62g raw):					
Fried	42	1 ea	82	7	6
Roasted	44	1 ea	76	12	9
Stewed	46	1 ea	78	6	4
Thigh, meat & skin (94g raw; 120g w/bone):					
Batter-fried	86	1 ea	238	8	20
Flour-fried	62	1 ea	162	12	20
Roasted	62	1 ea	153	9	13
Stewed	68	1 ea	158	4	7
Thigh, meat only (69g raw):					
Fried	52	1 ea	113	10	11
Roasted	52	1 ea	109	10	11
Stewed	55	1 ea	107	6	7
Wing, meat & skin (49g raw; 90g w/bone):					
Batter-fried	49	1 ea	159	6	10
Flour-fried	32	1 ea	103	10	10
Wing, meat only (29g raw):					
Fried	20	1 ea	42	10	4
Roasted	21	1 ea	43	10	4
Chicken gizzard	22	1 ea	34	8	3
Chicken liver	20	1 ea	30	21	6
DUCK, domestic:					
roasted, meat & skin	85	3 oz	286	6	17
GOOSE, domestic:					
roasted, meat & skin	85	3 oz	259	7	17
TURKEY:					
Breast meat, seasoned:					
Barbecued	28	1 oz	40	18	7
Hickory smoked	28	1 oz	35	20	7
Ground turkey, cooked	100	3.5 oz	229	11	25
Roasted:					
All types of meat	85	3 oz	145	6	8
Dark meat	85	3 oz	159	6	9
Light meat	85	3 oz	133	6	7
Frozen slices	85	3 oz	130	8	10
Frozen slices w/gravy	142	5 oz	95	21	20
Turkey patty, breaded, fried	64	1 ea	181	7	13

Meats: Sausages & Lunchmeats

	Wt (g)	Svg	Cal	Selenium (mcg) 100 Cal	Svg
Beef lunchmeat, thin sliced	28	1 oz	50	7	4
Bologna:					
Beef	23	1 pce	72	5	4
Beef and pork	28	1 oz	89	4	4
Cured pork	23	1 pce	57	7	4
Turkey	28	1 oz	56	5	3
Braunschweiger	57	2 oz	205	3	6
FRANKFURTER (hotdog):					
Beef, 8/pkg	57	1 ea	184	3	5
Beef and pork, 8/pkg	57	1 ea	183	3	5
Chicken, 10/pkg	45	1 ea	115	4	4
Turkey, 10/pkg	45	1 ea	102	6	6
HAM:					
Chopped, pkg	42	2 pce	98	6	6
Lunchmeat:					
Extra lean	57	2 oz	75	8	6
Regular	57	2 oz	103	6	6
Thin sliced	28	3 pce	37	13	5
Minced	21	1 pce	55	6	3

ESHA Research

	Wt (g)	Svg	Cal	Selenium (mcg) 100 Cal	Svg
Meats, Sausages & Lunchmeats, continued:					
Ham patty, cooked	60	1 ea	203	6	13
Ham salad spread	240	1 c	518	6	30
Luncheon meat, canned	21	1 pce	70	8	6
Luxury loaf	57	2 oz	80	9	7
Polish sausage	28	1 oz	92	5	5
Pork sausage:					
Cooked link	13	1 ea	48	4	2
Cooked patty	27	1 pce	100	4	4
Poultry spread	13	1 T	25	9	2
SALAMI:					
Beef salami	23	1 pce	58	8	5
Pork and beef	57	2 oz	143	5	7
Turkey salami	57	2 oz	111	6	7
Sausage (brown & serve)	13	1 ea	50	4	2
Smoked link sausage:					
Beef & pork	68	1 ea	229	4	9
Pork	68	1 ea	265	3	9
TURKEY (other):					
Breakfast sausage	28	1 oz	65	8	5
Ham	57	2 oz	73	6	5
Loaf, breast meat	43	2 pce	46	15	7
Pastrami	57	2 oz	74	7	5
Roll, light & dark	57	2 pce	84	10	8
Roll, light meat	57	2 pce	83	9	7
Summer sausage	28	1 oz	50	5	3

Mixed Dishes & Fast Foods

	Wt (g)	Svg	Cal	Selenium (mcg) 100 Cal	Svg
Beef & vegetable stew:					
Recipe	245	1 c	220	7	15
Canned	245	1 c	194	7	14
Beef, macaroni, tomato					
sauce, recipe	226	1 c	189	5	10
Beef pot pie, from frozen	234	1 ea	426	3	12
BURRITO:					
Bean	174	1 ea	322	7	22
Beef	177	1 ea	463	6	29
Beef and bean	175	1 ea	390	6	23
Deluxe combination	198	1 ea	424	5	23
Cheese souffle, recipe	112	1 c	221	7	15
Chicken a la king, recipe	245	1 c	470	2	9
Chicken & noodles, recipe	240	1 c	365	8	29
Chicken chow mein:					
Recipe	250	1 c	255	3	8
Canned	250	1 c	95	1	1
Chicken curry, recipe	337	1.5 c	305	5	14
Chicken pot pie, f/frozen	230	1 ea	430	7	32
Chicken salad w/celery	78	1/2 c	266	4	10
Chili with beans, canned	255	1 c	286	2	6
Cole slaw	120	1 c	84	2	2
Egg salad	183	1 c	438	9	40
ENCHILADA:					
Beef	120	1 ea	292	4	12
Cheese	120	1 ea	330	3	11
Chicken	120	1 ea	269	4	11
French toast, recipe	65	1 pce	123	8	10
LASAGNA:					
Recipe w/meat	245	1 pce	398	9	34
Recipe w/o meat	218	1 pce	316	4	12
Macaroni & cheese:					
Recipe	200	1 c	430	7	30
Canned	240	1 c	230	12	28
Macaroni salad w/o cheese	141	1 c	371	2	7
Manicotti, frozen entree	225	1 ea	271	3	9
Meat loaf	85	3 oz	198	7	14

	Wt (g)	Svg	Cal	Selenium (mcg) 100 Cal	Svg
Moussaka, lamb & eggplant	250	1 c	250	6	16
Pies, fried pastry, commercial:					
Apple	85	1 ea	255	2	5
Cherry	85	1 ea	250	2	4
PIZZA, cheese:					
Regular crust 1/8 of 15"	120	1 pce	290	7	20
Thick crust 1/2 of 10"	208	1 pce	519	7	36
Potato salad w/mayo & eggs	250	1 c	358	4	13
Quiche lorraine, 1/8 pie	176	1 pce	600	3	15
Ravioli, beef, canned	226	1 c	220	4	9
SANDWICHES, fast foods:					
Cheeseburger, 3 oz meat	112	1 ea	300	9	28
Cheeseburger, 4 oz meat	194	1 ea	524	5	24
Chicken patty sandwich	157	1 ea	436	5	23
English muffin, egg, cheese,					
and bacon	138	1 ea	360	3	12
Fish sandwich:					
Large, without cheese	170	1 ea	470	7	35
Regular with cheese	140	1 ea	420	8	35
Hamburger, 3 oz meat	98	1 ea	245	10	25
Hamburger, 4 oz meat	174	1 ea	445	6	28
Hotdog (frankfurter) and bun	85	1 ea	260	5	14
SANDWICHES on part whole wheat bread,					
except when stated as rye:					
Avocado, cheese,					
tomato, sprouts	195	1 ea	432	7	31
Bacon, lettuce, tomato	135	1 ea	327	8	26
Chicken salad sandwich	99.7	1 ea	294	12	35
Corned beef & swiss on rye	147	1 ea	429	6	26
Egg salad sandwich	111	1 ea	319	10	32
Grilled cheese	117	1 ea	393	7	29
Ham sandwich	122	1 ea	256	13	33
Ham on rye	116	1 ea	242	12	28
Ham salad sandwich	125	1 ea	339	9	30
Ham & American cheese	151	1 ea	363	10	37
Ham & swiss on rye	145	1 ea	350	9	33
Peanut butter & jam	100	1 ea	341	7	24
Patty melt on rye	177	1 ea	567	6	36
Reuben, grilled	233	1 ea	480	6	26
Roast beef	122	1 ea	280	10	28
Tuna salad sandwich	116	1 ea	303	17	50
Turkey sandwich	122	1 ea	271	10	27
Turkey ham	122	1 ea	253	11	28
Turkey ham on rye	116	1 ea	239	9	23
SPAGHETTI, pasta &					
tomato sauce w/cheese:					
Recipe	250	1 c	260	8	21
Canned	250	1 c	190	11	21
Spinach souffle	136	1 c	218	3	7
Taco, beef	78	1 ea	207	5	10
Taco, chicken	78	1 ea	172	5	8
Tostada:					
Beans & beef	192	1 ea	332	21	69
Beans & chicken	157	1 ea	249	27	67
Tuna noodle casserole, recipe	202	1 c	251	32	80
Tuna salad	205	1 c	383	37	140
Turkey pot pie f/frozen	233	1 ea	416	8	32
Veal Parmigiana, entree	205	7.25 oz	372	4	15

Nuts & Seeds

	Wt (g)	Svg	Cal	Selenium (mcg) 100 Cal	Svg
Almonds, dried, whole	142	1 c	837	1	7
Brazil nuts, dry	140	1 c	919	165	1518
Cashews, oil roasted	130	1 c	748	5	27-37
Chestnuts, roasted	143	1 c	350	1	5

	Wt (g)	Svg	Cal	Selenium (mcg) 100 Cal	Svg
Nuts & Seeds, continued:					
Coconut:					
Raw, grated	80	1 c	283	6	16
Dried unsweetened	78	1 c	515	6	29
Flaked, pkg, sweetened	74	1 c	351	4	14
Coconut milk, raw	240	1 c	552	2	9
Filberts (hazelnuts), whole	135	1 c	853	<1	4
MIXED NUTS w/peanuts (almonds, brazil nuts, cashews, filberts, peanuts, pecans):					
Dry roasted	137	1 c	814	4	32
MIXED NUTS w/o peanuts (cashews, almonds, brazil nuts, pecans, filberts):					
Oil roasted	144	1 c	886	3	28
PEANUTS:					
Dry roasted	146	1 c	855	1	11
Oil roasted	144	1 c	837	1	11
Peanut butter	32	2 T	188	1	3
Peanut flour, defatted	60	1 c	196	1	2
Pecans, dried, chopped	119	1 c	794	1	6-14
Pistachios, dried, shelled	128	1 c	739	2	13
Pumpkin seeds, roasted	64	1 c	285	1	4
Pumpkin/squash kernels:					
Dried kernels	138	1 c	747	1	10
Roasted kernels	227	1 c	1185	2	22
Sesame seeds:					
Dried kernels	150	1 c	882	1	12
Whole, dried	144	1 c	825	1	8
Soybeans, roasted	86	1/2 c	405	1	3.5
Sunflower seeds:					
Dried	144	1 c	821	13	103
Oil roasted	135	1 c	830	12	98
Sunflower seed butter	16	1 T	93	11	10
Tahini, sesame butter	15	1 T	91	7	6
Walnuts, chopped:					
Black	125	1 c	759	3	22
English	120	1 c	770	1	9

Soups, Sauces, Gravies

	Wt (g)	Svg	Cal	Selenium (mcg) 100 Cal	Svg
GRAVY:					
Beef, recipe	135	1/2 c	151	1	2
Beef, canned	233	1 c	124	2	2
Chicken, recipe	130	1/2 c	163	1	2
Chicken, canned	238	1 c	189	1	2
SAUCES (also see "Other"):					
Cheese sauce:					
Recipe	101	1/2 c	216	3	6
From mix w/milk	279	1 c	305	1	3
Hollandaise, recipe	160	1 c	867	4	33
Spaghetti sauce:					
Recipe	220	1 c	179	2	3
Canned	249	1 c	272	.7	2
Spaghetti sauce w/meat:					
Recipe	248	1 c	297	3	9
Canned	206	7/8 c	220	2	4
White sauce:					
Recipe, medium	250	1 c	395	2	8
From mix w/milk	264	1 c	240	2	5
SOUPS: soups are prepared from canned unless otherwise stated. RTS = Ready To Serve. For soups prepared with milk, assume whole milk.					
Celery, cream of	248	1 c	165	5	9
Chicken, cream of	248	1 c	191	4	9
Chicken rice	241	1 c	60	5	3

	Wt (g)	Svg	Cal	Selenium (mcg) 100 Cal	Svg
SOUPS, continued:					
Chicken rice, chunky, RTS	240	1 c	127	3	3
Chili beef	250	1 c	169	14	24
Clam chowder, New England	248	1 c	163	10	17
Mushroom, cream of, prep w/milk	248	1 c	205	4	9
Oyster stew	245	1 c	134	25	33
Pea soup	254	1 c	239	1	3
Potato, cream of	248	1 c	148	2	2
Tomato soup:					
Prep. with milk	248	1 c	160	5	9
Prep. with water	244	1 c	86	1	1

Vegetables & Legumes

	Wt (g)	Svg	Cal	Selenium (mcg) 100 Cal	Svg
Artichoke hearts, marinated	170	6 oz	168	.7	1.1
Asparagus, pieces:					
Fresh, uncooked	67	1/2 c	15	36	5
Cooked from fresh	90	1/2 c	23	16	4
Cooked from frozen	180	1 c	50	13	7
Canned, drained	121	1/2 c	16	7	1
BEANS: also see Garbanzo.					
Baked beans (dry white beans w/spices & sauce):					
Canned with franks	257	1 c	366	3	12
Canned w/pork & swt sce	253	1 c	282	4	12
Canned w/pork & tom sce	253	1 c	247	5	12
Black beans, ckd from dry	172	1 c	227	6	14
Great northern, ckd from dry	177	1 c	210	2	4
Green beans (snap):					
Cooked from fresh	125	1 c	44	3	1.4
Cooked from frozen	135	1 c	36	4	1.4
Canned, drained	135	1 c	26	3	.8
Canned w/liquid	240	1 c	36	2.5	.9
Kidney beans, all types:					
Ckd f/dry	177	1 c	225	2	4
Canned	256	1 c	208	2	3
Lima beans:					
Cooked from fresh	170	1 c	208	2	4
Ckd f/frozen, large	85	1/2 c	85	2	2
Ckd f/frzn, baby	90	1/2 c	94	2	2
Cooked from dry	188	1 c	217	2	5
Canned, drained	170	1 c	164	1	2
Canned with liquid	241	1 c	191	1	2
Navy, ckd from dry	182	1 c	259	2	5
Pinto, ckd from dry	171	1 c	235	5	11
Refried, canned	253	1 c	270	5	14
Yellow wax beans, see green beans.					
White beans, ckd from dry	179	1 c	253	1.5	3.8
Beets:					
Ckd from fresh, whole	100	2 ea	31	1.3	.4
Canned, diced, drained	85	1/2 c	27	.8	.2
Beet greens, ckd f/fresh, drained	144	1 c	40	4.3	1.7
Black-eyed peas: see Peas.					
BROCCOLI chopped:					
Fresh, uncooked	88	1 c	24	.7	.2
Cooked from fresh	156	1 c	44	.7	.312
Cooked from frozen	184	1 c	51	.7	.4
Broccoli, cauliflower, red pepper, ckd f/frzn	95	2/3 c	25	2	.4
Brussels sprouts:					
Cooked from fresh	156	1 c	60	18	11
Cooked from frozen	155	1 c	65	15	10
CABBAGE:					
Common, shredded	70	1 c	16	10	1.6
Common, ckd from fresh	150	1 c	32	8	2.6

Vegetables & Legumes, continued:

	Wt (g)	Svg	Cal	Selenium 100 Cal	Selenium Svg
CABBAGE, continued:					
Red, fresh, shredded	70	1 c	19	27	5
Red, ckd from fresh	75	1/2 c	16	25	4
Savoy, shredded	70	1 c	20	13	3
Savoy, ckd from fresh	145	1 c	35	8	3
CARROTS:					
Fresh, 7.5" x 1-1/8" dm.	72	1 ea	31	3	1.0
Fresh, grated	55	1/2 c	24	3	.8
Cooked f/fresh, slices	78	1/2 c	35	3	.9
Cooked f/frozen	73	1/2 c	26	4	.9
Canned, drained	73	1/2 c	17	6	.9
Carrot juice	123	1/2 c	49	2	.9
Cauliflower:					
Fresh, uncooked	50	1/2 c	12	3	.4
Cooked from fresh	62	1/2 c	15	3	.4
Cooked from frozen	180	1 c	34	3	1
Celeriac (celery root) cooked	100	3.5 oz	25	32	8
Celery:					
Fresh, outer stalk, 7.5"	40	1 ea	6	15	.88
Cooked from fresh, diced	150	1 c	27	6	1.65
CORN, kernels:					
Fresh, uncooked	77	1/2 c	66	.5	.3
Cooked from fresh	82	1/2 c	89	.4	.3
Cooked from frozen	82	1/2 c	67	.6	.4
Canned, drained	82	1/2 c	66	.5	.3
Canned with liquid	128	1/2 c	79	.4	.4
Canned, vacuum pack	210	1 c	166	.6	1.1
Corn, cream style, canned	128	1/2 c	93	.7	.6
Cucumber, fresh slices w/peel	28	7 pce	4	90	3
Dandelion greens:					
Fresh, chopped	55	1 c	25	1.8	.4
Cooked from fresh	105	1 c	35	1.8	.6
Eggplant, ckd from fresh	160	1 c	45	3	1
Escarole (curly endive) chopped	50	1 c	9	32	3
Garbanzos, cooked f/dry	164	1 c	269	2	6
LETTUCE, chopped:					
Butterhead	56	1 c	7.3	6	.5
Iceberg	56	1 c	7.3	3	.2
Loose leaf	56	1 c	10	2	.2
Romaine	56	1 c	9	3	.2
Kohlrabi:					
Fresh slices	140	1 c	38	30	11
Cooked from fresh	165	1 c	48	28	13
Leeks, chopped, ckd f/fresh	52	1/2 c	24	13	3
Lentils, ckd from dry	198	1 c	231	3	7
Mushrooms:					
Fresh, sliced (1 = 18g)	35	1/2 c	9	45	4
Cooked from fresh	78	1/2 c	21	43	9
Canned, drained	78	1/2 c	19	42	8
Onions:					
Fresh, chopped	160	1 c	61	5	3
Cooked from fresh	105	1/2 c	46	3	1.45
Dehydrated flakes	14	1/4 c	45	3	1
Onions, spring, chopped	50	1/2 c	16	4	.6
Parsley:					
Fresh chopped	30	1/2 c	10	1.5	.2
Freeze dried	1.4	1/4 c	4	2	.1
Parsnips:					
Fresh slices	133	1 c	100	3	3
Cooked from fresh	156	1 c	125	3	4
PEAS:					
Black-eyed peas:					
Cooked from fresh	165	1 c	160	19.4	31
Cooked from frozen	170	1 c	224	14.2	32
Cooked from dry	171	1 c	198	23.7	47
Canned	240	1 c	184	26.1	48

	Wt (g)	Svg	Cal	Selenium 100 Cal	Selenium Svg
PEAS, continued:					
Green peas:					
Fresh, uncooked	145	1 c	118	6.1	7.25
Cooked from fresh	160	1 c	134	6.6	9
Cooked from frozen	80	1/2 c	63	5.1	3
Canned, drained	85	1/2 c	59	5.8	3
Canned with liquid	124	1/2 c	61	6.6	4
Green peas, edible-pods:					
Fresh, uncooked	145	1 c	61	7.5	5
Cooked from fresh	160	1 c	67	5.2	4
Cooked from frozen	80	1/2 c	42	4.0	2
Split peas, ckd f/dry	196	1 c	231	8.2	19
Peas & carrots:					
Cooked f/frozen	80	1/2 c	38	5.9	2
Canned	128	1/2 c	48	5.4	3
Peas, sprouted:					
Fresh sprouts	120	1 c	154	5.7	9
Cooked from fresh	100	3.5 oz	118	4.9	6
PEPPERS, HOT, green/red (1 pod ≈ 45g):					
Chopped, fresh	75	1/2 c	30	2.0	.6
Jalapeno, chopped, cnd	68	1/2 c	17	3.2	.5
Green chili, canned	68	1/2 c	17	1.4	.2
PEPPERS, SWEET, green/red (1 pod ≈ 74g):					
Fresh, chopped	50	1/2 c	14	3.8	.53
Ckd f/fresh, chopped	68	1/2 c	19	2.6	.5
POTATOES:					
Baked:					
Flesh & skin	202	1 ea	220	.8	2
Flesh only	156	1 ea	145	1.0	1.4
Boiled, flesh only	135	1 ea	116	1.0	1.2
Cooked f/frozen	70	1 ea	46	1.3	.6
Cottage fried, f/frozen	50	10 ea	109	.4	.5
French fried f/frozen:					
Fried in oil	50	10 ea	158	.3	.5
Oven heated	50	10 ea	111	.4	.5
Hash brown, f/frozen	156	1 c	340	.4	1.4
Mashed:					
Prepared with milk	210	1 c	162	1.2	2
Prepared f/instant	215	1 c	239	.9	2.1
Canned, 1" diam.	70	2 ea	42	1.5	0.6
Dehydrated flakes	45	1 c	159	1.3	2
Potato puffs/tater tots heated f/frzn	62	1/2 c	138	.4	0.6
Potato dishes:					
Au gratin, recipe	245	1 c	322	.7	2
Au gratin, from mix	245	1 c	228	.5	1
Scalloped, recipe	245	1 c	210	1.0	2
Scalloped, from mix	245	1 c	228	.5	1
Potato chips, 1 oz ≈ 14 chips	28	14 ea	148	1.5	2
Potato pancakes	76	1 ea	237	.2	.6
Pumpkin, mashed:					
Cooked from fresh	245	1 c	50	4.9	2.5
Canned (can be a mixture w/squash)	123	1/2 c	42	3.9	2
Soybeans, ckd from dry	172	1 c	298	1.0	3
Soybeans, mature, sprouted:					
Fresh sprouts	35	1/2 c	45	.9	.4
Steamed	94	1 c	76	.9	.7

SOYBEAN PRODUCTS: see tofu in this section; miso in Other; roasted soybeans in Nuts & Seeds; soy milk in Dairy; soy flour in Grains.

	Wt (g)	Svg	Cal	Selenium 100 Cal	Selenium Svg
Spinach:					
Fresh, chopped	56	1 c	12.3	6.6	.8
Cooked from fresh	180	1 c	41	6.6	3
Cooked f/frozen, leaf	190	1 c	53	5.6	3
Canned, drained	214	1 c	50	3.0	1.5

	Wt (g)	Svg	Cal	Selenium (mcg) 100 Cal	Svg

SQUASH, SUMMER:

	Wt (g)	Svg	Cal	100 Cal	Svg
Crookneck, fresh slices	130	1 c	24	8.1	.2
Crookneck, ckd from fresh	180	1 c	36	6.9	2.5
Scallop, fresh slices	130	1 c	24	8.1	.2
Scallop, slices, ckd from fresh	90	1/2 c	14	8.3	1
Zucchini, fresh slices	130	1 c	19	10.3	.2
Zucchini, ckd from fresh	180	1 c	29	5.2	1.5

SQUASH, WINTER:

	Wt (g)	Svg	Cal	100 Cal	Svg
Acorn (Danish), baked	245	1 c	137	2.3	3
Acorn (Danish), boiled	245	1 c	83	3.8	3
Butternut, baked	245	1 c	99	3.2	3
Butternut, ckd f/frozen	240	1 c	94	3.2	3
Hubbard, baked	240	1 c	120	3.2	4
Hubbard, boiled	236	1 c	70	3.6	2.5
Spaghetti, baked/boiled	155	1 c	45	2.5	1
Succotash:					
Cooked from fresh	192	1 c	222	.5	1
Cooked from frozen	170	1 c	158	.5	1
Sweet potatoes:					
Baked, whole	114	1 ea	118	.6	.7
Boiled, peeled	151	1 ea	160	.4	.7
Candied, recipe	105	1 pce	144	.5	.7
Canned, mashed	128	1/2 c	129	.6	.7
Canned, vacuum pack	255	1 c	233	.6	1.5
Tofu, firm	126	1/2 c	183	1.9	3.5
Tofu, regular	124	1/2 c	94	1.9	1.8

TOMATOES:

	Wt (g)	Svg	Cal	100 Cal	Svg
Fresh, whole, 2.6" dm	123	1 ea	26	2.3	.6
Fresh, chopped	180	1 c	38	2.4	.9
Cooked from fresh	240	1 c	65	2.2	1.4
Canned, whole	240	1 c	47	2.3	1.1
Tomato products:					
Tomato juice	244	1 c	42	1.1	.5
Tomato paste	262	1 c	220	.6	1.4
Tomato puree	250	1 c	102	2.0	.2
Tomato sauce	245	1 c	74	2.0	1.5
Turnips, fresh cubes	130	1 c	35	2.6	.9
Turnips, cooked cubes	78	1/2 c	14	3.9	.5
Turnip greens:					
Cooked from fresh	144	1 c	29	4.5	1.3
Cooked from frozen	82	1/2 c	24	3.0	.7

VEGETABLES, MIXED (corn, peas, carrots, green beans, lima beans):

	Wt (g)	Svg	Cal	100 Cal	Svg
Cooked from frozen	182	1 c	107	3.4	4
Canned, drained	163	1 c	77	4.3	3
Vegetable juice cocktail	242	1 c	46	1.0	0.5

Zucchini: see Squash, summer.

Other

Cooking ingredients, condiments, fats, flavorings, other beverages, spices, etc.

	Wt (g)	Svg	Cal	100 Cal	Svg
Beer (12 fl oz = 1.5 c):					
Regular	356	1.5 c	146	3	4.5
Light	354	1.5 c	100	4	3.6
Butter	14	1 T	100	.1	.1

CANDY & CANDY BARS:

	Wt (g)	Svg	Cal	100 Cal	Svg
Caramel, plain/chocolate	28	1 oz	115	1	1
Chocolate covered:					
Almonds	165	1 c	935	.4	4
Coconut	28	1 oz	133	1	2
Peanuts	170	1 c	954	5	44
Raisins	187	1 c	733	2	14
Fondant (candy corn)	28	1 oz	105	1	1
Fudge, chocolate	28	1 oz	115	1	1

CANDY, continued:

	Wt (g)	Svg	Cal	100 Cal	Svg
Fudge, choc w/nuts	28	1 oz	114	1	1
Milk chocolate, plain	28	1 oz	145	1	1
Milk chocolate, with almonds	28	1 oz	150	1	1
Milk chocolate, with peanuts	28	1 oz	155	1	2
Sugar coated almonds	28	7 ea	146	.5	1
Carob flour	103	1 c	185	4	7
Catsup	15	1 T	16	6	1
Chili sauce, red pepper	31	2 T	6	4	0.2

CHOCOLATE:

	Wt (g)	Svg	Cal	100 Cal	Svg
Baking chocolate, unsweetened	28	1 oz	145	1	1
Choc. chips, semi-sweet	170	1 c	860	.1	7
Dark, sweet	28	1 oz	150	.1	.1
Syrup, thin type	38	2 T	85	2	1
Cocoa powder	86	1 c	224	8	18
Coffee, brewed	240	1 c	2	5	.1
Gelatin, dry, plain, env.	7	1 ea	25	8	2
Gelatin salad dessert	120	1/2 c	70	3	2
Grape drink	250	1 c	112	2	2.0
Honey	21	1 T	65	.3	.2
Hummous or Humous	246	1 c	420	1	6
Lemonade, prep f/frzn	248	1 c	100	.1	.1
Limeade, prep f/frzn	247	1 c	102	.1	.1
Jam or preserves	20	1 T	54	1	.4
Jelly, assorted	18	1 T	49	1	.4
Margarine:					
Regular, 80% fat	14	1 T	100	.1	.14
Spread, 60% fat	14	1 T	75	.1	.1
Imitation, soft, 40% fat	14.2	1 T	50	.1	.07
Mayonnaise	13.8	1 T	100	.8	.8
Miso (soybean product)	138	1/2 c	284	1	2
Molasses:					
Blackstrap	40	2 T	85	30	25
Light	20	1 T	43	30	13
Mustard, prepared	125	1/2 c	94	31	29
Olive oil	216	1 c	1909	<.1	.4
Olives:					
Green, pitted	39	10 ea	45	1	.4
Black, pitted, large	45	10 ea	52	.8	.4
Pickles:					
Sweet, medium	35	1 ea	41	.4	.15
Dill pickle	65	1 ea	11.7	6	.7
Fresh pack, slices	30	4 pce	20	2	.3
Pickle relish, sweet	15	1 T	20	1	.2
Sugar, brown	220	1 c	820	<1	2
Sunflower oil	14	1 T	125	2.0	2.5
Syrup:					
Corn syrup, dark	328	1 c	944	3	32
Corn syrup, light	328	1 c	912	4	40
Pancake syrup	84	1/4 c	244	.6	2

SPICES:

	Wt (g)	Svg	Cal	100 Cal	Svg
Caraway seed	6.7	1 T	22	3	.6
Chili powder	7.5	1 T	24	9	.2
Cinnamon	6.8	1 T	17.7	6	1
Garlic, cloves	12	4 ea	17.9	17	3
Garlic powder	8.4	1 T	28	19	5
Oregano, ground	4.5	1 T	14	1	.1
Paprika	6.9	1 T	20	4	.8
Pepper, black	6.4	1 T	16.3	2	.3
Pepper, cayenne/red	5.3	1 T	17	1	.2
Salt	16.5	1 T	0	High	.5
Tea, brewed	240	1 c	2	8	.15
Wine:					
Rred	118	1/2 c	85	.7	.6
Rosé	118	1/2 c	84	.4	.4
White	118	1/2 c	79	.4	.4
Yeast, brewer's	8	1 T	25	14	3.4
Yeast, dry active	30	4 T	80	3	2.6

ESHA Research

Sodium

Sodium is the major element in regulating fluid balance in the body, and its function is balanced by potassium. Both these elements dissolve in water and are called electrolytes.

The movement of fluids in and out of the cells is largely a function of the relative concentration of the electrolytes on either side of the cell membrane. Primarily, the sodium ion is on the outside of the cell and the potassium ion is on the inside. As you might suspect, there should generally be similar intakes of both minerals to continue this balance. Only under extreme conditions of heavy long term sweating (such as athletic tri-athelons) would you need more sodium to replace the sodium lost through sweat. Sodium is also also involved with the generation of nerve impulse, acid-base balance, and the metabolism of carbohydrates and protein.

There is currently great concern about excess sodium intake in connection with high blood pressure or hypertension. The only findings that seem to hold up strongly, however, are that some people are sensitive to sodium, and those people can reduce their high blood pressure by reducing salt intake. People with normal blood pressure do not seem to be affected by large salt intakes.

Most people are usually unaware of the sodium they consume every day. Because processed foods contain large amounts of salt (sodium chloride), and there are so many processed foods, Americans consume far more sodium than they need (see "Food for Thought" at the beginning of this section for some of the food additives that contain sodium as well).

It will be interesting to see the developments from the increased use of salt substitutes, which are primarily potassium-based rather than sodium-based, thus creating perhaps another imbalance if over-consumed.

Estimated Safe and Adequate Amounts for Adults

USA: Adults: Approximately 2,000-2,400 mg, with a *minimum* requirement of 500 mg.
   The *minimum* recommended intake of 500 mg a day is substantially exceeded by the average diets in the U.S. For a variety of reasons, a Food and Nutrition Board committee recently recommended that daily intakes of sodium chloride be limited to 6g (2.4 g or 2,400 mg of sodium) or less. The National Dietary Goals recommend less than 5 grams of salt a day. And since salt is about 39% sodium, that translates to about 2 grams or 2000 mg of sodium per day

Canada: Adults: 9 mg per kilogram of body weight, per day.

Food Sources

Most of the sodium in our diet (75%) is from salt added during processing and manufacturing. Studies have found that only 10% of the salt we consume comes from the natural content of foods, and 15% comes from salt added at the table or during cooking

You will notice in the list of foods that canned goods and processed goods are quite high in sodium. These amounts *will change* in the future as manufacturers respond to the consumer interest in reducing the sodium content in foods. Many sources of sodium do not taste salty because the sodium may be part of compounds added to the food to stabilize, preserve, or color it.

Baked Goods

Because recipes vary, and manufacturers are changing formulations to reduce sodium, these values should be considered close approximations.

	Wt (g)	Svg	Cal	Sodium (mg) 100 Cal	Svg
Apple crisp, 3" x 3"	78	1 pce	146	45	66
Bagel, plain or egg, 3.5" dm.	68	1 ea	180	167	300
Biscuits, prepared:					
Homemade	28	1 ea	100	195	195
From mix	28	1 ea	94	278	262
From refrig dough	20	1 ea	65	382	249
BREADS:					
Banana nut, 1/2" slice	50	1 pce	161	71	115
Boston brown, cnd, 1/2"	45	1 pce	95	119	113
Cornbread: see Muffins.					
Cracked wheat	25	1 pce	65	163	106
French, 5" x 2.5" x 1"	35	1 pce	100	203	203
Mixed grain	25	1 pce	65	163	106
Oatmeal bread	25	1 pce	65	191	124
Pita pocket, 6.5" dm.	60	1 ea	165	205	339
Pumpernickel, 5' x 4" x 3/8"	32	1 pce	80	221	177
Raisin bread	25	1 pce	68	135	92
Rye, light 5" x 3.5" x 7/16"	25	1 pce	65	269	175
Wheat (white &					
whole wheat flour)	28	1 pce	72	210	151
White bread	28	1 pce	75	192	144
Whole wheat bread	28	1 pce	70	257	180
Bread crumbs:					
Soft	45	1 c	120	193	231
Dry grated	100	1 c	390	189	736
Brownies, frosted w/nuts	25	1 ea	100	55	55
CAKES, pce = 1/16th cake					
unless stated otherwise.					
(Cupcakes ≈ 42 g):					
Angel food, 1/12 cake	53	1 pce	125	215	269
Boston cream pie, 1/8	120	1 pce	260	87	225
Carrot cake & cream					
cheese frosting, 2.5"x 3"	112	1 pce	406	36	146
Cheesecake:					
Recipe, 1/12	92	1 pce	278	73	204
From mix, 1/8	103	1 pce	300	122	366
Chocolate, choc frosting	69	1 pce	235	77	181
Coffee cake f/mix, 2.4" x 2.8"	72	1 pce	230	135	310
Fruitcake, dark, 2/3" arc	43	1 pce	165	41	67
Gingerbread, 1/9 of 8" square	63	1 pce	174	110	192
Pound cake, 1/2" slice:	30	1 pce	115	94	108
Sheet Cake:					
Plain	86	1 pce	315	82	258
White frosting	121	1 pce	445	62	275
Snack cake, cream filled:					
Chocolate, eg Ding Dongs	28	1 ea	105	100	105
Sponge cake, eg Twinkies	42	1 ea	155	100	155
Sponge cake, 1/12 tube	66	1 pce	194	108	210
White cake:					
Chocolate frosting	77	1 pce	291	60	176
Coconut frosting	70	1 pce	270	66	177
White frosting	71	1 pce	260	68	176
Yellow cake, choc frosting:					
From mix	69	1 pce	235	67	157
Commercial	69	1 pce	245	78	192
Cheese puffs, (Cheetos)	28	1 oz	158	218	344
Cherry cobbler, 3" x 3"	129	1 pce	199	156	311
Chips: see corn & tortilla chips in this section; potato chips are in Vegetable section.					

	Wt (g)	Svg	Cal	Sodium (mg) 100 Cal	Svg
COOKIES:					
Animal cookies, box ≈ 1 oz	28	27 ea	120	94	113
Chocolate chip:					
Homemade	40	4 ea	185	44	82
Commercial	42	4 ea	180	78	140
Fr/chilled dough	48	4 ea	225	77	173
Butter cookies	25	5 ea	115	91	105
Fig bars	56	4 ea	210	86	180
Gingersnaps, homemade	7	1 ea	34	59	20
Lady fingers	44	4 ea	158	32	50
Oatmeal raisin	52	4 ea	245	60	148
Peanut butter, home	48	4 ea	245	58	142
Sandwich type, all	40	4 ea	195	97	189
Shortbread cookies	32	4 ea	155	79	123
Snickerdoodles	20	1 ea	110	84	93
Sugar f/ refrigerator dough	48	4 ea	235	111	261
Vanilla wafers	40	10 ea	185	81	150
Corn chips	28	1 oz	155	150	233
CRACKERS:					
Cheese crackers	10	10 ea	50	224	112
Cheese, peanut butter fill	30	4 ea	150	225	338
Graham crackers	14	2 ea	60	143	86
Melba toast, plain	5	1 pce	20	220	44
Oyster crackers, 10 = 7g	28	1 oz	120	297	356
Round, like Ritz	9	3 ea	45	200	90
Rye wafers, whole grain	14	2 ea	55	209	115
Saltines	12	4 ea	50	310	155
Sesame crackers	12	4 ea	60	180	108
Soda crackers	17	6 ea	75	249	187
Wheat crackers, thin	8	4 ea	35	197	69
Whole wheat crackers	8	2 ea	35	169	59
Cream puff, custard filled	110	1 ea	280	44	122
Crepe, without filling	27	1 ea	46.6	247	115
Croissant, 4.5" x 4" x 2"	57	1 ea	235	192	452
Croutons, seasoned					
(dry bread cubes)	30	1 c	111	359	399
Danish pastry:					
Plain	57	1 ea	220	99	218
With fruit	65	1 ea	235	99	233
Doughnut:					
Cake type, medium	50	1 ea	210	91	192
Yeast raised, plain	60	1 ea	235	94	222
Eclair, custard filled, choc icing	94	1 ea	262	39	101
English muffin:					
Plain, enriched	57	1 ea	140	270	378
Sourdough	56	1 ea	129	197	254
MUFFINS:					
Blueberry, recipe	45	1 ea	135	147	198
Blueberry, fr/mix	45	1 ea	140	161	225
Bran, wheat, recipe	45	1 ea	125	151	189
Bran, wheat, fr/mix	45	1 ea	140	275	385
Cornmeal, recipe	45	1 ea	145	117	169
Cornmeal, fr/mix	45	1 ea	145	201	291
Pancakes, 4" diam:					
Buckwheat, f/mix	27	1 ea	55	227	125
Plain, homemade	27	1 ea	60	192	115
Plain, f/mix	27	1 ea	60	267	160
Whole wheat, f/mix	52	1 ea	94	160	150
Peach cobbler, 3" x 3"	130	1 pce	130	238	309
PIES: pce is 1/6 of 9" pie.					
Apple pie	158	1 pce	405	118	476
Banana cream	198	1 pce	319	92	294
Blueberry pie	158	1 pce	380	111	423
Cherry pie	158	1 pce	410	117	480
Chocolate cream	175	1 pce	311	137	427
Coconut cream	172	1 pce	343	132	452
Coconut custard	165	1 pce	384	112	430
Cream, commercial	152	1 pce	455	81	369

187

Sodium (Na) mg — Values are for edible portion of foods

	Wt (g)	Svg	Cal	Sodium (mg) 100 Cal	Svg
Baked Goods, continued:					
PIES, continued:					
Custard	152	1 pce	293	114	333
Lemon meringue	140	1 pce	355	111	395
Mincemeat	160	1 pce	395	84	330
Peach pie	158	1 pce	405	104	423
Pecan pie	138	1 pce	583	52	304
Pumpkin pie	200	1 pce	367	92	338
Strawberry chiffon, recipe	162	1 pce	372	70	259
Poptart-like pastry	54	1 ea	210	118	248
Pretzels, dutch twists	16	1 ea	65	397	258
Pretzels, thin twists	60	10 ea	240	403	966
ROLLS:					
Cinnamon bun, small	50	1 ea	158	123	195
Dinner roll, 2.5" x 2"	28	1 ea	85	182	155
Hamburger bun	45	1 ea	129	210	271
Hard roll, white	50	1 ea	155	202	313
Hotdog bun	40	1 ea	115	210	241
Rye roll, light	28	1 ea	76	258	196
Rye roll, dark	28	1 ea	79	322	254
Submarine (hoagie) roll	135	1 ea	400	171	683
Whole wheat roll	35	1 ea	88	247	217
Stuffing, w/enr bread:					
Prep from dry	70	1/2 c	250	251	626
Stove Top stuffing, prep	108	1/2 c	176	359	632
Tortillas:					
Corn, enr, 6" dm, fried	30	1 ea	87	18	16
Flour, 10.5" dm.	56.8	1 ea	168	128	215
Flour, 8" dm.	35.4	1 ea	105	128	134
Tortilla chips, Doritos:	28	1 oz	139	101	140
Nacho flavor	28	1 oz	139	77	107
Taco flavor	28	1 oz	140	136	191
Waffles:					
Homemade, 9" dm.	75	1 ea	245	182	445
From mix, 9" dm.	75	1 ea	205	251	515
Frozen, 4" dm.	35	1 ea	98	247	242

Beverages

See milk beverages in Dairy & juices in Fruits or Vegetables.

	Wt (g)	Svg	Cal	100 Cal	Svg
Alcoholic:					
Beer (12 fl oz = 1.5 cup)	356	1.5 c	146	13	19
Beer, light	354	1.5 c	100	10	10
Bloody mary (5 fl oz)	148	1 ea	116	286	332
Tom collins (7.5 fl oz)	222	1 ea	121	32	39
Wine, average all	118	1/2 c	80-85	7	6
CARBONATED (1.5 c = 12 fl oz):					
Club soda	355	1.5 c	0	75	75
Cola, regular	370	1.5 c	151	10	15
Cream soda	371	1.5 c	191	23	43
Diet soda pop, avg	355	1.5 c	2	1600	32
Fruit flavored soda	372	1.5 c	170	28	48
Ginger ale	366	1.5 c	124	20	25
Grape soda	372	1.5 c	161	35	57
Lemon-lime soda	368	1.5 c	149	28	41
Orange drink	372	1.5 c	177	26	46
"Pepper" type soda	368	1.5 c	151	25	38
Root beer	370	1.5 c	152	32	49
Clam & tomato juice, 5.5 fluid oz can	166	.7 c	77	862	664
COFFEE, brewed	240	1 c	2	271	5
Coffee, instant, prepared:					
Cafe francais	185	3/4 c	58	41	24
Cafe vienna	187	3/4 c	64	147	94
Swiss mocha	188	3/4 c	51	71	36
Cappuchino	192	3/4 c	62	168	104

	Wt (g)	Svg	Cal	100 Cal	Svg
Fruit punch drink	247	1 c	113	10	11
Gatorade	230	1 c	39	315	123
Grape drink	250	1 c	112	14	16
Lemonade, f/frozen	248	1 c	100	2	2
Limeade, f/frozen	247	1 c	102	<1	<1
Pineapple grapefruit drink	250	1 c	117	29	34
Pineapple orange drink	250	1 c	125	7	9
Tea, brewed	240	1 c	2	337	7
Water:					
Tap water (varies)	237	1 c	0	7	7
Perrier, 6.5 fl oz	192	1 ea	0	3	3
Poland Springs	237	1 c	0	1	1
Tonic (quinine)	366	1.5 c	125	12	15

Dairy & Dairy Products

	Wt (g)	Svg	Cal	100 Cal	Svg
CHEESE (1.5" cube ≈ 1 oz):					
American processed	28	1 oz	106	383	406
American cheese food	28	1 oz	94	291	274
American cheese food, jar	28	1 oz	93	362	337
American cheese spread	28	1 oz	82	465	381
Blue cheese	28	1 oz	100	396	396
Brick cheese	28	1 oz	105	151	159
Brie cheese	28	1 oz	95	187	178
Camembert	28	1 oz	85	278	236
Caraway	28	1 oz	107	183	196
Cheddar cheese	28	1 oz	114	154	176
Cheshire	28	1 oz	110	180	198
Colby	28	1 oz	112	153	171
Cottage cheese:					
Lowfat 1%	226	1 c	164	560	918
Lowfat 2%	226	1 c	205	448	918
Creamed, large curd	225	1 c	235	388	911
Creamed, small curd	210	1 c	215	395	850
Dry curd	145	1 c	123	15	19
Cream cheese, 1 T = 15g	28	1 oz	99	85	84
Edam cheese	28	1 oz	101	271	274
Feta cheese	28	1 oz	75	421	316
Gjetost	28	1 oz	132	129	170
Gorgonzola	28	1 oz	111	462	513
Gouda cheese	28	1 oz	101	230	232
Gruyere	28	1 oz	117	81	95
Liederkranz	28	1 oz	87	448	390
Limburger	28	1 oz	93	244	227
Monterey jack	28	1 oz	106	143	152
Mozzarella, low moisture:					
Part skim	28	1 oz	80	188	150
Whole milk	28	1 oz	90	132	119
Muenster	28	1 oz	104	171	178
Neufchatel	28	1 oz	74	153	113
Parmesan, grated (1 T = 5g)	28	1 oz	129	409	528
Pimento processed	28	1 oz	106	382	405
Port du salut	28	1 oz	100	151	151
Provolone	28	1 oz	100	248	248
Ricotta cheese:					
With part skim	246	1 c	340	90	307
With whole milk	246	1 c	428	48	207
Romano, grated (1 T = 5g)	28	1 oz	110	309	340
Roquefort	28	1 oz	105	489	513
Swiss cheese	28	1 oz	107	69	74
Swiss cheese food	28	1 oz	92	478	440
Swiss processed cheese	28	1 oz	95	408	388
Tilsit cheese	28	1 oz	96	222	213
CREAM, SWEET, fluid:					
Coffee or table	15	1 T	30	20	6
Half and half	15	1 T	20	30	6
Light whipping cream	239	1 c	699	12	82
Heavy whipping cream	238	1 c	821	11	89

	Wt (g)	Svg	Cal	Sodium (mg) 100 Cal	Svg
CREAM, SWEET, whipped:					
Heavy cream, unsweetened	119	1 c	410	11	45
Pressurized	60	1 c	154	51	78
CREAM, SOUR, dairy:					
Cultured, dairy	14	1 T	30	25	8
Half & half, dairy	15	1 T	20	30	6
Cream, sour, Imitation, non-dairy	14	1 T	29	48	14
CREAM SUBSTITUTES (non-dairy):					
Coffee whitener, liquid	15	1 T	20	60	12
Coffee whitener, powder	2	1 tsp	11	36	4
Dessert toppings:					
Dessert powder, dry mix	43	1.5 oz	245	21	52
Frozen (e.g. Coolwhip)	75	1 c	239	8	19
Pressurized, non-dairy	70	1 c	185	23	43
Kefir beverage	233	1 c	160	31	50
MILK (cow):					
Skim	245	1 c	86	147	126
Lowfat 1%	244	1 c	102	121	123
Lowfat 2%	244	1 c	121	101	122
Whole (3.3% fat)	244	1 c	150	80	120
Buttermilk	245	1 c	99	260	257
Canned:					
Skim, evap	255	1 c	200	147	293
Whole, evap	252	1 c	340	79	267
Sweetened, condensed	306	1 c	982	40	389
Dry, nonfat, instant	68	1 c	244	153	373
Milk (other):					
Goat milk	244	1 c	168	73	122
Human breast milk	246	1 c	171	25	42
Soybean milk	240	1 c	79	38	30
MILK BEVERAGES & mixes:					
Carob flavor mix:					
Powder	12	3 T	45	27	12
Prep w/whole milk	256	1 c	195	68	132
Chocolate milk, commercial:					
Lowfat 1%	250	1 c	160	95	152
Lowfat 2%	250	1 c	180	84	151
Whole (3.3% fat)	250	1 c	210	71	149
Chocolate flavor mix, to be mixed w/water:					
Powder (includes dry milk)	28	1 oz	100	139	139
Drink, prepared	206	3/4 c	100	139	139
Chocolate flavor mix, to be mixed w/milk:					
Powder	21.6	3/4 oz	75	60	45
Drink, w/whole milk	266	1 c	226	73	165
Eggnog, commercial	254	1 c	342	40	138
Instant Breakfast, dry	37	1 env	130	128	166
Malted milk, w/whole milk:					
Chocolate flavored	265	1 c	229	75	172
Natural flavored	265	1 c	237	94	223
Milkshakes, 10 fl oz = 1.25 c:					
Chocolate	283	1.25 c	360	76	273
Strawberry	283	1.25 c	319	73	234
Vanilla	283	1.25 c	314	74	232
MILK DESSERTS:					
Ice cream, vanilla:					
Regular	133	1 c	269	43	116
Rich	148	1 c	349	31	108
Soft serve	173	1 c	377	41	153
Ice milk, vanilla, regular	131	1 c	184	57	105
Ice milk, soft serve	175	1 c	223	73	163
MILK DESSERTS, continued:					
PUDDINGS, prepared (5 oz can ≈ 1/2 c):					
Assorted flavors, low cal	130	1/2 ea	69	213	147
Chocolate:					
Fr/mix, instant	260	1 c	310	284	880
Cooked fr/mix	260	1 c	300	111	334
Canned, 5 oz can	142	1 can	205	139	285
Coconut, f/instant	149	1/2 c	184	192	354
Lemon, f/instant	149	1/2 c	178	217	387
Rice:					
Fr/mix, instant	149	1/2 c	175	90	158
Cooked fr/mix	132	1/2 c	155	90	140
Tapioca:					
Cooked fr/mix	130	1/2 c	145	105	152
Canned, 5 oz can	142	1 can	160	158	252
Vanilla:					
Fr/instant-mix	130	1/2 c	150	250	375
Cooked fr/mix	130	1/2 c	145	123	178
Canned, 5 oz can	142	1 can	220	139	305
Pudding Pops:					
Chocolate & choc fudge	57	1 ea	99	105	104
All other flavors	57	1 ea	94	67	63
Sherbet	193	1 c	270	33	88
Yogurt, frozen (average)	174	1 c	220	41	90
YOGURT:					
Lowfat, plain	227	1 c	144	110	159
Lowfat with fruit	227	1 c	231	54	125
Lowfat, coffee or vanilla	227	1 c	193	78	150
Nonfat, plain	227	1 c	127	136	173
Whole, plain	227	1 c	138	75	104
Yogurt cheese, recipe	208	1 c	222	84	186

Eggs

	Wt (g)	Svg	Cal	Sodium (mg) 100 Cal	Svg
Eggs, chicken:					
Whole, cooked or raw	50	1 ea	77.5	80	62
White, cooked or raw	33.4	1 ea	17	323	55
Yolk, cooked or raw	16.6	1 ea	59	12	7.14
Egg substitute, frozen	60	1/2 c	96	125	120
Egg substitute, liquid	251	1 c	211	210	444

Fats, Oils & Salad Dressings

	Wt (g)	Svg	Cal	Sodium (mg) 100 Cal	Svg
FATS:					
Butter, regular	14.2	1 T	100	116	116
Bacon fat	14	1 T	126	111	140
Cooking fat, veg shortening	205	1 c	1812	0	0
Lard	13	1 T	115	<1	<1
Margarine:					
Hard, 80% fat	14	1 T	100	133	133
Soft, 80% fat	14	1 T	100	153	153
Spread, 60% fat	14	1 T	75	185	139
Soft, 40% fat (imitation)	14	1 T	50	272	136
Mayonnaise, imitation	15	1 T	35	214	75
Mayonnaise	14	1 T	100	80	80
OILS, all	14	1 T	125	0	0
SALAD DRESSINGS (Brands vary. Check label)					
Blue cheese	15	1 T	75	219	164
Ceasar's	92	1/2 c	408	211	860
French	16	1 T	85	221	188
French, low cal	16	1 T	24	1250	300
Italian	15	1 T	75	96	72
Italian, low cal	15	1 T	5	2720	136
Mayonnaise type	15	1 T	58	181	105
Mayo type, low cal	15	1 T	35	215	75
Ranch style	15	1 T	54	120	65

ESHA Research

	Wt (g)	Svg	Cal	Sodium (mg) 100 Cal	Svg
Fats, Oils & Salad Dressings, continued:					
Russian salad dressing	15	1 T	76	176	133
1000 island	16	1 T	60	183	110
1000 island, low cal	15	1 T	25	612	153
Vinegar & oil	16	1 T	70	0	0
Tartar sauce	14	1 T	74	246	182

Fruits & Fruit Juices

	Wt (g)	Svg	Cal	Sodium (mg) 100 Cal	Svg
Acerola juice, fresh	242	1 c	51	14	7
Apple, 2.75" dm. w/peel	138	1 ea	80	1	1
Apple juice:					
From frozen	239	1 c	111	15	17
Cnd, bottled	248	1 c	116	6	7
Applesauce, unswtnd	244	1 c	106	5	5
APRICOTS:					
Fresh, whole, pitted	106	3 ea	51	2	1
Cnd halves, juice pack	84	3 ea	40	8	3
Cnd halves, heavy syrup	85	3 ea	70	4	3
Dried halves	35	10 ea	83	4	4
Apricot nectar, canned	251	1 c	141	6	8
Avocado, whole:					
California	173	1 ea	305	7	21
Florida	304	1 ea	340	4	14
Banana, 8.75", 176g w/ peel	114	1 ea	105	1	1
Blackberries:					
Fresh	144	1 c	74	0	0
Canned	256	1 c	236	3	7
Blackberry juice, fresh	250	1 c	93	3	3
Blueberries:					
Fresh	145	1 c	82	11	9
Canned	256	1 c	225	4	9
Boysenberries:					
Frozen, unthawed	132	1 c	66	3	2
Canned	256	1 c	225	4	9
Cantaloupe or Casaba: see Melon.					
Cherries, sour:					
Frozen	155	1 c	72	1	1
Canned	244	1 c	90	19	17
CHERRIES, SWEET, pitted:					
Whole, fresh	68	10 ea	49	0	0
Canned w/liquid	257	1 c	213	3	7
Cranberries, fresh	95	1 c	46	2	1
Cranberry juice cocktail	253	1 c	145	3	5
Cranberry apple juice	253	1 c	169	3	5
Cranberry sauce, canned	277	1 c	419	19	80
Dates, pitted	83	10 ea	228	1	2
Figs:					
Fresh, medium	50	1 ea	37	3	1
Dried	187	10 ea	477	4	21
Fruit cocktail, canned:					
Juice pack	248	1 c	115	9	10
Heavy syrup	255	1 c	185	8	15
Gooseberries:					
Fresh berries	150	1 c	67	1	1
Canned with liquid	252	1 c	185	3	6
GRAPEFRUIT (half = 241g whole):					
Fresh, pink/red	123	1 half	37	0	0
Fresh, white	118	1 half	39	0	0
Canned sections	254	1 c	152	3	4
Grapefruit juice:					
Fresh	247	1 c	96	2	2
Fr/frzn concentrate	247	1 c	102	2	2
Cnd, unsweetened	247	1 c	93	2	2
Cnd, sweetened	250	1 c	115	4	5

	Wt (g)	Svg	Cal	Sodium (mg) 100 Cal	Svg
Grapes:					
Fresh, all	50-57	10 ea	35-38	3	1
Cnd, hvy syrup	256	1 c	187	7	14
Grape juice:					
From frozen	250	1 c	128	4	5
Bottled/canned	253	1 c	155	5	8
Guava sauce, cooked	238	1 c	87	10	9
Honeydew: see Melon.					
Kiwi fruit	76	1 ea	46	9	4
Lemon juice:					
Fresh	15	1 T	4	3	<1
Bottled	15	1 T	5	60	3
Lime juice:					
Fresh	15	1 T	4	4	<1
Bottled	15	1 T	3	78	2
Loganberries, fresh/frzn	147	1 c	70	1	1
Mandarin oranges, canned	252	1 c	155	10	15
Mango, fresh slices	165	1 c	108	3	3
MELON: see also Watermelon.					
Cantaloupe cubes	160	1 c	57	25	14
Casaba cubes	170	1 c	45	44	20
Honeydew cubes	170	1 c	60	28	17
Frozen melon balls, mixed	173	1 c	55	96	53
ORANGE, 2-5/8" dm					
(180g w/peel)	131	1 ea	60	<1	<1
Orange juice:					
Fresh squeezed	248	1 c	111	2	2
Prep f/frozen	249	1 c	110	2	2
Canned, unswtnd	249	1 c	105	5	5
Orange grapefruit jce, cnd	247	1 c	105	7	7
Papaya (454g w/refuse)	304	1 ea	117	7	8
Papaya nectar, canned	250	1 c	142	10	14
Passion fruit	18	1 ea	18	28	5
Passion fruit juice:					
Purple	247	1 c	126	6	7
Yellow	247	1 c	149	10	15
PEACHES:					
Fresh, 2.5" dm.	87	1 ea	37	0	0
Canned, juice pack	77	1 half	34	9	3
Canned, heavy syrup	81	1 half	60	8	5
Peach nectar, canned	249	1 c	134	7	10
PEARS:					
Fresh, Bartlett (18g whole)	166	1 ea	98	1	1
Canned, juice pack	77	1 half	38	8	3
Canned, heavy syrup	79	1 half	59	7	4
Pear nectar, canned	250	1 c	149	5	8
PINEAPPLE:					
Fresh chunks	155	1 c	76	3	2
Cnd pieces, juice pack	250	1 c	150	2	3
Cnd pieces, heavy syrup	255	1 c	199	2	3
Pineapple juice	250	1 c	135	2	3
Plantain:					
Fresh slices	148	1 c	181	3	6
Ckd slices	154	1 c	179	4	8
PLUMS:					
Fresh, 2-1/8" dm	66	1 ea	36	1	<1
Canned, juice pack	95	3 ea	55	2	1
Canned, heavy syrup	110	3 ea	98	21	21
Pricklypear, fresh	103	1 ea	42	14	6
Prunes, dried	84	10 ea	201	1	3
Prune juice, bottled	256	1 c	181	6	10
Raisins, unpacked measure	145	1 c	435	4	17
Raspberries:					
Fresh	123	1 c	60	0	0
Canned w/liquid	256	1 c	234	4	9

	Wt (g)	Svg	Cal	Sodium (mg) 100 Cal	Svg
Fruits & Fruit Juices, continued:					
Rhubarb, fresh dices	122	1 c	26	19	5
STRAWBERRIES:					
Fresh berries	149	1 c	45	4	2
Frozen, unthawed	149	1 c	52	6	3
Tangerine, fresh	84	1 ea	37	3	1
Watermelon pieces	160	1 c	50	6	3

Grains & Grain Products
Cereals, Grains, Flours, Noodles, Pasta, Popcorn

	Wt (g)	Svg	Cal	Sodium (mg) 100 Cal	Svg
Amaranth grain, dry	195	1 c	729	5	40
Barley:					
Pearled, cooked	157	1 c	193	2	4.71
Whole, cooked	200	1 c	200	<1	1
Bran: see Oat, Rice, Wheat.					
Buckwheat flour:					
Dark	98	1 c	338	<1	1
Light	98	1 c	340	<1	1
Bulgar wheat, cooked	182	1 c	151	6	9.1
CEREALS, COLD (Ready to Eat):					
Sodium content varies. Check label.					
CEREALS, HOT, cooked					
w/o added salt:					
Corn grits, cooked, enriched	242	1 c	145	0	0
Cream of Rice	244	1 c	126	2	2
Cream of Wheat	244	1 c	140	4	5
Farina, cooked	233	1 c	116	0	0
Malt-O-Meal	240	1 c	122	2	2
Maypo	180	3/4 c	128	5	6
Oatmeal, rolled oats					
(Instant, Quick, Reg)	234	1 c	145	2	2.34
Oatmeal, fortified instant fr/packet:					
Plain	177	3/4 c	104	274	285
Flavors averaged	164	3/4 c	160	159	254
Ralston cereal	253	1 c	134	3	4
Roman Meal	181	3/4 c	111	2	2
Wheatena	243	1 c	135	4	5
Whole wheat cereal	242	1 c	150	1	1
Corn flour:					
Regular	117	1 c	422	1	5.85
Masa Harina, enr	110	1 c	416	1	5.7
Cornmeal:					
Degermed, enr, dry	138	1 c	505	1	4.14
Whole unbolted, dry	122	1 c	442	10	43
MACARONI, cooked without salt:					
Regular, enriched	140	1 c	197	1	1.4
Whole wheat	140	1 c	174	3	4.2
Millet, cooked	120	1/2 c	143	2	2.4
NOODLES:					
Chow mein, dry	45	1 c	237	83	197
Egg, cooked	160	1 c	213	5	11.2
Ramen noodles:					
Beef	227	1 c	225	384	865
Chicken	227	1 c	202	365	738
Oriental	227	1 c	207	400	829
Spinach noodles, ckd	140	1 c	182	11	20
Oat bran, 1T ≈ 6g	94	1 c	132	3	3.76
Oats, rolled:					
Dry	81	1 c	311	1	3.24
Cooked: see Oatmeal under Cereals.					
PASTA: see Macaroni, Noodles, Spaghetti.					
Popcorn, ckd in oil, salted	11	1 c	55	156	86

	Wt (g)	Svg	Cal	Sodium (mg) 100 Cal	Svg
RICE, cooked:					
Brown rice	195	1 c	217	4	9.75
White, converted	175	1 c	200	3	5.3
White, regular	205	1 c	264	2	4
White, instant	165	1 c	162	3	5
Wild rice	164	1 c	166	3	4.92
Rice bran	83	1 c	262	2	4.1
Rye flour:					
Dark	128	1 c	415	<1	1.3
Light	102	1 c	361	<1	2
SOY FLOUR:					
Low fat, stirred	44	1/2 c	163	5	8
Defatted, stirred	50	1/2 c	164	6	10
Full fat, stirred, raw	42	1/2 c	182	3	5
Spaghetti noodles, cooked:					
Enriched, w/o salt	140	1 c	197	1	1.4
Enriched, with salt	140	1 c	197	71	140
Whole wheat spaghetti	140	1 c	174	2	4.2
Tapioca, dry	152	1 c	518	.3	1.5
WHEAT:					
Bran	30	1/2 c	65	.9	.6
FLOURS:					
All-purpose, white, unsifted	125	1 c	455	.5	2.5
Cake flour, sifted	96	1 c	348	1	2
Gluten flour	140	1 c	529	1	3
Self-rising flour	125	1 c	442	359	1587
Semolina flour	167	1 c	601	.3	1.7
Whole wheat flour	120	1 c	407	1	6
Wheat germ:					
Raw	100	1 c	360	3	12
Toasted	113	1 c	432	1	5
Wheat, rolled, cooked	240	1 c	142	1	2
Whole grain wheat					
(wheat berries) cooked	50	1/3 c	28	.7	.2
Whole wheat, sprouted	108	1 c	214	8	17.3

Meats: Fish & Shellfish

	Wt (g)	Svg	Cal	Sodium (mg) 100 Cal	Svg
Abalone, fried	85	3 oz	161	312	502
Anchovies cnd in oil, drained	45	11 ea	95	1738	1651
Bass, freshwater, baked/brld	100	3.5 oz	125	60	75
Bluefish:					
Baked/broiled	100	3.5 oz	159	48	77
Fried in crumbs	100	3.5 oz	205	33	67
Steamed	100	3.5 oz	148	34	50
Carp, baked/brld	100	3.5 oz	162	39	63
Catfish, cornmeal fried	100	3.5 oz	229	122	280
Caviar (1T = 16g)	100	6.25 T	252	595	1500
CLAMS:					
Breaded, fried, small	188	20 ea	379	180	684
Canned, drained	160	1 c	236	76	179
Canned w/liquid, minced	183	1 can	145	202	293
Steamed, meat only	90	20 ea	133	75	100
Clam nectar, canned	240	1 c	6	8600	516
COD:					
Baked, broiled, poached	100	3.5 oz	105	74	78
Batter fried	100	3.5 oz	199	50	100
Smoked	100	3.5 oz	79	1481	1170
CRAB, cooked meat:					
Alaska King crab, leg	134	1 ea	129	1113	1436
Blue crab meat:					
Cooked, unpacked	135	1 c	138	272	376
Canned, unpacked	135	1 c	133	338	450
Dungeness crab, cooked	101	3/4 c	85	352	299

ESHA Research

Fish & Shellfish, continued:

	Wt (g)	Svg	Cal	Sodium (mg) 100 Cal	Svg
Crab, imitation fr/surimi	85	3 oz	87	822	715
Crab cakes, recipe	60	1 ea	93	213	198
Crayfish, ckd, moist heat	85	3 oz	97	60	58
Croaker, breaded, fried fillet	87	1 ea	192	158	303
Dolfinfish, (Mahi Mahi), raw	100	3.5 oz	85	104	88
Eel:					
Baked or broiled	100	3.5 oz	236	28	65
Smoked	100	3.5 oz	330	242	798
Fish cakes, fried:	100	3.5 oz	193	259	500
Fish sticks, heated f/frzn	57	2 ea	155	214	332
HADDOCK:					
Baked, broiled, poached	85	3 oz	95	78	74
Breaded, fried	85	3 oz	175	70	123
Smoked	100	3.5 oz	116	658	763
HALIBUT:					
Baked or broiled	85	35 oz	119	50	59
Smoked	100	3.5 oz	224	214	480
Herring:					
Baked/broiled	100	3.5 oz	203	57	115
Canned in oil	100	3.5 oz	208	224	466
Smoked (kippered)	100	3.5 oz	217	423	918
Pickled (1 pce ≈ 15 g)	100	3.5 oz	262	332	870
Lobster meat ckd, moist heat	145	1 c	142	388	551
Mackerel:					
Baked/broiled, Atlantic	100	3.5 oz	262	32	83
Canned, Jack	361	1 can	563	243	1368
Mullet, baked/brld	85	3 oz	127	48	61
Mussels, Blue, steamed	85	3 oz	147	213	313
Ocean Perch, baked/brld	100	3.5 oz	121	79	96
OYSTERS:					
Raw, Eastern	248	1 c	170	163	277
Raw, Pacific	248	1 c	200	131	262
Breaded, fried, medium	88	6 ea	173	212	367
Simmered	100	3.5 oz	137	164	224
Perch, freshwater, breaded, fried	85	3 oz	185	75	138
Pike, baked/broiled, Northern	100	3.5 oz	113	43	49
Pollock, baked/broiled	100	3.5 oz	99	99	98
Rockfish, baked/broiled	100	3.5 oz	121	64	77
Roe, raw, mixed species	28	1 oz	39	156	61
SALMON, cooked:					
Baked broiled, avg	85	3 oz	183	31	56
Smoked, Chinook	85	3 oz	99	673	666
Cnd, Atlantic, small can	220	1 can	281	31	87
Cnd, Pink, drained, #1 can	454	1 can	631	398	2514
Sardines, canned:					
Atlantic, 2 ≈ 24g	92	1 can	192	242	465
Pacific, 1 ≈ 38g	100	3.5 oz	208	243	505
Scallops:					
Breaded, fried	93	6 ea	200	215	431
Steamed	100	3.5 oz	113	174	197
Scallops, imitation f/surimi	85	3 oz	84	805	676
Sea Bass, baked/broiled	85	3 oz	82	71	58
Seatrout/Steelhead, ckd	100	3.5 oz	131	51	67
Shark, batter-fried	85	3 oz	194	53	103
SHRIMP:					
Boiled, 2 large ≈ 11g	100	3.5 oz	99	226	224
Breaded, fried, 2 large ≈ 15g	90	12 ea	218	142	310
Canned, drained	128	1 c	154	140	216
Canned w/liquid	100	3.5 oz	102	2255	2300
Shrimp, imitation f/surimi	85	3 oz	86	697	599
Smelt, Rainbow, cooked	85	3 oz	106	61	65
Snapper, baked/broiled	100	3.5 oz	128	45	57
SOLE (also Flounder):					
Baked/broiled	85	3 oz	99	90	89
Batter-fried	85	3 oz	250	78	194

	Wt (g)	Svg	Cal	Sodium (mg) 100 Cal	Svg
Sole, breaded, fried	100	3.5 oz	188	106	200
Squid, flour-fried	85	3 oz	149	174	260
Surimi, processed walleye pollock (also see Imitation crab, scallops, shrimp):	85	3 oz	84	144	122
Swordfish, baked/broiled	100	3.5 oz	155	74	115
Trout, baked/broiled	85	3 oz	129	22	29
TUNA:					
Cooked from fresh, Bluefin	85	3 oz	157	27	43
Canned, drained (No. 1/2 can):					
Light, canned in oil	171	1 can	339	179	606
Light, water pack	165	1 can	216	272	588

Meats:

Beef, Pork, Lamb, Rabbit, Veal, Venison

	Wt (g)	Svg	Cal	Sodium (mg) 100 Cal	Svg
BEEF:					
Breakfast strips, cured beef	34	3 strips	153	501	766
Chuck blade, pot roasted:					
Lean and fat (5.4 oz raw)	85	3 oz	325	16	53
Lean only	85	3 oz	230	26	60
Ground beef, avg ckd:					
Extra lean	85	3 oz	215	27	59
Lean	85	3 oz	231	28	65
Regular	85	3 oz	250	28	70
Frozen patty, broiled	85	3 oz	240	27	66
Rib, choice, oven roasted:					
Lean & fat (5 oz raw)	85	3 oz	324	17	54
Lean only	85	3 oz	204	31	63
Round steak, broiled, choice:					
Lean & fat (4.5 oz raw)	85	3 oz	233	22	51
Lean only	85	3 oz	165	33	54
Round, tip, roasted, all grades:					
Lean & fat	85	3 oz	213	25	53
Lean only	85	3 oz	162	34	55
Sirloin steak, broiled, all grades:					
Lean & fat	85	3 oz	238	23	54
Lean only	85	3 oz	172	33	57
T-bone steak, broiled, choice:					
Lean & fat	85	3 oz	276	18	51
Lean only	85	3 oz	182	31	56
Variety meats:					
Heart, simmered pieces	85	3 oz	140	39	54
Liver, fried	85	3 oz	184	49	90
Tongue	85	3 oz	241	21	51
Dried beef, cured (6 to 7 pce)	28	1 oz	47	2094	984
Corned beef, canned	85	3 oz	213	401	855
HAM: see Pork, cured; Lunchmeats; & Turkey Ham.					
LAMB:					
Arm chop, braised:					
Lean & fat	70	1 ea	244	21	51
Lean only	55	1 ea	152	27	41
Loin chop, broiled:					
Lean & fat	64	1 ea	201	24	49
Lean only	46	1 ea	100	39	39
Cutlet, lean, ckd avg	85	3 oz	175	37	64
Leg of lamb:					
Lean & fat	85	3 oz	219	26	56
Lean only	85	3 oz	162	36	58
Rib roast:					
Lean & fat	85	3 oz	305	20	62
Lean only	85	3 oz	197	35	69
Shoulder roast:					
Lean & fat	85	3 oz	235	24	56
Lean only	85	3 oz	173	33	58

	Wt (g)	Svg	Cal	Sodium (mg) 100 Cal	Svg
PORK:					
Bacon, cooked:					
Regular	19	3 pce	109	278	303
Canadian style	47	2 pce	86	836	719
Breakfast strips	34	3 pce	156	458	714
Center loin chop:					
Braised, lean & fat	75	1 ea	266	14	38
Braised, lean only	61	1 ea	166	20	33
Broiled, lean & fat	82	1 ea	284	21	61
Broiled, lean only	72	1 ea	166	34	56
Pan-fried, lean & fat	89	1 ea	333	19	64
Pan-fried, lean only	67	1 ea	178	32	57
Roasted, lean & fat	88	1 ea	268	21	56
Roasted, lean only	72	1 ea	180	29	52
Center rib chop:					
Braised, lean & fat	67	1 ea	246	13	32
Braised, lean only	53	1 ea	147	19	28
Broiled, lean & fat	77	1 ea	264	18	47
Broiled, lean only	63	1 ea	162	26	42
Pan-fried, lean & fat	88	1 ea	343	12	40
Pan-fried, lean only	62	1 ea	160	19	31
Roasted, lean & fat	79	1 ea	252	14	35
Roasted, lean only	66	1 ea	162	19	30
Pork roast, leg:					
Lean & fat	85	3 oz	250	20	50
Lean only	85	3 oz	187	29	54
Pork roast, average:					
Lean & fat	85	3 oz	265	17	45
Lean only	85	3 oz	206	24	49
Shoulder, braised:					
Lean & fat	85	3 oz	293	25	74
Lean only	67	2.4 oz	166	41	68
Spareribs, ckd f/1 lb raw	177	6.25 oz	703	23	165
Pork heart, braised	145	1 c	214	24	51
Pork liver, braised	85	3 oz	141	30	42
PORK, CURED (HAM):					
Roasted, lean & fat	85	3 oz	207	487	1009
Roasted, lean only	85	3 oz	133	848	1128
Canned, roasted	85	3 oz	140	649	908
RABBIT, roasted	85	3 oz	175	18	31
VEAL (calf):					
Cutlet, lean, cooked avg	85	3 oz	166	46	76
Rib, roasted	85	3 oz	151	54	82
Liver, pan-fried	85	3 oz	208	54	112
VENISON (deer) roasted	85	3 oz	134	34	46

Meats: Poultry

	Wt (g)	Svg	Cal	Sodium (mg) 100 Cal	Svg
CHICKEN:					
All types:					
Fried	140	1 c	307	41	127
Roasted	140	1 c	266	45	120
Stewed	140	1 c	248	40	98
Canned, boned, w/broth	142	5 oz	235	304	714
Dark meat only:					
Fried	85	3 oz	203	41	83
Roasted	85	3 oz	174	45	79
Stewed	85	3 oz	163	39	63
Light meat only:					
Fried	85	3 oz	163	43	69
Roasted	85	3 oz	147	45	66
Stewed	85	3 oz	135	41	55
Breast*, meat & skin:					
Batter-fried	140	1 ea	364	106	385
Flour-fried	98	1 ea	218	34	74
Roasted	98	1 ea	193	36	69
Stewed	110	1 ea	202	34	68

	Wt (g)	Svg	Cal	Sodium (mg) 100 Cal	Svg
CHICKEN, continued:					
Breast*, meat only:					
Fried	86	1 ea	161	42	68
Roasted	86	1 ea	142	45	64
Stewed	95	1 ea	144	41	59
*2 pieces per bird					
Drumstick, meat & skin:					
Batter-fried	72	1 ea	193	101	194
Flour-fried	49	1 ea	120	37	44
Roasted	52	1 ea	112	42	47
Stewed	57	1 ea	116	37	43
Drumstick, meat only:					
Fried	42	1 ea	82	49	40
Stewed	46	1 ea	78	47	37
Roasted	44	1 ea	76	55	42
Thigh, meat & skin:					
Batter-fried	86	1 ea	238	104	248
Flour-fried	62	1 ea	162	34	55
Roasted	62	1 ea	153	34	52
Stewed	68	1 ea	158	31	49
Thigh, meat only:					
Fried	52	1 ea	113	43	49
Roasted	52	1 ea	109	42	46
Stewed	55	1 ea	107	38	41
Wing, meat & skin:					
Batter-fried	49	1 ea	159	99	157
Flour-fried	32	1 ea	103	24	25
Roasted	34	1 ea	99	28	28
Stewed	40	1 ea	100	27	27
Wing, meat only:					
Roasted	21	1 ea	43	44	19
Fried	20	1 ea	42	43	18
Stewed	24	1 ea	42	42	18
Variety meats, simmered:					
Gizzard	22	1 ea	34	44	15
Heart	3.3	1 ea	6	33	2
Liver	20	1 ea	30	33	10
DUCK, domestic, roasted:					
Meat & skin	85	3 oz	286	18	51
Meat only	85	3 oz	171	32	55
GOOSE, domestic, roasted:					
Meat & skin	85	3 oz	259	23	60
Meat only	85	3 oz	202	32	64
TURKEY:					
Breast meat:					
Barbecued	28	1 oz	40	390	156
Hickory smoked	28	1 oz	35	594	208
Ground turkey, cooked	100	3.5 oz	229	36	83
Roasted:					
All types	85	3 oz	144	42	60
Dark meat	85	3 oz	159	42	67
White meat	85	3 oz	133	41	54
Frozen slices:					
Roasted	85	3 oz	130	445	578
With gravy	142	5 oz	95	828	787
Turkey patty, breaded, fried	64	1 ea	181	283	512

Meats: Sausages & Lunchmeats

	Wt (g)	Svg	Cal	Sodium (mg) 100 Cal	Svg
Barbecue loaf, pork/ beef	23	1 pce	40	768	307
Beef lunchmeat:					
Loaf (roll)	28	1 oz	87	433	377
Thin sliced	28	1 oz	50	816	408
Beerwurst, (beer salami):					
Beef salami	23	1 pce	75	285	214
Pork salami	23	1 pce	55	518	285

ESHA Research

Sodium (Na) mg

Values are for edible portion of foods

Meats: Sausages & Lunchmeats, continued:

Food	Wt (g)	Svg	Cal	Sodium 100 Cal	Sodium Svg
Berliner	23	1 pce	53	562	298
Bologna:					
Beef	23	1 pce	72	319	230
Beef & pork	28	1 oz	89	325	289
Cured pork	23	1 pce	57	477	272
Turkey	28	1 oz	56	443	248
Braunschweiger	57	2 oz	205	318	652
Brotwurst, link	70	1 ea	226	344	778
Cheesefurter (cheese smoki)	43	1 ea	141	330	465
Chicken roll, light meat	57	2 oz	90	368	331
Corned beef loaf, jellied	28	1 oz	46	639	294
Dutch brand loaf	28	1 oz	68	521	354
FRANKFURTER (hotdogs):					
Beef, 8 per pkg	57	1 ea	184	317	584
Beef & pork, 8 per pkg	57	1 ea	183	349	639
Chicken, 10 per pkg	45	1 ea	115	536	616
Turkey, 10 per pkg	45	1 ea	102	445	454
HAM, cured pork:					
Chopped, pkg	42	2 pce	98	585	573
Lunchmeat slices:					
Regular	57	2 oz	103	724	746
Extra lean (3 slices ≈ 1 oz)	57	2 oz	75	1080	810
Minced	21	1 pce	55	475	261
Ham & cheese roll/loaf	57	2 oz	147	518	762
Ham salad spread	80	1/4 c	130	422	547
Ham patty, cooked	60	1 ea	203	311	632
Italian sausage link, ckd	67	1 ea	216	286	618
Keilbasa	26	1 pce	81	346	280
Knockwurst, link	68	1 ea	209	329	687
Liverwurst, pork	18	1 pce	59	93	55
Luncheon meat, canned	21	1 pce	70	387	271
Luncheon sausage, beef/pork	23	1 pce	60	453	272
Luxury loaf	57	2 oz	80	868	694
Mortadella	15	1 pce	47	398	187
Olive loaf	57	2 oz	133	633	842
Pastrami:					
Beef, cured	57	2 oz	198	351	695
Turkey	57	2 oz	74	769	569
Peppered loaf	28	1 oz	42	1029	432
Pepperoni, small slice	22	4 pce	109	412	449
Pickle & pimento loaf	57	2 oz	149	528	787
Polish sausage	28	1 oz	92	270	248
PORK SAUSAGE, cooked:					
Link	13	1 ea	48	350	168
Patty	27	1 pce	100	349	349
Brown & serve links	13	1 ea	50	210	105
Salami:					
Beef	23	1 pce	58	459	266
Dry, beef & pork	20	2 pce	85	438	372
Pork & beef	57	2 oz	143	422	604
Turkey	57	2 oz	111	482	535
Sandwich spread, pork & beef	15	1 T	35	434	152
Smoked link sausage:					
Beef & pork	68	1 ea	229	280	642
Pork	68	1 ea	265	385	1020
Turkey	28	1 oz	55	398	219
Turkey, breakfast	28	1 oz	65	294	191
Summer sausage, turkey	28	1 oz	50	608	304
Turkey ham	57	2 oz	73	751	548
Turkey loaf, breast meat	43	2 pce	46	1322	608
Turkey roll, lunchmeat:					
Light & dark	57	2 oz	84	395	332
Light meat	57	2 oz	83	334	277
Turkey sandwich spread	13	1 T	25	196	49
Vienna sausage, canned	16	1 ea	45	338	152

Mixed Foods & Fast Foods

Food	Wt (g)	Svg	Cal	Sodium 100 Cal	Sodium Svg
Beef & vegetable stew:					
Recipe	245	1 c	220	133	292
Canned	245	1 c	194	511	992
Beef, macaroni, tomato sce, recipe	226	1 c	189	515	974
Beef pot pie, f/frzn	234	1 ea	426	257	1093
BURRITOS:					
Bean	174	1 ea	322	320	1030
Beef	177	1 ea	463	83	382
Beef & bean	175	1 ea	390	132	516
Deluxe combination	198	1 ea	424	127	537
Cheese soufflé, recipe	112	1 c	221	195	431
Chicken & noodles, recipe	240	1 c	365	164	600
Chicken a la king, recipe	245	1 c	470	162	760
Chicken chow mein, recipe	250	1 c	255	282	718
Chicken curry, recipe	337	1.5 c	305	318	970
Chicken pot pie, f/frzn	230	1 ea	430	211	907
Chicken salad w/celery	78	1/2 c	266	75	199
Chili w/beans, canned	255	1 c	286	465	1330
Chop suey, beef & pork	250	1 c	300	351	1053
Cole slaw	120	1 c	84	33	28
Corn dog	111	1 ea	330	379	1252
Corn pudding	250	1 c	271	51	138
Corned beef hash, canned	220	1 c	382	354	1354
Egg salad	183	1 c	438	98	428
ENCHILADAS:					
Beef	120	1 ea	292	54	157
Cheese	120	1 ea	330	94	311
Chicken	120	1 ea	269	59	160
French toast, recipe	65	1 pce	123	154	189
LASAGNA:					
Recipe w/meat	245	1 pce	398	197	783
Recipe w/o meat	218	1 pce	316	241	760
Frozen entree	205	1 pce	275	352	967
Macaroni & cheese:					
Recipe	200	1 c	430	253	1086
Canned	240	1 c	230	317	730
Macaroni salad, no cheese	141	1 c	371	85	315
Manicotti, frozen entree	225	1 ea	271	293	795
Meat loaf:					
Beef and 1/3 pork	87	1 pce	212	185	392
Beef only	87	1 pce	193	176	340
Moussaka, lamb & eggplant	250	1 c	250	194	485
Pie, fried pastry:					
Apple	85	1 ea	255	128	326
Cherry	85	1 ea	250	148	371
PIZZA, cheese:					
Regular crust 1/8 of 15"	120	1 pce	290	241	699
Thick crust 1/2 of 10"	208	1 pce	519	230	1193
Potato salad w/mayo & eggs	250	1 c	358	370	1323
Quiche Lorraine, 1/8	176	1 pce	600	109	653
Ravioli, beef, canned	226	1 c	220	409	900
SANDWICHES, fast food:					
Cheeseburger, 3 oz patty	112	1 ea	300	224	672
Cheeseburger, 4 oz patty	194	1 ea	524	234	1224
Chicken patty sandwich	157	1 ea	436	627	2732
English muffin with egg, cheese, & bacon	138	1 ea	360	231	832
Fish sandwich:					
Regular w/cheese	140	1 ea	420	159	667
Large, w/o cheese	170	1 ea	470	132	621
Hamburger, 3 oz patty	98	1 ea	245	189	463
Hamburger, 4 oz patty	174	1 ea	445	171	763
Hotdog (frankfurter) & bun	85	1 ea	260	287	745

ESHA Research

194

	Wt (g)	Svg	Cal	Sodium (mg) 100 Cal	Sodium (mg) Svg
Mixed Foods & Fast Foods, continued:					
SANDWICHES on part whole wheat,					
except when stated as rye:					
Avocado, cheese,					
tomato and sprouts	195	1 ea	432	120	518
Bacon, lettuce, tomato	135	1 ea	327	202	661
Chicken salad sandwich	100	1 ea	294	141	415
Corned beef & swiss on rye	147	1 ea	429	244	1045
Egg salad sandwich	111	1 ea	319	145	461
Grilled cheese	117	1 ea	393	297	1169
Ham sandwich	122	1 ea	256	474	1213
Ham on rye	116	1 ea	242	521	1261
Ham salad sandwich	125	1 ea	339	266	901
Ham & swiss on rye	145	1 ea	350	382	1336
Ham & cheese	151	1 ea	363	447	1624
Patty melt, grilled	177	1 ea	567	163	923
Peanut butter & jam	100	1 ea	341	122	417
Reuben, grilled	233	1 ea	480	342	1642
Roast beef sandwich	122	1 ea	280	275	771
Tuna salad sandwich					
Turkey sandwich	122	1 ea	271	430	1165
Turkey ham	122	1 ea	253	371	938
SPAGHETTI, pasta & tomato sauce & cheese:					
Homemade	250	1 c	260	367	955
Canned	250	1 c	190	503	955
Spinach soufflé, recipe	136	1 c	218	350	763
Taco:					
Beef	78	1 ea	207	68	141
Chicken	78	1 ea	172	84	145
TOSTADAS:					
Beans and beef	192	1 ea	332	145	483
Beans and chicken	157	1 ea	249	190	474
Refried beans	157	1 ea	212	292	618
Tuna noodle casserole, recipe	202	1 c	251	346	869
Tuna salad	205	1 c	383	215	824
Turkey pot pie, frozen	233	1 ea	416	240	1000
Veal Parmigiana, frzn entree	205	7.25 oz	372	379	1410
Waldorf salad	142	1 c	424	58	246

Nuts, Seeds & Products

	Wt (g)	Svg	Cal	Sodium (mg) 100 Cal	Sodium (mg) Svg
ALMONDS, whole:					
Dried, whole	142	1 c	837	2	15
Dry roasted, salted	138	1 c	810	133	1076
Dry roasted, unsalted	138	1 c	810	2	15
Oil roasted, salted	157	1 c	970	126	1223
Oil roasted, unsalted	157	1 c	970	2	16
Almond butter	16	1 T	101	2	2
Brazil nuts, dried, unsalted	140	1 c	919	<1	2
CASHEWS:					
Dry roasted, salted	137	1 c	787	111	877
Dry roasted, unsalted	137	1 c	787	3	21
Oil roasted, salted	130	1 c	748	109	814
Oil roasted, unsalted	130	1 c	748	3	22
Cashew butter	16	1 T	94	2	2
Chestnuts, roasted	143	1 c	350	1	3
COCONUT:					
Fresh, grated	80	1 c	283	6	16
Flaked, sweetened:					
Canned	77	1 c	341	4	15
Packaged	74	1 c	351	54	189
Shredded, sweetened	93	1 c	466	52	244
Dried unsweetened	78	1 c	515	6	29
Coconut cream, raw	240	1 c	792	1	10
Coconut milk, raw	240	1 c	552	7	36

	Wt (g)	Svg	Cal	Sodium (mg) 100 Cal	Sodium (mg) Svg
Coconut water, raw	240	1 c	46	553	252
Filberts (hazelnuts), whole	135	1 c	853	<1	4
Macadamias:					
Oil roasted, salted	134	1 c	962	36	348
Oil roasted, unsalted	134	1 c	962	1	9
MIXED NUTS with peanuts					
(cashews, peanuts, brazil nuts,					
filberts, almonds, pecans):					
Dry roasted, salted	137	1 c	814	113	917
Dry roasted, unsalted	137	1 c	814	2	16
Oil roasted, salted	142	1 c	876	106	926
Oil roasted, unsalted	142	1 c	876	2	16
MIXED NUTS without peanuts					
(cashews, almonds, brazil nuts,					
pecans, filberts):					
Oil roasted, salted	144	1 c	886	114	1008
Oil roasted, unsalted	144	1 c	886	2	16
PEANUTS:					
Dry roasted, salted	146	1 c	855	139	1187
Dry roasted, unsalted	146	1 c	855	1	9
Oil roasted, salted	144	1 c	837	74	624
Oil roasted, unsalted	144	1 c	837	1	9
Peanut butter:	16	1 T	95	79	75
Regular, salted	32	2 T	188	81	153
Unsalted	32	2 T	188	3	5
Peanut flour, defatted	60	1 c	196	55	108
PECANS:					
Dried, chopped	119	1 c	794	<1	1
Dry roasted, salted	28	1/4 c	187	118	221
Dry roasted, unsalted	28	1/4 c	187	<1	<1
Oil roasted, salted	110	1 c	754	110	832
Oil roasted, unsalted	110	1 c	754	<1	1
Pine nuts, dried:					
Pignola	28	1 oz	146	1	1
Pinyon	28	1 oz	161	12	20
Pistachio nuts:					
Dried, no shells	128	1 c	739	1	8
Dry roasted, salted	128	1 c	776	129	998
Pumpkin/squash seed kernels:					
Dried, unsalted	138	1 c	747	3	25
Roasted, salted	227	1 c	1185	110	1305
Roasted, unsalted	227	1 c	1185	3	41
Pumpkin/squash seeds, whole:					
Roasted, salted	64	1 c	285	129	368
Roasted, unsalted	64	1 c	285	4	12
Sesame flour, low to high fat	28	1 oz	95-149	8-12	12
Sesame seeds, dried:					
Kernels	38	1/4 c	221	7	15
Whole	36	1/4 c	206	2	4
Soybeans, roasted:					
Salted	86	1/2 c	405	35	140
Unsalted	86	1/2 c	405	1	4
Sunflower seeds:					
Dried kernels	36	1/2 c	205	<1	1
Oil roasted, salted	68	1/2 c	415	98	407
Oil roasted, unsalted	68	1/2 c	415	<1	2
Tahini, sesame butter:					
From roasted kernels	15	1 T	89	12	11
From unroasted kernels	15	1 T	91	5	5
Walnuts, chopped:					
Black	125	1 c	759	<1	1
English	120	1 c	770	2	12

	Wt (g)	Svg	Cal	Sodium (mg) 100 Cal	Svg

Soups, Sauces & Gravies

GRAVIES:

	Wt (g)	Svg	Cal	100 Cal	Svg
Beef gravy:					
Recipe	135	1/2 c	151	260	392
Canned	233	1 c	124	1052	1305
From dry mix	258	1 c	75	1435	1076
Brown, from packet	24.8	1 ea	85	1428	1214
Chicken gravy:					
Homemade	130	1/2 c	163	238	388
Canned	238	1 c	189	728	1375
From dry mix	260	1 c	85	1334	1134
Mushroom, f/dry	258	1 c	70	2003	1402
Mushroom, canned	238	1 c	120	1131	1357
Onion, f/dry mix	261	1 c	78	1295	1013
Pork, f/dry mix	258	1 c	76	1625	1235
Turkey, f/dry mix	261	1 c	87	1722	1498

SAUCES:

	Wt (g)	Svg	Cal	100 Cal	Svg
Au Jus, prep f/dry	246	1 c	32	1299	964
Cheese sce, recipe	101	1/2 c	216	250	539
Curry sce, recipe	115	1/2 c	115	528	607
Hollandaise sauce:					
Recipe	160	1 c	867	220	1905
From mix					
W/milk & butter	255	1 c	703	161	1134
W/water	259	1 c	240	652	1564
Hot chili, red pepper	31	2 T	6	50	3
Mushroom, f/dry w/milk	267	1 c	228	672	1533
Sour cream sce, f/dry					
w/milk	314	1 c	509	198	1007
Spaghetti sauce, plain:					
Homemade	220	1 c	179	503	900
Canned	249	1 c	272	454	1236
Stroganoff sauce:					
Dry mix only	55.6	2 oz	195	1155	2252
Prep w/milk & water	296	1 c	271	675	1829
Sweet & sour sce, f/dry	313	1 c	294	265	779
White sauce:					
Recipe	250	1 c	395	225	888
F/dry w/milk	264	1 c	240	332	797

SOUPS: Soups are prep fr/canned unless otherwise stated. RTS = Ready To Serve. For soups prep w/milk, assume whole milk.

	Wt (g)	Svg	Cal	100 Cal	Svg
Asparagus, cream of, prep w/milk	248	1 c	165	545	900
Bean w/bacon	253	1 c	173	491	850
Beef bouillon:					
Canned, condensed	240	1 c	16	5125	1640
Cube prep w/water	241	1 c	8	14400	1152
Beef noodle w/water	244	1 c	84	988	830
Celery, cream of:					
Prepared with milk	248	1 c	165	490	809
Condensed, undiluted	252	1 c	100	880	830
Cheese soup:					
Prepared with milk	251	1 c	230	352	810
Condensed, undiluted	258	1 c	260	577	1500
Chicken, cream of:					
Prepared with milk	248	1 c	191	455	870
Condensed, undiluted	251	1 c	220	736	1620
Chicken bouillon:					
Canned, condensed	244	1 c	39	3846	1500
Cube prep w/water	243	1 c	13	6092	792
CHICKEN NOODLE soup	241	1 c	75	1160	870
Chicken rice w/water	241	1 c	60	1333	800

	Wt (g)	Svg	Cal	100 Cal	Svg
Clam chowder, New England prep w/milk	248	1 c	163	571	930
Mushroom, cream of:					
Prep with milk	248	1 c	205	429	880
Condensed, undiluted	251	1 c	200	810	1640
Split pea w/ham & bacon prep w/water	253	1 c	150	533	800
Tomato soup:					
Prepared w/milk	248	1 c	160	456	730
Condensed, undiluted	252	1 c	180	744	1340
Vegetable beef	244	1 c	79	949	750

SOUP: Campbell's Special Request.
1/3 less salt

	Wt (g)	Svg	Cal	100 Cal	Svg
Bean w/bacon	253	1 c	120	450	540
Chicken, cream of, w/milk	248	1 c	185	313	580
Chicken noodle	241	1 c	70	800	560
Chicken w/rice	241	1 c	60	867	520
Mushroom, cream of, w/milk	248	1 c	175	337	590
Tomato soup w/milk	248	1 c	165	321	530
Vegetable soup	241	1 c	80	650	520
Vegetable beef	244	1 c	70	686	480

SOUP: Campbell's Low Sodium.
Serving size varies. Check label.

	Wt (g)	Svg	Cal	100 Cal	Svg
Chicken broth	298	1 can	40	175	70
Chicken with noodles	305	1 can	160	53	85
Chunky chicken veg	305	1 can	240	40	95
Mushroom, cream of	298	1 can	190	32	60
Split pea	305	1 can	240	10	25
Tomato w/tomato pces	298	1 can	180	22	40

Sweets & Sweeteners

	Wt (g)	Svg	Cal	100 Cal	Svg
Butterscotch topping	50	3 T	156	71	111
CANDY/CANDY BARS:					
Almonds, sugar coated	28	7 ea	146	1	1
Caramel, plain/choc	28	1 oz	115	56	64
Chocolate kisses	28	6 pce	154	16	25
Chocolate covered:					
Almonds	165	1 c	935	1	12
Coconut candy	28	1 oz	133	5	7
Mints	28	1 oz	116	45	52
Peanuts	170	1 c	954	11	102
Raisins	187	1 c	733	4	26
Divinity with nuts	20	1 pce	80	24	19
English toffee bar	32	1 ea	220	41	90
Fondant (candy corn)	28	1 oz	105	54	57
Fudge, chocolate	28	1 oz	115	47	54
Fudge, vanilla	28	1 oz	118	42	50
Gum drops	28	1 oz	98	10	10
Hard candy, all	28	1 oz	109	6	7
Jelly beans	28	1 oz	104	7	7
Kit Kat bar	43	1 ea	210	18	38
Krackle candy bar	34	1 ea	179	27	49
M&M's plain candies	48	1 pkg	237	17	41
M&M's peanut candies	47	1 pkg	240	12	29
Malted milk balls	28	14 pce	135	21	28
Mars bar	50	1 ea	240	35	85
Milk chocolate:					
Plain	28	1 oz	145	16	23
With almonds	28	1 oz	150	15	23
With peanuts	28	1 oz	155	12	19
With rice cereal	28	1 oz	140	33	46
Milky Way	60	1 ea	260	54	140
Mr. Goodbar	47	1 ea	250	8	21
Reese's peanut butter cup	45	2 ea	240	38	92
Snickers , 2.2 oz bar`	61	1 ea	290	59	170
Caramel topping	50	3 T	155	98	152

CHOCOLATE:	Wt (g)	Svg	Cal	Sodium (mg) 100 Cal	Svg
Baking, unsweetened	28	1 oz	145	.7	1
Bittersweet	28	1 oz	141	1	2
Chips, semi-sweet	170	1 c	860	3	24
Dark, sweet	28	1 oz	150	3	5
Hot fudge topping	38	2 T	129	33	42
Syrup, thin type	38	2 T	85	36	31
Custard, baked:					
Recipe	265	1 c	305	69	209
Prepared from mix	143	1/2 c	161	136	219
Gelatin salad (dessert)	120	1/2 c	70	79	55
Granola bar	28	1 ea	127	67	85
Honey	21	1 T	65	2	1
Hot fudge topping	300	1 c	1020	33	336
Icing (cake):					
Prepared from mix	39	2.5 T	167	54	91
Canned, average all	39	2.5 T	160	53	84
Jam or preserves	20	1 T	54	4	2
Jelly, assorted	18	1 T	49	8	4
Marmalade	20	1 T	52	8	4
Marshmallows	28	4 ea	90	28	25
Molasses:					
Blackstrap	40	2 T	85	45	38
Light	20	1 T	43	7	3
Popsicle, (3 oz fluid)	95	1 ea	70	16	11
Sugar:					
Brown	220	1 c	820	12	97
White, granulated	200	1 c	770	1	5
Syrup:					
Corn syrup, light	328	1 c	912	16	148
Corn syrup, dark	328	1 c	944	24	223
Maple syrup	20	1 T	50	6	3
Pancake syrup	84	1/2 c	244	16	38

Vegetables & Legumes

	Wt (g)	Svg	Cal	Sodium (mg) 100 Cal	Svg
Alfalfa sprouts	33	1 c	10	20	2
Amaranth leaf:					
Fresh, chopped	28	1 c	7.4	81	6
Ckd from fresh, boiled	132	1 c	28	96	27
Artichoke globe, cooked	120	1 ea	60	190	114
Artichoke hearts:					
Ckd fr/ frozen (pkg)	240	9 oz	108	118	127
Marinated, jar	170	6 oz	168	536	900
ASPARAGUS, pieces:					
Fresh pieces	67	1/2 c	15	7	1
Ckd from fresh	90	1/2 c	23	18	4
Ckd from frozen	180	1 c	50	14	7
Canned, drained	121	1/2 c	16	1775	284
Canned with liquid	122	1/2 c	17	2500	425
Bamboo shoots, sliced:					
Ckd from fresh	120	1 c	15	33	5
Canned slices	131	1 c	25	36	9
Bean sprouts:					
Fresh sprouts	104	1 c	31	19	6
Boiled fr/ fresh, drained	124	1 c	26	46	12
Ckd fr/ fresh, stir fried	124	1 c	62	23	14
Canned, drained	125	1 c	16	88	14
BEANS: see also Garbanzo, Lentils, Soybeans.					
Baked beans (dry white beans w/spice & sauce):					
Home prepared	253	1 c	382	279	1068
Canned, plain/veg	254	1 c	235	429	1008
Canned w/franks	257	1 c	366	302	1105
Canned w/pork	253	1 c	268	391	1048
Canned w/pork, swt sce	253	1 c	282	301	849
Canned w/pork, tomato sce	253	1 c	247	451	1113

	Wt (g)	Svg	Cal	Sodium (mg) 100 Cal	Svg
BEANS, continued:					
Black beans, ckd f/dry	172	1 c	227	<1	1
Broadbeans:					
Ckd f/fresh veg	100	3.5 oz	56	73	41
Ckd f/dry legume	170	1 c	186	4	8
Canned (cooked fr/ dry)	256	1 c	183	634	1161
Great northern (legume):					
Ckd from dry	177	1 c	210	2	4
Canned with liquid	262	1 c	300	4	11
Green beans (snap):					
Fresh, uncooked	110	1 c	34	18	6
Ckd from fresh	125	1 c	44	9	4
Ckd from frozen	135	1 c	36	47	17
Canned, with liquid	240	1 c	36	2453	883
Canned, drained:					
Regular	135	1 c	26	1304	339
Low sodium	135	1 c	26	12	3
Hyacinth beans:					
Ckd from fresh veg	87	1 c	43	5	2
Ckd from dry legume	194	1 c	228	6	13
Kidney beans:					
Ckd from dry	177	1 c	225	2	4
Canned with liquid	256	1 c	208	427	889
Lima beans:					
Ckd from fresh	170	1 c	208	14	29
Ckd f/ frozen, baby	90	1/2 c	94	28	26
Ckd f/ frozen, large	170	1 c	170	53	90
Ckd from dry, average	185	1 c	223	2	5
Canned, drained	170	1 c	164	245	402
Canned w/ liquid, large	241	1 c	191	424	809
Navy beans:					
Ckd from dry	182	1 c	259	1	2
Canned	262	1 c	296	396	1173
Pinto beans:					
Ckd from dry	171	1 c	235	1	3
Canned	240	1 c	186	536	998
Refried beans, canned	253	1 c	270	397	1071
White beans, cooked fr/ dry	179	1 c	253	2	4
Winged beans:					
Ckd fr/ raw veg	62	1 c	23	13	3
Ckd fr/ dry legume	172	1 c	252	9	22
Yambean, ckd f/fresh veg	100	1 c	46	13	6
Yardlong beans, fresh pods:					
Ckd fr/ fresh veg	104	1 c	49	8	4
Ckd fr/ dry legume	171	1 c	202	4	9
Yellow wax: see green beans.					
Bean sprouts (Mung beans):					
Fresh sprouts	104	1 c	31.2	19	6
Boiled, drained	124	1 c	26	46	12
BEETS:					
Ckd from fresh, diced	85	1/2 c	26	162	42
Canned dices w/liquid:					
Regular	123	1/2 c	36	900	324
Low sodium	123	1/2 c	36	158	57
Canned dices, drained:					
Regular	85	1/2 c	27	863	233
Low sodium	85	1/2 c	27	289	78
Canned, pickled, slices	114	1/2 c	74	407	301
Beet greens:					
Fresh pieces	19	1/2 c	4	950	38
Ckd fr/fresh, drained	144	1 c	40	865	346
Borage:					
Fresh	44	1/2 c	9	389	35
Ckd from fresh	100	3.5 oz	25	352	88
BROCCOLI, chopped:					
Fresh, uncooked	88	1 c	24	100	24
Ckd from fresh	156	1 c	44	91	40
Ckd from frozen	184	1 c	51	86	44

197

	Wt (g)	Svg	Cal	Sodium (mg) 100 Cal	Svg
Vegetables & Legumes, continued:					
BROCCOLI, ckd fr/frozen, continued:					
With cheese sce	142	1/2 c	166	401	665
With hollandaise sce	95	1/2 c	105	111	117
Brussels sprouts, 1 ≈ 21g:					
Ckd from fresh	156	1 c	60	28	17
Ckd from frozen	155	1 c	65	55	36
CABBAGE:					
Common, fresh	70	1 c	16	75	12
Common, ckd	150	1 c	32	91	29
Bok choy, fresh	70	1 c	9	5	<1
Bok choy, ckd	170	1 c	20	285	57
Pe-tsai, fresh	76	1 c	11	64	7
Pe-tsai, ckd	119	1 c	16	69	11
Savoy, fresh	70	1 c	20	100	20
Savoy, ckd	145	1 c	35	97	34
CARROTS:					
Fresh (7.5" x 1-1/8")	72	1 ea	31	81	25
Fresh, grated	55	1/2 c	24	79	19
Ckd fr/frozen, slices	73	1/2 c	26	165	43
Canned, drained	73	1/2 c	17	1035	176
Canned, drained, diet	73	1/2 c	17	176	30
Canned with liquid	123	1/2 c	28	1061	297
Carrot juice	123	1/2 c	49	73	36
Cauliflower:					
Fresh, uncooked	50	1/2 c	12	58	7
Ckd from fresh	62	1/2 c	15	27	4
Ckd from frozen	180	1 c	34	97	33
Celery:					
Fresh stalk, 7.5" ≈ 40g	40	1 ea	6.4	547	35
Fresh, chopped	60	1/2 c	9.6	552	53
Ckd, diced	150	1 c	27	500	135
Celery root, cooked	100	3.5 oz	25	244	61
Chard, Swiss:					
Fresh, uncooked	36	1 c	6.8	1121	77
Ckd from fresh	175	1 c	35	894	313
Collards:					
Fresh, uncooked	36	1 c	11.2	64	7.2
Ckd from fresh	128	1 c	35	59	20.5
Ckd from frozen	170	1 c	63	135	85
CORN, kernels:					
Fresh kernels	77	1/2 c	66	18	12
Ckd from fresh	82	1/2 c	89	16	14
Ckd from frozen	82	1/2 c	67	6	4
Canned, regular					
Drained	82	1/2 c	66	288	190
With liquid	128	1/2 c	79	410	324
Vacuum pack	210	1 c	166	345	572
Canned, low sodium:					
Drained	82	1/2 c	66	6	4
With liquid	128	1/2 c	79	5	4
Vacuum pack	210	1 c	166	4	6
Corn, cream style, cnd:					
Regular	128	1/2 c	93	392	365
Low sodium	128	1/2 c	93	4	4
Cucumber slices, w/peel	28	7 pce	4	15	1
Dandelion greens:					
Fresh, chopped	55	1 c	25	168	42
Ckd from fresh	105	1 c	35	131	46
Dock (sorrel) greens:					
Fresh	133	1 c	29	17	5
Ckd from fresh	100	1 c	20	15	3
Eggplant cubes, ckd	160	1 c	45	11	5
Endive, chopped	25	1/2 c	4	150	6
Escarole, chopped	50	1 c	8.5	129	11
Garbanzo beans (chickpeas):					
Ckd from dry	164	1 c	269	4	11
Canned	240	1 c	285	252	718

	Wt (g)	Svg	Cal	Sodium (mg) 100 Cal	Svg
Garden cress:					
Fresh	25	1/2 c	8	50	4
Ckd from fresh	135	1 c	31	35	11
Jerusalem artichoke, fresh	150	1 c	114	5	6
Jicama	100	3.5 oz	20	130	26
Kale, chopped:					
Fresh	67	1 c	33	88	29
Ckd from fresh	130	1 c	42	72	30
Ckd from frozen	130	1 c	39	51	20
Kohlrabi:					
Fresh slices	140	1 c	38	74	28
Ckd from fresh	165	1 c	48	71	34
Leeks, ckd fr/ fresh slc	52	1/2 c	24	21	5
Lentils, ckd from dry	198	1 c	231	2	5
Lentils, sprouted:					
Fresh sprouts	77	1 c	81	10	8
Stir fried	100	3.5 oz	101	9	9
LETTUCE, fresh, chopped:					
Butterhead	56	1 c	7.3	38	3
Iceberg	56	1 c	7.3	69	5
Loose leaf	56	1 c	10	50	5
Romaine	56	1 c	9	50	4
Lotus root slices, ckd fr/ fresh	89	10 ea	59	68	40
Mushrooms:					
Fresh (1 average = 18g)	35	1/2 c	9	16	1
Ckd from fresh	78	1/2 c	21	8	2
Canned, drained	78	1/2 c	19	1775	332
Mustard greens:					
Fresh, uncooked	56	1 c	15	96	14
Ckd from fresh	140	1 c	21	105	22
Ckd from frozen	150	1 c	29	133	38
Okra, cooked:					
From fresh, pods	85	8 ea	27.2	15	4
From frozen, slices	92	1/2 c	34	9	3
ONIONS, chopped:					
Fresh	160	1 c	60.8	8	4.8
Ckd from fresh	105	1/2 c	46	7	3
Ckd from frozen	105	1/2 c	30	43	13
Onion, dehydrated flakes	14	1/2 c	45	7	3
Onion, Spring, chopped	50	1/2 c	16	50	8
Onion rings, from frozen	20	2 ea	81	93	75
Parsley:					
Fresh, chopped	30	1/2 c	10	120	12
Freeze dried	1.4	1/2 c	4	125	5
Parsnips, ckd from fresh	156	1 c	125	13	16
PEAS:					
Black-eyed peas:					
Ckd from fresh	165	1 c	160	4	7
Ckd from frozen	170	1 c	224	4	9
Ckd from dry	171	1 c	198	3	6
Canned (legume)	240	1 c	184	390	718
Green peas:					
Fresh	145	1 c	118	6	7
Ckd from fresh	160	1 c	134	3	4
Ckd from frozen	80	1/2 c	63	111	70
Cooked from dry	196	1 c	231	2	4
Canned, drained:					
Regular	85	1/2 c	59	315	186
Low sodium	85	1/2 c	59	3	2
Canned with liquid:					
Regular	124	1/2 c	61	557	340
Low salt	124	1/2 c	61	4	3
Green peas, edible-pods:					
Fresh	145	1 c	61	10	6
Ckd from fresh	160	1 c	67	9	6
Ckd from frozen	80	1/2 c	42	10	4
Split peas, ckd from dry	196	1 c	231	2	4

	Wt (g)	Svg	Cal	Sodium (mg) 100 Cal	Svg
Vegetables & Legumes, continued:					
Peas, sprouted, mature:					
Fresh sprouts	120	1 c	154	16	24
Ckd from fresh	100	3.5 oz	118	3	3
Peas & carrots:					
Frozen	70	1/2 c	37	149	55
Canned w/liquid:					
Regular	128	1/2 c	48	692	332
Low sodium	128	1/2 c	48	10	5
PEPPERS, HOT chili,					
green or red, chopped:					
Fresh (1 pod=45g)	75	1/2 c	30	17	5
Canned, Hot Green	68	1/2 c	17	56	10
Canned, Jalapeño	68	1/2 c	17	5853	995
PEPPERS, SWEET,					
green or red, chopped:					
Fresh (1 pod=74 g)	50	1/2 c	14	7	1
Cooked from fresh	68	1/2 c	19	5	1
Poi, two finger	240	1 c	269	10	28
POTATOES:					
Baked in oven:					
Flesh and skin	202	1 ea	220	7	16
Flesh only	156	1 ea	145	6	8
Potato skin	58	1 ea	115	10	12
Boiled, flesh only					
Ckd without skin	135	1 ea	116	6	7
Ckd with skin	136	1 ea	119	5	6
Canned, 1" diam.	70	2 ea	42	338	142
Cooked fr/ frzn, small	70	1 ea	46	30	14
Dehydrated flakes	200	1 c	722	38	275
Cottage fried, fr/ frzn	50	10 strips	109	21	23
French fried, fr/ frzn:					
Cooked in oil	50	10 strips	158	68	108
Oven heated	50	10 strips	111	14	15
Hash brown, fr/ frzn	156	1 c	340	16	53
Mashed, prepared:					
W/milk & salt	210	1 c	162	393	636
From instant	215	1 c	239	307	733
Potato puff (tater tot) f/fzn	62	1/2 c	138	335	462
Potato dishes:					
Au gratin, recipe	245	1 c	322	330	1064
Au gratin, fr/ dry mix	245	1 c	228	472	1076
Scalloped, recipe	245	1 c	210	391	821
Scalloped, fr/ dry mix	245	1 c	228	366	835
Potato chips:					
Plain	28	14 ea	148	90	133
Sour cream & onion	28	1 oz	153	164	251
Potato flour	179	1 c	628	10	61
Potato pancakes	76	1 ea	237	164	388
Pumpkin, mashed:					
Ckd from fresh	245	1 c	50	6	3
Canned	123	1/2 c	42	14	6
Rutabaga:					
Fresh cubes	140	1 c	51	55	28
Cooked fr/ fresh	85	1/2 c	29	52	15
Salsify, cooked slices	135	1 c	92	23	21
Sauerkraut, cnd w/ liquid	236	1 c	44	3548	1561
SEAWEED:					
Agar, fresh	28	1 oz	7.4	35	3
Agar, dried	28	1 oz	87	33	29
Irish moss, fresh	28	1 oz	13.9	137	19
Kelp, fresh	28	1 oz	12.2	541	66
Lavar, fresh	28	1 oz	9.9	137	14
Wakame, fresh	28	1 oz	12.8	1930	247
Spirulina, dried	28	1 oz	82.2	361	297
Soybeans, ckd f/dry	172	1 c	298	<1	1

	Wt (g)	Svg	Cal	Sodium (mg) 100 Cal	Svg
Soybeans, sprouted:					
Fresh sprouts	35	1/2 c	45	11	5
Steamed	94	1 c	76	13	10
SOYBEAN PRODUCTS: see tofu this section; miso in Other; roasted soybeans in Nuts & Seeds; soy milk in Dairy; soy flour in Grains.					
SPINACH:					
Fresh, chopped	56	1 c	12	358	44
Ckd f/fresh, chopped	180	1 c	41	307	126
Ckd f/frozen, leaf	190	1 c	53	306	163
Canned, drained:					
Regular	214	1 c	50	1366	683
Low sodium	214	1 c	50	116	58
SQUASH, SUMMER, sliced:					
Crookneck, fresh	130	1 c	24	8	2
Crookneck, cooked	180	1 c	36	6	2
Scallop, fresh	130	1 c	24	8	2
Scallop, cooked	90	1/2 c	14	7	1
Zucchini, fresh	130	1 c	19	16	3
Zucchini, cooked	180	1 c	29	19	5
SQUASH, WINTER, mashed:					
Acorn, baked	245	1 c	137	8	11
Acorn, boiled	245	1 c	83	7	6
Butternut, baked	245	1 c	99	8	8
Butternut, ckd f/frzn	240	1 c	94	4	4
Hubbard, baked	240	1 c	120	15	19
Hubbard, boiled	236	1 c	70	17	12
Spaghetti, baked/boiled	155	1 c	45	62	28
Succotash:					
Ckd from fresh	192	1 c	222	14	32
Ckd from frozen	170	1 c	158	49	77
Canned w/liquid	255	1 c	161	350	564
Sweet potatoes, 5"x2":					
Baked in skin, flesh only	114	1 ea	118	10	12
Boiled without skin	151	1 ea	160	13	20
Canned, mashed	128	1/2 c	129	74	95
Vacuum pack, mashed	255	1 c	233	58	136
Candied, 2.5" x 2" pce	105	1 pce	144	51	73
Taro chips	23	10 ea	110	77	85
Taro, fresh slices	104	1 c	112	11	12
Taro, ckd fr/ fresh slices	132	1 c	187	11	20
Tofu (soybean curd):					
Regular, raw	124	1/2 c	94	10	9
Firm, raw	126	1/2 c	183	9	17
TOMATOES:					
Fresh, whole, 2-3/5" dm	123	1 ea	26	42	11
Fresh, chopped	180	1 c	38	42	16
Ckd from fresh	240	1 c	65	40	26
Canned, whole	240	1 c	47	830	390
Tomato juice, canned:					
Regular	244	1 c	41.5	2123	881
No salt added	244	1 c	41.5	58	24
Tomato products:					
Paste, canned	262	1 c	220	77	170
Puree, canned	250	1 c	102	48	49
Sauce, canned	245	1 c	74	2001	1481
Turnips:					
Fresh cubes	130	1 c	35	251	88
Ckd fr/ fresh	78	1/2 c	14	279	39
Turnip greens:					
Ckd from fresh	144	1 c	29	141	41
Ckd, from frozen	82	1/2 c	24	50	12
Vegetable juice cocktail	242	1 c	46	1920	883

ESHA Research

	Wt (g)	Svg	Cal	Sodium (mg) 100 Cal	Svg
VEGETABLE COMBINATIONS,					
cooked from frozen:					
Broccoli, carrots, pasta	95	2/3 c	88	299	263
Broccoli, carrots					
& water chestnuts	91	2/3 c	32	69	22
Broccoli, cauliflower					
& red pepper	95	2/3 c	25	72	18
Broccoli, & water chestnuts	95	1/2 c	33	648	214
Cantonese stir fry	95	1/2 c	53	896	475
Chinese stir fry	95	1/2 c	31	1623	503
Green beans & spaetzle	95	1/2 c	108	389	420
Japanese vegetables	95	1/2 c	29	1852	537
Mexicana vegetables	95	1/2 c	125	371	464
Mixed vegetables (corn, peas, carrots,					
green beans, lima beans):					
Ckd from frozen	182	1 c	107	60	64
Canned, drained	163	1 c	77	316	243
New England vegetables	95	1/2 c	128	363	464
Peas, carrots, onions	91	1/2 c	54	111	60
Peas, carrots, onions, pasta	95	1/2 c	122	366	446
Peas, cauliflower, cream sce	95	1/2 c	118	321	379
Peas & mushrooms	95	1/2 c	73	295	215
Peas & onions	95	1/2 c	71	439	312
Peas, onions, carrots, butter sce	71	2/3 c	100	458	458
Peas, onions, cheese sce	142	1/2 c	165	316	522
Peas, pasta, corn, cream sce	95	1/2 c	132	211	279
Peas, pasta, mushrooms,					
& cream sauce	95	1/2 c	129	352	454
Peas, potatoes, cream sce	76	1/2 c	140	346	485
Spinach & water chestnuts	95	1/2 c	29	886	257
Water chestnuts, cnd slices	70	1/2 c	35	17	6
Watercress, fresh	17	1/2 c	2	350	7
Yams:					
Hawaii mtn, steamed	145	1 c	119	15	17
Orange: see Sweet Potato.					
White cubes, ckd f/ fresh	136	1 c	158	7	11
Zucchini, see Squash.					

Other

Cooking Ingredients, Condiments, Flavorings, Spices, etc.

	Wt (g)	Svg	Cal	Sodium (mg) 100 Cal	Svg
Bac-O-Bits, General Mills	8	1 T	33	573	189
Baking chocolate	28	1 oz	145	1	1
Baking powder:					
#1 w/monocalcium					
phosphate monohydrate	3	1 tsp	5	6580	329
#2 w/monocalcium phosphate					
monohydrate calcium sulfate	2.9	1 tsp	5	5800	290
#3 straight phosphate	3.8	1 tsp	5	6240	312
#4 low sodium	4.3	1 tsp	5	t	t
Baking soda	3	1 tsp	0	821	821
Barbecue sauce	16	1 T	10	1280	128
Carob flour	103	1 c	185	19	36
Catsup	17	1 T	18	1122	202
Chili sauce, tomato base	273	1 c	284	1286	3653
Cocoa powder	86	1 c	224	8	17
Cornstarch	8	1 T	20	3	.5
Enchilada dip, Fritos	28	1 oz	35	483	169
Falafel (2-1/4" patty = 17g)	17	1 patty	57	88	50
Gelatin, dry, envelope	7	1 ea	25	24	6
Ginger root slices, raw	11	5 pce	8	13	1
Horseradish, prepared	15	1 T	6	233	14
Hummous, Humous	246	1 c	420	143	599
Jalapeño bean dip, Fritos	28	1 oz	33	494	163
Miso (soybean product)	138	1/2 c	284	1772	5032
Mustard, prepared	125	1/2 c	94	1665	1565
Olives:					
Green, pitted	39	10 ea	45	2080	936
Ripe, large	45	10 ea	52	754	392

	Wt (g)	Svg	Cal	Sodium (mg) 100 Cal	Svg
Pectin (Sure gel) pkg	12	1/4 ea	37	8	3
Pickles:					
Dill, medium, 3.75" x 1.25"	65	1 ea	11.7	7120	833
Fresh pack, slices	35	4 pce	41	490	201
Sweet, small	15	1 ea	20	1645	329
Pickle relish, sweet	15	1 T	20	535	107
Salsa:					
Picante by Tostitos	85	6 T	40	1200	480
From recipe	81	6 T	35	240	84
Soy sauce:					
Regular (wheat & soy)	18	1 T	9	11433	1029
Low sodium (wheat & soy)	18	1 T	9	3333	300
Tamari (soy)	18	1 T	11	9136	1005
Fr/hydrolized veg. protein	18	1 T	7	14628	1024
SPICES:					
Allspice	6	1 T	16	31	5
Basil, dried	4.5	1 T	11	18	2
Celery seed	2.0	1 tsp	8.5	41	4
Cinnamon	2.3	1 tsp	5.9	10	1
Cloves, ground	6.6	1 T	21	76	16
Chili powder	7.5	1 T	24	317	76
Coriander, fresh	4	1/2 c	.08	1375	1
Garlic, cloves	12	4 ea	17.9	11	2
Garlic, powder	8.4	1 T	28	7	2
Cumin seed	6	1 T	22	45	10
Curry powder	6.3	1 T	18	17	3
Dill weed, dried	3.1	1 T	8	75	6
Fenugreek seed	11.1	1 T	36	19	7
Ginger, ground	5.4	1 T	19	11	2
Mace, ground	5.3	1 T	25	16	4
Onion powder	6.5	1 T	15	20	3
Oregano, ground	4.5	1 T	14	7	1
Paprika	6.9	1 T	20	12	2
Pepper, black	6.4	1 T	16.3	18	3
Pepper, red, cayenne	5.3	1 T	17	41	7
Pepper, white	7.1	1 T	21	<1	<1
Poppyseed	8.8	1 T	47	4	2
Poultry seasoning	3.7	1 T	11	9	1
Pumpkin pie spice	5.6	1 T	19	16	3
Rosemary, dried	3.3	1 T	11	18	2
Sage, ground	2	1 T	6	0	0
Salt	16.5	1 T	0	6396	6396
Salt substitute, Morton:					
Regular	6	1 T	0	t	t
Light	6	1 tsp	0	1100	1100
Tarragon, ground	4.8	1 T	14	21	3
Thyme, ground	4.3	1 T	12	17	2
Turmeric, ground	6.8	1 T	24	13	3
Vinegar, cider	15	1 T	1.8	7	<1
Teriyaki sauce	18	1 T	15	4600	690
Tobasco sauce	15	1 T	1.6	61	101
Yeast:					
Brewer's yeast	8	1 T	25	40	10
Dry active (pkt=7.5g)	7.5	1 T	20	19	10

Zinc

Zinc is necessary for the body's metabolism of protein, carbohydrates, fats, and alcohol. Over 70 enzymes require zinc as a cofactor, and it is found throughout the body. The highest concentrations are in the muscle, bones, skin, eyes, male reproductive organs, and kidneys. Saliva and pancreatic fluids also contain large quantities of zinc.

Zinc is needed for the synthesis of proteins, and DNA and RNA; immune reactions; the hormone insulin; growth and repair of tissues; the utilization of vitamin A; and for the cell's ability to produce and dispose of carbon dioxide. It is very important to growth in children, the development of the fetus (therefore for pregnancy), healing of wounds, ability to taste foods, and the making of sperm.

Deficiencies result in loss of appetite, poor growth, hypogonadism, slow healing of wounds, and decreased ability to taste foods.

Nutrient Losses

Milling and processing, such as wheat into flour and peanuts to peanut butter, result in significant losses of zinc. When dried beans are soaked in water and cooked, about 5 percent to 10 percent of the zinc is leached into the water.

Phytate and fiber are prevalent in plant foods and will bind with zinc and reduce its availability to the body. But even with the presence of phytate and fiber, whole grain foods provide more zinc than refined foods (with less fiber and phytate). Yeast breads have higher zinc availability because the yeast produces enzymes that destroy phytate.

Recommended Daily Amounts for Adults

USA: Men, 15 mg; Women, 12 mg; (Pregnant, 15 mg; Nursing: the first 6 months, 19 mg; over 6 months, 16 mg)
Canada: Men, 9 mg; Women, 8 mg (Pregnant, 9-10 mg; Nursing, 14 mg)

There is evidence that small amounts of zinc are more efficiently absorbed than large amounts. The body also appears to adjust absorption -- persons in poor zinc status absorb more efficiently than those in good status.

Toxicity

Zinc does not occur in natural diets at levels that will be toxic. Moderately elevated zinc levels due to supplementation of the U.S. diet is not uncommon, however, and does cause concern. Copper status can be impaired by intakes of 25 mg of zinc a day, and 80 mg a day can cause a decline of high density lipoproteins. Chronic supplementation of more than 15 mg per day is not recommended without medical supervision.

Food Sources

As a general rule of thumb, zinc tends to follow protein in foods. Therefore, seafoods, meats, whole grains, and legumes are good sources. Vegetables and fruits are lesser sources, and the amount they have can be influenced by fertilizer and soil conditions. Zinc levels have been shown to vary according to genetic breeding of some plants, and considerably higher levels of both zinc and copper are in the peel rather than in the flesh of some fruits. Shellfish are higher in zinc concentrations than white fish, and darker poultry meat has more than light meat.

ESHA Research

	Wt (g)	Svg	Cal	Zinc (mg) 100 Cal	Svg

Baked Goods
Breads, Cakes, Cookies, Crackers, Muffins, Pancakes, Pastries, Pies, Rolls

Item	Wt (g)	Svg	Cal	100 Cal	Svg
Apple crisp	78	1 pce	146	.1	.088
Bagel, 3.5" dm, plain/egg	68	1 ea	180	.3	.612
Biscuits:					
Homemade	28	1 ea	100	.2	.153
From mix	28	1 ea	94	.2	.179
From refrig dough	20	1 ea	65	.1	.094
BREADS:					
Boston brown, canned	45	1 pce	95	.4	.350
Cornbread muffin, avg	45	1 ea	145	.2	.325
Cracked wheat	25	1 pce	65	.5	.350
French, 5" x 2.5" x 1"	35	1 pce	100	.2	.221
Mixed grain bread	25	1 pce	65	.5	.300
Oatmeal bread	25	1 pce	65	.4	.245
Pita pocket, 6.5" dm.	60	1 ea	165	.3	.501
Pumpernickel, 5" x 4" x 3/8"	32	1 pce	80	.5	.400
Raisin bread	25	1 pce	68	.2	.155
Rye, light 5" x 3.5" x 7/16"	25	1 pce	65	.6	.380
Wheat (white & whole wheat flour)	28	1 pce	72	.4	.294
White bread	28	1 pce	75	.2	.173
Whole wheat bread	28	1 pce	70	.7	.500
Bread pudding w/raisins	165	1 c	349	.2	.863
Breadsticks, 4" x 1/2" dm.	100	10 ea	384	.1	.570
Brownies w/frosting & nuts	25	1 ea	100	.4	.36
CAKES: pce = 1/16th cake (3" x 3") unless otherwise stated. Cupcakes ≈ 42 grams.					
Angel food, 1/12 tube cake	53	1 pce	125	.1	.070
Boston cream pie, 1/8	120	1 pce	260	.1	.230
Carrot, cream cheese frosting	112	1 pce	406	.1	.45
Cheesecake:					
From recipe, 1/12 cake	92	1 pce	278	.1	.386
From mix, 1/8	103	1 pce	300	.1	.427
Chocolate, chocolate frosting	69	1 pce	235	.2	.530
Coffeecake, f/mix, 2.4" x 2.8"	72	1 pce	230	.3	.619
Fruitcake, dark, 2/3" arc	43	1 pce	165	.1	.215
Gingerbread, 1/9 of 8" sq.	63	1 pce	174	.35	.610
Pound cake, commercial	29	1 pce	110	.1	.11
Sheet cake, 3" x 3":					
Plain cake	86	1 pce	315	.1	.306
White frosting	121	1 pce	445	.1	.322
Snack cake, cream filled:					
Chocolate, like Ding-dongs	28	1 ea	105	.2	.172
Sponge cake, like Twinkies	42	1 ea	155	.1	.210
Sponge cake, 1/12	66	1 pce	194	.2	.475
White, chocolate frosting	77	1 pce	291	.1	.323
White, coconut/white frosting	70	1 pce	270	.1	.212
Yellow, chocolate frosting, avg	69	1 pce	240	.1	.206
Cherry crisp, 3" x 3"	138	1 pce	157	.1	.145
Chips: see Corn and Tortilla in this section; see vegetable section for Potato chips.					
COOKIES:					
Chocolate chip:					
Recipe	40	4 ea	185	.1	.220
Refrig dough	48	4 ea	225	.1	.240
Commercial	42	4 ea	180	.2	.304
Fig bars	56	4 ea	210	.2	.358
Lady fingers	44	4 ea	158	.4	.576
Oatmeal raisin	52	4 ea	245	.2	.530
Peanut butter, homemade	48	4 ea	245	.1	.360
Sandwich type, all	40	4 ea	195	.1	.214
Sugar, from refrigerator dough	48	4 ea	235	.1	.240

Item	Wt (g)	Svg	Cal	100 Cal	Svg
Corn chips (Fritos)	28	1 oz	155	.3	.440
CRACKERS:					
Armenian cracker bread	28	4 pce	117	.8	.900
Rye wafers, whole grain	14	2 ea	55	2.9	1.60
Sesame	12	4 ea	60	.2	.125
Wheat crackers, thin	8	4 ea	35	.7	.240
Whole wheat	8	2 ea	35	.7	.233
Cream puff, custard filled	110	1 ea	280	.2	.624
Croissant, 4.5" x 4" x 2"	57	1 ea	235	.1	.322
Danish pastry:					
Plain pastry	57	1 ea	220	.2	.479
With fruit	65	1 ea	235	.2	.546
Eclair, custard filled, choc icing	94	1 ea	262	.2	.546
English muffin, plain/sourdough	57	1 ea	135	.3	.410
MUFFINS:					
Blueberry:					
Recipe	45	1 ea	135	.2	.290
From mix	45	1 ea	140	.2	.210
Bran, wheat:					
Recipe	45	1 ea	125	.3	.370
From mix	45	1 ea	140	.7	.950
Cornmeal:					
Recipe	45	1 ea	145	.2	.310
From mix	45	1 ea	145	.2	.340
Pancakes:					
Plain, 4" recipe/mix	27	1 ea	60	.4	.226
Buckwheat, 4" dm, mix	27	1 ea	55	.9	.500
Whole wheat, 5" dm.	52	1 ea	94	.6	.519
PIES: piece is 1/6 of 9" pie unless otherwise stated.					
Apple pie	158	1 pce	405	.1	.267
Banana cream, commercial	152	1 pce	333	.3	.873
Chocolate cream	175	1 pce	311	.2	.743
Coconut cream	172	1 pce	343	.2	.823
Coconut custard	165	1 pce	384	.3	1.21
Cream, commercial	152	1 pce	455	.2	.785
Custard pie	152	1 pce	293	.3	.792
Lemon meringue	140	1 pce	355	.1	.510
Peach pie	158	1 pce	405	.1	.352
Pecan pie	138	1 pce	583	.3	1.47
Pumpkin pie	200	1 pce	367	.3	.993
Poptart-type toaster pastry	54	1 ea	210	.1	.313
Pretzel, dutch twist	16	1 ea	65	.3	.173
Pretzel, thin twists	60	10 ea	240	.2	.419
ROLLS:					
Cinnamon bun, small	50	1 ea	158	.3	.452
Dinner roll, 2.5" x 2"	28	1 ea	85	.3	.223
Hamburger bun	45	1 ea	129	.3	.408
Hard roll, white	50	1 ea	155	.3	.438
Hotdog bun	40	1 ea	115	.3	.363
Rye roll, light	28	1 ea	76	.6	.426
Rye roll, dark	28	1 ea	79	.3	.274
Submarine roll (hoagie)	135	1 ea	400	.3	1.17
Whole wheat roll	35	1 ea	88	.7	.580
Tortillas:					
Corn, 6" diam, fried	30	1 ea	87	.3	.300
Flour, 10.5" diam.	57	1 ea	168	.3	.432
Flour, 8" diam.	35	1 ea	105	.3	.269
Tortilla chips, all kinds	28	1 oz	139	.3	.420
Waffles:					
Homemade, 7" dm.	75	1 ea	245	.3	.652
Prep f/mix, 7" dm.	75	1 ea	205	.3	.515
Frozen, 4" dm.	35	1 ea	98	.3	.288

Dairy & Dairy Products

	Wt (g)	Svg	Cal	Zinc (mg) 100 Cal	Zinc (mg) Svg
CHEESE [1.5" cube ≈ 1 oz]:					
American processed cheese	28	1 oz	106	.9	.933
American cheese food	28	1 oz	94	.9	.850
American cheese spread	28	1 oz	82	1.0	.780
Blue cheese	28	1 oz	100	.8	.750
Brick cheese	28	1 oz	105	.7	.734
Brie cheese	28	1 oz	95	.7	.700
Camembert	28	1 oz	85	.8	.675
Caraway	28	1 oz	107	.8	.882
Cheddar cheese	28	1 oz	114	.8	.924
Cheshire	28	1 oz	110	.7	.800
Colby	28	1 oz	112	.8	.870
Cottage cheese:					
Lowfat 1%	226	1 c	164	.5	.860
Lowfat 2%	226	1 c	205	.5	.950
Creamed, lrg curd	225	1 c	235	.3	.800
Creamed, sm curd	210	1 c	215	.4	.802
Cream cheese	28	1 oz	99	.3	.325
Edam cheese	28	1 oz	101	1.0	1.06
Feta cheese	28	1 oz	75	1.1	.813
Fontina	28	1 oz	110	.9	.990
Gjetost	28	1 oz	132	.7	.946
Gouda	28	1 oz	101	1.1	1.10
Gruyere	28	1 oz	117	.9	1.00
Liederkranz	28	1 oz	87	.8	.700
Limburger	28	1 oz	93	.6	.600
Monterey jack	28	1 oz	106	.8	.846
Mozzarella, low moisure:					
Part skim	28	1 oz	80	1.0	.825
Whole milk	28	1 oz	90	1.0	.895
Muenster	28	1 oz	104	.8	.843
Parmesan, grated (1 T = 5g)	28	1 oz	129	.8	1.00
Pimento, processed	28	1 oz	106	.8	.840
Port du salut	28	1 oz	100	.8	.800
Provolone	28	1 oz	100	.9	.889
Ricotta, part skim	246	1 c	340	1.0	3.29
Ricotta, whole milk	246	1 c	428	.7	2.85
Romano, grated (1 T = 5g)	28	1 oz	128	.9	1.20
Roquefort	28	1 oz	105	.5	.570
Swiss cheese	28	1 oz	107	1.0	1.10
CREAM, SWEET, fluid:					
Coffee or table	240	1 c	469	.1	.649
Half and half	242	1 c	315	.4	1.23
Light whipping cream	239	1 c	699	.1	.600
Heavy whipping cream	238	1 c	821	.1	.550
CREAM, SWEET, whipped:					
Heavy cream, unsweetened	119	1 c	410	.1	.275
Pressurized	60	1 c	154	.1	.220
CREAM, SOUR, cultured, dairy	230	1 c	493	.1	.690
Cream, sour, Imitation (non-dairy)	230	1 c	479	0	0
CREAM SUBSTITUTES, non-dairy:					
Coffee whitener, powder	94	1 c	514	.1	.480
Coffee whitener, liquid	120	1/2 c	163	<.1	.020
Dessert Toppings, non-dairy,					
Frozen (like Coolwhip)	75	1 c	239	<.1	.029
Kefir beverage	233	1 c	160	.6	.900
MILK (cow):					
Skim milk	245	1 c	86	1.1	.915
Lowfat 1%	244	1 c	102	.9	.963
Lowfat 2%	244	1 c	121	.8	.963
Whole (3.3% fat)	244	1 c	150	.6	.930
Buttermilk	245	1 c	99	1.0	1.03
Canned, evap, skim	255	1 c	200	1.1	2.18

	Wt (g)	Svg	Cal	Zinc (mg) 100 Cal	Zinc (mg) Svg
MILK (cow) continued:					
Canned, evap, whole	252	1 c	340	.6	1.94
Dry, nonfat instant	68	1 c	244	1.3	3.06
MILK (other):					
Goat milk	244	1 c	168	.4	.730
Human breast milk	246	1 c	171	.2	.420
Soy milk	240	1 c	79	.7	.540
MILK BEVERAGES & mixes:					
Chocolate milk, commercial:					
Lowfat 1%	250	1 c	160	.6	1.02
Lowfat 2%	250	1 c	180	.5	.910
Whole (3.3%)	250	1 c	210	.5	1.02
Chocolate flavored mix, to be mixed with water:					
Powder (includes dry milk)	28	1 oz	100	1.3	1.26
Drink, prepared	206	3/4 c	100	1.3	1.26
Chocolate flavored mix, to be mixed with milk:					
Powder	21.6	3/4 oz	75	.4	.330
Drink, prep w/wh milk	266	1 c	226	.6	1.26
Eggnog, commercial	254	1 c	342	.3	1.17
Instant Breakfast, dry	37	1 env	130	2.3	3.00
Malted milk, prep w/whole milk:					
Chocolate flavor	265	1 c	229	.5	1.09
Natural flavor	265	1 c	237	.5	1.14
Milkshake (10 fl oz = 1.25 c):					
Chocolate	283	1.25 c	360	.3	1.15
Strawberry	283	1.25 c	319	.3	1.00
Vanilla	283	1.25 c	314	.3	1.01
MILK DESSERTS:					
Custard, baked:					
Recipe	265	1 c	305	.5	1.53
Prep from mix	143	1/2 c	161	.4	.645
Ice cream, vanilla:					
Regular	133	1 c	269	.5	1.41
Soft serve	173	1 c	377	.5	1.99
Rich	148	1 c	349	.3	1.21
Ice milk, vanilla:					
Regular	131	1 c	184	.3	.550
Soft serve, 3% fat	175	1 c	223	.4	.860
PUDDINGS, prepared (5 oz can ≈.55c)					
Chocolate:					
From mix-ckd or instant	260	1 c	305	.4	1.18
Canned	142	1 can	205	.3	.700
Coconut, f/instant	149	1/2 c	184	.3	.474
Lemon, f/instant	149	1/2 c	178	.3	.480
Rice, ckd/instant mix	141	1/2 c	165	.4	.577
Tapioca pudding, ckd f/mix	130	1/2 c	145	.3	.500
Tapioca, canned	142	1 can	160	.4	.700
Vanilla, ckd/instant mix	130	1/2 c	148	.3	.500
Vanilla, canned	142	1 can	220	.3	.700
Pudding Pops:					
Banana/butterscotch/van	57	1 ea	94	.3	.245
Chocolate/choc fudge	57	1 ea	99	.4	.355
Sherbet	193	1 c	270	.5	1.33
Yogurt, frozen, avg	174	1 c	220	.5	1.12
YOGURT:					
Lowfat, plain	227	1 c	144	1.4	2.02
Lowfat, with fruit	227	1 c	231	.7	1.68
Lowfat, coffee or vanilla	227	1 c	193	1.0	1.88
Nonfat	227	1 c	127	1.7	2.20
Whole	227	1 c	138	1.0	1.34
Yogurt cheese, recipe	208	1 c	222	1.7	3.72

ESHA Research

Zinc (Zn) mg

	Wt (g)	Svg	Cal	Zinc (mg) 100 Cal	Svg
Eggs					
Egg, chicken, raw/cooked:					
Whole egg	50	1 ea	77.5	.7	.55
White only	33.4	1 ea	17	0	0
Yolk only	16.6	1 ea	59	.9	.54
Egg substitutes					
(check label, products vary):					
Frozen	60	1/4 c	96	.6	.590
Liquid	251	1 c	211	1.5	3.26
Fruit & Fruit Juices					
APPLE, w/peel, 2.75" dm.	138	1 ea	80	.1	.050
APRICOTS:					
Fresh, pitted	106	3 ea	51	.5	.280
Canned, juice pack	248	1 c	119	.2	.270
Canned, heavy syrup	258	1 c	214	.1	.270
Dried apricots	35	10 ea	83	.3	.260
Apricot nectar, canned	251	1 c	141	.2	.230
Avocado, whole:					
California	173	1 ea	305	.2	.730
Florida	304	1 ea	340	.4	1.28
Banana, 8.75", 176g w/peel	114	1 ea	105	.2	.190
Blackberries:					
Fresh berries	144	1 c	74	.5	.390
Frozen, unthawed	151	1 c	97	.4	.370
Canned	256	1 c	236	.2	.470
Blueberries:					
Fresh	145	1 c	82	.2	.160
Canned	256	1 c	225	.1	.170
Boysenberries:					
Frozen	132	1 c	66	.4	.290
Canned	256	1 c	225	.2	.480
Cantaloupe: see Melon.					
Cassava, fresh	100	3.5 oz	120	.8	.980
Cherries, sour:					
Frozen	155	1 c	72	.2	.160
Canned	244	1 c	90	.2	.170
CHERRIES, SWEET:					
Fresh, pitted	68	10 ea	49	.1	.040
Canned w/liquid	257	1 c	213	.1	.260
Cranberries, fresh, whole	95	1 c	46	.3	.120
Cranberry juice cocktail	253	1 c	145	.1	.177
Currants:					
Black, fresh	112	1 c	71	.4	.300
Red or white, fresh	112	1 c	63	.4	.260
Zante, dried	144	1 c	407	.2	.940
Dates, whole, pitted	83	10 ea	228	.1	.242
Figs, fresh, medium	50	1 ea	37	.2	.070
Figs, dried	187	10 ea	477	.2	.940
Fruit cocktail, canned:					
Juice pack	248	1 c	115	.2	.210
Heavy syrup	255	1 c	185	.1	.210
Gooseberries:					
Fresh berries	150	1 c	67	.3	.180
Canned with liquid	252	1 c	185	.2	.280
GRAPEFRUIT (half = 241g w/refuse):					
Pink or red half	123	1 ea	37	.2	.090
White half	118	1 ea	39	.2	.080
Canned sections	254	1 c	152	.1	.21
Grapefruit juice, canned	247	1 c	93	.2	.220
Guava, fresh	90	1 ea	45	.5	.210
Lemon juice, bottled	244	1 c	52	.3	.150
Lime juice, bottled	246	1 c	50	.3	.150
Loganberries, frozen	147	1 c	80	.6	.500
Lychees, canned	100	3.5 oz	68	.3	.200
Mandarin oranges, canned	252	1 c	155	<.1	.075
Mango, fresh slices	165	1 c	108	.2	.260
Melon, also see Watermelon:					
Cantaloupe cubes	160	1 c	57	.4	.256
Frozen melon balls, mixed	173	1 c	55	.5	.290
Mixed fruit, dried	293	11 oz	712	.2	1.47
ORANGE, avg (180g w/refuse)	131	1 ea	60	.1	.090
Orange juice:					
Fresh juice	248	1 c	111	.1	.124
Prep from frozen	249	1 c	110	.1	.128
Frozen conc, 6 oz can	213	3/4 c	339	.1	.383
Canned, unsweetened	249	1 c	105	.2	.174
Orange grapefruit juice, cnd	247	1 c	105	.2	.180
Papaya (454g w/ refuse)	304	1 ea	117	.2	.220
Papaya nectar, canned	250	1 c	142	.3	.380
PEACHES:					
Fresh, 2.5" diam	87	1 ea	37	.3	.120
Canned, juice pack	77	1 half	34	.3	.085
Canned, heavy syrup	81	1 half	60	.1	.070
Peach nectar, canned	249	1 c	134	.1	.200
PEARS:					
Fresh, Bartlett (180g w/ refuse)	166	1 ea	98	.2	.200
Canned, juice pack	77	1 half	38	.2	.069
Canned, heavy syrup	79	1 half	59	.1	.063
Pear nectar, canned	250	1 c	149	.1	.160
Persimmon, (Japanese)	168	1 ea	118	.2	.180
PINEAPPLE:					
Fresh, chunks	155	1 3	76	.2	.120
Canned pieces (1 ring = 58g):					
Juice pack	250	1 c	150	.2	.250
Heavy syrup	255	1 c	199	.2	.306
Pineapple juice	250	1 c	135	.2	.283
Plantain, fresh slices	148	1 c	181	.1	.270
Plantain, cooked slices	154	1 c	179	.1	.210
PLUMS:					
Fresh, med, 2-1/8" dm.	66	1 ea	36	.2	.06
Canned, juice pack	95	3 ea	55	.2	.11
Canned, heavy syrup	110	3 ea	98	.1	.08
Prunes, dried, pitted	84	10 ea	201	.2	.445
Prune juice, bottled	256	1 c	181	.3	.538
Raisins, dark, unpacked	145	1 c	435	.1	.464
Raspberries:					
Fresh berries	123	1 c	60	.9	.566
Frozen, thawed	250	1 c	255	.2	.450
Rhubarb:					
Fresh, diced	122	1 c	26	.5	.130
Cooked with sugar	240	1 c	279	.1	.192
STRAWBERRIES:					
Fresh berries	149	1 c	45	.4	.194
Frozen, unsweetened	149	1 c	52	.4	.190
Frozen, thawed, swtnd	255	1 c	245	.1	.153
Tangerine	84	1 ea	37	1.0	.380
Watermelon cubes	160	1 c	50	.2	.112
Grains & Grain Products					
Cereals, Flour, Grains, Noodles, Pasta, and Popcorn					
Amaranth grain, dry	195	1 c	729	.85	6.20
Barley, cooked:					
Whole	200	1 c	200	.6	1.16
Pearled	157	1 c	193	.7	1.29

	Wt (g)	Svg	Cal	Zinc (mg) 100 Cal	Svg
Grains & Grain Products, continued:					
Bran: see Oat, Rice, Wheat.					
Buckwheat flour, light	98	1 c	340	.8	2.56
Buckwheat flour, dark	98	1 c	338	.8	2.65
Bulgar wheat, cooked	182	1 c	151	.7	1.04
CEREALS, COLD (Ready to eat) Cereals are often fortified with zinc. Check the label.					
CEREALS, HOT (cooked):					
Corn grits, cooked, enriched	242	1 c	145	.1	.169
Cream of Rice	244	1 c	126	.3	390
Cream of Wheat, cooked	244	1 c	140	.2	.347
Farina	233	1 c	116	.1	.163
Malt-O-Meal	240	1 c	122	.1	.168
Maypo cereal	180	3/4 c	128	.9	1.12
Oatmeal, from rolled oats (regular, quick, instant)	234	1 c	145	.8	1.15
Oatmeal, fortified, instant, prepared from packet:					
Plain	177	3/4 c	104	1.0	1.00
With bran & raisin	195	3/4 c	158	.9	1.35
Other flavors, avg	164	3/4 c	160	.6	1.00
Ralston cereal	253	1 c	134	1.1	1.42
Roman Meal	181	3/4 c	111	1.2	1.34
Wheatena	243	1 c	135	1.2	1.68
Whole wheat	242	1 c	150	.8	1.16
Corn flour	117	1 c	422	.5	2.02
Corn flour, Masa Harina, enr	114	1 c	416	.5	2.03
Cornmeal:					
Degermed, dry	138	1 c	505	.2	.994
Bolted, nearly whole	122	1 c	441	.5	2.22
FLOUR: see specific grain, nut or vegetable.					
Macaroni, cooked:					
Enriched	140	1 c	197	.4	.742
Vegatable, enriched	134	1 c	172	.3	.59
Whole wheat	140	1 c	174	.6	1.13
Millet, cooked	120	1/2 c	143	.8	1.10
Noodles, cooked:					
Egg noodles, enriched	160	1 c	213	.5	.992
Spinach noodles	140	1 c	182	.8	1.51
Oat bran, 1T ≈ 6 g	94	1 c	132	2.2	2.92
Oats, rolled, dry, uncooked	81	1 c	311	.8	2.49
PASTA: see Macaroni, Noodles, Spaghetti.					
Popcorn, ckd in oil, salted	11	1 c	55	.5	.285
RICE, cooked:					
Brown rice	195	1 c	217	.6	1.23
White, regular	205	1 c	264	.4	.943
White, converted	175	1 c	200	.3	.542
White, instant	165	1 c	162	.2	.396
Wild rice	164	1 c	166	1.3	2.20
Rye flour:					
Dark	128	1 c	415	1.7	7.19
Light	102	1 c	361	.5	1.79
Soy flour, stirred					
Low fat	44	1/2 c	163	.32	.52
Defatted	50	1/2 c	164	.75	1.23
Full fat, raw	42	1/2 c	182	.92	1.67
Spaghetti noodles, ckd:					
Enriched	140	1 c	197	.4	.742
Whole wheat spaghetti	140	1 c	174	.7	1.14
WHEAT:					
Wheat bran	30	1/2 c	65	3.4	2.18
Wheat flours:					
All purpose, white, unsifted	125	1 c	455	.2	.875
Cake flour, sifted	96	1 c	348	.2	.595

	Wt (g)	Svg	Cal	Zinc (mg) 100 Cal	Svg
WHEAT flour, continued:					
Self-rising	125	1 c	442	.2	.775
Semolina	167	1 c	601	.3	.175
Whole wheat	120	1 c	407	.9	3.52
Wheat germ:					
Raw	100	1 c	360	3.4	12.3
Toasted	113	1 c	432	4.4	18.9
Wheat, rolled, cooked	240	1 c	142	.9	1.22
Wheat, rolled, dry	85	1 c	289	.9	2.5
Whole grain (wheatberries) cooked	50	1/3 c	28	.9	.244
Whole wheat, sprouted	108	1 c	214	.8	1.78

Meats: Fish & Shellfish

	Wt (g)	Svg	Cal	Zinc (mg) 100 Cal	Svg
Abalone, fried	85	3 oz	161	.5	.800
Anchovies cnd in oil, drained	45	11 ea	95	1.2	1.10
Bass, freshwater, baked/broiled	100	3.5 oz	125	.6	.700
Bluefish:					
Baked/broiled	100	3.5 oz	159	.7	1.04
Fried in crumbs	100	3.5 oz	205	.4	.900
Carp, baked/broiled	100	3.5 oz	162	1.2	1.90
Catfish, cornmeal fried	100	3.5 oz	229	.5	1.20
CLAMS, meat only:					
Canned, drained	160	1 c	236	1.9	4.37
Minced w/liquid, small can	183	1 can	145	.4	.561
Breaded, fried, small	188	20 ea	379	.7	2.74
Steamed meat	90	20 ea	133	1.8	2.46
Clam nectar, canned	240	1 c	6	4.0	.240
Cod, Atlantic:					
Broiled/baked/poached	100	3.5 oz	105	.6	.580
Batter fried	100	3.5 oz	199	.3	.500
Smoked	100	3.5 oz	79	.5	.380
CRAB meat, cooked:					
Alaska King crab leg	134	1 ea	129	7.9	10.2
Blue crab, unpacked measure					
Cooked	135	1 c	138	4.1	5.70
Canned	135	1 c	133	4.1	5.42
Dungeness meat, ckd	101	3/4 c	85	5.1	4.33
Crab, imitation fr/surimi	85	3 oz	87	.3	.250
Crab cakes fr/recipe	60	1 ea	93	2.6	2.46
Crayfish, ckd, moist heat	85	3 oz	97	1.5	1.42
Eel, baked/broiled	100	3.5 oz	236	.9	2.08
Eel, smoked	100	3.5 oz	330	.2	.70
Fish cakes, fried:					
Homemade	100	3.5 oz	172	.3	.480
From frozen	100	3.5 oz	213	.2	.400
Fish sticks, frzn, heated	57	2 ea	155	.2	.380
HADDOCK:					
Baked/broiled/poached	85	3 oz	95	.4	.410
Breaded, fried	85	3 oz	175	.5	.850
Smoked	100	3.5 oz	116	.4	.500
HALIBUT, baked/broiled	85	3.5 oz	119	.4	.450
Herring:					
Baked/broiled	100	3.5 oz	203	.6	1.27
Canned w/liquid	100	3.5 oz	208	.8	1.72
Canned w/tomato sce	100	3.5 oz	173	.9	1.60
Smoked, kippered	100	3.5 oz	217	.6	1.36
Pickled, 1 pce = 15g	100	3.5 oz	262	.2	.53
Lobster meat, cooked	145	1 c	142	3.0	4.23
Mackerel:					
Baked/broiled, Atlantic	100	3.5 oz	262	.4	.94
Canned, Jack, 1 tall can	361	1 can	563	.7	3.68

ESHA Research

Fish & Shellfish, continued:	Wt (g)	Svg	Cal	Zinc (mg) 100 Cal	Zinc (mg) Svg
Mullet, baked/broiled	85	3 oz	127	.6	.750
Ocean perch:					
Baked/broiled	100	3.5 oz	121	.5	.610
Breaded, fried	85	3 oz	185	.2	.410
Octopus, raw	100	3.5 oz	82	2.0	1.68
OYSTER:					
Raw, Eastern	248	1 c	170	133	226
Raw, Pacific	248	1 c	200	21	41.2
Simmered, Eastern	100	3.5 oz	137	133	182
Breaded, fried, med, Eastern	88	6 ea	173	44	76.7
Perch, baked/broiled	92	2 ea	108	1.2	1.32
Pike, baked/broiled, Northern	100	3.5 oz	113	.8	.860
Pollock:					
Baked/broiled, mixed	100	3.5 oz	99	.5	.536
Baked/broiled, Walleye	100	3.5 oz	113	.5	.600
Rockfish, baked/broiled	100	3.5 oz	121	.4	.530
SALMON:					
Average, baked/brld	85	3 oz	183	.2	.430
Chinook, smoked	85	3 oz	99	.3	.260
Coho, steamed/poached	100	3.5 oz	185	.3	.520
Sockeye, baked/brld	100	3.5 oz	216	.2	.510
Canned, Atlantic, small can	220	1 can	281	.6	1.58
Canned, Pink, 1# can	454	1 can	631	.7	4.19
Canned, Sockeye, 1# can	369	1 can	566	.7	3.75
Sardines, cnd, drained:					
Atlantic, 2 ≈ 24g	92	1 can	192	.6	1.21
Pacific, 1 ≈ 38g	100	3.5 oz	178	.8	1.40
Scallops:					
Breaded, fried	93	6 ea	200	.5	.986
Steamed	100	3.5 oz	113	1.0	1.16
Sea Bass, baked/brld	100	3.5 oz	124	.4	.520
Seatrout/Steelhead, ckd	100	3.5 oz	131	.4	.520
Shad:					
Baked w/bacon	100	3.5 oz	201	.1	.295
Batter-fried	85	3 oz	194	.2	.410
SHRIMP:					
Boiled, 2 large ≈ 11g	100	3.5 oz	99	1.6	1.56
Breaded, fried (2 large ≈ 15g)	90	12 ea	218	.6	1.24
Canned, drained	128	1 c	154	1.0	1.61
Canned w/liquid	100	3.5 oz	102	2.3	2.30
Smelt, Rainbow, ckd	85	3 oz	106	1.7	1.80
Snapper, baked/brld	100	3.5 oz	128	.3	.440
SOLE (Flounder):					
Baked/broiled	85	3 oz	99	.5	.530
Breaded, fried	100	3.5 oz	188	.2	.453
Batter-fried	85	3 oz	250	.2	.450
Squid, flour-fried	85	3 oz	149	1.0	1.50
Sturgeon:					
Cooked	85	3 oz	115	.4	.460
Smoked	85	3 oz	147	.4	.658
Surimi, processed walleye (Alaska) pollock; see imitation crab.					
Swordfish, baked/brld	100	3.5 oz	155	.9	1.47
Trout, baked/brld	85	3 oz	129	.9	1.18
TUNA:					
Light, canned, drained (No. 1/2 can):					
Canned in oil	171	1 can	339	.5	1.54
Water pack	165	1 can	216	.6	1.30
Bluefin, ckd f/fresh	85	3 oz	157	.4	.650

Meats

Beef, Ham, Pork, Frog legs, Rabbit, Venison, and Veal

	Wt (g)	Svg	Cal	Zinc (mg) 100 Cal	Zinc (mg) Svg
BEEF:					
Breakfast strips					
(cured beef), cooked	34	3 pce	153	1.4	2.17
Chuck blade, pot roasted, all grades:					
Lean & fat (5.4 oz raw)	85	3 oz	325	2.0	6.66
Lean only	85	3 oz	230	3.8	8.73
Ground beef, average baked, broiled, fried:					
Extra lean (17% fat, raw)	85	3 oz	215	2.1	4.59
Lean (20.7% fat, raw)	85	3 oz	231	1.9	4.44
Regular (26.6% fat, raw)	85	3 oz	250	1.7	4.29
Frzn patty, brld (23% fat raw)	85	3 oz	240	1.9	4.59
Rib, choice, roasted:					
Lean & fat (5 oz raw)	85	3 oz	324	1.4	4.40
Lean only	85	3 oz	204	2.9	5.90
Round steak, choice, brld:					
Lean & fat (4.5 oz raw)	85	3 oz	233	1.5	3.51
Lean only	85	3 oz	165	2.4	3.97
Round tip, all grades, rstd:					
Lean & fat	85	3 oz	213	2.5	5.41
Lean only	85	3 oz	162	3.7	6.01
Sirloin steak, all grades, broiled (11.3 oz raw = 8.2 oz ckd, lean & fat; 6.9 oz lean. Cooked values follow):					
Lean & fat	85	3 oz	238	1.6	3.91
Lean only	85	3 oz	172	2.6	4.44
T-bone steak, choice, broiled (16 oz raw = 9.7 oz ckd, lean & fat; 7.4 oz lean. Cooked values follow):					
Lean & fat	85	3 oz	276	1.4	3.79
Lean only	85	3 oz	182	2.5	4.59
Variety meats:					
Corned beef, canned	85	3 oz	213	1.4	3.03
Dried beef, cured (6 - 7 pces)	28	1 oz	47	3.2	1.49
Heart, simmered	85	3 oz	140	1.9	2.66
Liver, fried	85	3 oz	184	2.5	4.63
Tongue, cooked	85	3 oz	241	1.7	4.08
HAM: see Pork, cured; Turkey ham; and Lunchmeat section.					
LAMB:					
Arm chop, braised (5.6 oz w/bone, raw):					
Lean & fat	70	1 chop	244	1.8	4.28
Lean only	55	1 chop	152	2.6	3.98
Loin chop, broiled (4.2 oz w/bone, raw):					
Lean & fat	64	1 chop	201	1.1	2.22
Lean only	46	1 chop	100	1.9	1.91
Cutlet, lean, ckd average	85	3 oz	175	2.6	4.48
Leg of lamb, roasted:					
Lean & fat	85	3 oz	219	1.7	3.74
Lean only	85	3 oz	162	2.6	4.20
Rib roast:					
Lean & fat	85	3 oz	305	1.0	2.96
Lean only	85	3 oz	197	1.9	3.80
Shoulder roast:					
Lean & fat	85	3 oz	235	1.9	4.44
Lean only	85	3 oz	173	3.0	5.14
Lamb liver, pan-fried	85	3 oz	202	2.4	4.79
PORK:					
Bacon, cooked:					
Regular	19	3 pce	109	.6	.620
Canadian style	47	2 pce	86	.9	.790
Breakfast strips	34	3 pce	156	.8	1.25
Blade chop:					
Braised, lean & fat	67	1 ea	275	.9	2.58
Braised, lean only	50	1 ea	156	1.6	2.47
Broiled, lean & fat	77	1 ea	303	.8	2.35
Broiled, lean only	59	1 ea	117	1.9	2.24

	Wt (g)	Svg	Cal	Zinc (mg) 100 Cal	Svg
PORK, Blade chop, continued:					
Pan-fried, lean & fat	89	1 ea	368	.7	2.41
Pan-fried, lean only	62	1 ea	175	1.3	2.28
Roasted, lean & fat	88	1 ea	321	.8	2.63
Roasted, lean only	71	1 ea	198	1.3	2.55
Center loin chop:					
Braised, lean & fat	75	1 ea	266	.7	1.85
Braised, lean only	61	1 ea	166	1.1	1.78
Broiled, lean & fat	87	1 ea	275	.6	1.68
Broiled, lean only	72	1 ea	166	1.0	1.61
Pan-fried, lean & fat	89	1 ea	333	.5	1.74
Pan-fried, lean only	67	1 ea	178	.9	1.61
Roasted, lean & fat	88	1 ea	268	.7	1.80
Roasted, lean only	72	1 ea	180	1.0	1.71
Center rib chop:					
Braised, lean & fat	67	1 ea	246	.6	1.57
Braised, lean only	53	1 ea	147	1.0	1.49
Broiled, lean & fat	77	1 ea	264	.6	1.56
Broiled, lean only	63	1 ea	162	.9	1.50
Pan-fried, lean & fat	88	1 ea	343	.4	1.43
Pan-fried, lean only	62	1 ea	160	.8	1.28
Roasted, lean & fat	79	1 ea	252	.6	1.55
Roasted, lean only	66	1 ea	162	.9	1.47
Leg of pork, roasted:					
Lean & fat	85	3 oz	250	1.0	2.43
Lean only	85	3 oz	187	1.5	2.77
Pork roast, average loin & rib:					
Lean & fat	85	3 oz	265	.6	1.70
Lean only	85	3 oz	206	.9	1.92
Shoulder, braised (yield from 6.8 oz raw w/bone & skin):					
Lean & fat	85	3 oz	293	1.2	3.43
Lean only	67	2.4 oz	166	2.0	3.33
Spareribs, from 1 lb raw	177	6.25 oz	703	1.2	8.14
Pork heart	145	1 c	214	2.1	4.48
Pork liver	85	3 oz	141	4.0	5.71
PORK, CURED — HAM, also see bacon under Pork:					
Roasted, lean & fat	85	3 oz	207	1.0	1.97
Roasted, lean only	85	3 oz	133	1.6	2.19
Canned, roasted	85	3 oz	140	1.4	1.97
RABBIT, roasted	85	3 oz	131	1.2	1.51
VEAL (calf):					
Cutlet, lean, ckd avg	85	3 oz	166	2.6	4.33
Rib, roasted	85	3 oz	151	2.5	3.81
Heart, braised	85	3 oz	134	1.7	2.34
Liver, pan-fried	85	3 oz	208	3.2	6.69
VENISON (deer) roasted	85	3 oz	134	1.7	2.34

Meats: Poultry

CHICKEN: A 3 lb chicken ≈ 1.45 lbs raw; 1.1 lbs cooked.

	Wt (g)	Svg	Cal	Zinc (mg) 100 Cal	Svg
All types:					
Fried	140	1 c	307	1.0	3.13
Roasted	140	1 c	266	1.1	2.94
Stewed	140	1 c	248	1.1	2.79
Canned, boned w/broth	142	5 oz	235	.9	2.13
Dark meat:					
Fried	85	3 oz	203	1.2	2.47
Roasted	85	3 oz	174	1.4	2.38
Stewed	85	3 oz	163	1.4	2.26
Light meat:					
Fried	85	3 oz	163	.7	1.08
Roasted	85	3 oz	147	.7	1.05
Stewed	85	3 oz	135	.7	1.01

	Wt (g)	Svg	Cal	Zinc (mg) 100 Cal	Svg
CHICKEN, continued:					
Breast*, meat & skin (145g raw; 181g raw w/bone):					
Batter-fried	140	1 ea	364	.4	1.33
Flour-fried	98	1 ea	218	.5	1.07
Roasted	98	1 ea	193	.5	1.00
Stewed	110	1 ea	202	.5	1.06
Breast*, meat only (118g raw):					
Fried	86	1 ea	161	.6	.930
Roasted	86	1 ea	142	.6	.860
Stewed	95	1 ea	144	.6	.920
*2 pieces per bird					
Drumstick, meat & skin (73g raw; 110g raw w/bone):					
Batter-fried	72	1 ea	193	.9	1.67
Flour-fried	49	1 ea	120	1.2	1.42
Roasted	52	1 ea	112	1.3	1.49
Stewed	57	1 ea	116	1.3	1.51
Drumstick, meat (62g raw):					
Fried	42	1 ea	82	1.6	1.35
Roasted	44	1 ea	76	1.8	1.40
Stewed	46	1 ea	78	1.8	1.39
Thigh, meat & skin (94g raw; 120g raw w/bone):					
Batter-fried	86	1 ea	238	.7	1.75
Flour-fried	62	1 ea	162	1.0	1.56
Roasted	62	1 ea	153	1.0	1.46
Stewed	68	1 ea	158	1.0	1.53
Thigh, meat (69g raw):					
Fried	52	1 ea	113	1.3	1.45
Roasted	52	1 ea	109	1.2	1.34
Stewed	55	1 ea	107	1.3	1.42
Wing, meat & skin (49g raw; 90g raw w/bone):					
Batter-fried	49	1 ea	159	.4	.670
Flour-fried	32	1 ea	103	.5	.560
Roasted	34	1 ea	99	.6	.62
Stewed	40	1 ea	100	.7	.650
Wing, meat only (29g raw):					
Fried	20	1 ea	42	1.0	.420
Roasted	21	1 ea	43	1.0	.450
Stewed	24	1 ea	43	1.1	.490
Chicken gizzard	22	1 ea	34	2.8	.963
Chicken heart	3.3	1 ea	6	4.0	.240
Chicken liver	20	1 ea	30	2.9	.867
Chicken roll, light meat	57	2 pce	90	.5	.410
DUCK, domestic, roasted:					
Meat & skin	85	3 oz	286	.6	1.58
Meat only	85	3 oz	171	1.3	2.21
GOOSE, domestic, roasted:					
Meat & skin	85	3 oz	259	.7	1.76
Meat only	85	3 oz	202	1.1	2.30
TURKEY:					
Breast meat, seasoned:					
Barbecued	28	1 oz	40	.9	.350
Hickory smoked	28	1 oz	35	.9	.300
Ground, cooked	100	3.5 oz	229	1.2	2.86
Roasted:					
All types	140	1 c	238	1.8	4.34
Dark meat only	85	3 oz	159	2.4	3.80
Light meat only	85	3 oz	133	1.3	1.73
Frozen slices w/gravy	142	5 oz	95	1.0	.994
Frozen slices	85	3 oz	130	1.8	2.37
Breaded, fried patty	64	1 ea	181	.8	1.50
Turkey gizzard	67	1 ea	109	2.6	2.79
Turkey heart	16	1 ea	28	3.0	.843
Turkey liver	75	1 ea	127	1.8	2.32

ESHA Research

Zinc (Zn) mg

Values are for edible portion of foods

Meats: Sausages & Lunchmeats

Food	Wt (g)	Svg	Cal	Zinc 100 Cal	Zinc Svg
Barbecue loaf, pork & beef	23	1 pce	40	1.4	.570
Beef lunchmeat:					
Loaf or roll	28	1 oz	87	.8	.720
Thin sliced	28	1 oz	50	2.3	1.13
Beerwurst (beer salami):					
Beef salami	23	1 pce	75	.8	.610
Pork salami	23	1 pce	55	.7	.400
Berliner sausage	23	1 pce	53	1.1	.570
BOLOGNA:					
Beef bologna	23	1 pce	72	.6	.460
Beef and pork	28	1 oz	89	.6	.550
Cured pork	23	1 pce	57	.8	.470
Turkey	28	1 oz	56	.9	.492
Braunschweiger	18	1 pce	65	.8	.510
Brotwurst, link	70	1 ea	226	.7	1.47
Cheesefurter (cheese smoki)	43	1 ea	141	.7	.970
Chicken roll, light meat	57	2 oz	90	.5	.410
Corned beef loaf, jellied	28	1 oz	46	2.3	1.08
Dutch brand loaf	28	1 oz	68	.7	.490
FRANKFURTER (hotdog):					
Beef, 8/pkg	57	1 ea	184	.7	1.21
Beef & pork, 8/pkg	57	1 ea	183	.6	1.05
Chicken, 10/pkg	45	1 ea	115	.9	1.00
Turkey, 10/pkg	45	1 ea	102	1.0	1.00
HAM, lunchmeat:					
Extra lean	57	2 oz	75	1.5	1.09
Regular	57	2 oz	103	1.2	1.21
Thin slices, (3 ≈ 1 oz)	28	3 pce	37	1.5	.550
Ham, chopped, packaged	42	2 pce	98	.8	.769
Ham, minced	21	1 pce	55	.7	.400
Ham patty, cooked	60	1 ea	203	.6	1.13
Ham & cheese roll/loaf	57	2 oz	147	.8	1.13
Ham salad spread	240	1 c	518	.5	2.64
Italian sausage link, cooked	67	1 ea	216	.7	1.59
Keilbasa	26	1 pce	81	.6	.520
Knockwurst, link	68	1 ea	209	.5	1.13
Liverwurst, pork	18	1 pce	59	.8	.468
Luncheon meat, canned	21	1 pce	70	.4	.310
Luncheon sausage, beef & pork	23	1 pce	60	.9	.560
Luxury loaf	57	2 oz	80	2.2	1.73
Mortadella	15	1 pce	47	.7	.320
Olive loaf	57	2 oz	133	.6	.780
Pastrami:					
Beef, cured	57	2 oz	198	.6	1.21
Turkey, cured	57	2 oz	74	2.0	1.46
Peppered loaf	28	1 pce	42	2.2	.920
Pepperoni sausage, small slice	22	4 pce	109	.5	.550
Pickle & pimento loaf	57	2 oz	149	.5	.790
Polish sausage	28	1 oz	92	.6	.550
Pork sausage:					
Link, cooked	13	1 ea	48	.7	.330
Patty, cooked	27	1 pce	100	.7	.680
Brown & serve, links	13	1 ea	50	.3	.150
Poultry sandwich spread	13	1 T	25	1.0	.250
SALAMI:					
Beef	23	1 pce	58	.8	.490
Pork and beef	57	2 oz	143	.8	1.21
Turkey	57	2 oz	111	1.1	1.25
Salami, dry, beef & pork	20	2 pce	85	.8	.640
Smoked link sausage:					
Beef & pork	68	1 ea	229	.6	1.44
Pork link	68	1 ea	265	.7	1.92

Food	Wt (g)	Svg	Cal	Zinc 100 Cal	Zinc Svg
TURKEY lunchmeats (other):					
Smoked turkey sausage	28	1 oz	55	1.3	.710
Summer sausage	23	1 pce	80	.6	.470
Breakfast sausage	28	1 oz	65	1.5	.970
Turkey ham	57	2 oz	73	2.2	1.58
Turkey roll, light & dark	57	2 oz	84	1.3	1.13
Turkey roll, liight meat	57	2 oz	83	1.1	.880
Turkey summer sausage	28	1 oz	50	1.4	.720
Vienna sausage, canned	16	1 can	45	.6	.260

Mixed Dishes & Fast Foods

Food	Wt (g)	Svg	Cal	Zinc 100 Cal	Zinc Svg
Beef & vegetable stew:					
Recipe	245	1 c	220	2.4	5.29
Canned	245	1 c	194	2.2	4.23
Beef, macaroni, tomato					
sauce, recipe	226	1 c	189	1.1	2.07
Beef pot pie, f/frzn	234	1 ea	426	.6	2.64
BURRITO:					
Bean burrito	174	1 ea	322	.7	2.37
Beef burrito	177	1 ea	463	1.3	5.80
Beef and bean	175	1 ea	390	.8	3.30
Deluxe combination	198	1 ea	424	.9	3.91
Cheese soufflé, recipe	112	1 c	221	.6	1.35
Chicken a la king, recipe	245	1 c	470	.4	1.80
Chicken chow mein:					
Recipe	250	1 c	255	.8	2.12
Canned	250	1 c	95	1.4	1.30
Chicken egg roll	100	1 ea	242	.2	.400
Chicken & noodles, recipe	240	1 c	365	.6	2.14
Chicken pot pie, f/frozen	230	1 ea	430	.3	1.22
Chili w/beans, canned	255	1 c	286	1.8	5.10
Cole slaw	120	1 c	84	.3	.240
Chicken salad w/celery	78	1/2 c	266	.3	.804
Chop suey w/beef & pork	250	1 c	300	1.2	3.58
Corn dog	111	1 ea	330	.4	1.44
Corn fritter, recipe	45	1 ea	116	.3	.295
Corn pudding	250	1 c	271	.5	1.26
Corned beef hash, canned	220	1 c	382	1.1	4.38
Egg salad	183	1 c	438	.5	2.24
ENCHILADA:					
Beef enchilada	120	1 ea	292	.8	2.25
Cheese enchilada	120	1 ea	330	.5	1.50
Chicken enchilada	120	1 ea	269	.4	1.21
French toast, recipe	65	1 pce	123	.4	.474
LASAGNA:					
Recipe, w/meat	245	1 pce	398	.8	3.23
Recipe, w/o meat	218	1 pce	316	.6	1.93
Frozen entree	205	1 pce	275	.5	1.25
Macaroni & cheese:					
Recipe	200	1 c	430	.3	1.20
Canned	240	1 c	230	.5	1.20
Macaroni salad, no cheese	141	1 c	371	.1	.335
Manicotti:					
Meat & tomato sce	233	1 ea	320	.6	1.78
Frozen entree	225	1 ea	271	.7	2.00
Meat loaf:					
Beef only	87	1 pce	193	1.8	3.50
Beef & 1/3 pork	87	1 pce	212	1.3	2.86
Moussaka, (lamb & eggplant)	250	1 c	250	1.3	3.29
Pies, fried, commercial:					
Apple pie	85	1 ea	255	.1	.144
Cherry pie	85	1 ea	250	.1	.150
PIZZA, cheese:					
Thick crust 1/2 of 10"	208	1 pce	519	.5	2.66
Regular crust 1/8 of 15"	120	1 pce	290	.6	1.81

ESHA Research

Mixed Foods & Fast Foods, continued:	Wt (g)	Svg	Cal	Zinc (mg) 100 Cal	Svg
Potato salad w/mayo & eggs	250	1 c	358	.2	.780
Quiche Lorraine, 1/8 pie	176	1 pce	600	.3	1.95
Ravioli, beef, frzn w/sce	28	2 ea	33	.5	.170
Ravioli, canned	226	1 c	220	.6	1.37
SANDWICHES, Fast food:					
Cheeseburger, 3 oz patty	112	1 ea	300	.8	2.53
Cheeseburger, 4 oz patty	194	1 ea	524	1.0	5.27
Chicken patty sandwich	157	1 ea	436	.2	1.00
English muffin, egg, cheese, bacon	138	1 ea	360	.5	1.86
Fish sandwich:					
Regular w/cheese	140	1 ea	420	.2	.952
Large, w/o cheese	170	1 ea	470	.2	.884
Hamburger, 3 oz patty	98	1 ea	245	.8	2.00
Hamburger, 4 oz patty	174	1 ea	445	1.1	5.01
Hotdog (frankfurter) w/bun	85	1 ea	260	.5	1.19
Roast beef w/bun	150	1 ea	345	1.1	3.66
SANDWICHES on part whole wheat bread unless stated as rye.					
Avocado, cheese, sprouts, tomato	195	1 ea	432	.4	1.87
Bacon, lettuce, tomato	135	1 ea	327	.4	1.30
Chicken salad sandwich	100	1 ea	294	.3	.998
Corned beef & swiss on rye	147	1 ea	429	1.0	4.37
Egg salad sandwich	111	1 ea	319	.4	1.16
Grilled cheese	117	1 ea	393	.6	2.49
Ham sandwich	122	1 ea	256	.7	1.74
Ham on rye	116	1 ea	242	.1	1.91
Ham & cheese	151	1 ea	363	.7	2.69
Ham & swiss on rye	145	1 ea	350	.9	3.03
Ham salad sandwich	125	1 ea	339	.4	1.26
Patty melt on rye	177	1 ea	567	1.2	6.63
Peanut butter & jam	100	1 ea	341	.4	1.30
Reuben sandwich, grilled	233	1 ea	480	.9	4.55
Roast beef sandwich	122	1 ea	280	1.0	2.87
Tuna salad sandwich	116	1 ea	303	.3	.893
Turkey sandwich	122	1 ea	271	.5	1.24
Turkey ham sandwich	122	1 ea	253	.9	2.20
SPAGHETTI (pasta, tomato sauce & cheese):					
Homemade	250	1 c	260	.5	1.30
Canned	250	1 c	190	.6	1.12
SPAGHETTI (pasta, tomato sauce & meat):					
Homemade	248	1 c	330	.7	2.45
Canned	250	1 c	260	.9	2.39
Spinach soufflé	136	1 c	218	.6	1.29
Taco, beef	78	1 ea	207	1.4	2.89
Taco, chicken	78	1 ea	172	.8	1.34
TOSTADA:					
W/refried beans	157	1 ea	212	.7	1.55
W/beans & beef	192	1 ea	332	1.1	3.57
W/beans & chicken	157	1 ea	249	.8	1.94
Tuna noodle casserole, recipe	202	1 c	251	.4	.966
Tuna salad	205	1 c	383	.3	1.15
Turkey pot pie, f/frzn	233	1 ea	416	.4	1.50
Veal Parmigiana, frzn entree	205	7.25 oz	372	1.1	3.97
Waldorf salad	142	1 c	424	.2	.690

Nuts & Seeds

	Wt (g)	Svg	Cal	Zinc (mg) 100 Cal	Svg
ALMONDS, dried, whole	142	1 c	837	.5	4.15
Almond butter	16	1 T	101	.5	.488
Brazil nuts, dry (≈7)	28	1 oz	186	.7	1.30
Cashews, oil roasted	130	1 c	748	.8	6.18

	Wt (g)	Svg	Cal	Zinc (mg) 100 Cal	Svg
Cashew butter	16	1 T	94	.9	.830
Chestnuts, roasted	143	1 c	350	.2	.815
Coconut:					
Fresh, grated	80	1 c	283	.3	.880
Packaged, flaked, sweet	74	1 c	351	.4	1.30
Dried, unsweetened	78	1 c	515	.3	1.57
Coconut milk, canned	226	1 c	445	.3	1.27
Coconut water, raw	240	1 c	46	.5	.240
Filberts (hazelnuts), whole	135	1 c	853	.4	3.24
Macadamias:					
Dried	134	1 c	940	.2	2.29
Oil roasted	134	1 c	962	.2	1.47
MIXED NUTS w/peanuts (cashews, pnuts, brazil nuts, filberts, almonds, pecans):					
Dry roasted	137	1 c	814	.6	5.21
Oil roasted	142	1 c	876	.8	7.22
MIXED NUTS w/o peanuts (cashews, almonds, brazil nuts, pecans, filberts):					
Oil roasted	144	1 c	886	.8	6.71
PEANUTS:					
Dry roasted	146	1 c	855	.6	4.83
Oil roasted	144	1 c	837	1.1	9.60
Peanut butter	32	2 T	190	.4	.802
Peanut flour, defatted	60	1 c	196	1.6	3.06
Pecans, dried, chopped	119	1 c	794	.8	6.51
Pine nuts, dried pignola/pinyon	28	1 oz	154	.8	1.22
Pistachio, dried, unshelled	128	1 c	739	.2	1.72
Pumpkin/squash seeds:					
Dry kernels	138	1 c	747	1.4	10.3
Roasted kernels	227	1 c	1185	1.4	16.9
Whole seeds, roasted	64	1 c	285	2.3	6.59
Sesame flour:					
High fat	28	1 oz	149	2.0	3.03
Low fat	28	1 oz	95	3.0	2.84
Part defatted	28	1 oz	109	2.8	3.04
Sesame seeds:					
Dried kernels	150	1 c	882	1.7	15.4
Whole, dried	36	1/4 c	206	1.4	2.80
Soybeans, roasted	86	1/2 c	405	.7	2.7
Sunflower seeds:					
Dried	36	1/4 c	205	.9	1.82
Oil roasted	34	1/4 c	208	.8	1.76
Tahini (sesame butter)	15	1 T	91	1.7	1.57
Walnuts, chopped:					
Black	125	1 c	759	.6	4.28
English	120	1 c	770	.4	3.28

Soups, Sauces, Gravies

	Wt (g)	Svg	Cal	Zinc (mg) 100 Cal	Svg
GRAVIES:					
Beef, canned	233	1 c	124	1.9	2.33
Chicken gravy:					
From dry mix	260	1 c	85	.4	.320
Canned	238	1 c	189	1.0	1.91
Mushroom gravy:					
From dry mix	258	1 c	70	.5	.328
Canned	238	1 c	120	1.4	1.66
Onion gravy:					
Prepared from dry	261	1 c	78	.4	.287
Canned	241	1 c	57	1.1	.612
SAUCES:					
Cheese sce f/mix w/milk	279	1 c	305	.3	.950
Hollandaise	160	1 c	867	.3	2.39

ESHA Research

Zinc (Zn) mg

Values are for edible portion of foods

	Wt (g)	Svg	Cal	Zinc (mg) 100 Cal	Svg
Soups, Sauces, Gravies, continued:					
Spaghetti sauce, plain:					
Homemade	220	1 c	179	.3	.579
Canned	249	1 c	272	.2	.530
Spaghetti sauce w/meat:					
Homemade	248	1 c	297	.7	2.11
Canned	206	.8 c	220	.5	1.05
White sauce:					
Home recipe, med.	250	1 c	395	.3	1.05
From mix with milk	264	1 c	240	.5	1.15
SOUPS: soups are prep. from canned unless otherwise stated. RTS = Ready To Serve. For soups prepared with milk, assume whole milk.					
Bean w/bacon, w/water	253	1 c	173	.6	1.03
Beef bouillon	240	1 c	16	3.8	.600
Beef soup, chunky, RTS	240	1 c	171	1.5	2.64
Beef noodle	244	1 c	84	1.8	1.54
Celery, cream of:					
Prep with milk	248	1 c	165	.1	.196
Prep with water	244	1 c	90	.2	.151
Cheese, prep w/milk	251	1 c	230	.3	.688
Chicken broth, w/water	244	1 c	39	.6	.249
Chicken soup, chunky, RTS	251	1 c	178	.6	1.00
Chicken, cream of:					
Prepared with milk	248	1 c	191	.4	.675
Prepared with water	244	1 c	115	.5	.627
Condensed, undiluted	251	1 c	233	.5	1.26
Chicken gumbo	244	1 c	56	.7	.376
Chicken noodle:					
Prep with water	241	1 c	75	.7	.550
From dry	252	1 c	53	.4	.199
Chicken & rice	241	1 c	60	.4	.263
Chicken vegetable:					
Prep with water	241	1 c	74	.5	.366
From dry	251	1 c	49	.4	.208
Chicken vegetable, chunky, RTS	240	1 c	167	.2	.366
Chili beef soup	250	1 c	169	.8	1.40
Clam chowder, tom. base:					
Manhattan style	244	1 c	78	1.3	.976
Manhattan, chunky, RTS	240	1 c	133	1.3	1.68
Clam chowder, New England	248	1 c	163	.8	1.30
Minestrone soup	241	1 c	80	.9	.735
Mushroom, cream of:					
Prep with water	244	1 c	130	.5	.593
Prep with milk	248	1 c	205	.3	.640
Condensed, undiluted	251	1 c	257	.5	1.19
Oyster stew, prep w/milk	245	1 c	134	7.7	10.3
Potato, cream of:					
Prep with milk	248	1 c	148	.5	.675
Prep with water	244	1 c	73	.9	.630
Shrimp, cream of, w/milk	248	1 c	165	.5	.799
Split pea & ham	253	1 c	189	.7	1.32
Split pea , prep from dry	255	1 c	133	.4	.591
Tomato beef noodle	244	1 c	140	.5	.752
Tomato, cream of:					
Prep with milk	248	1 c	160	.2	.290
Prep with water	244	1 c	86	.3	.244
From dry	265	1 c	102	.2	.209
Tomato rice soup	247	1 c	120	.4	.514
Tomato noodle	244	1 c	69	.8	.583
Vegetable, from dry	253	1 c	55	.3	.167
Turkey soup	241	1 c	74	.8	.612
Turkey noodle, chunky, RTS	236	1 c	136	3.5	4.79
Vegetable, chunky, RTS	240	1 c	122	2.6	3.12
Vegetable beef:					
Prep with water	244	1 c	79	2.5	2.00
From dry	253	1 c	53	.5	.270
Vegetarian vegetable	241	1 c	70	.7	.460

Vegetables & Legumes

	Wt (g)	Svg	Cal	Zinc (mg) 100 Cal	Svg
Alfalfa sprouts	33	1 c	10	3.0	.304
Amaranth leaves:					
Fresh, chopped	28	1 c	7	3.6	.255
Boiled	132	1 c	28	4.3	1.21
Artichoke, globe, ckd (300g whole)	120	1 ea	60	1.0	.588
Artichoke hearts:					
Cooked f/frozen	240	9 oz	108	.8	.864
Marinated	170	6 oz	168	.3	.540
Asparagus, pieces:					
Fresh, uncooked	67	1/2 c	15	3.1	.469
Ckd from fresh	90	1/2 c	23	1.9	.432
Ckd from frozen	180	1 c	50	2.0	1.01
Canned, drained	121	1/2 c	16	3.0	.484
Canned w/liquid	122	1/2 c	17	3.4	.573
Bamboo shoots:					
Cooked from fresh	120	1 c	15	2.1	.319
Canned	131	1 c	25	1.2	.300
BEANS: see also garbanzo, lentils, soybeans.					
Baked (dry White beans with spices & sauce):					
Home prepared	253	1 c	382	.5	1.84
Canned, plain/vegetarian	254	1 c	235	1.5	3.55
Canned w/frankfurters	257	1 c	366	1.3	4.79
Canned with pork	253	1 c	268	1.4	3.69
Canned w/pork, swt sce	253	1 c	282	1.3	3.80
Canned w/pork, tom sce	253	1 c	247	1.1	2.60
Black beans, ckd f/dry	172	1 c	227	.8	1.92
Broadbeans, ckd f/dry	170	1 c	186	.9	1.72
Great northern, ckd f/dry	177	1 c	210	.7	1.55
Green (snap) beans:					
Fresh, uncooked	110	1 c	34	.8	.260
Ckd from fresh	125	1 c	44	1.0	.450
Ckd from frozen	135	1 c	36	2.3	.840
Canned, drained	135	1 c	26	1.5	.392
Canned w/liquid	240	1 c	36	1.3	.480
Hyacinth, ckd f/dry	194	1 c	228	2.4	5.53
Kidney beans, all:					
Ckd from dry	177	1 c	225	.8	1.89
Canned w/liquid	256	1 c	208	.7	1.41
Lima beans:					
Ckd from fresh	170	1 c	208	.6	1.34
Ckd from frozen:					
Large	85	1/2 c	85	.4	.370
Baby	90	1/2 c	94	.5	.500
Ckd from dry	188	1 c	217	.8	1.79
Canned, drained	170	1 c	164	1.0	1.60
Canned w/liquid	241	1 c	191	.8	1.57
Navy, ckd from dry	182	1 c	259	.7	1.93
Pinto beans:					
Cooked from dry	171	1 c	235	.8	1.85
Canned	240	1 c	186	1.0	1.66
Refried beans, canned	253	1 c	270	1.3	3.45
White beans, ckd f/dry	179	1 c	253	.8	1.96
Winged beans, ckd f/dry	172	1 c	252	1.0	2.48
Yardlong beans, ckd f/dry	171	1 c	202	.9	1.84
Yellow wax: see green beans.					
Bean sprouts (Mung beans):					
Fresh sprouts	104	1 c	31	1.4	.426
Boiled, drained	124	1 c	26	2.2	.580
Stir fried	124	1 c	62	1.8	1.12
BEETS:					
Ckd f/fresh, whole	100	2 ea	31	.8	.250
Canned dices, drained	85	1/2 c	27	.7	.180
Canned dices w/liquid	123	1/2 c	36	.8	.283
Canned, pickled, slices	114	1/2 c	74	.4	.296

ESHA Research

210

Vegetables & Legumes, continued:	Wt (g)	Svg	Cal	Zinc (mg) 100 Cal	Svg		Wt (g)	Svg	Cal	Zinc (mg) 100 Cal	Svg
						Kohlrabi:					
Beet greens, ckd fresh, drained	144	1 c	40	1.8	.720	Fresh slices	140	1 c	38	.8	.322
BROCCOLI, chopped:						Cooked f/fresh	165	1 c	48	.7	.322
Fresh, uncooked	88	1 c	24	1.5	.360	Leeks, chopped, fresh	104	1 c	32	.5	.165
Ckd f/fresh	156	1 c	44	1.3	.592	Lentils, ckd f/dry	198	1 c	231	1.1	2.50
Ckd f/frozen	184	1 c	51	1.1	.560	Lentils, sprouted:					
W/cheese sce	142	1/2 c	166	.2	.370	Fresh sprouts	77	1 c	81	1.4	1.16
W/hollandaise sce	95	1/2 c	105	.3	.290	Stir fried	100	3.5 oz	101	1.6	1.60
Brussels sprouts:											
Ckd f/fresh (1 sprout = 21g)	156	1 c	60	.8	.500	LETTUCE, chopped:					
Ckd f/frozen	155	1 c	65	.8	.550	Butterhead	56	1 c	7.3	2.0	.144
						Iceberg	56	1 c	7.3	1.8	.123
CABBAGES, chopped:						Loose leaf	56	1 c	10	1.8	.185
Common, fresh	70	1 c	16	.8	.120	Romaine	56	1 c	9	2.1	.185
Common, cooked	150	1 c	32	.8	.240	Mushroom:					
Bok choy, fresh	70	1 c	9	3.2	.288	Fresh slices (1 avg ≈ 18g)	35	1/2 c	9	3.4	.300
Bok choy, cooked	170	1 c	20	2.2	.432	Cooked from fresh	78	1/2 c	21	3.2	.679
Pe-tsai, fresh	76	1 c	11	1.5	.170	Canned, drained	78	1/2 c	19	3.0	.562
Pe-tsai, cooked	119	1 c	16	1.4	.220	Mustard greens:					
Red, fresh	70	1 c	19	.8	.150	Fresh, chopped	56	1 c	15	1.0	.144
Red, cooked	150	1 c	32	.7	.220	Ckd f/fresh	140	1 c	21	1.4	.300
Savoy, fresh	70	1 c	20	1.3	.255	Ckd f/frozen	150	1 c	29	1.1	.300
Savoy, cooked	145	1 c	35	.8	.263	Okra:					
						Ckd f/fresh pods	85	8 ea	27	1.7	.468
CARROTS:						Ckd f/frozen slices	92	1/2 c	34	1.7	.570
Fresh (7.5" x 1-1/8" dm.)	72	1 ea	31	.5	.14	Onion:					
Ckd f/fresh, sliced	78	1/2 c	35	.7	.234	Fresh, chopped	160	1 c	61	.5	.304
Ckd f/frozen, sliced	73	1/2 c	26	.7	.175	Ckd f/fresh, chopped	105	1/2 c	46	.5	.22
Canned, drained	73	1/2 c	17	1.1	.190	Dehydrated flakes	14	1/4 c	45	.6	.260
Canned w/liquid	123	1/2 c	28	1.3	.357	Spring chopped, all	50	1/2 c	16	1.3	.20
Carrot juice	123	1/2 c	49	.5	.221	Parsley, fresh, chopped	30	1/2 c	10	2.2	.220
Cauliflower:						Parsnips:					
Fresh, raw	50	1/2 c	12	.8	.090	Fresh slices	133	1 c	100	.8	.785
Ckd fr/fresh	124	1 c	30	1.0	.298	Ckd f/fresh	156	1 c	125	.3	.400
Ckd fr/frozen	180	1 c	34	.7	.234						
Celeriac (celery root) cooked	100	3.5 oz	25	1.2	.310	PEAS:					
Celery stalk (7.5" stalk = 40g)	40	1 ea	6	.9	.052	Black-eyed peas:					
Celery, cooked dices	150	1 c	27	.8	.21	Ckd f/fresh	165	1 c	160	1.1	1.7
Chard, Swiss:						Ckd f/frozen	170	1 c	224	1.1	2.42
Fresh, chopped	36	1 c	7	2.3	.163	Ckd f/dry	171	1 c	198	1.1	2.20
Ckd fr/fresh	175	1 c	35	1.7	.589	Canned	240	1 c	184	.9	1.68
Collards:						Green peas:					
Fresh, chopped	36	1 c	11.2	.4	.047	Fresh, uncooked	145	1 c	118	1.5	1.80
Ckd fr/fresh	128	1 c	35	.4	.141	Ckd f/fresh	160	1 c	134	1.4	1.90
Ckd fr/frozen	170	1 c	63	.7	.460	Ckd f/frozen	80	1/2 c	63	1.2	.750
						Canned, drained	85	1/2 c	59	1.0	.600
CORN, kernels:						Canned w/liquid	124	1/2 c	61	1.4	.860
Fresh uncooked	77	1/2 c	66	.5	.347	Green peas, edible-pods:					
Cooked from fresh	82	1/2 c	89	.4	.394	Fresh, uncooked	145	1 c	61	1.0	.590
Cooked from frozen	82	1/2 c	67	.4	.290	Ckd f/fresh	160	1 c	67	.9	.600
Canned, drained	82	1/2 c	66	.6	.381	Ckd f/frozen	80	1/2 c	42	.9	.390
Canned with liquid	128	1/2 c	79	.6	.460	Split peas, ckd f/dry	196	1 c	231	.8	1.96
Canned, vacuum pack	210	1 c	166	.6	.966	Peas, sprouted:					
Corn, creamed, canned	128	1/2 c	93	.7	.678	Fresh sprouts	120	1 c	154	.8	1.26
Cucumber, 8" x 2" dm	301	1 ea	39	1.8	.690	Cooked from fresh	100	3.5 oz	118	.7	.780
Dandelion greens:						Peas & carrots, ckd f/frzn	80	1/2 c	38	.9	.360
Fresh chopped	55	1 c	25	2.5	.620						
Ckd f/fresh	105	1 c	35	2.3	.800	PEPPERS, HOT, chili peppers:					
Eggplant, ckd cubes	160	1 c	45	.5	.240	Fresh, chopped	75	1/2 c	30	.8	.225
Endive, fresh, chopped	25	1/2 c	4	5.0	.200	Canned, Jalapeno, chopped	68	1/2 c	17	.8	.130
Escarole, chopped, curly endive	50	1 c	9	4.4	.395						
Garbanzo beans:						PEPPERS, SWEET, green/red, chopped:					
Cooked f/dry	164	1 c	269	1.0	2.51	Fresh	50	1/2 c	14	.4	.06
Canned w/liquid	240	1 c	285	1.0	2.53	Ckd from fresh	68	1/2 c	19	.4	.082
Garden cress:						Poi, two finger	240	1 c	269	.8	2.04
Fresh, chopped	25	1/2 c	8	15.0	1.20						
Ckd f/fresh	135	1 c	31	3.2	1.00	POTOTOES:					
Kale, chopped:	67	1 c	33	.9	.295	Baked in oven, 4.75" x 2.3":					
Ckd f/fresh	130	1 c	42	.8	.312	Flesh & skin	202	1 ea	220	.3	.650
Ckd f/frozen	130	1 c	39	.6	.234	Flesh only	156	1 ea	145	.3	.450
						Potato skin	58	1 ea	115	.2	.280

Vegetables & Legumes, continued:	Wt (g)	Svg	Cal	Zinc (mg) 100 Cal	Zinc (mg) Svg
POTATOES, continued:					
Boiled, flesh only:					
Cooked w/o skin	135	1 ea	116	.3	.370
Cooked in skin	136	1 ea	119	.3	.410
Ckd f/frzn, small	70	1 ea	46	.4	.175
Canned, 1" dm.	70	2 ea	42	.5	.200
Cottage fries, f/frzn	50	10 ea	109	.2	.210
Dehydrated flakes	200	1 c	722	.2	1.72
French fried					
from frozen:					
Cooked in oil	50	10 ea	158	.1	.190
Oven heated	50	10 ea	111	.2	.210
Hash brown:					
Homemade	156	1 c	163	.1	.234
From frozen	156	1 c	340	.1	.500
Mashed:					
Prep w/milk	210	1 c	162	.4	.600
From instant	215	1 c	239	.2	.509
Potato puffs (tater tots),					
heated from frozen	62	1/2 c	138	.1	.190
Potato dishes, prepared:					
Au gratin, recipe	245	1 c	322	.5	1.69
Au gratin, from mix	245	1 c	228	.3	.588
Scalloped, recipe	245	1 c	210	.5	.980
Scalloped, from mix	245	1 c	228	.3	.613
Potato chips	28	14 ea	148	.2	.300
Potato pancakes	76	1 ea	237	.3	.680
Pumpkin:					
Ckd f/fresh, mashed	245	1 c	50	.9	.450
Canned	123	1/2 c	42	.5	.209
Radish, red	45	10 ea	7	1.8	.130
Radish seeds, sprouted	38	1 c	16	1.4	.213
Rutabaga, fresh cubes	140	1 c	51	.9	.480
Rutabaga, cooked	85	1/2 c	29	.9	.260
Sauerkraut, cnd w/liquid	236	1 c	44	1.0	.440
Seaweed:					
Irish moss, fresh	28	1 oz	14	4.0	.553
Kelp, fresh	28	1 oz	12	2.9	.349
Lavar, fresh	28	1 oz	10	3.0	.298
Soybeans, ckd f/dry	172	1 c	298	.7	1.98
Soybeans, sprouted:					
Fresh sprouts	35	1/2 c	45	.9	.41
Steamed	94	1 c	76	1.3	.98
Stir fried	100	1 c	125	1.7	2.10
SOYBEAN PRODUCTS: see tofu in this section; miso, natto, tempeh in Other; roasted soybeans in Nuts & Seeds; soy milk in Dairy; soy flour in Grains.					
Spinach:					
Fresh, chopped	56	1 c	12	2.4	.297
Ckd from fresh	180	1 c	41	3.3	1.37
Ckd from frozen, leaf	190	1 c	53	2.5	1.33
Canned, drained	214	1 c	50	2.0	.990
SQUASH, SUMMER, sliced:					
Crookneck, fresh	130	1 c	24	1.6	.380
Crookneck, ckd f/fresh	180	1 c	36	2.0	.710
Scallop, fresh	130	1 c	24	1.6	.380
Scallop, ckd f/fresh	90	1/2 c	14	1.6	.220
Zucchini, fresh	130	1 c	19	1.4	.260
Zucchini, ckd f/fresh	180	1 c	29	1.1	.324
SQUASH, WINTER, mashed:					
Acorn, baked	245	1 c	137	.3	.418
Acorn, boiled	245	1 c	83	.3	.270
Butternut, baked	245	1 c	99	.3	.319
Butternut, ckd f/frzn	240	1 c	94	.3	.288
Hubbard, baked	240	1 c	120	.3	.360

	Wt (g)	Svg	Cal	Zinc (mg) 100 Cal	Zinc (mg) Svg
SQUASH, WINTER, continued:					
Hubbard, boiled	236	1 c	70	.3	.220
Spaghetti, baked/boiled	155	1 c	45	.7	.310
Succotash:					
Ckd f/frzn	170	1 c	158	.5	.760
Canned w/liquid	255	1 c	161	.8	1.28
Sweet potato (whole ≈ 5" x 2"):					
Baked in skin, flesh only	114	1 ea	118	.3	.330
Boiled w/o skin, flesh only	151	1 ea	160	.3	.400
Canned, mashed	128	1/2 c	129	.2	.270
Vacuum pack, mashed	255	1 c	233	.2	.460
Candied, recipe, 2.5" x 2"	105	1 pce	144	.1	.160
Taro, ckd slices	132	1 c	187	.1	.240
Taro chips	23	10 ea	110	.3	.293
Tofu, soybean curd					
Firm, raw	126	1/2 c	183	1.1	1.98
Regular, raw	124	1/2 c	94	1.1	1.00
TOMATOES:					
Fresh, 2-3/5" dm	123	1 ea	26	.4	.111
Fresh, chopped	180	1 c	38	.4	.162
Cooked from fresh	240	1 c	65	.4	.264
Canned, whole	240	1 c	47	.8	.380
Tomato juice, canned	244	1 c	42	.8	.342
Tomato paste, canned	262	1 c	220	1.0	2.10
Tomato puree, canned	250	1 c	102	.5	.540
Tomato sauce, canned	245	1 c	74	.8	.600
Turnips, cubes:					
Fresh cubes	130	1 c	35	.4	.156
Ckd from fresh	156	1 c	28	.6	.166
Turnip greens:					
Ckd from fresh	144	1 c	29	1.0	.290
Ckd from frozen	82	1/2 c	24	1.4	.340
Vegetable juice cocktail	242	1 c	46	1.1	.484
VEGETABLES, COMBINATIONS,					
cooked from frozen:					
Broccoli, mixed with:					
Carrots, pasta	95	2/3 c	88	.3	.300
Carrots, water chestnuts	91	2/3 c	32	.8	.258
Cauliflower & red pepper	95	2/3 c	25	.9	.230
Water chestnuts	95	1/2 c	33	.8	.273
Cantonese stir fry	95	1/2 c	53	.5	.257
Green beans & spaetzle	95	1/2 c	108	.2	.258
Japanese style	95	1/2 c	29	.8	.230
Mixed vegetables (corn, peas, carrots, green beans and lima beans):					
Ckd from frozen	182	1 c	107	.8	.892
Canned, drained	163	1 c	77	.9	.668
Peas, mixed with:					
Cauliflower	95	1/2 c	118	.3	.397
Mushrooms	95	1/2 c	73	.8	.550
Onions	95	1/2 c	71	.7	.485
Onions, pasta	95	1/2 c	122	.4	.530
Onions, cheese sauce	142	1/2 c	165	.5	.750
Potatoes, cream sauce	76	1/2 c	140	.2	.260
Rice, mushrooms	66	2/3 c	108	.3	.340
Pasta, corn, cream sce	95	1/2 c	132	.5	.610
Spinach & water chestnuts	95	1/2 c	29	.8	.223
Water chestnuts, cnd slices	70	1/2 c	35	.8	.270
Yams, orange: see Sweet potatoes.					
Yams, white, cooked cubes	136	1 c	158	.2	.272
Zucchini: see Squash, summer.					

	Wt (g)	Svg	Cal	Zinc (mg) 100 Cal	Svg
Other					
Cooking Ingredients, condiments, flavorings, other beverages, spices, sweeteners, etc.					
Barbecue sauce	250	1 c	160	.3	.417
BEVERAGES, other than milk beverages, fruit juices, or vegetable juices:					
Beer (12 fl oz = 1.5 c):					
Regular	356	12 fl oz	146	<.1	.071
Light	354	12 fl oz	100	.1	.106
Carbonated (12 fl oz = 1/5 c):					
Club soda	355	1.5 c	0	.36	.360
Cream soda	371	1.5 c	191	.1	.240
Cola beverages	355	1.5 c	2	14	.280
Fruit flavored soda pop	372	1.5 c	170	.2	.272
Ginger ale	366	1.5 c	124	.1	.183
Grape soda	372	1.5 c	161	.2	.260
Lemon-lime soda	368	1.5 c	149	.1	.180
Orange drink soda	372	1.5 c	177	.2	.380
"Pepper" type soda	368	1.5 c	151	<.1	.150
Root beer	370	1.5 c	152	.2	.260
Coffee, brewed	240	1 c	2	4.0	.080
Fruit punch, canned	253	1 c	118	.3	.313
Lemonade, f/frzn conc	248	1 c	100	<.1	.050
Pineapple grapefruit drink	250	1 c	117	.1	.150
Tea, brewed	240	1 c	2	2.5	.050
Wine:					
Red	118	1/2 c	85	.1	.110
Rose	118	1/2 c	84	.1	.073
White, dry	118	1/2 c	79	.1	.078
CANDY and CANDY BARS:					
Almonds, sugar coated	28	7 ea	146	.4	.526
Caramel, plain/choc	28	1 oz	115	.1	.153
Chocolate covered:					
Almonds	165	1 c	935	.5	4.21
Coconut	28	1 oz	133	.3	.404
Peanuts	170	1 c	954	.5	4.46
Mints	28	1 oz	116	.2	.220
Raisins	187	1 c	733	.2	1.25
Chocolate kisses	28	6 pce	154	.2	.360
Fudge, chocolate, plain	28	1 oz	115	.1	.155
Fudge, chocolate, w/ nuts	28	1 oz	114	.2	.245
Fudge, vanilla, w/nuts	28	1 oz	122	.1	.179
Granola bar	28	1 ea	127	.8	1.00
Kit Kat bar	43	1 ea	210	.2	.430
Krackle bar	34	1 ea	179	.2	.370
M&M's plain candies	48	1 pkg	237	.2	.571
M&M's peanut candies	47	1 pkg	240	.3	.661
Mars bar	50	1 ea	240	.2	.585
Milk chocolate, plain	28	1 oz	145	.3	.367
Milk chocolate w/ almonds	28	1 oz	150	.3	.482
Milk chocolate w/ peanuts	28	1 oz	155	.4	.676
Milk chocolate w/ rice cereal	28	1 oz	140	.2	.294
Carob flour	103	1 c	185	.5	.940
Catsup	245	1 c	255	.2	.56
Chili sauce, tom base	273	1 c	284	.3	.740
Cocoa powder	86	1 c	224	2.9	6.53
CHOCOLATE:					
Baking, unsweetened	28	1 oz	145	.7	1.01
Bittersweet	28	1 oz	141	.8	1.13
Chips, semi-sweet	170	1 c	860	.3	2.39
Dark, sweet	28	1 oz	150	.3	.425
Fudge topping	38	2 T	125	.3	.388
Hot fudge topping	300	1 c	1020	.3	3.10
Syrup, thin	38	2 T	85	.5	.394

	Wt (g)	Svg	Cal	Zinc (mg) 100 Cal	Svg
Dill pickle	65	1 ea	12	1.5	.176
Falafel, (2.25" patty)	17	1 patty	57	<.1	.26
Honey	339	1 c	1030	<.1	.400
Horseradish, prepared	15	1 T	6	2.9	.176
Hummous (Humous)	246	1 c	420	.6	2.70
Miso (soybean product)	138	1/2 c	284	1.6	4.58
Mustard, prepared	125	1/2 c	94	.8	.788
Natto (soybean product)	88	1/2 c	187	1.4	2.67
Pickle relish	245	1 c	320	<.1	.163
Salsa:					
Picante by Tostitos	85	6 T	40	2.3	.900
Home recipe	108	1/2 c	46	.3	.116
Soy sauce:					
Regular (wheat & soy) shoyu	18	1 T	9	.8	.07
Tamari (soy)	18	1 T	11	.7	.08
From hydrolyzed veg protein	18	1 T	7	.9	.06
SPICES:					
Basil, dried	4.5	1 T	11	2.4	.262
Caraway seed	6.7	1 T	22	1.7	.372
Cardamom	5.8	1 T	18	2.4	.433
Chili powder	7.5	1 T	24	.8	.203
Cloves, ground	6.6	1 T	21	1.3	.280
Coriander seed	5	1 T	15	1.6	.235
Cumin seed	6	1 T	22	1.3	.288
Curry powder	6.3	1 T	18	1.4	.260
Fenugreek seed	11.1	1 T	36	.8	.280
Garlic cloves	12	4 ea	17.9	5.9	1.06
Garlic powder	8.4	1 T	28	.8	.221
Ginger, ground	5.4	1 T	19	1.3	.250
Ginger root slices, raw	11	5 pce	8	2.8	.220
Nutmeg, ground	7	1 T	37	.4	.150
Onion powder	6.5	1 T	15	1.0	.150
Oregano, ground	4.5	1 T	14	1.4	.200
Paprika	6.9	1 T	20	1.4	.280
Poppyseed	8.8	1 T	47	1.9	.900
Tarragon, ground	4.8	1 T	14	1.3	.187
Thyme, ground	4.3	1 T	12	2.2	.266
Turmeric, ground	6.8	1 T	24	1.3	.300
Tempeh (soybean product)	83	1/2 c	165	.9	1.5
Tobasco sauce	18	1 T	1.6	1.0	.016
Vinegar, cider	240	1 c	29	1.0	.280
Wheat germ oil	13.6	1 T	120	.4	.520
Yeast:					
Brewer's yeast	8	1 T	25	2.5	.632
Dry active yeast	30	4 T	80	2.3	1.80

ESHA Research

Dietary Fiber

Dietary fiber is a collective term for many components that are not normally broken down by the digestive tract and help pass food through the intestine. It is extremely important to the normal functioning of the digestive tract, plus:

* High fiber foods, because of their ability to absorb water, provide a feeling of fullness and satisfaction in a meal.

* A high fiber diet is automatically *high in nutrients* and *low in calories and fat*. High fiber breads (whole grains, bran added, nuts added etc.) have more nutrients than low fiber breads (white breads) and often fewer calories. A high fiber diet has a larger share of plant foods (fruits and vegetables). Fiber supplements may add fiber to the diet, but they do not supply the other benefit of such a diet -- the increased nutrients.

* Fiber absorbs water, increasing and softening the stools and preventing constipation. This alone can prevent the formation of hemorrhoids, varicose veins, hiatus hernia, and diverticulosis.

* Because fiber speeds the passage of food through the digestive tract, it reduces calorie absorption and may reduce the exposure of tissue to any cancer-causing agents in the food.

* Some forms of fiber (soluble, like oat bran and pectin) also lower blood cholesterol levels, either by binding to certain lipids, like cholesterol, or by other mechanisms.

Scientists realized the importance of fiber from studies in underdeveloped countries, where people were free of certain diseases that are common in "more developed" countries. Lack of fiber in the diet has since been connected to diverticular disease in the intestine; colon and rectal cancer; hiatus hernia; constipation; varicose veins; hemorrhoids; appendicitis; diabetes mellitus; gall bladder disease; high blood serum cholesterol; heart disease; breast cancer; and problems with being overweight.

"The consumption of diets rich in plant foods (and therefore fiber) is inversely related to the incidence of cardiovascular disease, colon cancer, and diabetes. Because an increase in dietary fiber consumption is almost invariably associated with a change in other dietary constituents, it is difficult to establish a clear relationship with dietary fiber alone. One plausible mechanism for an anticarcinogenic effect is the rapid passage of the digestive mass through the colon, thereby reducing the possibility that potential carcinogens have an opportunity to interact with the mucosal surface. In addition, the increased mass of the softer stool may dilute carcinogens." (443)

Nutrient Losses

Processing foods removes much of the fiber from many of the basic ingredients. The milling of whole grains into flour, which takes out many of the nutrients, takes out most of the dietary fiber as well. Therefore, baked goods using all-purpose enriched white flour, are also low in fiber.

Recommended Amounts

As of this writing there are a variety of unofficial recommendations for dietary fiber. The National Cancer Institute suggests 20 to 35 grams of dietary fiber daily (depending on body size) from a variety of foods. Scientists in the field are presenting recommendations ranging from 1 gram to 2.5 grams of dietary fiber per 100 calories consumed. Depending on the version you have, The Food Processor nutrition system has used 1 or 2 grams of dietary fiber for every 100 calories consumed, since its initial release in 1984.

The current typical dietary fiber intake of Americans is about 12g/day. It is recommended that this amount be increased -- not by adding fiber concentrates to the diet, but by consumption of fruits, vegetables, whole grain cereals and legumes (beans). These foods will also provide valuable vitamins and minerals to the diet. (Your body might appreciate your doing it gradually as it adjusts to the modification).

Food Sources

Good sources of dietary fiber are fruits, vegetables, legumes (beans and peas), nuts, seeds, whole grains, and whole grain products. There is no dietary fiber in meats or dairy products. There is very little dietary fiber in white flour or baked goods made with white flour.

It is Important to know

Nutrient values for dietary fiber are emerging data. Many different analytical techniques are used to measure various components of fiber, and these methods are undergoing their own review of accuracy in the scientific community. Expect additional research and refinement of analytical techniques to modify the data in the future. We also envision that we will be reporting insoluble and soluble fiber separately in the future, when more information is available.

Dietary fiber values in this table are either an estimate of the total dietary fiber (a specific analytical technique), or a combination of analytical results for the various components of fiber within a food. A specific analytical technique known as the neutral detergent fiber method -- NDF -- measures the insoluble components. It is combined with data on pectin, when available, for a measure of total dietary fiber. Little data exists for gums and mucilages at this time.

Sources of information are from the most recent Provisional tables from the USDA Human Nutrition Information Service in Hyattsville, Maryland; Southgate, from England; many scientific journal articles; and extensive unpublished information from the USDA . In some cases there is wide variation in reported data. Part of this is due to the variance between analyzed samples; part may be due to the analytical technique used.

Dietary fiber is not crude fiber. Dietary fiber includes cellulose, hemi-cellulose, lignin, pectin, gums, mucilages, and algal polysaccharides; where crude fiber (a nutritionally obsolete term) contains only portions of the first three components stated. Older information tables on fiber are for crude fiber, and should not be confused with "Dietary" fiber.

Crude fiber is a measure of the residue of plant food that remains after a harsh chemical procedure. Techniques for analyzing dietary fiber attempt to measure the actual fiber that remains after digestion in the body. These techniques are less harsh. Generally if a food item has 1 gram of crude fiber, it will have approximately 2 to 3 grams of dietary fiber, but there is no consistent quantitative relationship between the two.

ESHA Research

Baked Goods
Breads, Cakes, Cookies, Crackers, Muffins, Pancakes,
Pastries, Pies, Rolls, and some desserts.

	Wt (g)	Svg	Cal	Fiber(g) 100 Cal	Fiber(g) Svg
Apple crisp, 3" x 3"	78	1 pce	146	1.0	1.4
Biscuit, average, from mix	28	1 ea	94	.5	.5
BREADS:					
Banana nut, 1/2" slice	50	1 pce	161	.7	1.2
Boston brown, canned, 1/2" slice	45	1 pce	95	2.2	2.1
Cornbread, muffin	45	1 ea	145	1.1	1.6
Cracked wheat	25	1 pce	65	2.0	1.3
French, 5" x 2.5"	35	1 pce	100	7.3	7.3
Mixed grain	25	1 pce	65	2.4	1.6
Oatmeal bread	25	1 pce	65	1.5	1.0
Pita pocket, 6.5" dm.	60	1 ea	165	.6	1.0
Pumpernickel, 5" x 4" x 3/8"	32	1 pce	80	2.4	1.9
Rye, light, 4.75" x 3.75" x 7/16"	25	1 pce	65	2.5	1.7
Vienna	25	1 pce	70	1.1	.8
Wheat (white & whole wheat flour)	28	1 pce	72	1.4	1.0
White bread	28	1 pce	75	.7	.5
Whole wheat bread	28	1 pce	70	3.0	2.1

CAKES: pce = 1/16th cake (≈3"x3") unless stated
otherwise; cupcake ≈ 42 g.

	Wt (g)	Svg	Cal	Fiber(g) 100 Cal	Fiber(g) Svg
Boston cream pie, 1/8 cake	120	1 pce	260	.4	1.0
Carrot cake, cream cheese frosting	112	1 pce	406	.3	1.2
Cheesecake, f/recipe, 1/12	92	1 pce	278	.4	1.2
Chocolate, choc. frosting	69	1 pce	235	.3	.74
Coffee cake, f/mix	72	1 pce	230	.7	1.5
Fruitcake, dark, 2/3" arc	43	1 pce	165	1.0	1.6
Gingerbread: 3" x 3" pce	63	1 pce	174	1.0	1.8
Sheet cake:					
Plain	86	1 pce	315	.1	.34
W/white frosting	121	1 pce	445	.1	.34
White, coconut frosting	70	1 pce	270	.5	1.3
Yellow, chocolate frosting, avg	69	1 pce	240	.1	.35
Cherry crisp, 3" x 3" pce	138	1 pce	157	.8	1.3

Chips: See corn & tortilla this section; see Vegetables
for potato chips.

COOKIES:	Wt (g)	Svg	Cal	Fiber(g) 100 Cal	Fiber(g) Svg
Chocolate chip:					
Commercial	42	4 ea	180	.3	.5
Homemade	40	4 ea	185	.5	1.0
From refrig dough	48	4 ea	225	.4	1.0
Fig bars	56	4 ea	210	1.2	2.6
Oatmeal raisin	52	4 ea	245	.6	1.5
Peanut butter, homemade	48	4 ea	245	.4	.9
Sandwich type, all	40	4 ea	195	.2	.4
Sugar, from refrig dough	48	4 ea	235	<.1	.2
Corn chips	28	1 oz	155	.8	1.3

CRACKERS:	Wt (g)	Svg	Cal	Fiber(g) 100 Cal	Fiber(g) Svg
Armenian cracker bread	28	4 pce	117	3.1	3.7
Cheese crackers, small	10	10 ea	50	<.1	.02
Cheese w/peanut butter fill	30	4 ea	150	.2	.33
Graham crackers	14	2 ea	60	.6	.35
Round, like Ritz	9	3 ea	45	.2	.11
Rye, whole grain	14	2 ea	55	4.0	2.2
Saltines	12	4 ea	50	.5	.3
Whole wheat crackers	8	2 ea	35	2.4	.8
Croissant, 4.5" x 4" x 2"	57	1 ea	235	.5	1.1
Danish pastry:					
Plain danish	57	1 ea	220	.2	.7
With fruit	65	1 ea	235	.5	1.2

Doughnuts:	Wt (g)	Svg	Cal	Fiber(g) 100 Cal	Fiber(g) Svg
Cake type, medium	50	1 ea	210	.3	.6
Yeast raised, plain	60	1 ea	235	.3	.7
Yeast raised, jelly filling	65	1 ea	226	.3	.8
English muffin, plain/sourdough	57	1 ea	135	1.1	1.5
MUFFINS:					
Blueberry	45	1 ea	138	1.2	1.6
Bran, average (wheat or oat)	45	1 ea	133	2.1	2.8
Cornmeal	45	1 ea	145	1.1	1.6
Pancakes:					
Buckwheat, 4" dm.	58	1 ea	55	2.4	1.3
Plain or buttermilk, 4" dm.	52	1 ea	60	.7	.4
Whole wheat, 5" dm.	52	1 ea	94	2.6	2.4
Peach crisp, 3" x 3"	139	1 pce	166	1.1	1.9

PIES: piece = 1/6 of 9" pie.

	Wt (g)	Svg	Cal	Fiber(g) 100 Cal	Fiber(g) Svg
Apple pie	158	1 pce	405	.7	2.8
Banana cream, recipe	198	1 pce	319	.5	1.7
Blueberry	158	1 pce	380	1.0	3.7
Cherry	158	1 pce	410	.5	1.9
Chocolate cream	175	1 pce	311	1.1	3.5
Coconut cream	172	1 pce	343	.7	2.5
Coconut custard	165	1 pce	384	.6	2.4
Cream pie, commercial	152	1 pce	455	.1	.6
Custard pie	152	1 pce	293	.2	.6
Lemon meringue	140	1 pce	355	.5	1.7
Mincemeat	160	1 pce	395	1.2	4.7
Peach pie	158	1 pce	405	.5	2.1
Pecan pie	138	1 pce	583	.8	4.8
Pumpkin	200	1 pce	367	1.5	5.4
Pretzels, thin twists	60	10 ea	240	.7	1.7

ROLLS:	Wt (g)	Svg	Cal	Fiber(g) 100 Cal	Fiber(g) Svg
Cinnamon bun, small	50	1 ea	158	1.0	1.6
Hamburger bun	45	1 ea	129	.9	1.2
Hard roll, white	50	1 ea	155	.5	.8
Hot dog bun	40	1 ea	115	.9	1.1
Dinner roll, 2.5" x 2"	28	1 ea	85	.9	.8
Rye roll, dark	28	1 ea	79	2.4	1.9
Rye roll, light	28	1 ea	76	2.4	1.9
Submarine (Hoagie) roll	135	1 ea	400	.6	2.2
Whole wheat roll	35	1 ea	88	2.8	2.5
Stuffing, prep. w/enr bread:					
From dry	140	1 c	500	.8	4.0
Stove Top	108	1/2 c	176	1.8	3.1
Taco shell	13.6	1 ea	59	1.8	1.1
Tortillas:					
Corn, 6" dm, fried	30	1 ea	87	2.8	2.4
Flour, 10.5" dm.	56.8	1 ea	168	1.0	1.6
Flour, 8" dm.	35.4	1 ea	105	1.0	1.0
Tortilla chips, all	28	1 oz	139	.8	1.1
Waffles, prepared:					
Homemade, 7" diam.	75	1 ea	245	.4	1.1
From mix, 7" diam.	75	1 ea	205	.5	1.1
Frozen, 4" diam.	35	1 ea	98	.9	.8

Dairy & Dairy Products
There is no dietary fiber in Dairy foods, except from
added ingredients, or non-dairy products like soy.

Milk Beverages:	Wt (g)	Svg	Cal	Fiber(g) 100 Cal	Fiber(g) Svg
Carob flavor mix, w/whole milk	256	1 c	195	.9	1.7
Chocolate milk/Hot Cocoa drink:					
Lowfat 2%	250	1 c	180	2.1	3.0
Whole (3.3% fat)	250	1 c	210	1.8	3.0
Soy milk	120	1/2 c	39	1.1	.43

Fruits & Fruit Juices

Food	Wt (g)	Svg	Cal	Fiber/100 Cal	Fiber/Svg
Acerola juice, fresh	242	1 c	51	2.0	1.0
APPLE, 2.75" diam:					
with skin	138	1 ea	80	4.3	3.0
without skin	128	1 ea	72	3.8	2.7
Applesauce, unsweetened	244	1 c	106	3.8	4.0
APRICOTS:					
Fresh, pitted	106	3 ea	51	4.0	2.0
Canned, juice pack	85	3 ea	40	3.3	1.3
Canned, heavy syrup	85	3 ea	70	1.8	1.3
Dried halves	35	10 ea	83	3.3	2.7
Apricot nectar, canned:	251	1 c	141	1.1	1.5
Avocado, fresh, whole:					
California	173	1 ea	305	5.4	16.6
Florida	304	1 ea	340	8.6	29.2
Bananas, 8.75" (176g whole)	114	1 ea	105	2.2	2.3
Blackberries:					
Fresh berries	144	1 c	74	13.1	9.7
Frozen, unthawed	151	1 c	97	8.0	7.8
Canned	256	1 c	236	5.5	13.0
Blueberries:					
Fresh berries	145	1 c	82	4.8	3.9
Frozen, unsweetened	155	1 c	78	4.6	3.6
Frozen, sweetened, thawed	230	1 c	185	2.9	5.4
Canned	256	1 c	225	2.7	6.0
Boysenberries:					
Frozen berries	132	1 c	66	7.8	5.2
Canned	256	1 c	225	4.7	10.5
Cantaloupe, Casaba: see Melon.					
Cherries, sour:					
Frozen, unthawed	155	1 c	72	2.9	2.1
Canned	244	1 c	90	3.0	2.7
CHERRIES, SWEET:					
Fresh, pitted	68	10 ea	49	2.1	1.0
Frozen, thawed	259	1 c	232	.9	2.1
Canned with liquid	257	1 c	213	.7	1.4
Cranberries, whole, raw	95	1 c	46	8.7	4.0
Cranberry juice cocktail	253	1 c	145	.5	.8
Cranberry sauce, canned	277	1 c	419	1.4	5.9
Currants:					
Fresh, Black	112	1 c	71	13.7	9.7
Fresh, Red/white	112	1 c	63	7.7	4.8
Dried, Zante	144	1 c	407	2.3	9.4
Dates, pitted	83	10 ea	228	3.0	6.8
Figs:					
Fresh, medium	50	1 ea	37	5.0	2.0
Canned in heavy syrup	85	3 ea	75	3.4	2.6
Dried figs	18	1 ea	48	4.3	21.0
Fruit cocktail, canned:					
Juice pack	248	1 c	115	2.2	2.6
Heavy syrup	255	1 c	185	1.4	2.5
Gooseberries:					
Fresh berries	150	1 c	67	7.2	4.8
Canned w/liquid	252	1 c	185	3.3	6.1
Grapes, fresh:					
Thompson, seedless	160	1 c	114	1.6	1.9
Tokay/Emperor	57	10 ea	40	2.4	.9
Grape juice:					
Bottled, canned	253	1 c	155	.8	1.3
Prep f/frozen	250	1 c	128	1.0	1.3
GRAPEFRUIT (half = 241g w/refuse):					
Pink or red half	123	1 ea	37	4.3	1.6
White half	118	1 ea	39	3.9	1.5
Canned	254	1 c	152	1.8	2.8
Grapefruit juice, fresh	247	1 c	96	1.0	1.0
Guava, raw	90	1 ea	45	11.8	5.3

Food	Wt (g)	Svg	Cal	Fiber/100 Cal	Fiber/Svg
Honeydew melon: see Melon.					
Kiwi fruit	76	1 ea	46	5.6	2.6
Lemon juice:					
Fresh juice	244	1 c	60	1.4	.9
Frozen, std strength	244	1 c	54	1.5	.8
Lime juice:					
Fresh juice	246	1 c	65	1.5	1.0
Bottled	246	1 c	50	1.7	.9
Loganberries, fresh/frzn	150	1 c	90	10	9.2
Mandarin oranges, canned	252	1 c	155	2.7	4.3
Mango, fresh slices	165	1 c	108	5.4	5.8
Melon: also see Watermelon.					
Cantaloupe, cubes	160	1 c	57	2.8	1.6
Casaba, cubes	170	1 c	45	3.8	1.7
Honeydew, cubes	170	1 c	60	3.1	1.8
Frozen melon balls, mixed	173	1 c	55	3.3	1.8
Nectarines (1 med = 1 c slc)	136	1 ea	67	4.0	2.7
ORANGE, 2-5/8" 180g w/peel	131	1 ea	60	5.2	3.1
Orange juice:					
Fresh	248	1 c	111	0.9	1.0
Prep f/frozen	249	1 c	110	.5	.5
Papaya (454g w/refuse)	304	1 ea	117	4.4	5.2
Papaya nectar, canned	250	1 c	142	1.1	1.5
PEACHES:					
Fresh, 2.5" diam.	87	1 ea	37	4.1	1.5
Frozen slices, thawed	250	1 c	235	1.8	4.1
Canned half:					
Juice pack	77	1 ea	34	2.6	.9
Heavy syrup	81	1 ea	60	1.5	.9
Canned slices:					
Juice pack	248	1 c	109	2.8	3.0
Heavy syrup	256	1 c	190	1.6	3.0
Peach nectar, canned	249	1 c	134	1.1	1.5
PEARS:					
Fresh, Bartlett	166	1 ea	98	4.7	4.6
Cnd half, juice pack	77	1 ea	38	2.9	1.1
Cnd half, heavy syrup	79	1 ea	59	1.9	1.1
Pear nectar, canned	250	1 c	149	1.2	1.8
Pineapple:					
Fresh chunks	155	1 c	76	3.0	2.3
Cnd pieces, juice pack	250	1 c	150	1.7	2.5
Cnd pieces, heavy syrup	255	1 c	199	1.2	2.4
Pineapple juice:					
Prep f/frozen	250	1 c	129	.6	.8
Canned, unsweetened	250	1 c	140	.5	.8
Plantain:					
Fresh slices	148	1 c	181	3.6	6.5
Ckd slices	154	1 c	179	3.7	6.6
PLUMS:					
Fresh, 2-1/8" diam.	66	1 ea	36	3.6	1.3
Cnd, juice pack	95	3 ea	55	3.1	1.7
Cnd, heavy syrup	110	3 ea	98	1.7	1.7
Prunes, dried	84	10 ea	201	4.0	8.0+
Prune juice, bottled	256	1 c	181	1.5	2.8
Raisins, dark, unpacked	145	1 c	435	2.0	8.5
Raspberries:					
Fresh berries	123	1 c	60	12.8	7.7
Frozen, thawed	250	1 c	255	4.5	11.5
Canned w/liquid	256	1 c	234	4.6	10.8
Rhubarb:					
Fresh, diced	122	1 c	26	12.2	3.2
Cooked w/sugar	240	1 c	279	1.9	5.3
STRAWBERRIES:					
Fresh berries	149	1 c	45	8.6	3.9
Frozen measure	149	1 c	52	7.4	3.9
Frozen, thawed, sweetened	255	1 c	245	3.1	7.6

ESHA Research

	Wt (g)	Svg	Cal	Fiber (g) 100 Cal	Svg
Fruits & Fruit Juices, continued:					
Tangerines, fresh	84	1 ea	37	4.6	1.7
Tangerine juice:					
Prep f/frozen	241	1 c	110	.9	1.0
Canned, sweetened	249	1 c	125	.6	.8
Watermelon, diced pieces	160	1 c	50	1.3	.6

Grains & Grain Products
Cereals, Grains, Flours, Noodles, Pasta, Popcorn

	Wt (g)	Svg	Cal	Fiber (g) 100 Cal	Svg
Barley, cooked:					
Pearled	157	1 c	193	2.3	4.4
Whole	200	1 c	200	2.3	4.6
Bran: see Oat, Rice, Wheat.					
Buckwheat flour:					
Dark	98	1 c	338	2.4	8.0
Light	98	1 c	340	1.8	6.0
Bulgar wheat, cooked	182	1 c	151	7.0	10.6
CEREALS, COLD (Ready To Eat): Check label.					
Values vary. Some cereals have fiber added.					
CEREALS, HOT (Cooked): Check label.					
Some cereals have fiber added.					
Corn grits	242	1 c	145	3.1	4.5
Cream of rice	244	1 c	126	1.1	1.4
Cream of wheat	244	1 c	140	2.4	3.4
Farina	233	1 c	116	2.8	3.2
Malt-O-Meal	240	1 c	122	2.5	3.0
Maypo	180	3/4 c	128	1.7	2.2
Oatmeal, prepared:					
Rolled oats (reg, quick, inst)	234	1 c	145	2.8	4.1
Oatmeal, instant, fortified:					
Plain, from packet	177	3/4 c	104	2.8	2.9
With bran & raisins	195	7/8 c	158	2.1	3.3
Flavored, average	164	3/4 c	160	1.9	3.0
Ralston cereal	253	1 c	134	2.5	3.3
Roman Meal	181	3/4 c	111	5.0	5.6
Wheatena	243	1 c	135	2.4	3.3
Whole wheat cereal	242	1 c	150	2.6	3.9
Corn flour & Masa Harina, enr.	117	1 c	422	3.5	14.7
Cornmeal, dry:					
Degermed, enriched	138	1 c	505	2.0	10.2
Bolted, nearly whole	122	1 c	441	2.6	11.6
Corn grits: see Cereals, Hot.					
Macaroni:					
Cooked, enriched	140	1 c	197	1.1	2.2
Cooked whole wheat	140	1 c	174	1.0	1.8
Cooked vegetable, enriched	134	1 c	172	1.0	1.75
Millet, cooked from dry	120	1/2 c	143	.9	1.3
NOODLES:					
Chow mein, dry	45	1 c	237	.8	1.8
Egg noodles, ckd	160	1 c	213	1.6	3.5
Spinach noodles, ckd	140	1 c	182	1.8	3.2
Oat bran, 1T ≈ 6 g	94	1 c	132	12.4	16.4
Oats, rolled:					
Dry	81	1 c	311	2.8	8.8
Cooked: see cereals, hot.					
Popcorn:					
Air popped, plain	8	1 c	30	4.3	1.3
Popped in oil	11	1 c	55	2.2	1.2
RICE, cooked:					
Brown rice	190	1 c	217	1.5	3.3
White, converted	175	1 c	200	.6	1.1
White, instant	165	1 c	162	.8	1.3
Wild rice	164	1 c	166	2.4	4.0
Rice bran	83	1 oz	262	8.3	21.7

	Wt (g)	Svg	Cal	Fiber (g) 100 Cal	Svg
Rye flour, dark	128	1 c	415	4.3	17.7
Rye flour, light	102	1 c	361	3.9	14
Soy flour:					
Full fat, raw	85	1 c	368	2.9	10.8
Low fat	88	1 c	326	3.7	12.0
Spaghetti noodles, cooked:					
Enriched	140	1 c	197	1.1	2.2
Whole wheat spaghetti	140	1 c	174	2.3	4.5
Tapioca, dry	152	1 c	518	.3	1.7
WHEAT:					
Bran, dry	30	1/2 c	65	11.7	7.6
FLOUR, unbleached:					
All-purpose, white, unsifted	125	1 c	455	.7	3.4
Cake, sifted	96	1 c	348	.8	2.7
Gluten	140	1 c	529	.2	1.2
Self-rising	125	1 c	442	.7	3.3
Semolina	167	1 c	601	1.0	3.9
Whole wheat flour	120	1 c	407	3.7	15.1
Wheat germ:					
Raw	100	1 c	360	3.2	11.6
Toasted	113	1 c	432	3.6	15.6
Wheat, rolled, cooked	240	1 c	142	3.9	5.5
Wheat, rolled, dry	85	1 c	289	3.8	11.0
Wheat, sprouted	108	1 c	214	4.2	8.9
Whole grain (wheat berries) ckd	50	1/3 c	28	4.3	1.2

Meats
There is no fiber in meat.

Mixed Dishes & Fast Foods

	Wt (g)	Svg	Cal	Fiber (g) 100 Cal	Svg
Beef, macaroni, tom sce, recipe	226	1 c	189	1.2	2.3
Beef & vegetable stew:					
Recipe	245	1 c	220	1.5	3.4
Canned	245	1 c	194	.7	1.3
Beef pot pie, f/frzn	234	1 ea	426	.2	.9
Brunswick stew	250	1 c	186	3.7	6.8
BURRITO:					
Bean burrito	174	1 ea	322	2.6	8.2
Beef burrito	177	1 ea	463	.3	1.6
Beef and bean	175	1 ea	390	1.3	5.0
Deluxe combination	198	1 ea	424	1.2	4.9
Chicken a la king, recipe	245	1 c	470	.3	1.3
Chicken chow mein:					
Recipe	250	1 c	255	1.6	4.1
Canned	250	1 c	95	5.3	5.0
Chicken & noodles, recipe	240	1 c	365	.3	1.2
Chicken pot pie, f/frzn	230	1 ea	430	.4	1.7
Chili w/beans, cnd	255	1 c	286	2.9	8.2
Chop suey w/beef & pork	250	1 c	300	.8	2.4
Cole slaw	120	1 c	84	2.9	2.4
Corn fritter, recipe	45	1 ea	116	1.6	1.8
Corn pudding	250	1 c	271	3.5	9.4
Corned beef hash, cnd	220	1 c	382	.3	1.2
ENCHILADA:					
Beef enchilada	120	1 ea	292	.7	2.0
Cheese enchilada	120	1 ea	330	.6	2.0
Chicken enchilida	120	1 ea	269	.7	2.0
French toast, recipe	65	1 pce	123	.6	.8
LASAGNA:					
Recipe, w/meat	245	1 pce	398	.4	1.6
Recipe w/o meat	218	1 pce	316	.5	1.6
Macaroni & cheese:					
Recipe	200	1 c	430	.3	1.2
Canned	240	1 c	230	.6	1.4

Mixed Foods & Fast Foods, continued:	Wt (g)	Svg	Cal	Fiber (g) 100 Cal	Svg
Macaroni salad w/o cheese	141	1 c	371	.4	1.3
Manicotti, meat & tom sce	233	1 ea	320	.4	1.1
Moussaka, lamb & eggplant	250	1 c	250	2.2	5.6
Pie, fried, commercial:					
Apple pie	85	1 ea	255	.6	1.5
Cherry	85	1 ea	250	.4	1.0
PIZZA, cheese:					
Regular crust, 1/8 of 15"	120	1 pce	290	.7	2.2
Thick crust, 1/2 of 10"	208	1 pce	519	.8	4.2
Potato salad w/mayo & eggs	250	1 c	358	1.0	3.7
Quiche Lorraine, 1/8 pie	176	1 pce	600	.2	1.0
Ravioli, beef, canned	226	1 c	220	.9	2.0
SANDWICHES, fast foods:					
Cheeseburger, 3 oz meat	112	1 ea	300	.5	1.4
Cheeseburger, 4 oz meat	194	1 ea	524	.4	2.3
Chicken patty sandwich	157	1 ea	436	.3	1.4
English muffin w/egg,					
cheese, bacon	138	1 ea	360	.4	1.5
Fish sandwich:					
Large, w/o cheese	170	1 ea	470	.3	1.4
Regular w/cheese	140	1 ea	420	.3	1.3
Hamburger, 3 oz meat	98	1 ea	245	.5	1.3
Hamburger, 4 oz meat	174	1 ea	445	.3	1.4
Hotdog (frankfurter & bun)	85	1 ea	260	.5	1.2
SANDWICHES (on part whole wheat bread, except when stated as rye):					
Avocado, cheese, tomato,					
& sprouts	195	1 ea	432	1.9	8.2
Bacon, lettuce & tomato	135	1 ea	327	1.0	3.3
Chicken salad sandwich	99.7	1 ea	294	.9	2.7
Corned beef & swiss on rye	147	1 ea	429	1.2	5.1
Egg salad sandwich	111	1 ea	319	.8	2.5
Grilled cheese	117	1 ea	393	.6	2.5
Ham sandwich	122	1 ea	256	1.0	2.5
Ham & cheese	151	1 ea	363	.7	2.5
Ham & swiss on rye	145	1 ea	350	1.4	5.1
Ham salad sandwich	125	1 ea	339	.8	2.6
Patty melt on rye	177	1 ea	567	.9	5.0
Peanut butter & jam	100	1 ea	341	1.4	4.7
Reuben sandwich, on rye	233	1 ea	480	1.4	6.5
Tuna salad sandwich	116	1 ea	303	1.0	3.1
Turkey sandwich	122	1 ea	271	.9	2.5
Turkey ham	122	1 ea	253	1.0	2.5
Turkey ham on rye	116	1 ea	239	2.1	5.0
SPAGHETTI, pasta & tomato sauce w/cheese:					
Homemade	250	1 c	260	1.0	2.5
Canned	250	1 c	190	1.3	2.5
Spinach soufflé	136	1 c	218	1.7	3.8
Stuffed cabbage rolls	228	8 oz	218	1.6	3.5
Stuffed green pepper	172	1 ea	217	.7	1.4
Taco:					
Beef	78	1 ea	207	.5	1.1
Chicken	78	1 ea	172	.7	1.1
TOSTADA:					
Refried beans	157	1 ea	212	3.3	6.9
Beans & beef	192	1 ea	332	1.2	4.0
Beans & chicken	157	1 ea	249	1.5	3.7
Tuna salad, fr/recipe	205	1 c	383	.3	1.1
Turkey pot pie, fr/frozen	233	1 ea	416	.2	.9
Waldorf salad	142	1 c	424	.9	3.6

Nuts & Seeds

	Wt (g)	Svg	Cal	Fiber (g) 100 Cal	Svg
ALMONDS, dried, whole	142	1 c	837	2.1	17.4
Almond butter, plain	16	1 T	101	1.4	1.4
Brazil nuts, dried	140	1 c	919	1.4	12.5
Cashews, dry or oil rstd	137	1 c	748-787	1.1	8.1
Cashew butter	16	1 T	94	1.0	1.0
Chestnuts, roasted	143	1 c	350	5.3	18.5
COCONUT:					
Raw, grated	80	1 c	283	3.1	8.9
Dried, unsweetened	78	1 c	515	2.4	12.2
Sweetened, shredded, pkg	93	1 c	466	2.7	12.6
Coconut milk, raw	240	1 c	552	.2	1.2
Filberts/hazelnuts, whole	135	1 c	853	1.4	11.6
Macadamias, average	134	1 c	950	.7	7.0
MIXED NUTS (cashews, peanuts, brazil nuts, filberts, almonds, pecans):					
Dry or oil roasted	137	1 c	814-876	1.5	12.5
MIXED NUTS, w/o peanuts (cashews, almonds, brazil nuts, pecans, filberts)					
Oil roasted	144	1 c	886	.4	3.2
PEANUTS, dry or oil rstd, avg	144	1 c	846	1.5	12.8
Peanut butter	32	2 T	188	1.3	2.4
Peanut flour, defatted	60	1 c	196	4.1	8.1
Pecans, dried, chopped	119	1 c	794	1.0	7.7
Pine nuts, dried:					
Pignola	28	1 c	146	.3	.5
Pinyon	28	1 oz	161	1.3	2.1
Pistachios (shelled), dried	128	1 c	739	1.9	13.8
Pumpkin seed kernels:					
Roasted	227	1 c	1185	.8	9.0
Dry kernels	138	1 c	747	1.0	7.5
Pumpkin seeds, roasted	64	1 c	285	8.1	23.0
Sesame flour, lowfat	28	1 oz	95	1.5	1.4
Sesame seeds:					
Kernels, dried	150	1 c	882	1.2	10.3
Whole dried	144	1 c	825	1.7	14.0
Soybeans, roasted	86	1.2 c	405	1.2	5
SUNFLOWER seeds:					
Oil roasted	34	1/4 c	208	.9	1.8
Dried	144	1 c	821	1.2	10.0
Tahini (sesame butter)	15	1 T	91	1.6	1.5
Walnuts, chopped:					
Black	125	1 c	759	.9	7.0
English	120	1 c	770	.9	7.1

Soups, Sauces & Gravies

	Wt (g)	Svg	Cal	Fiber (g) 100 Cal	Svg
Spaghetti sauce, plain:					
Homemade	220	1 c	179	1.5	2.7
Canned	249	1 c	272	1.1	3.0
SOUPS: Soups are canned unless otherwise stated. RTS = Ready To Serve. For soups prepared w/milk, assume whole milk.					
Bean & 'frank', w/water	250	1 c	187	1.1	2.0
Bean & bacon	253	1 c	173	1.5	2.6
Bean & ham, chunky, RTS	243	1 c	231	.9	2.0
Beef, chunky, RTS	240	1 c	171	.5	.8
Black bean soup	247	1 c	116	1.7	2.0
Chili beef	250	1 c	169	.9	1.5
Chicken noodle soup	241	1 c	75	1.0	.7
Clam chowder:					
New England style	248	1 c	163	.6	1.0
Manhattan style	244	1 c	78	1.5	1.2

ESHA Research

	Wt (g)	Svg	Cal	Fiber (g) 100 Cal	Fiber (g) Svg
Soups, Sauces & Gravies, continued:					
Potato soup w/water	244	1 c	73	1.2	.9
Green pea w/water	250	1 c	85	1.4	1.2
Lentil & ham soup, RTS	248	1 c	140	1.4	1.9
Minestrone soup	241	1 c	80	1.2	1.0
Minestrone, chunky, RTS	240	1 c	127	.8	1.0
Split pea soup w/ham	253	1 c	189	.7	1.3
Split pea soup, chunky, RTS	240	1 c	184	.9	1.6
Tomato rice	247	1 c	120	.6	.8
Turkey soup	236	1 c	136	.7	1.0
Vegetable beef soup	244	1 c	79	1.0	.8
Vegetarian vegetable soup	241	1 c	70	2.3	1.6

Vegetables & Legumes

	Wt (g)	Svg	Cal	Fiber (g) 100 Cal	Fiber (g) Svg
Alfafa sprouts	33	1 c	10	12.0	1.2
Amaranth leaves:					
Raw, chopped	28	1 c	7	2	.14
Boiled	132	1 c	28	8.4	2.3
Artichoke:					
Globe (300g whole)	120	1 ea	60	16.2	9.7
Hearts, ckd f/frozen	240	9 oz	108	16.8	18.1
Hearts, marinated, jar	170	6 oz	168	6.4	10.7
Asparagus, pieces:					
Fresh, raw	67	1/2 c	15	6.8	1.0
Ckd f/frozen	180	1 c	50	6.4	3.2
Canned, drained	121	1/2 c	16	12.1	1.94
Bamboo shoots:					
Ckd slices	120	1 c	15	20.0	3.0
Canned slices	131	1 c	25	13.0	3.3
BEANS (see also Garbanzo,					
Lentils, Soybeans):					
Baked beans (dry white					
beans w/spices & sauce):					
Home prepared	253	1 c	382	3.7	14.0
Canned, plain/vegetarian	254	1 c	235	6.0	14.0
Canned w/franks	257	1 c	366	4.8	17.5
Canned w/pork	253	1 c	268	5.2	14.0
Canned, pork & swt sce	253	1 c	282	4.9	13.9
Canned, pork & tom sce	253	1 c	247	5.6	13.9
Black beans, cooked	172	1 c	227	6.8	15.4
Broadbeans, ckd f/dry	170	1 c	186	1.0	1.9
Great northern, ckd f/dry	177	1 c	210	5.1	10.7
Green beans (snap):					
Fresh	110	1 c	34	7.4	2.5
Ckd from fresh	125	1 c	44	7.4	3.3
Ckd from frozen	135	1 c	36	11.6	4.2
Canned, drained	135	1 c	26	6.9	1.8
Canned w/liquid	240	1 c	36	5.3	1.9
Kidney beans, all types:					
Ckd from dry	177	1 c	225	6.7	15
Canned with liquid	256	1 c	208	9.1	19
Lima beans:					
Ckd fr/fresh	170	1 c	208	7.7	16
Ckd fr/frozen, baby, small	90	1/2 c	94	8.1	7.7
Ckd fr/frozen, large	85	1/2 c	85	5.4	4.6
Ckd fr/dry	188	1 c	217	8.3	18
Canned, drained	170	1 c	164	8.8	14.5
Canned w/liquid	241	1 c	191	8.6	16.5
Navy beans, ckd f/dry	182	1 c	259	6.2	16
Pinto beans, ckd f/dry	171	1 c	235	8.3	19.5
Refried beans, canned	253	1 c	270	8.1	22
Yellow wax: see green beans.					
Bean sprouts (Mung beans):					
Fresh	104	1 c	31	8.7	2.7
Ckd f/fresh, drained	124	1 c	26	9.6	2.5
Canned, drained	125	1 c	16	11.9	1.9

	Wt (g)	Svg	Cal	Fiber (g) 100 Cal	Fiber (g) Svg
Beets:					
Ckd from fresh, whole	100	2 ea	31	6.1	1.9
Canned dices	85	1/2 c	27	6.6	1.8
Pickled slices	114	1/2 c	74	3.2	2.4
Beet greens, ckd f/fresh	144	1 c	40	7.4	3.0
BROCCOLI, chopped:					
Fresh, uncooked	88	1 c	24	12.1	2.9
Ckd from fresh	156	1 c	44	10.7	4.7
Ckd from frozen	184	1 c	51	10.5	5.4
Brussels sprouts:					
Ckd from fresh	156	1 c	60	11.2	6.7
Ckd from frozen	155	1 c	65	10.3	6.7
CABBAGES:					
Bok choy, shredded	70	1 c	9	13.3	1.2
Bok choy, ckd f/fresh	170	1 c	20	13.6	2.7
Common, shredded	70	1 c	16	10.5	1.7
Common, ckd f/fresh	150	1 c	32	10.9	3.5
Pe-tsai, fresh, shredded	76	1 c	11	10.9	1.2
Pe-tsai, ckd f/fresh	119	1 c	16	11.9	1.9
Red cabbage, shredded	70	1 c	19	8.4	1.6
Red cabbage, ckd f/fresh	75	1/2 c	16	9.4	1.5
Savoy, shredded	70	1 c	20	8.4	1.7
Savoy, ckd f/fresh	145	1 c	35	9.7	3.4
CARROTS:					
Fresh (7.5 " x 1-1/8")	72	1 ea	31	7.4	2.3
Fresh, grated	55	1/2 c	24	7.3	1.8
Ckd slices fr/fresh	78	1/2 c	35	7.8	2.7
Ckd slices fr/frozen	73	1/2 c	26	10.4	2.7
Canned, drained	73	1/2 c	17	6.6	1.1
Canned w/liquid	123	1/2 c	28	5.0	1.4
Carrot juice	123	1/2 c	49	3.4	1.7
Cauliflower:					
Fresh pieces	50	1/2 c	12	10.6	1.3
Ckd from fresh	62	1/2 c	15	9.1	1.4
Ckd from frozen	180	1 c	34	10.0	3.4
Celeriac (celery root), ckd	100	3.5 oz	25	19.6	4.9
Celery, chopped:					
Fresh (7.5" stalk = 40g)	60	1/2 c	10	10.4	1.0
Ckd from fresh	150	1 c	27	8.9	2.4
Chard, Swiss:					
Fresh, chopped	36	1 c	7	14.3	1.0
Cooked	175	1 c	35	10.9	3.8
Chayote:					
Fresh	132	1 c	32	2.9	.9
Boiled, drained	160	1 c	38	2.4	.9
Collards:					
Fresh, chopped	36	1 c	11	7.3	.8
Ckd from fresh	128	1 c	35	11.7	4.1
Ckd from frozen	170	1 c	63	7.6	4.8
CORN, kernels:					
Fresh, uncooked	77	1/2 c	66	5.0	3.3
Ckd from fresh	82	1/2 c	89	4.8	4.3
Ckd from frozen	82	1/2 c	67	5.1	3.4
Cnd, drained	82	1/2 c	66	1.7	1.2
Cnd with liquid	128	1/2 c	79	1.5	1.2
Cnd, vacuum pack	210	1 c	166	1.8	3.0
Corn, canned, creamed	128	1/2 c	93	1.7	1.6
Cucumber, whole, 8" x 2+"	301	1 ea	39	8.2	3.2
Dandelion greens:					
Fresh chopped	55	1 c	25	3.8	.9
Cooked	105	1 c	35	3.6	1.3
Dock (sorrel), fresh	133	1 c	29	3.7	1.1
Eggplant, ckd cubes	160	1 c	45	13.4	6.0
Garbanzo beans, ckd f/dry	164	1 c	269	4.1	11.1
Garden cress:					
Fresh, chopped	25	1/2 c	8	3.5	.28
Cooked from fresh	135	1 c	31	3.9	1.2
Jerusalem artichoke, fresh	150	1 c	114	1.7	2.0

	Wt (g)	Svg	Cal	Fiber (g) 100 Cal	Fiber (g) Svg
Vegetables & Legumes, continued:					
Jicama	120	1 c	49	2.2	1.1
Kale, chopped:					
Fresh chopped	67	1 c	33	11.8	3.9
Ckd from fresh	130	1 c	42	8.5	3.6
Ckd from frozen	130	1 c	39	8.3	3.2
Kohlrabi, slices:					
Fresh slices	140	1 c	38	5.3	2.0
Ckd from fresh	165	1 c	48	4.8	2.3
Lambquarters:					
Fresh chopped	56	1 c	24	5.4	1.3
Ckd f/fresh	180	1 c	58	6.2	3.6
Leeks, chopped:					
Fresh, chopped	104	1 c	63	3.9	2.5
Cooked	52	1/2 c	24	7.1	1.7
Lentils, ckd f/dry	198	1 c	231	4.3	10.0
Lentils, sprouted:					
Fresh, uncooked	77	1 c	81	3.9	3.1
Stir-fried	100	3.5 oz	101	3.8	3.9
LETTUCE, chopped:					
Butterhead	56	1 c	7.3	10	.73
Iceberg	56	1 c	7.3	10	.73
Loose leaf	56	1 c	10	7.7	.8
Romaine	56	1 c	9	10.5	.95
Lotus root slices, ckd	89	10 ea	59	1.9	1.1
Mushrooms:					
Fresh slices (1 avg = 18g)	35	1/2 c	9	7.2	.63
Ckd from fresh	78	1/2 c	21	8.2	1.7
Canned, drained	78	1/2 c	19	8.8	1.6
Mushrooms, Shiitake:					
Cooked, pieces	145	1 c	80	7.2	5.8
Dried	15	4 ea	44	10.8	4.8
Mustard greens:					
Fresh, chopped	56	1 c	15	10.3	1.5
Ckd from fresh	140	1 c	21	13.8	2.9
Ckd from frozen	150	1 c	29	10.5	3.0
Okra, cooked:					
From fresh, pods	85	8 ea	27	6.9	1.9
From frozen, slices	92	1/2 c	34	6.8	2.3
ONIONS:					
Fresh, chopped	160	1 c	61	4.3	2.6
Ckd from fresh	105	1/2 c	46	3.3	1.5
Ckd from frozen	105	1/2 c	30	5.1	1.5
Dehydrated flakes	14	1/4 c	45	2.8	1.2
Parsley:					
Fresh, chopped	30	1/2 c	10	16.4	1.6
Freeze dried	1.4	1/4 c	4	27.3	1.1
Parsnips, sliced:					
Fresh slices	133	1 c	100	6.0	6.0
Cooked	156	1 c	125	4.4	5.5
PEAS:					
Black-eyed peas:					
Ckd from fresh	165	1 c	160	7.6	12.2
Ckd from frozen	170	1 c	224	6.3	14.0
Ckd from dry	171	1 c	198	10.4	20.5
Canned	240	1 c	184	9.2	17.0
Green peas:					
Fresh, uncooked	145	1 c	118	6.0	7.1
Ckd from fresh	160	1 c	134	5.7	7.7
Ckd from frozen	80	1/2 c	63	5.7	3.6
Canned, drained	85	1/2 c	59	6.7	4.0
Green peas, edible-pods:					
Fresh	145	1 c	61	6.7	4.1
Ckd from fresh	160	1 c	67	6.7	4.5
Ckd from frozen	80	1/2 c	42	6.2	2.6
Split peas, green, ckd f/dry	196	1 c	231	4.2	9.8
Peas & carrots:					
Ckd from frozen	80	1/2 c	38	8.7	3.3
Canned, solids & liquid	128	1/2 c	48	8.7	4.14
Peas, sprouted:					
Fresh peas	120	1 c	154	2.8	4.3
Cooked	100	3.5 oz	118	3.7	4.3
PEPPER, HOT, green/red (1 pod=45 g):					
Fresh, chopped	75	1/2 c	30	4.0	1.2
Canned, chopped	68	1/2 c	17	6.4	1.1
PEPPER, SWEET, green/red (1 pod=74 g):					
Fresh, chopped	50	1/2 c	14	5.7	.8
Cooked from fresh	68	1/2 c	19	4.7	.9
Poi, two finger	240	1 c	269	2.0	5.5
POTATOES:					
Baked, in oven:					
Flesh & skin	202	1 ea	220	2.1	4.7
Flesh only	156	1 ea	145	1.7	2.4
Potato skin	58	1 ea	115	2.0	2.3
Boiled f/fresh, peeled	136	1 ea	118	1.7	2.0
Ckd f/frozen, small	70	1 ea	46	1.8	.8
Canned, 1" diam.	70	2 ea	42	1.9	.8
Cottage fried, f/frozen	50	10 ea	109	0.9	1.0
French fries, f/frozen	50	10 ea	158	.6	1.0
Hash browns, ckd f/frzn	156	1 c	340	.9	3.1
Mashed, prep w/milk	210	1 c	162	1.8	3.0
Potato puffs (tater tots)					
frozen, heated	62	1/2 c	138	.9	1.2
Potatoes au gratin, recipe	245	1 c	322	1.4	4.4
Potatoes, scalloped, recipe	245	1 c	210	1.2	2.5
Potato chips (1 oz≈14 chips)	28	14 ea	148	.9	1.4
Potato flour	179	1 c	628	.3	2.0
Potato pancakes	76	1 ea	237	.6	1.5
Pumpkin:					
Cooked from fresh	245	1 c	50	8.6	4.3
Canned	123	1/2 c	42	5.4	2.3
Radish seeds, sprouted	38	1 c	16	6.3	1.0
Rutabaga, cubes:					
Fresh cubes	140	1 c	51	4.1	2.1
Ckd f/fresh	85	1/2 c	29	4.5	1.3
Salsify, cooked slices	135	1 c	92	3.3	3.0
Sauerkraut, cnd w/liquid	236	1 c	44	10.0	4.4
Seaweed:					
Agar, dried	28	1 oz	87	1.7	1.5
Spirulina, dried	28	1 oz	82	1.3	1.0
Wakame, fresh	28	1 oz	13	9.3	1.2
Soybeans, ckd f/dry	172	1 c	298	1.7	6.0
Soybeans, sprouted:					
Fresh, uncooked	35	1/2 c	45	1.8	.8
Steamed	94	1 c	76	2.4	1.8

SOYBEAN PRODUCTS: see tofu this section; miso, natto, tempeh in Other; roasted soybeans in Nuts & Seeds; soy milk in Dairy, soy flour in Grains.

	Wt (g)	Svg	Cal	Fiber (g) 100 Cal	Fiber (g) Svg
Spinach:					
Fresh, chopped	56	1 c	12	15.2	1.9
Ckd from fresh	180	1 c	41	10.9	4.5
Ckd from frozen	190	1 c	53	9.4	5.0
Canned, drained	214	1 c	50	11.6	5.8
SQUASH, SUMMER:					
Crookneck, fresh slices	130	1 c	24	7.7	1.8
Crookneck, ckd f/fresh	180	1 c	36	8.3	3.0
Scallop, fresh slices	130	1 c	24	5.8	1.4
Scallop, ckd f/fresh	90	1/2 c	14	9.9	1.4
Zucchini, fresh slices	130	1 c	19	7.0	1.3
Zucchini, ckd f/fresh	180	1 c	29	12.4	3.6

ESHA Research

	Wt (g)	Svg	Cal	Fiber (g) 100 Cal	Fiber (g) Svg
Vegetables & Legumes, continued:					
SQUASH, WINTER:					
Acorn, baked	245	1 c	137	5.0	6.9
Acorn, boiled	245	1 c	83	7.1	5.9
Butternut, baked	245	1 c	99	6.9	6.9
Butternut, ckd f/frozen:	240	1 c	94	7.1	6.7
Hubbard, baked	240	1 c	120	5.3	6.4
Hubbard, boiled	236	1 c	70	9.4	6.6
Spaghetti, baked/boiled	155	1 c	45	9.6	4.3
Succotash:					
Ckd from fresh	192	1 c	222	4.4	9.7
Ckd from frozen	170	1 c	158	5.6	8.8
Sweet potato:					
Baked in skin, peeled	114	1 ea	118	2.9	3.4
Boiled, flesh only	151	1 ea	160	2.8	4.5
Canned regular, mashed	128	1/2 c	129	2.1	2.8
Vacuum pack, mashed	255	1 c	233	2.4	5.5
Candied, recipe	105	1 pce	144	1.4	2.0
Tofu, raw (soybean curd):					
Firm	126	1/2 c	183	1.2	2.2
Regular	124	1/2 c	94	1.6	1.5
TOMATO:					
Fresh, 2.6" diam.	123	1 ea	26	6.9	1.8
Fresh, chopped	180	1 c	38	6.8	2.6
Ckd from fresh	240	1 c	65	6.2	4.0
Tomato products, canned:					
Tomato juice	244	1 c	42	4.6	1.9
Tomato paste	262	1 c	220	5.1	11.2
Tomato puree	250	1 c	102	5.6	5.8
Tomato sauce	245	1 c	74	5.0	3.7
Turnips, cubed:					
Fresh cubes	130	1 c	35	7.7	2.7
Ckd from fresh	78	1/2 c	14	11.7	1.6
Turnip greens:					
Ckd from fresh	144	1 c	29	15.4	4.5
Ckd from frozen	82	1/2 c	24	14.6	3.5
Vegetable juice cocktail	242	1 c	46	3.3	1.5
Vegetables, mixed (corn, peas, carrots, green beans, lima beans):					
Ckd from frozen	182	1 c	107	6.7	7.2
Canned, drained	163	1 c	77	8.2	6.3
Water chestnuts:					
Fresh slices	62	1/2 c	66	2.9	1.9
Canned slices	70	1/2 c	35	4.4	1.5
Yams:					
Hawaii mtn, steamed	145	1 c	119	3.7	4.4
Orange: see sweet potato					
White, cooked cubes	136	1 c	158	2.5	3.9

Other

Beverages (other), cooking ingredients, condiments, flavorings, spices, sweets, misc.

	Wt (g)	Svg	Cal	Fiber (g) 100 Cal	Fiber (g) Svg
Beer (12 fl oz = 1.5 c):					
Regular	356	1.5 c	146	1.2	1.8
Light	354	1.5 c	100	1.2	1.2
Bloody mary, 5 fl oz drink	148	1 ea	116	.7	.9
Barbecue sauce	250	1 c	160	.9	1.5
CANDY and CANDY BARS:					
Almonds, sugar coated	28	7 ea	146	1.7	2.4
Chocolate covered almonds	165	1 c	935	1.9	17.5
Chocolate covered peanuts	170	1 c	954	1.2	11.0
Chocolate fudge	28	1 oz	115	1.3	1.5
Granola bar	28	1 ea	127	1.1	1.4
KRACKLE	34	1 ea	179	.5	.8
M & M's peanut candies	47.3	1 pkg	240	.6	1.4
MARS bar	50	1 ea	240	.4	.9

	Wt (g)	Svg	Cal	Fiber (g) 100 Cal	Fiber (g) Svg
MR. GOODBAR	47	1 ea	250	.6	1.5
REESE'S peanut butter cup	45	2 ea	240	.7	1.7
SNICKERS, 2.2 oz	61.2	1 bar	290	.5	1.5
Catsup	245	1 c	255	1.5	4.4
Carob flour	103	1 c	185	18.3	33.8
Chili sauce:					
Tomato base	273	1 c	284	1.2	3.5
Hot, red pepper	31	2 T	6	41.7	2.5
CHOCOLATE:					
Baking, unsweetened	28	1 oz	145	3.0	4.4
Bittersweet	28	1 oz	141	.6	.9
Chips, semi-sweet	170	1 c	860	.6	5.4
Dark, sweet	28	1 oz	150	.6	.9
Hot fudge topping	300	1 c	1020	.7	7.2
Syrup, thin type	38	2 T	85	1.1	.9
Cocoa powder	86	1 c	224	11.4	25.6
Coffee, brewed	240	1 c	<2	0	0
Dill pickle	65	1 ea	11.7	6.8	.8
Falafel (2-1/4" patty = 17g)	17	1 patty	57	.3	.18
Gelatin, plain, dry, env	7	1 ea	25	4.0	1.0
Grape drink, canned	250	1 c	112	.2	.25
Hummous, Humous	246	1 c	420	3.3	14
Lemonade/Limeade, fr/frzn	248	1 c	100	.3	.29
Miso (soybean product)	138	1/2 c	284	1.5	4.2
Natto, (soybean, product)	88	1/2 c	187	1.3	2.5
Olives, without pits:					
Green	39	10 ea	45	3.0	1.4
Ripe, large	45	10 ea	52	3.5	1.8
Pickle relish	245	1 c	320	1.3	4.1
Salad dressings:					
1000 island low cal	245	1 c	389	.7	2.9
French dressing	250	1 c	1321	.1	1.9
Salsa:					
Picante by Tostitos	85	6 T	40	3.4	1.4
Salsa, homemade	108	1/2 c	46	3.7	3.0
SPICES:					
Allspice	6	1 T	16	8.1	1.3
Basil, dried	4.5	1 T	11	7.3	.8
Caraway seed	6.7	1 T	22	6.1	1.4
Chili powder	7.5	1 T	24	10.7	2.6
Cinnamon	6.8	1 T	18	9.4	1.7
Coriander seed	5	1 T	15	9.7	1.5
Curry powder	6.3	1 T	18	11.6	2.1
Fenugreek seed	11.1	1 T	36	3.1	1.1
Paprika	6.9	1 T	20	7.2	1.4
Pepper, black	6.4	1 T	16	9.8	1.6
Pepper, cayenne/red	5.3	1 T	17	14.1	2.4
Pumpkin pie spice	5.6	1 T	19	4.4	.8
Thyme, ground	4.3	1 T	12	6.7	.8
Tempeh (soybean product)	83	1/2 c	165	3.0	5
Wine, all	118	1/2 c	85-90	0	0
Yeast:					
Active dry, pkg	7	1 ea	20	11.1	2.2
Brewer's	8	1 T	25	10.1	2.5

Cholesterol

Cholesterol is of great interest among the health conscious, and is somewhat falsely maligned. It is a waxy, fatlike substance classified as a lipid (fat), and is essential for the body. It is incorporated in the bile acids, cell membrances, and sex hormones. The brain is the area of richest concentration in the body, with every cell sheathed in cholesterol (remember that when bragging of extremely low cholesterol levels).

Although essential to bodily function, cholesterol is not an essential nutrient. The liver manufactures most of the cholesterol which passes through our systems, about 1,000 mg per day. The typical American diet provides another 400-500 mg.

The recent cholesterol furor is not due to its beneficial uses, of course, but rather its connection with coronary artery disease. This has evolved into a discussion of LDL (Low Density Lipoprotein) versus HDL (High Density Lipoprotein) cholesterol, and led many people to look for dietary sources of these. The designations of LDL and HDL refer, however, to the type of protein carriers involved in the transport of cholesterol in human blood, not to cholesterol in food. All food cholesterol is the same and is simply referred to as cholesterol. The fractions of blood cholesterol (LDL & HDL), on the other hand, have quite different functions and effects on the body.

LDL brings cholesterol into the bloodstream to be used for cell building, and it is the LDL ("bad" cholesterol) that can leave residues of cholesterol on artery walls.

HDL collects cholesterol residues and transports them to the liver for reprocessing or excretion. High levels of HDL ("good" cholesterol) seem to work to keep arteries clear of deposits and reduce the risk of coronary artery disease.

Recommendations for all Adults
by the National Cholesterol Education Program
Blood serum levels of Cholesterol

	LDL (mg/dl)	Total Cholesterol (mg/dl)
Desireable	under 130	under 200
Borderline/Average	130-159	200-239
Undesireable	160 or more	240 or more

Average HDL levels in American adults range between 45 and 65 mg, with women averaging higher levels than men. Values below 35 signal coronary risk, while levels above 70 may be protective. Therefore, in evaluating coronary risk, more than just the total blood cholesterol level must be considered.

There are ways to boost the levels of protective HDL cholesterol. The most important is exercise. Prolonged, intense and frequent exercise is most effective, but even moderate activity conveys benefits over being totally sedentary. Research work at Stanford University suggests that reducing body fat -- whether by increasing exercise or by reducing calorie intake -- increases HDL levels. HDL levels will also rise to the levels seen in non-smoking, when one stops smoking. Anarobic steroids should be avoided for many reasons, including the fact that they also depress levels of HDL.

The most dramatic changes in blood cholesterol profiles, however, are achieved by changing the LDL fraction. The major portion of cholesterol circulating in the bloodstream is produced by the liver. So, avoiding cholesterol from foods is not as important as other factors in achieving lower blood levels of total cholesterol, and especially LDL cholesterol. Since dietary saturated fat stimulates the production of cholesterol by the liver, reducing saturated fat intake may be of far greater importance than reducing dietary cholesterol.

The Cholesterol-Saturated Fat Index (CSI)

CSI = (1.01 x grams of saturated fat) plus (0.05 x mg of cholesterol)

The CSI indicates the concentrations of cholesterol and saturated fat in foods. Generally speaking, the lower the CSI number, the better the diet is. Researchers are recommending a daily CSI value of about 16 for women and about 22 for men, related to body size and general nutrient needs. To keep the CSI down, it is more effective to concentrate on avoiding saturated fat than to be overly preoccupied with cholesterol itself. The Cholesterol-Saturated Fat Index (CSI) was initially reported in the May 31 1986 issue of *Lancet*, and further discussed in the June 1989 issue of the *Journal of the American Dietetic Association*. The Food Processor®II system has included this indicator as a basic part of the program since it was released in 1987.

Lowering Blood Cholesterol

Dietary factors -- the main factors which contribute to a lowering of blood cholesterol levels are to:

1. Reduce dietary saturated fat (palm oil, palm kernal oil, coconut oil, meat fat, butterfat)
2. Increase dietary fiber, especially soluble fiber (legumes, oats, barley, fruits & vegetables).
3. Lose weight if obese (reduce your percent of body fat)
4. Reduce dietary cholesterol   (when you act on numbers 1 and 2, you will automatically reduce cholesterol).

Non-Dietary Factors -- these factors also have strong beneficial effects on the blood cholesterol profile:

1. Stop smoking
2. Increase exercise
3. Use no anabolic steroids

It is important to remember that many foods which contain small amounts of cholesterol also contain *nutrients of great value*. You will more accurately modify the diet for maximum health benefits *if you lower saturated fat intake*, rather than cholesterol. Select lean meats, skim milk, and other low-fat dairy products, and avoid large quantities of foods that are heavily fried.

Losses in Foods

Cholesterol is relatively stable in foods, so there is minimal loss in cooking.

Recommended Amounts for Adults

There is no official recommendation because cholesterol is not an essential nutrient (the body makes it). Since the National Dietary Goals have recommended that less than 300 mg be consumed per day, that is used as a semi-official benchmark. Many others are recommending a lower amount, but current information suggests there should be concern about saturated fat, rather than the cholesterol alone.

Food Sources

Cholesterol in foods is found only in products of animal origin: meat, fish, poultry, eggs, and dairy products. Not being produced by plant cells, cholesterol will not be found in foods of plant origin (grains, vegetables, fruits, nuts, legumes).

Cholesterol (mg)

	Wt (g)	Svg	Cal	Choles (mg) 100 Cal	Choles (mg) Svg

Baked Goods

Breads, Cakes, Cookies, Crackers, Muffins, Pancakes, Pies, Rolls, some desserts.

	Wt (g)	Svg	Cal	100 Cal	Svg
Bagel, egg, 3.5" dm.	68	1 ea	180	24	44
Bagel, plain, 3.5" dm.	68	1 ea	180	0	0
Biscuit, average	28	1 ea	95	0	0-t
BREAD:					
Banana nut, 1/2" slice	50	1 pce	161	20	32
Boston brown, cnd, 1/2"	45	1 pce	95	3	3
Cornbread muffin	45	1 ea	145	22	32
Breads, w/o egg, average	28	1 pce	72	0	0
Bread pudding w/raisins	165	1 c	349	41	142
Brownies, frosted w/nuts	25	1	100	15	15
CAKES: pce = 1/16th cake (≈ 3"x3") unless stated.					
(Cupcake ≈ 42g)					
Carrot cake, crm cheese frosting	112	1 pce	406	13	57
Cheesecake:					
Homemade, 1/12	92	1 pce	278	61	170
From mix, 1/8	103	1 pce	300	10	30
Chocolate, choc frosting	69	1 pce	235	15	34
Coffeecake, 2.4" x 2.8"	72	1 pce	230	20	47
Fruitcake, dark, 2/3" arc	43	1 pce	165	12	20
Gingerbread, 1/9 of 8":					
Homemade	110	1 pce	351	9	30
From mix	63	1 pce	174	<1	1
Pound cake, 1/2" slice:					
From mix	30	1 pce	120	27	32
Commercial	29	1 pce	110	58	64
Sheet cake:					
Plain	86	1 pce	315	19	61
White frosting	121	1 pce	445	16	70
Snack cake, cream filled:					
Chocolate, like Ding-Dongs	28	1 ea	105	14	15
Sponge, like Twinkies	42	1 ea	155	5	7
Sponge cake, 1/12	66	1 pce	194	71	137
White w/frosting, commercial, avg	73	1 pce	274	1	3
Yellow, choc frosting	69	1 pce	245	15	36
Cheese puffs (Cheetos)	28	1 oz	158	3	5
Cherry cobbler, 3" x 3"	129	1 pce	199	1	1
COOKIES:					
Animal cookies	28	27 ea	120	<1	<1
Butter cookies	25	5 ea	115	3	4
Chocolate chip:					
Homemade	40	4 ea	185	10	18
From refrig dough	48	4 ea	225	10	22
Commercial	42	4 ea	180	3	5
Fig bars	56	4 ea	210	13	27
Lady fingers	44	4 ea	158	99	157
Oatmeal raisin	52	4 ea	245	1	2
Peanut butter, homemade	48	4 ea	245	9	22
Sandwich cookies	40	4 ea	195	3	5
Shortbread, commercial	32	4 ea	155	17	27
Snickerdoodle	20	1 ea	110	11	13
Sugar, f/refrig dough	48	4 ea	235	12	29
Vanilla wafers	40	10 ea	185	14	25
CRACKERS:					
Cheese crackers	10	10 ea	50	12	6
Cheese, pnut butter filling	30	4 ea	150	3	4
Graham crackers	14	2 ea	60	0	0
Saltines	12	4 ea	50	8	4
Cream puff, custard filling	110	1 ea	280	81	228
Crepe (without filling)	27	1 ea	47	87	41
Croissant, 4.5"x 4" x 2"	57	1 ea	235	6	13

	Wt (g)	Svg	Cal	100 Cal	Svg
Eclair, custard filled, choc icing	94	1 ea	262	64	167
Danish pastry:					
Plain pastry	57	1 ea	220	22	49
With fruit	65	1 ea	235	24	56
Doughnuts:					
Cake type, medium	50	1 ea	210	10	20
Yeast raised	60	1 ea	235	9	21
MUFFINS:					
Blueberry, f/mix	45	1 ea	140	32	45
Bran, wheat or oat:					
From recipe	45	1 ea	125	19	24
From mix	45	1 ea	140	20	28
Cornmeal, f/recipe	45	1 ea	145	16	23
Cornmeal, f/mix	45	1 ea	145	29	42
Pancakes:					
Buckwheat f/mix, 4"	27	1 ea	55	36	20
Plain, recipe, 4"	27	1 ea	60	27	16
Plain, f/mix, 4"	27	1 ea	60	27	16
Whole wheat, 5"	52	1 ea	94	32	30
Peach cobbler, 3 x 3"	130	1 pce	130	1	1
PIES: piece = 1/6 of 9" pie.					
Banana cream, recipe	198	1 pce	319	5	15
Boston cream, 1/8	120	1 pce	260	8	20
Chocolate cream	175	1 pce	311	5	15
Coconut cream	172	1 pce	343	4	15
Coconut custard	165	1 pce	384	48	183
Cream, commercial	152	1 pce	455	2	8
Custard pie	152	1 pce	293	51	148
Lemon meringue	140	1 pce	355	39	137
Pecan pie	138	1 pce	583	23	137
Pumpkin pie	200	1 pce	367	30	109
Strawberry chiffon, recipe	162	1 pce	372	11	41
Popovers	51	1 ea	96	71	68
ROLLS:					
Dinner roll, 2.5" x 2":					
Homemade	35	1 ea	120	10	12
Commercial	28	1 ea	85	<1	<1
Whole wheat roll	35	1 ea	88	3	3
Stuffing, from enriched bread:					
Moist, with egg	203	1 c	420	16	67
Stove Top stuffing	108	1/2 c	176	12	21
Waffles:					
Frozen, 4" dm.	35	1 ea	98	20	20
Homemade, 7" dm.	75	1 ea	245	42	102
From mix, 7" dm.	75	1 ea	205	29	59

Dairy & Dairy Products

	Wt (g)	Svg	Cal	100 Cal	Svg
CHEESE (1.5" cube = about 1 oz):					
American, processed	28	1 oz	106	25	27
American cheese food	28	1 oz	94	19	18
American cheese spread	28	1 oz	82	20	16
Blue cheese	28	1 oz	100	21	21
Brick cheese	28	1 oz	105	26	27
Brie cheese	28	1 oz	95	29	28
Camembert	28	1 oz	85	24	20
Cheddar cheese	28	1 oz	114	26	30
Cheshire	28	1 oz	110	26	29
Colby cheese	28	1 oz	112	24	27
Cottage cheese:					
Lowfat 1%	226	1 c	164	6	10
Lowfat 2%	226	1 c	205	9	19
Creamed, large curd	225	1 c	235	14	34
Creamed, small curd	210	1 c	215	14	31
Creamed, w/fruit	226	1 c	279	9	25
Dry curd	145	1 c	123	8	10

ESHA Research

Dairy & Dairy Products, continued:	Wt (g)	Svg	Cal	Choles (mg) 100 Cal	Choles (mg) Svg
CHEESE, continued:					
Cream cheese (1 T = 15 g)	28	1.9 T	99	31	31
Edam cheese	28	1 oz	101	25	25
Feta cheese	28	1 oz	75	33	25
Fontina	28	1 oz	110	30	33
Gorgonzola	28	1 oz	111	23	25
Gouda cheese	28	1 oz	101	32	32
Gruyere	28	1 oz	117	26	31
Gjetost	28	1 oz	132	19	25
Liederkranz	28	1 oz	87	24	21
Limburger	28	1 oz	93	28	26
Monterey jack	28	1 oz	106	25	26
Mozzarella, low moisture:					
Part skim	28	1 oz	80	19	15
Whole milk	28	1 oz	90	28	25
Muenster	28	1 oz	104	26	27
Neufchatel	28	1 oz	74	30	22
Parmesan grated (1T=5g)	28	1 oz	129	17	22
Pimento, processed	28	1 oz	106	25	27
Port du salut	28	1 oz	100	35	35
Provolone	28	1 oz	100	20	20
Ricotta, part skim milk	246	1 c	340	22	76
Ricotta, whole milk	246	1 c	428	29	124
Romano, grated	28	1 oz	128	23	29
Roquefort	28	1 oz	105	25	26
Swiss cheese	28	1 oz	107	24	26
Swiss cheese food	28	1 oz	92	25	23
Swiss, processed	28	1 oz	95	25	24
Tilsit cheese	28	1 oz	96	30	29
CREAM, SWEET, fluid					
Table or coffee	15	1 T	30	33	10
Half and half	15	1 T	20	28	6
Light whipping	239	1 c	699	38	265
Heavy whipping	238	1 c	821	40	326
CREAM, SWEET, whipped:					
Heavy cream, unsweetened	119	1 c	410	40	163
Pressurized	60	1 c	154	30	46
CREAM, SOUR, dairy:					
Cultured	14	1 T	30	21	6
Half & half, dairy (12% fat)	15	1 T	20	30	6
Cream, sour, Imitation, non-dairy	14	1 T	29	0	0
CREAM SUBSTITUTES, non-dairy					
Coffee whitener:					
Liquid, frozen	15	1 T	20	0	0
Powdered	2	1 tsp	11	0	0
Dessert Topping, non-dairy					
Frozen, like Coolwhip	75	1 c	239	0	0
Dry, powder only	43	1.5 oz	245	0	0
Pressurized topping	70	1 c	185	0	0
Kefir beverage	233	1 c	160	6	10
MILK (cow):					
Skim	245	1 c	86	5	4
Lowfat 1%	244	1 c	102	10	10
Lowfat 2%	244	1 c	121	18	22
Whole (3.3% fat)	244	1 c	150	22	33
Buttermilk (< 1% fat)	245	1 c	99	9	9
Canned:					
Skim, evap	255	1 c	200	5	10
Whole, evap (7.6% fat)	252	1 c	340	22	74
Sweet, condensed (8.7% fat)	306	1 c	982	11	104
Dried Instant nonfat, env.	91	1 c	326	4	12
Dry buttermilk	120	1 c	464	18	83
Milk (other):					
Goat milk	244	1 c	168	17	28
Human breast milk	246	1 c	171	20	34
Soy milk	240	1 c	79	0	0

MILK BEVERAGES & mixes:	Wt (g)	Svg	Cal	Choles (mg) 100 Cal	Choles (mg) Svg
Carob flavored mix,					
with milk	256	1 c	195	17	33
Chocolate milk, commercial:					
Lowfat 1%	250	1 c	160	4	7
Lowfat 2%	250	1 c	180	9	17
Whole (3.3% fat)	250	1 c	210	15	31
Chocolate flavored mix:					
To be mixed w/water:					
Powder (includes dry milk)	28	1 oz	100	1	1
Drink, prepared	206	3/4 c	100	1	1
To be mixed w/milk:					
Powder	21.6	3/4 oz	75	0	0
Drink, w/whole milk	266	1 c	226	15	33
Eggnog, commercial	254	1 c	342	44	149
Instant Breakfast, dry	37	1 env	130	0	0
Malted milk, prepared					
with whole milk:					
Chocolate flavor	265	1 c	229	15	34
Natural malt flavor	265	1 c	237	16	37
Milkshakes,10 fl oz = 1.25 c:					
Chocolate	283	1.25 c	360	10	37
Strawberry	283	1.25 c	319	10	31
Vanilla	283	1.25 c	314	10	32
MILK DESSERTS:					
Custard:					
Baked	265	1 c	305	70	213
From mix	143	1/2 c	161	50	80
Ice cream, vanilla:					
Regular	133	1 c	269	22	59
Rich	148	1 c	349	25	88
Soft serve	173	1 c	377	41	153
Ice milk, vanilla:					
Regular (about 4% fat)	131	1 c	184	10	18
Soft serve (3% fat)	175	1 c	223	6	13
PUDDINGS, prepared					
(5oz can = .55c):					
Assorted, low calorie	130	1/2 ea	69	3	2
Chocolate:					
Fr/mix, ckd, instant	260	1 c	305	10	29
Canned	142	1 can	205	<1	1
Coconut, f/instant	149	1/2 c	184	9	17
Lemon, f/instant	149	1/2 c	178	10	17
Rice, from mix,					
avg ckd/inst	141	1/2 c	165	10	16
Tapioca, ckd f/mix	130	1/2 c	145	10	15
Tapioca, canned	142	1 can	160	1	1
Vanilla pudding:					
From recipe	255	1 c	285	13	29
Fr/mix, ckd/instant	130	1/2 c	148	10	15
Canned	142	1 can	220	<1	1
Pudding pops, average	57	1 ea	94	1	1
Sherbet (2% fat)	193	1 c	270	5	14
Yogurt, frozen (avg)	174	1 c	220	6	14
YOGURT:					
Lowfat:					
Plain	227	1 c	144	10	14
With fruit	227	1 c	231	4	10
Coffee/vanilla	227	1 c	193	6	11
Nonfat	227	1 c	127	3	4
Whole	227	1 c	138	21	30
Yogurt cheese, recipe	208	1 c	222	10	23

ESHA Research

Eggs
(cholesterol is in the yolk):

	Wt (g)	Svg	Cal	Choles 100 Cal	Choles Svg
Chicken egg, large					
Whole, raw or cooked	50	1 ea	77.5	272	211
Duck egg, raw	70	1 ea	130	476	619
Egg substitute (brands vary, check label):					
Frozen	60	1/4 c	96	1	1
Liquid	251	1 c	211	1	3
Quail egg	9	1 ea	14	543	76
Turkey egg	79	1 ea	135	546	737

Fats, Oils & Salad Dressings
(Animal only. There is no cholesterol in vegetables or vegetable oils)

	Wt (g)	Svg	Cal	Choles 100 Cal	Choles Svg
Bacon fat	14	1 T	126	67	84
Beef fat (tallow, drippings)	12.8	1 T	115	12	14
Butter	14.2	1 T	100	31	31
Butter oil (ghee)	13	1 T	112	31	34
Chicken fat, melted	13	1 T	115	10	11
Cod liver oil	14	1 T	126	63	80
Duck fat	205	1 c	1846	11	205
Goose Fat	205	1 c	1846	11	205
Lard (pork fat)	205	1 c	1849	11	195
Mutton tallow	205	1 c	1846	11	209
Mayonnaise:					
Regular	14	1 T	100	8	8
Low cal, imitation	15	1 T	35	11	4
SALAD DRESSINGS:					
1000 island	16	1 T	60	7	4
1000 island, low cal	15	1 T	25	7	2
Blue cheese	15	1 T	75	4	3
Ceasar's	92	1/2 c	408	27	110
Italian, low cal	240	1 c	130	11	14
Mayo type	15	1 T	58	7	4
Mayo type, low cal	15	1 T	35	10	4
Ranch style	14.9	1 T	54	11	6
Russian	15.3	1 T	76	13	10
Tartar sauce	14	1 T	74	5	4
Turkey fat	13	1 T	115	11	13

Grains & Grain Products
Grains do not contain cholesterol. It is present from egg, meat or dairy ingredients.

	Wt (g)	Svg	Cal	Choles 100 Cal	Choles Svg
Chow mein noodles, dry	45	1 c	237	0	0
Egg noodles, cooked	160	1 c	213	23	50

Meats: Fish & Shellfish

	Wt (g)	Svg	Cal	Choles 100 Cal	Choles Svg
Abalone, fried	85	3 oz	161	50	80
Anchovies, cnd, drained	45	11 ea	95	33	31
Bass, freshwater,					
baked/broiled	100	3.5 oz	125	64	80
Bluefish:					
Baked/broiled	100	3.5 oz	159	40	63
Fried in crumbs	100	3.5 oz	205	29	60
Carp, baked/broiled	100	3.5 oz	162	52	84
Catfish, cornmeal fried	100	3.5 oz	229	35	81
Caviar (1 T = 16g)	100	3.5 oz	252	230	580

	Wt (g)	Svg	Cal	Choles 100 Cal	Choles Svg
CLAMS, meat only:					
Canned, drained	160	1 can	236	45	107
Canned w/liquid,					
minced, small can	183	1 can	145	45	65
Breaded, fried, small	188	20 ea	379	30	115
Steamed	100	3.5 oz	148	45	67
Clam nectar, cnd	240	1 c	6	0	0
COD:					
Baked/broiled	100	3.5 oz	105	52	55
Batter- fried	100	3.5 oz	199	28	55
Cnd w/liquid, 11 oz can	312	1 can	327	52	171
Smoked	100	3.5 oz	79	63	50
CRAB:					
Alaska King, leg, ckd	134	1 ea	129	56	72
Blue Crab, whole:					
Cooked, unpacked	135	1 c	138	98	135
Canned, unpacked	135	1 c	133	90	120
Dungeness, cooked	101	3/4 c	85	75	64
Crab, imitation f/surimi	85	3 oz	87	20	17
Crab cakes, recipe	60	1 ea	93	97	90
Crayfish, ckd, moist heat	85	3 oz	97	156	151
Croaker, breaded, fried fillet	87	1 ea	192	38	73
Dolfinfish (Mahi Mahi), raw	100	3.5 oz	85	86	73
Eel:					
Baked/broiled	100	3.5 oz	236	68	161
Smoked	100	3.5 oz	330	21	70
Fish cakes:					
Fried f/frozen	100	3.5 oz	213	42	90
Heated f/frozen	57	2 ea	155	41	64
Gefiltefish, sweet,					
commercial	42	1 pce	35	34	12
Grouper, cooked	100	3.5 oz	118	40	47
Haddock:					
Baked/broiled	85	3 oz	95	66	63
Breaded, fried	85	3 oz	175	31	55
Smoked	100	3.5 oz	116	66	77
HALIBUT:					
Baked/broiled	85	35 oz	119	29	35
Smoked	100	3.5 oz	224	45	100
Herring:					
Baked/broiled	100	3.5 oz	203	38	77
Canned in oil	100	3.5 oz	208	47	97
Smoked, kippered	100	3.5 oz	217	37	80
Pickled (1 pce = 15g)	100	3.5 oz	262	5	13
Lobster meat, cooked	145	1 c	142	73	104
Mackerel:					
Baked/brld, Atlantic	100	3.5 oz	262	29	75
Baked/brld, Spanish	100	3.5 oz	158	46	73
Canned, Jack, #1 can	361	1 can	563	51	285
Mullet, baked/broiled	85	3 oz	127	43	54
Mussels, Blue,					
steamed meat	85	3 oz	147	33	48
Ocean perch:					
Baked/broiled	100	3.5 oz	121	45	54
Breaded, fried	85	3 oz	185	25	46
Octopus, raw	100	3.5 oz	82	59	48
Orange roughy, raw	85	3 oz	107	16	17
OYSTERS:					
Raw, Eastern	248	1 c	170	80	136
Raw, Pacific	248	1 c	200	68	136
Breaded, fried, Eastern	88	6 ea	173	42	72
Simmered, Eastern	100	3.5 oz	137	387	530
Perch, freshwater,					
baked/broiled	92	2 ea	108	98	106
Pike, Northern,					
baked/broiled	100	3.5 oz	113	44	50

ESHA Research

Cholesterol (mg)

	Wt (g)	Svg	Cal	Choles (mg) 100 Cal	Choles (mg) Svg
Fish & Shellfish, continued:					
Pollock:					
Baked/broiled, avg	100	3.5 oz	99	71	70
Baked/broiled, Walleye	100	3.5 oz	113	85	96
Pompano, baked/brld	100	3.5 oz	211	30	64
Rockfish, baked/brld	100	3.5 oz	121	36	44
Roe, raw, mixed species	28	1 oz	39	269	105
SALMON, cooked:					
Mixed species, bkd/brld	85	3 oz	183	40	74
Chinook, smoked	85	3 oz	99	20	20
Sockeye, ckd f/fresh	100	3.5 oz	216	40	87
Canned:					
Atlantic, small can	220	1 can	281	47	132
Chum, #1 can	369	1 can	521	28	144
Pink, drained, #1 can	454	1 can	631	36	230
Sockeye, drained, #1 can	369	1 can	566	28	161
Sardines, cnd, drained:					
Atlantic (2 ≈ 24g)	92	1 can	192	68	131
Pacific (1 ≈ 38g)	100	3.5 oz	178	34	61
Scallops:					
Breaded, fried	93	6 ea	200	28	57
Steamed	100	3.5 oz	113	39	44
Imitation f/surimi	85	3 oz	84	21	18
Seatrout (Steelhead), cooked	100	3.5 oz	131	87	114
Sea bass, baked/broiled	100	3.5 oz	124	43	53
Shad, baked w/bacon	100	3.5 oz	201	30	60
Shark, batter-fried	85	3 oz	194	26	50
SHRIMP:					
Boiled, 2 lrg ≈ 11g	100	3.5 oz	99	197	195
Breaded, fried, 2 large ≈ 15g	90	12 ea	218	73	159
Canned, drained	128	1 c	154	144	222
Canned, with liquid	100	3.5 oz	102	144	147
Shrimp, imitation fr/surimi	85	3 oz	86	36	31
Smelt, Rainbow, ckd	85	3 oz	106	72	76
Snapper, baked/brld	100	3.5 oz	128	37	47
Sole (flounder):					
Baked/broiled	85	3 oz	99	59	58
Batter-fried	85	3 oz	250	18	45
Breaded, fried	100	3.5 oz	188	28	53
Squid, flour-fried	85	3 oz	149	148	221
Surimi, processed walleye pollock (also see imitation crab, scallops, shrimp, lobster)	100	3.5 oz	99	30	30
Swordfish, baked/broiled	100	3.5 oz	155	32	50
Trout, baked/broiled	85	3 oz	129	48	62
TUNA:					
Ckd f/fresh, Bluefin	85	3 oz	157	27	42
Canned, drained, #1/2 can:					
Light, canned in oil	171	1 can	339	9	30
Light, water pack	165	1 can	216	43	93

Meats
Beef, Pork, Ham, Rabbit, Frog legs, Venison, Veal

	Wt (g)	Svg	Cal	Choles (mg) 100 Cal	Choles (mg) Svg
BEEF:					
Breakfast strips	34	3 ea	153	26	40
Chuck roast, pot roasted:					
Lean & fat (5.4 oz raw)	85	3 oz	325	27	87
Lean only	85	3 oz	230	39	90
GROUND BEEF, cooked avg:					
Extra lean (17% fat, raw)	85	3 oz	215	33	70
Lean (20.7% fat, raw)	85	3 oz	231	30	70
Regular (26.6% fat, raw)	85	3 oz	250	30	75
Frzn patty (23% fat, raw)	85	3 oz	240	33	80

	Wt (g)	Svg	Cal	Choles (mg) 100 Cal	Choles (mg) Svg
BEEF, continued:					
Rib, oven rstd, choice:					
Lean & fat (5 oz raw)	85	3 oz	324	22	72
Lean only	85	3 oz	204	33	68
Round steak, brld, choice:					
Lean & fat (4.5 oz raw)	85	3 oz	233	30	71
Lean only	85	3 oz	165	42	70
Round tip, oven rstd, all:					
Lean & fat	85	3 oz	213	33	70
Lean only	85	3 oz	162	43	69
Sirloin steak, broiled, all:					
Lean & fat	85	3 oz	238	28	67
Lean only	85	3 oz	172	38	65
T-bone steak, brld, choice:					
Lean & fat	85	3 oz	276	26	71
Lean only	85	3 oz	182	37	68
Variety meats:					
Brains, pan fried	85	3 oz	167	1016	1696
Heart, simmered	85	3 oz	140	117	164
Kidney, ckd	85	3 oz	122	270	329
Liver, fried	85	3 oz	184	223	410
Tongue, ckd	85	3 oz	241	38	91
Corned beef, canned	85	3 oz	213	34	73
Dried beef, cured, 6-7 pces:	28	1 oz	47	98	46
FROG LEGS, raw meat only	100	3.5 oz	73	68	50
HAM: see Pork, Cured (also Lunchmeats, turkey ham)					
LAMB:					
Arm chop, braised:					
Lean & fat	70	1 ea	244	34	84
Lean only	55	1 ea	152	43	66
Loin chop, broiled:					
Lean & fat	80	1 ea	201	32	64
Lean only	64	1 ea	100	44	44
Leg of lamb, roasted:					
Lean & fat	85	3 oz	219	36	79
Lean only	85	3 oz	162	47	76
Cutlet, lean, ckd avg	85	3 oz	175	45	78
Rib roast:					
Lean & fat	85	3 oz	305	27	82
Lean only	85	3 oz	197	38	74
Shoulder roast:					
Lean & fat	85	3 oz	235	33	78
Lean only	85	3 oz	173	43	74
Lamb liver	85	3 oz	202	207	419
PORK:					
Bacon, cooked:					
Regular	19	3 pce	109	15	16
Canadian style	47	2 pce	86	31	27
Breakfast strips	34	3 pce	156	23	36
Blade chop:					
Braised, lean & fat	67	1 ea	275	26	72
Braised, lean only	50	1 ea	156	37	57
Pan-fried, lean & fat	89	1 ea	368	23	85
Pan-fried, lean only	62	1 ea	175	34	60
Roasted, lean & fat	88	1 ea	321	25	79
Roasted, lean only	71	1 ea	198	32	63
Center loin chop:					
Braised, lean & fat	75	1 ea	266	30	81
Braised, lean only	61	1 ea	166	41	68
Broiled, lean & fat	82	1 ea	284	30	84
Broiled, lean only	72	1 ea	166	43	71
Pan-fried, lean & fat	89	1 ea	334	28	92
Pan-fried, lean only	67	1 ea	178	40	71
Roasted, lean & fat	88	1 ea	268	30	80
Roasted, lean only	72	1 ea	180	38	68

PORK, continued / Meats: Poultry

	Wt (g)	Svg	Cal	Choles (mg) 100 Cal	Choles (mg) Svg
PORK, continued:					
Center rib chop:					
Braised, lean & fat	67	1 ea	246	26	64
Braised, lean only	53	1 ea	147	35	51
Broiled, lean & fat	77	1 ea	264	27	72
Broiled, lean only	63	1 ea	162	36	59
Pan-fried, lean & fat	88	1 ea	343	22	74
Pan-fried, lean only	62	1 ea	160	31	50
Roasted, lean & fat	79	1 ea	252	25	64
Roasted, lean only	66	1 ea	162	32	52
Pork roast, leg:					
Lean & fat	85	3 oz	250	32	79
Lean only	85	3 oz	187	43	80
Pork roast, loin & rib:					
Lean & fat	85	1 pce	265	28	73
Lean only	85	1 pce	206	35	72
Shoulder, braised:					
Lean & fat	85	3 oz	293	32	93
Lean only	67	2.4 oz	166	46	76
Spareribs, ckd f/1 lb raw	177	6.25 oz	703	30	214
Pork heart, braised	85	3 oz	125	150	188
Pork liver, braised	85	3 oz	141	214	302
PORK, CURED - HAM (also see Bacon under Pork):					
Roasted, lean & fat	85	3 oz	207	26	53
Roasted, lean only	85	3 oz	133	35	47
Canned, roasted	85	3 oz	140	25	35
RABBIT, roasted	85	3 oz	175	31	55
VEAL (calf):					
Cutlet, braised/broiled	100	3.5 oz	181	55	100
Heart, cooked	85	3 oz	158	95	150
Liver, pan-fried	85	3 oz	208	135	280
Rib roast	85	3 oz	151	64	97
VENISON (deer) roasted	85	3 oz	134	71	95

Meats: Poultry

	Wt (g)	Svg	Cal	Choles (mg) 100 Cal	Choles (mg) Svg
CHICKEN:					
All types meat:					
Fried	140	1 c	307	43	131
Roasted	140	1 c	266	47	125
Stewed	140	1 c	248	47	116
Canned, boned w/broth	142	5 oz	235	37	88
Dark meat:					
Fried	85	3 oz	203	40	82
Roasted	85	3 oz	174	45	79
Stewed	85	3 oz	163	46	75
Light meat:					
Fried	85	3 oz	163	47	76
Roasted	85	3 oz	147	49	72
Stewed	85	3 oz	135	48	65
Breast*, meat & skin:					
Batter-fried	140	1 ea	364	33	119
Flour-fried	98	1 ea	218	40	88
Roasted	98	1 ea	193	43	83
Stewed	110	1 ea	202	41	83
Breast* meat only:					
Fried	86	1 ea	161	48	78
Stewed	95	1 ea	144	51	73
Roasted	86	1 ea	142	51	73
***2 pieces per bird**					
Drumstick, meat & skin:					
Batter-fried	72	1 ea	193	32	62
Flour-fried	49	1 ea	120	37	44
Roasted	52	1 ea	112	43	48
Stewed	57	1 ea	116	41	48

CHICKEN, continued (right column)

	Wt (g)	Svg	Cal	Choles (mg) 100 Cal	Choles (mg) Svg
CHICKEN, continued:					
Drumstick, meat only:					
Fried	42	1 ea	82	49	40
Roasted	44	1 ea	76	54	41
Stewed	46	1 ea	78	51	40
Thigh, meat & skin:					
Batter-fried	86	1 ea	238	34	80
Flour-fried	62	1 ea	162	37	60
Roasted	62	1 ea	153	38	58
Stewed	68	1 ea	158	36	57
Thigh meat only:					
Fried	52	1 ea	113	47	53
Roasted	52	1 ea	109	45	49
Wing, meat & skin:					
Batter-fried	49	1 ea	159	25	39
Flour-fried	32	1 ea	103	25	26
Roasted	34	1 ea	99	29	29
Stewed	40	1 ea	100	28	28
Wing meat only:					
Fried	20	1 ea	42	40	17
Roasted	21	1 ea	43	42	18
Stewed	24	1 ea	43	42	18
Variety meats:					
Gizzard, simmered	22	1 ea	34	125	43
Heart, simmered	3.3	1 ea	6	133	8
Liver, simmered	20	1 ea	30	420	126
DUCK, domestic:					
Roasted, meat & skin	85	3 oz	286	25	71
Roasted, meat only	85	3 oz	171	44	76
GOOSE, domestic:					
Roasted, meat & skin	85	3 oz	259	30	78
Roasted, meat only	85	3 oz	202	40	82
TURKEY:					
Ground turkey, cooked	100	3.5 oz	229	30	69
Roasted:					
All types	140	1 c	238	45	107
Dark meat	85	3 oz	159	45	72
White meat	85	3 oz	133	44	59
Frozen slices	85	3 oz	145	44	64
Frozen, slices w/gravy	142	5 oz	95	27	26
Breast meat:					
Barbecued,	28	1 oz	40	40	16
Hickory smoked	28	1 oz	35	37	13
Turkey gizzard	67	1 ea	109	142	155
Turkey heart	16	1 ea	28	129	36
Turkey liver	75	1 ea	127	369	469

Meats: Sausages & Lunchmeats

	Wt (g)	Svg	Cal	Choles (mg) 100 Cal	Choles (mg) Svg
Barbecue loaf, pork & beef	23	1 pce	40	23	9
Beef lunchmeat:					
Beef loaf	28	1 oz	87	21	18
Thin sliced	28	1 oz	50	24	12
Beerwurst (beer salami):					
Beef salami	23	1 pce	75	17	13
Pork salami	23	1 pce	55	24	13
Berliner sausage	23	1 pce	53	21	11
BOLOGNA:					
Beef bologna	23	1 pce	72	18	13
Beef and pork	28	1 oz	89	18	16
Cured pork	23	1 pce	57	25	14
Turkey (pce = 1 oz)	28	1 oz	56	50	28
Braunschweiger	18	1 pce	65	43	28
Brotwurst, link	70	1 ea	226	19	44
Cheesefurter (cheese smoki)	43	1 ea	141	21	29

Cholesterol (mg)   Values are for edible portion of foods

	Wt (g)	Svg	Cal	Choles (mg) 100 Cal	Choles (mg) Svg
Meats, Sausages & Lunchmeats, continued:					
Chicken roll	57	2 pce	90	31	28
Corned beef loaf, jellied	28	1 oz	46	26	12
Dutch brand loaf	28	1 oz	68	19	13
FRANKFURTER (hotdog):					
Beef, 8/pkg	57	1 ea	184	15	27
Beef & pork, 8/pkg	57	1 ea	183	16	29
Chicken, 10/pkg	45	1 ea	115	39	45
Turkey, 10/pkg	45	1 ea	102	38	39
Ham, chopped, pkg	42	2 pce	98	21	21
Ham lunchmeat:					
Extra lean	57	2 oz	75	35	27
Regular	57	2 oz	103	31	32
Thin sliced	28	3 pce	37	35	13
Ham, minced	21	1 pce	55	27	15
Ham salad spread	240	1 c	518	17	88
Ham & cheese loaf	57	2 oz	147	22	33
Italian sausage link, ckd	67	1 ea	216	24	52
Keilbasa	26	1 pce	81	21	17
Knockwurst, link	68	1 ea	209	19	39
Liverwurst, pork	18	1 pce	59	47	28
Luncheon meat, canned	21	1 pce	70	19	13
Luncheon sausage:					
Beef & pork	23	1 pce	60	25	15
Luxury loaf	57	2 oz	80	25	20
Mortadella	15	1 pce	47	17	8
Olive loaf	57	2 oz	133	17	22
Pastrami:					
Beef, cured	57	2 oz	198	27	53
Turkey, cured	57	2 oz	74	41	30
Peppered loaf	28	1 oz	42	31	13
Pepperoni, small slice	22	4 pce	109	7	8
Pickle & pimento loaf	57	2 oz	149	14	21
Polish sausage	28	1 oz	92	22	20
Pork sausage:					
Cooked link	13	1 ea	48	23	11
Cooked patty	27	1 pce	100	22	22
Raw, 2 oz patty	57	1 ea	238	16	39
Poultry sandwich spread	13	1 T	25	16	4
SALAMI:					
Beef salami	23	1 pce	58	24	14
Beef & pork, dry	20	2 pce	85	19	16
Pork & beef	57	2 pce	143	26	37
Turkey	57	2 pce	111	41	46
Sandwich spread, pork & beef	15	1 T	35	17	6
Sausage links, brown & serve	13	1 ea	50	18	9
Smoked link sausage:					
Beef & pork	68	1 ea	229	21	48
Pork link	68	1 ea	265	17	46
Turkey	28	1 oz	55	35	19
Summer sausage:					
Beef, pork	23	1 oz	80	20	16
Turkey	28	1 oz	50	44	22
TURKEY lunchmeats (other):					
Breakfast sausage	28	1 oz	65	35	23
Ham	57	2 oz	73	44	32
Loaf, breast meat	43	2 pce	46	38	17
Roll, light & dark	57	2 pce	84	37	31
Roll, light meat	57	2 pce	83	29	24
Vienna sausage, canned	16	1 ea	45	18	8

Mixed Dishes & Fast Foods

	Wt (g)	Svg	Cal	Choles (mg) 100 Cal	Choles (mg) Svg
Beef, macaroni, tomato sauce, recipe	226	1 c	189	12	22
Beef & vegetable stew:					
Recipe	245	1 c	220	32	71
Canned	245	1 c	194	8	15
Beef pot pie, f/frzn	234	1 ea	426	10	41
BURRITO:					
Bean burrito	174	1 ea	322	5	15
Beef burrito	177	1 ea	463	19	89
Beef and bean	175	1 ea	390	13	52
Deluxe combination	198	1 ea	424	14	58
Cheese soufflé, recipe	112	1 c	221	132	291
Chicken & noodles, recipe	240	1 c	365	28	103
Chicken a la king, recipe	245	1 c	470	47	221
Chicken chow mein:					
Recipe	250	1 c	255	29	75
Canned	250	1 c	95	8	8
Chicken curry, recipe	337	1.5 c	305	21	64
Chicken pot pie, f/frzn	230	1 ea	430	9	40
Chicken salad w/celery	78	1/2 c	266	18	48
Chili w/beans, canned	255	1 c	286	15	43
Chop suey w/beef & pork	250	1 c	300	23	68
Corn dog	111	1 ea	330	11	37
Corn fritter, recipe	45	1 ea	116	29	34
Corn pudding	250	1 c	271	85	230
Cole slaw	120	1 c	84	11	10
Corned beef hash, canned	220	1 c	382	35	132
Egg salad	183	1 c	438	144	629
ENCHILADA:					
Beef enchilada	120	1 ea	292	13	38
Cheese enchilada	120	1 ea	330	13	43
Chicken enchilada	120	1 ea	269	14	39
French toast, recipe	65	1 pce	123	59	73
LASAGNA, recipe:					
With meat	245	1 pce	398	14	56
Without meat	218	1 pce	316	10	30
Macaroni & cheese:					
Recipe	200	1 c	430	10	44
Canned	240	1 c	230	10	24
Macaroni salad, no cheese	141	1 c	371	7	24
Manicotti, frozen entree	225	1 ea	271	26	70
Meat loaf:					
Beef only	87	1 pce	193	51	98
Beef and 1/3 pork	87	1 pce	212	46	97
Moussaka (lamb & eggplant)	250	1 c	250	57	143
Pies, fried, commercial:					
Apple, fried pastry	85	1 ea	255	5	14
Cherry, fried pastry	85	1 ea	250	5	13
PIZZA, cheese:					
Regular crust, 1/8 of 15"	120	1 pce	290	19	56
Thick crust, 1/2 of 10"	208	1 pce	519	15	77
Potato salad w/mayo & eggs	250	1 c	358	47	170
Quiche Lorraine, 1/8	176	1 pce	600	48	285
Ravioli, beef, canned	226	1 c	220	6	12
SANDWICHES, fast foods:					
Cheeseburger, 3 oz patty	112	1 ea	300	15	44
Cheeseburger, 4 oz patty	194	1 ea	524	20	104
Chicken patty sandwich	157	1 ea	436	16	68
English muffin, egg, cheese, bacon	138	1 ea	360	59	213

	Wt (g)	Svg	Cal	Choles (mg) 100 Cal	Choles (mg) Svg		Wt (g)	Svg	Cal	Choles (mg) 100 Cal	Choles (mg) Svg
Mixed Dishes & Fast Foods, continued:						**GRAVY,** continued:					
SANDWICHES, fast foods, continued:						Onion, from dry	261	1 c	78	0	0
Fish sandwich:						Pork, from dry	258	1 c	76	4	3
Regular, w/cheese	140	1 ea	420	13	56	Turkey gravy:					
Large, w/o cheese	170	1 ea	470	19	90	Canned	238	1 c	122	4	5
Hamburger, 3 oz patty	98	1 ea	245	13	32	From dry mix	261	1 c	87	3	3
Hamburger, 4 oz patty	174	1 ea	445	16	71						
Hotdog (frankfurter) w/bun	85	1 ea	260	9	23	**SAUCES** (also see "Other"):					
						Au Jus, canned	238	1 c	38	3	1
SANDWICHES on part whole wheat,						Au Jus, f/dry mix	246	1 c	32	8	2.5
unless stated as on rye.						Bearnaise, recipe	73	1/2 c	441	58	254
Avocado, cheese, tomato,						Bechamel, recipe	145	1/2 c	141	22	32
sprouts	195	1 ea	432	7	32	Bordelaise, recipe	233	1/2 c	197	16	31
Bacon, lettuce, tomato	135	1 ea	327	7	21	Cheese sauce,					
Chicken salad	100	1 ea	294	9	25	fr/mix w/milk	279	1 c	305	17	53
Corned beef & swiss on rye	147	1 ea	429	20	85	Hollandaise:					
Egg salad	111	1 ea	319	51	163	Recipe	160	1 c	867	92	799
Grilled cheese	117	1 ea	393	14	55	Mix w/milk & butter	255	1 c	703	27	189
Ham sandwich	122	1 ea	256	11	29	Mix w/water	259	1 c	240	22	52
Ham & cheese	151	1 ea	363	16	56	Mornay, recipe	172	1/2 c	380	49	186
Ham on rye	116	1 ea	242	12	29	Spaghetti sauce w/meat:					
Ham & swiss on rye	145	1 ea	350	16	55	Homemade	248	1 c	297	11	34
Ham salad	125	1 ea	339	8	27	Canned	206	7/8 c	220	8	17
Patty melt on rye	177	1 ea	567	19	107	Stroganoff sauce from mix					
Reuben, grilled	233	1 ea	480	18	85	w/milk & water	296	1 c	271	14	38
Roast beef	122	1 ea	280	11	30	White sauce:					
Tuna salad	116	1 ea	303	8	25	Recipe. medium	250	1 c	395	8	32
Turkey sandwich	122	1 ea	271	11	29	Dry mix w/milk	264	1 c	240	14	34
Turkey ham	122	1 ea	253	14	35						
Turkey ham & cheese	151	1 ea	361	17	62	**SOUPS:** soups are from canned unless otherwise					
Turkey ham on rye	116	1 ea	239	14	35	stated. RTS = Ready To Serve; for soups prep.					
Turkey ham & cheese on rye	145	1 ea	347	18	62	w/milk, assume whole milk.					
						Bean & 'frank' w/water	250	1 c	187	6	12
SPAGHETTI pasta, tomato						Bean & bacon:					
sauce & cheese:						Prep w/water	253	1 c	173	2	3
Homemade	250	1 c	260	3	8	From dry mix	265	1 c	105	3	3
Canned	250	1 c	190	2	3	Bean & ham, chunky, RTS	243	1 c	231	10	22
SPAGHETTI pasta, tomato						Beef, chunky, RTS	240	1 c	171	8	14
sauce & meat:						Beef bouillon, all	241	1 c	8-16	2-5	<1-1
Homemade	248	1 c	330	27	89	Beef noodle:					
Canned	250	1 c	260	9	23	Prep w/water	244	1 c	84	6	5
Spinach soufflé	136	1 c	218	84	184	From dry w/water	251	1 c	41	5	2
Taco, beef	78	1 ea	207	21	45	Celery, cream of:					
Taco, chicken	78	1 ea	172	26	45	Prep w/milk	248	1 c	165	19	32
						Prep w/water	244	1 c	90	17	15
TOSTADA:						Cheese soup:					
W/beans & beef	192	1 ea	332	19	62	Prep w/milk	251	1 c	230	21	48
W/beans & chicken	157	1 ea	249	21	53	Prep w/water	247	1 c	155	19	30
W/refried beans	157	1 ea	212	7	15	Chicken, chunky, RTS	251	1 c	178	17	30
Tuna noodle casserole	202	1 c	251	21	52	Chicken, cream of:					
Tuna salad	205	1 c	383	7	27	Prep w/milk	248	1 c	191	14	27
Turkey pot pie, frozen	233	1 ea	416	5	20	Prep w/water	244	1 c	115	9	10
Veal Parmigiana, frzn entree	205	7.25 oz	372	30	113	From dry mix	261	1 c	107	3	3
Waldorf salad	142	1 c	424	5	22	Chicken bouillon, all	244	1 c	21-39	3-5	1
						Chicken Noodle:					
Soups, Sauces & Gravies						Canned	241	1 c	75	9	7
						Canned, chunky, RTS	240	1 c	114	16	18
						From dry	252	1 c	53	6	3
GRAVY:						From packet	188	3/4 c	40	5	2
Beef, recipe	135	1/2 c	151	6	9	Chicken & rice:					
Beef, canned	233	1 c	124	6	7	Prep w/water	241	1 c	60	12	7
Brown gravy fr/dry mix	258	1 c	75	3	2.6	From dry mix	253	1 c	60	5	3
Chicken gravy:						Chicken & rice,					
Homemade	130	1/2 c	163	7	11	chunky, RTS	240	1 c	127	9	12
From dry mix	260	1 c	85	4	3	Chicken vegetable:					
Canned	238	1 c	189	3	5	From dry	251	1 c	49	6	3
Mushroom gravy:						Chunky, RTS	240	1 c	167	10	17
From dry w/milk	267	1 c	228	15	34	Chili beef soup	250	1 c	169	7	12
From dry w/water	258	1 c	70	1	1						

	Wt (g)	Svg	Cal	Choles (mg) 100 Cal	Svg
Soups, continued:					
Clam chowder:					
New England w/milk	248	1 c	163	13	22
Manhatten w/water	244	1 c	78	3	2
Lentil & ham, RTS	248	1 c	140	5	7
Minestrone:					
Canned or fr/dry	241	1 c	80	4	3
Chunky, RTS	240	1 c	127	4	5
Mushroom, cream of:					
Prep w/milk	248	1 c	205	10	20
Prep w/water	244	1 c	130	2	2
Prep fr/dry	253	1 c	96	1	1
Potato, cream of:					
Prep w/milk	248	1 c	148	15	22
Prep w/water	244	1 c	73	7	5
Onion soup, packet:					
From dry w/water	184	1 pkt	21	0	0
Dry, unprepared	39	1 pkt	115	2	2
Oxtail soup, from dry	253	1 c	71	4	3
Oyster stew:					
Prep w/milk	245	1 c	134	24	32
Prep w/water	241	1 c	59	24	14
Shrimp, cream of:					
Prep w/milk	248	1 c	165	21	35
Prep w/water	244	1 c	90	19	17
Split pea w/ham:					
Prep w/water	253	1 c	189	4	8
From dry	255	1 c	133	2	3
Split pea, chunky, RTS	240	1 c	184	4	7
Tomato, prep w/milk	248	1 c	160	11	17
Tomato beef noodle prep w/water	244	1 c	140	4	5
Tomato rice w/water	247	1 c	120	2	2
Turkey, chunky, RTS	236	1 c	136	7	9
Turkey noodle w/water	244	1 c	69	7	5
Turkey vegetable w/water	241	1 c	74	3	2
Vegetable beef w/water	244	1 c	79	6	5
Vegetable beef f/dry	253	1 c	53	2	1

Vegetables & Legumes
There is no cholesterol in vegetables or legumes.

Other
There is no cholesterol in chocolate, only in added ingredients such as milk.

CANDY & CANDY BARS:					
Caramel, plain/choc	28	1 oz	115	1	1
Fudge, chocolate	28	1 oz	115	1	1
Fudge, vanilla	28	1 oz	118	8	10
Milky Way bar	60	1 ea	260	5	14
Milk chocolate	28	1 oz	145	4	6
Milk chocolate w/almonds	28	1 oz	150	3	5
Milk chocolate w/peanuts	28	1 oz	155	2	3
Milk chocolate w/rice cereal	28	1 oz	140	4	6
Mr. Goodbar	47	1 ea	250	3	7
Reese's peanut butter cup	45	2 ea	240	1	3

Sources & References

The following sources of food nutrient data were compiled for The Food Processor® II nutrition system, and the ESHA nutrient database. This document -- The Food Finder -- is a compilation from this ongoing research effort.

(1) Composition of Foods—Dairy and Egg Products...Raw, Processed, Prepared; Linda P. Posati and Martha L. Orr; U.S. Department of Agriculture Handbook 8-1, 1976.

(2) Composition of Foods—Spices and Herbs... Raw, Processed, Prepared; Anne C. Marsh, Mary K. Moss and Elizabeth W. Murphy; U.S. Department of Agriculture Handbook No. 8-2, 1977.

(3) Composition of Foods—Baby Foods...Raw, Processed, Prepared; S.E. Gebhardt, R. Cutrufelli and R.H. Matthews; U.S. Department of Agriculture Handbook No. 8-3, 1978.

(4) Composition of Foods—Fats and Oils...Raw, Processed, Prepared; James B. Reeves, III and John L. Weihrauch; U.S. Department of Agriculture Handbook No.8-4, 1979.

(5) Composition of Foods—Poultry Products... Raw, Processed, Prepared; Linda P. Posati; U.S. Department of Agriculture. Handbook No. 8-5, 1979.

(6) Composition of Foods—Soups, Sauces and Gravies...Raw, Processed, Prepared; Anne C. Marsh; U.S. Department of Agriculture Handbook No. 8-6, 1980.

(7) Composition of Foods—Sausages and Luncheon Meats...Raw, Processed, Prepared; Martha Richardson, Linda P. Posati and Barbara A.Anderson; U.S. Department of Agriculture Handbook No. 8-7, 1980.

(8) Composition of Foods—Breakfast Cereals... Raw, Processed, Prepared; Judith S. Douglas, Ruth H. Matthews and Frank N. Hepburn; U.S. Department of Agriculture Handbook No. 8-8, 1982.

(9) Composition of Foods—Fruits and Fruit Juices...Raw, Processed, Prepared; Susan E. Gebhardt, Rena Cutrufelli and Ruth H. Matthews; U.S. Department of Agriculture Handbook No. 8-9, 1982.

(10) Composition of Foods—Pork and Pork Products... Raw, Processed, Prepared; Barbara A. Anderson; U.S. Department of Agriculture Handbook No. 8-10, 1983.

(11) Composition of Foods: Vegetables and Vegetable Products...Raw, Processed, Prepared; David B. Haytowitz and Ruth H. Matthews; US Department of Agriculture Handbook No. 8-11, August 1984.

(12) Composition of Foods: Nut and Seed Products...Raw, Processed, Prepared; Marie A. McCarthy and Ruth H. Matthews; US Department of Agriculture Handbook No. 8-12, September 1984.

(13) Composition of Foods: Beverages...Raw, Processed, Prepared; Rena Cutrufelli and Ruth H. Matthews; US Department of Agriculture Handbook No. 8-14, May 1986.

(14) Nutritive Value of American Foods in Common Units; Catherine F. Adams; U.S. Department of Agriculture Handbook No. 456, 1975.

(15) Composition of Foods...Raw, Processed, Prepared; Bernice K. Watt and Annabel L. Merrill; U.S. Department of Agriculture Handbook No. 8, revised 1963, reprinted 1975.

(16) Provisional Table on the Content of Omega 3 Fatty Acids and Other Fat Components in Selected Foods; Jacob Exler and John L. Weihrauch; Nutrient Data Research Branch, Nutrition Monitoring Division, US Department of Agriculture, Human Nutrition Information Service, February 1986.

(17) Provisional Table on the Content of Fast Foods; Anna C. Marsh and John L. Weihrauch with technical assistance of Carolina E. Szymanski; Nutrient Data Research Branch, Nutrition Monitoring Division, USDA, Human Nutrition Information Service, 1984.

(18) Provisional Table on the Fatty Acid and Cholesterol Content of Selected Foods; John L. Weihrauch; Nutrient Data Research Branch, Nutrition Monitoring Division, USDA, Human Nutrition Information Service, October 1984.

(19) Conversations, Correspondence, and Provisional data from the professional staff members at the U.S. Department of Agriculture, Human Information Service: Many thanks to Ruth Matthews, Susan Gephardt, Betty Perloff, Dennis Drake, David Haytowitz, Jacob Exler, John Weihrach, Pamela Pehrsson, Jean Stewart, and others; 1984 through 1988.

(20) Nutritive Values of Foods; Susan E. Gephardt and Ruth H. Matthews; US Department of Agriculture; Human Nutrition Information Service; Home and Garden Bulletin No. 72, revised 1985, released 1986.

(21) Provisional Table on Percent Retention of Nutrients in Food Preparation; Sharon L. Garland under direction of Ruth Matthews; Nutrient Data Research Group, Consumer Nutrition Center, Human Nutrition Information Service, USDA.

(22) Iron Content of Food; Jacob Exler; Home Economics Research Report No.45; Human Nutrition Information Service, USDA, 1983.

(23) Recommended Dietary Allowances; National Academy of Sciences; National Research Council; 9th revised edition, 1980.

(24) Nutritive Value of Foods; Catherine F. Adams and Martha Richardson; U.S.Department of Agriculture Home and Garden Bulletin No. 72, revised 1981.

(25) The Composition of Foods (McCance and Widdowson's); A.A. Paul and D.A.T. Southgate; fourth revised and extended edition of Medical Research Council's Special Report No. 297; Her Majesty's Stationery Office, Elsevier/North-Holland Biomedical Press.

(26) Geigy Scientific Tables, vol.1: Units of Measurement, Body Fluids, Composition of the Body, Nutrition; C. Lentner, editor; 8th revised and enlarged edition; published by CIBA-GEIGY Ltd, 1981.

(27) Pantothenic Acid, Vitamin B6 and Vitamin B12 in Foods; Martha L. Orr; U.S. Department of Agriculture Home Economics Research Report No.36, 1969.

(28) Nutrient Value of Some Common Foods; published by authority of the Minister of National Health and Welfare, Canada; revised 1979.

(29) Nutrients in Foods; Gilbert A. Leveille, Mary Ellen Zabik and Karen J. Morgan; Published by The Nutrition Guild; Cambridge, Mass., 1983.

(30) Comparisons of Methods for Calculating Retentions of Nutrients in Cooked Foods; Elizabeth W. Murphy, Patricia E. Criner and Bruce C. Gray; Journal Agric. Food Chemistry; v.23:1153, 1975.

(31) Composition of Foods Used in Far Eastern Countries; W.W. Leung, R.K. Pecot and B.K. Watt; U.S. Department of Agriculture Handbook No. 34, 1952.

(32) The Sodium Content of Your Food; Anne C. Marsh, Ruth N. Klippstein and Sybil D. Kaplan; U.S. Department of Agriculture Home and Garden Bulletin No. 233, 1980.

(33) Vitamin B6 Components in Fresh and Dried Vegetables; Marilyn M. Polansky; Journal of Am. Dietetic Assoc.; v.54:118, 1969.

(34) Nutrients in Cooked and Frozen Vegetables; L.J. Teply, PhD and P.H. Derse; Journal of Am. Dietetic Assoc., 1958.

ESHA Research

(35) Thiamin, Riboflavin, and Vitamin B6 Contents of Selected Foods as Served; Michael H. Dong, MPH, MPA; Evelyn L.McGown, PhD; Bruce W. Schwenneker and Howerde W. Sauberlich, PhD; Journal of Am. Dietetic Assoc.; v.76:156, 1980.

(36) Vitamin B6 in Reheated, Held, and Freshly Cooked Turkey Breast; Pamela P. Engler and Jane A. Bowers, PhD; Journal of Am. Dietetic Assoc; v.67:42, 1975.

(37) Vitamin B6 Components in Fruits and Nuts; Marilyn M. Polansky and Elizabeth W. Murphy; Journal of Am. Dietetic Assoc.;v.48:109, 1966

(38) B-Vitamin Retention in Meat During Storage and Preparation; Pamela P. Engler and Jane A. Bowers, PhD; Journal of Am. Dietetic Assoc.; v.69:253, 1976.

(39) Vitamin B6 and Niacin Contents of Broiler Meat of Different Strains, Sexes, and Production Regions; Catharina Y.W. Ang; Journal of Food Science; v.45:898, 1980.

(40) Nutrient Composition of Selected Wheats and Wheat Products; Marilyn M. Polansky and Edward W. Toepfer; Cereal Chemistry v.46:664, 1969.

(41) Lesser Known Vitamins in Foods; Mervyn G. Hardinge, MD, DR, PH, PhD and Hulda Crooks; Journal of Am. Dietetic Assoc.; v.38:240, 1961.

(42) The Vitamin Content of Wheat, Flour and Bread; W.K. Calhoun, W.G. Bechtel and W.B. Bradley; Cereal Chemistry; v.35:350, 1958.

(43) Nutritional Value of Dehydrated Foods; Miriam H. Thomas and Doris H. Calloway, PhD; Journal of Am. Dietetic Assoc.; v.39:105, 1961.

(44) Pantothenic Acid Content of 75 Processed and Cooked Foods; Joan H. Walsh, PhD, RD, Bonita W. Wyse, PhD, RD and R. Gaurth Hansen, PhD; Journal of Am. Dietetic Assoc.; v.78:140, 1981.

(45) Pantothenic Acid and Vitamin B6 in Beef; Bernadine H. Meyer, PhD, Mary A. Mysinger and Lois A. Wodarski; Journal of Am. Dietetic Assoc.; v. 54:122, 1969.

(46) Retention of the B Vitamins in Beef and Lamb After Stewing. III. Pantothenic Acid; Sylvia Cover, Esther M. Dilsaver and Rene M. Hays; Journal of Am. Dietetic Assoc; v.23:696, 1947.

(47) Vitamin B12...Microbiological Assay Methods and Distribution in Selected Foods; Harold Lichtenstein, Aram Beloian and Elizabeth W. Murphy; U.S. Department of Agriculture Home Economics Research Report No. 13, 1961

(48) Vitamin B12, E, and D Content of Raw and Cooked Beef; M.R. Bennink and K. Ono; Journal of Food Science; v.47:1786, 1982.

(49) The Vitamin B12 Content of Meals and Items of Diet; J.F. Adams, Fiona McEwan and Alison Wilson; British Journal of Nutrition; v.29:65, 1973.

(50) Vitamin B6 Components in Some Meats, Fish, Dairy Products, and Commercial Infant Formulas; Marilyn M. Polansky and Edward W. Toepfer; Journal of Agric. Food Chemistry; v.17:1394, 1969.

(51) Nutritive Value of Canned Foods...Vitamin B6, Folic Acid, Beta-Carotene, Ascorbic Acid, Thiamine, Riboflavin, and Niacin Content and Proximate Composition; L.J. Teply, P.H. Derse and C.H. Krieger; Agriculture and Food Chemistry; v.1:1204, 1953.

(52) Folacin in Selected Foods; Betty P. Perloff and Ritva R. Burtrum; Journal of Am. Dietetic Assoc.; v.70:161, 1977.

(53) Folic Acid Content of Foods...Micro- biological Assay by Standardized Methods and Compilation of Data from Literature; Edward W. Toepfer, Elizabeth G. Zook and Martha L. Orr; U.S. Department of Agriculture Handbook No. 29, 1951.

(54) Folacin in Wheat and Selected Foods; Susan Butterfield and Doris H. Calloway, PhD; Journal of Am. Dietetic Assoc.; v.60:310, 1972.

(55) Folate Distribution in Fruit Juices; Faye M. Dong and Susan M. Oace, PhD, RD, Journal of Am. Dietetic Assoc.; v.62:162, 1973.

(56) The Free and Total Folate Activity in Foods Available on the Canadian Market; K. Hoppner, B. Lampi and D.E. Perrin; Journal Inst. Canadian Technol. Aliment. v.5, n.2:60, 1972.

(57) Folic Acid Activity in Puerto Rican Foods; Rafael Santini, Jr., PhD, Maj. Florence M. Berger, AMSC, Gloria Berdasco, Capt. Thomas W. Sheehy, MC, Josefina Aviles and Ivonne Davila; Journal of Am. Dietetic Assoc.; v.41:562, 1962.

(58) Thermal Destruction of Folacin in Microwave and Conventional Heating; R.G. Cooper, Tung-Shan Chen, PhD and M.A. King; Journal of Am. Dietetic Assoc.; v.73:406, 1978.

(59) Thermal Destruction of Folacin: Effect of Ascorbic Acid, Oxygen and Temperature; Tung- Shan Chen, PhD and R.G. Cooper; Journal of Food Science; v.44:715, 1979.

(60) Folacin Content of Tea; Tung-Shan Chen, PhD, Cliff K.F. Lui and Christine H. Smith, PhD, RD; Journal of Am. Dietetic Assoc.; v.82:627, 1983.

(61) Folacin Activity of Frozen Convenience Foods; K. Hoppner, B. Lampi and D.E. Perrin; Journal of Am. Dietetic Assoc.; v.63:536, 1973.

(62) Folacin Content of Supplemental Foods for Pregnancy; Shirley W. Thenen, PhD; Journal of Am. Dietetic Assoc.; v.80:237, 1982.

(63) Effect of Processing on Provitamin A in Vegetables; J.P. Sweeney and A.C. Marsh; Journal of Am. Dietetic Assoc.; v.59:238, 1971.

(64) Vitamin E Content of Foods; P.J. McLaughlin, PhD and John L. Weihrauch; Journal of Am. Dietetic Assoc.; v.75:647, 1979.

(65) Vitamin E Adequacy of Vegetable Oils; J.G. Bieri, PhD andR. Poukka Evars, DVM; Journal of Am. Dietetic Assoc.; v.66:134, 1975.

(66) Tocopherol Contents of Vegetables and Fruits; V.H. Booth and M.P. Bradford; British Journal of Nutrition; v.17:575, 1963.

(67) Tocopherols in Foods and Fats; Hal T. Slover; Lipids; v.6,n.5:291, 1971.

(68) Alpha-Tocopherol Content of Cereal Grains and Processed Cereals; David C. Herting and Emma-Jane E. Drury; Journal of Agric. Food Chemistry; v.17:785, 1971.

(69) The Effect of Cooking on Alpha-Tocopherol in Vegetables; V.H. Booth and M. P. Bradford; Zeitschrift Vitaminforsch; v.33:276, 1963.

(70) Alpha-Tocopherol Content of Foods; R.H. Bunnel, PhD, J. Keating, BS, A. Quaresimo, BS and G.K. Parman, BS; The American Journal of Clinical Nutrition; v.17:1, 1965.

(71) Cheese Products: Protein, Moisture, Fat, and Acceptance; Pamela J. Nystrom, Joyce G. Ostrander, PhD and Charlene S. Martinsen, RD; Journal of Am. Dietetic Assoc.; v.65:40, 1974.

(72) Composition of Native American Fruits in the Pacific Northwest; Patrick B. Keely, RD, Charlene S. Martinsen, PhD, RD, Eugene S. Hunn, PhD and Helen H. Norton; Journal of Am. Dietetic Assoc.; v.81:568, 1982.

(73) Nutrients in Fresh Peeled Oranges and Grapefruit from California and Arizona; James A. Staroscik, Felipe U. Gregorio, Jr. and Samuel K. Reeder, PhD; Journal of Am. Dietetic Assoc.; v.77:567, 1980.

(74) Nutrients in California Lemons and Oranges; John J. Birdsall, Philip H. Derse and Lester J. Teply; Journal of Am. Dietetic Assoc.; v.38:555, 1961.

(75) Mineral Composition of Fruits: Nitrogen, Calcium, Magnesium, Phosphorus, Potassium, Aluminum, Boron, Copper, Iron, Manganese, and Sodium; Elizabeth G. Zook and Joanna Lehmann; Journal of Am. Dietetic Assoc.; v.52:225, 1968.

(76) Mineral Composition of Fruits: Edible Yield, Total Solids, and Ash of 30 Fresh Fruits; Elizabeth G. Zook; Journal of Am. Dietetic Assoc.; v.52:218, 1968.

(77) Mineral Content of Dairy Products: II. Cheeses; N.P. Wong, PhD, D.E. LaCroix and John A. Alford, PhD; Journal of Am. Dietetic Assoc.; v.72:608, 1978.

(78) Major Mineral Elements in Dairy Products; Ruth M. Feeley, Patricia E. Criner, Elizabeth W. Murphy and Edward W. Toepfer, PhD; Journal of Am. Dietetic Assoc.; v.61:505, 1972.

(79) Mineral Contents of Selected Pre-Prepared Foods Sampled in a Hospital Food Service Line; Cecilia Leung, PhD., RD, Helen H. Koehler and Margaret M. Hard; Journal of Am. Dietetic Assoc.; v.80:530, 1982.

(80) Mineral Elements in Fresh Vegetables from Different Geographic Areas; Homer Hopkins and Jacob Eisen; Agriculture and Food Chemistry; v.7,n.9:643, 1959.

(81) Minerals and Proximate Composition of Organ Meats; H.T. Hopkins, E.W. Murphy and D.P. Smith; Journal of Am. Dietetic Assoc.; v.38:344, 1961.

(82) Essential Mineral Elements in Peanuts and Peanut Butter; Leila C.A. Galvao, Anthony Lopez and Harriet L. Williams; Journal of Food Science; v.41:1305, 1976.

(83) Comprehensive Evaluation of Fatty Acids in Foods: I. Dairy Products; Linda A. Posati, John E. Kinsella, PhD and Bernice K. Watt, PhD, RD; Journal of Am. Dietetic Assoc.; v.66:482, 1975.

(84) Comprehensive Evaluation of Fatty Acids in Foods: II. Beef Products; Barbara A. Anderson, John A. Kinsella, PhD and Bernice K. Watt, PhD, RD; Journal of Am. Dietetic Assoc.; v.67:35, 1975.

(85) Comprehensive Evaluation of Fatty Acids in Foods: III. Eggs and Egg Products; Linda P. Posati, John E. Kinsella, PhD and Bernice K. Watt, PhD, RD; Journal of Am. Dietetic Assoc.; v.67:111, 1975.

(86) Comprehensive Evaluation of Fatty Acids in Foods: IV. Nuts, Peanuts and Soups; Geraldine A. Fristrom, PhD, Bernice C. Stewart, PhD, John L. Weibrauch and Linda P. Positi; Journal of Am. Dietetic Assoc.; v.67:351, 1975.

(87) Comprehensive Evaluation of Fatty Acids in Foods: V. Unhydroge-nated Fats and Oils; Carol A. Brignoli, PhD, John E. Kinsella, PhD. and John L. Weibrauch; Journal of Am. Dietetic Assoc.; v.68:224, 1976.

(88) Comprehensive Evaluation of Fatty Acids in Foods: VI. Cereal Products; John L. Weihrauch, John E. Kinsella, PhD and Bernice K. Watt, PhD, RD; Journal of Am. Dietetic Assoc.; v.68:335, 1976.

(89) Comprehensive Evaluation of Fatty Acids in Foods: VII. Pork Products; Barbara A. Anderson; Journal of Am. Dietetic Assoc.; v.69:44, 1976.

(90) Comprehensive Evaluation of Fatty Acids in Foods: VIII. Finfish; Jacob Exler, PhD and John L. Weihrauch; Journal of Am. Dietetic Assoc.; v.69:244, 1976.

(91) Total Fat and Fatty Acid Composition of Commercially Available Chocolate Candies; P.C. Ahn, N. Kassim, RD and P.V.J. Hegarty, PhD; Journal of Am. Dietetic Assoc.; v.79:552, 1981.

(92) Effects of Four Cooking Methods on the Proximate, Mineral and Fatty Acid Composi- tion of Fish Fillets; K.L. Gall, W.S, Otwell, J.S. Koburger and H. Appledorf; Journal of Food Science; v.48:1068, 1983

(93) Copper Content of Foods; Jean T. Pennington PhD and Doris H. Calloway, PhD, RD; Journal of American Dietetic Association; v.63:143, 1973.

(94) Copper Content of Some Low-Copper Foods; Liesbeth Hook and Ira K. Brandt, MD; Journal of Am. Dietetic Assoc.; 202, 1966.

(95) Copper Content of Tea; Mary A. Kenney, PhD and Saroja Thimaya; Journal of American Dietetic Association; v.82:509, 1983.

(96) Copper in Chinese Food Materials; William H. Adolph and T'Ung-Pi Chou; Chinese Journal of Physiology; v.VII, n.3:185, 1933

(97) Zinc and Copper Content of Foods Used in Vegetarian Diets; Jeanne H. Freeland-Graves PhD, RD, M. Lavone Ebangit and Pamela W. Bodzy; Journal of Am. Dietetic Association; v.77:648, 1980.

(98) Zinc Content of Selected Foods; Kathryn A. Haeflein and Arlette I. Rasmussen, PhD, RD; Journal; of Am. Dietetic Assoc.; v.70:610, 1977.

(99) Zinc Content of Selected Foods; Jeanne H. Freeland, PhD and Robert J. Cousins, PhD; Journal of Am. Dietetic Assoc.; v.68:526, 1976.

(100) Provisional Tables on the Zinc Content of Foods; Elizabeth W. Murphy, Barbara W. Willis and Bernice K. Watt, PhD, RD; Journal of Am. Dietetic Assoc.; v.66:345, 1975.

(101) The Lead, Zinc and Copper Content of Foods; Doris Larkin, Margaret Page, J.C. Bartletand Ross A. Chapman; Food Research; 212,1954.

(102) The Zinc and Copper Content of Seeds and Nuts; Kenneth G.D. Allen, Leslie M. Klevay and Hugh L. Springer; Nutrition Reports International; v.12:75, 1975.

(103) Magnesium Content of Selected Foods; J.L. Greger, S. Marhefka and A.H. Geissler; Journal of Food Science;v.43:1610, 1978.

(104) Magnesium Content of Accessory Foods; Gweneth Y. Nelson and Mary R. Gram, PhD; Journal of Am. Dietetic Assoc.; v.38:437, 1960.

(105) Yields, Proximate Composition and Mineral Content of Finfish and Shellfish; Jane E. Anthony, Pauline N. Hadgis, Rhonda S. Milam Gudrun A. Herzfeld, L. Janette Taper and S.J. Richey; Journal of Food Science; v.48:313, 1983.

(106) Composition of the Edible Portion of Raw (Fresh or Frozen) Crustaceans, Finfish, and Mollusks. I. Protein, Fat, Moisture, Ash, Carbo-hydrate, Energy Value, and Cholestrol; Virginia D. Sidwell, Pauline R. Foncannon, Nancy S. Moore and James C. Bonnet; Marine Fisheries Review; v.36:21, 1974.

(107) Chemical and Nutritive Values of Several Fresh and Canned Finfish, Crustaceans, and Mollusks. Part I: Proximate Composition, Calcium, and Phosphorus; Virgina D. Sidwell, James C. Bonnet, and Elizabeth G. Zook; Marine Fisheries Review; v.35:16, 1973.

(108) Mineral and Proximate Composition of Pacific Coast Fish; Dennis T. Gordon and G. Louis Roberts; Journal of Agric. Food Chemistry; v.25:1262, 1977.

(109) Nutritive Value of Fish. I. Nicotinic Acid, Riboflavin, Vitamin B12, and Amino Acids of Various Salt-Water Species; A.E. Teeri, M.E. Loughlin and D. Josselyn; Food Research; v.22:145, 1957.

(110) Comparison of Nutrients in Raw, Commercial- ly Breaded and Hand-Breaded Shrimp; I.H. Ahamad, R.M. Rao, J.A. Liuzzo and M.A. Kahn; Journal of Food Science; v.48:307, 1983.

(111) Vitamin Retention During Home Drying of Vegetables and Fruits; Zoe A. Holmes, Lorraine Miller, Margaret Edwards and Eva Benson; Home Economics Research Journal; v.7:261, 1979.

(112) The Content of Nine Mineral Elements in Raw and Cooked Mature Dry Legumes; Christine R. Meiners, Nellie L. Derise, Herbert C. Lau, Michael G. Crews, S.J. Richey and Elizabeth W. Murphy; Journal of Agric. Food Chemistry; v.24:1126, 1976.

(113) Nutrient Composition of Selected Wheats and Wheat Products. VI. Distribution of Manganese, Copper, Nickel, Zinc, Magnesium, Lead, Tin, Cadmium, Chromium, and Selenium as Determined by Atomic Absorption Spectro- oscopy and Colorimetry; Elizabeth G. Zook, F. Ella Greene and E.R. Morris; Cereal Chemistry; v.47:720, 1970.

(114) Composition and Utilization of Milled Barley Products. IV. Mineral Components; D.J. Liu, G.S. Robbins and Y. Pomeranz; Cereal Chemistry; v.51:309, 1974.

(115) Nutrient Content and Edible Yield of Selected Cuts of Cooked Pork; Wayne A. Johnson and Dorothy E. Deethardt; Journal of Food Science; v.48:1352, 1983.

(116) Nutrient Composition of Historical Canned Food Samples; J.A. Dudek and E.R. Elkins, Jr.; Journal of Food Science; v.48:654, 1983.

(117) Ascorbic Acid Retention in Frozen Corn; Irene R. Payne, PhD; Journal of Am. Dietetic Assoc.; v.51:344, 1967.

(118) Composition of Raw and Cooked Potato Peel and Flesh: Amino Acid Content; E.A. Talley R.B. Toma and P.H. Orr; Journal of Food Science; v.48:1360, 1983.

(119) Cholesterol Index of Foods; Donald B. Zilversmit, PhD; Journal of Am. Dietetic Assoc.; v.74:562, 1979.

(120) Cholesterol Content of Foods; Ruth M. Feel- ey, Patricia E. Criner and Bernice K. Watt, PhD, RD; Journal of Am. Dietetic Assoc.; v.61:134, 1972.

(121) Fiber Contents of Selected Raw and Process Vegetables, Fruits and Fruit Juices as Served; J. Zyren, R.R. Elkins, J.A. Dudek and R.E. Hagen; Journal of Food Science; v.48:600, 1983.

(122) Cellulose, Hemicellulose and Lignin Content of Raw and Cooked Processed Vegetables; Josefa Herranz, Concepcion Vida-Valverde and Enrique Rojas-Hidalgo; Journal of Food Science; v.48:274, 1983.

(123) Dietary Fiber in Spanish Fruits; Concepcion Vidal-Valverde, Josefa Herranz, Inmaculada Blanco and Enrique Rojas-Hidalgo; Journal of Food Science; v.47:1840, 1982.

(124) Selenium Content of Foods; V.C. Morris and O.A. Levander; Journal of Nutrition; v.100:1383, 1970.

(125) Selenium Content of Canadian Foods; D. Arthur; Canadian Inst. Food Science Technology Journal; v.5:165, 1972.

(126) Selenium and Chromium in Human Nutrition; Orville A. Levander, PhD; Journal of Am. Dietetic Assoc.; v.66:338, 1975.

(127) Selenium Content of Selected Foods; Helen W. Lane, PhD, RD, Barbara J. Taylor, Elizabeth Stool, RD, Diane Servance and Doris C. Warren PhD; Journal of Am. Dietetic Assoc.; v.82:24, 1983.

(128) Trace Nutrients. Selenium in British Food; Janet Thorn, Jean Robertson, D.H. Buss and N.C. Bunton; British Journal of Nutrition; v.39:391, 1978.

(129) Selenium Content of Food Consumed by Canadians; J.N. Thompson, Paula Erdody and Dorothy C. Smith; Journal of Nutrition; v.105:224, 1975.

(130) Ubersicht Spurenelemente in Lebensmittein; D. Schlettwein-Guell and S. Mommen-Straub; VIII. Selen. Internat. Z. Vit. Ern. Forschung; 42:607, 1972.

(131) Uber Selengehalte Pflanzlicher, Tierischer Dnderer Stoffe. 2. Mitteilung: V.W. Oelschlager and K.H. Menke; Selen-und Schwefelgehalte in Nahrungmittein; Zeitschrift fur Ernahrungswissen; v.9:216, 1968.

(132) Effect of Cooking on Selenium Content of Foods; D.J. Higgs, V.C. Morris and O.A. Levander; Journal of Agric. Food Chemistry; v.20:678, 1972.

(133) Toxicity of Selenium in Brazil Nuts to Rats; Ivan S. Palmer, Ailene Herr and Terri Nelson; Journal of Food Sciences; v.47:1595, 1982.

(134) Determination of Selenium in Biological Materials; R.C. Ewan, C.A. Baumann and A.L. Pope; Journal of Agric. Food Chemistry; v.16:212, 1968.

(135) Nutrients in Raw vs. Cooked Globe Artichokes; Bessie B. Cook, PhD and Saroja Sundaram; Journal of Am. Dietetic Assoc.; v.42:231, 1963.

(136) Trace Minerals in Commercially Prepared Baby Foods; Susan B. Deeming, PhD, RD and Charles W. Weber, PhD; Journal of Am. Dietetic Assoc.; v.75:149, 1979.

(137) Metallocalorie Ratios for Copper, Iron, and Zinc in Fruits and Vegetables; Phyllis E. Johnson, Cynthia Straus and Gary W. Evans; Nutrition Reports International; v.15:469, 1977.

(138) Fibre vs. Phylate as Determinant of the Availability of Calcium, Zinc, and Iron of Breadstuffs; John G. Reinhold, Faramarz Ismail-Beigi and Bahram Faradji; Nutrition Reports International; v.12:75, 1975.

(139) Proximate and Mineral Content of Fast Foods; H. Appledorf and L.S. Kelly; Journal of Am. Dietetic Assoc.; v.74:35, 1979.

(140) Thiamin Content of Freshly Prepared and Leftover Italian Spaghetti Served in a University Cafeteria Foodservice; M.A. Kahn, B.P. Klein and F.V. Lee; Journal of Food Science; v.47:2093, 1982.

(141) Effect of Marination Upon Mineral Content and Tenderness of Beef; Paula M. Howat, Lucille M. Sievert, Pamela J. Myers, Kenneth L. Koonce and Thomas D. Bidner; Journal of Food Science; v.48:662, 1983.

(142) Salt Content of Selected Snack Foods; Mahmood A. Khan and Judith A. Martin; Journal of Food Science; v.48:656, 1983.

(143) Simultaneous Determination of Thiamin and Riboflavin in Selected Foods by High- Performance Liquid Chromatography; J.K. Fellman, W.E. Artz, P.D. Tassinari, C.L. Cole and J. Augustin; Journal of Food Science; v.47:2048, 1982.

(144) Fast Foods: A Perspective on Their Nutritional Impact; B.M. Shannon and S.C. Parks; Journal of Am. Dietetic Assoc.; v.76:242, 1980.

(145) Vitamins in Frozen Convenience Dinners and Pot Pies; E. DeRitter, M. Osadca, J. Scheiner and J. Keating; Journal of Am. Dietetic Assoc.; v.64:391, 1974.

(146) Tocopherols in Canned Entrees and Vended Sandwiches; H.H. Koehler, H.C. Lee and M. Jacobson; Journal of Am. Dietetic Assoc.; v.70:616, 1977.

(147) Expression of Nutrient Allowances per 1000 Kilocalories; R. Gaurth Hansen, PhD and Bonita W. Wyse, PhD, RD; Journal of Am. Dietetic Assoc.; v.76:233, 1980.

(148) Analysis of Dietary Data: An Interactive Computer Method for Storage and Retrieval; Jelia Witschi, RD, Holly Kowaloff, Saul Bloom and Warner Slack, MD; Journal of Am. Dietetic Assoc.; v.78:609, 1981.

(149) Computerized Nutrient Data Bases: I. Comparison of Nutrient Analysis Systems; Loretta W. Hoover, PhD, RD; Journal of Am. Dietetic Assoc.; v.82:501, 1983.

(150) Computerized Nutrient Data Bases; II. Development of Model for Appraisal of Nutrient Data Base System Capabilities; Loretta W. Hoover, PhD, RD and Betty P. Perloff; Journal of Am. Dietetic Assoc.; v.85:506, 1982.

(151) Bowes and Church's Food Values of Portions Commonly Used: J.A.T. Pennington and H.N. Church; Harper & Row, NY, 13th revised edition, 1980.

(152) The Complete Book of Vitamins; staff of Prevention (r) Magazine; compiled and prepared by C. Gerras; Rodale Press, Emmaeus, PA. 1977.

(153) Tocopherols and Fatty Acids in American Diets: The Recommended Allowance for Vitamin E; J.G. Bier and R.P. Evarts; Journal of the Am. Dietetic Assoc.; v.62:147-151, 1973.

(154) Comparisons of Methods for Calculating Retentions of Nutrients in Cooked Foods; E.W. Murphy, P.E. Criner, B.C. Gray; Journal of Agric. and Food Chemistry, v.23:6, p.1153, 1975.

(155) Nutrition Almanac; John D. Kirschmann; McGraw-Hill Book Company, NY, revised edition 1979.

(156) You Are What You Eat; V.H. Lindlahr; Lancer Books, Inc., NY, 1972.

(157) Fatty Acid Content of Franchise Chicken Dinners; W.P. Donovan and H. Appledorf; Journal of Food Science; v.37:961, 1972.

(158) Nutritional Analysis of Foods from Fast Food Chains; Howard Appledorf; Food Technology; April 1974.

(159) The A.M.A.'s Nutrients in Processed Foods— Vitamins and Minerals; Publishing Sciences Group, Inc., Acton, Mass. 1974.

(160) Understanding Nutrition; E.N. Whitney and W.M.N. Hamilton; West Publishing Company, NY, second edition 1981, and third edition 1987.

(161) Diet for a Small Planet; Frances M. Lappe; Ballantine Books, NY, revised edition, 1975.

(162) The Save Your Life Diet; David Reuben, MD; Random House, 1975.

(163) Everything You Wanted to Know About Nutrition; David Reuben, MD; Simon Schuster, NY, 1978.

(164) The Computer Diet; V. Antonelli; M. Evans Co., Inc., NY, 1973.

(165) Fundamentals of Normal Nutrition; C.H. Robinson; MacMillan Publishing Co., NY, third edition, 1977.

(166) Normal and Therapeutic Nutrition; C.H. Robinson and M.R. Lawler; MacMillan Publishing Co., NY, fifteenth edition, 1977.

(167) Introduction to Nutrition; H. Fleck; MacMillan Publishing Co. Inc., NY, third edition, 1976.

(168) The Brand Name Nutrition Counter; J. Carper; Bantam Books, Inc. 1975.

(169) The Joy of Cooking; I.S. Rombauer and M.R. Becker; Bobbs-Merill Co., NY, revised edition, 1975.

(170) The World Encyclopedia of Food; L.P. Coyle, Facts on File, Inc., NY, 1982.

(171) Information Resources for Food and Human Nutrition; R.C. Frank; Journal of Am. Dietetic Associations; v.80-344, 1983.

(172) The Sodium and Potassium Content of Selected Vegetables; Anne C. Marsh and Percilla C. Koons, RD; Journal of Am. Dietetic Assoc.; v.83:24, 1983.

(173) Comprehensive Evaluation of Fatty Acids in Foods: XIII. Sausages and Luncheon Meats; Barbara A. Anderson; Journal of Am. Dietetic Assoc.; v.72:48, 1978.

(174) Comprehensive Evaluation of Fatty Acids in Foods: XII. Shellfish; Jacob Exler, and John L. Weihrauch; Journal of Am. Dietetic Assoc.; v.71:518, 1977.

(175) Comprehensive Evaluation of Fatty Acids in Foods: XI. Leguminous Seeds; Jacob Exler, PhD, Remedios M. Avena, PhD and John L. Weihrauch; Journal of Am. Dietetic Assoc.; v.71:412, 1977.

(176) Comprehensive Evaluation of Fatty Acids in Foods: X. Lamb and Veal; Barbara A. Anderson, Geraldine A. Fristrom, PhD and John L. Weihrauch; Journal of Am. Dietetic Assoc.; v.70:53, 1977.

(177) Folate Content of Various Nigerian Foods; Razia S. Huq, Joseph A. Abalaka and Winnie L. Stafford; Journal of the Science of Food and Agriculture; v.34:404, 1983.

(178) Vitamin E Content of Feedstuffs Determined by High-Performance Liquid Chromatographic Fluorescence; Winifred M. Cort, Thelma S. Vicente, Edward H. Waysek and Beverly D. Williams; Journal Agric. Food Chemistry; v.31:1330, 1983.

(179) Effects of Processing on the Sodium, Potassium, Calcium, Phosphorus Content in Foods; C. Jane Wyatt and K. Ronan; Journal Agric. Food Chemistry; v.31:415, 1983.

(180) Tocopherol Content of Some Southeast Asian Foods; John K. Candlish; Journal Agric. Food Chemistry; v.31:168, 1983.

(181) Dietary Fiber Content of Different Cereal Cereal Products in Norway; W. Frlich and B. Hestangen; Cereal Chemistry; v.60:82, 1983.

(182) Mineral and Protein Contents in Hard Red Winter Wheat Flours; Y. Pomeranz and E. Dikeman; Cereal Chemistry; v.60:80, 1983.

(183) Nutrient Composition of Stone Fruit (Prunus spp.) Cultivars: Apricots, Cherry, Necta rine, Peach and Plum; Ron B.H. Wills, Frances M. Scriven and Heather Greenfield; Journal of the Science of Food and Agri- culture; v.34:1383, 1983.

(184) The Nutritional Composition of British Bread—A Nationwide Study; Robert W. Wenlock, Lorna M. Sivell, Richard T. King, David Scuffam and Robert A. Wiggins; Journal of the Science of Food and Agric; v.34:1302, 1983.

(185) Composition of Fruit, Vegetable and Cereal Dietary Fibre; Warren D. Holloway; Journal of the Science of Food and Agriculture; v.43:1236. 1983.

(186) Nutrient Composition of Taro (Colocasia esculenta) Cultivars from the Papus New Guinea Highlands; Ron B.H. Wills, Jessie S.K. Lim, Heather Greenfield and Tim Bayliss-Smith; Journal of the Science of Food and Agriculture; v.34:1137, 1983.

(187) The Folacin Contents of Foods as Measured by a Radiometric Microbiologic Method; Marianne F. Chen, Janice W. Hill and Patricia A. McIntyre; Journal of Nutrition; v.113:2192, 1983.

(188) The Discrepancy Between Normal Folate Intakes and the Folate RDA; Bales, Black, Phillips, Wright and Southgate; Human Nutrition; Applied Nutrition; v.36:422, 1982.

(189) Studies on the Response of Lactobacillus casei to Folate Vitamin in Foods; D.R. Phillips and A.J.A. Wright; British Journal of Nutrition; v.49:181, 1983.

(190) Analysis of Arsenic and Selenium in Marine Raw Materials; G. Lunde, Journal of the Science of Food and Agriculture; v.21:242, 1970.

(191) Sprouting of Seeds and Nutrient Composition of Seed and Sprouts; J.R. Fordham, C.E. Wells and L.H. Chen; Journal of Food Science; v.40:552, 1975.

(192) Nutrients in Seeds and Sprouts of Alfalfa, Lentils, Mung Beans and Soybeans; Anne M. Kylen and Rolland M. McCready; Journal of Food Science; v.40:1008, 1975.

(193) Effects of Heating Methods on Vitamin Retention in Six Fresh or Frozen Prepared Food Products; Catharina Y.W. Ang, Charlotte M. Chang, A.E. Frey and G.E. Livingston; Journal of Food Science; v.40:997, 1975.

(194) Effect of Cooking on Vegetable Fiber; V. Mathee' and H. Appledorf; Journal of Food Science; v.48, 1978.

(195) Tocopherols in Nuts; G. Lambertsen, H. Myklestad and O.R. Braekkan; Journal of the Science of Food and Agriculture; v.13:617, 1962.

(196) Effect of Baking and Frying on Nutritive Value of Potatoes. Nitrogenous Constituents: Rathy Ponnam palam and Nell I. Mondy; Journalof Food Science; v.48:1613, 1983.

(197) Effect of Baking and Frying on Nutritive Value of Potatoes: Minerals; Nell I. Mondy and Rathy Ponnampalam; Journal of Food Science; v.48:1475, 1983.

(198) Effect of Age, Sex and Strain on the Fatty Acid Composition of Goose Muscle and Depot Fats; D.W. Friend, J.K.G. Kramer and A. Fortin; Journal of Food Science; v.48:1442, 1983.

(199) Physical, Chemical, Nutritional and Sensory Properties of Corn-Based Fortified Food Products; Maria L. Tonella, Manuel Sanchez and Maria G. Salazar; Journal of Food Science; v.48:1637, 1983.

(200) Essential Elements in Oysters (Crassostrea virginica) as Affected by Processing Method; Anthony Lopez, D.R. Ward and H.L. Williams; Journal of Food Sciences; v.48:1680, 1983.

(201) Measurement of Glycosylated Vitamin B6 in Foods; Hossein Kabir, James Leklem and Lorraine T. Miller; Journal of Food Science v.48:14_42, 1983.

(202) Functional Properties of Wheat-Bean Composite Flours; S.S. Seshpande, P.D. Rangnekar, S.K. Sathe, and D.K. Salunkhe; Journal of Food Science; v.48:1659, 1983.

(203) Selenium and Acute Alcoholism; Sudhir K. Dutta, MD, Pamela A. Miller, BS, Lynn B. Greenberg, MS and Orville A. Levander, PhD; v.38:713, 1983.

(204) Effect of Oral Contraceptive Agents on Vitamin and Mineral Requirements; Valerie J. Thorp; Journal of Am. Dietetic Assoc.; v.76:581, 1980.

(205) Food Manufacture and Nutrition; Arnold E. Bender; Nutrition Reviews Supplement; January 1982.

(206) The Use, Dietary Significance and Production of Fruit; G.R. Wadsworth; Journal of Human Nutrition; v.32:27, 1975.

(207) Protein, Fat and Mineral Analyses of Franchise Chicken Dinners; W.P. Donovan and H. Appledorf; Journal of Food Science; v.38:79, 1973.

(208) Bioavailability of Vitamin B-6 from Wheat Bread in Humans; James E. Leklem, Lorraine T. Miller, Anne D. Perera and Diane E. Peffers; Journal of Nutrition; v.110:1829, 1980.

(209) Pantothenic Acid in Foods; E.G. Zook, M.J. MacArthur, and E.W. Toepfer; U.S. Department of Agriculture Handbook No. 97, 1956.

(210) Enzyme Neutral Detergent Fiber Analysis of Selected Commercial and Home-Prepared Foods; Wen-Li John Jwuang and Mary E. Zabik; Journal of Food Science; v.44:924, 1979.

(211) Comparison of Two Radioassay Methods for Cyonalcobalamin in Seafoods; Robert A. Beck; Journal of Food Science; v.44:1077, 1979.

(212) Losses of Vitamins and Trace Minerals Resulting from Processing and Preservation of Foods; Henry A. Schroeder, MD; The American Journal of Clinical Nutrition; 562, 1971.

(213) The Distribution of Folic Acid Active Compounds in Individual Foods; Rafael Santini, PhD, Carol Brewster, BS and C.E. Butterworth, Jr., MD; The American Journal of Clinical Nutrition; v.14:205, 1964.

(214) Beneficial Physiologic Action of Beans; E.W. Hellendoorn, PhD; Journal of Am. Dietetic Assoc.; v.69:248, 1976.

(215) Unpublished Data on Vitamin B-6, Selenium, and Other Nutrients in Foods, 1983; Professor James E. Leklem, Dept. of Foods and Nutrition, Oregon State University.

(216) Selenium in Crops in the United States in Relation to Selenium-Responsive Diseases in Animals; J. Kubota, W.H. Allaway, D.L. Carter, E.E. Cary and V.A. Lazar; Journal of Agric. and Food Chemistry; v.15:448,1967.

(217) The B Vitamin Content of Grapes, Musts, and Wines; Alice P. Hall, Lisa Brinner, Maynard A. Amerine, and Agnes Fay Morgan; Food Research; v.21:362, 1956.

(218) Composition of Typical Mexican Foods; Rene Cravioto, Ernest E. Lockhart, Richmond K. Anderson, Francisco ce P. Miranda, and Robert S. Harris, et. al; Journal of Nutrition; v.29:317, 1945.

(219) The Value of Dairy Products in Nutrition; Royal A. Sullivan, Evelyn Bloom and Joan Jarmol; Journal of Nutrition; 1942-45.

(220) Thermal Destruction of Folacin: Effect of pH and Buffer Ions; B. Paine-Wilson and T.S. Chen; Journal of Food Science; v.44:717, 1979.

(221) Pyridoxine, Inositol and Vitamin K Contents of Germinated Pulses; Food Research; v.20: 545, 1965.

(222) Nutrient Content of Selected Baby Foods; Ruth H. Matthews and Martha Y. Workman; Journal of Am. Dietetic Assoc.; v.72:27, 1978.

(223) Folic Acid Content of Canned Garbanzo Beans; K.C. Lin, B.S. Luh and B.S. Schwei-gert; Journal of Food Science; v.40:562, 1975.

(224) Tocopherol Contents of Nine Vegetable Frying Oils and Their Changes Under Simulated Deep- Fat Frying Conditions; E. Yuki and Y. Ishikawa; Journal of Am. Oil Chem. Society; v.53:673, 1976.

(225) Nutrients in Vegetarian Foods; Delores D. Truesdell, Eleanor N. Whitney, PhD, RD, and Phyllis B. Acosta, Dr. P.J., RD; Journal of Am. Dietetic Assoc.; v.84:28, 1984.

(226) Comprehensive Evaluation of Fatty Acids in Foods; IX. Fowl; Geraldine A. Fristrom, PhD, and John L. Weibranch; Journal of Am. Dietetic Assoc.; v.69:517, 1976.

(227) A Guide to Calculating Intakes of Dietary Fiber; D.A.T. Southgate, PhD, Barbara Bailey, Edna Collison and Ann F. Walker, MS; Journal of Human Nutrition; v.30:303, 1976.

(228) Gamma Tocopherols: Metabolism, Biological Activity and Significance in Human Vitamin E Nutrition; John C. Bieri, PhD, and R. Poukka Evarts, DVM; The American Journal of Clinical Nutrition; v.27:980, 1974.

(229) Vitamin E Content of Infant Formulas and Cereals; Martha W. Dicks-Bushnell, PhD and Karen C. Davis, BS; The American Journal of Clinical Nutrition; v.20:262, 1967.

(230) Effects of Processing on the Dietary Fiber Content of Wheat Bran, Pureed Green Beans and Carrots; N.E. Anderson and F.M. Clydes- dale; Journal of Food Science; v.45:1533, 1980.

(231) Studies of Convenience Foods, III Packaged Dry and Canned Entrees; Marion Jacobson and Helen H. Koehler; Washington Agricultural Esperiment Station, Circular 569.

(232) Dietary Fiber—in Historical Perspective; Franklin C. Bing, PhD; Journal of Am. Dietetic Assoc.; v.69:498, 1976.

(233) Dietary Fiber; M.A. Eastwood and R. Passmore; The Lancet, p.202, July 1983.

(234) Armour Food Company

(235) Oscar Mayer Food Corporation, and Louis Rich, a division of Oscar Mayer.

(236) Carnation Healthcare Services; Carnation Inc.

(237) Artichoke Industries, Inc.

(238) LaChoy Food Products, a division of Beatrice Foods.

(239) Rosarita Foods, a division of Beatrice Foods.

(240) Del Monte Corporation

(241) Taco Bell, Inc.

(242) General Foods, Inc.

(243) Sunshine Biscuits, Inc.

(244) Peter Paul Cadbury, Inc.

(245) Land O'Lakes, Inc.

(246) Heublein, Inc.

(247) Libby, McNeil, Libby, Inc.

(248) Keebler Company.

(249) General Mills, Inc.

(250) The Nestle Company.

(251) Golden Grain Macaroni Co.

(252) Pillsbury Co.

(253) Campbell Soup Company.

(254) Kentucky Fried Chicken Corporation, Nutritional information 1983, 1986.

(255) The Quaker Oats Company

(256) Van de Kamp's Frozen Foods, a division of General Host Corporation.

(257) Kraft, Inc.

(258) A Nutritional Analysis of Food Served at McDonald's Restaurants, McDonald's Corporation, Oak Brook Illinois. Nutrition Analysis by Raltech Scientific Services, Inc., 1984, 1986, 1988.

(259) Hunt-Wesson Foods.

(260) MM/Mars, a division of Mars, Incorporated, Hackettstown, New Jersey. MM/Mars Guide to the Nutritional Value of Selected Snack Foods (c) 1983 MARS, Inc.

(261) Loma Linda Foods, Inc.

(262) Worthington Foods, Inc.

(263) Bordon, Inc.

(264) Hershey Foods Corporation, Hershey, Pennsylvania 17033, 1984.

(265) Nutrient Content of Pasta Products; Judith Spungen Douglas and Ruth H. Matthews; Cereal Foods World; vol. 27:558, November 1982.

(266) Evaluation of the Nutrient Composition of Wheat. III. Minerals; K.R. Davis, M.S., Louis J. Peters, Robert F. Cain, Ph.D., Duane LeTourneau, Ph.D., and James McGinnis,Ph.D.; Cereal Foods World; vol.29:246, April 1984.

(267) Variability of the Vitamin Content in Wheat; K.R. Davis, M.S., L.J. Peters, and D. LeTourneau, Ph.D.; Cereal Foods World; vol.29:364, June 1984.

(268) Nutritional Composition of Corn and Flour Tortillas; Guadalupe Saldana and Harold E. Brown; Journal of Food Science; vol.49: 1202, 1984.

(269) Minerals in Selected Variety Breads Commercially Produced in Four Major U.S. Cities; Gur Ranhotra, Janette Gelroth, Frances Novak, and Ruth Matthews; Journal of Food Science; vol.50:365, 1985.

(270) Measuring Dietary Fiber in Human Foods; Judith A.Marlett and Joan G. Chesters; Journal of Food Science; vol.50:410, 1985.

(271) Effects of Cooking in Solutions of Varying pH on the Dietary Fiber Components of Vegetables; Laura M. Brandt, Melissa A. Jeltema, Mary E. Zabik, and Brian D. Jeltema; Journal of Food Science; vol.49:900, 1984.

(272) Influence of Initial Riboflavin Content on Retention in Pasta During Photodegradation and Cooking; E.M. Furuya and J.J. Warthesen Journal of Food Science; vol. 49:984, 1984.

(273) Mineral and Vitamin Contents of Seeds and Sprouts of Newly Available Small-Seeded Soybeans and Market Samples of Mung beans; Aminah Abdullah and Ruth E. Baldwin; Journal of Food Science; vol. 49:656, 1984.

(274) Vitamin B-6 Content of Selected Foods Served in Dining Halls; Lou Anne Reiter, MS, and Judy A. Driskell, PhD, RD; Journal of The American Dietetic Association; vol.85:1625, December 1985.

(275) Vitamin A Activity of Selected Fruits;   C.D. Johnson,   R.R. Eitenmiller, PhD, D.A. Lillard, PhD, and M. Rao, PhD; Journal of The American Dietetic Associaton; vol.85:1627, December 1985.

(276) Yields and Nutrient Content of Selected Fresh Fruits; L. Janette Taper, PhD, Deborah A. McNeill, PhD, and S.J. Ritchey, PhD; Journal of The American Dietetic Association; vol.85:718, June 1985.

(277) Mineral Analyses of Vegetarian, Health, and Conventional Foods: Magnesium, Zinc, Copper, and Manganese Content; Deborah A. McNeill, PhD, Perveen S. Ali, MS, and Young S. Song, MS; Journal of The American Dietetic Association; vol. 85:569, May 1985.

(278) Selenium in Foods Purchased or Produced in South Dakota; Oscar E. Olson and Ivan S. Palmer; Journal of Food Science; vol.49, 1984.

(279) Assessment and Improvment of Selenium Composition Data; Deborah G. Lurie, MS, RD, Anita Schubert, MS,Joanne Holden, MD, and Wayne R. Wolf, PhD; US Department of Agriculture Nutrient Composition Lab in Beltsville, MD; Abstract at Society for Nutrition Education Meeting in Washington D.C.; July 7-9, 1986.

(280) Nutrient Content of Sprouted Wheat and Selected Legumes; Jorg Augustin, C.L. Cole, J.K. Fellman, R.H. Matthews, P.D. Tassinari, and H. Woo; Cereal Foods World; vol. 28:358, June 1983.

(281) Nutrient Profiles of Selected Snack Foods; News: Chocolate Manufacturers Association of the USA; New York, NY.

(282) Kellogg's Ready-To-Eat Cereals: Nutritive Values and Product Information; Kellogg Company Consumer Affairs Department; Battle Creek, Michigan, 1986.

(283) Selenium and Vitamins A, E, and C: Nutrients with Cancer Prevention Properties; Ronald Ross Watson, PhD and Tina K. Leonard; Journ- al of The American Dietetic Association; vol.86:505, April 1986.

(284) Proximate Composition, Cholesterol, and Calcium Content in Mechanically Separated Fish Flesh From Three Species of the Gadidae Family; Judith Krzynowek, Denise Peton, and Kate Wiggin; Journal of Food Science; vol.49:1182, 1984.

(285) Composition of Red Delicious Apples; C.D. Johnson, R. R. Eitenmiller, J. B. Jones, Jr., V. N. M. Rao, and S.E. Gebhardt; Journal of Food Science; vol.49:952, 1984.

(286) Yield of Chicken Parts: Proximate Composition and Mineral Content; Christine Meiners, PhD, M.G. Crews, PhD, and S.J. Ritchey, PhD; Journal of The American Dietetic Association; vol.81:435, 1982.

(287) Salt Substitutes and Medical Potassium Sources: Risks and Benefits; Danielle Riccardella, MS, and Johanna Dwyer, DSc., RD; Journal of The American Dietetic Association; vol.85:471, April 1985.

(288) New Carrot Excels in Carotene and Flavor; The Journal of The American Dietetic Association; From Agricultural Research, 32 (May): 12, 1984.

(289) Discussions with and Information from James E. Leklem, PhD, Oregon State University, regarding Vitamin B6, 1986, 1987.

(290) Proximate Components in Selected Variety Breads Commercially Produced in Major U.S. Cities; Gur Ranhotra, Janette Gelroth, Frances Novak, Faustina Bohannon, and Ruth Matthews; Journal of Food Science; vol.49:642, 1984.

(291) Contents and Retention of Nutrients in Extra Lean, Lean and Regular Ground Beef; K. Ono, B.W. Berry, and E. Paroczay; Journal of Food Science; vol.50:701, 1985.

(292) Content of Zinc in Selected Muscles from Beef, Pork, and Lamb; B. R. Schricker, D. D. Miller, and J. R. Stouffer; Journal of Food Science; vol.47:1020, 1982.

(293) Tocopherols and Tocotrienols in Finnish Foods: Meat and Meat Products; Vieno Piironen, Eeva-Liisa Syvaoja, Pertti Varo, Kari Salminen, and Pekka Koivistoinen; Journal of Agric. Food Chemistry; vol.33:1218, 1985.

(294) Nutrient Composition of Retail Ground Beef; Joanne M. Holden, Elaine Lanza, and Wayne R. Wolf; Journal of Agricultural Food Chemistry; vol.34:302, 1986.

(295) Tocopherols and Tocotrienols in Finnish Foods: Vegetables, Fruits, and Berries; Vieno Piironen, Eeva-Liisa Syvaoja, Pertii Varo, Kari Salminen, and Pekka Koivistoinen; Journal of Agricultural Food Chemistry; vol.34:742, 1986.

(296) Lipid Content and Fatty Acid Composition of Indica and Japonica Types of Nonglutinous Brown Rice; Hirokadzu Taira and Wan-Lai Chang; Journal of Agric. Food Chemistry; vol.34: 542, 1986.

(297) Studies on Dietary Fiber. 3. Improved Procedures for Analysis of Dietary Fiber; Olof Theander and Eric A. Westerlund; Journal of Agric. Food Chemist_ry; vol.34:330, 1986.

(298) Effect of Processing on Available Carbohydrates in Legumes; Sudesh Jood, Usha Mehta, and Randhir Singh; Journal of Agric. Food Chemistry; vol.34:417, 1986.

(299) The Composition of Commercially Important Fish Taken from New England Waters. II. Proximate Analysis of Butterfish, Flounder, Pollock, and Hake, and Their Seasonal Variation; Journal of Food Science; vol.27,1962.

(300) Ascorbic Acid and Vitamin A Activity in Selected Vegetables from Different Geographical Areas of the United States; B. P. Klein and A. K. Perry; Journal of Food Science; vol.47:941, 1982.

(301) Nutrient Composition of Cantaloupe and Honeydew Melons; R. R. Eitenmiller, C. D. Johnson, W. D. Bryan, D. B. Warren, and S. E. Gebhardt; Journal of Food Science; vol.50:136, 1985.

(302) Mineral and Vitamin Content of Goat's Milk; Wajih N. Sawaya, Ph.D., Jehangir K. Khalil, Ph.D., and Abdullah F. Al-Shalhat, B.Sc.; The Journal of The American Dietetic Association; vol.84:433, April 1984.

(303) Vitamin E Composition of Some Seed Oils as Determined by High-Performance Liquid Chromatography with Fluorometric Detection; A. J. Speek, J. Schriver, and W.H P. Schreurs; Journal of Food Science; vol.50:121, 1985.

(304) Measuring Dietary Fiber in Human Foods; Judith A. Marlett and Joan G. Chesters; Journal of Food Science; vol.50:410, 1985.

(305) Beef Patty Composition: Effects of Fat Content and Cooking Method; Bradford W. Berry, Ph.D., and Kathleen Leddy; The American Dietetic Association; vol. 84:654, June 1984.

(306) Metal Content of Wines; C. S. Ough, E. A. Crowell, and J. Benz; Journal of Food Science; vol.47:825, 1982.

(307) Vitamin Retention During Preparation and Holding of Mashed Potatoes Made from Commercially Dehydrated Flakes and Granules; Jorg Augustin, G.A. Marousek, W.E. Artz, and B.C. Swanson; Journal of Food Science; vol.47:274, 1981.

(308) Cholesterol Content of Raw and Cooked Beef Longissimus Muscles with Different Degrees of Marbling; Ki Soon Rhee, Thayne R. Dutson Gary C. Smith, Robert L. Hostetler, and Raymond Reiser; Journal of Food Science; vol. 47:716, 1982.

(309) Effect of Home and Industrial Processing on Protein Quality of Baby Foods and Breakfast Cereals; M. A. Kahn and B. O. Eggum; Journal of Science Food Agriculture; vol.30:369, 1979.

(310) Energy Value of Foods - Basis and Derivation; US Department of Agriculture Handbook 74; Al Merrilland B. K. Watt; 1973.

(311) Consumer and Food Economic Institute; Food, Home and Garden Institute; vol.228:64, 1979.

(312) Trace Elements - A Selective Survey; J. G. Reinhold; Clinical Chemistry; vol.21:476, 1975.

(313) Composition of Hawaii Fruits; Hawaii Agriculture Exp. Bulletin; N. S. Wenkam, and C. D. Miller; No. 135:87, 1965.

(314) Effect of Cooking on Vegetable Fiber; V. Matther and H. Appledorf; Journal of Food Science; vol.43:1344, 1978.

(315) Lipid Content and Fatty Acid Profiles of Various Deep-Fat Fried Foods; L. M. Smith, A. J. Clifford, R. K. Creveling and C. L. Hamblin; Journal of American Oil Chemists Society (JAOCS); vol.62:996, June 1985.

(316) Dietary Fiber: Analysis and Food Sources; D.A.T. Southgate; American Journal of Clinical Nutrition; vol. 31:5107, 1978.

(317) Effect of Processing on Dietary Fiber Content of Bran, Beans, and Carrots; Anderson and Clydesdede; Journal of Food Science; vol.45:1538, 1980.

(318) Effect of Cooking on Nutrient Retention of Legumes; David B. Haytowitz, and Ruth H. Matthews; Cereal Foods World; vol.28:362, June 1983.

(319) Nutrient Composition of Fresh Retail Pork; M. Moss, J. M. Holden, K. Ono, R. Cross, H. Slover, B. Berry, E. Lanza, R. Thompson, W. Wolf, J. Vanderslice, H. Johnson, and K. Stewart; Journal of Food Sciences; vol.48:1767, 1983.

(320) Protein Fortification of Cookies; K. Lorenz; Cereal Foods World; vol.28:449, August 1983.

(321) Evaluation of the Nutrient Composition of Wheat. II. Proximate Analysis, Thiamin, Riboflavin, Niacin, and Pyridoxine; K.R. Davis, R.F. Cain, D. Le Tourneau, L.J. Peters, and J. McGinnis; Cereal Chemistry; vol.58:116, 1981.

(322) The Vitamins of Triticale, Wheat, and Rye; P. Michela and K. Lorenz; Cereal Chemistry; vol.53:853, 1976.

(323) Bioavailability of Magnesium in Cereal-Based Foods; Gur S. Ranhotra, Ph.D.; Cereal Foods World; vol.28:349, June 1983.

(324) Nutritional Profile of Corn and Flour Tortillas; Gur S. Ranhotra, Ph.D.; Cereal Foods World; vol.30:703, October 1985.

(325) Retention of Selected B Vitamins in Cooked Pasta Products; G.S. Ranhotra, J.A. Gelroth F. A. Novak, and R. H. Matthews; Cereal Chemistry; vol.62:476, 1985.

(326) Nutritional Composition of Bagels Commercially Produced in the United States; C.S. Ranhotra, J.A. Geiroth, and F.A. Novak; Cereal Foods World; vol.30:209, March 1985.

(327) B Vitamins in Selected Variety Breads Commercially Produced in Major US Cities; Gur Ranhotra, Janette Gelroth, Frances Novak, and Ruth Matthews; Journal of Food Science; vol.50:1174, 1985.

(328) A New Look at Dietary Fiber; Martin A. Eastwood, MD, and Reginald Passmore, MD; Nutrition Today; September/October 1984.

(329) Selenium in Wheats and Commercial Wheat Flours; K. Lorenz; Cereal Chemistry; vol.55:287, 1978.

(330) Selenium Content of Foods Grown or Sold in Ohio; A. L. Moxon, and D. L. Palmquist; Ohio Report on Research and Development; Ohio Agricultural Research and Development Center; Jan./Feb. 1980.

(331) Alaska King Crab: Fatty Acid Composition, Carotenoid Index and Proximate Analysis; Richard A. Krzeczkowski, R. D. Tenney and C. Kelley; Journal of Food Science; vol.36:604, 1971.

(332) California Walnuts and Light Foods; Tom Payne; Cereal Foods World; vol.30:215, March; 1985.

(333) Nutrient Composition of Fresh Retail Pork; M. Moss, J. M. Holden, K. Ono, R. Cross, H. Slover, B. Berry, E. Lanza, R. Thompson, W. Wolf, J. Vanderslice, H. Johnson, and K. Stewart; Journal of Food Science; vol. 48:1767, 1983.

(334) Nutritive Value of Selected Variety Breads and Pastas; Gur S. Ranhotra, PhD, Janette A. Gelroth, Frances A. Novak, M. Ann Bock, PhD, RD, Gay L. Winterringer, PhD, RD, and Ruth H. Matthews, RD; Journal of The American Dietetic Association; vol.84:324, March 1984.

(335) The Content of Nine Mineral Elements in Raw and Cooked Mature Dry Legumes; Christine R. Meiners, Nellie L. Derise Herbert C. Lau, Michael G. Crews, S.J. Ritchey, and Elizabeth W. Murphy; Journal of Agric. Food Chemistry; vol.24:1126, 1976.

(336) Proximate Composition and Yield of Raw and Cooked Mature Dry Legumes; Christine R. Meiners, Nellie L. Derise, Herbert C. Lao, S.J. Ritchey, and Elizabeth W. Murphy; Journal of Agric. Food Chemistry; vol.24: 1122, 1976.

(337) Content of Selected Nutrients in Raw, Cooked, and Processed Legumes; David B. Haytowitz, Anne C. Marsh, and Ruth H. Matthews; Food Technology; March 1981.

(338) Proximate and Mineral Content of Selected Baked Products; Catherine McQuilkin Tarone, and Ruth H. Matthews; Cereal Foods World; vol.27:308, 1982.

(339) Food Values of Portions Commonly Used; Jean A. T. Pennington and Helen Nichols Church; Harper and Row Pub., New York, 1985.

(340) Effect of Germination on the Nutritive Value of Legumes; J. Vanderstoep; Food Technology; 35(3): 83.

(341) The Alaska Dietary Survey 1956-1961; Christine A. Heller, PhD, and Edward M. Scott, PhD; Public Health Service Publication No. 999-AH-2; Arctic Health Research Center; Anchorage, Alaska.

(342) Food Composition Table from Unpublished Manuscript; Christine Heller, Nutritionist; Arctic Health Research Center; US Dept. HEW, Anchorage, Alaska.

(343) Potato Chip/Snack Food Association Nutrition Service; Salted Snacks; Sampling Dates and Sources of Laboratory Composites for Data Submitted to Nutrient Data Banks; Laguna Beach, CA; 1986.

(344) Ross Medical Nutritional System; Product Information; Enteral Nutrition; Ross Laboratories; Columbus, OH; June 1986.

(345) Vivonex T.E.N. (Total Enteral Nutrition; Norwich Eaton Pharmaceutiacals, Inc.; Norwich, NY; 1986.

(346) Nutritional Analysis of Food Served at McDonald's Restaurants; Hazleton Laboratories America, Inc.; Chemical and Biomedical Sciences Division; McDonald's Corp.; 1986.

(347) Nutritional facts about California Ripe olives, California Olive Industry, October 1986.

(348) Enteral Nutrition Formulary Composition Chart; San Jose State Univ; Dept. of Nutrition and Food Services; Produced under an Educational Grant fron NUTREX Corporation; Sunnyvale, CA; 1986.

(349) The Enteral Nutritional Management System Product Handbook; Mead Johnson Nutritional Division; Evansville, IN; 1985.

(350) Provisional Tables on the Content of Omega-3 Fatty Acids and Other Fat Components of Selected Foods; Frank N. Hepburn; Jacob Exler; and John L. Weihrauch; Journal of the American Dietetic Association; vol.86:788; June 1986.

(351) Mineral Content of Foods and Total Diets: The Selected Minerals in Foods Survey, 1982 to 1984; Jean A.T. Pennington, Ph.D., R.D.; Barbara E. Young; Dennis B. Wilson; Roger D. Johnson; and John E. Vanderveen, Ph.D; Journal of the American Dietetic Association; vol.86:878; July 1986.

(352) Vitamin E in Foods from High and Low Linoleic Acid Diets; J. Lehmann, M.S., and J.T. Judd, Ph.D.; Journal of the American Dietetic Association; vol.86:1208; September 1986.

(353) A Critical Review of Food Fiber Analysis and Data; Elaine Lanza, Ph.D., and Ritva R. Butrum, Ph.D.; Journal of the American Dietetic Association; vol.86:734; June 1986.

(354) Folacin Values in Food; Eddie M. Lancaster, M.S., R.D.; Cornelia H. Boone, M.S., R.D.; Mary D. Brooks, M.S., R.D.; and Mary A. Smith, Ph.D.,R.D.; University of Tennessee Center for the Health Sciences, Child Development Center; 1985.

(355) Chemistry and Analysis of Soluble Dietary Fiber; Alfred Olson; Gregory M. Gray; and Meichen Chiu; Food Technology; vol.41:71; February, 1987.

(356) Vitamin E and Fatty Acid Composition of Human Milk; Lennart Jansson, MD; Bjorn Akesson, MD; and Lars Holmberg, MD; The American Journal of Clinical Nutrition; vol. 34:8; January, 1981.

(357) Crude Protein, Minerals, and Total Carotenoids in Sweet Potatoes; David H. Picha; Journal of Food Science; vol. 50:1768; 1985.

(358) Dietary Fiber and Other Constituents of Some Tongan Foods; W.D. Holloway; J.A. Monro; J.C. Gurnsey; E.W. Pomare; and N.H. Stace; Journal of Food Science; vol. 50:1756; 1985.

(359) Elemental Analysis of Fruit and Vegetables From Tonga; J.A. Monro; W.D. Holloway; and J. Lee; Journal of Food Science; vol.51:522; 1986.

(360) Distribution of Selenium in Human Milk; J.A. Milner, PhD.; L. Sherman, MS; and M.F. Picciano, PhD; American Journal of Clinical Nutrition; vol. 45:617; 1987.

(361) Dietary Selenium Intake and Selenium Concentrations of Plasma, Erythrocytes, and Breast Milk in Pregnant and Post-partum Lactating and Nonlactating Women; Orville A. Levander, PhD; Phylis B. Moser, PhD; and Virginia C. Morris, MS; American Journal of Clinical Nutrition; vol. 46:694; October, 1987.

(362) Selenium Content of Processed Soybeans; C.M. Weaver; J. Davis; H.S. Marks; and R.K. Sensmeier; Journal of Food Science; vol. 53:300; 1988.

(363) Nutritional, Fatty Acid, and Oil Characteristics of Pumpkin and Melon Seeds; Evangelos S. Lazos; Journal of Food Science; vol. 51:1382; 1986.(364) Vitamin B12 Activity in Miso and Tempeh; Delores D. Truesdell; Nancy R. Green; and Phyllis B. Acosta; Journal of Food Science; vol. 52:493; 1987.

(365) Proximate Composition, Mineral Content, and Fatty Acids of Catfish (Ictalurus punctatus, Rafinesque) for Different Seasons and Cooking Methods; P.A. Mustafa and D.M. Medeiros; Journal of Food Science; vol. 50:585; 1985.

(366) Ocean Pout (Macrozoarces americanus): Nutrient Analysis and Utilization; Sudip N. Jhaveri; Pavlos A. Karakoltsidis; Soliman Y.K. Shenouda; and Spiros M. Constantinides; Journal of Food Science; vol. 50:719; 1985.

(367) Proximate and Amino Acid Composition of the Rice and Muscle of Selected Marine Species; Mariko Iwasaki and Rokuro Harada; Journal of Food Science; vol. 50:1585; 1985.

(368) Contents and Retention of Nutrients in Extra Lean, Lean and Regular Ground Beef; K. Onio; B.W. Berry; and E. Paroczay; Journal of Food Science; vol. 50:701; 1985.

(369) The Folate in Human Milk; Jack M. Cooperman, PhD; Harry S. Dweek, MD; Leonard J. Newman, MD; Charles Garbarino, MD; and Rafael Lopez, MD; The American Journal of Clinical Nutrition; vol. 36:576; October, 1982.

(370) Polyunsaturated Fatty Acids and Fat in Fish Flesh for Selecting Species for Health Benefits; Thomas L. Hearn; Sandra A. Sgoutas; James A. Hearn; and Demetrios S. Sgoutas; Journal of Food Science; vol. 52:1209; 1987.

(371) Lipid Content and Fatty Acid Composition of Buckwheat Seed; G. Mazza; Cereal Chemistry; vol. 65:122; March/April, 1988.

(372) Stability of Polyunsaturated Fatty Acids After Microwave Cooking of Fish; Thomas L. Hearn; Sandra A. Sgoutas; Demetrios S. Sgoutas; and James A. Hearn; Journal of Food Science; vol. 52:1430; 1987.

(373) Soluble and Total Dietary Fiber in White Bread; Gur Ranhotra; and Janette Gelroth; American Association of Cereal Chemists, Inc; vol. 65:155; 1988.

(374) Soluble and Insoluble Fiber in Soda Crackers; Gur Ranhotra and Janette Gelroth; American Association of Cereal Chemists, Inc; vol. 65:159; 1988.

(375) Dietary Fiber and Resistant Starch; H.N. Englyst, PhD; H. Trowell, MD; D.A.T. Southgate, PhD; and J.H. Cummings, MSc; American Journal of Clinical Nutrition; vol. 46:873; 1987.

(376) Workshop 1 — Component Analysis of Fiber in Food; Chairman: Peter J. Van Soest, MD; The American Journal of Clinical Nutrition; vol. 31:S75; October, 1978.

(377) Dietary Fiber in the Reduction of Colon Cancer Risk; Peter Greenwald, MD, Dr, PH; Elaine Lanza, PhD; and Gerald A. Eddy, DVM, PhD; Journal of the American Dietetic Association; vol. 87:1178; September, 1987.

(378) Dietary Fiber: Classification, Chemical Analyses, and Food Sources; Joanne L. Slavin, PhD, RD; Journal of the American Dietetic Association; vol. 87:1164; September, 1987.

(379) Pantothenic Acid Content of Human Milk; L. Johnston; L. Vaughan; and H.M. Fox; The American Journal of Clinical Nutrition; vol. 34:2205; October, 1981.

(380) Thiamine Partitioning and Retention in Cooked Rice and Pasta Products; H.T. Vandrasek and J.J. Warthesen; Cereal Chemistry; 64:116; 1987.

(381) Nutritional Profile of Corn and Flour Tortillas; Gur S. Ranhotra, PhD; Cereal Foods World; vol. 30:704; October, 1985.

(382) Nutrient Composition, Protein Quality, and Sensory Properties of Thirty-Six Cultivars of Dry Beans (Phaseolus vulgaris L.) H.H. Koehler, Ch'iung-Hsia Chang; Genevieve Scheier, and D.W. Burke; Journal of Food Science; vol. 52:1336; 1987.

(383) Vitamin and Proximate Composition of Fast-Food Fried Chicken; Jane A. Bowers, PhD; Jean A. Craig, MS; Tammy Tucker, MS; Joanne M. Holden, MS; and Linda P. Posati, MS; Journal of the American Dietetic Association; vol.87:736; June, 1987.

(384) Chemical Forms of Iron, Calcium, Magnesium and Zinc in Black, Oolong, Green and Instant Black Tea; Lauren S. Jackson and Ken Lee; Journal of Food Science; vol. 53:181; 1988.

(385) Moisture, Fat and Cholesterol Content of Some Raw, Barbecued and Cooked Organ Meats of Beef and Mutton; Faisal A. Mustafa; Journal of Food Science; vol. 53:270; 1988.

(386) Effects of Chemical Preservatives on Storage and Nutrient Composition of Soybean Curd; Anna Miskovsky and Martha B. Stone; Journal of Food Science; vol. 52:1535; 1987.

(387) Loss of Vitamin C in Vegetables During the Foodservice Cycle; Beth L. Carlson, PhD; and Mary H. Tabacchi, PhD, RD; Journal of the American Dietetic Association; vol. 88:65; January, 1988.

(388) Fatty Acid Composition of Mature Human Milk of Egyptian and American Women; Marlene W. Borschel, RD, PhD; Robert G. Elkin, PhD; Avanelle Kirksey, PhD; Jon A. Story, PhD; Osman Galal, MD, PhD; Gail G. Harrison, PhD; and Norge W. Jerome, PhD; American Journal of Clinical Nutrition; vol. 44:330; 1986.

(389) Contents and Retentions of Sodium and Other Minerals in Pasta Cooked in Unsalted or Salted Water; J.A. Albrecht; E.H. Asp; and I.M. Buzzard; Cereal Chemistry; vol. 64:106; March/April, 1987.

(390) Nutrient Composition Protein Quality, and Sensory Properties of Thirty-Six Cultivars of Dry Beans (Phaseolus vulgaris L.); H.H. Koehler; Ch'lung-Hsia Chang; Genevieve Scheier; and D.W. Burke; Journal of Food Science; vol. 52:1335; 1987.

(391) Mineral Content of Market Samples of Fluid Whole Milk; Jean A.T. Pennington, PhD, RD; Dennis B. Wilson; Barbara E. Young; Roger D. Johnson; and John E. Vanderveen, PhD; Journal of the American Dietetic Association; vol. 87:1036; August, 1987.

(392) HPLC Determination of Carotenoids in Fruits and Vegetables in the United States; Janice L. Bureau; and Rodney J. Bushway; Journal of Food Science; vol. 51:128; 1986.

(393) Selenium in Foods Produced and Consumed in Greece; Michael S. Bratakos; Theodore F. Zafiropoulos; Panayiotis A. Siskos; and Panayiotis V. Ioannou; Journal of Food Science; vol. 52:817; 1987.

(394) Selenium Content of Foods Purchased or Produced in Ohio; Jean T. Snook, PhD; Donna Kinsey; Donald L. Palmquist, PhD; James P. DeLany, PhD; Virginia M. Vivian, PhD, RD; and Alvin L. Moxon, PhD; Journal of the American Dietetic Association; vol. 87:744; June, 1987.

(395) Selenium Content of a Core Group of Foods Based on a Critical Evaluation of Published Analytical Data; Anita Schubert, MS; Joanne M. Holden, MS; and Wayne R. Wolf, PhD; Journal of the American Dietetic Association; vol. 87:285; March, 1987.

(396) Heinz Baby Food and Juices; Idamarie Laquatra, PhD, RD; Heinz USA; Pittsburgh, PA; 1988.

(397) Nutrient Composition of Some Fresh and Cooked Retail Cuts of Veal; K. Ono; B.W. Berry; and L.W. Douglass; Journal of Food Science; vol. 51(5):1352; 1986.

(398) Dietary Fiber and Bread: Intake, Enrichment, Determination, and Influence on Colonic Function; Hans Georg Becker; Dr. Werner Steller; Prof. Dr. Walter Feldheim; Dr.Elizabeth Wisker; Wolfgang Kulikowski; Dr.Peter Suckow; Prof. Dr. Fredrich Meuser; and Prof. Dr. Wilfried Seibel; Cereal Foods World; vol. 31(4):306; April, 1986.

(399) Soluble and Insoluble Plant Fiber in Selected Cereals and Vegetables; Wen-Ju Lin Chen, PhD; and James W. Anderson, MD; The American Journal of Clinical Nutrition; vol. 34:1077; June, 1981.

(400) Dietary Fiber Intake in The US Population; Elaine Lanza, PhD; D. Yvonne Jones, PhD; Gladys Block, PhD; and Larry Kessler, ScD; American Journal of Clinical Nutrition; vol. 46(5):790; November, 1987.

(401) Phytate and Zinc Contents of Coffees, Cocoas, and Teas; B.F. Harland; and D. Oberleas; Journal of Food Science; vol. 50:832; 1985.

(402) Fiber Analysis Tables; Prepared by D.A.T. Southgate B. Bailey, E. Collinson and A.F. Walker. From: J. Human Nutrition 30:303, 1976; The Journal of Clinical Nutrition; vol. 31:S281; October, 1978.

(403) Food Fiber Choices for Diabetic Diets; Eugenio Del Toma, MD, PhD; Aldo Clementi, MD, MS; Marcello Marcelli, MD, MS; Marsilio Cappelloni, RT; and Claudia Lintas, PhD; The American Journal of Clinical Nutrition; vol. 47:243; February, 1988.

(404) Dietary Fibers: Their Definition and Nutritional Properties; Peter J. Van Soest, PhD; The American Journal of Clinical Nutrition; vol. 31:S12; October, 1978.

(405) Composition of Foods—Legumes and Legume Products...Raw, Processed, Prepared; David B. Haytowitz; and Ruth H. Matthews; US Department of Agriculture Handbook No. 8-16; December, 1986.

(406) Composition of Foods—Finfish and Shellfish Products...Raw, Processed, Prepared; Jacob Exler; US Department of Agriculture Handbook No. 8-15; September, 1987.

(407) Composition of Foods—Beef Products...Raw, Processed, Prepared; Barbara A. Anderson; Jeanne L. Lauderdale; and Margaret Hoke; US Department of Agriculture Handbook No. 8-13; August, 1986.

(408) Provisional, unpublished data on Dietary fiber from USDA, HNIS, Hyattsville, Maryland. Ruth Mathews and Pamela Pehrsson, July 1988.

(409) Taco Bell, Inc, revised information, fall 1987.

(410) Oat bran data from USDA, HNIS, Hyattsville, MD, spring, 1988 (David Haytowitz, Jacob Exler, John Weihrach, Pamela Pehrsson.)

(411) Ambrosia Chocolate Company, Milwaukie, Wisconsin, February 1988.

(412) Macdonald's Corporation, revised information, spring, 1988.

(413) Arby's, Inc.

(414) Burger King Nutrition Guide; Burger King Corporation, 1986, 1989.

(415) International Dairy Queen, Inc.

(416) Domino's Pizza, Inc. 1988.

(417) Hardee's Food Systems, Inc.

(418) Jack in the Box Restaurants; Foodmakers, Inc. 1986.

(419) Pizza Hut, Inc.

(420) Rax Products Nutritional Information; Rax Corporate Center, Ohio, 1987, 1989.

(421) Taco Bell, Inc.., revised 1989.

(422) Wendy's International, Inc., 1986, 1989.

(423) Nutritional Information for White Castle Products, 1985, 1989.

(424) Long John Silver's Restaurants; Department of Nutrition & Food Science, University of Kentucky.

(425) The Coca Cola Company.

(426) The Shasta Beverage Company, 1988.

(427) Riverside Food Facts; vol. II:4; 1988.

(428) Composition of Foods: Fast Foods...Raw, Processed, Prepared; Lynn E. Dickey and John L. Weihrauch; U.S. Department of Agriculture Handbook No. 8-21, 1988.

(429) Lea and Perrins, Inc., Summer 1989.

(430) Pillsbury Consumer Products Nutrition Information; Winter 1989.

(431) Dannon Company, 1989.

(432) General Mills, Inc., 1989.

(433) Composition of Foods: Lamb, Veal, and Game Products...Raw, Processed, Prepared; Marjorie L. Clements; Lynn E. Dickey; Jacob Exler; and I. Margaret Hoke; U.S. Department of Agriculture Handbook No. 8-17, 1989.

(434) Food Composition and Nutrition Tables 1986/87; S.W. Souci; W. Fachmann; and H. Kraut; Wissenschaftliche Verlagsgesellschaft mbH Stuttgart, Germany, 1986; third revised and completed edition.

(435) Canola Oil, Properties and Performance; M. Vaisey-Genser & N.A. Michael Eskin; published by the Canola Council, 1987.

(436)   Recommended Nutrient Intakes for Canadians; Bureau of
Nutritional Sciences, Food Directorate, Health Protection Branch, Dept
of National Health & Welfare; 1983.

(437)   Vitamin K Content of Foods, June 1986; U. S. Department of
Agriculture Provisional Tables, Washington DC.

(438)   Phytate and Zinc Contents of Coffees, Cocoas, and Teas; B.F.
Harland; and D. Oberleas; Journal of Food Science; vol. 50:832; 1985.

(439)   Fiber Analysis Tables; Prepared by D.A.T. Southgate B. Bailey,
E. Collinson and A.F. Walker. From: J. Human Nutrition 30:303, 1976;
The Journal of Clinical Nutrition; vol. 31:S281; October, 1978.

(440)   Food Fiber Choices for Diabetic Diets; Eugenio Del Toma, MD,
PhD; Aldo Clementi, MD, MS; Marcello Marcelli, MD, MS; Marsilio
Cappelloni, RT; and Claudia Lintas, PhD; The American Journal of
Clinical Nutrition; vol. 47:243; February, 1988.

(441)   Woodstock Research Center; Morton Salt Division of Morton
Thiokol, Inc.; Woodstock, Illinois, 1989.

(442)   Nutritional Data on the Composition of Eggs, 1989, from
Hazelton Labs, Wisconsin; and the USDA, HNIS, Hyattsville, Maryland,
1989.

(443)   Recommended Dietary Allowances, 10th edition; National
Research Council, Food & Nutrition Board, Commission on Life Sci-
ences, Subcommittee on the Tenth Edition opf the RDAs; Washinton D.
C. 1989.

(444)   Nutritional elements in U.sS. diets: results from the Total Diet
Study, 1982-86; J.A.T. Pennington, B.E. Young, and D.B. Wilson;
Journal of American Dietatic Association, v. 89:659-664, 1989.

(445)   Selenium-related endemic diseases and the daily selenium
requirement of humans; G. Yang, K. Ge, J. Chen, and X. Chen; World
Review Nutr. Diet. v. 55:98--152; 1988.

(446)   The concentrations of copper and zinc in human milk. A
longitudinal study; E. Vuori and P. Kuitunen; Acta Paediatrics Scand.,
v. 68:33-37, 1979.

(447)   Studies in human lactation: zinc, copper, manganesse, and
chromium in human milk in the first month of lactation; R.K. Chandra; J.
Am. Medical Association, v. 252:1443-1446, 1985.

(448)   Apparent absorption and retention of Ca, Cu, Mg, Mn, and Zn
from a diet containing bran; R.J. Apgar Schwartz and E.M. Wien; Am.
J. Clinical Nutrition, v. 43:444-455, 1986.

(449)   Pantothenic acid content of human milk; L. Johnston, L.
Vaughan, and H.M. Fox; Am. J. Clinical Nutrition, v. 34:2205-2209, 1981.

(450   On the requirements of ascorbic acid in man: steady -state
turnover and body pool in smokers; A. Kallner, D. Hartmann, and D.H.
Hornig, 1981; Am. J. of Clinical Nutrition v. 34: 1347-1355, 1981.

(451)   Composition of Foods: Cereal Grains and Pasta, USDA
Handbook 8-18, pre-printed release from the USDA, Human Nutrition
Information Service, 1989.

(452)   USDA Supplemental nutriiton information, providing modifica-
tions to previous printed data in various editions of the USDA Handbook
8 series, 1989.

1488